MATHEMATICS HL & SL
WITH HL OPTIONS

MATHEMATICS HL & SL
WITH HL OPTIONS

Ideal for the
INTERNATIONAL BACCALAUREATE DIPLOMA

◆

PETER SMYTHE

MATHEMATICS PUBLISHING PTY. LIMITED
2003

National Library of Australia Catalogue-in-Publication Data

Smythe, Peter, 1943- .
    Mathematics HL & SL with HL options : Ideal for the International Baccalaureate Diploma.
    For secondary school students.
    ISBN 0 9750831 0 4.
    1. Mathematics - Textbooks. 2. International Baccalaureate.
    I. Title.
    510

First Published 2003

Copyright © Mathematics Publishing Pty Ltd 2003. No part of this publication may be reproduced, stored in a retrieval system, or transmitted in any form, or by any means, without the prior permission of the publishers.

While every care has been taken to trace and acknowledge copyright, the publishers tender their apologies for any accidental infringement or unconscious copying that has occured or where copyright has proved untraceable. In any case where it is considered copying has occurred, the publishers would be pleased to come to a suitable arrangement with the rightful owner.

This publication has been developed by the author in his personal capacity independently of the International Baccalaureate Organisation and this publication is not endorsed by that organisation.

Published by Mathematics Publishing Pty. Limited
PO Box 926, Glenelg, South Australia 5045
Email address: info@mathematicspublishing.com
Visit our website: mathematicspublishing.com

Cover design by David Blaiklock

Printed at Hyde Park Press
4 Deacon Avenue, Richmond, South Australia 5033

DEDICATION

For Olive

ACKNOWLEDGEMENTS

I would like to pay tribute to the following people for the enormous help given to me in creating this text. Dr Barbara Possingham a physicist, mathematician and gifted teacher whose masterly critical editing of the text added important substance that will be helpful for both teachers and students. Graham Nancarrow, for his experience and advice in book production and whose commitment and dedication to the overall project has been invaluable. I feel honoured that such talented individuals have supported me in this endeavour. Trevor Grant, a life-time friend and teaching colleague, for writing my biography; Paul Hassing for his creative contribution; Michael Pelling for his immeasurable support; Zita for sharing my passion for mathematics and for her total devotion to the project; and the naturally talented and gifted Fontella who has been supportive from the outset.

Finally, but certainly not least, my wife Jenny, for her tolerance of my obsession with mathematics and for her encouragement and total dedication in bringing this text to fruition.

PREFACE

This text has been specifically designed to provide a comprehensive approach to the teaching of the International Baccalaureate Mathematics Higher and Standard Level Courses. For ease of reference, the major topics have been generally presented in a single chapter and the order of the chapters has been chosen to provide a suitable teaching order. There is, however, ample opportunity for individual teachers to choose the order which best suits their teaching style. It should be noted that each chapter can presume knowledge of any and all previous chapters.

Chapters 1-17 contain all of the material common to both the Higher Level and Standard Level courses. Material which is required for the Higher Level course but not the Standard Level course is headed **Higher Level** and placed in shaded boxes to emphasise the fact that it is not required for the Standard Level course. Chapters 18-24 contain all material relevant to the Higher Level Core. Chapters 24 and 25 contain material needed for the Higher Level Options of Sets, Relations & Groups and Analysis & Approximation.

In each chapter the information relevant to the topic is discussed and several examples providing various approaches to the solutions are given. If the material in any Higher Level section is not specifically required by the Syllabus, it is designated as "Optional". These few sections have been included either because they could be useful in a prescribed section which follows or because they could be of interest. As an example of the first type, knowledge of the formulae for the sum and product of the roots of a quadratic (Chapter 19.5) is extremely useful when it comes to finding a quadratic factor of a real polynomial when given a single non-real zero (Chapter 20.2). As an example of the second type, Chapter 11 contains a section which discusses vector methods for finding the volumes of parallepipeds and tetrahedra. Knowledge of these sections is not absolutely necessary for a student to achieve a high grade for the course.

The exercises provided with each section have been carefully graded from the relatively easy to the more difficult. Whenever a given exercise is accompanied by an asterisk (*), it is considered to be too challenging for some students but suitable for extending others. There are more exercises provided in this text than can possibly be done by most students within the teaching time allotted. Answers to the odd-numbered questions have been provided. Where it was considered helpful, the answers to a few even-numbered questions have been added.

This text has not been written with the graphic display calculator (GDC) specifically in mind. However, there are countless examples where the use of this technology would be a definite advantage. It is left to the individual teacher to encourage sensible use of a GDC. Students should finally be able to decide for themselves when it is reasonable to call upon the technology and when a 'by hands' approach is more appropriate. It is most important that students understand the underlying mathematics and do not simply call upon a GDC to provide an answer when they have no idea of how that answer is produced.

Peter Smythe

CONTENTS

PART 1
COMMON CORE WITH HIGHER LEVEL EXTENSIONS

CHAPTER 1   PROPERTIES OF REAL NUMBERS
1.1	Subsets of the Real Numbers	3
1.2	Order Properties and the Modulus Function	7
1.3	Roots of Real Numbers and Surds	11
1.4	Irrational Equations	16
1.5	Exponents	17
1.6	Exponential and Logarithmic Functions	20

CHAPTER 2   COORDINATE GEOMETRY IN 2 DIMENSIONS
2.1	The General Form of the Equation of a Straight Line	31
2.2	The Distance of a Point from a Line	33
2.3	The General Quadratic Function	36
2.4	The General Solution to a Quadratic Equation	41
2.5	The Roots of a Quadratic Equation	43
2.6	Transformations of Graphs in the Plane	47
2.7	Ratio of Division of a Line Segment (Optional)	52
2.8	Circles in the Cartesian Plane	55

CHAPTER 3   SOLVING TRIANGLES
3.1	The Sine Rule	61
3.2	The Cosine Rule	68

CHAPTER 4   TRIGONOMETRY
4.1	The Sine, Cosine and Tangent Functions	73
4.2	The Graphs of the Trigonometric Functions	77
4.3	The Radian Measure of an Angle and the Double Angle Formulae	79
4.4	Graphs of Circular Functions of the Form $y = a \sin x$	83
4.5	Graphs of Circular Functions of the Form $y = \sin bx$	84
4.6	Graphs of Circular Functions of the Form $y = \sin(x-c)$	86
4.7	Graphs of Circular Functions of the Form $y = \sin x + d$	88
4.8	Mensuration of the Circle	90
4.9	Trigonometric Equations	94
4.10	The Six Circular Functions, their Graphs and Identities	99
4.11	Addition Formulae	104
4.12	Addition Formulae for the Tangent Function	107
4.13	Duplication and Half-angle Formulae	110
4.14	Functions of the Form $a\sin\theta + b\cos\theta$	115
4.15	Converting Sums and Differences to Products (Optional)	117
4.16	Converting Products to Sums or Differences (Optional)	120

Chapter 5   Relations and Functions
5.1   Relations and Functions - Domain and Range   123
5.2   The Composition of Functions   129
5.3   Inverse Functions   131

Chapter 6   Sequences and Series
6.1   Sequences   141
6.2   Arithmetic Sequences (Arithmetic Progressions)   142
6.3   Arithmetic Series and Sigma Notation   146
6.4   Geometric Sequences (Geometric Progressions)   150
6.5   Geometric Series   153
6.6   Compound Interest   155
6.7   Infinite Geometric Series   158

Chapter 7   Statistics 1
7.1   Frequency Tables and Frequency Histograms   163
7.2   Measures of Central Tendency - Mean, Median, Mode   168
7.3   Cumulative Frequency - Quartiles and Percentiles   172
7.4   Measures of Dispersion - Discrete Data   178
7.5   The Mean and Variance for Grouped Data   181

Chapter 8   Counting Techniques and the Binomial Theorem
8.1   The Product Principle   185
8.2   Factorial Notation   187
8.3   Permutations   188
8.4   Combinations and Partitions   192
8.5   The Binomial Theorem   199

Chapter 9   Matrices and Linear Equations
9.1   Matrix Addition   205
9.2   Matrix Multiplication   208
9.3   The Determinant of a (Square) Matrix   213
9.4   Matrix Algebra   218
9.5   Systems of Linear Equations   227
9.6   Geometrical Interpretation of Solutions (Optional)   235

Chapter 10   Differential Calculus 1
10.1   Limits   241
10.2   Gradient   245
10.3   The Derived Function   248
10.4   The Derivative of x^n 251
10.5   The Second Derivative - Motion of a Body on a Straight Line   255
10.6   Points of Increase, Points of Decrease and Stationary Points   259
10.7   Extreme Points   265
10.8   Optimisation   271

Chapter 11   Vector Geometry
11.1   Addition of Vectors - The Zero Vector - Scalar Multiples   277
11.2   Position Vectors and the Ratio Formula (Optional)   281

11.3	Vectors in Cartesian 2-Space	284
11.4	The Scalar Product of Two Vectors - Orthogonal Projection	288
11.5	The Vector Equation of a Line in Cartesian 2-Space	294
11.6	Vectors in Cartesian 3-Space	298
11.7	The Vector Product of Two Vectors	301
11.8	Volumes of Parallelepipeds and Tetrahedra (Optional)	308
11.9	The Equations of a Line	311
11.10	The Equations of a Plane	314
11.11	The Angles between Lines and Planes	320
11.12	The Intersections of Lines and Planes	322
11.13	Shortest Distance between Points, Lines and Planes	328

CHAPTER 12   DIFFERENTIAL CALCULUS 2

12.1	Derivatives of Composite Functions - The Chain Rule	337
12.2	The Product Rule	341
12.3	The Quotient Rule	343
12.4	Implicit Differentiation	347
12.5	Related Rates	351
12.6	The Graphs of Rational Functions	354
12.7	Graphs of Rational Functions with and without the Use of a GDC	356

CHAPTER 13   PROBABILITY

13.1	Sets	361
13.2	Elementary Probability Theory	366
13.3	Sum and Product Laws	371
13.4	Conditional Probability and Bayes' Theorem	375
13.5	Binomial Probabilities	379

CHAPTER 14   STATISTICS 2

14.1	Discrete Probability Distributions	385
14.2	The Mean (or Expected Value) of a Distribution	389
14.3	Variance and Standard Deviation	392
14.4	The Binomial Random Variable	397
14.5	The Normal Probability Distribution	400
14.6	An Introduction to Sampling	407
14.7	Interval Estimates and Confidence	412
14.8	Significance Testing	417
14.9	Contingency Tables	421

CHAPTER 15   INTEGRAL CALCULUS

15.1	Antiderivatives and the Integral Notation	427
15.2	The Use of the Chain Rule to Find $\int (ax+b)^n \, dx \ (n \neq -1)$	431
15.3	Integration by Substitution	434
15.4	The Definite Integral and Area	437

CHAPTER 16   TRIGONOMETRIC CALCULUS

16.1	Limits	447
16.2	The Derivatives of $\sin x$, $\cos x$ and $\tan x$	449

16.3	Integration of sin x, cos x and $1/cos^2 x$	456
16.4	The Derivatives of csc x, sec x and cot x	459
16.5	Further Integration of Trigonometric Functions	461

CHAPTER 17   EXPONENTIAL AND LOGARITHMIC FUNCTIONS

17.1	The Exponential Function a^x	465
17.2	Differentiation of Logarithmic Functions	470
17.3	Integration of Exponential Functions	475
17.4	Further Integration of Exponential Functions	477
17.5	Integration of Functions of the Form $\dfrac{1}{ax+b}$	479
17.6	Integration of Functions which can be Written in the Form $\dfrac{f'(x)}{f(x)}$	481

PART 2
HIGHER LEVEL CORE

CHAPTER 18   MATHEMATICAL INDUCTION

18.1	The Principle of Mathematical Induction	487
18.2	Making and Proving Conjectures	495

CHAPTER 19   POLYNOMIALS

19.1	Addition, Multiplication and the Division Process	497
19.2	The Remainder and Factor Theorems	502
19.3	Contracted (Synthetic) Division	506
19.4	Polynomial Equations with Integer Coefficients	511
19.5	Relations between the Zeros and Coefficients of a Quadratic Polynomial (Optional)	512

CHAPTER 20   COMPLEX NUMBERS

20.1	Addition, Multiplication and Division	517
20.2	Zeros of a Polynomial with Real Coefficients	524
20.3	Geometrical Representation of a Complex Number - Modulus	526
20.4	Argument	533
20.5	The Polar Form of a Complex Number	535
20.6	De Moivre's Theorem	539

CHAPTER 21   PROBABILITY DENSITY FUNCTIONS

21.1	Mean and Variance	547
21.2	Median and Mode	551

CHAPTER 22   INVERSE TRIGONOMETRIC FUNCTIONS

22.1	The Inverse Sine, Inverse Cosine and Inverse Tangent Functions	555
22.2	The Derivatives of the Inverse Trigonometric Functions	559
22.3	Integrals Involving Inverse Trigonometric Functions	562

CHAPTER 23 FURTHER INTEGRATION
23.1 Integration by Substitution — 565
23.2 Integration by Parts — 568
23.3 Partial Fractions — 571
23.4 Differential Equations — 576
23.5 Equations Reducible to Separable Form-Homogeneous Equations — 582

CHAPTER 24 MATRICES AND TRANSFORMATIONS IN THE PLANE
24.1 Linear Transformations in the Plane — 587
24.2 Rotations, Reflections, Stretches, Dilations and Shears — 592
24.3 The Image of the Curve $y = f(x)$ — 602

PART 3
HIGHER LEVEL OPTIONS

CHAPTER 25 SETS, RELATIONS AND GROUPS
25.1 Sets — 609
25.2 Cartesian Products and Binary Relations — 613
25.3 Equivalence Relations — 617
25.4 Matrices of Relations — 620
25.5 Functions - Injections, Surjections and Bijections — 623
25.6 Binary Operators and Closure — 628
25.7 The Associative Law — 630
25.8 The Identity Element — 631
25.9 Inverse Elements — 633
25.10 Residue Classes — 635
25.11 Permutations — 638
25.12 Cyclic Notation (Optional) — 641
25.13 Groups — 643
25.14 Properties of a Group — 646
25.15 Subgroups, Cyclic Groups and Isomorphism — 648

CHAPTER 26 ANALYSIS AND APPROXIMATION
26.1 Infinite Series - Tests for Convergence — 659
26.2 The Ratio and nth-Root Tests — 665
26.3 Improper Integrals and the Integral Test — 668
26.4 Alternating Series and Absolute Convergence — 671
26.5 Power Series — 675
26.6 The Mean Value Theorem — 679
26.7 Taylor and Maclaurin Series — 685
26.8 Differentiation of Integration of Power Series — 694
26.9 The Definite Integral as a Limit of a Sum — 698
26.10 Numerical Integration — 702
26.11 Numerical Solutions of Equations — 709

PART 4

Answers — 723

PART 1
COMMON CORE WITH HIGHER LEVEL EXTENSIONS

1 Properties of Real Numbers

1.1 Subsets of the Real Numbers

The development of the real number system has taken thousands of years. This development has been driven by man's need to solve problems. Initially all that was required were the small positive integers; enough to count the sheep in a flock for example. Later developments in commerce, construction of cities, navigation, measurement of time and warfare required quite advanced mathematical skills. Knowledge of numbers and the skills to manipulate them became invaluable.

The Natural Numbers

From the time we first learn to speak, we are introduced to the world of mathematics with the ***natural numbers*** 0, 1, 2, 3, ... , sometimes called the ***counting numbers***. We denote the set of all natural numbers by $\mathbb{N}$. We now take for granted the fact that whenever we add or multiply any two natural numbers, the result is always another natural number.

$$\mathbb{N} = \{\, 0, 1, 2, 3, \ldots \,\}$$
$\mathbb{N}$ is closed under addition and multiplication

The Integers

It does not take long to realise that if we try to subtract one natural number from another we do not always obtain another natural number. For example, if the temperature (in degrees) is 4 and falls a further 6, the temperature is –2 which is not a natural number. To include such problems in the list of those we may solve with the available mathematics, we need to introduce the negative integers –1, –2, –3, ... When this is done, we obtain the set of ***integers*** denoted by $\mathbb{Z}$. As before, it is obvious that the sum and product of any two integers are integers.

$$\mathbb{Z} = \{\, \ldots, -3, -2, -1, 0, 1, 2, 3, \ldots \,\}$$
$\mathbb{Z}$ is closed under addition and multiplication

The Rational Numbers

If we attempt to divide one integer by another we do not always obtain another integer. For example, if we divide 5 apples equally between 2 people, each person does not receive an integral number of apples. We need the ***rational numbers*** to cope with such situations.

Chapter 1

Definition   A rational number is one which *can be expressed* in the form $\dfrac{p}{q}$ where p and q are integers and $q \neq 0$.

We denote the set of all rational numbers by $\mathbb{Q}$. It is obvious that the sum and product of two rational numbers are rational numbers.

$$\mathbb{Q} = \left\{ \dfrac{p}{q} \mid p, q \in \mathbb{Z}, q \neq 0 \right\}$$

$\mathbb{Q}$ is closed under addition and multiplication

The numbers $\dfrac{3}{4}, -\dfrac{2}{3}, 6 = \dfrac{6}{1}, 0.\overline{3} = \dfrac{1}{3}, \sqrt{4} = \dfrac{2}{1}$ are all rational, whereas the numbers $\sqrt{2}, \dfrac{2}{\sqrt{3}}, 4\sqrt{5} - 6, \pi$ are not rational. They are said to be *irrational*.

Higher Level

All rational numbers and many irrational numbers are called *algebraic*. Any root of an equation of the form $a_n x^n + a_{n-1} x^{n-1} + \cdots + a_2 x^2 + a_1 x + a_0 = 0$ where $a_0, a_1, a_2, \cdots, a_n$ are integers, is called algebraic. Numbers which are not algebraic are called *transcendental*. The most familiar transcendental number is π, and in this course you will be introduced to a second transcendental number, denoted by e. The value of e is approximately 2.71828.

Joseph Liouville was the first to prove that certain numbers are transcendental. He found that numbers of the form $\dfrac{1}{n} + \dfrac{1}{n^2} + \dfrac{1}{n^6} + \dfrac{1}{n^{24}} + \dfrac{1}{n^{120}} + \cdots$ where n is a number greater than 1 and the exponents 1, 2, 6, 24, 120 are 1, 1×2, 1×2×3, 1×2×3×4, 1×2×3×4×5, etc., are transcendental. The particular case when $n = 10$ is $\dfrac{1}{10} + \dfrac{1}{10^2} + \dfrac{1}{10^6} + \dfrac{1}{10^{24}} + \cdots = 0.110001000000000000000001\ldots\ldots$.

To prove a given number is transcendental is a difficult task, well beyond the requirements of this course. However, to prove a given number is irrational can be a much easier task.

Theorem   The number $\sqrt{2}$ is irrational.

Proof   (The method of *reductio ad absurdum*.)

Properties of Real Numbers

> Suppose that $\sqrt{2}$ is rational. i.e., $\sqrt{2} = \dfrac{p}{q}$ where p and q are integers.
>
> Since we may always reduce a fraction to its lowest terms, we may also assume that p and q are relatively prime, i.e., they have no common positive integer factor other than 1.
>
> Then $p^2 = 2q^2$ and so p^2 is even which makes p even.
> Let $p = 2r$ where r is an integer.
> Then $p^2 = 4r^2 = 2q^2$ which means that $q^2 = 2r^2$.
> Thus q^2 is even and so q is even.
> But this is impossible since p and q do not have 2 as a common factor.
> Therefore $\sqrt{2}$ cannot be rational and so must be irrational.

All rational numbers can be expressed as either a ***finite*** or a ***recurring*** decimal (or as both: $1.4\overline{9} = 1.5\,!!$). For example, $\dfrac{1}{2} = 0.5$ is a finite decimal and $\dfrac{1}{3} = 0.\overline{3}$ is a recurring decimal. An irrational number cannot be expressed in this way.

Example   Express the recurring decimal $0.2\overline{45}$ in the form $\dfrac{p}{q}$ where p and q are integers and $q \neq 0$.

Let $x = 0.2\overline{45}$.
Then $100x = 24.5\overline{45}$.
By subtraction we obtain $99x = 24.3$.
Hence $x = 0.2\overline{45} = \dfrac{243}{990} = \dfrac{27}{110}$, as required.

The Real Numbers

The rational and irrational numbers together make up the set of ***real*** numbers, denoted by $\mathbb{R}$. The sets $\mathbb{N}$, $\mathbb{Z}$ and $\mathbb{Q}$ are all subsets of $\mathbb{R}$. In fact, $\mathbb{N} \subset \mathbb{Z} \subset \mathbb{Q} \subset \mathbb{R}$.

The one-to-one correspondence between the real numbers and the points on the number line is familiar to us all. Corresponding to each real number there is exactly one point on the line: corresponding to each point on the line there is exactly one real number.

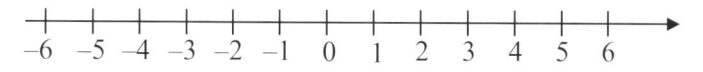

Chapter 1

Throughout this book the following notation for various subsets of $\mathbb{R}$ will be used. In each case a and b are real numbers with $a < b$.

1. $\mathbb{R} =]-\infty, \infty[$
   $(-\infty, \infty)$

2. $x < a$ is denoted by $]-\infty, a[$
   $(-\infty, a)$

3. $x > a$ is denoted by $]a, \infty[$
   (a, ∞)

4. $x \leq a$ is denoted by $]-\infty, a]$
   $(-\infty, a]$

5. $x \geq a$ is denoted by $[a, \infty[$
   $[a, \infty)$

6. $a < x < b$ is denoted by $]a, b[$
   (a, b)

7. $a \leq x \leq b$ is denoted by $[a, b]$
   $[a, b]$

8. $a < x \leq b$ is denoted by $]a, b]$
   $[a, b)$

9. $a \leq x < b$ is denoted by $[a, b[$

Exercise 1.1

1. Determine which of the following numbers are (i) integers; (ii) rational; (iii) irrational.

   $-2, \dfrac{3}{7}, \sqrt{5}, \dfrac{1}{\pi}, \sqrt{441}, \sqrt{6\tfrac{1}{4}}, 0.8, 0.\bar{8}, 1.\bar{9}, \dfrac{1}{\sqrt{3}-1}, \dfrac{4-\sqrt{32}}{\sqrt{18}-3}, \pi^2.$

2. Determine which of the following equations have roots which are (i) integral; (ii) rational; (iii) irrational; (iv) non-real.

   (a) $x^2 - 3x - 10 = 0$;
   (b) $6x^2 + 5x - 6 = 0$;
   (c) $2x^2 - 4x + 3 = 0$;
   (d) $2x^2 - 4x - 3 = 0$;
   (e) $6x^2 - 7x - 10 = 0$;
   (f) $x^2 - 2\sqrt{5}x + 5 = 0$;
   (g) $x^2 - (2+\sqrt{3})x + 2\sqrt{3} = 0$;
   (h) $x^2 - (\pi - 3)x - 3\pi = 0$.

3. Write each of the following numbers in the form $\dfrac{p}{q}$ where p, q are integers:

   $0.04, 1.\bar{2}, \sqrt{12.25}, 0.\overline{23}, 0.1234, 0.\overline{1234}, 0.\overline{285714}, 0.\bar{9}, 1.32169.$

Properties of Real Numbers

1.2   Order Properties and the Modulus Function

If a and b are any two real numbers, then either $a < b$ or $b < a$ or $a = b$.

The sum and product of any two positive real numbers are both positive.

If $a < 0$ then $(-a) > 0$.

If $a > b$ then $(a - b) > 0$. This means that the point on the number line corresponding to a is to the right of the point corresponding to b.

The most elementary rules for inequalities are:

(1)   If $a < b$ and c is any real number then $a + c < b + c$. That is we may add (or subtract) any real number to (or from) both sides of an inequality.

(2)   (i)   If $a < b$ and $c > 0$, then $ac < bc$.

(ii)   If $a < b$ and $c < 0$, then $ac > bc$.

That is we may multiply (or divide) both sides of an inequality by a positive real number, but when we multiply (or divide) both sides of an inequality by a negative real number we must change the direction of the inequality.

(3)   (i)   If $0 < a < b$ then $0 < a^2 < b^2$.

(ii)   If $a < b < 0$ then $0 < b^2 < a^2$.

That is we may square both sides of an inequality if both sides are positive. However if both sides are negative we may square both sides but we must reverse the direction of the inequality and then both sides become positive. If one side is positive and the other negative we cannot square both sides. For example, $-2 < 3$ and $(-2)^2 < 3^2$, but $-3 < 2$ and $(-3)^2 > 2^2$.

(4)   (i)   If $0 < a < b$ then $0 < \dfrac{1}{b} < \dfrac{1}{a}$.

(ii)   If $a < b < 0$ then $\dfrac{1}{b} < \dfrac{1}{a} < 0$.

That is we may take the reciprocal of both sides of an inequality only if both sides have the same sign and, in each possible case, we must reverse the direction of the inequality.

7

Chapter 1

Example Solve the inequality $3x + 2 \leq 5x + 6$.

$3x + 2 \leq 5x + 6$ $\Rightarrow$ $3x \leq 5x + 4$ (subtracting 2 from both sides)
$\Rightarrow$ $-2x \leq 4$ (subtracting 5x from both sides)
$\Rightarrow$ $x \geq -2$. (dividing both sides by –2)

Example Solve the inequality $\dfrac{3x-4}{x+1} \geq 3$.

$\dfrac{3x-4}{x+1} \geq 3$ $\Rightarrow$ $\dfrac{3x-4}{x+1} - 3 \geq 0$

$\Rightarrow$ $\dfrac{3x - 4 - 3(x+1)}{x+1} \geq 0$

$\Rightarrow$ $\dfrac{-7}{x+1} \geq 0$

$\Rightarrow$ $x + 1 < 0$

$\Rightarrow$ $x < -1$.

Higher Level

Example Solve the inequality $0 < \dfrac{3}{x+1} < \dfrac{2}{x}$.

$0 < \dfrac{3}{x+1} < \dfrac{2}{x}$ $\Rightarrow$ $0 < \dfrac{x}{2} < \dfrac{x+1}{3}$, $x > 0$, $x + 1 > 0$

$\Rightarrow$ $0 < 3x < 2x + 2$, $x > 0$

$\Rightarrow$ $0 < x < 2$.

Example Solve the inequality $\dfrac{2}{x+2} < \dfrac{1}{x+3} < 0$.

$\dfrac{2}{x+2} < \dfrac{1}{x+3} < 0$ $\Rightarrow$ $x + 3 < \dfrac{x+2}{2} < 0$, $x + 3 < 0$, $x + 2 < 0$

$\Rightarrow$ $2x + 6 < x + 2 < 0$, $x < -3$

$\Rightarrow$ $x + 4 < 0$, $x < -3$

$\Rightarrow$ $x < -4$.

Properties of Real Numbers

The Modulus Function

Sometimes we order numbers according to their size. We denote the size or absolute value of the real number x by $|x|$, called the ***modulus of x***.

Definition $\quad |x| = \begin{cases} x & \text{if } x \geq 0 \\ -x & \text{if } x < 0. \end{cases}$

The graph of $y = |x|$ is as follows:

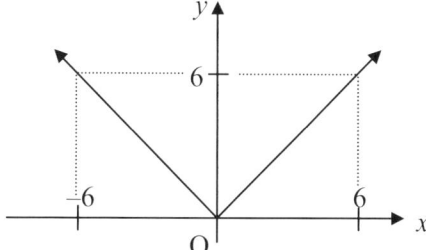

Since the symbol $\sqrt{}$ is used to mean any non-negative square root, we may also define the modulus of the real number x by:

$$|x| = \sqrt{x^2}.$$

Geometrically $|x|$ measures the distance from the point representing the real number x on the number line to the origin, the point representing 0. Thus $|x| < 2$ implies that x is within 2 units of the origin, i.e., $-2 < x < 2$; and $|x| > 2$ implies that x is further than 2 units from the origin, i.e., $x > 2$ or $x < -2$.

In general, if a is any positive number,
$\quad |x| < a \implies -a < x < a$,
and $\quad |x| > a \implies x > a$ or $x < -a$.

Higher Level

The following relations are true for all real numbers a and b:

(1) $|-a| = |a|$
(2) $-|a| \leq a \leq |a|$
(3) $|ab| = |a||b|$ and $|a/b| = |a|/|b|$ $(b \neq 0)$
(4) $|a+b| \leq |a| + |b|$
(5) $|a-b| \geq |a| - |b|$

Chapter 1

Example Solve the inequality $|3x-2|<1$.

$$|3x-2|<1 \Rightarrow -1<3x-2<1$$
$$\Rightarrow 1<3x<3$$
$$\Rightarrow \tfrac{1}{3}<x<1.$$

Example Solve the inequality $\left|\dfrac{2x-1}{x+1}\right|<2$.

$$\left|\dfrac{2x-1}{x+1}\right|<2 \Rightarrow |2x-1|<2|x+1|$$

(Square both sides since both are non-negative.)
$$4x^2-4x+1 < 4(x^2+2x+1)$$
$$-4x+1 < 8x+4$$
$$12x > -3$$
$$x > -\tfrac{1}{4}.$$

Exercise 1.2

1. Find the values of x which satisfy each of the following inequalities:
   (a) $3x-1<11$;
   (b) $1-x\geq 7$;
   (c) $2(3x+2)>5$;
   (d) $1-3(x+2)\leq 10$;
   (e) $\dfrac{x+3}{2}>\dfrac{1-x}{3}$;
   (f) $\dfrac{3x+1}{2}>\dfrac{2x-3}{5}$;
   (g) $\dfrac{6}{x-2}>0$;
   (h) $\dfrac{-1}{3-2x}>0$;
   (i) $\dfrac{x+3}{x-2}<1$;
   (j) $\dfrac{2x+3}{x-1}<2$;
   (k) $\dfrac{5x-1}{2x+1}<\dfrac{5}{2}$;
   (l) $\dfrac{2x-3}{4-3x}>-\dfrac{2}{3}$.

2. Find the values of x which satisfy each of the following inequalities:
   (a) $|x|\leq 3$;
   (b) $|2x|>6$;
   (c) $|3x+2|<8$;
   (d) $|6-x|\leq 4$;
   (e) $|3x+2|+2<0$;
   (f) $|5x+4|>12$;
   (g) $\left|\dfrac{x}{x+2}\right|<1$;
   (h) $2+|x-2|>0$;
   (i) $1-|x+3|<8$;
   (j) $3-|2x+5|>9$;
   (k) $\left|\dfrac{3-2x}{x+1}\right|\geq 2$;
   (l) $\left|\dfrac{4x-3}{2x+1}\right|\leq 2$.

Higher Level

3. Solve each of the following inequalities:
   (a) $0 < \dfrac{5}{x+1} < \dfrac{6}{x}$;
   (b) $\dfrac{3}{x+1} < \dfrac{2}{x+5} < 0$;
   (c) $0 < \dfrac{5}{x-2} < \dfrac{3}{2x+1}$;
   (d) $\dfrac{2}{x+1} < \dfrac{5}{2x+1} < 0$.

4. Decide whether each of the following statements is true or false. If the statement is false, give an example to confirm that it is false.
   (a) $x < a \Rightarrow 2 - x < 2 - a$;
   (b) $x^2 > 4 \Rightarrow x > 2$;
   (c) $x > 2 \Rightarrow x^2 > 4$;
   (d) $x^2 < 4 \Rightarrow x < 2$;
   (e) $x < 2 \Rightarrow x^2 < 4$;
   (f) $x < y \Rightarrow \dfrac{1}{x} > \dfrac{1}{y}$;
   (g) $x < y \Rightarrow \sqrt{x} < \sqrt{y}$;
   (h) $x < y < 0 \Rightarrow \dfrac{1}{x} > \dfrac{1}{y}$;
   (i) $x < y \Rightarrow |x| < |y|$;
   (j) $|x| < |y| \Rightarrow x < y$.

1.3 Roots of Real Numbers and Surds

Square Roots

If $a^2 = b$ then b is **the square of a** and a is **a square root of b**. We say **a** square root of b since both $2^2 = 4$ and $(-2)^2 = 4$, and so 4 has (at least) two square roots, 2 and -2. We write $\sqrt{4} = 2$ to mean that the positive square root of 4 is 2, and we write $-\sqrt{4} = -2$ to mean that the negative square root of 4 is -2.

Thus:
1. The symbol $\sqrt{}$ denotes the positive square root only.
2. Every positive number has two square roots, one of which is positive, and the other negative.
3. The square root of zero is zero.
4. Negative real numbers have no real square roots.

Cube Roots

If $a^3 = b$ then b is **the cube of a** and a is **the cube root of b**. This time we say **the** cube root of b since $2^3 = 8$ but $(-2)^3 = -8$.

Chapter 1

Thus:
1. No definition is needed to specify which cube root is meant since a real number has one and only one cube root in the real number system.
2. Every positive number has one and only one cube root which is positive.
3. The cube root of zero is zero.
4. Every negative number has one and only one cube root which is negative.

Nth Roots

In general, if *n* is an **even** positive integer then in the real number system,
(a) if $a > 0$, *a* has two *n*th roots, one positive and one negative.
(b) if $a = 0$, *a* has one *n*th root which is zero.
(c) if $a < 0$, *a* has no real *n*th roots.

If *n* is an **odd** positive integer (not 1) then in the real number system,
(a) if $a > 0$, *a* has one *n*th root which is positive.
(b) if $a = 0$, *a* has one *n*th root which is zero.
(c) if $a < 0$, *a* has one *n*th root which is negative.

If *n* is a positive integer other than 1 or 2 the radical sign $\sqrt[n]{\ }$ is used to indicate an *n*th root of a real number. If $n = 2$ just the radical $\sqrt{\ }$ is used.

Surds

A surd is an irrational number expressed with a radical (root) sign.

$\sqrt{2},\ 2+\sqrt{7},\ \sqrt[3]{4},\ \dfrac{3}{\sqrt{3}+1}$ are true surds.

$\sqrt{4},\ \sqrt{6\tfrac{1}{4}},\ 1+\sqrt[3]{8},\ \sqrt{0.64},\ \dfrac{7}{\sqrt[4]{16}}$ are written in surd form but are not true surds since they can be written without the radical as 2, 2.5, 3, 0.8 and 3.5 respectively.

The rules used to manipulate surds should be already familiar to the reader. These rules may be summarised as follows:

(1) Product Rule $\quad \sqrt{a} \times \sqrt{b} = \sqrt{ab}$ for $a \geq 0,\ b \geq 0$

e.g., $\sqrt{5} \times \sqrt{x} = \sqrt{5x},\quad \sqrt{2} \times \sqrt{18} = \sqrt{36} = 6$

(2) Converse of Product Rule $\quad \sqrt{ab} = \sqrt{a} \times \sqrt{b}$ for $a \geq 0,\ b \geq 0$

e.g., $\sqrt{5a} = \sqrt{5} \times \sqrt{a},\quad \sqrt{24} = \sqrt{4} \times \sqrt{6} = 2\sqrt{6}$

(3) Quotient Rule $\quad \dfrac{\sqrt{a}}{\sqrt{b}} = \sqrt{\dfrac{a}{b}}$ for $a \geq 0,\ b > 0$

e.g., $\dfrac{\sqrt{2}}{\sqrt{5}} = \sqrt{\dfrac{2}{5}} = \sqrt{0.4},\quad \dfrac{\sqrt{28}}{\sqrt{7}} = \sqrt{\dfrac{28}{7}} = \sqrt{4} = 2$

Properties of Real Numbers

Definition If a, b, c and d are rational with b and d not both squares of rational numbers, then $a\sqrt{b}+c\sqrt{d}$ is a binomial surd. The **conjugate** of the binomial surd $a\sqrt{b}+c\sqrt{d}$ is $a\sqrt{b}-c\sqrt{d}$.

Note that the product of a binomial surd and its conjugate is rational:
$$(a\sqrt{b}+c\sqrt{d})(a\sqrt{b}-c\sqrt{d}) = a^2 b - c^2 d.$$

We often use this result to simplify surd expressions, particularly the quotient of binomial surds.

Example Express $5\sqrt{3}$ as an entire surd.

$$5\sqrt{3} = \sqrt{25} \times \sqrt{3} = \sqrt{75}.$$

Example Express $\sqrt{252}$ in the form $a\sqrt{b}$ where a and b are integers and a is as large as possible.

$$\sqrt{252} = \sqrt{9 \times 28} = \sqrt{9 \times 4 \times 7} = 3 \times 2 \times \sqrt{7} = 6\sqrt{7}.$$

(Factorise the number, choosing the largest possible perfect square that you can *easily* find as a first factor. In this example, if you notice immediately that 36 is a factor of 252 then this slightly shortens the working.)

Example Expand $(\sqrt{5}+2\sqrt{2})(\sqrt{5}-2\sqrt{2})$.

$$(\sqrt{5}+2\sqrt{2})(\sqrt{5}-2\sqrt{2}) = (\sqrt{5})^2 - (2\sqrt{2})^2 = 5 - 8 = -3.$$

Rationalisation

If a simple surd such as $\sqrt{3}$ is multiplied by itself, the result is rational: $\sqrt{3} \times \sqrt{3} = 3$. If a binomial surd is multiplied by its conjugate, the result is also rational. We make use of these results when simplifying certain surd expressions by 'rationalising the denominator'.

Example Express $\dfrac{\sqrt{2}}{\sqrt{5}}$ with a rational denominator.

$$\frac{\sqrt{2}}{\sqrt{5}} = \frac{\sqrt{2}}{\sqrt{5}} \times \frac{\sqrt{5}}{\sqrt{5}} = \frac{\sqrt{10}}{5}, \text{ as required.}$$

Chapter 1

Example Express $\dfrac{3}{3\sqrt{2}-2\sqrt{3}}$ with a rational denominator and in simplest form.

$$\dfrac{3}{3\sqrt{2}-2\sqrt{3}} = \dfrac{3}{3\sqrt{2}-2\sqrt{3}} \times \dfrac{3\sqrt{2}+2\sqrt{3}}{3\sqrt{2}+2\sqrt{3}}$$
$$= \dfrac{3(3\sqrt{2}+2\sqrt{3})}{18-12}$$
$$= \dfrac{3\sqrt{2}+2\sqrt{3}}{2}.$$

Example Find the rational numbers a and b such that $(a+b\sqrt{2})(2+3\sqrt{2})=1$.

Method 1: $\quad (a+b\sqrt{2})(2+3\sqrt{2})=1$
$\Rightarrow \quad 2a+6b+\sqrt{2}(3a+2b)=1=1+0\sqrt{2}$
$\Rightarrow \quad$ Thus $2a+6b=1$ and $3a+2b=0$ since $\sqrt{2}(3a+2b)$ must be zero if a and b are rational. Solving these equations gives $a=-\tfrac{1}{7}$ and $b=\tfrac{3}{14}$.

Method 2: $\quad (a+b\sqrt{2})(2+3\sqrt{2})=1 \Rightarrow a+b\sqrt{2} = \dfrac{1}{2+3\sqrt{2}}$
$$= \dfrac{2-3\sqrt{2}}{4-18}$$
$$= -\tfrac{1}{7}+\tfrac{3}{14}\sqrt{2}.$$

Thus $a=-\tfrac{1}{7}$ and $b=\tfrac{3}{14}$.

Exercise 1.3

1. Write each of the following as a single surd:
   (a) $\sqrt{3}\times\sqrt{5}$; (b) $\sqrt{xy}\times\sqrt{y}$; (c) $\sqrt{3}\times\sqrt{5}\times\sqrt{6}$; (d) $\sqrt[3]{4}\times\sqrt[3]{5}$.

2. Simplify each of the following:
   (a) $\sqrt{6}\times\sqrt{2}$; (b) $\sqrt{15}\times\sqrt{3}$; (c) $\dfrac{\sqrt{15}}{\sqrt{5}}$;
   (d) $\dfrac{\sqrt{12}}{\sqrt{108}}$; (e) $\dfrac{\sqrt{12}\times\sqrt{18}}{\sqrt{8}\times\sqrt{27}}$; (f) $\dfrac{\sqrt{40}\times\sqrt{42}}{\sqrt{72}\times\sqrt{28}}$.

Properties of Real Numbers

3. Write each of the following in the form $a\sqrt{n}$ where a is rational and n is the smallest possible positive integer:
   (a) $\sqrt{18}$;
   (b) $\sqrt{200}$;
   (c) $\sqrt{75}$;
   (d) $\sqrt{72}$;
   (e) $\sqrt{450}$;
   (f) $\sqrt{0.98}$.

4. Simplify each of the following:
   (a) $\sqrt{18} + \sqrt{50} - \sqrt{98}$;
   (b) $4\sqrt{20} - \sqrt{45} - \sqrt{80}$;
   (c) $\sqrt{75} \times \sqrt{12}$;
   (d) $4\sqrt{12} - \sqrt{125} - \sqrt{192}$.

5. Simplify each of the following:
   (a) $2\sqrt{5} \times 3\sqrt{5}$;
   (b) $(3\sqrt{7})^2$;
   (c) $(11\sqrt{6})^2$;
   (d) $\sqrt{3}(2\sqrt{3} - \sqrt{5})$;
   (e) $\sqrt{10}(2\sqrt{5} - 5\sqrt{2})$;
   (f) $\sqrt{12}(2\sqrt{3} - \sqrt{6})$.

6. Expand each of the following:
   (a) $(\sqrt{3} - \sqrt{2})(\sqrt{3} + \sqrt{2})$;
   (b) $(2\sqrt{5} + 3)(2\sqrt{5} - 3)$;
   (c) $(\sqrt{x} - \sqrt{5})(\sqrt{x} + \sqrt{5})$;
   (d) $(a\sqrt{a} + b\sqrt{b})(a\sqrt{a} - b\sqrt{b})$;
   (e) $(\sqrt{7} - \sqrt{2})^2$;
   (f) $(3\sqrt{2} + 2\sqrt{3})^2$;
   (g) $(\sqrt{x-4} + 2)^2$;
   (h) $(\sqrt{\ell + m} + \sqrt{\ell - m})^2$.

7. Express each of the following with rational denominator and in simplest form:
   (a) $\dfrac{1}{\sqrt{3}}$;
   (b) $\dfrac{4}{\sqrt{6}}$;
   (c) $\dfrac{\sqrt{5}}{2\sqrt{15}}$;
   (d) $\dfrac{1}{\sqrt{2}-1}$;
   (e) $\dfrac{6}{\sqrt{5}+\sqrt{2}}$;
   (f) $\dfrac{6}{3\sqrt{2}+2\sqrt{3}}$;
   (g) $\dfrac{\sqrt{3}+\sqrt{2}}{\sqrt{3}-\sqrt{2}}$;
   (h) $\dfrac{\sqrt{5}}{6-2\sqrt{5}}$;
   (i) $\dfrac{\sqrt{2}-\sqrt{3}}{\sqrt{3}+2\sqrt{2}}$.

8. Solve each of the following equations, expressing each answer in simplest surd form:
   (a) $2x - \sqrt{3} + 2(1 - x\sqrt{3}) = 5 + 2\sqrt{3}$;
   (b) $2\sqrt{2} - x - (x\sqrt{2} - 2) = 3$;
   (c) $\dfrac{1+2x}{2x-1} = \sqrt{2}$;
   (d) $\dfrac{x-\sqrt{3}}{2x+\sqrt{3}} = \sqrt{3}$.

9. Find the values of the rational numbers a and b for which
   (a) $(a + b\sqrt{3})(5 - 2\sqrt{3}) = 1$;
   (b) $(a + 2\sqrt{3})(4 - b\sqrt{3}) = 4$.

15

Chapter 1

Higher Level

1.4 Irrational Equations

Consider the equation $x - 2 = 4$. Clearly $x = 6$ is the only solution. But if we square both sides we obtain the equation

$$x^2 - 4x + 4 = 16 \quad \ldots\ldots\ldots\ldots (1)$$
$$x^2 - 4x - 12 = 0$$
$$(x - 6)(x + 2) = 0$$

Thus $x = 6$ or $x = -2$.

But we know that $x = -2$ is **not** a solution of our original equation. In fact this so-called '*false solution*' was introduced when we squared both sides of our original equation. This comes about because $(-4)^2 = 16$ just as $4^2 = 16$. Thus the equation $x - 2 = -4$ leads, on squaring, to the same equation (1) as before. The process of squaring both sides is used in solving irrational equations, i.e., equations involving surd expressions. It is therefore *always* necessary to check any solutions obtained.

Example   Solve the equation $\sqrt{2x + 10} = 9 - \sqrt{2x + 55}$.

Square both sides: $2x + 10 = 81 - 18\sqrt{2x + 55} + 2x + 55$
$\Rightarrow$ $18\sqrt{2x + 55} = 126$
$\Rightarrow$ $\sqrt{2x + 55} = 7$.
Square both sides again: $2x + 55 = 49$
$\Rightarrow$ $x = -3$.

Check: If $x = -3$, LHS $= \sqrt{2x + 10} = \sqrt{4} = 2$, and
RHS $= 9 - \sqrt{2x + 55} = 9 - \sqrt{49} = 9 - 7 = 2$, which checks.
Therefore $x = -3$ is a solution.

Example   Solve the equation $\sqrt{3x + 4} - \sqrt{8 - x} = 2$.

Transpose one surd: $\sqrt{3x + 4} = 2 + \sqrt{8 - x}$.
Square both sides: $3x + 4 = 4 + 4\sqrt{8 - x} + 8 - x$
$\Rightarrow$ $4x - 8 = 4\sqrt{8 - x}$
$\Rightarrow$ $x - 2 = \sqrt{8 - x}$.

Properties of Real Numbers

Square both sides again:
$$x^2 - 4x + 4 = 8 - x$$
$$\Rightarrow \quad x^2 - 3x - 4 = 0$$
$$\Rightarrow \quad (x-4)(x+1) = 0.$$

Whence $x = 4$ or $x = -1$.

Check:
If $x = 4$, $\sqrt{3x+4} - \sqrt{8-x} = \sqrt{16} - \sqrt{4} = 4 - 2 = 2$, which checks.
If $x = -1$, $\sqrt{3x+4} - \sqrt{8-x} = \sqrt{1} - \sqrt{9} = 1 - 3 = -2$, which does not check.
Therefore $x = 4$, only.

Exercise 1.4

1. Solve each of the following equations, checking carefully all solutions obtained:
   (a) $\sqrt{x+4} = 4$;
   (b) $\sqrt{3x-2} = 7$;
   (c) $\sqrt{3x-2} = \sqrt{1-2x}$;
   (d) $4\sqrt{1-x} = 3\sqrt{1-2x}$;
   (e) $3\sqrt{x+1} - 2\sqrt{3x-1} = 0$;
   (f) $\sqrt{6x-1} - 3\sqrt{1-x} = 0$.

2. Solve the following equations:
   (a) $\sqrt{x^2+9} = \sqrt{2x^2-7}$;
   (b) $\sqrt{2x^2-6x} = x-3$;
   (c) $\sqrt{2x^2-1} = \sqrt{3x+4}$;
   (d) $\sqrt{2x+1} = 1 - \sqrt{x}$.

3. Solve the following equations:
   (a) $\sqrt{x+2} + \sqrt{x+18} = 8$;
   (b) $\sqrt{2x-2} + \sqrt{2x+1} = \sqrt{4x+3}$;
   (c) $\sqrt{16x+1} - 3 = 2\sqrt{3x-5}$;
   (d) $\sqrt{x+3} + \sqrt{4x+1} = 2$;
   (e) $\sqrt{x+3} - 2\sqrt{x+1} = \sqrt{x+2}$;
   (f) $\sqrt{x} + \sqrt{5x-4} = \sqrt{3x+1}$.

1.5 Exponents

The rules used to manipulate exponential expressions should already be familiar to the reader. These rules may be summarised as follows:

(1) Product Rule $\qquad (a^m)(a^n) = a^{m+n}$

(2) Quotient Rule $\qquad \dfrac{a^m}{a^n} = a^{m-n} \quad (a \neq 0)$

(3) Power of a Power Rule $\qquad (a^m)^n = a^{mn}$

(4) Power of a Product Rule $\qquad (ab)^m = a^m b^m$

Chapter 1

(5)   Power of a Quotient Rule $\left(\dfrac{a}{b}\right)^m = \dfrac{a^m}{b^m}$ $(b \neq 0)$.

For the quotient rule to hold when $m = n$ we must agree that $\dfrac{a^m}{a^m} = a^{m-m} = a^0$, $(a \neq 0)$. Then it is natural to define $a^0 = 1$ $(a \neq 0)$.

Now $\dfrac{1}{a^n} = \dfrac{a^0}{a^n} = a^{0-n} = a^{-n}$ $(a \neq 0)$, which gives meaning to negative exponents.

Rational Exponents

If the product rule for exponents is to hold, then $\left(a^{1/2}\right)\left(a^{1/2}\right) = a^{\frac{1}{2}+\frac{1}{2}} = a^1 = a$. But we know that $\left(\sqrt{a}\right)\left(\sqrt{a}\right) = a$ $(a \geq 0)$. Thus $a^{1/2}$ is a square root of a (provided $a \geq 0$). To avoid any confusion, take $a^{1/2} = \sqrt{a}$ $(a \geq 0)$, i.e., $a^{1/2}$ is taken to be the positive square root of a $(a > 0)$.

More generally, if n is a positive integer, $a^{1/n}$ is defined by $a^{1/n} = \sqrt[n]{a}$.
Note that if n is *even*, a must be non-negative; if n is *odd*, a may be any real number.
Next we define $a^{m/n} = \left(a^{1/n}\right)^m = \left(\sqrt[n]{a}\right)^m$ which exists for all a if n is *odd* and exists for $a \geq 0$ if n is *even*.

Example   Evaluate $27^{5/3}$.

$$27^{5/3} = \left(\sqrt[3]{27}\right)^5 = 3^5 = 243.$$

Example   Evaluate $0.064^{-4/3}$.

$$0.064^{-4/3} = \left(\sqrt[3]{0.064}\right)^{-4} = 0.4^{-4} = \left(\dfrac{5}{2}\right)^4 = \dfrac{625}{16}.$$

Example   Solve the following equations:
(a) $8^{1-x} = 4^{2x+3}$;   (b) $a^{3/2} = 8$;   (c) $a^{2/3} = 16$.

(a) $8^{1-x} = 4^{2x+3}$
Here we may express each side as a power of 2.
$\left(2^3\right)^{1-x} = \left(2^2\right)^{2x+3}$
$\Rightarrow$ $2^{3-3x} = 2^{4x+6}$
$\Rightarrow$ $3 - 3x = 4x + 6$
$\Rightarrow$ $x = -3/7$.

Properties of Real Numbers

(b) $\qquad a^{3/2} = 8$
$\Rightarrow \qquad a^{1/2} = \sqrt[3]{8} = 2$
$\Rightarrow \qquad a = 4.$

(c) $\qquad a^{2/3} = 16$
$\Rightarrow \qquad a^{1/3} = \pm 4$
$\Rightarrow \qquad a = \pm 64.$

Exercise 1.5

1. Express each of the following in simpler form:
   (a) $(a^2)(a^4)$;
   (b) $(b^3)(b^4)(b^2)$;
   (c) $(x^2)(x^{-2})$;
   (d) $\dfrac{a^3}{a}$;
   (e) $\dfrac{a^2}{a^3}$;
   (f) $\dfrac{(x^{-3})(y^{-2})}{(x^{-4})(y^{-1})}$.

2. Write each of the following without brackets or negative exponents:
   (a) $(x^2)^3$;
   (b) $(x^{-2})^3$;
   (c) $(x^{-1})^{-2}$;
   (d) $(x^2 y^3)^2$;
   (e) $(3a^{-2}b)^2$;
   (f) $(x^2 - y)^2$;
   (g) $\left(\dfrac{x^2}{y^3}\right)^{-1}$;
   (h) $\left(\dfrac{2x^{-2}}{y^{-3}}\right)^3$.

3. Write each of the following without exponents:
   (a) $4^{3/2}$;
   (b) $16^{3/4}$;
   (c) $9^{-1/2}$;
   (d) $0.04^{3/2}$;
   (e) $0.01^{-3/2}$;
   (f) $(x^2)^{1/2}$.

4. Solve each of the following equations:
   (a) $2^{x-1} = 16$;
   (b) $3^x = \dfrac{1}{81}$;
   (c) $9^{1-x} = 1$;
   (d) $5^{x-1} = 0.008$;
   (e) $4^x = \dfrac{1}{8}$;
   (f) $2^{2-x} = 0.25$.

5. Solve each of the following equations:
   (a) $a^2 = 4$;
   (b) $a^3 = 27$;
   (c) $a^{1/2} = -4$;
   (d) $a^{3/2} = 27$;
   (e) $a^{4/3} = 16$;
   (f) $a^{2/3} = 0.25$;
   (g) $a^{3/5} = 64$;
   (h) $5a^3 = 0.04$;
   (i) $3a^{4/3} = \dfrac{1}{27}$.

19

Chapter 1

Higher Level

6. Solve each of the following equations:
   - (a) $x - 5x^{1/2} + 4 = 0$;
   - (b) $2^{x/2} + 4(2^{-x/2}) = 5$;
   - (c) $9^x - 10(3^x) + 9 = 0$;
   - (d) $2^{4x+3} - 33(2^{2x-1}) + 1 = 0$;
   - (e) $x^{4/3} - 9x^{2/3} + 8 = 0$;
   - (f) $x^3 + 19x^{3/2} = 216$;
   - (g) $8x^{3/2} - 27x^{-3/2} = 215$;
   - (h) $36x^{2/3} + 36x^{-2/3} = 97$.

1.6 Exponential and Logarithmic Functions

Functions of the form $f(x) = a^x$ $(a > 0, \ a \neq 1)$, are called **exponential functions**. If $a > 1$ the graph of $f(x) = a^x$ has the form:

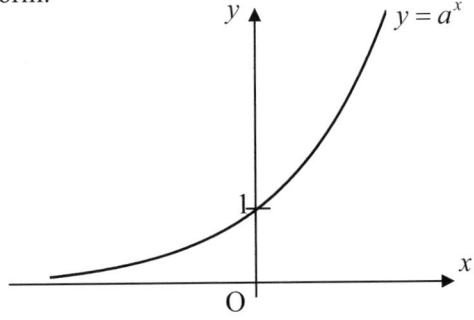

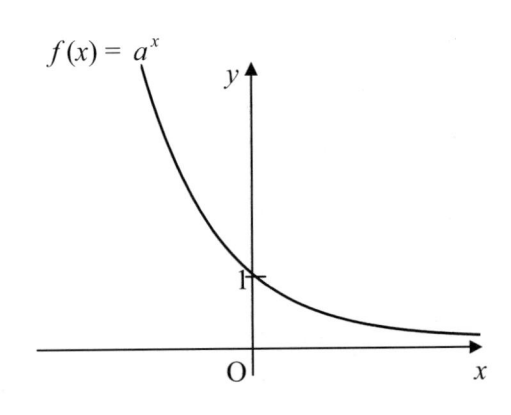

This is a function with domain $\left]-\infty, \infty\right[$, i.e., $\mathbb{R}$, and range $\left]0, \infty\right[$, i.e., $\mathbb{R}^+$. The x-axis is an asymptote and the y-intercept is 1.

If $0 < a < 1$ the graph of the function $f(x) = a^x$ has the form:

20

Properties of Real Numbers

This is a function with domain $]-\infty, \infty[$, i.e., $\mathbb{R}$, and range $]0, \infty[$, i.e., $\mathbb{R}^+$. The x-axis is an asymptote and the y-intercept is 1.

The second graph may be obtained from the graph of $y = b^x$ where $b = \dfrac{1}{a} > 1$ by reflecting in the y-axis.

These functions often occur in nature.

Since the function $y = a^x$ ($a > 0$ or $0 < a < 1$) is also one-to-one (no horizontal line crosses the graph in more than one place), its inverse is also a function.

The inverse of the function $y = a^x$ has an equation $x = a^y$. When we make y the subject of the formula (i.e., write y as a function of x), we write $y = \log_a x$, i.e., y is the logarithm of x in base a. The inverses of exponential functions are logarithmic functions.

The graph of a logarithmic function is found by reflecting the graph of the corresponding exponential function in the line $y = x$.

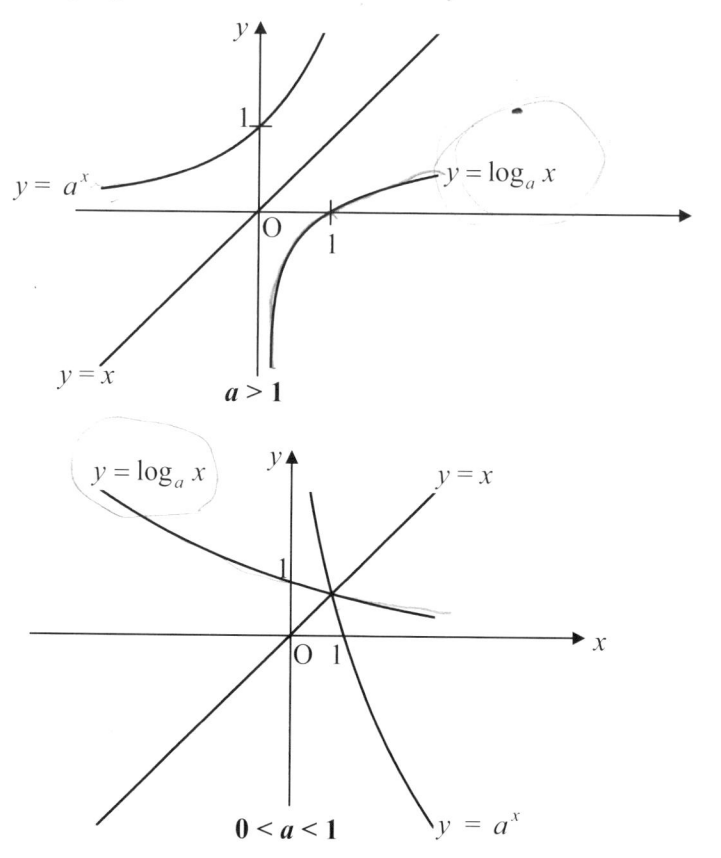

Chapter 1

The domain of $y = \log_a x$ is $]0, \infty[$, i.e., $\mathbb{R}^+$, and the range is $]-\infty, \infty[$, i.e., $\mathbb{R}$.

Each exponential expression has a corresponding logarithmic expression.

The relationship is $\quad y = a^x \Leftrightarrow x = \log_a y$.

Thus we may write $y = a^{\log_a y}$, and so the logarithm of any positive number y in base a is equal to the exponent needed to express y as a power of a.

For example, $10^2 = 100$, and $\log_{10} 100 = 2$; $\log_{10} 2 \approx 0.301$, and $10^{0.301} \approx 2$.

Example (a) Write $4^3 = 64$ in logarithmic form.
(b) Write $\log_m p = q$ in exponential form.

(a) $4^3 = 64 \quad \Rightarrow \quad \log_4 64 = 3$.
(b) $\log_m p = q \quad \Rightarrow \quad p = m^q$.

Example Find the numerical value of $\log_3 \sqrt[3]{9}$.

Let $y = \log_3 \sqrt[3]{9}$, then $3^y = \sqrt[3]{9} = 3^{2/3} \Rightarrow y = \tfrac{2}{3} \Rightarrow \log_3 \sqrt[3]{9} = \tfrac{2}{3}$.

The Laws of Logarithms

1. **The Multiplication Law** $\qquad \log_a(mn) = \log_a m + \log_a n$

   Proof If $x = \log_a m$, $y = \log_a n$ and $z = \log_a(mn)$ then $m = a^x$, $n = a^y$ and $mn = a^z$.
   Now $mn = a^x a^y = a^{x+y} = a^z \Rightarrow z = x + y$.
   Thus $\log_a(mn) = \log_a m + \log_a n$, as required.

2. **The Division Law** $\qquad \log_a\left(\dfrac{m}{n}\right) = \log_a m - \log_a n$

   Proof If $x = \log_a m$, $y = \log_a n$ and $z = \log_a\left(\dfrac{m}{n}\right)$ then $m = a^x$, $n = a^y$ and $\dfrac{m}{n} = a^z$.

Now $\dfrac{m}{n} = \dfrac{a^x}{a^y} = a^{x-y} = a^z \Rightarrow z = x - y$.

Thus $\log_a\left(\dfrac{m}{n}\right) = \log_a m - \log_a n$, as required.

3. **The Power Law** $\qquad \log_a(m^p) = p\log_a m$

   Proof If $x = \log_a m$ and $z = \log_a(m^p)$ then $m = a^x$, $m^p = a^z$.

   Now $m^p = (a^x)^p = a^{px} = a^z \Rightarrow z = px$.

   Thus $\log_a(m^p) = p\log_a m$, as required.

4. **The Change of Base Law** $\qquad \log_a b = \dfrac{\log_c b}{\log_c a}$

   Proof If $x = \log_a b$ then $a^x = b$.

   Take logarithms in base c of both sides: $\log_c(a^x) = \log_c b$.

   This gives $x\log_c a = \log_c b \Rightarrow x = \dfrac{\log_c b}{\log_c a}$.

   Thus $\log_a b = \dfrac{\log_c b}{\log_c a}$, as required.

Note: The change of base rule is very useful since all logarithmic calculations (even using (some) graphics calculators) are performed either in base 10, or in base e (e ≈ 2.71828). In this text we will denote $\log_{10} x$ by $\log x$ and $\log_e x$ by $\ln x$ (the 'natural' logarithm of x). The student should be aware that some texts may use $\log x$ for the natural logarithm of x and not $\ln x$. In any such text, a base-10 logarithm must indicate the base, although $\lg x$ may sometimes be used.

Example If $y = 3x^2$, find a linear expression connecting $\log x$ and $\log y$.

Since $y = 3x^2$, then $\log y = \log(3x^2) = \log 3 + \log(x^2) = \log 3 + 2\log x$.

Example Write an expression equivalent to $\log y = 3 - 2\log x$ without using logarithms.

$\log y = 3 - 2\log x = \log 1000 - \log(x^2) = \log\left(\dfrac{1000}{x^2}\right) \Rightarrow y = \dfrac{1000}{x^2}$.

Chapter 1

Example Write an expression equivalent to $\log y = 2 + 0.301x$ without using logarithms.

$\log y = 2 + 0.301x$ is equivalent to $y = 10^{(2+0.301x)}$, and in fact this answer will do. However a 'neater' answer can be found as follows:
$\log y = 2 + 0.301x \approx \log 100 + x\log 2 = \log 100 + \log(2^x) = \log((100)2^x)$
Thus $y \approx (100)2^x$.

Example Evaluate $\log_5 50$.

As 50 cannot be expressed as a simple power of 5, a change of base can be used here.
$$\log_5 50 = \frac{\log 50}{\log 5} = 2.43.$$

Example Solve the equations (a) $5^x = 0.4$; (b) $5^x = 0.4$.

(a) Since $0.04 = \frac{1}{25} = 5^{-2}$ we do not need logarithms.

Here $5^x = 5^{-2}$ and so $x = -2$.

(b) 0.4 cannot (easily) be written as a power of 5 and so logarithms should be used.
$$5^x = 0.4 \implies x\log 5 = \log 0.4 \implies x = \frac{\log 0.4}{\log 5} = -0.569.$$

Example Solve the equation $2^{3x} = 3^{2x-1}$.

Since $2^{3x} = 3^{2x-1}$, then $3x\log 2 = (2x-1)\log 3$.
Thus $3x\log 2 = 2x\log 3 - \log 3$
$2x\log 3 - 3x\log 2 = \log 3$
$x(2\log 3 - 3\log 2) = \log 3$.
This gives $x = \frac{\log 3}{2\log 3 - 3\log 2} = 9.33$.

Growth and Decay

There are many situations encountered in real life in which the exponential function provides the most accurate mathematical model.

Properties of Real Numbers

Example The amount, $A(t)$ gram, of radioactive material in a sample after t years is given by $A(t) = 80\left(2^{-t/100}\right)$.

(a) Find the amount of material in the original sample.
(b) Calculate the half-life of the material. [The *half-life* is the time taken for half of the original material to decay.]
(c) Calculate the time taken for the material to decay to 1 gram.

(a) The original amount of material present is
$A(0) = 80\left(2^0\right) = 80$ gram.

(b) For the half-life, $A(t) = 40$.
$\Rightarrow 40 = 80\left(2^{-t/100}\right)$
$\Rightarrow \dfrac{1}{2} = 2^{-t/100}$
$\Rightarrow 2 = 2^{t/100}$
$\Rightarrow \dfrac{t}{100} = 1$
$\Rightarrow t = 100$.
Therefore the half-life is 100 years.

(c) $A(t) = 1 \Rightarrow 80\left(2^{-t/100}\right) = 1$
$\Rightarrow 2^{-t/100} = \dfrac{1}{80}$
$\Rightarrow 2^{t/100} = 80$
$\Rightarrow \dfrac{t}{100}\log 2 = \log 80$
$\Rightarrow t = \dfrac{100\log 80}{\log 2} = 632.$

Therefore it will take 632 years for the material to decay to 1 gram.

Higher Level

The Empirical Determination of Formulae

Example Values of two related quantities are measured in the laboratory. The collected data is given in the following table:

x	0.8	2.4	3.8	5.2	8.5
y	78	8.7	3.5	1.8	0.69

Draw the graph of the values of log y against those of log x, and use this graph to find a relationship between x and y which does not use logarithms.

From the data found in the laboratory, we obtain the following table of corresponding values of log x and log y:

log x	−0.10	0.38	0.58	0.72	0.93
log y	1.89	0.94	0.54	0.26	−0.16

The graph of log y against log x is as follows:

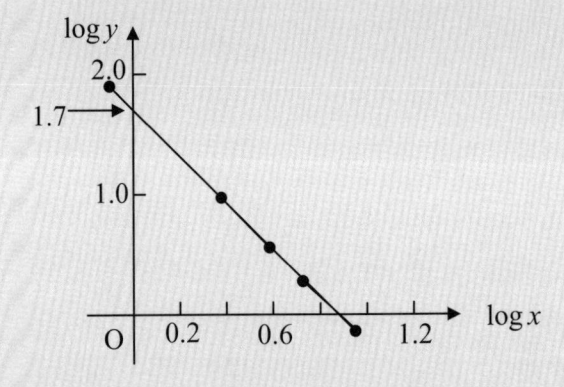

Since the graph of log y against log x is a straight line, log y and log x can be connected by a relationship of the form

$$\log y = m \log x + c$$

where m is the gradient and c is the vertical axis intercept.

From the graph $m = \dfrac{1.89 + 0.16}{-0.10 - 0.93} = \dfrac{2.05}{-1.03} \approx -2$ and $c = 1.7$.

Therefore $\log y = -2\log x + 1.7 \approx \log(x^{-2}) + \log 50 = \log\left(\dfrac{50}{x^2}\right)$.

Thus the required relationship is $y = \dfrac{50}{x^2}$.

Exercise 1.6

1. Write each of the following in logarithmic form:
   (a) $3^2 = 9$;
   (b) $2^{-2} = \dfrac{1}{4}$;
   (c) $p = q^3$;
   (d) $\left(\dfrac{1}{2}\right)^x = y$;
   (e) $k^3 = \ell$;
   (f) $5^{-p} = q$.

2. Write each of the following in exponential form:
   (a) $\log_3 81 = 4$;
   (b) $\log_2\left(\dfrac{1}{2}\right) = -1$;
   (c) $\log_p q = -2$;
   (d) $\log_5 t = u$;
   (e) $\log_{1/2} y = z$;
   (f) $\log_b c = d$.

3. Find the numerical value of each of the following:
   (a) $\log_2 32$;
   (b) $\log_4 8$;
   (c) $\log_6 7$;
   (d) $\log_5 \sqrt{125}$;
   (e) $\log_5 0.008$;
   (f) $\log_9 10$.

4. Find a linear expression connecting $\log x$ and $\log y$ in each of the following:
   (a) $y = 7x^3$;
   (b) $y = 10x^{-2}$;
   (c) $y = ax^t$.

5. Find a relationship between x and y which does not use logarithms in each of the following:
   (a) $\log y = \log 2 + 3\log x$;
   (b) $\log y = \log 5 - \dfrac{1}{2}\log x$;
   (c) $\log y = 2 + 3\log x$;
   (d) $\log y = 0.699 + \dfrac{1}{4}\log x$.

6. Solve each of the following equations:
   (a) $6^x = 3$;
   (b) $8^x = 0.5$;
   (c) $9^{x-2} = 1$;
   (d) $0.3^x = 4$;
   (e) $3^{2x+4} = 5^{x-2}$;
   (f) $16^{1-2x} = 0.125$.

7. The amount, $A(t)$ gram, of radioactive material in a sample after t years is given by the following formulae. In each case find the time taken for half of the material to decay (i.e., find the half-life).
   (a) $A(t) = 250 \times 2^{-t/10}$;
   (b) $A(t) = 50 \times 10^{-t/20}$;
   (c) $A(t) = A(0) \times 5^{-t/25}$;
   (d) $A(t) = A(0) \times e^{-2t/75}$.

Chapter 1

8. The weight, $W(t)$ gram, of bacteria in a certain culture t hours after it was established is given by $W(t) = 0.10 \times e^{t/10}$. Find the time taken for the amount of bacteria to double.

9. The speed, $V(t)$, of a certain chemical reaction at t °C is given by $V(t) = V(0) \times 5^{t/30}$. At what temperature will the speed of reaction be twice that at 0°C?

10. The number, $N(t)$, of bacteria present in a culture t minutes after it is established is given by $N(t) = 500e^{t/100}$. Find the time taken for the number of bacteria in the culture to double.

11. The population of a city, $P(n)$, n years after the population was p is given by $P(n) = p\left(e^{n/30}\right)$. Find
    (a) the time taken for the population to double ;
    (b) the time taken for the population to reach 1 million from an original population of 10 000.

12. If W_0 gram of radioactive substance decays to $W(t)$ gram in t years and k is the half-life of the material, then it is known that $W(t) = W_0 \times 2^{-t/k}$. Find the half-life of the substance, 50 gram of which decays to 49 gram in ten years.

13. The speed, $S(t)$ m s^{-1}, at which a man falls t seconds after jumping from a plane is given by $S(t) = 48\left(1 - 2^{-0.3t}\right)$. After how long is the man falling at
    (a)   24 m s^{-1} ;   (b)   30 m s^{-1} ;   (c)   45 m s^{-1} ?

Higher Level

14. The temperature of hot coffee in a container, $T(t)$ °C, t minutes after it is placed in a room whose ambient temperature is A°C, is given by $T(t) = A + Be^{-t/k}$. Five minutes after the coffee with an initial temperature of 100°C is placed in a room, the temperature of the coffee is 60°C and five minutes after that the temperature is 40°C.
    (a)   Find A, B and k.
    (b)   Find the time taken for the temperature of the coffee to reach 30°C.
    (c)   Find the temperature of the coffee 20 minutes after the coffee was placed in the room.
    (d)   Find the time taken for the coffee to reach a temperature which is within 1°C of the ambient room temperature.

Properties of Real Numbers

15. The times of oscillation for simple pendulums of various lengths were found by experiment, and the following results obtained:

Length (ℓ cm)	100	80	65	49	30	20	10
Time (t s)	2.00	1.79	1.61	1.40	1.10	0.89	0.63

    Draw a graph of log t against log ℓ and find a formula connecting t and ℓ which does not contain logarithms.

16. If a given mass of gas is compressed, or expanded, suddenly, Boyle's Law (pV = constant) does not hold. The following table shows how pressure and volume were connected in such a case:

V	100	80	60	40	20
p	14	19.1	28.6	50.5	133

    Draw the graph of log p against log V to find the formula which holds in this case.

17. Values of x and corresponding values of y are given in the following table:

x	2	4	6	8	10
y	18.4	42.2	68.7	97.0	127

    By drawing the graph of log y against log x, show that $y = kx^n$ where k and n are real numbers. Use your graph to find the values of k and n.

18. Two variables x and y are thought to satisfy a relationship of the form $y = A(b^{-x})$ where A and b are constants. Some values of x and the corresponding values of y are found by experiment, and are given in the following table.

x	1	2	3	4	5	6
y	184	67.7	24.9	9.16	3.37	1.24

    By plotting a graph of log y against x, confirm the suspected relationship and find the values of A and b.

Chapter 1

Required Outcomes

After completing this chapter, a student should be able to
- convert a repeating decimal to rational form.
- solve inequalities which can be reduced to linear form, including those involving the modulus function.
- simplify expressions containing surds.
- solve irrational equations which simplify to linear or quadratic form. **(HL)**
- solve simple exponential equations.
- convert a number from logarithmic form to exponential form.
- change the base of any logarithm.
- determine the relationship between two variables from a graph. **(HL)**

2 Coordinate Geometry in 2 Dimensions

2.1 The General Form of the Equation of a Straight Line

Whenever A, B and C are constants with A and B not both zero, the graph of the equation $Ax + By + C = 0$ is a straight line.

This form of the equation of a straight line is called the ***general form***.

Now $By = -Ax - C$ and so $y = -\dfrac{A}{B}x - \dfrac{C}{B}$ $(B \neq 0)$.

This equation is now in the form $y = mx + b$ (called the ***slope-intercept*** form), and so we see that the ***gradient*** of the line $Ax + By + C = 0$ is $-\dfrac{A}{B} = -\dfrac{\text{coefficient of } x}{\text{coefficient of } y}$.

We are able to use this information to enable us to **write down** the equation of any straight line in its general form once we know its gradient and any point on it.

Example   Find the equation of the straight line which passes through the point $(2, -3)$ and has gradient $-\tfrac{1}{3}$.

Since the gradient is $-\tfrac{1}{3}$ we may choose $A = 1$ and $B = 3$, and so the equation of the line is $x + 3y + C = 0$.
But the point $(2, -3)$ lies on the line and so $x + 3y = 2 + 3(-3) = -7$.
Thus the required equation is $x + 3y = -7$ or $x + 3y + 7 = 0$.

Note: With practice, all the required steps can be carried out at once.

Example   Write down the equation of the straight line with gradient $-\tfrac{3}{2}$ and passing through the point $(4, 1)$.

The required equation is $3x + 2y = 14$. $(3 \times 4 + 2 \times 1 = 14)$

Example   Find the equation of the straight line passing through $A(-2, 5)$ and $B(4, 17)$.

The gradient of $(AB) = \dfrac{17 - 5}{4 + 2} = 2$.   (Let $A = 2$ and $B = -1$.)
The equation is $2x - y = -9$. $(2 \times -2 - 5 = -9$ or $2 \times 4 - 17 = -9)$

Chapter 2

Exercise 2.1

1. Write down the gradient of each of the following straight lines:
   (a) $5x + 4y + 3 = 0$;
   (b) $3x + y = 10$;
   (c) $4x - 3y + 2 = 0$;
   (d) $3x + 5 = 2y$;
   (e) $3y = 6x - 5$;
   (f) $11x + 5y = 9$;
   (g) $5x - 7y + 35 = 0$;
   (h) $9y + 5x = 27$;
   (i) $y = 7 - 3x$;
   (j) $5x + 8 = 0$;
   (k) $7 - 8y = 0$;
   (l) $3(x - 2) + 7(2 - y) = 21$.

2. Write down the equation of the straight line passing through the given point and with the given gradient in each of the following:
   (a) $(4, 3), -\frac{3}{4}$;
   (b) $(5, -2), -3$;
   (c) $(-2, -3), \frac{1}{2}$;
   (d) $(7, -1), 4$;
   (e) $(0, 0), 4\frac{1}{2}$;
   (f) $(6, -6), -3\frac{1}{3}$.

3. Find the equation of the straight line passing through each of the following pairs of points:
   (a) $(2, 1), (5, 3)$;
   (b) $(-1, -4), (7, 4)$;
   (c) $(4, 0), (0, 3)$;
   (d) $(-2, 3), (-2, -7)$;
   (e) $(-4, 1), (5, 1)$;
   (f) $(a, b), (b, a), a \neq b$.

4. Find the equation of the line through the given point and parallel to the given line in each of the following:
   (a) $(6, 2), 4x - 3y = 0$;
   (b) $(-2, 3), 2x + 5y = 10$;
   (c) $(-3, -4), 7x + 3y = 0$;
   (d) $(4, -5), y = 5 - 4x$.

5. Find the equation of the line through the given point and perpendicular to the given line in each of the following:
   (a) $(3, 2), 2x + 3y = 7$;
   (b) $(-1, -3), y = 3x + 5$;
   (c) $(0, -3), 5x - 4y = 20$;
   (d) $(7, 2), 5y - 3x = 2$.

6. Find the coordinates of the point of intersection of the lines (AB) and (CD) where $A = (7, -2), B = (0, 4), C = (5, 7)$ and $D = (-9, -2)$.

7. Find the coordinates of the orthocentre of the triangle ABC in which $A = (0, 4), B = (2, 9)$ and $C = (6, 3)$.
   [The **orthocentre** of a triangle is the point of intersection of the three altitudes.]

8. Find the coordinates of the circumcentre of the triangle ABC in which $A = (1, 0), B = (5, 0)$ and $C = (3, 4)$.
   [The *circumcentre* of a triangle is the point of intersection of the three perpendicular bisectors of the sides.]

Coordinate Geometry in 2 Dimensions

9. Find the coordinates of the centroid of the triangle ABC in which A = (−6, −3), B = (3, −7) and C = (0, 6).

[The *centroid* of a triangle is the point of intersection of the three medians – the lines joining each vertex to the mid-point of the opposite side.]

10. The triangle ABC has vertices A(1, −1), B(1, 3) and C(5, 7). Determine the coordinates of the orthocentre, the circumcentre and the centroid of the triangle, and show that these points are collinear. [In fact the orthocentre, circumcentre and centroid of *every* triangle are collinear.]

Higher Level

2.2 The Distance of a Point from a Line

Consider the point $P(x_1, y_1)$ which does not lie on the straight line with equation $Ax + By + C = 0$ (A, B not both zero). Let $Q(x, y)$ be that point on the line which is closest to P, and let the length of the line segment [PQ] be d.

Theorem   The shortest distance from the point $P(x_1, y_1)$ to the straight line $Ax + By + C = 0$ is given by $d = \dfrac{|Ax_1 + By_1 + C|}{\sqrt{A^2 + B^2}}$.

Proof

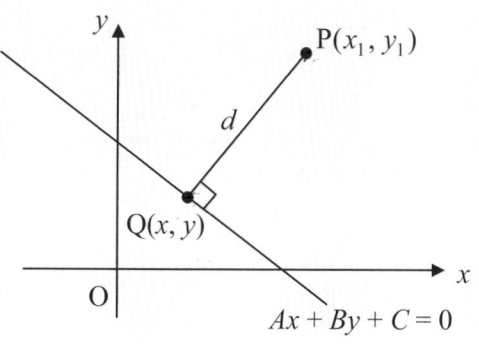

The gradient of the given line is $-\dfrac{A}{B}$, and so the gradient of (PQ) is $\dfrac{B}{A}$.

Thus $\dfrac{y - y_1}{x - x_1} = \dfrac{B}{A}$.

Let $x - x_1 = Ak$ and $y - y_1 = Bk$ for some real number k. ....(*)

Now $\begin{aligned}d^2 &= (x - x_1)^2 + (y - y_1)^2 \\ &= A^2 k^2 + B^2 k^2 \\ &= k^2(A^2 + B^2).\end{aligned}$

Thus $d = |k|\sqrt{A^2 + B^2}$.

Chapter 2

From (*), $x = x_1 + Ak$ and $y = y_1 + Bk$, and Q(x, y) lies on $Ax + By + C = 0$.

Hence $A(x_1 + Ak) + B(y_1 + Bk) + C = 0$ giving $k = \dfrac{-(Ax_1 + By_1 + C)}{A^2 + B^2}$.

Therefore $d = \dfrac{|-(Ax_1 + By_1 + C)|}{A^2 + B^2}\sqrt{A^2 + B^2} = \dfrac{|Ax_1 + By_1 + C|}{\sqrt{A^2 + B^2}}$.

Example Find the shortest distance from the point (–3, 4) to the straight line $3x + y = 5$.

The required distance is $\dfrac{|3(-3) + 4 - 5|}{\sqrt{3^2 + 1^2}} = \dfrac{10}{\sqrt{10}} = \sqrt{10}$.

Example Find the area of the triangle ABC in which A = (1, 2), B = (7, 5) and C = (2, 11).

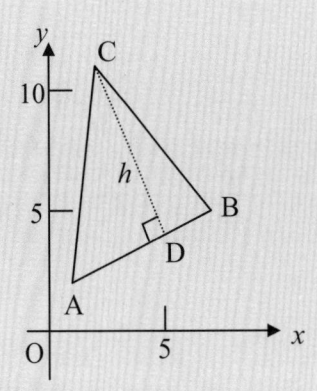

The gradient of (AB) = $\dfrac{5 - 2}{7 - 1} = \dfrac{1}{2}$, and so its equation is $x - 2y = -3$. The altitude from C to (AB) is the shortest distance from C to (AB) with length $h = \dfrac{|2 - 22 + 3|}{\sqrt{1^2 + 2^2}} = \dfrac{17}{\sqrt{5}}$.

Also AB = $\sqrt{6^2 + 3^2} = 3\sqrt{5}$, and the area of $\triangle ABC = \tfrac{1}{2} AB \times h = \tfrac{1}{2} \times 3\sqrt{5} \times \dfrac{17}{\sqrt{5}} = 25\tfrac{1}{2}$.

Example Find the value of p if the shortest distance from the point (2, –1) to the straight line $px + 4y = 12$ is 2.

The distance from (2, –1) to $px + 4y = 12$ is $\dfrac{|2p - 4 - 12|}{\sqrt{p^2 + 16}} = 2$.

Therefore $|2p - 16| = 2\sqrt{p^2 + 16}$ so that $|p - 8| = \sqrt{p^2 + 16}$.

Squaring both sides gives $p^2 - 16p + 64 = p^2 + 16$ or $p = 3$.

Example Find the equations of the two bisectors of the angles between the lines $x + y = 5$ and $x - 7y + 11 = 0$.

If P(x, y) lies on a bisector of an angle between the given lines, then P is equidistant from these lines.

Thus $\dfrac{|x+y-5|}{\sqrt{2}} = \dfrac{|x-7y+11|}{5\sqrt{2}} \Rightarrow 5|x+y-5| = |x-7y+11|$.

This gives $5(x + y - 5) = \pm(x - 7y + 11)$ or $x + 3y = 9$ and $3x - y = 7$ which are the required bisectors.

Exercise 2.2

1. Find the shortest distance from the origin to each of the following lines:
   (a) $3x + 4y = 15$; (b) $5x - 12y = 26$; (c) $8x + 15y + 34 = 0$;
   (d) $x + 2y = 5$; (e) $y = 2x + 10$; (f) $y = 5 - 3x$.

2. In each of the following, find the shortest distance from the given point to the given line:
   (a) $(3, 1), 2x - 3y = 16$; (b) $(3, 2), x + 2y = 27$;
   (c) $(1, -2), 3x + 4y = 5$; (d) $(-3, -4), x - 4y = 3$;
   (e) $(2, 6), 2x - 5y = 3$; (f) $(-3, 2), 6x - 2y + 11 = 0$;
   (g) $(7, -2), x + 5 = 0$; (h) $(3, -5), y = 3x + 2$;
   (i) $(a, b), lx + my + n = 0$; (j) $(p, q), qx + py = pq$.

3. Find the distance between each of the following pairs of parallel lines:
   (a) $3x + 5y = 1, 3x + 5y + 9 = 0$; (b) $7x - 3y = 5, 7x - 3y = 12$;
   (c) $3x - 4y = 3, 3x - 4y + 12 = 0$; (d) $2x + 2y = 9, x + y + 9 = 0$;
   (e) $7x - y = 6, y = 7x - 31$; (f) $x - 2y = 10, 2x - 4y + 1 = 0$.

4. Show that the lines $4x - 3y = 15$, $12x + 5y + 39 = 0$ and $x = -3$ are equidistant from the origin.

5. Calculate the area of the triangle
   (a) with vertices $(0, 0), (5, 6)$ and $(-3, 5)$;
   (b) with vertices $(5, 4), (3, -2)$ and $(7, 1)$;
   (c) with vertices $(-2, 3), (-3, -2)$ and $(7, -4)$;
   (d) whose sides have equations $2y + 3 = 0, 2y + x = 4$ and $3y - x = 1$;
   (e) whose sides have equations $y = x + 3, 8x + y = 39$ and $x + 2y = 3$.

6. Find the distance, r, of $(-3, 7)$ from the line $3x + 4y = 9$, and then find the coordinates of the two points on the y-axis which are r units from $3x + 4y = 9$.

7. Find the value of k if the straight line $2x - 3y = k$ is a tangent to the circle of radius 5 and centre $(3, -1)$.

Chapter 2

> 8. Find the value of p for which the shortest distance from the point $(2, -1)$ to the straight line $px + y = 5$ is 2.
>
> 9. Find the values of p for which the line $px + y = 9$ is a tangent to the circle with centre $(3, -2)$ and radius $\sqrt{5}$.
>
> 10. Find the equations of the bisectors of the angles between lines $y = 8x - 7$ and $4x + 7y = 11$.
>
> *11. The equation of the side AB of the triangle ABC is $x + y = 8$, that of side AC is $x - y = -4$, and that of BC is $7x + y = 44$.
>   (a) Write down three inequalities in x and y if $P(x, y)$ lies *inside* the triangle ABC.
>   (b) Find the equations of the interior bisectors of the angles A and B of the triangle.
>   (c) Find the coordinates of the incentre of the triangle.
>   [The incentre of a triangle is the point of intersection of the interior bisectors of the angles of the triangle.]

2.3 The General Quadratic Function

The function $f : x \mapsto ax^2 + bx + c$ where a, b and c are constants and $a \neq 0$ is a *quadratic* function. The graph of such a function is always a parabola. The turning point of the graph of a quadratic is called its **vertex**, and the vertical line through the vertex is called the **axis** of the parabola.

Completing the Square

An algebraic process which proves to be very useful in discussions of quadratic functions is called "*completing the square*". The process involves writing the expression $ax^2 + bx + c$ in the form $a(x - h)^2 + k$. The following examples illustrate the method.

Example   Find the values of h and k for which $x^2 + 4x - 5 = (x - h)^2 + k$ for all values of x.

$$\begin{aligned}
x^2 + 4x - 5 &= x^2 + 4x + \quad - 5 \quad \text{[leaving a temporary space for a constant to be added so that the first 3 terms form a } perfect\ square] \\
&= x^2 + 4x + 4 - 5 - 4 \quad [x^2 + 4x + 4 = (x + 2)^2] \\
&= (x + 2)^2 - 9 \Rightarrow h = -2 \text{ and } k = -9.
\end{aligned}$$

36

Coordinate Geometry in 2 Dimensions

Example   Find the values of h and k for which $2x^2 - 3x + 1 = 2(x-h)^2 + k$.

$$\begin{aligned} 2x^2 - 3x + 1 &= 2(x^2 - \tfrac{3}{2}x + \quad) + 1 \quad \text{[preparing to complete the square]} \\ &= 2\left(x^2 - \tfrac{3}{2}x + (\tfrac{3}{4})^2\right) + 1 - 2(\tfrac{3}{4})^2 \\ &= 2\left(x - \tfrac{3}{4}\right)^2 - \tfrac{1}{8}. \end{aligned}$$

Therefore $h = \tfrac{3}{4}$ and $k = -\tfrac{1}{8}$.

Example   Complete the square for the quadratic $6 - x - \tfrac{1}{2}x^2$.

$$\begin{aligned} 6 - x - \tfrac{1}{2}x^2 &= -\tfrac{1}{2}(x^2 + 2x + \quad) + 6 \\ &= -\tfrac{1}{2}(x^2 + 2x + 1) + 6 + \tfrac{1}{2} \\ &= -\tfrac{1}{2}(x+1)^2 + 6\tfrac{1}{2}. \end{aligned}$$

The Coordinates of the Vertex and the Equation of the Axis

Consider
$$\begin{aligned} y &= f(x) \\ &= ax^2 + bx + c \qquad (a \ne 0) \\ &= a\left(x^2 + \frac{b}{a}x \quad\right) + c \qquad \text{\{Preparing to complete the square.\}} \\ &= a\left(x^2 + \frac{b}{a}x + \frac{b^2}{4a^2}\right) + c - \frac{b^2}{4a} \\ &= a\left(x + \frac{b}{2a}\right)^2 + \frac{4ac - b^2}{4a} \\ &= a\left(x + \frac{b}{2a}\right)^2 - \frac{\Delta}{4a} \qquad \text{where } \Delta = b^2 - 4ac. \end{aligned}$$

[Δ is called the *discriminant* of the quadratic.]

If $a > 0$, $a\left(x + \dfrac{b}{2a}\right)^2 \ge 0$ and so $y \ge -\dfrac{\Delta}{4a}$ with equality when $x = -\dfrac{b}{2a}$.

Then the graph of $y = ax^2 + bx + c$ will *open at the top*, and its vertex will have coordinates $\left(-\dfrac{b}{2a}, -\dfrac{\Delta}{4a}\right)$. The equation of the axis of symmetry will be $x = -\dfrac{b}{2a}$.

In this case the graph will be of the form:

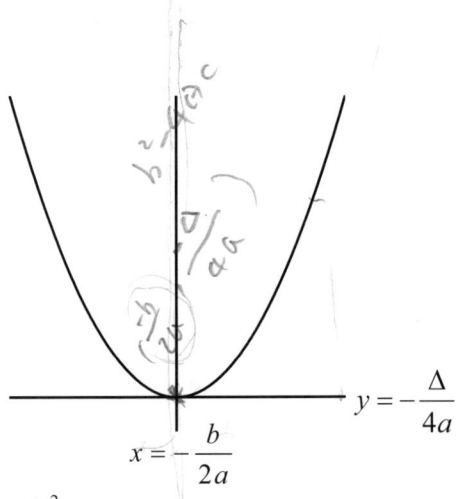

Chapter 2

If $a < 0$, $a\left(x + \dfrac{b}{2a}\right)^2 \leq 0$ and so $y \leq -\dfrac{\Delta}{4a}$ with equality when $x = -\dfrac{b}{2a}$.

Then the graph of $y = ax^2 + bx + c$ will *open at the bottom*, and its vertex will have coordinates $\left(-\dfrac{b}{2a}, -\dfrac{\Delta}{4a}\right)$. The axis of symmetry will again be $x = -\dfrac{b}{2a}$.

In this case the graph will be of the form:

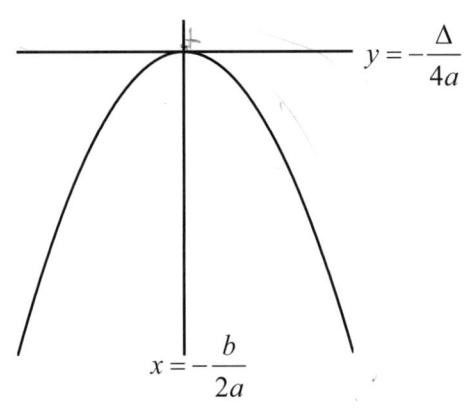

Example Find the coordinates of the vertex of the graph of $y = 2x^2 + 3x - 5$ and the equation of its axis of symmetry.

Method 1 (Completing the square.)
$$\begin{aligned} y &= 2x^2 + 3x - 5 \\ &= 2\left(x^2 + \tfrac{3}{2}x\phantom{+\tfrac{9}{16}}\right) - 5 \\ &= 2\left(x^2 + \tfrac{3}{2}x + \tfrac{9}{16}\right) - 5 - \tfrac{9}{8} \\ &= 2\left(x + \tfrac{3}{4}\right)^2 - \tfrac{49}{8} \end{aligned}$$
The vertex is $\left(-\tfrac{3}{4}, -\tfrac{49}{8}\right)$, and the axis is $x = -\tfrac{3}{4}$.

Method 2   The x-coordinate of the vertex is $-\frac{b}{2a}=-\frac{3}{4}$, and the y-coordinate of the vertex is $2\left(-\frac{3}{4}\right)^2+3\left(-\frac{3}{4}\right)-5=-\frac{49}{8}$.
The results are the same as with method 1.

Method 3 $-\frac{b}{2a}=-\frac{3}{4}$, and $\Delta=b^2-4ac=9-4(2)(-5)=49$.
The vertex is $\left(-\frac{b}{2a},-\frac{\Delta}{4a}\right)=\left(-\frac{3}{4},-\frac{49}{8}\right)$ giving the same results as in the previous methods.

Method 4   A graphic display calculator may be used to find the minimum value of $y=2x^2+3x-5$. The coordinates of the vertex obtained are (–0.75, –6.125) which give the same results as with all the previous methods.

Example   The sum of two numbers is 10. What is the least value of half the sum of their squares?

Let the numbers be x and $10-x$. Then half the sum of their squares is given by
$$S = \tfrac{1}{2}\left[x^2+(10-x)^2\right]$$
$$= \tfrac{1}{2}\left[x^2+100-20x+x^2\right]$$
$$= x^2-10x+50$$
$$= (x-5)^2+25.$$

Thus the least value of half the sum of the squares of the two numbers is 25.

Sketching Graphs of Quadratic Functions

A graphic display calculator can clearly be used to determine the graph of a quadratic function. However students should be able to use simple algebraic techniques which are not only easy to perform but are more informative.

The graph of a quadratic function can be sketched by drawing the axis of symmetry, plotting the vertex and y-intercept, and then using the symmetry property. If the x-intercepts can also be found by factorising the quadratic, the task is made even easier.

Example   Sketch the graph of the quadratic $y=x^2+2x-3$.

The x-coordinate of the vertex is $-\frac{b}{2a}=-1$.
The y-coordinate of the vertex is $(-1)^2-2-3=-4$.
The axis is $x=-1$ and the y-intercept is at (0, –3) which has an image of (–2,–3) under reflection in the axis.

Chapter 2

Since $x^2 + 2x - 3 = (x+3)(x-1)$, the x-intercepts are –3 and 1.
The graph is as follows:

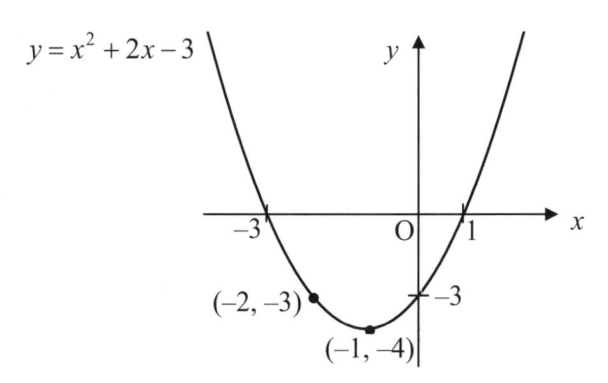

Example   Sketch the graph of $y = 3 - 2x - 2x^2$.

$y = 3 - 2x - 2x^2$
$= -2(x^2 + x\quad) + 3$
$= -2(x^2 + x + \frac{1}{4}) + 3 + \frac{1}{2}$
$= -2(x + \frac{1}{2})^2 + 3\frac{1}{2}$.

The vertex is $(-\frac{1}{2}, 3\frac{1}{2})$; the axis is $x = -\frac{1}{2}$; the y-intercept is 3.
The graph is as follows:

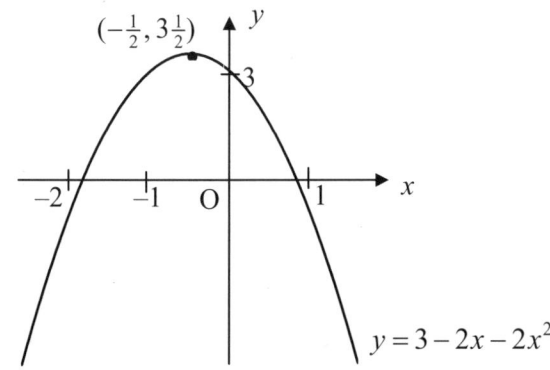

Exercise 2.3

1. For the graph of each of the following functions find the coordinates of the vertex and the equation of the axis by "completing the square":
   (a) $y = x^2 + 2x - 3$;
   (b) $y = x^2 - 4x + 3$;
   (c) $y = 2 - 2x - x^2$;
   (d) $y = 5 + 4x - x^2$;

(e) $y = x^2 + x - 4$; (f) $y = x^2 - 3x + 2$;
(g) $y = 5x - x^2$; (h) $y = 3 - 3x - x^2$;
(i) $y = 3x^2 - 3x - 4$; (j) $y = 6 - 2x - 3x^2$.

2. Sketch the graph of each of the following functions:
(a) $y = x^2 - 6x + 8$; (b) $y = 2 - x - x^2$;
(c) $y = 2x^2 + 4x + 3$; (d) $y = 3 - 4x - 2x^2$;
(e) $y = -3x^2 + 2x - 1$; (f) $y = \frac{1}{2}x^2 + x - 2$;
(g) $y = \frac{1}{3}x^2 - 2x + 1$; (h) $y = -\frac{1}{2}x^2 + x$;
(i) $y = 2x^2 + \frac{1}{2}x - 1$; (j) $y = -\frac{1}{3}x^2 - x + 2$.

3. The perimeter of a rectangle is 36 cm. What is its greatest possible area?

4. The square of a number is subtracted from the number. What is the maximum value of the result?

5. One hundred metres of fencing is used to form 3 sides of a rectangular enclosure with the wall of a shed providing the fourth side. Find the largest area that the enclosure may have.

6. A farmer has 2 km of fencing with which he wishes to enclose a rectangular field. What is the largest number of hectares he can enclose?

2.4 The General Solution to a Quadratic Equation

Consider the equation $ax^2 + bx + c = 0$ ($a \neq 0$).

Then
$$x^2 + \frac{b}{a}x = -\frac{c}{a}$$

$$x^2 + \frac{b}{a}x + \frac{b^2}{4a^2} = -\frac{c}{a} + \frac{b^2}{4a^2}$$

$$\left(x + \frac{b}{2a}\right)^2 = \frac{b^2 - 4ac}{4a^2} = \frac{\Delta}{4a^2}.$$

Thus
$$x + \frac{b}{2a} = \pm\frac{\sqrt{\Delta}}{2a}$$

and
$$x = \frac{-b \pm \sqrt{\Delta}}{2a} \quad \text{where } \Delta = b^2 - 4ac.$$

This is called *the general solution* of the quadratic equation $ax^2 + bx + c = 0$.

Chapter 2

Example   Solve the equations   (a) $2x^2 - 3x - 2 = 0$;
(b) $2x^2 - 3x - 1 = 0$.

(a)   Since the quadratic is easily factorised we do not need the general solution.
$2x^2 - 3x - 2 = (2x + 1)(x - 2) = 0$ when $x = -\frac{1}{2}$ or $x = 2$.

Note that the general solution may be used.
Here $a = 2$, $b = -3$, $c = -2$ and so $\Delta = 9 + 16 = 25$.
Therefore the solution is $x = \dfrac{3 \pm \sqrt{25}}{4} = \dfrac{3 \pm 5}{4} = 2$ or $-\frac{1}{2}$.

A graphic display calculator may also be used.

(b)   This quadratic does not factorise easily so we use the general solution. Here $a = 2$, $b = -3$, $c = -1$ and so $\Delta = 9 + 8 = 17$.
Therefore the solution is $x = \dfrac{3 \pm \sqrt{17}}{4}$.

When solving quadratic equations without the use of a graphic display calculator look to see if the given quadratic can be *easily* factorised. If so, factorise it and solve. If not, use the general solution.

Exercise 2.4

1. Solve each of the following equations by
   (1) factorising the given quadratic ;   (2) using the general solution.
   (a) $x^2 + 5x + 6 = 0$;   (b) $x^2 + 7x + 12 = 0$;
   (c) $x^2 + 9x + 20 = 0$;   (d) $x^2 - 7x + 10 = 0$;
   (e) $x^2 - 6x + 5 = 0$;   (f) $x^2 + x - 2 = 0$;
   (g) $x^2 + 5x - 14 = 0$;   (h) $x^2 + 4x - 21 = 0$;
   (i) $x^2 - 3x - 10 = 0$;   (j) $x^2 - x - 2 = 0$.

2. Solve the following equations:
   (a) $4x^2 + 8x + 3 = 0$;   (b) $6x^2 + 7x + 2 = 0$;
   (c) $2x^2 - 7x + 3 = 0$;   (d) $3x^2 - 14x + 8 = 0$;
   (e) $x^2 - 2x - 2 = 0$;   (f) $2x^2 - 3x - 1 = 0$;
   (g) $6x^2 + 5x - 6 = 0$;   (h) $5x^2 - x - 2 = 0$;
   (i) $6x^2 + 7x - 10 = 0$;   (j) $5 - 21x + 4x^2 = 0$.

3. Solve the following equations:
   (a) $x^2 - 9x = 0$;
   (b) $4x^2 = 25$;
   (c) $2x^2 + 6x + 4 = 0$;
   (d) $3x^2 + 12x + 9 = 0$;
   (e) $2x^2 - 3x + 1 = 0$;
   (f) $x^2 - 10x + 20 = 0$;
   (g) $5x^2 - x - 1 = 0$;
   (h) $2x^2 - 3x - 1 = 0$;
   (i) $2 - 6x - x^2 = 0$;
   (j) $18 - 9x - 2x^2 = 0$.

Higher Level

4. Solve the following equations:
   (a) $4x^4 - 13x^2 + 9 = 0$;
   (b) $3x^4 - 14x^2 + 8 = 0$;
   (c) $8x^6 + 7x^3 - 1 = 0$;
   (d) $x - 3x^{1/2} - 4 = 0$;
   (e) $x^{2/3} + x^{1/3} - 6 = 0$;
   (f) $(3x^2 - x)^2 - 12(3x^2 - x) + 20 = 0$.

2.5 The Roots of a Quadratic Equation

We have already seen that the roots of the quadratic equation $ax^2 + bx + c = 0$ $(a \neq 0)$ are given by $x = \dfrac{-b \pm \sqrt{b^2 - 4ac}}{2a}$.

We define the **discriminant** Δ to be $\Delta = b^2 - 4ac$.

If $\Delta > 0$ the equation has **two distinct real roots** given by $x = \dfrac{-b + \sqrt{\Delta}}{2a}$ and $x = \dfrac{-b - \sqrt{\Delta}}{2a}$.

If $\Delta = 0$ the equation has **only one real root (or two coincident roots)** given by $x = -\dfrac{b}{2a}$.

If $\Delta < 0$ the equation has **no real root** since a negative real number cannot have a real square root.

Geometrically the graph of $y = ax^2 + bx + c$ $(a \neq 0)$ meets the x-axis
(1) in two distinct points if $\Delta > 0$;
(2) in exactly one point if $\Delta = 0$;
(3) in no point at all if $\Delta < 0$.

Chapter 2

If **a > 0** we have:

$\Delta > 0$ $\Delta = 0$ $\Delta < 0$

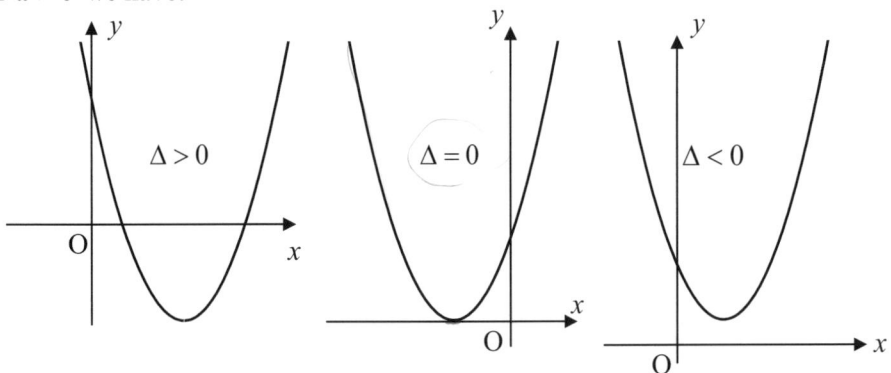

If **a < 0** we have:

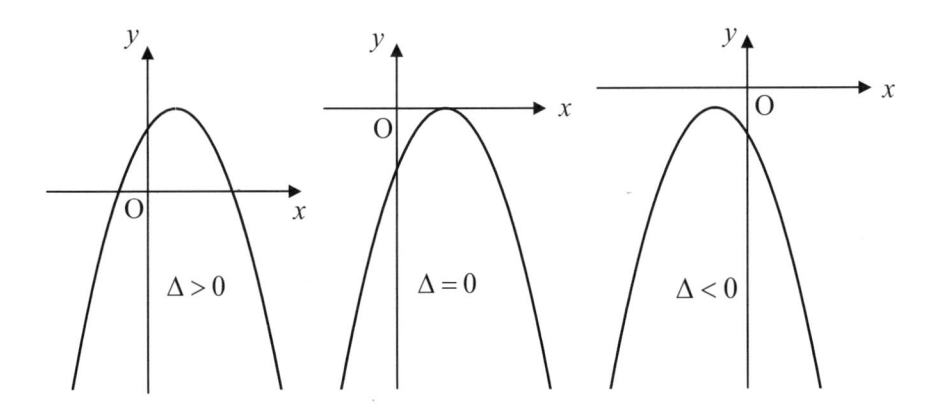

$\Delta > 0$ $\Delta = 0$ $\Delta < 0$

If $a > 0$ and $\Delta < 0$, the graph of $y = ax^2 + bx + c$ lies entirely above the *x*-axis. In this case the quadratic $ax^2 + bx + c$ is said to be ***positive definite***.

If $a < 0$ and $\Delta < 0$, the graph of $y = ax^2 + bx + c$ lies entirely below the *x*-axis. In this case the quadratic $ax^2 + bx + c$ is said to be ***negative definite***.

Example   Find the number of real roots that each of the following equations possesses: (a) $2x^2 + 5x - 4 = 0$;   (b) $2x^2 + 5x + 4 = 0$.

(a)   The quadratic $2x^2 + 5x - 4$ has discriminant
$\Delta = 5^2 - 4(2)(-4) = 57 > 0$.
Thus the given equation has two real (distinct) roots.

(b)   The quadratic $2x^2 + 5x + 4$ has discriminant $\Delta = 25 - 32 < 0$.
Thus the given equation has no real roots.

Coordinate Geometry in 2 Dimensions

Example Show that the quadratic $y = 2x^2 - x + 1$ is positive definite and find the minimum value of y.

The discriminant of the given quadratic is $\Delta = (-1)^2 - 4(2)(1) = -7 < 0$ and $a = 2 > 0$. Therefore the quadratic is positive definite.

To find the minimum value of y we may use a graphic display calculator or use any one of several algebraic methods as follows:

Method 1 $y = 2x^2 - x + 1$
$= 2\left(x^2 - \tfrac{1}{2}x + \left(\tfrac{1}{4}\right)^2\right) - 2\left(\tfrac{1}{4}\right)^2 + 1$
$= 2\left(x - \tfrac{1}{4}\right)^2 + \tfrac{7}{8}$.
Minimum $y = \tfrac{7}{8}$.

Method 2 The x-coordinate of the vertex $= -\tfrac{b}{2a} = \tfrac{1}{4}$ and so the y-coordinate of the vertex $= 2\left(\tfrac{1}{4}\right)^2 - \tfrac{1}{4} + 1 = \tfrac{7}{8}$.
Minimum $y = \tfrac{7}{8}$.

Method 3 The y-coordinate of the vertex $= -\tfrac{\Delta}{4a} = \tfrac{7}{8}$.
Minimum $y = \tfrac{7}{8}$.

Higher Level

Example Show that $\dfrac{22}{3x^2 - 4x + 5}$ is positive for all real values of x and find its greatest value.

The discriminant of the quadratic $3x^2 - 4x + 5$ is $\Delta = 16 - 60 = -44 < 0$ and since $a = 3 > 0$, the quadratic is positive definite.
Hence $\dfrac{22}{3x^2 - 4x + 5}$ is positive for all real values of x as required.

Now $3x^2 - 4x + 5$ has a minimum value of $-\dfrac{\Delta}{4a} = \dfrac{11}{3}$ and so

$\dfrac{22}{3x^2 - 4x + 5}$ has a maximum value of $\dfrac{22}{11/3} = 6$.

Chapter 2

Exercise 2.5

1. In each of the following find the value of the discriminant and so decide whether the equation has one, two or no real solutions. If the equation has any real solutions, find them.
   (a) $x^2 + 3x - 5 = 0$;
   (b) $2x^2 - x - 4 = 0$;
   (c) $2x^2 - 2x + 3 = 0$;
   (d) $3x^2 + 2x - 2 = 0$;
   (e) $4x^2 - 12x + 9 = 0$;
   (f) $5x^2 + 16x + 3 = 0$;
   (g) $3x^2 + 6x + 2 = 0$;
   (h) $3x^2 + 5x + 2 = 0$;
   (i) $25 - 20x + 4x^2 = 0$;
   (j) $6x^2 - 7x + 3 = 0$.

2. In each of the following, find the values of k for which the equation has only one real root:
   (a) $3x^2 - 4x + k = 0$;
   (b) $x^2 + kx + 1 = 0$;
   (c) $x^2 + kx + (k + 3) = 0$;
   (d) $2kx^2 + kx + 2 = 0$;
   (e) $kx^2 + (k + 1)x + 1 = 0$;
   (f) $kx^2 + 12x + (k + 9) = 0$.

3. Show that the values of the following functions have the same sign for all real values of x:
   (a) $x^2 + x + 1$;
   (b) $x^2 - 2x + 2$;
   (c) $4x^2 - x + 1$;
   (d) $x - 2x^2 - 2$;
   (e) $4x - 3x^2 - 6$;
   (f) $5x - 4x^2 - 2$.

4. Prove that of all rectangles with a given perimeter, the square has the largest area.

5. Triangle ABC is right-angled at B. The lengths of AB and BC are 4 cm and 7 cm respectively. The points P, Q and R lie on AB, AC and BC respectively such that PQRB is a rectangle. If BR = x cm, show that the area of the rectangle PQRB is $(4x - \frac{4}{7}x^2)$ cm². Hence find the positions of P, Q and R for which this area is a maximum. What is the maximum area?

6. A piece of wire 1 m long is cut into two pieces. If one piece is bent to form a circle and the other piece bent to form a square, find the maximum and minimum values of the sum of the areas of the circle and square.

7. A farmer wishes to use an existing fence as one side of a rectangular enclosure and has 1.2 km of fencing to use for the other three sides. What is the maximum number of hectares that can be enclosed?

Higher Level

8. (a) Show that $x^2 + 2x + 3$ is positive definite.

   (b) Find the values of a and b if $\dfrac{5x^2 + 10x + 17}{x^2 + 2x + 3} = a + \dfrac{b}{x^2 + 2x + 3}$ for all real values of x.

   (c) Prove that $5 < \dfrac{5x^2 + 10x + 17}{x^2 + 2x + 3} \leq 6$ for all real values of x.

9. Find the largest value of m and the smallest value of M for which
$$m < \dfrac{14}{2x^2 + 5x + 4} \leq M.$$

10. Find the largest value of m and the smallest value of M for which
$$m \leq \dfrac{9}{2x - 1 - 4x^2} < M.$$

*11. Triangle ABC is such that AB + AC = 10 cm and BC = 6 cm. If AB = x cm and D is the foot of the altitude from A to BC, show that

   (a) $AD^2 = -\dfrac{16}{9}(x^2 - 10x + 16)$;

   (b) the maximum area of the triangle occurs when the triangle is isosceles.

   Find the maximum area of the triangle ABC.

2.6 Transformations of Graphs in the Plane

Vertical Shift

The graph of $y = f(x)$ is given a vertical translation of q units.

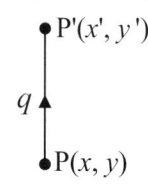

If P(x, y) maps onto P'(x', y'), the transformation equations are $x' = x$ and $y' = y + q$. Thus the equation of the curve, $y = f(x)$, becomes $y' - q = f(x')$ which is the condition that P'(x', y') lies on the curve $y = f(x) + q$.

Chapter 2

Horizontal Shift

The graph of $y = f(x)$ is given a horizontal translation of p units.

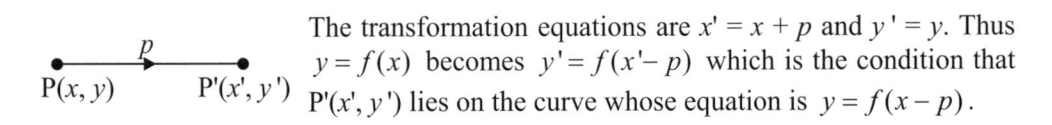

The transformation equations are $x' = x + p$ and $y' = y$. Thus $y = f(x)$ becomes $y' = f(x' - p)$ which is the condition that $P'(x', y')$ lies on the curve whose equation is $y = f(x - p)$.

Translation

The graph of $y = f(x)$ is given a translation with vector $\begin{pmatrix} p \\ q \end{pmatrix}$.

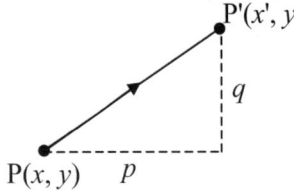

This is a combination of vertical and horizontal shifts. The equations are $x' = x + p$ and $y' = y + q$. Thus $y = f(x)$ becomes $y' - q = f(x' - p)$ which is the condition that $P'(x', y')$ lies on the curve whose equation is $y = f(x - p) + q$.

Vertical Stretch

Here, all points on the curve $y = f(x)$ are translated parallel to the y-axis so that the point $P(x, y)$ maps to the point $P'(x, hy)$. The number h is called the *stretch constant*.

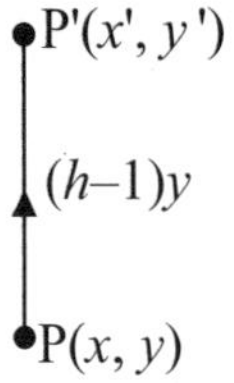

The transformation equations are $x' = x$ and $y' = hy$. Thus $y = f(x)$ becomes $y'/h = f(x')$ or $y' = hf(x')$ which is the condition that $P'(x', y')$ lies on the curve whose equation is $y = hf(x)$.

The distance of translation is proportional to the distance of P from the x-axis.

Horizontal Stretch

Here, all points on the curve $y = f(x)$ are translated parallel to the x-axis so that the point $P(x, y)$ maps to the point $P'(kx, y)$. The number k is called the *stretch constant*.

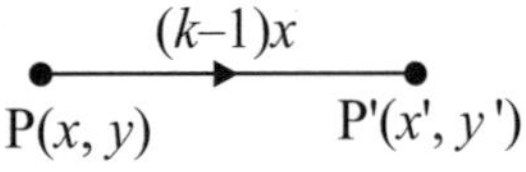

The transformation equations are $x' = kx$ and $y' = y$. Thus $y = f(x)$ becomes $y' = f(x'/k)$ which is the condition that $P'(x', y')$ lies on the curve whose equation is $y = f(x/k)$.

The distance of translation is proportional to the distance of P from the y-axis.

Reflection in the x-axis

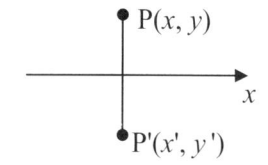

The equations of the transformation are $x' = x$ and $y' = -y$ and so the curve $y = f(x)$ becomes $-y' = f(x')$ which is the condition that P'(x', y') lies on the curve with equation is $y = -f(x)$.

Reflection in the y-axis

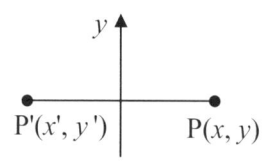

The equations of the transformation are $x' = -x$ and $y' = y$ and so the curve $y = f(x)$ becomes $y' = f(-x')$ which is the condition that P'(x', y') lies on the curve with equation is $y = f(-x)$.

The results of the above transformations on the graph of $y = x^2$ are summarised in the following table:

Transformation	Image of $y = x^2$	Graph
Vertical shift of q units	$y = x^2 + q$	
Horizontal shift of p units	$y = (x-p)^2$	
Translation – vector $\begin{pmatrix} p \\ q \end{pmatrix}$	$y = (x-p)^2 + q$	

Vertical stretch with stretch constant h	$y = hx^2$	
Horizontal stretch with stretch constant k	$y = (x/k)^2$	
Reflection in the x-axis	$y = -x^2$	
Reflection in the y-axis	$y = (-x)^2 = x^2$	$y = x^2$ is symmetrical about the y-axis and is therefore its own image under a reflection in the y-axis.

Example Describe the transformation(s) under which $y = x^2$ maps onto $y = 5 - 2(x+3)^2$.

One possible order of transformations would be a shift of 3 units to the left, followed by a stretch parallel to the y-axis with stretch constant 2, a reflection in the x-axis and an upward vertical shift of 5 units.

Example The diagram at the top of the next page shows the graph of the curve $y = f(x)$. Sketch, on separate axes, the graph of each of the following curves:
(a) $y = f(x+2)$;
(b) $y = f(x+1) - 2$;
(c) $y = f(1-x)$;
(d) $y = 1 - f(x)$.

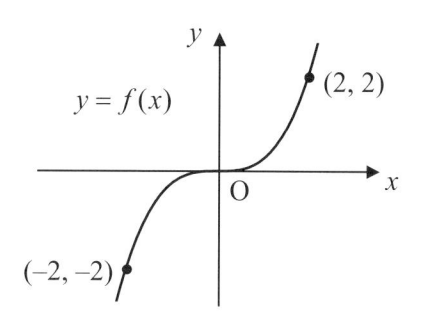

Coordinate Geometry in 2 Dimensions

$y = f(x)$ with points $(2, 2)$ and $(-2, -2)$

(a) $y = f(x+2)$

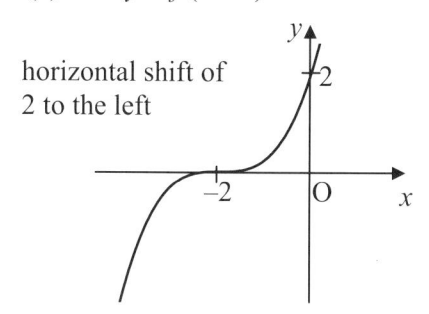

horizontal shift of 2 to the left

(b) $y = f(x+1) - 2$

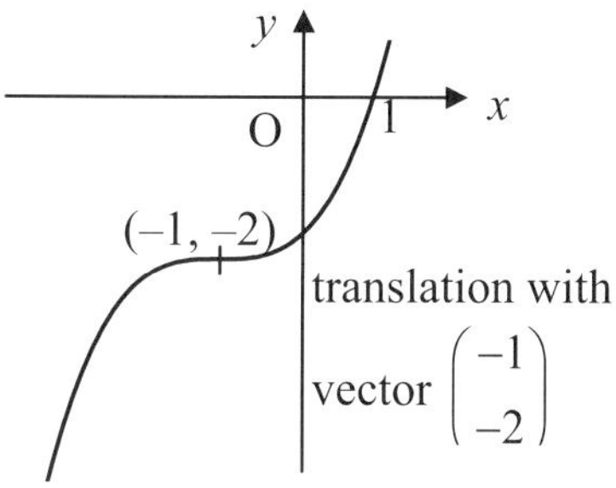

translation with vector $\begin{pmatrix} -1 \\ -2 \end{pmatrix}$

(c) $y = f(1-x)$

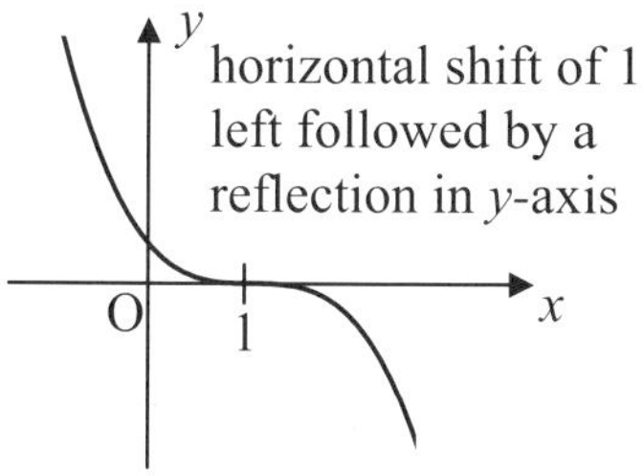

horizontal shift of 1 left followed by a reflection in y-axis

(d) $y = 1 - f(x)$

reflection in the x-axis followed by a vertical shift of 1 upwards

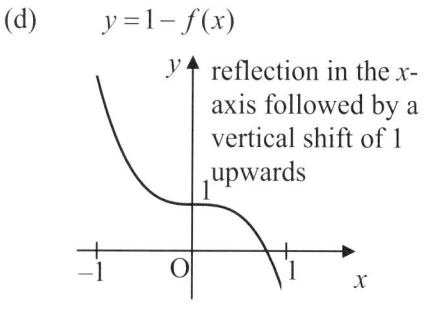

Exercise 2.6

1. Sketch the graph of each function and the graph of $y = x^2$ on the same set of coordinate axes in each case:

   (a) $y = x^2 + 2$;
   (b) $y = x^2 - 3$;
   (c) $y = (x-2)^2$;
   (d) $y = (x+1)^2$;
   (e) $y = (x-2)^2 - 3$;
   (f) $y = (x+2)^2 + 1$;
   (g) $y = 2x^2$;
   (h) $y = x^2/9$;
   (i) $y = 2(x-2)^2 - 3$;
   (j) $y = 1 - 3(x+1)^2$;
   (k) $y = (2x-3)^2 + 4$;
   (l) $3 - (3x-2)^2$.

51

Chapter 2

2. Sketch graphs of the functions $y = f(x)$ and $y = F(x)$ for each of the following:

(a) $f(x) = x^2$, $F(x) = f(x+2) - 3$;   (b) $f(x) = 2^x$, $F(x) = f(x) + 2$;

(c) $f(x) = \dfrac{1}{x}$, $F(x) = f(x-2) + 1$;   (d) $f(x) = x^3$, $F(x) = 2f(x) - 1$;

(e) $f(x) = \log x$, $F(x) = 2f(1-x)$;   (f) $f(x) = |x|$, $F(x) = 2f(x) - 3$.

3. (a) The graph of $f(x) = x^2$ is translated with vector $\begin{pmatrix} 2 \\ -3 \end{pmatrix}$ and the resulting graph is then reflected in the x-axis. Find the equation of the final image.

(b) The graph of $f(x) = x^2$ is stretched parallel to the y-axis with stretch constant 2 and the result is then shifted horizontally a distance of 3 units to the right. If the image after these transformations is then reflected in the y-axis, find the equation of the final curve.

(c) Describe the transformation(s) under which the graph of $y = f(x)$ maps onto $y = 2f(x-1) + 3$.

(d) Describe the transformation(s) under which the graph of $y = f(x)$ maps onto $y = 1 - f(\tfrac{1}{2}x + 2)$.

Higher Level (Optional)

2.7 Ratio of Division of a Line Segment

Let the point $P(x, y)$ divide the line segment joining $P_1(x_1, y_1)$ to $P_2(x_2, y_2)$ in the ratio $m : n$.

Let us consider for the moment the case where P divides $[P_1P_2]$ internally.

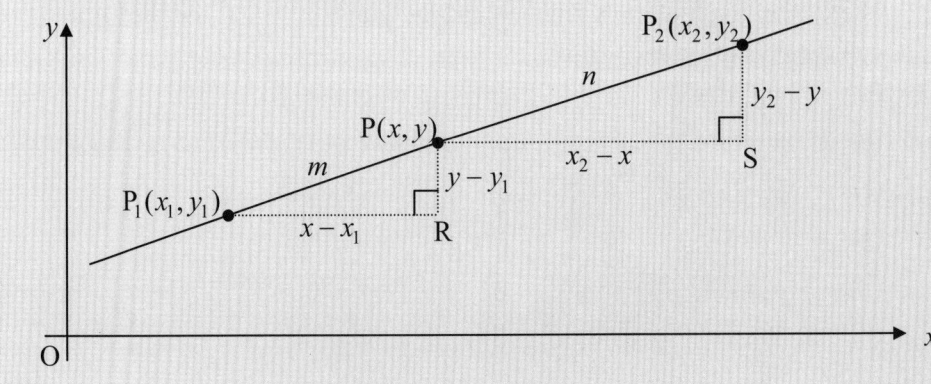

Triangles P_1PR and PP_2S are similar and so $\dfrac{P_1P}{PP_2} = \dfrac{m}{n} = \dfrac{x - x_1}{x_2 - x} = \dfrac{y - y_1}{y_2 - y}$.

Thus $mx_2 - mx = nx - nx_1$ and $my_2 - my = ny - ny_1$.

i.e. $x = \dfrac{mx_2 + nx_1}{m + n}$ and $y = \dfrac{my_2 + ny_1}{m + n}$.

Therefore the coordinates of P are $\left(\dfrac{mx_2 + nx_1}{m + n}, \dfrac{my_2 + ny_1}{m + n} \right)$.

This argument assumed that P divided $[P_1P_2]$ *internally* in the ratio $m:n$.

Let us now consider the case where P divides $[P_1P_2]$ *externally* in the ratio $m:n$.

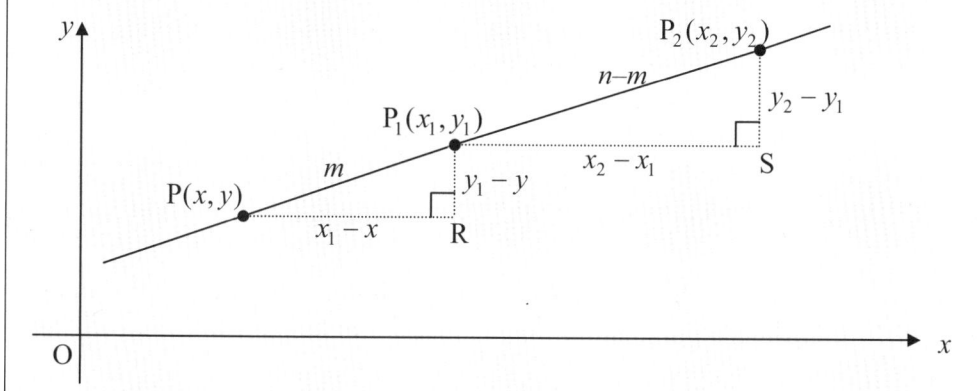

Triangles PRP_1 and P_1SP_2 are similar and so $\dfrac{PP_1}{P_1P_2} = \dfrac{m}{n - m} = \dfrac{x_1 - x}{x_2 - x_1} = \dfrac{y_1 - y}{y_2 - y_1}$.

Thus $mx_2 - mx_1 = (n - m)(x_1 - x)$ and $my_2 - my_1 = (n - m)(y_1 - y)$.

i.e. $x = \dfrac{mx_2 - nx_1}{m - n}$ and $y = \dfrac{my_2 - ny_1}{m - n}$.

Thus the coordinates of P are $\left(\dfrac{mx_2 - nx_1}{m - n}, \dfrac{my_2 - ny_1}{m - n} \right)$.

Example The points A and B have coordinates (2, 3) and (−4, 7) respectively. Find the coordinates of the points P and Q which divide the line segment [AB] internally and externally in the ratio 2 : 1.

Here $m = 2$, $n = 1$, $P = \left(\dfrac{2 \times (-4) + 1 \times 2}{2 + 1}, \dfrac{2 \times 7 + 1 \times 3}{2 + 1} \right) = \left(-2, \dfrac{17}{3} \right)$ and

$Q = \left(\dfrac{2 \times (-4) - 1 \times 2}{2 - 1}, \dfrac{2 \times 7 - 1 \times 3}{2 - 1} \right) = (-10, 11)$.

Chapter 2

Exercise 2.7

1. Given points A(4, 5) and B(−2, 8), find the coordinates of the points P and Q on the line (AB) such that P, Q divide [AB] internally and externally in the ratio   (a) 1 : 2 ;   (b) 2 : 3 ;   (c) 5 : 4 ;   (d) 3 : 1.

2. A(2, 3), B(−4, 0) and C(6, −5) are three points in Cartesian 2-space. Find the coordinates of the point
    (a)   P which divides [AB] internally in the ratio 2 : 1 ;
    (b)   Q which divides [BC] internally in the ratio 3 : 2 ;
    (c)   R which divides [AC] externally in the ratio 3 : 1.
    Show that P, Q, R are collinear and find the ratio in which Q divides [PR].

3. A(1, 4), B(−2, −2) and C are three points in Cartesian 2-space.
    (a)   Find the coordinates of the point P which divides [AB] externally in the ratio 2 : 1.
    (b)   Find the coordinates of C if the point Q(7, 1) divides BC externally in the ratio 3 : 1.
    (c)   Find the coordinates of the point R which divides [AC] externally in the ratio 6 : 1.
    (d)   Show that the points P, Q and R are collinear.

4. Given A(3, 5) and B(−2, −4), find the coordinates of the point P on the line (AB) which divides [AB] in the ratio $k : 1$. Hence, or otherwise, find the ratio in which [AB] is divided by
    (a)   the x-axis ;   (b)   the y-axis ;   (c)   the line $y = 7$;
    (d)   the line $x = 1$;   (e)   the line $y = x$;   (f)   the line $x + 2y = 0$.

5. The coordinates of A, B and C are (−6, −1), (0, 2) and (−4, −2) respectively. If AP : PB = 2 : 1, AQ : QB = 2 : 1, Q external to [AB] and AR : RC = 3 : 2, R external to [AC], find the coordinates of P, Q and R, and show that (PC) is parallel to (QR).

6. The point P lies on the straight line through A(3, 1) and B(−1, 4). Find the ratio in which P divides [AB] if P also lies on the line $x − 2y + 2 = 0$.

7. The point P lies on the line through A(−5, 5) and B(−9, 3). If P is the point where the line (AB) meets the line $4x + 3y = 6$, find the ratio AP : PB.

8. The circle with centre $C_1(3, 1)$ and radius $\sqrt{5}$ and the circle with centre $C_2(9, −2)$ and radius $2\sqrt{5}$ touch externally. Find the coordinates of P, the point of contact of the circles and the equation of the common tangent at P.

54

2.8 Circles in the Cartesian Plane

Definition The path traced by a point which moves so that a given condition is satisfied at all times, is called the *locus* of the point.

A circle is the locus of all points in a plane which are equidistant from a fixed point in that plane. For example, the locus of all points in the Cartesian plane each r units from the point $A(h, k)$, is a circle with centre A and radius r.

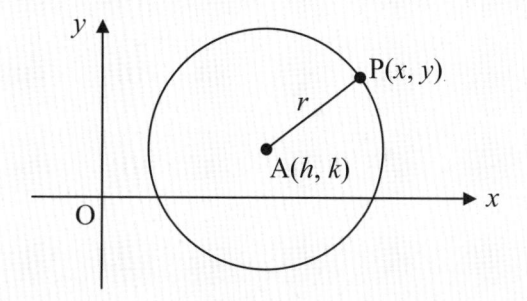

For all points $P(x, y)$ which lie on this circle the distance $AP = r$ or $AP^2 = r^2$.
Thus $(x - h)^2 + (y - k)^2 = r^2$ which is the equation of the circle.

Example Find the equation of the circle with centre $(3, -2)$ and radius 3.

The required equation is $(x - 3)^2 + (y + 2)^2 = 9$.

Example Find the equation of the circle with points $A(2, 3)$ and $B(4, -1)$ at opposite ends of a diameter.

The centre C is the mid-point of [AB].
Thus $C = (3, 1)$.
The radius r is equal to half the length of the diameter [AB].
Thus $r = \frac{1}{2}\sqrt{(4-2)^2 + (-1-3)^2} = \frac{1}{2}\sqrt{20} = \sqrt{5}$.
The equation of the circle is then $(x - 3)^2 + (y - 1)^2 = 5$.

Note that the equation $(x - h)^2 + (y - k)^2 = r^2$ can be written in the form
$$x^2 + y^2 - 2hx - 2ky + h^2 + k^2 - r^2 = 0$$
or
$$x^2 + y^2 - 2hx - 2ky + c = 0. \quad (*)$$

If the equation of a circle is given in this form, we can find the coordinates of its centre and the length of its radius by completing the square.

Chapter 2

Example   Find the coordinates of the centre and the length of the radius of the circle $x^2 + y^2 - 2x + 6y - 6 = 0$.

$$x^2 + y^2 - 2x + 6y - 6 = 0$$
$$x^2 - 2x + 1 + y^2 + 6y + 9 = 6 + 1 + 9$$
$$(x-1)^2 + (y+3)^2 = 16.$$

Thus the centre = $(1, -3)$ and the length of the radius = 4.

From equation (*) it can be seen that the centre has coordinates given by

$\left(-\frac{1}{2} \text{ of the coefficient of } x, \ -\frac{1}{2} \text{ of the coefficient of } y\right)$,

and the length of the radius is given by

$$r = \sqrt{-c + h^2 + k^2} = \sqrt{-c + (x\text{-coord. of centre})^2 + (y\text{-coord. of centre})^2}.$$

Example   Find the centre and radius of the circle $x^2 + y^2 + 2x - 5y = 5$.

$$C = \left(-\tfrac{1}{2} \times 2, -\tfrac{1}{2} \times -5\right) = \left(-1, \tfrac{5}{2}\right); \quad r = \sqrt{5 + (-1)^2 + \left(\tfrac{5}{2}\right)^2} = \sqrt{\tfrac{49}{4}} = 3\tfrac{1}{2}.$$

Example   Find the centre and radius of the circle $3x^2 + 3y^2 - 6x + 2y = 5$.

The equation must first be written in the form of (*), i.e., with the coefficients of x^2 and y^2 both equal to +1.

$$3x^2 + 3y^2 - 6x + 2y = 5 \quad\Rightarrow\quad x^2 + y^2 - 2x + \tfrac{2}{3}y = \tfrac{5}{3}.$$

Therefore the centre is $\left(1, -\tfrac{1}{3}\right)$ and the radius is $\sqrt{\tfrac{5}{3} + 1 + \tfrac{1}{9}} = \tfrac{5}{3}$.

Touching Circles

Two circles touch ***externally*** if the distance between their centres is equal to the sum of the lengths of their radii and touch ***internally*** if the distance between their centres is equal to the difference between the lengths of their radii.

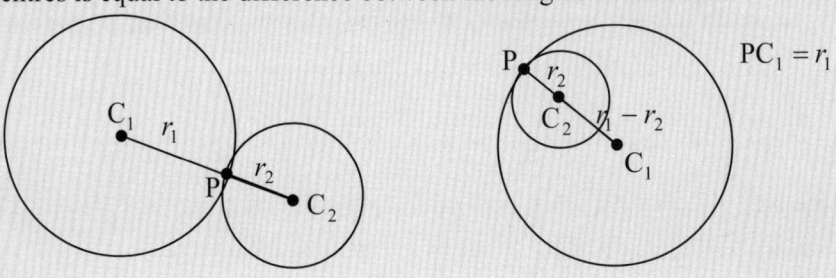

The circles touch externally if
$C_1C_2 = r_1 + r_2$ and then P divides
$[C_1C_2]$ *internally* in the ratio $r_1 : r_2$.

The circles touch internally if
$C_1C_2 = r_1 - r_2$ and then P divides
$[C_1C_2]$ *externally* in the ratio $r_1 : r_2$.

Example Show that the circles
$$x^2 + y^2 - 2x + 4y + 3 = 0 \text{ and } x^2 + y^2 - 8x - 2y + 9 = 0$$
touch. Find the coordinates of the point of contact P and the equation of the common tangent at P.

$C_1 = (1, -2)$, $r_1 = \sqrt{-3 + 1 + 4} = \sqrt{2}$, $C_2 = (4, 1)$, $r_2 = \sqrt{-9 + 16 + 1} = 2\sqrt{2}$
Now $C_1C_2 = \sqrt{9 + 9} = 3\sqrt{2} = r_1 + r_2$ and so the circles touch (externally).
P divides $[C_1C_2]$ internally in the ratio $r_1 : r_2 = 1 : 2$.
Therefore, P = (2, -1).
The gradient of $[C_1C_2] = \dfrac{1+2}{4-1} = 1$ and so the gradient of the tangent is -1.
Hence the equation of the common tangent at P is $x + y = 1$.

Tangents to Circles

The line $Ax + By + C = 0$ is a tangent to the circle $(x - h)^2 + (y - k)^2 = r^2$ if and only if the distance from the centre (h, k) to the line is exactly equal to the length of the radius r. Thus $Ax + By + C = 0$ is a tangent to the circle $(x - h)^2 + (y - k)^2 = r^2$ if and only if $\dfrac{|Ah + Bk + C|}{\sqrt{A^2 + B^2}} = r$.

If $\dfrac{|Ah + Bk + C|}{\sqrt{A^2 + B^2}} < r$ the line cuts the circle (in two places) and if $\dfrac{|Ah + Bk + C|}{\sqrt{A^2 + B^2}} > r$ the line does not meet the circle at all.

Example Show that the line $y = x - 6$ is a tangent to the circle with equation
$$x^2 + y^2 - 4x + 4y + 6 = 0$$
and find the coordinates of the point of contact.

The centre of the circle is C(2, -2) and the radius is $r = \sqrt{-6 + 4 + 4} = \sqrt{2}$.
The distance from the centre to the line is $\dfrac{|2 + 2 - 6|}{\sqrt{2}} = \sqrt{2} = r$ and so the line is a tangent to the circle.

Chapter 2

The radius through the point of contact is perpendicular to the tangent and passes through the centre of the circle. The equation of this radius is $x + y = 0$. The point of contact is therefore the point of intersection of the lines $y = x - 6$ and $x + y = 0$, i.e., the point $(3, -3)$.

Since the quadratic equation which results when we attempt to find the coordinates of any point of intersection of the line and the circle can have only one real root if the line is a tangent, the following alternative method can always be used. This method has the advantage of proving tangency and providing the point of contact at the same time.

For intersection of the line $y = x - 6$ and circle $x^2 + y^2 - 4x + 4y + 6 = 0$,

$$x^2 + (x-6)^2 - 4x + 4(x-6) + 6 = 0$$
$$\Rightarrow 2x^2 - 12x + 18 = 0$$
$$\Rightarrow x^2 - 6x + 9 = 0$$
$$\Rightarrow (x-3)^2 = 0.$$

Since $x = 3$ is the only solution, the line is a tangent to the circle at $(3, -3)$.

Exercise 2.8

1. Find the equation of the circle defined by
   (a) centre at (3, 1), radius 4 ;
   (b) centre at (–1, –2), radius 5 ;
   (c) centre at (–3, –2), radius 1.5 ;
   (d) centre at (–3, 0), radius $2\sqrt{2}$;
   (e) ends of a diameter at (–2, 5) and (6, –1) ;
   (f) ends of a diameter at (4, –3) and (6, –5) ;
   (g) centre at (–3, 2) and touching the x-axis ;
   (h) centre at (–5, –4) and touching the y-axis ;
   (i) radius $3\sqrt{5}$ and concentric with $x^2 + y^2 - 4x + 3y = 9$;
   (j) radius 2.5 and concentric with $x^2 + y^2 + 8y + 1 = 0$;
   (k) centre at (3, 1) and passing through (–1, –1) ;
   (l) centre at (–1, 0) and passing through (2, –4) ;
   (m) centre at (4, –1) and touching $5x - 12y = 6$;
   (n) centre at (–5, –3) and touching $x + 2y + 1 = 0$.

2. Find the coordinates of the centre and the length of the radius of each of the following circles:
   (a) $x^2 + y^2 - 6x - 8y = 11$;   (b) $x^2 + y^2 + 4x - 2y = 5$;
   (c) $x^2 + y^2 + 4x + 8y = 1$;   (d) $x^2 + y^2 + 4y = 4$;

(e) $2x^2 + 2y^2 - 4x + 8y + 1 = 0$; (f) $x^2 + y^2 - 3x + 5y + 1 = 0$;
(g) $3x^2 + 3y^2 - 6x + 4y = 3$; (h) $5x^2 + 5y^2 + 4y = 0$.

3. Show that the following pairs of circles touch. Determine whether the contact is internal or external. Find the coordinates of the point of intersection and the equation of the common tangent at this point.
   (a) $x^2 + y^2 - 4x - 2y = 0$ and $x^2 + y^2 - 10x - 14y + 54 = 0$;
   (b) $x^2 + y^2 - 6x - 8y = 0$ and $x^2 + y^2 + 14x - 8y = 160$;
   (c) $x^2 + y^2 - 8x - 2y = 23$ and $x^2 + y^2 - 14x + 39 = 0$;
   (d) $2x^2 + 2y^2 - 2x + y = 5$ and $8x^2 + 8y^2 - 36x - 52y + 85 = 0$.

4. The circle C_1 has equation $x^2 + y^2 + 2x - 6y + 1 = 0$ and C_2 is a circle with centre (3, 6). Find the equation of C_2 when the circles touch
   (a) internally ; (b) externally.

5. Show that in each of the following the line is a tangent to the circle. In each case find the coordinates of the point of contact.
   (a) $x + y = 3$, $x^2 + y^2 + 2x + 4y = 13$;
   (b) $2x - y = 16$, $x^2 + y^2 - 10x + 2y + 21 = 0$;
   (c) $x + 3y = 20$, $x^2 + y^2 + 4x - 8y + 10 = 0$;
   (d) $3x + 4y = 20$, $x^2 + y^2 - 2x + 4y = 20$.

6. Prove that the length of the tangent from the external point $P(x_1, y_1)$ to the circle $x^2 + y^2 - 2hx - 2ky + c = 0$ is $\sqrt{x_1^2 + y_1^2 - 2hx_1 - 2ky_1 + c}$. Hence find the length of the tangent from
   (a) (2, 3) to the circle $x^2 + y^2 + 6x + 10y + 30 = 0$;
   (b) (−3, 4) to the circle $x^2 + y^2 - 4x - 6y + 2 = 0$;
   (c) (2, 7) to the circle $x^2 + y^2 - 5x + 4y = 4$;
   (d) $(5, 3\frac{1}{2})$ to the circle $4x^2 + 4y^2 - 8x - 4y + 1 = 0$.

7. Find the values of k if
   (a) $2x + 3y = k$ is a tangent to the circle $x^2 + y^2 = 13$;
   (b) $x - 2y = k$ is a tangent to the circle $x^2 + y^2 - 8x = 4$;
   (c) $kx - y = 3k + 4$ is a tangent to the circle $x^2 + y^2 - 4x - 2y + 4 = 0$.
   In each case find the coordinates of the point of contact.

Chapter 2

8. Show that the locus of the point P(x, y) in each of the following is a circle. Find the coordinates of the centre of this circle and the length of its radius.
    (a) The sum of the squares of the distances from P to (–3, 2) and (7, 2) is 58.
    (b) The distance from P to (2, 0) is twice the distance from P to (5, 3).

9. The distance of P(x, y) from A(4, 1) is twice its distance from B(–8, 2). Show that the locus of P is a circle and that the points which divide [AB] internally and externally in the ratio 2 : 1 are extremities of a diameter of the circle.

10. The distance of P(x, y) from A(–2, –3) is three times its distance from B(6, 5). Show that the locus of the point P is a circle and find the coordinates of its centre and the length of its radius. Show also that the points which divide [AB] internally and externally in the ratio 3 : 1 are extremities of a diameter of the circle.

Required Outcomes

After completing this chapter, a student should be able to
- find the equation of any straight line in 2-space.
- find the shortest distance between a point and a line in 2-space. **(HL)**
- calculate the area of any triangle in 2-space from the coordinates of the vertices. **(HL)**
- find the coordinates of the vertex of any quadratic function.
- sketch the graph of any quadratic function.
- solve any quadratic equation which has real solutions.
- determine whether or not a given quadratic is positive/negative definite.
- perform common transformations of curves, particularly the quadratic curve $y = x^2$.
- find the coordinates of a point which divides a line segment in a given ratio. **(HL)**
- find an equation for a circle given the coordinates of its centre and the length of its radius. **(HL)**
- determine the coordinates of the centre and the length of the radius of any circle from its equation. **(HL)**
- determine whether or not two circles touch. **(HL)**
- determine whether or not a given line is a tangent to a circle and if it is, find the point of contact. **(HL)**

3 Solving Triangles

3.1 The Sine Rule

We know from our work with the congruence of triangles that a triangle is completely defined if we are given
(1) the lengths of its sides ;
(2) the measures of any two angles and the length of one side ;
(3) the lengths of any two sides and the measure of the included angle.

To 'solve' a triangle, we are required to determine the lengths of all three sides and the measures of all three angles. This can clearly be done if we are given the three pieces of data in (1), (2) or (3) above.

We *may* also be able to solve a triangle when we are given the lengths of any two sides and the measure of a non-included angle.

We *cannot* determine the lengths of the sides of a triangle, however, if we are given only the measures of the three angles. For example, we can construct an equilateral triangle of any size we require.

Also, we are already able to solve right-angled triangles given the required amount of data (two pieces other than the right angle). The following 'Sine Rule' and 'Cosine Rule' will enable us to solve non-right-angled triangles given sufficient data.

Preliminary Theorems

The Triangle Inequality

In any triangle, the sum of the lengths of any two sides is greater than the length of the third side. This is simply a statement of the fact that 'the shortest distance between any two points in a plane is the length of the straight line joining the points'.

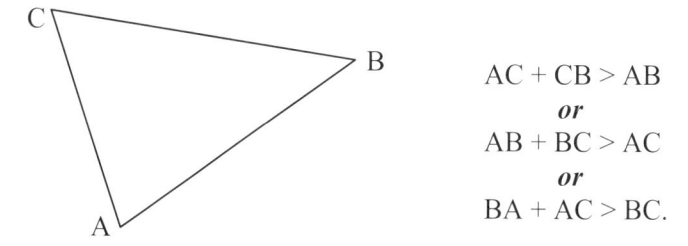

AC + CB > AB

or

AB + BC > AC

or

BA + AC > BC.

Chapter 3

Example In the triangle ABC, AB = x, AC = $2x - 1$ and BC = $2x + 1$.
Show that $x > 2$.

x must be positive and so BC is the longest side.
By the triangle inequality, AB + AC > BC and so $x + 2x - 1 > 2x + 1$ which gives $x > 2$, as required.

Notation: In all the work that follows, block letters, A, B, C, ..., will be used to represent the angles A, B, C, ..., (and their measures), and small letters, a, b, c, ..., will be used to represent the lengths of the sides opposite the angles A, B, C, ...

Theorem The area of a triangle ABC = $\frac{1}{2} ab \sin C$.

(That is, the area of a triangle is equal to half the product of the lengths of any two sides and the sine of the angle included by these sides.)

Proof

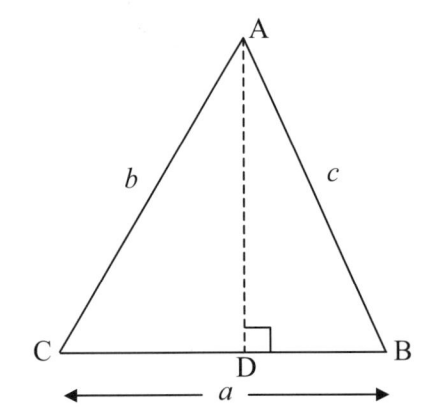

Draw (AD) perpendicular to (BC) to meet (BC) in D.

The area of the triangle ABC = $\frac{1}{2} a \times$ AD.

Now in triangle ADC, $\sin C = \dfrac{AD}{AC}$, and so AD = $b \sin C$.

Therefore the area of triangle ABC = $\frac{1}{2} ab \sin C$, as required.

Example Find the area of a triangular flower bed which has one angle of 30° and the sides about this angle of lengths 10 m and 12 m.

The required area = $\frac{1}{2}(10)(12) \sin 30° = 30 \text{ m}^2$.

Solving Triangles

Theorem   In any triangle ABC, $\dfrac{a}{\sin A} = \dfrac{b}{\sin B} = \dfrac{c}{\sin C}$.

(The Sine Rule)

Proof   The area of the triangle ABC $= \tfrac{1}{2} bc \sin A = \tfrac{1}{2} ac \sin B = \tfrac{1}{2} ab \sin C$.

Divide throughout by $\tfrac{1}{2} abc$: $\dfrac{\sin A}{a} = \dfrac{\sin B}{b} = \dfrac{\sin C}{c}$.

Taking reciprocals gives: $\dfrac{a}{\sin A} = \dfrac{b}{\sin B} = \dfrac{c}{\sin C}$, as required.

Notes on the Sine Rule

The Sine Rule is used to find the remaining sides and angles of a (non-right-angled) triangle if we are given
(1)   two angles and one side ;
(2)   two sides and a non-included angle.

In case (2), there may be two different triangles possible. The Sine Rule cannot show whether an angle is acute or obtuse since the sine of each is positive. For example, if we know that $\sin A = \tfrac{1}{2}$, we cannot be sure that $A = 30°$ since $\sin 150° = \tfrac{1}{2}$ also. An example of this 'ambiguous case' will be given in the following examples.

Example   In the triangle ABC, $a = 10$ cm, $A = 30°$, $B = 40°$. Find c.

$C = 110°$ (angle sum of a triangle is $180°$).

Also $\dfrac{c}{\sin C} = \dfrac{a}{\sin A}$ (Sine Rule).

Thus $c = \dfrac{10 \sin 110°}{\sin 30°} = 18.8$ cm.

Example   In the triangle ABC, $a = 8$ m, $b = 10$ m and $A = 30°$. Find the size of the angle C.

From the Sine Rule we have $\dfrac{\sin B}{b} = \dfrac{\sin A}{a}$ and so

$\sin B = \dfrac{10 \sin 30°}{8} = 0.625$.

Using a calculator we find that $\arcsin 0.625 = 38.7°$.
But $\sin 141.3° = 0.625$, also.

63

Chapter 3

Therefore   B = 38.7°   or   141°.
Then   C = 111°   or   8.68°.

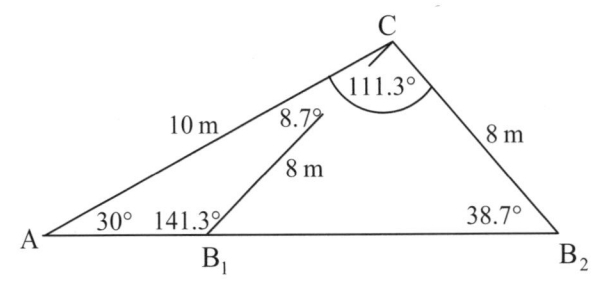

(This is the ambiguous case mentioned earlier.)

Example   From a point A, I observe that the angle of elevation of the top of a tree is 21°. I walk 10 m towards the foot of the tree to a point B, from where the angle of elevation of the top of the tree is 34°. Calculate the height of the tree above the level of observation.

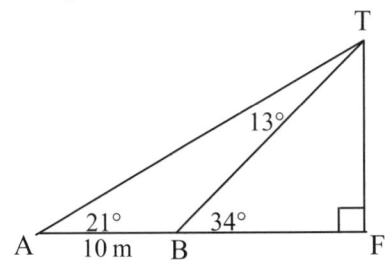

Angle ATB = 13° (exterior angle of a triangle is equal to the sum of the interior opposite angles.)

In the triangle ABT
$$\frac{AB}{\sin T} = \frac{BT}{\sin A}.$$

That is
$$BT = \frac{10 \sin 21°}{\sin 13°}.$$

In triangle BTF, $FT = BT \sin 34° = \dfrac{10 \sin 21° \sin 34°}{\sin 13°} = 8.91$.

Thus the height of the tree above the level of observation is 8.91 m.

Note: Since Pythagoras' theorem and simple definitions of sine and cosine can be used to solve any right-angled triangle, the Sine Rule is a waste of time in such cases. Also, the Sine Rule is not needed to solve an isosceles triangle as two congruent right-angled triangles can be created by joining the vertex to the mid-point of the base.

Solving Triangles

Example   Solve the triangle ABC in which $a = 15$ cm, $B = 35°$ and $C = 110°$.

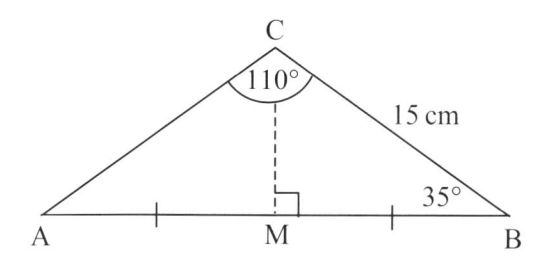

Firstly, $A = 35°$ (angle sum of a triangle is 180°). Therefore the triangle ABC is isosceles and so $b = 15$ cm. Join C to M, the midpoint of [AB]. Then, in the triangle ACM, (right-angled at M), AM = 15 cos 35° = 12.29 cm which gives $c = 2 \times 12.29 = 24.6$ cm.

Theorem   If d is the length of the diameter of the circumcircle of triangle ABC, then $d = \dfrac{a}{\sin A}$.

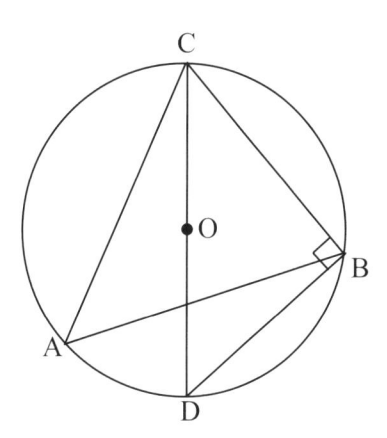

Proof   Let O be the centre of the circumcircle of the triangle ABC, and let CD be a diameter of length d. Then triangle BCD is right-angled at B and angle BDC = angle BAC (same segment). Now, in triangle BCD, $\sin D = \dfrac{BC}{d}$ and so $d = \dfrac{BC}{\sin D} = \dfrac{a}{\sin A}$.

Higher Level

Example   From a point A, the summits of two mountains, B and C, are on bearings of 024° and 058°, and the angles of elevation are 8.5° and 9.3° respectively. If B is 850 m above A, and C is due east of B, calculate

(a)   the horizontal distance from B to A ;
(b)   the height of C above A.

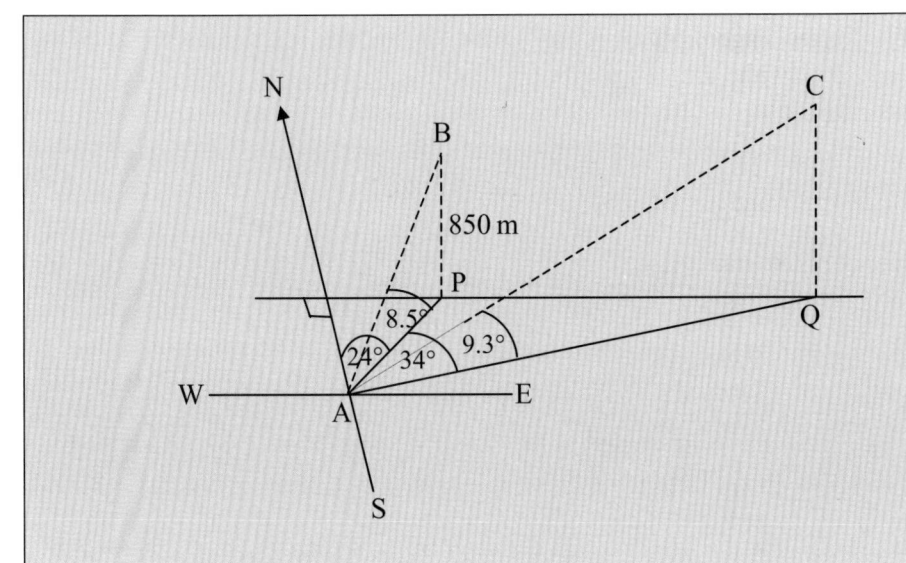

Chapter 3

(a) In triangle ABP, $\tan 8.5° = \dfrac{850}{AP} \Rightarrow AP = \dfrac{850}{\tan 8.5°} = 5690$.

Therefore the horizontal distance from A to B is 5690 m.

(b) In triangle APQ, $P = 114°$, $Q = 32°$ and $\dfrac{AQ}{\sin 114°} = \dfrac{AP}{\sin 32°}$.

$\Rightarrow AQ = \dfrac{AP \sin 114°}{\sin 32°} = \dfrac{850 \sin 114°}{\tan 8.5° \sin 32°}$.

In triangle ACQ, $\tan 9.3° = \dfrac{CQ}{AQ}$

$\Rightarrow CQ = AQ \tan 9.3° = \dfrac{850 \sin 114° \tan 9.3°}{\tan 8.5° \sin 32°} = 1610$.

Therefore C is 1610 m above A.

Exercise 3.1

[*Hint:* Never try to solve geometric type trigonometric problems without first drawing a neat sketch.]

1. In each of the following, solve the triangle ABC:
    (a) $b = 20$, $A = 120°$, $B = 30°$;
    (b) $a = 20$, $B = 125°$, $C = 30°$;
    (c) $b = 12$, $c = 20$, $C = 60°$;
    (d) $a = 15$, $c = 10$, $A = 30°$;
    (e) $a = 10$, $c = 15$, $A = 30°$.

Solving Triangles

2. The longer diagonal of a parallelogram is 36 cm long and forms angles of 44° and 26° with the sides. Determine the lengths of the sides of the parallelogram.

3. A landmark M is observed from two points A and B, 1.2 km apart. Angle BAM = 58° and angle ABM is 73°. Find the distance of M from A.

4. In a parallelogram ABCD, AC = 65 cm, AD = 30 cm and angle BAD = 46°. Calculate the measure of the angle BAC.

5. Point B is 500 m due east of point A on a straight east-west road. From A the bearing of a landmark is 064.5°, and from B the bearing is 032.5°. Find the distance of the landmark from the road.

6. Town B is 25 km from town A on a bearing of 085°. Town C is 30 km from A and its bearing from B is 190°. Find the bearing of A from C.

7. Points A and B are 60 m apart on one bank of a straight river. Point C is on the other bank, and the angles CAB and CBA are observed to be 46.4° and 67.6° respectively. Find the width of the river.

Higher Level

8. In a triangle ABC, the angle ACB is 31° and the ratio AB : AC = 2 : 3. If BC = 20 cm, calculate the lengths AB and AC.

9. A chord of a circle is 18 cm long and subtends an angle of 46° at the circumference. Find the length of a chord of this circle which subtends an angle at the circumference of 23°.

10. Calculate the area of the circumcircle of the triangle ABC in which A = 76°, B = 48° and c = 1.2 m.

11. In a triangle ABC, AC = 30 cm and BC = 40 cm. Find the greatest measure of the angle B. If B = 35°, solve the triangle.

12. In triangle ABC, BC = 56 cm, angle ABC = 102° and angle ACB = 29°. Point X lies on BC such that AX = 43 cm. Find the size of the angle AXC.

13. Consider the triangle ABC in which A, $0° < A < 90°$, a and b are given. Show that
    (a) no triangle exists if $a < b \sin A$;
    (b) exactly one triangle exists if $a = b \sin A$ or $a \geq b$;
    (c) two distinct triangles exist if $b \sin A < a < b$.
    Under what conditions is the triangle right angled?

Chapter 3

14. Four landmarks A, B, C, D are such that B is 4.8 km from A in a direction 152°, C's bearings from A and B are 193° and 234° respectively, and D bears 323° from C and is 6 km from A. Given that the angle ADC is acute, find the bearing of D from A and its distance from C.

15. Points P and Q lie in the same horizontal plane, P being 2500 m north of Q. The bearings of a mountain from these two points are 250° and 295° respectively. The angle of elevation of the summit from P is 20.4°. Find the height of the mountain.

16. Points A and B are on the same horizontal level as the foot of a wireless mast. Point A is due west of the mast and point B is 300 m from A in a direction of 118°. If the mast is north-east of B and subtends an angle of 13.3° at A, find the height of the mast.

3.2 The Cosine Rule

Theorem In any triangle ABC, $a^2 = b^2 + c^2 - 2bc\cos A$.

Proof Draw rectangular coordinate axes so that A coincides with the origin and [AB] coincides with the positive x-axis.

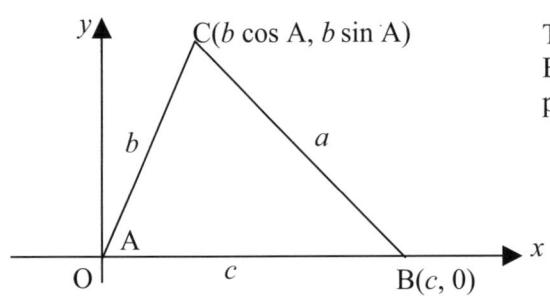

The coordinates of A are (0, 0), B is the point $(c, 0)$ and C is the point $(b\cos A, b\sin A)$.

$$\begin{aligned} BC^2 &= (b\cos A - c)^2 + (b\sin A - 0)^2 \\ &= b^2\cos^2 A - 2bc\cos A + c^2 + b^2\sin^2 A \\ &= b^2(\cos^2 A + \sin^2 A) + c^2 - 2bc\cos A. \quad [\cos^2 A + \sin^2 A = 1] \end{aligned}$$

Thus $a^2 = b^2 + c^2 - 2bc\cos A$, as required.

By symmetry $b^2 = a^2 + c^2 - 2ac\cos B$ and $c^2 = a^2 + b^2 - 2ab\cos C$.

Also by a re-arrangement of terms, $\cos A = \dfrac{b^2 + c^2 - a^2}{2bc}$, $\cos B = \dfrac{a^2 + c^2 - b^2}{2ac}$ and $\cos C = \dfrac{a^2 + b^2 - c^2}{2ab}$.

Notes on the Cosine Rule

The Cosine Rule is used to solve a (non-right-angled) triangle if we are given
(1)   the lengths of its three sides ;
(2)   the lengths of two sides and the measure of the included angle.

Since each of these sets of data completely determines the triangle, there can be no ambiguity in the use of the Cosine Rule. In any case, the cosine of an acute angle is positive and the cosine of an obtuse angle is negative.

If the lengths of the three sides are given, always find the largest angle (opposite the longest side) using the Cosine Rule first. This will ensure that if the Sine Rule is then used, any further angle found must be acute and any ambiguity is avoided.

As with the Sine Rule, the Cosine Rule is a waste of time with any right-angled or isosceles triangle.

Example   The sides of a triangle ABC are $a = 6$ cm, $b = 8$ cm and $c = 5$ cm. Find the angles of the triangle.

Angle B is the largest angle since it is opposite the longest side.

Also $\cos B = \dfrac{a^2 + c^2 - b^2}{2ac} = \dfrac{36 + 25 - 64}{60} = -\dfrac{1}{20}$.

Therefore $B = 92.9°$.

Using the Sine Rule now gives $\dfrac{\sin A}{a} = \dfrac{\sin B}{b} \Rightarrow \sin A = \dfrac{6 \sin B}{8}$, and so $A = 48.5°$ (A cannot be obtuse since B is the largest angle), and $C = 38.6°$.

Example   In a triangle ABC, $B = 60°$, $c = 12$ cm and $a = 7$ cm. Solve the triangle.

$b^2 = a^2 + c^2 - 2ac \cos B = 49 + 144 - 168 \cos 60°$, giving $b = 10.4$ cm.

The safest way to proceed now if the Sine Rule is used, is to find the measure of the angle A (not C, the largest angle, since this may be obtuse and the Sine Rule will not reveal it).

Then $\dfrac{\sin A}{a} = \dfrac{\sin B}{b} \Rightarrow \sin A = \dfrac{7 \sin 60°}{b} \Rightarrow A = 35.5° \Rightarrow C = 84.5°$.

Example   Calculate the area of triangle ABC if $a = 5$ cm, $b = 6$ cm and $c = 7$ cm.

Chapter 3

$$\cos C = \frac{a^2+b^2-c^2}{2ab} = \frac{25+36-49}{60} \Rightarrow C = \arccos 0.2 \text{ and the area of the}$$

triangle is then $\frac{1}{2}ab\sin C = 15\sin(\arccos 0.2) = 14.7 \text{ cm}^2$.

Higher Level

Example The bearing of a television tower on the top of a hill from a point A is 343°, and from a point B, on the same horizontal level as A, the bearing of the tower is 295°. From A, the angle of elevation of the top of the tower is 2.1° and from B, the angle of elevation is 2.7°. If the top of the tower is 850 m above the level of A and B, calculate the distance and bearing of B from A.

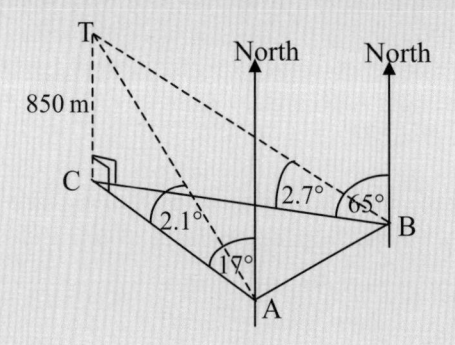

In triangle ACT, $\tan 2.1° = \dfrac{850}{AC} \Rightarrow AC = \dfrac{850}{\tan 2.1°}$.

In triangle BCT, $\tan 2.7° = \dfrac{850}{BC} \Rightarrow BC = \dfrac{850}{\tan 2.7°}$.

In triangle ABC,
$AB^2 = AC^2 + BC^2 - 2(AC)(BC)\cos 48° \Rightarrow AB = 17\,400$.

In triangle ABC, $\dfrac{\sin A}{BC} = \dfrac{\sin 48°}{AB} \Rightarrow \sin A = \dfrac{850\sin 48°}{AB\tan 2.7°} \Rightarrow A = 50.3°$.

Therefore B is 17.4 km from A on a bearing of 033.3°.

Exercise 3.2

1. In each of the following, solve the triangle ABC:
   (a) B = 60°, a = 12 cm, c = 15 cm ;
   (b) B = 117°, a = 3.4 m, c = 2.7 m ;
   (c) a = 17 cm, b = 21 cm, c = 34 cm ;
   (d) a = 5 m, b = 8 m, c = $5\sqrt{2}$ m.

Solving Triangles

2. A parallelogram has sides of lengths 5 cm and 6 cm, and one angle of 45.9°. Calculate the lengths of its diagonals.

3. Find the largest angle of the triangle whose sides have lengths
   (a)   4 cm, 6 cm, 7 cm ;   (b)   1.2 m, 85 cm, 77 cm.

4. In the triangle ABC, AB : AC = 5 : 7 and A = 60°. Find the ratio AB : BC and the measure of angle C.

5. Town A is 3 km from Town B in a direction of 031°, while Town C is 5.25 km from Town B in a direction of 057°. Calculate the distance between Towns A and C.

6. In the triangle ABC, BC = 15 cm, A = 75° and 3AC = 2AB. Calculate the length of AC and the measure of the angle B.

7. (a)   Find the angles of a triangle whose sides are in the ratio 3 : 5 : 7.
   (b)   Find the angles of a triangle whose sines are in the ratio 4 : 5 : 6.

8. The sides of a triangular field are 550 m, 680 m and 830 m. Calculate the area of the field in hectares.

Higher Level

9. A triangle has sides of lengths $(x + 1)$ cm, $(2x + 1)$ cm and $(2x + 3)$ cm.
   (a)   Show that $x > 1$.
   (b)   Find the angles of the triangle if $x = 10$.
   (c)   Find the value of x for which the triangle is right-angled.
   (d)   Find, in terms of x, the cosine of the largest angle, and hence find the value of x for which one angle of the triangle is 120°.
   (e)   Find the value of x for which one angle of the triangle is 60°.

10. Points O, A and B are at different levels. The bearing of A from O is 045° and its angle of elevation from O is 30°. The bearing of B from O is 310° and its angle of elevation from O is 60°. If A is 70 m higher than O, and B is 85 m higher than O, find the angle of elevation of B from A.

11. Points A and B are 5000 m apart. Hill, H, is on a bearing of 027.5° from A and 292° from B. If the angles of elevation of the top of the hill from A and B are 3.7° and 4.4°, respectively, find the height of the hill.

12. In a tetrahedron ABCD, B, C and D lie in a horizontal plane and AB is vertical. If AB = 15.5 cm, AC = 21.2 cm, AD = 24.8 cm and angle CBD = 75°, calculate the length of CD.

71

Chapter 3

Required Outcomes

After completing this chapter, a student should be able to
- apply the triangle inequality when appropriate.
- use the sine rule to solve a triangle which is not right-angled nor isosceles given two sides and a non-included angle or two angles and a side.
- find both triangles in the 'ambiguous case' involving the sine rule.
- use the cosine rule to solve a triangle which is not right-angled or isosceles given three sides or two sides and the included angle.
- calculate the area of a triangle either by using the formula $A = \frac{1}{2}bc\sin A$ directly or by using the cosine rule to first find an angle then applying this formula.

4 Trigonometry

4.1 The Sine, Cosine and Tangent Functions

Consider the **unit circle** – the circle in the xy-plane with centre at the origin and with a radius of 1 unit.

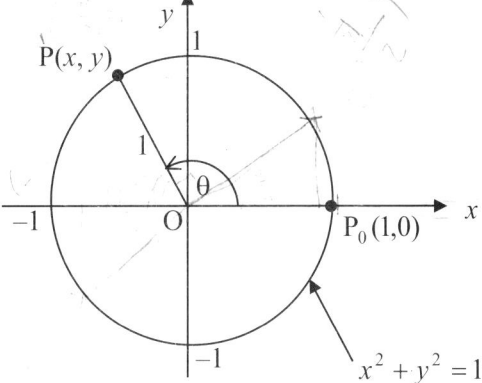

If $P(x, y)$ is any point on the unit circle, then $OP^2 = 1$.
Then $(x-0)^2 + (y-0)^2 = 1$ (distance formula), giving $x^2 + y^2 = 1$, which is the equation of the unit circle.
Let P_0 be the point $(1, 0)$. Rotate [OP] about O through angle θ with P initially at P_0.

Note: We consider θ to be positive if we rotate in an anticlockwise direction and negative if we rotate in a clockwise direction.

We define the cosine and sine of the angle θ to be the x- and y-coordinates respectively of the point P. That is $x = \cos\theta$ and $y = \sin\theta$. The tangent of the angle θ is defined as $\tan\theta = \dfrac{\sin\theta}{\cos\theta} = \dfrac{y}{x}$.

For example if $\theta = 180°$ (or $\theta = -180°$, etc.), P coincides with the point $(-1, 0)$. Thus $\sin 180° = 0$, $\cos 180° = -1$, $\tan 180° = 0/(-1) = 0$, $\sin(-180°) = 0$, $\cos(-180°) = -1$ and $\tan(-180°) = 0/(-1) = 0$.
Similarly if $\theta = 90°$ (or $\theta = -270°$, etc.), P coincides with the point $(0, 1)$. Thus $\sin 90° = 1$, $\cos 90° = 0$, $\tan 90° = 1/0$ which does not exist, $\sin(-270°) = 1$, $\cos(-270°) = 0$ and $\tan(-270°) = 1/0$ which does not exist.

73

Chapter 4

Since the equation of the unit circle is $x^2 + y^2 = 1$ and $x = \cos\theta$, $y = \sin\theta$, we have the very important relationship $\cos^2\theta + \sin^2\theta = 1$ which is true for **all** values of θ.

The Trigonometric Ratios of 0°, 30°, 45°, 60° and 90°

In order to establish the sine, cosine and tangent of angles of 0°, 30°, 45°, 60° and 90° we require the coordinates of the point P corresponding to each angle.

Consider the following diagrams of the unit circle.

Let P(x, y) correspond to $\theta = 60°$. Triangle OPP_0 is equilateral, so if M is the midpoint of $[OP_0]$ then $M = (\frac{1}{2}, 0)$ and [PM] is perpendicular to $[OP_0]$.

Thus $x = \frac{1}{2}$ and since $x^2 + y^2 = 1$ we find that $y = \frac{1}{2}\sqrt{3}$. Hence $P(\frac{1}{2}, \frac{1}{2}\sqrt{3})$ corresponds to $\theta = 60°$.

Let P(x, y) correspond to $\theta = 30°$. Let [PQ] which is perpendicular to $[OP_0]$, meet $[OP_0]$ at M and the unit circle at Q. Then triangle OPQ is equilateral and M is the mid-point of [PQ].

Thus $y = \frac{1}{2}$ and using symmetry and the above result we obtain $x = \frac{1}{2}\sqrt{3}$. Thus $P(\frac{1}{2}\sqrt{3}, \frac{1}{2})$ corresponds to $\theta = 30°$.

Let P(x, y) correspond to $\theta = 45°$. Let M lie on $[OP_0]$ such that [PM] is perpendicular to $[OP_0]$. Triangle OPM is isosceles since angles O and P are each equal to 45°. Thus $y = x$. From $x^2 + y^2 = 1$ we obtain $x = y = \frac{1}{2}\sqrt{2}$. Thus $P(\frac{1}{2}\sqrt{2}, \frac{1}{2}\sqrt{2})$ corresponds to $\theta = 45°$.

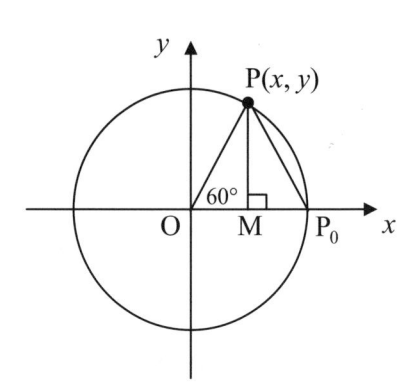

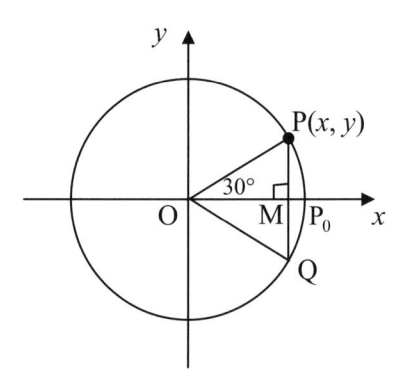

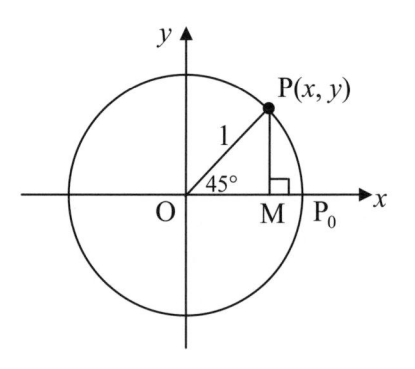

Trigonometry

We can now complete the following table, which could (should) be memorised.

θ	0°	30°	45°	60°	90°
sin θ	0	$\frac{1}{2}$	$\frac{1}{2}\sqrt{2}$	$\frac{1}{2}\sqrt{3}$	1
cos θ	1	$\frac{1}{2}\sqrt{3}$	$\frac{1}{2}\sqrt{2}$	$\frac{1}{2}$	0
tan θ	0	$\frac{1}{3}\sqrt{3}$	1	$\sqrt{3}$	—

Example Find, if possible, the three trigonometric ratios of 135° and 270°.

Using symmetry in the unit circle we find that θ = 135° corresponds to the point $(-\frac{1}{2}\sqrt{2}, \frac{1}{2}\sqrt{2})$.

Therefore $\sin 135° = y = \frac{1}{2}\sqrt{2}$, $\cos 135° = x = -\frac{1}{2}\sqrt{2}$ and $\tan 135° = \frac{y}{x} = -1$.

Clearly θ = 270° corresponds to the point (0, –1).
Therefore $\sin 270° = -1$, $\cos 270° = 0$ and $\tan 270°$ does not exist.

The coordinate plane is divided into 4 quadrants, numbered 1 to 4 as in the following diagram:

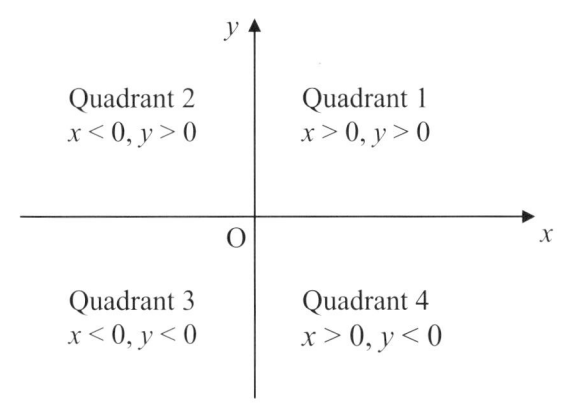

```
                      y ↑
                        |
      Quadrant 2        |    Quadrant 1
      x < 0, y > 0      |    x > 0, y > 0
                        |
    ––––––––––––––––––– O –––––––––––––––––→ x
                        |
      Quadrant 3        |    Quadrant 4
      x < 0, y < 0      |    x > 0, y < 0
                        |
```

In quadrant 1, both *x* and *y* are positive so all three trigonometric functions are positive here.

In quadrant 2, *x* is negative and *y* is positive so only sin θ is positive here.

In quadrant 3, *x* and *y* are both negative so only tan θ is positive here.

In quadrant 4, *x* is positive and *y* is negative so only cos θ is positive here.

Chapter 4

Exercise 4.1

1. Determine, if possible, the three trigonometric ratios of
   (a) $0°$;
   (b) $90°$;
   (c) $45°$;
   (d) $120°$;
   (e) $-210°$.

2. If P is a point on the unit circle and P_0 is the point $(1, 0)$, let θ be the angle POP_0. In which quadrant does P lie if
   (a) $\sin\theta$ is positive and $\tan\theta$ is negative;
   (b) $\cos\theta$ is negative and $\tan\theta$ is positive;
   (c) $\tan\theta$ is negative and $\sin\theta$ is negative;
   (d) $\cos\theta$ is negative and $\sin\theta$ is negative?

3. (a) Given $\sin\theta = 0.8$, find the possible values of $\cos\theta$ and $\tan\theta$.
   (b) Given $\cos\theta = 0.5$, find the possible values of $\sin\theta$ and $\tan\theta$.
   (c) Given $\tan\theta = -2$, find the possible values of $\sin\theta$ and $\cos\theta$.

4. For what points $P(x, y)$ on the unit circle is
   (a) $\sin\theta = 1$;
   (b) $\cos\theta = -0.8$;
   (c) $\tan\theta = 3$;
   (d) $\sin\theta = 0.2$;
   (e) $\cos\theta = -0.5$;
   (f) $\tan\theta = -5$.

5. Simplify the following expressions:
   (a) $\cos\theta \tan\theta$;
   (b) $\dfrac{\sin\theta}{\tan\theta}$;
   (c) $\dfrac{\cos\theta \tan\theta + \sin\theta}{\tan\theta}$;
   (d) $\tan\theta \left(\cos\theta + \dfrac{\sin\theta}{\tan\theta} \right)$;
   (e) $\dfrac{1}{\tan\theta}(\sin\theta + \cos\theta \tan\theta)$.

6. Which of the following points P lie on the unit circle?
   (a) $(0.6, 0.8)$;
   (b) $(-1, 0)$;
   (c) $(0.5, 0.5)$;
   (d) $(\tfrac{1}{2}\sqrt{2}, \tfrac{1}{2}\sqrt{2})$;
   (e) $(-0.28, 0.96)$;
   (f) $(-\tfrac{5}{13}, \tfrac{12}{13})$;
   (g) $(\tfrac{1}{\sqrt{5}}, -\tfrac{2}{\sqrt{5}})$;
   (h) $(-\tfrac{5}{9}, \tfrac{8}{9})$;
   (i) $(-\tfrac{15}{17}, -\tfrac{8}{17})$.

   Let θ be the angle subtended by the arc P_0P at the centre of the circle, where $P_0 = (1, 0)$. Determine, if possible, the values of $\sin\theta$, $\cos\theta$ and $\tan\theta$ in each case in which P lies on the unit circle.

7. Find, in each of the following, four possible values of θ for which
   (a) $\sin\theta = 0$;
   (b) $\cos\theta = 0$ or $\sin\theta = 0$;
   (c) $\tan\theta$ does not exist.

4.2 The Graphs of the Trigonometric Functions

The graphs of $y = \sin \theta$ and $y = \cos \theta$ for values of θ between $-360°$ and $360°$ are:

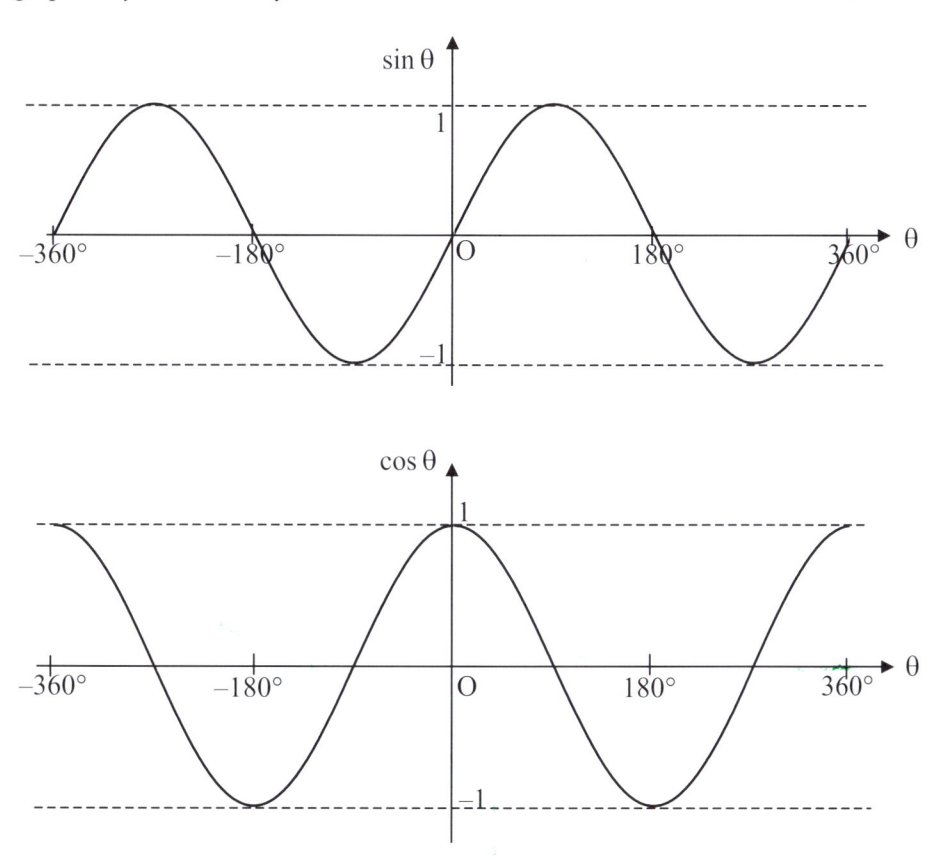

These graphs may be continued in either direction since the sine and cosine functions repeat their values every $360°$. This is due to the fact that the point on the unit circle corresponding to an angle θ also corresponds to angles $\theta + k(360°)$ for any integer k. Mathematically we say that the functions $\sin \theta$ and $\cos \theta$ are *periodic* with a period of $360°$.

Clearly each function has a *maximum* value of $+1$ and a *minimum* value of -1.

The *mean* value of each function is zero.

The maximum displacement from the mean position, called *the amplitude*, is 1.

Each function is *continuous* and each is defined for *all* values of θ.

Chapter 4

The graph of $y = \tan \theta$ has quite a different shape.

Firstly, $\tan \theta$ is undefined when $\cos \theta = 0$. This occurs when $P(\cos \theta, \sin \theta)$ has coordinates $(0, 1)$ or $(0, -1)$. The angles θ which correspond to either of these points are $\theta = (2k + 1)90°$, $k \in \mathbb{Z}$, i.e., all *odd* multiples of $90°$.

Secondly, the *period* of $\tan \theta$ is $180°$.

Thirdly, the tangent function has *no* maximum or minimum values.

The use of a graphic display calculator will confirm the following graph of $y = \tan \theta$ for values of θ between $-360°$ and $360°$:

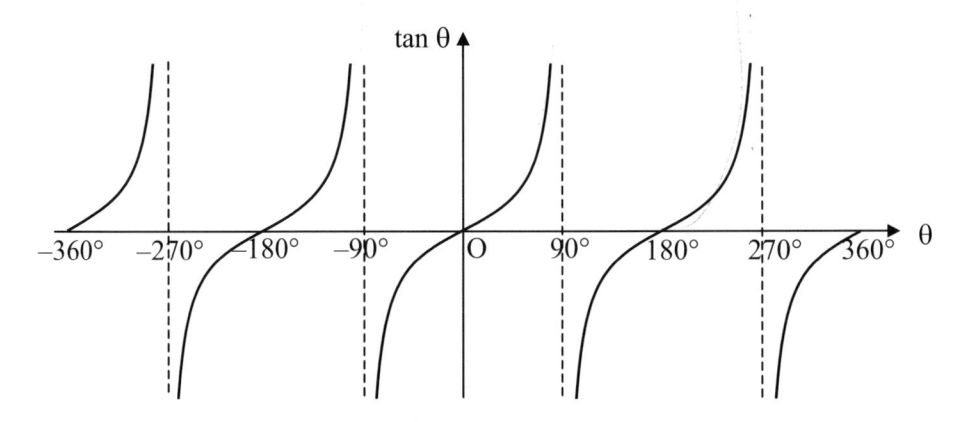

Exercise 4.2

1. Sketch the graph of $y = \sin \theta$ for $-180° \leq \theta \leq 180°$. Use the symmetry of your graph and a calculator to find all the values of θ, $-180° \leq \theta \leq 180°$, for which $\sin \theta = 0.4$.

2. Sketch the graph of $y = \cos \theta$ for $-360° \leq \theta \leq 360°$. Use the symmetry of your graph and a calculator to find all the values of θ, $-360° \leq \theta \leq 360°$, for which $\cos \theta = -0.2$.

3. Sketch the graph of $y = \tan \theta$ for $-270° < \theta < 270°$ ($\theta \neq -90°$ or $+90°$). Use the symmetry of your graph and a calculator to find all the values of θ in the given domain for which $\tan \theta = 2$.

4. By choosing a suitable transformation of the graph of $y = \sin \theta$ for $0° \leq \theta \leq 360°$, sketch the graph of $y = \sin(\theta + 90°)$ for $-90° \leq \theta \leq 270°$. Can you suggest a simpler expression for $\sin(\theta + 90°)$?

4.3 The Radian Measure of an Angle and the Double Angle Formulae

In plane geometry angles were measured in turns or degrees.
For example,

the measure of a right angle = $\frac{1}{4}$ turn = 90° ;

the measure of a straight angle = $\frac{1}{2}$ turn = 180° ;

the measure of a full turn = 360°.

It is convenient to introduce another unit for measuring angles – *the radian*.

An angle of 1 radian (1ᶜ) is the angle subtended by an arc of length 1 unit at the centre of the unit circle.

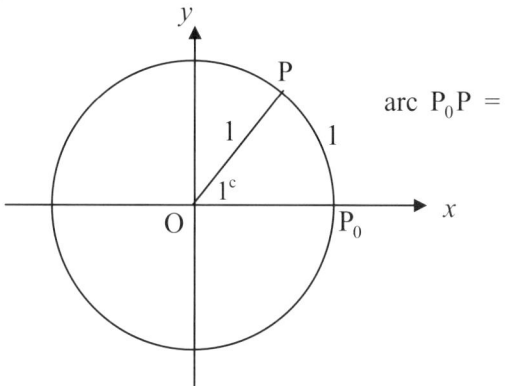

Note: 1ᶜ is also the size of the angle subtended at the centre of a circle of radius r by an arc of length r.

Now the angle subtended by the whole circumference of a unit circle at the centre is 360° and since the length of the circumference is 2π units we have

2π radians = 360 degrees   or π radians = 180 degrees.

Thus we have 1 radian = $\dfrac{180}{\pi}$ degrees ≈ 57.3 degrees.

Example   Convert 60° to radians and $\dfrac{7\pi}{6}$ radians to degrees.

60° = $\frac{1}{3}$ of 180° = $\frac{1}{3}\pi$ radians ;

$\dfrac{7\pi}{6}$ radians = $\dfrac{7}{6}$ × 180° = 210°.

Chapter 4

In all our work so far our trigonometric functions have mapped angles into real numbers. For example, the sine function maps an angle of 30° into the real number 0.5 since sin 30° = 0.5.

This is fine for surveyors, etc., but not always for scientists who often require the trigonometric functions (then called *circular functions*) to map real numbers into real numbers.

Therefore we define the circular functions of the **real number** x to be the trigonometric functions of the angle of x radians.

Thus if x is any real number, $\sin x = \sin x^c$, $\cos x = \cos x^c$ and $\tan x = \tan x^c$.

Example   Evaluate cos 2.

      (*Note:* cos 2 is not equal to cos 2°!)

      cos 2 = −0.416 (from a calculator in radian mode).

Double Angle Formulae

It can be shown (see Sections 4.11 and 4.13) that

    1. $\sin 2\theta = 2\sin\theta\cos\theta$
            and that
    2. $\cos 2\theta = \cos^2\theta - \sin^2\theta$.

[The proofs of these formulae are beyond the scope of the Standard Level course.]

Since $\sin^2\theta + \cos^2\theta = 1$, the second of the double angle formulae can be written in the forms

$$\cos 2\theta = \cos^2\theta - (1 - \cos^2\theta) = 2\cos^2\theta - 1$$

or $\cos 2\theta = (1 - \sin^2\theta) - \sin^2\theta = 1 - 2\sin^2\theta$.

The identities: $\sin 2\theta = 2\sin\theta\cos\theta$
                           $\cos 2\theta = \cos^2\theta - \sin^2\theta$
                               $= 2\cos^2\theta - 1$
                               $= 1 - 2\sin^2\theta$

are known as the "*double angle formulae*".

Trigonometry

Example   If $\sin A = \frac{4}{5}$ and $\frac{1}{2}\pi < A < \pi$, find the values of $\sin 2A, \cos 2A$ and $\tan 2A$ without the use of a calculator.

$\cos^2 A = 1 - \sin^2 A = 1 - \left(\frac{4}{5}\right)^2 = \frac{9}{25}$ and so $\cos A = \pm \frac{3}{5}$.

But A is in the second quadrant where cosines are negative so $\cos A = -\frac{3}{5}$.

Therefore $\sin 2A = 2 \sin A \cos A = 2\left(\frac{4}{5}\right)\left(-\frac{3}{5}\right) = -\frac{24}{25}$,

$\cos 2A = \cos^2 A - \sin^2 A = \left(-\frac{3}{5}\right)^2 - \left(\frac{4}{5}\right)^2 = -\frac{7}{25}$ and

$\tan 2A = \dfrac{\sin 2A}{\cos 2A} = \dfrac{-\frac{24}{25}}{-\frac{7}{25}} = \dfrac{24}{7}$.

Exercise 4.3

1. Convert each of the following radian measures to degree measure:
   (a) π^c ;   (b) $\frac{1}{2}\pi^c$;   (c) $\frac{2}{3}\pi^c$;   (d) $\frac{3}{5}\pi^c$;
   (e) $\frac{11}{9}\pi^c$;   (f) $\frac{5}{12}\pi^c$;   (g) 0.8^c ;   (h) 1.23^c.

2. Convert each of the following degree measures to radian measure:
   (a) $45°$;   (b) $150°$;   (c) $40°$;   (d) $105°$;
   (e) $195°$;   (f) $270°$;   (g) $32°$;   (h) $123°$.

3. Using the table of values found on page 75, and the symmetry of the unit circle, copy and complete the following table:

$\theta°$	0°	30°	45°	60°	90°	120°	135°	150°	180°
x^c									
$\sin x$									
$\cos x$									
$\tan x$									

4. Use your table in Question 3 to evaluate each of the following:
   (a) $\sin 30° + \cos 60°$;
   (b) $\tan 45° - \sin 30°$;
   (c) $\sin \frac{1}{4}\pi + \cos \frac{1}{4}\pi$;
   (d) $\sin \frac{1}{2}\pi + \cos \frac{1}{4}\pi$;
   (e) $\sin \frac{1}{6}\pi - \cos \frac{1}{2}\pi$;
   (f) $\sin^2 60° + \cos^2 30°$;
   (g) $\sin 150° - \cos 120°$;
   (h) $2\cos^2 \frac{2}{3}\pi - 1$;
   (i) $1 - 2\sin^2 \frac{5}{6}\pi$;
   (j) $\cos^2 \frac{3}{4}\pi - \sin^2 \frac{3}{4}\pi$;
   (k) $\sin^2 \frac{2}{3}\pi + \cos^2 \frac{2}{3}\pi$;
   (l) $\cos \frac{5}{6}\pi \cos \frac{2}{3}\pi + \sin \frac{5}{6}\pi \sin \frac{2}{3}\pi$.

81

Chapter 4

5. Verify the following for $\theta = \frac{1}{3}\pi$:
   (a) $\sin^2\theta + \cos^2\theta = 1$;
   (b) $\sin(\pi - \theta) = \sin\theta$.
   (c) $\sin 2\theta = 2\sin\theta\cos\theta$;
   (d) $\cos 2\theta = 1 - 2\sin^2\theta$;
   (e) $\tan 2\theta = \dfrac{2\tan\theta}{1 - \tan^2\theta}$;
   (f) $\cos 2\theta = \dfrac{1 - \tan^2\theta}{1 + \tan^2\theta}$.

   Are these also true for $\theta = \frac{3}{4}\pi$?

6. Evaluate each of the following without the use of a calculator.
   (a) $2\sin 15°\cos 15°$;
   (b) $2\sin 22\frac{1}{2}°\cos 22\frac{1}{2}°$;
   (c) $2\sin\frac{1}{12}\pi\cos\frac{1}{12}\pi$;
   (d) $\sin\frac{1}{8}\pi\cos\frac{1}{8}\pi$;
   (e) $\cos^2 30° - \sin^2 30°$;
   (f) $\sin^2 15° - \cos^2 15°$;
   (g) $2\cos^2\frac{1}{8}\pi - 1$;
   (h) $1 - 2\sin^2 15°$;
   (i) $2\sin^2\frac{1}{8}\pi - 1$.

7. Simplify each of the following:
   (a) $2\sin(x + 20°)\cos(x + 20°)$;
   (b) $\sin^2 2x - \cos^2 2x$;
   (c) $\cos^2 3x + \sin^2 3x$;
   (d) $1 - 2\sin^2(x + \frac{1}{4}\pi)$;
   (e) $1 - \sin^2\frac{1}{2}x$;
   (f) $(\sin x - \cos x)^2$.

8. Find the values of $\sin 2x$, $\cos 2x$ and $\tan 2x$ in each of the following without the use of a calculator:
   (a) $\sin x = \frac{3}{5}$;
   (b) $\cos x = \frac{12}{13}$;
   (c) $\tan x = -\frac{3}{4}$.

9. If $3\sin 2x = 2\cos x$, find the values of $\cos 2x$.

10. (a) If $\cos 2A = -\frac{1}{9}$ and $0 < A < \frac{1}{2}\pi$, find the values of $\sin A$, $\cos A$ and $\tan A$ without the use of a calculator.

    (b) If $\cos x = -\frac{7}{25}$, find $\sin\frac{1}{2}x$, $\cos\frac{1}{2}x$ and $\tan\frac{1}{2}x$ without the use of a calculator.

11. Simplify: (a) $\dfrac{\cos 2x + 1}{\cos x}$;
    (b) $\dfrac{\cos 2x - 1}{\sin 2x}$;
    (c) $\cos^4 x - \sin^4 x$.

82

Trigonometry

4.4 Graphs of Circular Functions of the Forms $y = a\sin x$

Multiplying the functions $\sin x$ and $\cos x$ by the positive number a simply changes the amplitude of each from 1 to a. The maximum value of $a\sin x$ (and $a\cos x$) is a and the minimum value of each is $-a$.

The graphs of $y = \sin x$, $y = 2\sin x$ and $y = 3\cos x$, $0 \leq x \leq 2\pi$, are as follows:

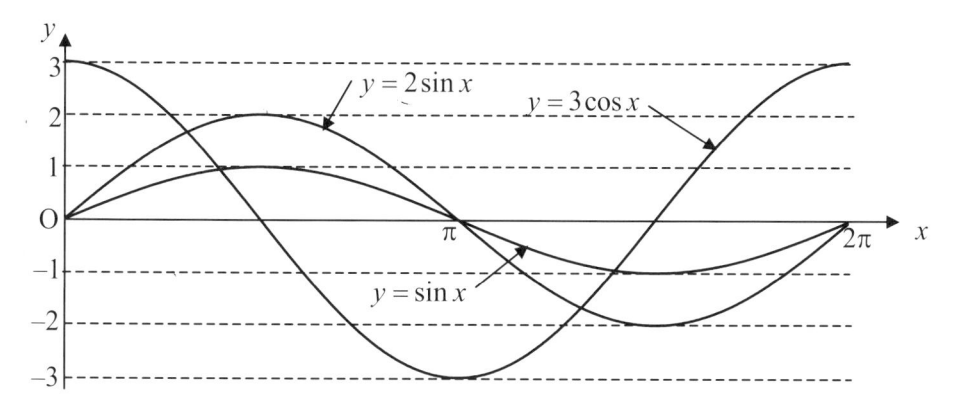

Multiplying the functions $\sin x$ and $\cos x$ by the negative number a first reflects the graphs in the x-axis and then changes the amplitude of each from 1 to $|a| = -a$.
The graphs of $y = \sin x$, $y = -\cos x$ and $y = -2\sin x$, $-\pi \leq x \leq \pi$, are as follows:

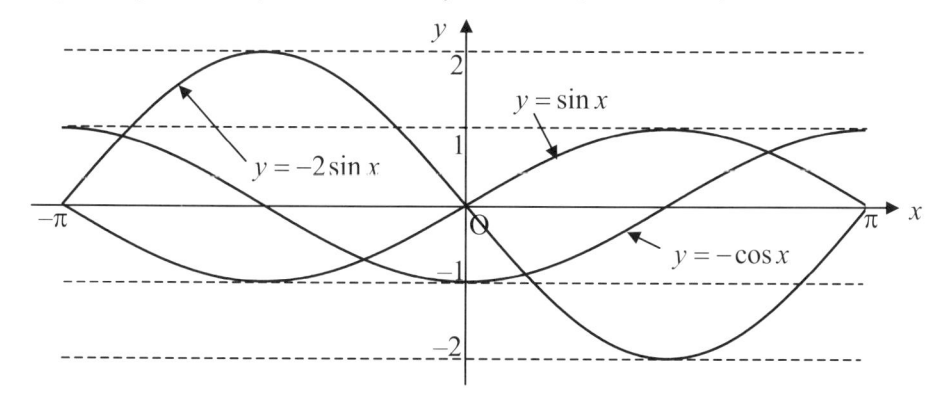

The corresponding effects on the graph of $y = \tan x$ are:
(1) $a > 0$ – a stretch parallel to the y-axis with scale factor a.
(2) $a < 0$ – a stretch parallel to the y-axis with scale factor $|a| = -a$ followed by a reflection in the x-axis (or y-axis in this case).

The graphs of $y = \tan x$, $y = 2\tan x$ and $y = -2\tan x$, $-\frac{1}{2}\pi < x < \frac{1}{2}\pi$, are:

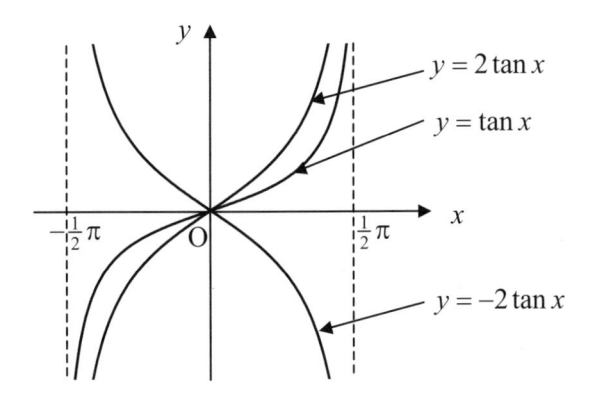

Chapter 4

Exercise 4.4

1. On the one set of coordinate axes, sketch the graphs of $y = \sin x$, $y = 2\sin x$ and $y = -\frac{1}{2}\sin x$, $0 \leq x \leq 2\pi$.

2. On the one set of coordinate axes, sketch the graphs of $y = \cos x$, $y = 3\cos x$ and $y = -2\cos x$, $-\pi \leq x \leq \pi$.

3. On the one set of coordinate axes, sketch the graphs of $y = \tan x$, $y = 3\tan x$ and $y = -\frac{1}{2}\tan x$, $-\frac{1}{2}\pi < x < \frac{1}{2}\pi$.

4.5 Graphs of Circular Functions of the Form $y = \sin bx$

Multiplying the arguments of the functions $y = \sin x$, $y = \cos x$ and $y = \tan x$ by the positive number b simply changes the period of the function.

Periodicity

Definition   A function $y = f(x)$ is **periodic with period p** if there exists a positive number p such that $f(x+p) = f(x)$ for all x in the domain of definition.

Example   Show that the function $f(x) = \sin 2x$ is periodic with period π.

$$\begin{aligned} f(x+\pi) &= \sin(2[x+\pi]) \\ &= \sin(2x + 2\pi) \\ &= \sin 2x \\ &= f(x), \text{ and so } f(x) \text{ is periodic with period } \pi. \end{aligned}$$

Trigonometry

Theorem If b is a positive number, $\sin bx$ and $\cos bx$ are each periodic with period $2\pi/b$, and $\tan bx$ is periodic with period π/b.

Proof Let $f(x) = \sin bx$, let $g(x) = \cos bx$ and let $h(x) = \tan bx$, $b > 0$.
Then $f(x + 2\pi/b) = \sin(b[x + 2\pi/b])$
$= \sin(bx + 2\pi)$
$= \sin bx$, and so $\sin bx$ is periodic with period $2\pi/b$;

and $g(x + 2\pi/b) = \cos(b[x + 2\pi/b])$
$= \cos(bx + 2\pi)$
$= \cos bx$, and so $\cos bx$ is periodic with period $2\pi/b$.

Also $h(x + \pi/b) = \tan(b[x + \pi/b])$
$= \tan(bx + \pi)$
$= \tan bx$, and so $\tan bx$ is periodic with period π/b.

Thus the periods of $\sin 2x$, $\cos \tfrac{1}{2}x$, $\tan 3x$ and $\sin \pi x$ are respectively $\pi \,(= 2\pi/2)$, $4\pi \,(= 2\pi/\tfrac{1}{2})$, $\pi/3$ and $2 \,(= 2\pi/\pi)$.

The graphs of $y = \sin x$, $y = \sin 2x$ and $y = \sin 3x$, $0 \le x \le \pi$, are:

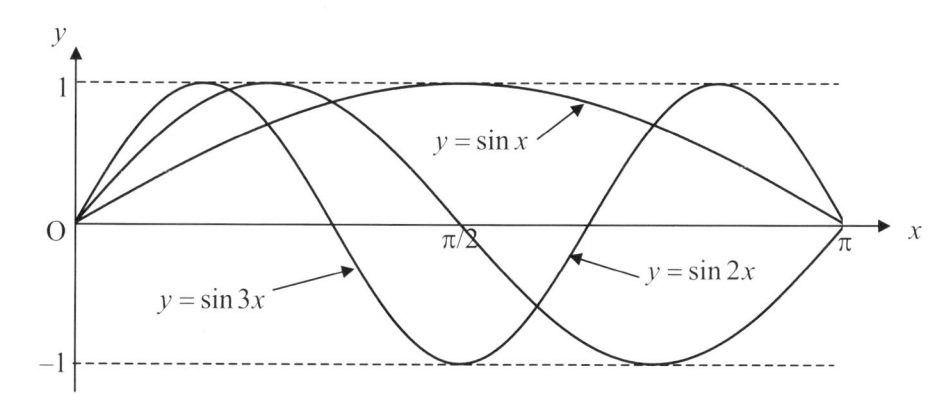

Multiplying the argument of the function by the negative number b, simply reflects the graph in the x-axis and changes the period from 2π for sine and cosine to $2\pi/|b|$ or from π for tangent to $\pi/|b|$.

85

Chapter 4

Exercise 4.5

1. On the one set of coordinate axes, sketch the graphs of $y = \sin 4x$, $y = \sin 2x$ and $y = \sin(-2x)$, $0 \le x \le \pi$.

2. On the one set of coordinate axes, sketch the graphs of $y = \cos 2x$, $y = \cos(-2x)$ and $y = \cos 3x$, $-\pi \le x \le \pi$.

3. On the one set of coordinate axes, sketch graphs of $y = \tan 2x$, $y = \tan\frac{1}{2}x$ and $y = \tan(-2x)$, $-\frac{1}{2}\pi < x < \frac{1}{2}\pi$.

4. Sketch the graph of each of the following functions:
   (a) $y = 2\sin 3x$, $0 \le x \le \pi$;
   (b) $y = 3\cos 2x$, $0 \le x \le \pi$;
   (c) $y = -\sin 3x$, $-\pi \le x \le \pi$;
   (d) $y = 4\sin(-2x)$, $-\pi \le x \le \pi$;
   (e) $y = -2\cos 4x$, $-\frac{1}{2}\pi \le x \le \frac{1}{2}\pi$;
   (f) $y = \frac{1}{2}\sin\frac{1}{2}x$, $-2\pi \le x \le 2\pi$.

5. Show that the function $f(x) = \sin 2x + 2\cos 5x$ is periodic with period 2π.

6. Show that the function $f(x) = \sin 3x + \sin 6x$ is periodic with period $2\pi/3$.

Higher Level

7. Find a value of p for each of the following functions given that each function is periodic with period p:
   (a) $f(x) = 3\sin 2x + \cos 2x$;
   (b) $f(x) = 3\cos 2x + \cos 4x$;
   (c) $f(x) = \tan x - \tan 3x$;
   (d) $f(x) = \sin 2x + \cos 3x$;
   (e) $f(x) = \sin 2x \cos 4x$;
   (f) $f(x) = \cos 3x \tan x$.

4.6 Graphs of Circular Functions of the Form $y = \sin(x - c)$

If P'(x', y') is the image of P(x, y) under a translation of c units to the right, c > 0, then $x' = x + c$ (or $x = x' - c$) and $y' = y$. Under this transformation, $y = f(x)$ becomes $y' = f(x' - c)$ which is the condition that P'(x', y') lies on the curve with equation $y = f(x - c)$. Thus under such a transformation the image of $y = f(x)$ is $y = f(x - c)$.

Trigonometry

Subtracting the positive number c from the argument simply translates the graph of the function c units to the right. Adding the positive number c to the argument simply translates the graph c units to the left.

The graphs $y = \sin x$, $y = \sin(x - \frac{1}{4}\pi)$ and $y = \sin(x + \frac{1}{4}\pi)$, $0 \le x \le 2\pi$ are:

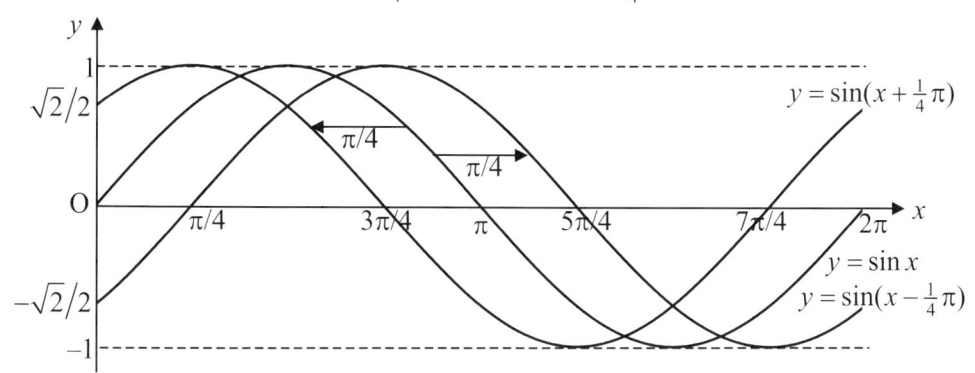

Note: In order to sketch graphs of trigonometric functions with arguments of the form $(bx + c)$, we must first express the argument in the form $b(x + c/b)$. This gives a function with period $2\pi/|b|$ shifted horizontally a distance $|c/b|$.

For example, the function $y = \sin(2x + \frac{1}{2}\pi) = \sin 2(x + \frac{1}{4}\pi)$ has period $\pi = 2\pi/2$ and its graph is that of $y = \sin 2x$ shifted $\pi/4$ to the left.

Exercise 4.6

1. On the one set of coordinate axes, sketch graphs of $y = \sin x$, $y = \sin(x - \frac{1}{2}\pi)$ and $y = \sin(x + \frac{1}{3}\pi)$, $-\pi \le x \le \pi$.

2. On the one set of coordinate axes, sketch graphs of $y = \cos x$, $y = \cos(x + \frac{1}{6}\pi)$ and $y = \cos(x - \frac{2}{3}\pi)$, $0 \le x \le 2\pi$.

3. Sketch the graph of each of the following:
   (a) $y = \sin(x - \pi)$, $0 \le x \le 2\pi$;
   (b) $y = \cos(x + \frac{1}{2}\pi)$, $-\pi \le x \le \pi$;
   (c) $y = 2\sin(x + \frac{1}{4}\pi)$, $0 \le x \le 2\pi$;
   (d) $y = 3\cos(x - 1)$, $0 \le x \le 2\pi$;
   (e) $y = \tan(x - \frac{1}{4}\pi)$, $-\frac{1}{4}\pi < x < \frac{3}{4}\pi$;
   (f) $y = \tan(x + \frac{1}{3}\pi)$, $-\frac{5}{6}\pi < x < \frac{1}{6}\pi$.

4. Sketch the graph of each of the following:
   (a) $y = \sin(2x - 2)$, $0 \le x \le \pi$;
   (b) $y = \cos(\pi - 3x)$, $-\pi \le x \le \pi$;
   (c) $y = 3\sin(2x + \frac{1}{2}\pi)$, $-\pi \le x \le \pi$;
   (d) $y = -2\sin(3x - 2)$, $0 \le x \le 3$.

87

Chapter 4

4.7 Graphs of Circular Functions of the Form $y = \sin x + d$

Adding the positive constant d to the functions simply translates their graphs d units upwards. Adding the negative constant d to the functions simply translates their graphs $|d| = -d$ units down.

The graphs of $y = \cos x$, $y = \cos x + 2$ and $y = \cos x - 1$ are:

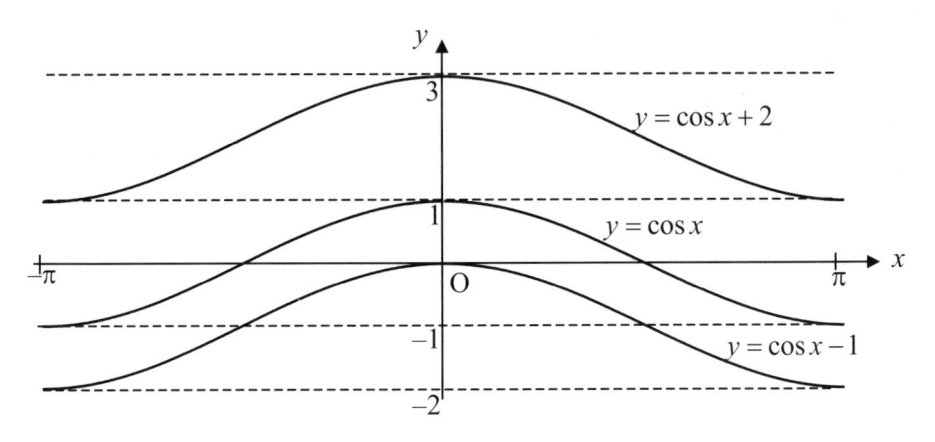

Example For the function $y = 2\sin 3x + 1$, find
(a) the period ; (b) the maximum and minimum values.
Sketch the graph of the function for $0 \leq x \leq \pi$.

(a) The period $= 2\pi/3$.
(b) The amplitude is 2 and so the maximum and minimum values of y are 1 ± 2, i.e., 3 and -1.

The graph of $y = 2\sin 3x + 1$ follows:

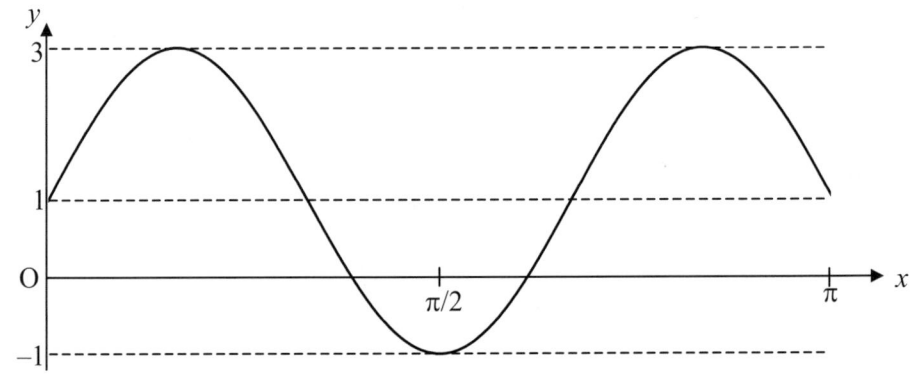

Trigonometry

Example   Sketch the graph of the function $y = 4 - 3\cos 2x$ for $-\pi \leq x \leq \pi$.

The period of this function is $2\pi/2 = \pi$ and the maximum and minimum values are 4 ± 3, i.e., 7 and 1.
Also to obtain the graph of this function we must first reflect the graph of $y = 3\cos 2x$ in the x-axis to give the graph of $y = -3\cos 2x$ as follows:

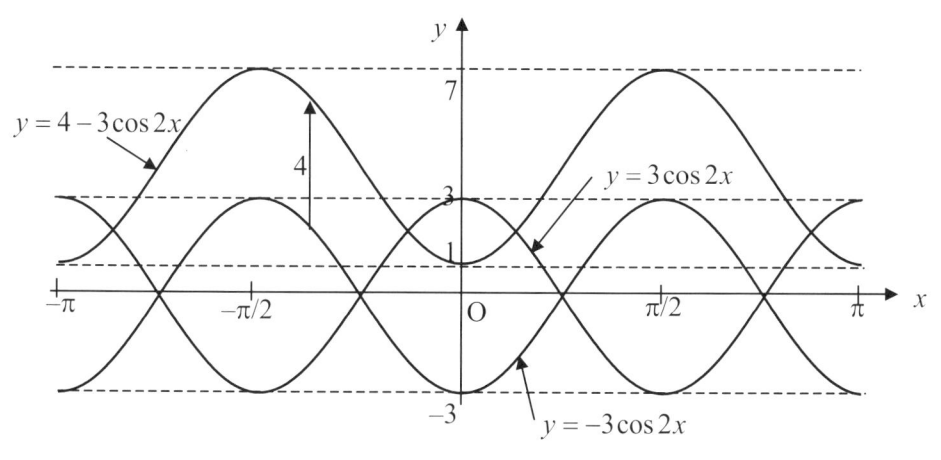

Exercise 4.7

1. On the one set of coordinate axes, sketch graphs of each of the following pairs of functions:
   (a) $y = \sin x$ and $y = 2\sin x - 1$, $0 \leq x \leq 2\pi$;
   (b) $y = \cos 3x$ and $y = 3\cos 3x - 2$, $0 \leq x \leq \pi$;
   (c) $y = 4\cos x$ and $y = 1 - 4\cos x$, $-\frac{1}{2}\pi \leq x \leq \frac{1}{2}\pi$;
   (d) $y = 3\sin \frac{1}{2}x$ and $y = 4 - 3\sin \frac{1}{2}x$, $-2\pi \leq x \leq 2\pi$.

2. Sketch graphs of each of the following functions for the domain indicated:
   (a) $y = 4\cos 3x - 3$, $0 \leq x \leq \pi$;
   (b) $y = 3\tan x + 2$, $-\frac{1}{2}\pi < x < \frac{1}{2}\pi$;
   (c) $y = 2 - \sin(x - \frac{1}{6}\pi)$, $-\pi \leq x \leq \pi$;
   (d) $y = 3\cos 2x + 2$, $0 \leq x \leq \pi$;
   (e) $y = \sin(2x - \frac{1}{2}\pi) - 1$, $0 \leq x \leq 2\pi$;
   (f) $y = 2 - 3\sin \pi x$, $-2 \leq x \leq 2$;
   (g) $y = 3 - 2\sin(2x + 1)$, $0 \leq x \leq 2\pi$;
   (h) $y = 4\cos(2 - 3x) + 1$, $0 \leq x \leq 4$.

Chapter 4

4.8 Mensuration of the Circle

The Length of an Arc of a Circle

Consider a circle of radius r. Let s be the length of the arc of this circle which subtends an angle of θ radians at the centre.

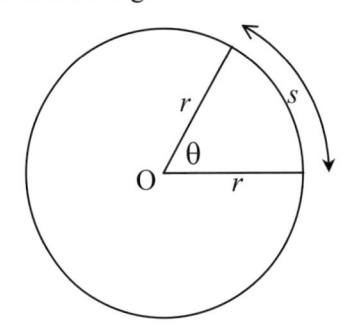

From the diagram,
$$\frac{\text{length of the arc}}{\text{circumference}} = \frac{\text{angle subtended by the arc}}{2\pi}.$$

Thus $\dfrac{s}{2\pi r} = \dfrac{\theta}{2\pi}$ which gives $s = r\theta$.

That is the length s of the arc of a circle of radius r which subtends an angle of θ radians at the centre of the circle is given by

$$s = r\theta.$$

Example Find the length of the arc which subtends an angle of $60°$ at the centre of a circle of radius 4 cm.

The required length $= 4 \times \frac{1}{3}\pi = \frac{4}{3}\pi$ cm.

Example Find the angle at the centre of the circle of radius 10 cm subtended by an arc of length 6 cm.

$s = r\theta$ and so $\theta = \dfrac{s}{r} = \dfrac{6}{10} = 0.6^c \approx 34.4°$.

The Area of a Sector of a Circle

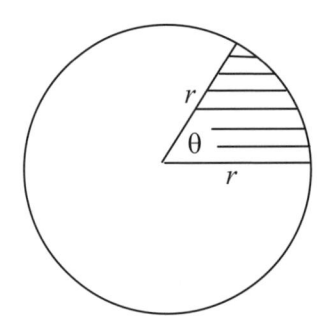

If θ radians is the angle of a sector of a circle of radius r, then

$$\frac{\text{area of sector}}{\text{area of circle}} = \frac{\text{angle of the sector}}{2\pi}.$$

Thus $A(\text{sector}) = \pi r^2 \times \dfrac{\theta}{2\pi} = \frac{1}{2}r^2\theta$.

Trigonometry

That is the area of the sector of a circle of radius r which subtends an angle of θ radians at the centre, is given by

$$A = \tfrac{1}{2}r^2\theta .$$

Example Calculate the area of the sector of a circle of radius 6 cm which subtends an angle of 75° at the centre.

The required area $= \tfrac{1}{2}r^2\theta = \tfrac{1}{2} \times 36 \times \tfrac{75}{180}\pi = 7.5\pi$ cm².

Example An arc of a circle of radius 8 cm is 12 cm in length. Find the area of the sector bounded by this arc and the radii at its ends.

$s = r\theta$ which gives $\theta = \dfrac{s}{r} = 1.5^c$.

Therefore the area of the sector $= \tfrac{1}{2} \times 8^2 \times 1.5 = 48$ cm².

The Area of a Segment of a Circle

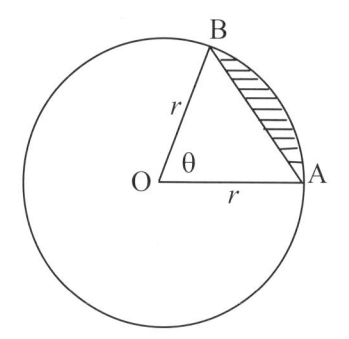

In the diagram, the area of the shaded segment is given by

Area of segment
= area of sector OAB − area of triangle OAB
$= \tfrac{1}{2}r^2\theta - \tfrac{1}{2}r^2 \sin\theta$.

That is the area of the segment of a circle of radius r which subtends an angle of θ radians at the centre, is given by

$$A = \tfrac{1}{2}r^2(\theta - \sin\theta) .$$

Example Find the area of the segment of the circle of radius 20 cm which subtends an angle of 72° at the centre.

The required area $= \tfrac{1}{2}(20)^2(\tfrac{2}{5}\pi - \sin\tfrac{2}{5}\pi)$ or $\tfrac{1}{2}(20)^2(\tfrac{2}{5}\pi - \sin 72°)$
$= 61.1$ cm².

Note: The first expression above is the one to be evaluated using a calculator when radian mode is used, and the second, when degree mode is used.

Chapter 4

Example Calculate the area of the segment of the circle of radius 4 cm if the arc of the segment has length 6 cm.

$$s = r\theta \text{ gives } \theta = \frac{s}{r} = 1.5^c.$$

Therefore the area of the segment = $\frac{1}{2}(16)(1.5 - \sin 1.5) = 4.02 \text{ cm}^2$, making sure that your calculator is in radian mode.

Exercise 4.8

1. Given that s is the length of the arc of a circle of radius r which subtends an angle of θ at the centre of the circle, find s when
   (a) $r = 10$ cm and $\theta = 30°$;
   (b) $r = 12$ cm and $\theta = 50°$;
   (c) $r = 1.2$ m and $\theta = 123°$;
   (d) $r = 65$ cm and $\theta = 36.5°$.

2. Given that s is the length of the arc of a circle of radius r which subtends an angle of θ at the centre of the circle, find:
   (a) θ when $s = 3$ cm and $r = 2$ cm ;
   (b) θ when $s = 1.0$ m and $r = 80$ cm ;
   (c) r when $s = 12.5$ cm and $\theta = 110°$;
   (d) r when $s = 95$ cm and $\theta = 56.5°$.

3. Chord PQ subtends an angle of 40° at O, the centre of a circle of radius 12 cm. Calculate:
   (a) the length of the chord PQ ;
   (b) the length of the minor arc PQ.

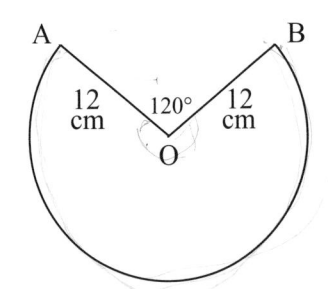

4. A right-circular cone is made by cutting a sector of a circle of angle 120° from a circle of radius 12 cm and joining the two radii OA, OB.
   Calculate:
   (a) the length of the major arc AB ;
   (b) the perpendicular height of the cone.

5. (a) Calculate the perimeter of a sector of a circle of radius 25 cm if the angle of the sector is 160°.
   (b) The sides of the sector formed by the radii in part (a) are joined together to form a right-circular cone. Find the base-radius of the cone and its perpendicular height.

Trigonometry

6. Two wheels of radii 10 cm and 5 cm have their centres, X and Y, 25 cm apart. The wheels are connected by a taut belt as shown in the following diagram. Find the length of the belt. [*Hint:* [XA] and [YB] are parallel.]

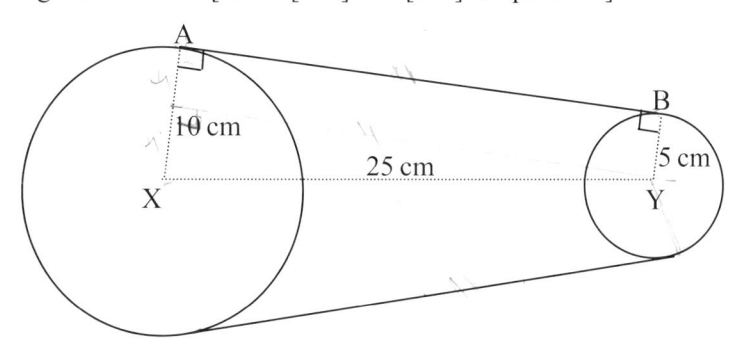

7. A circle of radius 2 m is drawn on a rectangle of dimensions 6 m by 3 m with the centre of the circle at the point of intersection of the diagonals of the rectangle. Find the area of overlap.

8. Given that s is the length of the arc of a circle of radius r which subtends an angle of θ at the centre of the circle, find the area of the sector and the segment defined by:
   (a) $r = 10$ cm and $\theta = 50°$;
   (b) $r = 2.6$ cm and $\theta = 48°$;
   (c) $s = 12.5$ cm and $\theta = 110°$;
   (d) $s = 1.05$ m and $\theta = 64.5°$.

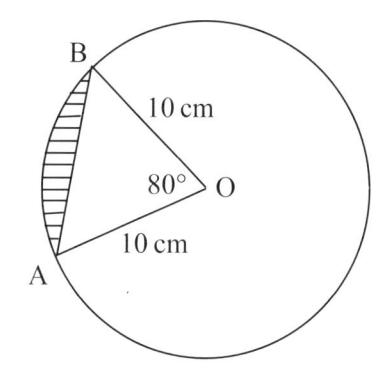

9. Chord AB subtends an angle of 80° at O, the centre of a circle of radius 10 cm. Calculate:
   (a) the length of the chord AB ;
   (b) the length of the minor arc AB ;
   (c) the area of the shaded segment.

10. Find the area of the segment of a circle of radius 12 cm cut off by a chord of length 10 cm.

11. Show that the curved surface area of a right-circular cone of base-radius r and slant-height s is given by $A = \pi rs$.

12. Two circles of radii 15 cm and 8 cm have their centres 17 cm apart. Find the area common to both circles.

93

Chapter 4

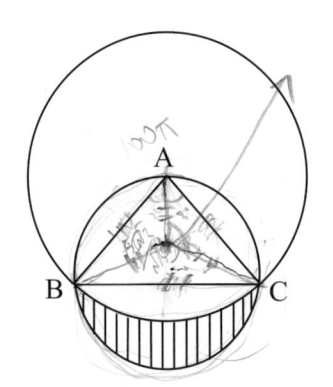

13. Consider two circles. One circumscribes an equilateral triangle ABC of side 10 cm, and the other has A as its centre and passes through both B and C.

Find the area shaded in the diagram.

14. Ten equal circles of radii 10 cm fit snugly into an equilateral triangle as shown in the diagram. Find:
   (a) the length of each side of the equilateral triangle;
   (b) the area of the equilateral triangle not covered by the circles.

4.9 Trigonometric Equations

The Solution of Equations Reducible to the Forms sin x = a, cos x = a, and tan x = b for $|a| \leq 1$ and $b \in \mathbb{R}$

Because of the periodic nature of the trigonometric functions, equations of the form sin x = a and cos x = a have an infinite number of solutions if $|a| \leq 1$. Equations of the form tan x = b have an infinite number of solutions for all real b.

Example   Solve the equation 2 cos x = 1.

$$2 \cos x = 1 \quad \Rightarrow \quad \cos x = \tfrac{1}{2}, \text{ and so } x = \tfrac{1}{3}\pi, \tfrac{5}{3}\pi \ (+2\pi k, k \in \mathbb{Z}).$$

This solution is known as the **general solution** of the given equation.

If we restrict our domain so that $0 \leq x \leq 2\pi$, there are only two solutions. They are $x = \tfrac{1}{3}\pi, \tfrac{5}{3}\pi$.

On the other hand, if we restrict our domain so that $-\pi \leq x \leq \pi$, the only solutions are $x = \pm\tfrac{1}{3}\pi$.

The following examples illustrate the general method.

Example   Solve the equation tan x + 1 = 0 for $0° \leq x \leq 360°$.

$\tan x + 1 = 0 \quad \Rightarrow \quad \tan x = -1$

The 'working angle' = 45°. We illustrate the possible quadrants for any solution with the following diagram:

The general solution is $x = 45° (+180°k, k \in \mathbb{Z})$.
[$\tan x$ has period 180°]
For $0° \leq x \leq 360°$, $x = 135°, 315°$.

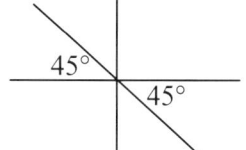

Example Solve the equation $4 \sin x + 3 = 0$ for $-\pi \leq x \leq \pi$.

$4 \sin x + 3 = 0 \Rightarrow \sin x = -0.75$
The 'working angle' is that value of x, $0 \leq x \leq \frac{1}{2}\pi$, such that $\sin x = 0.75$.
The working angle = 0.8481.
The corresponding diagram is:

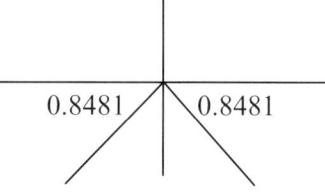

Therefore $x = \pi + 0.8481, 2\pi - 0.8481 (+2\pi k, k \in \mathbb{Z})$
$= 3.990, 5.435 (+2\pi k)$
$= -2.29, -0.848 (-\pi \leq x \leq \pi)$.

Example Solve the equation $2\cos^2 x + 3\cos x - 2 = 0$ for $0 \leq x \leq 2\pi$.

$2\cos^2 x + 3\cos x - 2 = 0$
$\Rightarrow \quad (2\cos x - 1)(\cos x + 2) = 0$
$\Rightarrow \quad \cos x = \frac{1}{2}$ or $\cos x = -2$.
But $\cos x \neq -2$ since $|\cos x| \leq 1$.
Therefore $\cos x = \frac{1}{2}$ only.
The working angle = $\frac{1}{3}\pi$.
The general solution is $x = \pm\frac{1}{3}\pi (+2\pi k, k \in \mathbb{Z})$.
Thus $x = \frac{1}{3}\pi, \frac{5}{3}\pi$ $(0 \leq x \leq 2\pi)$.

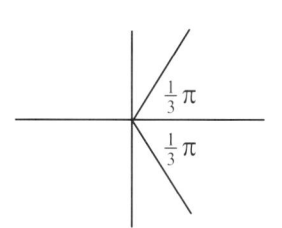

Example Solve the equation $2\cos^2 x - \sin x = 1$ for $-180° \leq x \leq 180°$.

$2\cos^2 x - \sin x = 1$
$\Rightarrow \quad 2(1 - \sin^2 x) - \sin x - 1 = 0$
$\Rightarrow \quad 2\sin^2 x + \sin x - 1 = 0$
$\Rightarrow \quad (2\sin x - 1)(\sin x + 1) = 0$
$\Rightarrow \quad \sin x = \frac{1}{2}$ or $\sin x = -1$.

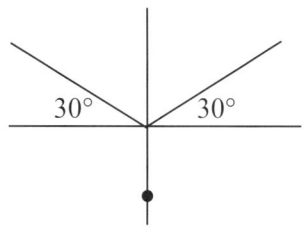

95

Chapter 4

The working angles are 30°, 90°.
The general solution is $x = 30°, 150°, -90°$ $(+360°k, k \in \mathbb{Z})$.
Thus $x = 30°, 150°, -90°$ $(-180° \leq x \leq 180°)$.

Example   Solve the equation $\sqrt{3} \cos x - \sin x = 0$ for $-\pi \leq x \leq \pi$.

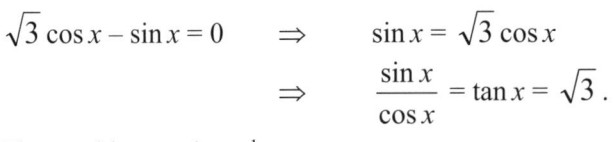

$$\sqrt{3} \cos x - \sin x = 0 \implies \sin x = \sqrt{3} \cos x$$
$$\implies \frac{\sin x}{\cos x} = \tan x = \sqrt{3}.$$

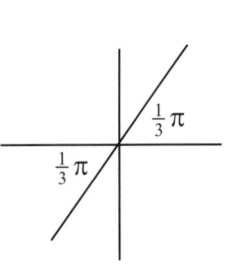

The working angle $= \frac{1}{3}\pi$.

The general solutions is $x = \frac{1}{3}\pi$ $(+\pi k, k \in \mathbb{Z})$.

Thus $x = \frac{1}{3}\pi, -\frac{2}{3}\pi$ $(-\pi \leq x \leq \pi)$.

Example   Solve the equation $\cos 2x - \cos x = 0$ for $0° \leq x \leq 360°$.

$$\cos 2x - \cos x = 0$$
$$\implies 2\cos^2 x - 1 - \cos x = 0$$
$$\implies 2\cos^2 x - \cos x - 1 = 0$$
$$\implies (2\cos x + 1)(\cos x - 1) = 0$$
$$\implies \cos x = -\tfrac{1}{2} \text{ or } \cos x = 1.$$

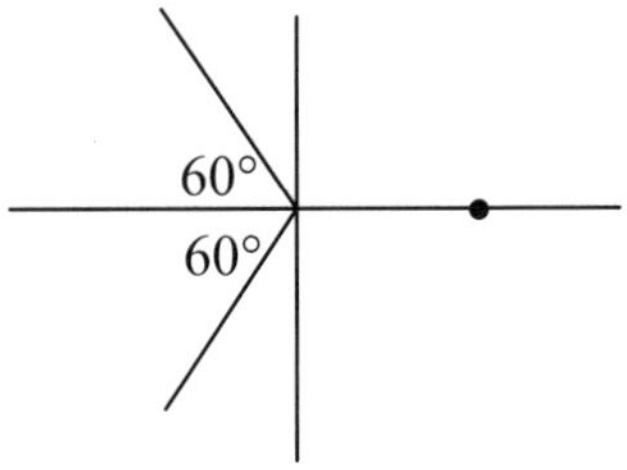

The working angles are 0° and 60°.

Therefore $x = 0°, 120°, 240°, 360°$ $(0° \leq x \leq 360°)$.

The Solution of Equations of the Form sin (x + α) = k

The method is best illustrated with examples.

Example   Solve the equation $2 \sin (x - 60°) = 1$ for $0° \leq x < 360°$.

$$2 \sin (x - 60°) = 1$$
$$\implies \sin (x - 60°) = \tfrac{1}{2}.$$

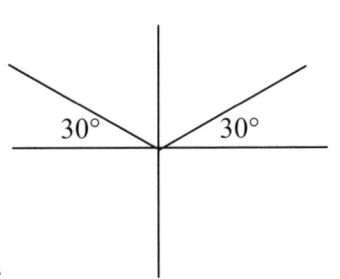

The working angle = 30°.
The general solution is
$x - 60° = 30°, 150°$ $(+360°k, k \in \mathbb{Z})$.
Thus $x = 90°, 210°$ $(+360°k)$,
and so $x = 90°, 210°$ $(0° \leq x < 360°)$.

96

Example   Solve the equation $3\cos(x - \frac{1}{6}\pi) + 1 = 0$ for $-\pi < x \le \pi$.

$$3\cos(x - \tfrac{1}{6}\pi) + 1 = 0$$
$\Rightarrow \quad \cos(x - \tfrac{1}{6}\pi) = -\tfrac{1}{3}.$

The working angle ≈ 1.231.
The general solution is
$x - \tfrac{1}{6}\pi = \pi \pm 1.231 \; (+2\pi k, \; k \in \mathbb{Z})$.
Thus $x = \tfrac{7}{6}\pi \pm 1.231 \; (+2\pi k)$, which gives $x = -1.40,\; 2.44 \; (-\pi < x \le \pi)$.

The Solution of Equations of the Form $\sin nx = k$

Once again the method is best illustrated with examples.

Example   Solve the equation $3\sin 3x + 1 = 0$ for $0 \le x \le \pi$.

$3\sin 3x + 1 = 0$
$\Rightarrow \quad \sin 3x = -\tfrac{1}{3}.$

The working angle = 0.3398.
Thus $3x = -0.3398,\; -\pi + 0.3398 \; (+2\pi k,\; k \in \mathbb{Z})$.
This gives the general solution
$x = -0.1133,\; -0.9339 \; (+\tfrac{2}{3}\pi k)$.
Thus $x = 1.16,\; 1.98 \; (0 \le x \le \pi)$.

Example   Solve the equation $\tan 4x = \sqrt{3}$ for $0° \le x \le 180°$.

$\tan 4x = \sqrt{3}$
The working angle is $60°$ and so
$4x = 60° \; (+180°k,\; k \in \mathbb{Z})$.
The general solution is $x = 15° \; (+45°k)$.
Thus $x = 15°,\; 60°,\; 105°,\; 150° \; (0° \le x \le 180°)$.

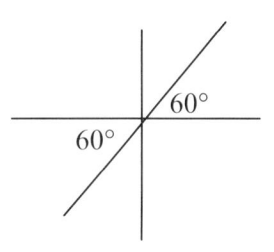

Example   Solve the equation $\dfrac{1}{\tan x} + \tan x = 4$ for $-\pi \le x \le \pi$.

$\dfrac{1}{\tan x} + \tan x = 4 \quad \Rightarrow \quad 1 + \tan^2 x = 4\tan x$

$\Rightarrow \quad \tan^2 x - 4\tan x + 1 = 0$

$\Rightarrow \quad \tan x = \dfrac{4 \pm \sqrt{16-4}}{2} = 2 \pm \sqrt{3}.$

97

Chapter 4

$\tan x = 2 + \sqrt{3} \Rightarrow x = 1.31, 1.309 - \pi = 1.31, -1.83$.

$\tan x = 2 - \sqrt{3} \Rightarrow x = 0.262, 0.2618 - \pi = 0.262, -2.88$.

Therefore $x = -2.88, -1.83, 0.262, 1.31 \ (-\pi \le x \le \pi)$.

Exercise 4.9

1. Find the values of x, $0 \le x < 2\pi$, for which
   (a) $2\cos x + 1 = 0$;
   (b) $2 \sin x = \sqrt{3}$;
   (c) $\tan x = -\frac{1}{2}$;
   (d) $3 \sin x + 4 \cos x = 0$;
   (e) $3 \sin x = 5$;
   (f) $4 \cos x + 1 = 0$;
   (g) $9\cos^2 x = 4$;
   (h) $\sin^2 x = \cos^2 x$;
   (i) $2 \sin^2 x + \sin x = 0$;
   (j) $2 \tan^2 x - 5 \tan x + 2 = 0$;
   (k) $3 \sin^2 x = 2(\cos x + 1)$;
   (l) $4 \tan^2 x - 1 = 3 \tan x$.

2. For each of the following equations, find the solutions in the given interval:
   (a) $\cos(x + 30°) = 1$, $\quad 0° \le x < 360°$;
   (b) $3 \tan(x - 1) = 4$, $\quad 0 \le x < 2\pi$;
   (c) $3 \sin(x + 1) = 2$, $\quad 0 \le x < 2\pi$;
   (d) $4 \cos(x - 42°) + 1 = 0$, $\quad -180° \le x \le 180°$;
   (e) $3 \sin(x - \frac{1}{2}\pi) = 2$, $\quad 0 \le x < 2\pi$;
   (f) $5 \tan(x + 123°) = 12$, $\quad 0° \le x < 360°$.

3. Solve the following equations for $0° \le x < 360°$:
   (a) $\sin 3x = 0.5$;
   (b) $\cos 2x + 1 = 0$;
   (c) $\sin 2x + 3 \cos 2x = 0$;
   (d) $\tan 3x = 2$;
   (e) $2 \cos(2x + 25°) + 1 = 0$;
   (f) $3 \sin(2x - 45°) = 2$;
   (g) $\sin 3x = 2 \cos 3x$;
   (h) $5 \cos 3x = 1$;
   (i) $\cos^2 2x + 2 \cos 2x = 0$;
   (j) $\sin^2 2x + \cos 2x + 1 = 0$.

4. For each of the following equations, find the solutions in the given interval:
   (a) $\cos x = 0.25$, $\quad 0 < x < 2\pi$;
   (b) $3\cos^2 4x - \cos 4x = 0$, $\quad -90° \le x \le 90°$;
   (c) $4 \tan(3x + 2) = 5$, $\quad 0 \le x \le \pi$;
   (d) $3 \sin(3 - 2x) + 1 = 0$, $\quad -\frac{1}{2}\pi \le x \le \frac{1}{2}\pi$;
   (e) $6\sin^2 2x - 5 \sin 2x + 1 = 0$, $\quad 0 \le x \le \pi$;
   (f) $\sin 3x + 2 \cos 3x = 0$, $\quad -3 < x < 3$;
   (g) $\tan 2x = \cos 2x$, $\quad 0° \le x \le 180°$;
   (h) $2 - 3 \tan 4x = 0$, $\quad 0° \le x \le 90°$;
   (i) $\sin^2 3x = \cos 3x - 1$, $\quad 0 \le x \le \pi$.

(j) $\sin^2 3x + \cos^2 3x = 1$, $\quad 0° \leq x \leq 180°$;
(k) $4\sin^3 3x - \sin 3x = 0$, $\quad 0° \leq x \leq 180°$;
(l) $\sin 2x = \tan x$, $\quad 0 \leq \pi \leq 2\pi$;
(m) $\cos 2x = 3 - 5\sin x$, $\quad 0° \leq x \leq 180°$.

Higher Level

4.10 The Six Circular Functions, their Graphs and Identities

If $P(x, y)$ is the point of the unit circle corresponding to angle θ, we have

the ***sine*** of angle θ: $\quad \sin\theta = y$;

the ***cosine*** of angle θ: $\quad \cos\theta = x$;

the ***tangent*** of angle θ: $\quad \tan\theta = \dfrac{\sin\theta}{\cos\theta} = \dfrac{y}{x}$ $(x \neq 0)$;

the ***cosecant*** of angle θ: $\quad \csc\theta = \dfrac{1}{\sin\theta} = \dfrac{1}{y}$ $(y \neq 0)$;

the ***secant*** of angle θ: $\quad \sec\theta = \dfrac{1}{\cos\theta} = \dfrac{1}{x}$ $(x \neq 0)$;

the ***cotangent*** of angle θ: $\quad \cot\theta = \dfrac{1}{\tan\theta} = \dfrac{\cos\theta}{\sin\theta} = \dfrac{x}{y}$ $(y \neq 0)$.

Further Graphs of the Trigonometric Functions

The function $y = \csc\theta$ is undefined when $\sin\theta = 0$. This occurs when $P(\cos\theta, \sin\theta)$ has coordinates $(1, 0)$ or $(-1, 0)$. The angles θ which correspond to either of these points are $\theta = k(180°)$, $k \in \mathbb{Z}$, i.e., ***all*** multiples of $180°$. Also, since $-1 \leq \sin\theta \leq 1$, the reciprocal function $\csc\theta \geq 1$ or $\csc\theta \leq -1$, i.e., the values of $\csc\theta$ cannot lie between -1 and $+1$.

The graph of $y = \csc\theta$ is at the top of the next page:

Chapter 4

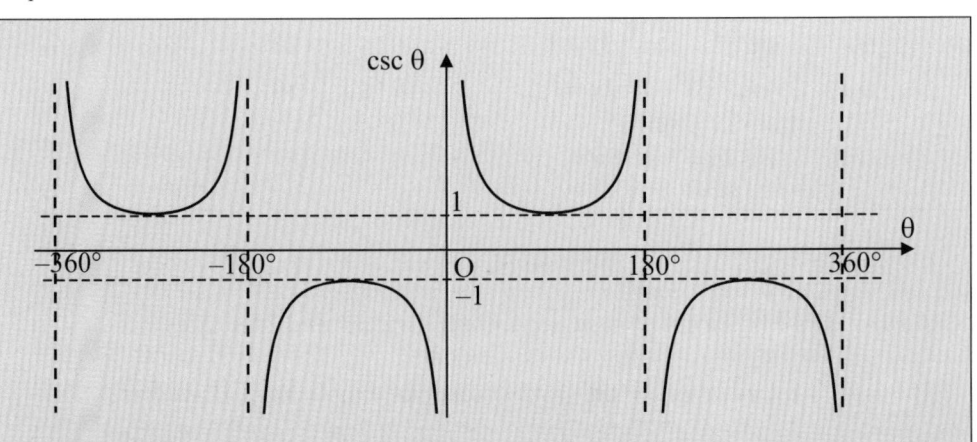

[Graph of csc θ]

The function $y = \sec\theta$ is undefined when $\cos\theta = 0$ and we have already seen that this occurs when θ is any **odd** multiple of 90°. As with csc θ, sec $\theta \geq 1$ or sec $\theta \leq -1$, i.e., the values of sec θ cannot lie between -1 and $+1$. The graph of $y = \sec\theta$ is as follows:

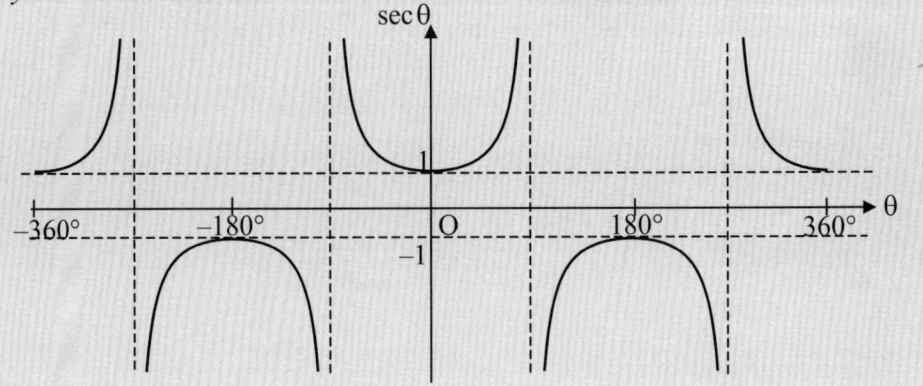

[Graph of sec θ]

The function $y = \cot\theta$ is undefined when $\sin\theta = 0$, i.e., when $\theta = k(180°)$, $k \in \mathbb{Z}$. The graph of $y = \cot\theta$, which can be confirmed by the use of a graphic display calculator, is as follows:

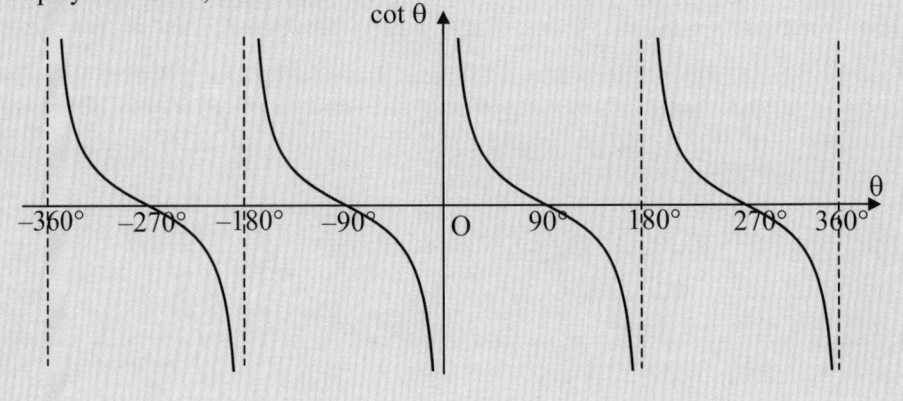

[Graph of cot θ]

Relationships between the Six Trigonometric Ratios

1. $\sin^2\theta + \cos^2\theta = 1$

    Proof $P(\cos\theta, \sin\theta)$ lies on the unit circle $x^2 + y^2 = 1$.
    Therefore $(\cos\theta)^2 + (\sin\theta)^2 = 1$ or $\sin^2\theta + \cos^2\theta = 1$.

2. $1 + \tan^2\theta = \sec^2\theta$ $(\cos\theta \neq 0)$

    Proof From 1, $\cos^2\theta + \sin^2\theta = 1$.
    Dividing both sides by $\cos^2\theta$ gives
    $$\frac{\cos^2\theta}{\cos^2\theta} + \frac{\sin^2\theta}{\cos^2\theta} = \frac{1}{\cos^2\theta}$$
    which simplifies to $1 + \tan^2\theta = \sec^2\theta$ $(\cos\theta \neq 0)$.

3. $1 + \cot^2\theta = \csc^2\theta$ $(\sin\theta \neq 0)$

    Proof From 1, $\sin^2\theta + \cos^2\theta = 1$.
    Dividing both sides by $\sin^2\theta$ gives
    $$\frac{\sin^2\theta}{\sin^2\theta} + \frac{\cos^2\theta}{\sin^2\theta} = \frac{1}{\sin^2\theta}$$
    which simplifies to $1 + \cot^2\theta = \csc^2\theta$ $(\sin\theta \neq 0)$.

Example Prove $\tan\theta + \cot\theta = \sec\theta \csc\theta$ whenever both sides have meaning.

$$\begin{aligned}
\text{L.H.S.} &= \tan\theta + \cot\theta \\
&= \frac{\sin\theta}{\cos\theta} + \frac{\cos\theta}{\sin\theta} \\
&= \frac{\sin^2\theta + \cos^2\theta}{\cos\theta\sin\theta} \\
&= \frac{1}{\cos\theta\sin\theta} \\
&= \sec\theta\csc\theta \\
&= \text{R.H.S.} \quad \text{provided } \sin\theta \neq 0 \text{ and } \cos\theta \neq 0, \text{ i.e., } \theta \neq \tfrac{1}{2}\pi k,\ k \in \mathbb{Z}.
\end{aligned}$$

Example Prove:
$$\sin^2 A - 4\cos^2 A + 1 = 2\sin^2 A - 3\cos^2 A = 3\sin^2 A - 2\cos^2 A - 1.$$

Chapter 4

$$\text{L.H.S.} = \sin^2 A - 4\cos^2 A + (\sin^2 A + \cos^2 A)$$
$$= 2\sin^2 A - 3\cos^2 A$$
$$= \text{Middle}$$
$$= 2\sin^2 A - 2\cos^2 A - \cos^2 A$$
$$= 2\sin^2 A - 2\cos^2 A - (1 - \sin^2 A)$$
$$= 3\sin^2 A - 2\cos^2 A - 1$$
$$= \text{R.H.S.}$$

Example Prove that $\dfrac{\cos^2 A}{1+\tan^2 A} - \dfrac{\sin^2 A}{1+\cot^2 A} = 1 - 2\sin^2 A$ whenever both sides have meaning.

$$\text{L.H.S.} = \dfrac{\cos^2 A}{\sec^2 A} - \dfrac{\sin^2 A}{\csc^2 A}$$
$$= \cos^4 A - \sin^4 A$$
$$= (\cos^2 A - \sin^2 A)(\cos^2 A + \sin^2 A)$$
$$= \cos^2 A - \sin^2 A$$
$$= (1 - \sin^2 A) - \sin^2 A$$
$$= 1 - 2\sin^2 A$$
$$= \text{R.H.S. provided } \tan A \text{ and } \cot A \text{ both exist, i.e., } \cos A \neq 0 \text{ and } \sin A \neq 0. \text{ Thus both sides have meaning provided } A \neq \tfrac{1}{2}\pi k, \; k \in \mathbb{Z}.$$

Exercise 4.10

1. Simplify the following:
    (a) $\sin^2 2A + \cos^2 2A$;
    (b) $1 + \tan^2 \tfrac{1}{4} A$;
    (c) $\sin^2 3 + \cos^2 3$;
    (d) $\cot^2 \theta + 1$;
    (e) $\cos^2 4A + \sin^2 4A$;
    (f) $\cos^2 1\tfrac{1}{2} + \sin^2 1\tfrac{1}{2}$;
    (g) $\dfrac{\sin\theta}{\csc\theta} + \dfrac{\cos\theta}{\sec\theta}$;
    (h) $(\sin A + \cos A)^2 + (\sin A - \cos A)^2$;
    (i) $1 - \sin^2 A$;
    (j) $1 - \cos^2 2B$;
    (k) $\sec^2 \theta - 1$;
    (l) $1 - \csc^2 A$.

2. Prove the following provided all expressions have meaning:
    (a) $\cos^2 \theta - \sin^2 \theta = 1 - 2\sin^2 \theta = 2\cos^2 \theta - 1$;
    (b) $2\tan^2 \theta - \sec^2 \theta = \tan^2 \theta - 1 = \sec^2 \theta - 2$;

(c) $2\cot^2\theta - 3\csc^2\theta = -2 - \csc^2\theta = -3 - \cot^2\theta$;
(d) $\sec A(\sin A - \cot A) = \tan A - \csc A$;
(e) $\sin A \sec A \cot A + \cos A \csc A \tan A = 2$;
(f) $\cot A(1 + \tan A) = 1 + \cot A$;
(g) $\cot^2 A \sec^2 A + \tan^2 A \csc^2 A = \sec^2 A + \csc^2 A$.

3. Prove the following provided each expression has meaning:
(a) $\sin^2 A - 4\cos^2 A = 3\sin^2 A - 2\cos^2 A - 2$;
(b) $2\csc^2 A + 3\cot^2 A = 6\cot^2 A - \csc^2 A + 3$;
(c) $2\tan^2 A - 3\sec^2 A = 1 - 4\sec^2 A + 3\tan^2 A$;
(d) $(\cos A - \sin A + 1)^2 = 2(1 + \cos A)(1 - \sin A)$.

4. Prove that the following statements are true whenever the functions involved have meaning:
(a) $\cos A \tan A = \sin A$;
(b) $\cos A \csc A = \cot A$;
(c) $\sec A \cos A + \cot^2 A = \csc^2 A$;
(d) $(\sin\theta + \cos\theta)^2 = 2\sin\theta\cos\theta + 1$;
(e) $\dfrac{1}{1-\sin^2\theta} = \sec^2\theta$;
(f) $\dfrac{\sin\theta}{1+\cos\theta} + \dfrac{1+\cos\theta}{\sin\theta} = 2\csc\theta$;
(g) $\dfrac{2\tan A}{1+\tan^2 A} = 2\sin A\cos A$;
(h) $\dfrac{2\cos A}{\sin A - \cos A} + 1 = \dfrac{\tan A + 1}{\tan A - 1}$;
(i) $\dfrac{\tan x}{1-\tan x} + \dfrac{\cot x}{1-\cot x} = -1$;
(j) $1 - \tan^4 x = 2\sec^2 x - \sec^4 x$;
(k) $1 + \dfrac{\cot^2 x}{1+\csc x} = \csc x$;
(l) $\dfrac{1+\sec A}{1-\sec A} = -(\cot A + \csc A)^2$.

103

Chapter 4

*5. Prove that the following statements are true whenever the functions involved have meaning:

(a) $(\sin A \cos B - \cos A \sin B)^2 + (\cos A \cos B + \sin A \sin B)^2 = 1$;

(b) $(\cos A + \sec A)(\sin A + \csc A) = 2 \sec A \csc A + \sin A \cos A$;

(c) $(1 + \cot A + \tan A)(1 - \cot A - \tan A) = 1 - \csc^2 A - \sec^2 A$;

(d) $\dfrac{\sec A + \tan A}{\sec A - \tan A} - \dfrac{\sec A - \tan A}{\sec A + \tan A} = 4 \tan A \sec A$;

(e) $\dfrac{(1 - \sin A)(1 - \cot A)}{(1 - \csc A)(1 - \tan A)} = \cos A$;

(f) $\dfrac{1 + \cot A}{1 + \tan A} - \dfrac{1 - \tan A}{1 - \cot A} = \tan A + \cot A$.

4.11 Addition Formulae

1. $\cos(A - B) = \cos A \cos B + \sin A \sin B$
2. $\cos(A + B) = \cos A \cos B - \sin A \sin B$
3. $\sin(A + B) = \sin A \cos B + \cos A \sin B$
4. $\sin(A - B) = \sin A \cos B - \cos A \sin B$

Proof [*Note:* The scalar product of two vectors required for this proof is defined in Chapter 11.]

1. Consider the points A $(\cos A, \sin A)$ and B $(\cos B, \sin B)$ on the unit circle.

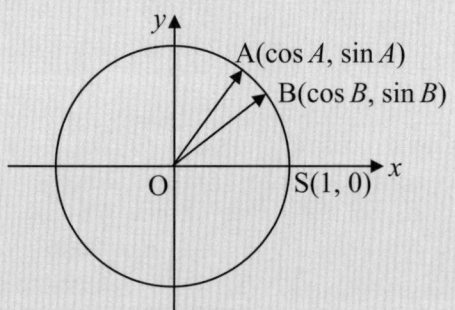

$A\hat{O}S = A$ and $B\hat{O}S = B$ and so $A\hat{O}B = A - B$.

$\cos(A - B) = \dfrac{\overrightarrow{OA} \cdot \overrightarrow{OB}}{|\overrightarrow{OA}||\overrightarrow{OB}|}$

$= \begin{pmatrix} \cos A \\ \sin A \end{pmatrix} \cdot \begin{pmatrix} \cos B \\ \sin B \end{pmatrix}$ since $|\overrightarrow{OA}| = |\overrightarrow{OB}| = 1$.

Thus $\cos(A - B) = \cos A \cos B + \sin A \sin B$.

Trigonometry

2. Replace B with $-B$ in 1.
$$\cos(A+B) = \cos A \cos(-B) + \sin A \sin(-B)$$
$$\cos(A+B) = \cos A \cos B - \sin A \sin B.$$

3. Replace A with $\frac{\pi}{2} + A$ in 2.
$$\cos(\tfrac{\pi}{2} + A + B) = \cos(\tfrac{\pi}{2} + A)\cos B - \sin(\tfrac{\pi}{2} + A)\sin B$$
$$-\sin(A+B) = -\sin A \cos B - \cos A \sin B$$
$$\sin(A+B) = \sin A \cos B + \cos A \sin B.$$

4. Replace B with $-B$ in 3.
$$\sin(A-B) = \sin A \cos(-B) + \cos A \sin(-B)$$
$$\sin(A-B) = \sin A \cos B - \cos A \sin B.$$

Example Verify, without the use of a calculator, the addition formula for $\sin(A+B)$ when $A = \pi/3$ and $B = \pi/6$.

$\sin(A+B) = \sin A \cos B + \cos A \sin B$
When $A = \pi/3$, $B = \pi/6$, LHS $= \sin(\pi/3 + \pi/6) = \sin(\pi/2) = 1$.
$\qquad$ RHS $= \sin(\pi/3)\cos(\pi/6) + \cos(\pi/3)\sin(\pi/6)$
$\qquad\qquad = \tfrac{1}{2}\sqrt{3} \times \tfrac{1}{2}\sqrt{3} + \tfrac{1}{2} \times \tfrac{1}{2}$
$\qquad\qquad = 1$ which verifies the formula.

Example Simplify $\tfrac{1}{2}\cos\theta + \tfrac{1}{2}\sqrt{3}\sin\theta$.

$\tfrac{1}{2}\cos\theta + \tfrac{1}{2}\sqrt{3}\sin\theta = \cos\theta \cos\tfrac{1}{3}\pi + \sin\theta \sin\tfrac{1}{3}\pi = \cos(\theta - \tfrac{1}{3}\pi)$.

Example If $\sin\alpha = \tfrac{3}{5}, 0 < \alpha < \tfrac{1}{2}\pi$ and $\sin\beta = \tfrac{4}{5}, \tfrac{1}{2}\pi < \beta < \pi$, evaluate $\cos(\alpha + \beta)$ without the use of a calculator.

$\sin\alpha = \tfrac{3}{5}, 0 < \alpha < \tfrac{1}{2}\pi$, so $\cos\alpha = \tfrac{4}{5}$; $\sin\beta = \tfrac{4}{5}, \tfrac{1}{2}\pi < \beta < \pi$, so $\cos\beta = -\tfrac{3}{5}$.
Thus $\cos(\alpha + \beta) = \cos\alpha \cos\beta - \sin\alpha \sin\beta$
$\qquad\qquad\qquad = \tfrac{4}{5} \times \left(-\tfrac{3}{5}\right) - \tfrac{3}{5} \times \tfrac{4}{5}$
$\qquad\qquad\qquad = -\tfrac{24}{25}$.

Example Find a value for $\sin\tfrac{7}{12}\pi$ in simplest surd form.

105

Chapter 4

$$\sin \tfrac{7}{12}\pi = \sin(\tfrac{1}{3}\pi + \tfrac{1}{4}\pi)$$
$$= \sin\tfrac{1}{3}\pi \cos\tfrac{1}{4}\pi + \cos\tfrac{1}{3}\pi \sin\tfrac{1}{4}\pi$$
$$= \tfrac{1}{2}\sqrt{3} \times \tfrac{1}{2}\sqrt{2} + \tfrac{1}{2} \times \tfrac{1}{2}\sqrt{2}$$
$$= \tfrac{1}{4}(\sqrt{6} + \sqrt{2}).$$

Example Verify that the expressions $\tan A + \cot B$ and $\dfrac{\cos(A-B)}{\cos A \sin B}$ are equal whenever both expressions have meaning.

$$\text{LHS} = \tan A + \cot B$$
$$= \frac{\sin A}{\cos A} + \frac{\cos B}{\sin B}$$
$$= \frac{\sin A \sin B + \cos A \cos B}{\cos A \sin B}$$
$$= \frac{\cos(A-B)}{\cos A \sin B}$$
$$= \text{RHS} \qquad \text{Q.E.D.}$$

Exercise 4.11

1. Verify the following identities using the addition formulae:
   (a) $\sin(\tfrac{1}{2}\pi + \theta) = \cos\theta$;
   (b) $\cos(\tfrac{1}{2}\pi - \theta) = \sin\theta$;
   (c) $\cos(\tfrac{1}{2}\pi + \theta) = -\sin\theta$;
   (d) $\sin(\pi + \theta) = -\sin\theta$;
   (e) $\cos(\pi - \theta) = -\cos\theta$;
   (f) $\sin(2\pi - \theta) = -\sin\theta$.

2. Without the use of a calculator verify the addition formula for
   (a) $\sin(\alpha - \beta)$ when $\alpha = \tfrac{2}{3}\pi, \beta = \tfrac{1}{3}\pi$;
   (b) $\sin(\alpha + \beta)$ when $\alpha = \tfrac{1}{2}\pi, \beta = \tfrac{1}{6}\pi$;
   (c) $\cos(\alpha + \beta)$ when $\alpha = \tfrac{3}{2}\pi, \beta = \tfrac{5}{6}\pi$;
   (d) $\cos(\alpha - \beta)$ when $\alpha = \pi, \beta = \tfrac{1}{3}\pi$.

3. Expand each of the following and simplify where possible:
   (a) $\sin(2x + y)$;
   (b) $\sin(3B + 40°)$;
   (c) $\sin 2(A + B)$;
   (d) $\sin(x - 2y)$;
   (e) $\sin(\theta - \tfrac{1}{6}\pi)$;
   (f) $\cos(x + \tfrac{1}{2}y)$;
   (g) $\cos 3(A + B)$;
   (h) $\cos(\tfrac{1}{6}\pi - A)$;
   (i) $\cos(\tfrac{1}{3}\pi + A)$;
   (j) $\cos(B - C)$;
   (k) $\sin(\pi - B)$;
   (l) $\cos(\tfrac{1}{3}\pi + 2A)$;
   (m) $\sin(\tfrac{2}{3}\pi - x)$;
   (n) $\sin 2x$;
   (o) $\cos 2x$.

4. Express each of the following as a single trigonometric function:
   (a) $\sin 50° \cos 40° + \cos 50° \sin 40°$; (b) $\sin\alpha \cos 40° - \cos\alpha \sin 40°$;
   (c) $\sin\frac{2}{3}\pi \cos\frac{1}{6}\pi - \cos\frac{2}{3}\pi \sin\frac{1}{6}\pi$; (d) $\cos 2A \cos A + \sin 2A \sin A$;
   (e) $\cos 55° \cos 35° - \sin 55° \sin 35°$; (f) $\sin x \cos\frac{1}{3}\pi - \cos x \sin\frac{1}{3}\pi$;
   (g) $\sin\frac{4}{15}\pi \cos\frac{1}{10}\pi - \cos\frac{4}{15}\pi \sin\frac{1}{10}\pi$; (h) $\frac{1}{2}\sin A - \frac{1}{2}\sqrt{3} \cos A$;
   (i) $\frac{1}{2}\sqrt{2} \sin B + \frac{1}{2}\sqrt{2} \cos B$; (j) $\cos^2 A - \sin^2 A$.

5. If α is an angle in quadrant 1 and β is an angle in quadrant 2 such that $\cos\alpha = \frac{3}{5}$ and $\sin\beta = \frac{3}{5}$, evaluate each of the following without the use of a calculator:
   (a) $\cos(\alpha + \beta)$; (b) $\sin(\alpha - \beta)$; (c) $\cos(\alpha - \beta)$.

6. If α is an angle in quadrant 2 and β is an angle in quadrant 3 such that $\sin\alpha = \frac{4}{5}$ and $\tan\beta = \frac{8}{15}$, evaluate each of the following without the use of a calculator:
   (a) $\cos(\alpha - \beta)$; (b) $\sin(\alpha + \beta)$; (c) $\cos(\alpha + \beta)$.

7. Find in simplest surd form the value of each of the following:
   (a) $\cos 105°$; (b) $\cos 75°$; (c) $\sin 15°$; (d) $\sin 75°$.

8. Find in simplest surd form the value if each of the following:
   (a) $\sin\frac{7}{12}\pi$; (b) $\cos\frac{13}{12}\pi$; (c) $\cos\frac{1}{12}\pi$; (d) $\cos\frac{23}{12}\pi$.

9. Verify that the following identities are true whenever the expressions have meaning:
   (a) $2\cos(\theta - \frac{1}{3}\pi) = \cos\theta + \sqrt{3}\sin\theta$; (b) $\sqrt{2}\cos(\theta + \frac{1}{4}\pi) = \cos\theta - \sin\theta$;
   (c) $\dfrac{\cos(\theta + \phi)}{\cos\theta \sin\phi} = \cot\phi - \tan\theta$; (d) $\dfrac{\sin(A + B)}{\cos A \sin B} = \tan A \cot B + 1$;
   (e) $\sin(\alpha + \beta) + \sin(\alpha - \beta) = 2\sin\alpha \cos\beta$;
   (f) $\sin(\alpha + \beta)\sin(\alpha - \beta) = \sin^2\alpha - \sin^2\beta$;
   (g) $\sin B + \cos A \sin(A - B) = \sin A \cos(A - B)$.

4.12 Addition Formulae for the Tangent Function

1. $\tan(A + B) = \dfrac{\tan A + \tan B}{1 - \tan A \tan B}$

Chapter 4

2. $\tan(A-B) = \dfrac{\tan A - \tan B}{1 + \tan A \tan B}$

Proof 1. $\tan(A+B) = \dfrac{\sin(A+B)}{\cos(A+B)}$

$= \dfrac{\sin A \cos B + \cos A \sin B}{\cos A \cos B - \sin A \sin B}$

$= \dfrac{\left(\dfrac{\sin A \cos B}{\cos A \cos B}\right) + \left(\dfrac{\cos A \sin B}{\cos A \cos B}\right)}{\left(\dfrac{\cos A \cos B}{\cos A \cos B}\right) - \left(\dfrac{\sin A \sin B}{\cos A \cos B}\right)}$

$= \dfrac{\tan A + \tan B}{1 - \tan A \tan B}.$

2. Put $-B$ for B in 1.

$\tan(A-B) = \dfrac{\tan A + \tan(-B)}{1 - \tan A \tan(-B)}$

$= \dfrac{\tan A - \tan B}{1 + \tan A \tan B}.$

Example If $\sin A = -\dfrac{5}{13}$, $\pi < A < \dfrac{3}{2}\pi$ and $\cos B = -\dfrac{3}{5}$, $\dfrac{1}{2}\pi < B < \pi$, find the value of $\tan(A-B)$ without the use of a calculator.

$\tan A = \dfrac{5}{12}$ and $\tan B = -\dfrac{4}{3}$

Therefore $\tan(A-B) = \dfrac{\tan A - \tan B}{1 + \tan A \tan B}$

$= \dfrac{\dfrac{5}{12} + \dfrac{4}{3}}{1 - \dfrac{5}{12} \times \dfrac{4}{3}}$

$= \dfrac{15 + 48}{36 - 20}$

$= \dfrac{63}{16}.$

Exercise 4.12

1. Verify the following identities using appropriate addition formulae:

 (a) $\tan(\pi + A) = \tan A$; (b) $\tan(\dfrac{1}{4}\pi - A) = \dfrac{1 - \tan A}{1 + \tan A}.$

108

2. Verify, without the use of a calculator, the addition formula for
   (a) $\tan(A - B)$ if $A = \frac{1}{3}\pi$ and $B = \frac{1}{6}\pi$;
   (b) $\tan(A + B)$ if $A = \frac{1}{3}\pi$ and $B = \frac{2}{3}\pi$.

3. Simplify each of the following:
   (a) $\dfrac{\tan 20° + \tan 25°}{1 - \tan 20° \tan 25°}$;
   (b) $\dfrac{1 + \tan A \tan B}{\tan A - \tan B}$;
   (c) $\dfrac{\tan \frac{5}{6}\pi - \tan \frac{2}{3}\pi}{1 + \tan \frac{5}{6}\pi \tan \frac{2}{3}\pi}$;
   (d) $\dfrac{\tan \frac{1}{4}\pi + \tan A}{1 - \tan \frac{1}{4}\pi \tan A}$;
   (e) $\dfrac{1 + \tan A}{1 - \tan A}$;
   (f) $\dfrac{2 \tan A}{1 - \tan^2 A}$.

4. If $\cos A = \frac{4}{5}$ where A is in the first quadrant and $\sin B = \frac{8}{17}$ where B is in the second quadrant, find the value of each of the following without the use of a calculator: (a) $\tan(A + B)$; (b) $\tan(A - B)$.

5. Verify that the following are true whenever the expressions have meaning:
   (a) $\tan(x + \alpha) = \dfrac{\sin x + \tan \alpha \cos x}{\cos x - \tan \alpha \sin x}$;
   (b) $\tan(\frac{1}{4}\pi - A) = \dfrac{\cos A - \sin A}{\cos A + \sin A} = \cot(\frac{1}{4}\pi + A)$;
   (c) $\dfrac{\tan A - \tan B}{\tan A + \cot B} = \tan B \tan(A - B)$;
   (d) $\dfrac{\sin x + \tan \alpha \cos x}{\cos x - \tan \alpha \sin x} = \tan(x + \alpha)$.

6. Find in simplest surd form the values of
   (a) $\tan \frac{5}{12}\pi$;
   (b) $\tan \frac{11}{12}\pi$.

7. Show that $\dfrac{\tan(x - y) + \tan y}{\tan(x - y) - \cot y} = -\tan x \tan y$ whenever all expressions have meaning.

8. Simplify $\dfrac{\tan 80° - \tan 35°}{\tan 80° + \cot 35°}$.

Chapter 4

4.13 Duplication and Half-angle Formulae

Duplication Formulae

1. $\sin 2A = 2\sin A \cos A$
2. $\cos 2A = \cos^2 A - \sin^2 A$
   $= 2\cos^2 A - 1$
   $= 1 - 2\sin^2 A$
3. $\tan 2A = \dfrac{2\tan A}{1-\tan^2 A}$

Proof 1. $\sin 2A = \sin(A+A)$
$= \sin A \cos A + \cos A \sin A$
$= 2\sin A \cos A.$

2. $\cos 2A = \cos(A+A)$
$= \cos A \cos A - \sin A \sin A$
$= \cos^2 A - \sin^2 A$
$= \cos^2 A - (1-\cos^2 A)$
$= 2\cos^2 A - 1$
$= 2(1-\sin^2 A) - 1$
$= 1 - 2\sin^2 A.$

3. $\tan 2A = \tan(A+A)$
$= \dfrac{\tan A + \tan A}{1-\tan A \tan A}$
$= \dfrac{2\tan A}{1-\tan^2 A}.$

Example If $\sin A = \tfrac{3}{5}$ and $\tfrac{1}{2}\pi < A < \pi$, find the values of $\sin 2A$ and $\cos 2A$ without the use of a calculator.

$\sin A = \tfrac{3}{5} \Rightarrow \cos A = -\tfrac{4}{5}$

Therefore, $\sin 2A = 2\sin A \cos A$ and $\cos 2A = 1 - 2\sin^2 A$
$= 2 \times \tfrac{3}{5} \times \left(-\tfrac{4}{5}\right)$ $= 1 - 2\left(\tfrac{9}{25}\right)$
$= -\tfrac{24}{25}.$ $= -\tfrac{7}{25}.$

Example Show that $\dfrac{\cos 2x - \sin 2x + 1}{\cos 2x + \sin 2x - 1} = \cot x$ whenever both expressions have meaning.

$$\dfrac{\cos 2x - \sin 2x + 1}{\cos 2x + \sin 2x - 1} = \dfrac{2\cos^2 x - 1 - 2\sin x \cos x + 1}{1 - 2\sin^2 x + 2\sin x \cos x - 1}$$

$$= \dfrac{2\cos^2 x - 2\sin x \cos x}{2\sin x \cos x - 2\sin^2 x}$$

$$= \dfrac{2\cos x(\cos x - \sin x)}{2\sin x(\cos x - \sin x)}$$

$$= \cot x.$$

Further Duplication Formulae

1. $\sin 2A = \dfrac{2\tan A}{1 + \tan^2 A}$ 2. $\cos 2A = \dfrac{1 - \tan^2 A}{1 + \tan^2 A}$

Proof 1. $\dfrac{2\tan A}{1 + \tan^2 A} = \dfrac{2\tan A \cos^2 A}{(1 + \tan^2 A)\cos^2 A}$

$$= \dfrac{2\sin A \cos A}{\cos^2 A + \sin^2 A}$$

$$= \sin 2A.$$

2. $\dfrac{1 - \tan^2 A}{1 + \tan^2 A} = \dfrac{(1 - \tan^2 A)\cos^2 A}{(1 + \tan^2 A)\cos^2 A}$

$$= \dfrac{\cos^2 A - \sin^2 A}{\cos^2 A + \sin^2 A}$$

$$= \cos 2A.$$

Example Find the value of $\tan x$ if $\sin 2x + 8\cos 2x = 3.2$.

Let $t = \tan x$ then $\sin 2x = \dfrac{2t}{1 + t^2}$ and $\cos 2x = \dfrac{1 - t^2}{1 + t^2}$.

Therefore $\dfrac{2t}{1 + t^2} + \dfrac{8(1 - t^2)}{1 + t^2} = \dfrac{16}{5}$

$\Rightarrow \quad 10t + 40(1 - t^2) = 16(1 + t^2)$

$\Rightarrow \quad 56t^2 - 10t - 24 = 0$

Chapter 4

$\Rightarrow \quad 28t^2 - 5t - 12 = 0$
$\Rightarrow \quad (4t - 3)(7t + 4) = 0$
Therefore $\tan x = t = \frac{3}{4}$ or $-\frac{4}{7}$.

Half-angle Formulae

1. $\sin \frac{1}{2} A = \pm \sqrt{\dfrac{1 - \cos A}{2}}$

2. $\cos \frac{1}{2} A = \pm \sqrt{\dfrac{1 + \cos A}{2}}$

3. $\tan \frac{1}{2} A = \pm \sqrt{\dfrac{1 - \cos A}{1 + \cos A}}$

Proof 1. $\cos A = 1 - 2 \sin^2 \frac{1}{2} A$ (duplication formula for cosine)

$\Rightarrow \quad \sin^2 \frac{1}{2} A = \dfrac{1 - \cos A}{2}$

$\Rightarrow \quad \sin \frac{1}{2} A = \pm \sqrt{\dfrac{1 - \cos A}{2}}.$

2. $\cos A = 2 \cos^2 \frac{1}{2} A - 1$ (duplication formula for cosine)

$\Rightarrow \quad \cos^2 \frac{1}{2} A = \dfrac{1 + \cos A}{2}$

$\Rightarrow \quad \cos \frac{1}{2} A = \pm \sqrt{\dfrac{1 + \cos A}{2}}.$

3. $\tan^2 \frac{1}{2} A = \dfrac{\sin^2 \frac{1}{2} A}{\cos^2 \frac{1}{2} A} = \dfrac{1 - \cos A}{1 + \cos A}$

$\Rightarrow \quad \tan \frac{1}{2} A = \pm \sqrt{\dfrac{1 - \cos A}{1 + \cos A}}.$

Note: Formulae 1 and 2 are sometimes very useful in the forms:
$\sin^2 A = \frac{1}{2} - \frac{1}{2} \cos 2A$ and $\cos^2 A = \frac{1}{2} + \frac{1}{2} \cos 2A$.

Example If $\cos A = -\frac{7}{25}$, find the values of $\sin \frac{1}{2} A$, $\cos \frac{1}{2} A$ and $\tan \frac{1}{2} A$.

$$\sin\tfrac{1}{2}A = \pm\sqrt{\frac{1-\cos A}{2}} = \pm\tfrac{4}{5};$$

$$\cos\tfrac{1}{2}A = \pm\sqrt{\frac{1+\cos A}{2}} = \pm\tfrac{3}{5};$$

$$\tan\tfrac{1}{2}A = \pm\sqrt{\frac{1-\cos A}{1+\cos A}} = \pm\tfrac{4}{3}.$$

Example If $\tan 2A = \tfrac{7}{24}$, $0 < A < \tfrac{1}{4}\pi$, find the value of $\tan A$.

Method 1: $\tan 2A = \dfrac{2\tan A}{1-\tan^2 A}$

$\Rightarrow\quad \dfrac{7}{24} = \dfrac{2\tan A}{1-\tan^2 A}$

$\Rightarrow\quad 7 - 7\tan^2 A = 48\tan A$

$\Rightarrow\quad 7\tan^2 A + 48\tan A - 7 = 0$

$\Rightarrow\quad (7\tan A - 1)(\tan A + 7) = 0$

$\Rightarrow\quad \tan A = \tfrac{1}{7}$ since $\tan A = -7$ is impossible for $0 < A < \tfrac{1}{4}\pi$.

Method 2: $\tan 2A = \tfrac{7}{24} \;\Rightarrow\; \cos 2A = \tfrac{24}{25}$

$\Rightarrow\quad \tan A = \sqrt{\dfrac{1-\cos 2A}{1+\cos 2A}}\quad (\tan A > 0)$

$\qquad\qquad = \sqrt{\dfrac{1/25}{49/25}}$

$\qquad\qquad = \tfrac{1}{7}.$

Method 3: Let O be the centre, and AB the diameter of a semi-circle of radius 25. Let C be a point on the semicircle such that the angle COB = 2A. Let CD be drawn perpendicular to OB to meet OB at D.

CD = 7
OD = 24
OC = OA = 25

Therefore $\tan A = \dfrac{CD}{AD} = \dfrac{7}{24+25} = \dfrac{1}{7}.$

Chapter 4

Example Prove that $\dfrac{1+\sin 2A}{\cos 2A} = \dfrac{\cos A+\sin A}{\cos A-\sin A}$ whenever both expressions have meaning.

$$\text{RHS} = \dfrac{\cos A+\sin A}{\cos A-\sin A} \times \dfrac{\cos A+\sin A}{\cos A+\sin A}$$

$$= \dfrac{\cos^2 A + 2\sin A\cos A + \sin^2 A}{\cos^2 A - \sin^2 A}$$

$$= \dfrac{1+\sin 2A}{\cos 2A}$$

= LHS, as required.

Exercise 4.13

1. Verify the formulae for
   (a) $\sin 2A$ when $A = \tfrac{1}{6}\pi$;
   (b) $\cos 2A$ when $A = \tfrac{1}{4}\pi$;
   (c) $\tan 2A$ when $A = \tfrac{1}{3}\pi$;
   (d) $\sin 2A$ when $\tan A = \sqrt{3}$;
   (e) $\cos 2A$ when $\tan A = 1$;
   (f) $\sin \tfrac{1}{2}A$ when $\cos A = \tfrac{1}{2}$;
   (g) $\cos \tfrac{1}{2}A$ when $\cos A = -1$;
   (h) $\tan \tfrac{1}{2}A$ when $\cos A = 0$.

2. Express each of the following in terms of the half angle (e.g. $\sin 4x = 2\sin 2x \cos 2x$):
   (a) $\sin 6A$;
   (b) $\sin A$;
   (c) $\sin 8A$;
   (d) $\sin 3x$;
   (e) $\cos 4x$;
   (f) $\cos x$;
   (g) $\tan 4x$;
   (h) $\tan \tfrac{1}{3}\pi$;
   (i) $\cos 50°$;
   (j) $\sin(\pi - 2x)$;
   (k) $\cos 2A - 1$;
   (l) $\tan(\tfrac{1}{2}\pi - A)$.

3. Evaluate each of the following without the use of a calculator:
   (a) $2\sin 15° \cos 15°$;
   (b) $\cos^2 22\tfrac{1}{2}° - \sin^2 22\tfrac{1}{2}°$;
   (c) $1 - 2\sin^2 \tfrac{1}{8}\pi$;
   (d) $\sin 75° \cos 75°$;
   (e) $\dfrac{2\tan 15°}{1 - \tan^2 15°}$;
   (f) $\dfrac{1 - \tan^2 22\tfrac{1}{2}°}{\tan 22\tfrac{1}{2}°}$;
   (g) $\cos^2 75° + \sin^2 75°$;
   (h) $\sin^3 \tfrac{1}{12}\pi \cos \tfrac{1}{12}\pi + \cos^3 \tfrac{1}{12}\pi \sin \tfrac{1}{12}\pi$;
   (i) $\sin^2 \tfrac{3}{8}\pi \cos^2 \tfrac{3}{8}\pi$;
   (j) $\cos^4 15° - \sin^4 15°$.

4. Simplify each of the following:
   (a) $\cos^2 \tfrac{1}{2}(A+B) - \sin^2 \tfrac{1}{2}(A+B)$;   (b) $\cos(\tfrac{1}{4}\pi - x)\sin(\tfrac{1}{4}\pi - x)$;
   (c) $\sin 2A \cos 2A \cos 4A$;   (d) $1 - \tan A \cot 2A$;
   (e) $\sin 3x - \cos 3x \tan x$;   (f) $\dfrac{1-\cos A}{1+\cos A}$;
   (g) $\dfrac{1-\cos 2A}{\sin A}$;   (h) $\dfrac{1-\sin 2x}{\cos x - \sin x}$.

5. (a) If $\cos A = -\tfrac{1}{3}$, find $\sin \tfrac{1}{2}A$ and $\cos \tfrac{1}{2}A$.
   (b) Find $\sin x$, $\cos x$ and $\tan x$ if $\cos 2x = \tfrac{1}{8}$.
   (c) Prove that $\tan \tfrac{1}{8}\pi = \sqrt{2} - 1$.
   (d) If A, B, C are acute angles such that $\tan A = 1$, $\tan B = 2$, $\tan C = 3$, show that $A + B + C = \pi$ without the use of a calculator.
   (e) If $5 \sin 2x = \cos x$, find the values of $\cos 2x$.
   (f) If $\tan A = \tfrac{1}{5}$, $\tan B = \tfrac{1}{239}$ with both A and B acute, prove that $4A - B = \tfrac{1}{4}\pi$.
   (g) In triangle ABC, $\tan A = -\tfrac{3}{4}$ and $\tan B = \tfrac{1}{3}$. Prove that the triangle is isosceles.

6. (a) Prove that the expressions $\dfrac{\tan 3A + \tan A}{\tan 3A - \tan A}$ and $2\cos 2A$ are equal whenever all expressions have meaning.

   (b) Prove that $\sec A + \tan A = \tan\left(\tfrac{1}{4}\pi + \tfrac{1}{2}A\right)$.

4.14 Functions of the Form $a \sin \theta + b \cos \theta$

The function $a \sin \theta + b \cos \theta$ can always be expressed in the form $k \sin(\theta + \alpha)$ or $k \cos(\theta + \alpha)$ where k and α are constants.

If $a \sin \theta + b \cos \theta = k \sin(\theta + \alpha) = k \sin \theta \cos \alpha + k \cos \theta \sin \alpha$, equating the coefficients of $\sin \theta$ and $\cos \theta$ gives $a = k \cos \alpha$ and $b = k \sin \alpha$.

Squaring and adding gives: $a^2 + b^2 = k^2(\cos^2 \alpha + \sin^2 \alpha) = k^2$.

This gives $k = \pm\sqrt{a^2 + b^2}$ but we are free to choose $k > 0$ so $k = \sqrt{a^2 + b^2}$ will do.

115

Then $\sin\alpha = b/k$, and $\cos\alpha = a/k$ or $\tan\alpha = b/a$. A suitable value of α can then be found.

Example Express $3\sin\theta - 4\cos\theta$ in the form $k\sin(\theta - \alpha)$ where $k > 0$ and $0 < \alpha < \frac{1}{2}\pi$, and hence find the maximum and minimum values of $3\sin\theta - 4\cos\theta$.

If $3\sin\theta - 4\cos\theta = k\sin(\theta - \alpha) = k\sin\theta\cos\alpha - k\cos\theta\sin\alpha$ then $k\cos\alpha = 3$ and $k\sin\alpha = 4$.
Choosing $k > 0$ we get $k = \sqrt{3^2 + 4^2} = 5$, then $\cos\alpha = \frac{3}{5}$ and $\sin\alpha = \frac{4}{5}$ which gives $\alpha = 0.927$ ($0 < \alpha < \frac{1}{2}\pi$).
Hence $3\sin\theta - 4\cos\theta = 5\sin(\theta - 0.927)$ and clearly the maximum and minimum values are 5 and −5 respectively.

Example Solve the equation $5\cos x + 12\sin x = 6.5$ for $0 \le x < 2\pi$.

Let $5\cos x + 12\sin x = A\cos(x - \alpha) = A\cos x\cos\alpha + A\sin x\sin\alpha$.
Firstly, $A = \sqrt{5^2 + 12^2} = 13$, then $A\cos\alpha = 5$ and $A\sin\alpha = 12$ giving $\alpha = 1.1760$.
We need to solve the equation $13\cos(x - \alpha) = 6.5$ or $\cos(x - \alpha) = \frac{1}{2}$.
$$x - \alpha = \pm\tfrac{1}{3}\pi \ (+2\pi k, k \in \mathbb{Z})$$
$$x = \alpha \pm \tfrac{1}{3}\pi \ (+2\pi k)$$
$$= 0.129, 2.22 \ (0 \le x < 2\pi).$$

Note: Clearly a graphic display calculator can be used here.

Exercise 4.14

1. Express each of the following functions in the form $A\sin(x + \alpha)$ where $A > 0$ and $0 < \alpha < \frac{1}{2}\pi$. State the maximum and minimum values of each function.
    (a) $2\sin x + 3\cos x$;   (b) $3\sin x + \cos x$;   (c) $5\cos x + 2\sin x$.

2. Express each of the following functions in one of the forms $A\sin(x \pm \alpha)$ or $A\cos(x \pm \alpha)$:
    (a) $\sin x - \cos x$;   (b) $\sqrt{3}\cos x + \sin x$;   (c) $3\cos x - 5\sin x$.

3. By expressing each of the following functions in the form $A\sin(x\pm\alpha)$ or $A\cos(x\pm\alpha)$, solve the equations for $0 \le x < 2\pi$.
   (a) $\cos x - 2\sin x = 1$;
   (b) $3\sin x - 5\cos x = 4$;
   (c) $5\sin x + \cos x = 3$;
   (d) $8\cos x + 15\sin x = 17$.

4. Solve each of the following equations for $0 \le \theta < 2\pi$:
   (a) $3\sin 2\theta - 2\cos 2\theta = 1$;
   (b) $4\cos 3\theta + \sin 3\theta = 2$;
   (c) $6\sin\theta + 3\cos\theta = \sqrt{5}$;
   (d) $\cos 2\theta + 3\sin 2\theta + 1 = 0$.

5. Solve each of the following equations for $0 \le x < 2\pi$
   (i) by expressing $\cos 2x$ and $\sin 2x$ in terms of $t = \tan x$;
   (ii) using an expression of the form $A\sin(2x\pm\alpha)$ or $A\cos(2x\pm\alpha)$.
   (a) $2\cos 2x + \sin 2x = 2$;
   (b) $2\sin 2x + \cos 2x = 0.4$;
   (c) $3\cos 2x - 2\sin 2x + 0.2 = 0$;
   (d) $2\sin 2x - 3\cos 2x = 0.5$.

6. The function $(1-2m)\cos x + m\sin x$ is written in the form $A\cos(x-\alpha)$ where $A > 0$.
   (a) Find the values of A and $\tan\alpha$ in terms of m.
   (b) Find the maximum value of $(1-2m)\cos x + m\sin x$ in terms of m.
   (c) Show that the maximum value must be greater than or equal to $\frac{1}{5}\sqrt{5}$ and find the value of m for which the maximum value occurs.

4.15 Converting Sums and Differences to Products (Optional)

The following identities can be used to express the sum or difference of two trigonometric functions as a product of two other trigonometric functions.

1. $\sin S + \sin T = 2\sin\frac{1}{2}(S+T)\cos\frac{1}{2}(S-T)$
2. $\sin S - \sin T = 2\cos\frac{1}{2}(S+T)\sin\frac{1}{2}(S-T)$
3. $\cos S + \cos T = 2\cos\frac{1}{2}(S+T)\cos\frac{1}{2}(S-T)$
4. $\cos S - \cos T = -2\sin\frac{1}{2}(S+T)\sin\frac{1}{2}(S-T)$

Proof $\sin(A+B) = \sin A\cos B + \cos A\sin B$ .... (i)
$\sin(A-B) = \sin A\cos B - \cos A\sin B$ .... (ii)
Adding (i) and (ii) gives:
$\sin(A+B) + \sin(A-B) = 2\sin A\cos B$ .... (iii)
Subtracting (ii) from (i) gives:
$\sin(A+B) - \sin(A-B) = 2\cos A\sin B$ .... (iv)

Chapter 4

Putting $S = A + B$ and $T = A - B$ gives $A = \frac{1}{2}(S + T)$ and $B = \frac{1}{2}(S - T)$.
Then (iii) becomes $\sin S + \sin T = 2 \sin \frac{1}{2}(S + T) \cos \frac{1}{2}(S - T)$ ......(1).
Also (iv) becomes $\sin S - \sin T = 2 \cos \frac{1}{2}(S + T) \sin \frac{1}{2}(S - T)$ ...... (2).

Similarly,
$\cos(A + B) = \cos A \cos B - \sin A \sin B$ .... (v)
$\cos(A - B) = \cos A \cos B + \sin A \sin B$ .... (vi)
Adding (v) and (vi) gives:
$\cos(A + B) + \cos(A - B) = 2 \cos A \cos B$ .... (vii)
Subtracting (vi) from (v) gives:
$\cos(A + B) - \cos(A - B) = -2 \sin A \sin B$ .... (viii)
Thus (vii) becomes $\cos S + \cos T = 2 \cos \frac{1}{2}(S + T) \cos \frac{1}{2}(S - T)$ ..... (3).
Also (viii) becomes $\cos S - \cos T = -2 \sin \frac{1}{2}(S + T) \sin \frac{1}{2}(S - T)$ ... (4).

Example Express $\cos(x + \frac{1}{3}\pi) + \cos(x - \frac{1}{3}\pi)$ as a product of two trigonometric functions and find the maximum value of $\cos(x + \frac{1}{3}\pi) + \cos(x - \frac{1}{3}\pi)$.

$\cos(x + \frac{1}{3}\pi) + \cos(x - \frac{1}{3}\pi)$
$= 2 \cos \frac{1}{2}(x + \frac{1}{3}\pi + x - \frac{1}{3}\pi) \cos \frac{1}{2}(x + \frac{1}{3}\pi - [x - \frac{1}{3}\pi])$
$= 2 \cos x \cos \frac{1}{3}\pi$
$= \cos x$.

Therefore the maximum value of $\cos(x + \frac{1}{3}\pi) + \cos(x - \frac{1}{3}\pi)$ is 1.

Example Prove that $\sin^2 A + \cos^2 B = 1 + \sin(A + B)\sin(A - B)$.

L.H.S. $= \frac{1}{2}(1 - \cos 2A) + \frac{1}{2}(1 + \cos 2B)$
$= 1 + \frac{1}{2}(\cos 2B - \cos 2A)$
$= 1 + \frac{1}{2}(-2 \sin \frac{1}{2}(2B + 2A) \sin \frac{1}{2}(2B - 2A))$
$= 1 + \frac{1}{2}(2 \sin(A + B) \sin(A - B))$
$= 1 + \sin(A + B) \sin(A - B)$
$=$ R.H.S.

Trigonometry

Exercise 4.15

1. (a) Verify the formula for $\sin S + \sin T$ when $S = \frac{2}{3}\pi$ and $T = \frac{1}{3}\pi$.
   (b) Verify the formula for $\sin S - \sin T$ when $S = \frac{5}{4}\pi$ and $T = \frac{3}{4}\pi$.
   (c) Verify the formula for $\cos S + \cos T$ when $S = \frac{7}{6}\pi$ and $T = \frac{1}{6}\pi$.

2. Express the following sums or differences as products:
   (a) $\sin 4x + \sin 2x$;
   (b) $\sin 3x - \sin x$;
   (c) $\cos 50° + \cos 30°$;
   (d) $\cos\theta - \cos 5\theta$;
   (e) $\sin \frac{3}{5}\pi - \sin \frac{1}{5}\pi$;
   (f) $\cos(x+y) + \cos(x-y)$;
   (g) $\sin 78° + \sin 38°$;
   (h) $\cos 3\alpha + \cos 5\alpha$;
   (i) $\cos 4\theta - \cos\theta$;
   (j) $\sin 50° - \sin\alpha$;
   (k) $\sin 54° - \sin 34°$;
   (l) $\cos(\frac{1}{2}\pi - x) + \cos(\frac{1}{2}\pi + x)$.

3. Prove the following identities true whenever the expressions have meaning:
   (a) $\dfrac{\sin 3x + \sin x}{\cos 3x + \cos x} = \tan 2x$;
   (b) $\dfrac{\cos x + \cos y}{\sin x - \sin y} = \cot\tfrac{1}{2}(x-y)$;
   (c) $(\cos x - \cos y)^2 + (\sin x - \sin y)^2 = 4\sin^2\tfrac{1}{2}(x-y)$;
   (d) $(\cos x + \cos y)^2 - (\sin x + \sin y)^2 = 4\cos(x+y)\cos^2\tfrac{1}{2}(x-y)$.

4. Give the exact value of each of the following:
   (a) $\dfrac{\cos(80° + x) - \cos(x - 40°)}{\sin(80° + x) + \sin(x - 40°)}$;
   (b) $\dfrac{\cos\frac{1}{12}\pi - \frac{1}{2}\sqrt{2}}{\frac{1}{2}\sqrt{2} - \sin\frac{1}{12}\pi}$.

5. Find the maximum value of each of the following functions:
   (a) $\sin A - \sin(\tfrac{1}{3}\pi - A)$;
   (b) $\cos(80° + A) - \cos(A - 40°)$.

6. Express each of the following in factors:
   (a) $\cos 3x + \cos 2x + \cos x$;
   (b) $\sin 7x + \sin 4x + \sin x$;
   (c) $\sin A + 2\sin 2A \cos A - \sin 5A$;
   (d) $\sin 5A - \sin 3A + \sin A$.

7. Evaluate each of the following without the use of a calculator:
   (a) $\sin^2 52\tfrac{1}{2}° - \sin^2 7\tfrac{1}{2}°$;
   (b) $\sin 140° + \sin 20° - \sin 80°$;
   (c) $\cos 85° + \cos 35° - \cos 25°$;
   (d) $\cos\frac{1}{30}\pi - \cos\frac{11}{30}\pi - \sin\frac{1}{5}\pi$.

119

Chapter 4

8. Solve the following equations for $0 \leq x < 2\pi$:
   (a) $\sin 4x + \sin 2x - \cos x = 0$;
   (b) $\sin 2x + \sin x = \cos 2x + \cos x$;
   (c) $\sin 3x = \sin 2x - \sin x$;
   (d) $\sin 5x + \cos 5x = \cos x - \sin x$.

4.16 Converting Products to Sums or Differences (Optional)

The following formulae are used to express the product of two trigonometric functions as the sum or difference of two other trigonometric functions.

1. $2\sin A \cos B = \sin(A+B) + \sin(A-B)$
2. $2\cos A \sin B = \sin(A+B) - \sin(A-B)$
3. $2\cos A \cos B = \cos(A+B) + \cos(A-B)$
4. $2\sin A \sin B = \cos(A-B) - \cos(A+B)$

Proof Each of the above formulae is proved during the proofs of the formulae converting sums and differences to products in Section 4.15.
These are formulae (iii), (iv), (vii) and (viii).

Example Evaluate $\sin 82\frac{1}{2}° \sin 37\frac{1}{2}°$ without the use of a calculator.

$$\sin 82\tfrac{1}{2}° \sin 37\tfrac{1}{2}° = \tfrac{1}{2}\cos(82\tfrac{1}{2}° - 37\tfrac{1}{2}°) - \tfrac{1}{2}\cos(82\tfrac{1}{2}° + 37\tfrac{1}{2}°)$$
$$= \tfrac{1}{2}\cos 45° - \tfrac{1}{2}\cos 120°$$
$$= \tfrac{1}{4}(\sqrt{2} + 1).$$

Exercise 4.16

1. Express the following as sums or differences of trigonometric functions:
   (a) $2\sin A \cos B$;   (b) $2\cos 2x \sin y$;
   (c) $2\cos 52° \cos 44°$;   (d) $2\sin 10° \sin 25°$;
   (e) $\cos 3x \cos x$;   (f) $\sin 6x \sin 4x$;
   (g) $\cos 3x \sin 2x$;   (h) $\sin 2x \cos 5x$.

2. Express the following as sums or differences of trigonometric functions of acute angles less than 45°:
   (a) $2\cos 54° \cos 26°$;   (b) $\cos 85° \sin 35°$;
   (c) $\sin 134° \cos 84°$;   (d) $\sin 112° \sin 82°$.

3. Prove that:
   (a) $2\sin 4x \cos 2x - 2\cos 5x \sin x = \sin 2x + \sin 4x$;
   (b) $\sin 2x \cos x + \cos 3x \sin 2x = \sin 4x \cos x$;
   (c) $\sin 4x \cos x - \cos 3x \sin 2x = 2\sin x \cos^2 x$;
   (d) $4\cos x \sin(\tfrac{1}{6}\pi + x)\sin(\tfrac{1}{6}\pi - x) = \cos 3x$.

4. Evaluate each of the following without the use of a calculator:
   (a) $2\sin 84° \cos 36° - \sin 48°$;
   (b) $\cos 38° \sin 22° + \tfrac{1}{2}\sin 164°$;
   (c) $4\cos 73° \sin 13° - 4\cos 58° \sin 28°$;
   (d) $4\cos 142° \cos 22° + 4\sin 112° \sin 52°$.

5. (a) If $\tan x = \tfrac{3}{4}$, $0 < x < \tfrac{1}{2}\pi$, find the value of $2\cos\tfrac{3}{2}x \cos\tfrac{1}{2}x$.
   (b) If $\tan 2x = -\tfrac{1}{2}$, $\tfrac{1}{4}\pi < x < \tfrac{1}{2}\pi$, find the value of $\sin 3x \cos x$.

6. Find the greatest and least values of each of the following functions:
   (a) $2\cos(2x + \tfrac{1}{6}\pi)\sin 2x$;   (b) $\sin(x - \tfrac{1}{4}\pi)\sin x$.

7. Solve each of the following equations for $0 \le x < 2\pi$:
   (a) $2\sin(x + \tfrac{1}{3}\pi)\sin x = 1$;   (b) $\sin(x - \tfrac{1}{6}\pi)\cos(x - \tfrac{1}{3}\pi) = \tfrac{1}{2}$.

8. Find the values of x, $0° \le x \le 360°$, for which
   (a) $\cos 2x \sin x = \cos 3x \sin 2x$;
   (b) $\sin 2x \sin 4x = \sin x \sin 5x$;
   (c) $5\sin x \sin(60° - x) = 1$;
   (d) $2\sin 3x \cos x = \sin 3x$;
   (e) $3\sin(x - 100°)\sin(x + 50°) + 1 = 0$;
   (f) $2\cos 2x \sin x = \sin 2x$;
   (g) $2\sin x \sin 3x = 1$;
   (h) $4\cos 3x \cos x = 1 + 2\cos 4x$.

121

Chapter 4

Required Outcomes

After completing this chapter, a student should be able to:
- quote the sine, cosine and tangent of angles of 0°, 30°, 45°, 60° and 90°.
- sketch graphs of the trigonometric functions of sine, cosine and tangent.
- convert from radians to degrees and vice versa.
- use the double angle formulae for $\sin 2\theta$ and $\cos 2\theta$ to simplify trigonometric expressions and/or solve equations.
- sketch the graphs of circular functions sine, cosine and tangent.
- calculate arc length, area of a sector and area of a segment of a circle.
- solve simple trigonometric equations.
- define and sketch the graphs of the functions cosecant, secant and cotangent. **(HL)**
- prove simple trigonometric identities. **(HL)**
- recall and use the addition formulae. **(HL)**
- recall and use the duplication and half-angle formulae. **(HL)**
- write $a\sin\theta + b\cos\theta$ in the forms $k\sin(\theta \pm \alpha)$ or $k\cos(\theta \pm \alpha)$. **(HL)**
- solve trigonometric equations using the formulae above. **(HL)**

5. Relations and Functions

5.1 Relations and Functions – Domain and Range

The concept of an ordered pair (x, y) where x is a member of one set, A, and y is a member of another set, B, is very common in mathematics. For the coordinates (x, y) of a point in the plane, $x \in \mathbb{R}$ and $y \in \mathbb{R}$. For the ordered pairs (x, y) for which $y = \sin x$, x could be any member of the set of all angles, and y any real number between –1 and 1 inclusive. The rational number p/q could be represented by the ordered pair (p, q) where p is any integer and q is any non-zero integer.

Relations

We are already familiar with the idea of a relation between x and y, the coordinates of any point in the Cartesian plane satisfying the equation, say, $y = 3x - 2$.

Definition   A *relation* is any set of ordered pairs.

$R = \{(x, y) \mid x \in \mathbb{R} \text{ and } y = 3x - 2\}$,
$S = \{(x, y) \mid x \in [-1, 1] \text{ and } x^2 + y^2 = 1\}$ and
$T = \{(x, y) \mid x \geq 0 \text{ and } y = 1 + x^2\}$ are examples of relations.

The set A of all possible values of x is called the ***domain*** of the relation and the set B of all possible values of y is called the ***range*** of the relation.

For the relation R above, the domain is the set of all real numbers $\mathbb{R}$, and the range is also $\mathbb{R}$.
For the relation S, the domain $= \{x \mid x \in [-1, 1]\} = [-1, 1]$ and the range is the same as the domain since $x^2 + y^2 = 1$ represents the unit circle.
For the relation T, the domain $= \{x \mid x \geq 0\} = [0, \infty[$ and the range $= \{y \mid y \geq 1\} = [1, \infty[$.

[***Note:*** The variable used in the definitions of the domain and range is a "***dummy***" variable since any symbol may be used. The sets $\{x \mid x \geq 0\}$, $\{y \mid y \geq 0\}$ and $\{z \mid z \geq 0\}$ are identical. The 'name' of the variable is irrelevant.]

Notation: If $x \in \mathbb{R}$ and $y \in \mathbb{R}$, then $(x, y) \in \mathbb{R}^2$. Thus the Cartesian plane is denoted by $\mathbb{R}^2$.

Chapter 5

Functions

Definition   A *function* is a relation in which no two different ordered pairs have the same first member.

Geometrically the graph of a function cannot be cut by any vertical line in more than one place.

More formally, a relation f with domain A and range B is a *function* if
- for every $x \in A$ there exists at least one $y \in B$ such that $(x, y) \in f$;
- if $(x, y_1) \in f$ and $(x, y_2) \in f$, then $y_1 = y_2$.

Example   Show that the relation $f = \{ (x, 3x - 2) \mid x \in \mathbb{R} \}$ is a function.

Consider any $x \in \mathbb{R}$. There exists (exactly) one value of $(3x - 2) \in \mathbb{R}$, and if $(x, y_1) \in f$ and $(x, y_2) \in f$, then $y_1 = 3x - 2$ and $y_2 = 3x - 2$ giving $y_1 = y_2$. Therefore f is a function.

Note: For subsets of the Cartesian plane, if a is any member of the domain, then the vertical line $x = a$ must cross the graph of a function in exactly one point.

Example   Show that the relation
$$f = \{ (x, y) \mid x \in [-1, 1], y \in [-1, 1], x^2 + y^2 = 1 \}$$
is not a function.

$\left(\frac{1}{2}, \frac{\sqrt{3}}{2}\right) \in f$ since $\frac{1}{2} \in [-1, 1]$, $\frac{\sqrt{3}}{2} \in [-1, 1]$ and $\left(\frac{1}{2}\right)^2 + \left(\frac{\sqrt{3}}{2}\right)^2 = 1$, and similarly $\left(\frac{1}{2}, -\frac{\sqrt{3}}{2}\right) \in f$, but $\frac{\sqrt{3}}{2} \neq -\frac{\sqrt{3}}{2}$.

Therefore f is not a function.

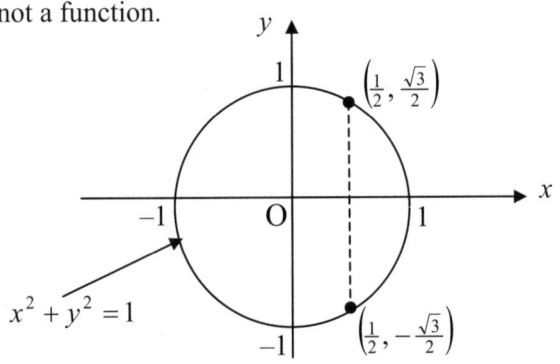

Clearly there is at least one vertical line which crosses the graph at more than one point.

Relations and Functions

Example Show that the relation $f = \{(x, y) \mid y = \dfrac{1}{x}, x \in \mathbb{R}\}$ is not a function.

Since there is no value of y for which $x = 0$, f is not a function.

Note that the relation $f = \{(x, y) \mid y = \dfrac{1}{x}, x \in \mathbb{R}, x \neq 0\}$ is a function since for any $x \in \mathbb{R}$, $x \neq 0$, there exists at least one value of $y \in \mathbb{R}$ for which $(x, y) \in f$, and if (x, y_1) and $(x, y_2) \in f$, then $y_1 = y_2$.

Finding the Domain of a Function

The domain of a function may be given explicitly when the function is defined or it may be *implied* by the expression used to define the function. The *implied domain* is the set of all real numbers for which the expression is defined.

For example, the function $f(x) = \dfrac{1}{x-1}$ has an implied domain which consists of all real numbers x except for $x = 1$ since division by zero is undefined.

The function $f(x) = \sqrt{x}$ is undefined for negative x and so the implied domain is the set of all non-negative values of x or $[0, \infty[$.

The function $f(x) = \log x$ is defined for all positive values of x and so the implied domain is $\mathbb{R}^+$.

The *range* of a function can be quite difficult to find, and perhaps it is best obtained from a graph of the function. A GDC is often the best tool to use here.

Example Find the domain and range of each of the following functions:

(a) $f(x) = \dfrac{x+2}{x+1}$;   (b) $f(x) = \dfrac{1}{\log x}$.

(a) $f(x) = \dfrac{x+2}{x+1}$ is not defined when $x = -1$ and so the domain of f is the set of all real numbers x except $x = -1$ or $]-\infty, -1[\cup]-1, \infty[$.

$y = f(x) = \dfrac{x+2}{x+1} = 1 + \dfrac{1}{x+1}$ and so $(y - 1)(x + 1) = 1$ which means that y cannot be equal to 1 {zero $\times (x + 1) \neq 1$} and so the range is all real numbers y except $y = 1$ or $]-\infty, 1[\cup]1, \infty[$.

125

Chapter 5

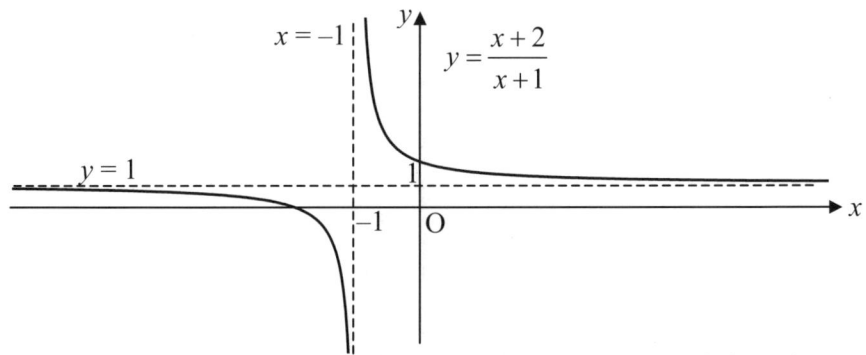

It is clear from the graph that the domain consists of all real x except $x = -1$ and the range consists of all real y except $y = 1$.

(b) $f(x) = \dfrac{1}{\log x}$ is not defined if $\log x$ is not defined or if $\log x = 0$.

Therefore the domain of f is the set of all positive real numbers x except for $x = 1$ or $]0, 1[\cup]1, \infty [$.

The following is a graph of $y = f(x) = \dfrac{1}{\log x}$:

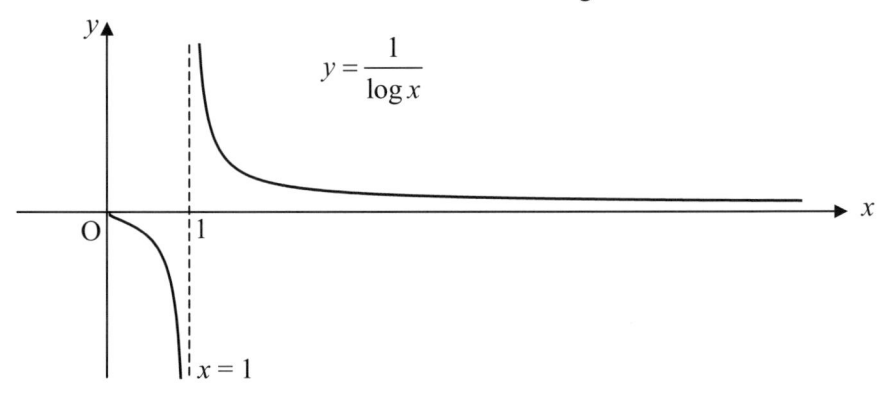

From the graph we see that the range is the set of all real numbers y except for $y = 0$ or $]-\infty, 0[\cup]0, \infty [$.

Notation

We often use the notation $f : A \to B$ to indicate that f is a function with domain A and range B. We interpret this as: "f is a function defined on A with values in B". Using Euler's notation, we usually write the statement "$(x, y) \in f$" as $y = f(x)$.

This notation suggests the fact that for each value of x belonging to the domain, there exists a unique value of $f(x)$ belonging to the range.

If the function $f(x)$ describes how the value of $f(x)$ is calculated for each value of x, then we write
$$f : x \mapsto f(x).$$

Thus if $f = \{ (x, 3x - 2) : x \in \mathbb{R} \}$, we write $f : \mathbb{R} \to \mathbb{R}$ or $f : x \mapsto 3x - 2$.

The function $f : A \to B$ can be pictured in the following way:

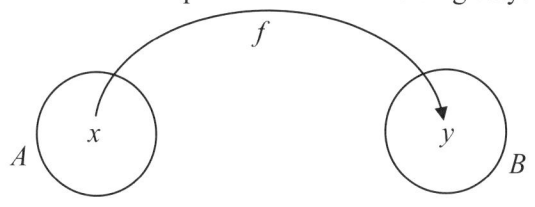

Definition Two functions $f_1 : A_1 \to B_1$ and $f_2 : A_2 \to B_2$ are said to be equal, $f_1 = f_2$, if $A_1 = A_2$, $B_1 = B_2$, and if $x \in A_1$, $f_1(x) = f_2(x)$.

Thus equal functions have the same domain and range, and have equal values for each element of the domain.

Example Consider the following functions from $\mathbb{R}$ to $\mathbb{R}$. State the range of each.

(a) $3x - 2$; (b) x^3; (c) $\cos x$; (d) 3^x.

(a) Range = $\mathbb{R}$. (b) Range = $\mathbb{R}$.
(c) Range = $[-1, 1]$. (d) Range = $\mathbb{R}^+$.

Exercise 5.1

1. Determine which of the following relations are functions:
   (a) $\{ (1, 4), (2, 7), (3, 10), (4, 13), (5, 16) \}$;
   (b) $\{ (4, 1), (5, 2), (6, 0), (7, 1), (8, 2), (9, 0) \}$;
   (c) $\{ (-3, 9), (-2, 4), (-1, 1), (0, 0), (1, 1), (2, 4), (3, 9) \}$;
   (d) $\{ (0, 2), (0, 4), (0, 6), (0, 8) \}$.

2. Determine which of the following relations are functions, giving reasons for your answers:
   (a) $R = \{ (x, y) \mid x \in \mathbb{R}, 3x - y = 2 \}$;
   (b) $S = \{ (x, y) \mid x \in \mathbb{R}, y = x^2 \}$;
   (c) $T = \{ (x, y) \mid x \in \mathbb{R}, (x - 1)(y - 1) = 1 \}$;
   (d) $U = \{ (x, y) \mid x \in \mathbb{R}^+, y = \log_2 x \}$;
   (e) $V = \{ (x, y) \mid x \in \mathbb{R}, y = \log_2 |x| \}$;
   (f) $W = \{ (x, y) \mid x \in \mathbb{Q}, xy = 4 \}$.

127

Chapter 5

3. State the range of each of the following functions:
   (a) $x \mapsto 5 - 4x$ with domain $\mathbb{R}$;
   (b) $x \mapsto \sqrt{x}$ with domain $[0, \infty[$;
   (c) $x \mapsto 2^x + 1$ with domain $\mathbb{R}$;
   (d) $x \mapsto \log x$ with domain $\mathbb{R}^+$;
   (e) $x \mapsto 3\sin x + 2$ with domain $\mathbb{R}$;
   (f) $x \mapsto \cos^2 x$ with domain $[0, \pi]$;
   (g) $x \mapsto \sqrt{9 - x^2}$ with domain $[-3, 3]$;
   (h) $x \mapsto 1/x$ with domain $\mathbb{R}^+$;
   (i) $x \mapsto \log\sqrt{1 - x^2}$ with domain $]-1, 0]$;
   (j) $x \mapsto 4/(4 + x^2)$ with domain $\mathbb{R}$;
   (k) $x \mapsto 1/(2x^2 - 4x + 3)$ with domain $\mathbb{R}$;
   (l) $x \mapsto \sqrt{4 - \dfrac{9}{x^2}}$ with domain $[\tfrac{3}{2}, \infty[$.

4. In each of the following, suggest a suitable domain and range to ensure that the relation is a function:
   (a) $\{(x, y) \mid y = 7 - 5x\}$;
   (b) $\{(x, y) \mid y = x^2\}$;
   (c) $\{(x, y) \mid y = \sqrt{x - 1}\}$;
   (d) $\{(x, y) \mid y = \sqrt{4 - x^2}\}$;
   (e) $\{(x, y) \mid y = 3^x\}$;
   (f) $\{(x, y) \mid y = \log x\}$;
   (g) $\{(x, y) \mid y = \log(x^2 - 1)\}$;
   (h) $\{(x, y) \mid y = 1/\sqrt{4 - x^2}\}$;
   (i) $\{(x, y) \mid y = \log(x^2 + 1)\}$;
   (j) $\{(x, y) \mid y = \sin x + 1\}$;
   (k) $\{(x, y) \mid y = \sqrt{\cos x}\}$;
   (l) $\{(x, y) \mid y = \tan x\}$.

5. In each of the following, state the 'largest' domain possible for the function and find the range corresponding to this domain:
   (a) $f : x \mapsto \sqrt{x^2 - 1}$;
   (b) $f : x \mapsto \dfrac{2}{x^2}$;
   (c) $f : x \mapsto \dfrac{1}{x - 2}$;
   (d) $f : x \mapsto \dfrac{x + 1}{x - 1}$;
   (e) $f : x \mapsto \dfrac{1}{(x - 1)(x - 2)}$;
   (f) $f : x \mapsto \dfrac{2}{2x^2 - 6x + 5}$.

Higher Level

6. In each of the following, give an example of a function which has the given domain and range:
   (a) Domain = $\mathbb{R}$, Range = $[-2, 2]$;
   (b) Domain = $\mathbb{R}$, Range = $]2, \infty[$;
   (c) Domain = $[-4, 4]$, Range = $[0, 2]$;
   (d) Domain = $]2, \infty[$, Range = $\mathbb{R}$;
   (e) Domain = $]-\infty, 0[\cup]0, \infty[$, Range = $\mathbb{R}^+$.

5.2 The Composition of Functions

Consider the functions $f : A \to B$ and $g : B \to C$, i.e., the range of f is the same as the domain of g. Then, the function $h : A \to C$, defined by $h(x) = g((f(x)))$ and sometimes written $h(x) = (g \circ f)(x)$, is called a *composition* of f and g.

The situation in diagrammatic form would be:

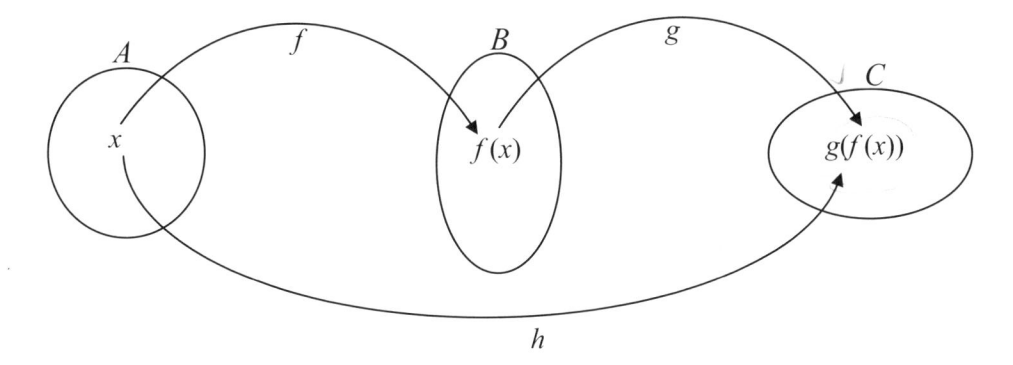

Example Given that $f = \{(1, 2), (2, 3), (3, 4)\}$ and $g = \{(2, 4), (3, 6), (4, 8)\}$, find the numerical values of $g(f(1))$, $g(f(2))$ and $g(f(3))$.

$g(f(1)) = g(2) = 4$; $g(f(2)) = g(3) = 6$; $g(f(3)) = g(4) = 8$.

Example For functions $f : \mathbb{R} \to \mathbb{R}$ defined by $f(x) = 2x + 1$ and $g : \mathbb{R} \to \mathbb{R}$ defined by $g(x) = 1 - x$, find the composite functions $f \circ g$ and $g \circ f$.

$(f \circ g)(x) = f(g(x)) = f(1 - x) = 2(1 - x) + 1 = 3 - 2x$, and
$(g \circ f)(x) = g(f(x)) = g(2x + 1) = 1 - (2x + 1) = -2x$.

Chapter 5

Example For functions $f: \mathbb{R} \to \mathbb{R}$, $g: \mathbb{R} \to \mathbb{R}$, defined by $f(x) = x+1$ and $g(x) = x^3$, find the composite functions $g \circ f$, $f \circ g$, $f \circ f$, $g \circ g$.

$(g \circ f)(x) = g(x+1) = (x+1)^3$; $(f \circ g)(x) = f(x^3) = x^3 + 1$;
$(f \circ f)(x) = f(x+1) = x + 2$; $(g \circ g)(x) = g(x^3) = (x^3)^3 = x^9$.

Note: We often use the shorthand notation f^2 for $f \circ f$. Thus, if $f(x) = 6 - x$, then $f^2(x) = f(f(x)) = f(6-x) = 6 - (6-x) = x$.

Exercise 5.2

1. For the functions $f = \{(1, 2), (2, 3), (3, 1)\}$, $g = \{(1, 1), (2, 3), (3, 2)\}$ and $h = \{(1, 3), (2, 2), (3, 1)\}$, find the numerical value of each of the following:
   (a) $(f \circ g)(1)$;
   (b) $(g \circ f)(2)$;
   (c) $(h \circ f)(3)$;
   (d) $(f \circ f)(2)$;
   (e) $(g \circ h)(1)$;
   (f) $(h \circ h)(3)$.

2. Given the functions f, g and h all with domain and range $\mathbb{R}$, defined by $f: x \mapsto x + 2$, $g: x \mapsto 3x - 1$ and $h: x \mapsto \dfrac{x+1}{3}$, find in simplest form the composite functions:
   (a) $f \circ g$;
   (b) $g \circ h$;
   (c) $h \circ g$;
   (d) h^2.

3. If $f(x) = \sqrt{x}$ and $g(x) = 2x - 1$, find the composite functions $f \circ g$ and $g \circ f$, and state the domain and range of each.

4. In each of the following find $f \circ g$ and $g \circ f$:
   (a) $f(x) = x + 1$, $g(x) = x - 1$;
   (b) $f(x) = 4x + 3$, $g(x) = \dfrac{x-3}{4}$;
   (c) $f(x) = 1 - \dfrac{1}{x}$, $g(x) = \dfrac{1}{1-x}$;
   (d) $f(x) = \dfrac{x-2}{x+1}$, $g(x) = \dfrac{x+2}{1-x}$.

Higher Level

5. Let f and g be functions each with domain $\mathbb{R}$ and range $\mathbb{R}$. Let $f(x) = 2x - 1$ and $g(x) = 2 - 3x$. Find a function $h: \mathbb{R} \to \mathbb{R}$ such that
   (a) $(f \circ h)(x) = x$;
   (b) $(h \circ g)(x) = 6x + 1$;
   (c) $(h \circ f)(x) = 6x + 1$;
   (d) $(g \circ h)(x) = 12x$.

6. Let f and g be functions each with domain $\mathbb{R}$ and range $\mathbb{R}$ defined by $f(x) = 3 - 2x$ and $g(x) = 2x + 3$. If h is a function with domain and range $\mathbb{R}$, find the value of
   (a) $h(1)$ if $f \circ h = h \circ f$;
   (b) $h(-3)$ if $g \circ h = h \circ g$.

7. Show that the domain and range of the function $f(x) = \dfrac{2x-5}{2(2x-1)}$ are the same, and find the function f^2.

*8. Let $f : \mathbb{R} \to \mathbb{R}^2$ and $g : \mathbb{R}^2 \to \mathbb{R}$ be functions defined by $f(x) = (2x+3, x)$ and $g(x, y) = \frac{1}{3}(x+y)$. Find $f \circ g$ and $g \circ f$.

*9. Find the domain of the function $f(x) = \dfrac{ax+b}{cx+d}$ ($c \ne 0$ and $ad - bc \ne 0$). Show that the range is equal to the domain provided $d = -a$, and then $f^2(x) = x$.

*10. Prove that the associative law holds for the composition of functions. That is, if $f : A \to B$, $g : C \to A$, and $h : D \to C$ are three functions, then
$$f \circ (g \circ h) = (f \circ g) \circ h.$$

5.3 Inverse Functions

Consider the function $f(x) = 2x + 1$ with domain $A = \{1, 2, 3, 4\}$ and range $B = \{3, 5, 7, 9\}$. Thus $f = \{(1, 3), (2, 5), (3, 7), (4, 9)\}$.

By interchanging the x- and y-coordinates of each element of f we form the *inverse function* of f, which is denoted f^{-1}. The inverse function has domain B and range A and can be written $f^{-1}(x) = \frac{1}{2}(x-1) = \{(3, 1), (5, 2), (7, 3), (9, 4)\}$.

The function and its inverse have the effect of 'undoing' each other.

If we denote the functions by 'machines' which take input and provide output, we may represent the function composition as follows:

$f : x \to \boxed{\text{multiply by 2}} \to 2x \to \boxed{\text{add 1}} \to 2x + 1$

$g : 2x + 1 \to \boxed{\text{subtract 1}} \to 2x \to \boxed{\text{divide by 2}} \to x$

Chapter 5

From this we see that $g(f(x)) = x$, i.e., function g has undone the effect of applying function f to the number x.

Note that function f undoes the effect of applying function g:

$g : x \to$ | subtract 1 | $\to x - 1 \to$ | divide by 2 | $\to \frac{1}{2}(x-1)$

$f : \frac{1}{2}(x-1) \to$ | multiply by 2 | $\to x - 1 \to$ | add 1 | $\to x$

Again we see that $f(g(x)) = x$ verifying that function f does indeed undo function g.

Definition Let f and g be two functions such that
- $f(g(x)) = x$ for every x in the domain of g, and
- $g(f(x)) = x$ for every x in the domain of f.

Then the function g is the *inverse* of the function f and we write $g = f^{-1}$.

Clearly from this definition, if function g is the inverse of function f, then function f is the inverse of function g. Thus the functions f and g are *inverses of each other*.

Definition The function $e(x) = x$ with domain and range equal to $\mathbb{R}$ is called the *identity* function.

Example Show that the functions $f(x) = 4x^3 + 2$ and $g(x) = \sqrt[3]{\dfrac{x-2}{4}}$ are inverses of each other.

First we note that the domain and range of both functions is $\mathbb{R}$, the set of all real numbers. We need to show that $f(g(x)) = x$ and $g(f(x)) = x$.

$$f(g(x)) = f\left(\sqrt[3]{\dfrac{x-2}{4}}\right) = 4\left(\sqrt[3]{\dfrac{x-2}{4}}\right)^3 + 2 = 4\left(\dfrac{x-2}{4}\right) + 2 = x - 2 + 2 = x \text{ and}$$

$$g(f(x)) = g(4x^3 + 2) = \sqrt[3]{\dfrac{4x^3 + 2 - 2}{4}} = \sqrt[3]{\dfrac{4x^3}{4}} = \sqrt[3]{x^3} = x \text{ which completes the proof.}$$

If (a, b) lies on the graph of f, then the point (b, a) lies on the graph of f^{-1} and vice versa. Thus the graph of f^{-1} is found by reflecting the graph of f in the line $y = x$.

Relations and Functions

Example Find the inverse of the function $f : \mathbb{R} \to \mathbb{R}$ defined by $f(x) = 3x - 2$ and sketch the graphs of both $y = f(x)$ and $y = f^{-1}(x)$ on the same set of coordinate axes.

To find the inverse of f we simply interchange x and y in the equation of f and then make y the subject of this new equation.

For the function f^{-1} we have $x = 3y - 2$ or $y = \frac{1}{3}(x + 2)$.

Thus $f^{-1}(x) = \frac{1}{3}(x + 2)$.

The graphs of the functions f and f^{-1} are as follows:

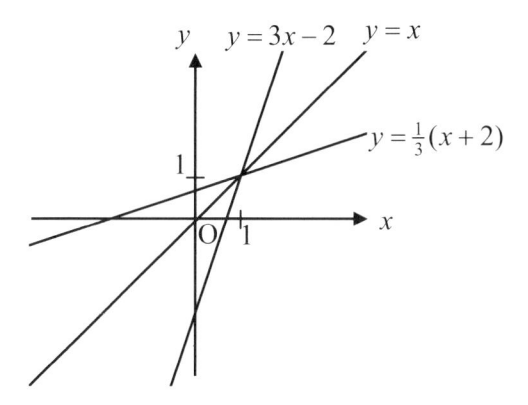

The Existence of an Inverse Function

A given function may not have an inverse function. For example, the function $f(x) = x^2, x \in \mathbb{R}$ does not have an inverse function since the relation $y^2 = x$ can have two differing elements $(1, 1)$ and $(1, -1)$ with the same first member and so cannot be a function.

Definition A function f is said to be **one-to-one** if no two different ordered pairs have the same second member.

More formally, a function f is said to be one-to-one if for all a and b belonging to its domain, $f(a) = f(b)$ implies that $a = b$.

Geometrically the graph of a one-to-one function cannot be crossed by a horizontal line in more than one place.

Example Show that the function $f(x) = 3x - 1$, $x \in \mathbb{R}$, is one-to-one.

Chapter 5

Let a and b be any real numbers such that $f(a) = f(b)$.
Then $3a - 1 = 3b - 1$ which gives $a = b$.
Thus f is one-to-one.

Theorem   A function f has an inverse function f^{-1} if and only if f is one-to-one.

Example   Determine whether or not each of the following functions has an inverse function:
(a) $f(x) = 3 - 2x$, $x \in \mathbb{R}$;   (b) $f(x) = x^2 - 2x$, $x \in \mathbb{R}$.
If the inverse function does not exist, give a domain which is as 'large' as possible so that an inverse function does exist.

(a)   Let a and b be any two real numbers such that $f(a) = f(b)$ where $f(x) = 3 - 2x$.
Then $3 - 2a = 3 - 2b$
$\Rightarrow$ $-2a = -2b$
$\Rightarrow$ $a = b$.
Therefore f is one-to-one and the inverse function exists.

(b)   Let a and b be any two real numbers such that $f(a) = f(b)$ where $f(x) = x^2 - 2x$.
Then $a^2 - 2a = b^2 - 2b$
$\Rightarrow$ $a^2 - 2a + 1 = b^2 - 2b + 1$
$\Rightarrow$ $(a-1)^2 = (b-1)^2$
$\Rightarrow$ $a - 1 = \pm(b - 1)$
$\Rightarrow$ $a = b$ or $a = 2 - b$.
Thus the function f is not one-to-one and the inverse function does not exist.

$f(x) = x^2 - 2x + 1 - 1 = (x-1)^2 - 1$

Thus f is one-to-one if $x \geq 1$ since $(a-1)^2 - 1 = (b-1)^2 - 1$ implies that $(a-1)^2 = (b-1)^2$ which in turn implies that $a - 1 = b - 1$ since both $a - 1 \geq 0$ and $b - 1 \geq 0$. Then a possible domain could be $[1, \infty[$. {It can be shown that f is also one-to-one if $x \leq 1$.}

The answers given can be verified by a graph of the function which follows at the top of the next page.

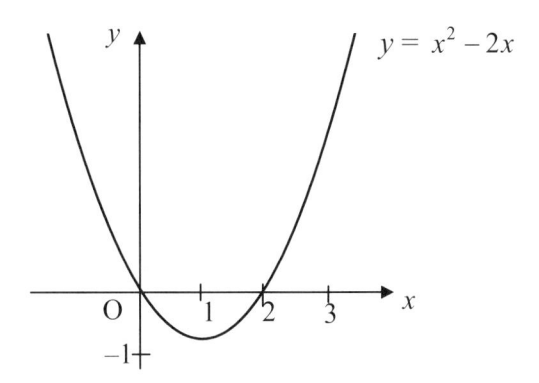

In part (b) of the previous example, all we need to do to prove that $f^{-1}(x)$ does not exist is to show that $f(2) = f(0)$ so that f is not one-to-one.

Finding the Inverse of a Function

To find the inverse of a function $y = f(x)$, follow the procedure below.

1. Test to see that f is one-to-one.
2. Write the function in the form $y = f(x)$.
3. Interchange x and y.
4. Make y the subject of this formula.
5. Replace y with $f^{-1}(x)$.
6. Check that the domain of f is the range of f^{-1} and that the domain of f^{-1} is the range of f.

Example   Find the inverse of the function $f : \mathbb{R} \to \mathbb{R}$ defined by $y = 6 - x$.

Let a and b be any two real numbers such that $f(a) = f(b)$.
Then $6 - a = 6 - b$ which gives $a = b$.
Therefore f is one-to-one.

For the function f^{-1}, $x = 6 - y$ or $y = 6 - x$, with domain and range both $\mathbb{R}$.
Thus $f^{-1}(x) = 6 - x$. [Function f is its own inverse.]

Example   Find the inverse of the function $f : x \mapsto \dfrac{3x+2}{x-1}$ with domain $x \neq 1$.

Let a and b be two real numbers not equal to 1 such that $\dfrac{3a+2}{a-1} = \dfrac{3b+2}{b-1}$.

Chapter 5

Then $(3a+2)(b-1) = (3b+2)(a-1)$
$\Rightarrow \quad 3ab - 3a + 2b - 2 = 3ab + 2a - 3b - 2$
$\Rightarrow \quad 5a = 5b$
$\Rightarrow \quad a = b$.
Therefore f is one-to-one.

For f^{-1} we have $\quad x = \dfrac{3y+2}{y-1}$

$\Rightarrow \quad x(y-1) = 3y+2$
$\Rightarrow \quad xy - x = 3y + 2$
$\Rightarrow \quad xy - 3y = x + 2$
$\Rightarrow \quad y(x-3) = x+2$
$\Rightarrow \quad y = \dfrac{x+2}{x-3} \quad (x \neq 3)$.

Thus $f^{-1}(x) = \dfrac{x+2}{x-3}$ with domain $x \neq 3$.

Now if $y = \dfrac{3x+2}{x-1} = 3 + \dfrac{5}{x-1}$ then $(y-3)(x-1) = 5$ and so $y \neq 3$.
Therefore the range of f is $x \neq 3$.

Similarly the domain of f^{-1} is $x \neq 3$ and since $y = \dfrac{x+2}{x-3} = 1 + \dfrac{5}{x-3}$ implies that $(y-1)(x-3) = 5$, then $y \neq 1$.

Thus f and f^{-1} are such that the domain of f is the range of f^{-1} and the domain of f^{-1} is the range of f. This completes the check.

Example Show that the function $f(x) = 2^{1-x}$ with domain $\mathbb{R}$ is one-to-one and find the inverse function $f^{-1}(x)$ stating its domain. Sketch the graphs of f and f^{-1} on the same set of coordinate axes.

Let a and b be two real numbers such that $f(a) = f(b)$.
Thus $2^{1-a} = 2^{1-b}$ which gives $1 - a = 1 - b$ or $a = b$.
Therefore f is one-to-one.

For the inverse function we have $x = 2^{1-y} \Rightarrow 1 - y = \log_2 x \Rightarrow y = 1 - \log_2 x$.
Thus $f^{-1}(x) = 1 - \log_2 x$ with domain $]0, \infty[$.

136

The graphs of f and f^{-1} are as follows:

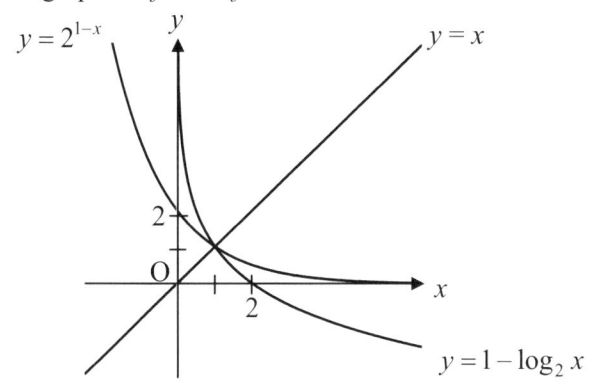

[Graph showing $y = 2^{1-x}$, $y = x$, and $y = 1 - \log_2 x$]

Note: The inverse of the function should not be confused with its reciprocal.
The reciprocal of $f(x)$ is $(f(x))^{-1} = \dfrac{1}{f(x)}$.

A diagrammatic relationship between inverses could be as follows:

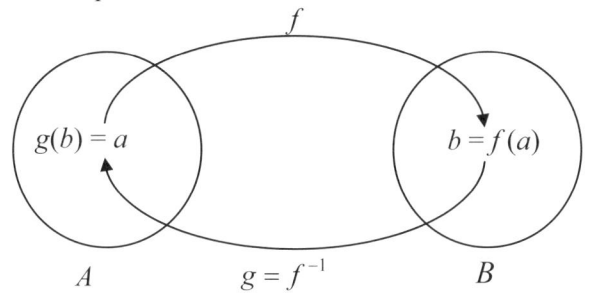

[Diagram showing sets A and B with $f: A \to B$, $b = f(a)$, and $g = f^{-1}$, $g(b) = a$]

The Inverse Trigonometric Functions

The function $f(x) = \sin x$ with domain $[-\tfrac{1}{2}\pi, \tfrac{1}{2}\pi]$ is one-to-one. Then an inverse function, $f^{-1}(x) = \arcsin x$, exists. The graphs of $y = \sin x$ and $y = \arcsin x$ are as follows:

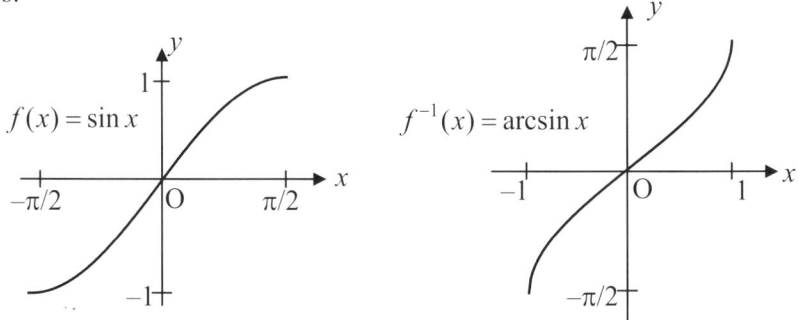

[Graph of $f(x) = \sin x$ on $[-\pi/2, \pi/2]$]

[Graph of $f^{-1}(x) = \arcsin x$]

137

Chapter 5

The function $f(x) = \cos x$ with domain $[0, \pi]$ is also one-to-one and so an inverse function, $f^{-1}(x) = \arccos x$, exists. The graphs of $y = \cos x$ and $y = \arccos x$ are as follows:

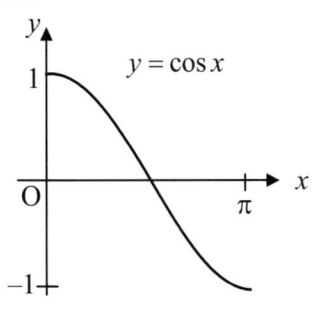

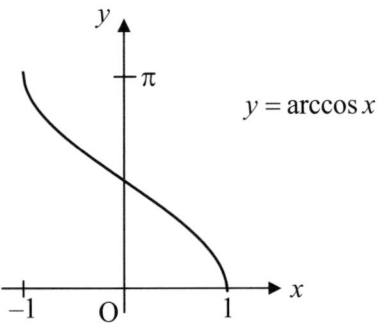

The function $f(x) = \tan x$ with domain $]-\frac{1}{2}\pi, \frac{1}{2}\pi[$ is also one-to-one and so an inverse function, $f^{-1}(x) = \arctan x$, exists. The graphs of $y = \tan x$ and $y = \arctan x$ are as follows:

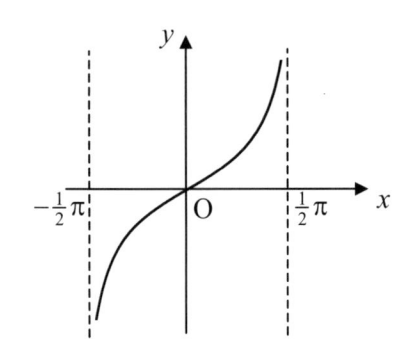

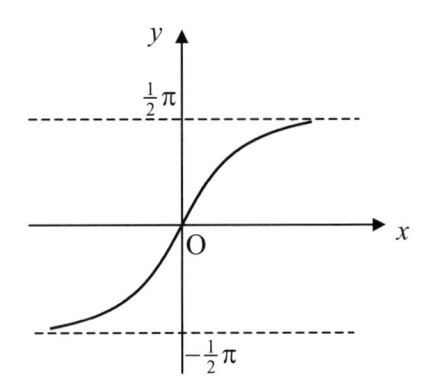

Exercise 5.3

1. Decide whether the following functions are one-to-one:
   (a) $f(x) = 2 - x$ with domain $\mathbb{R}$;
   (b) $f(x) = \sin x$ with domain $\mathbb{R}$;
   (c) $f(x) = x^3$ with domain $\mathbb{R}$;
   (d) $f(x) = \log x$ with domain $\mathbb{R}^+$;
   (e) $f(x) = \cos x$ with domain $[0, 2\pi]$;
   (f) $f(x) = 2/(1 + x^2)$ with domain $\mathbb{R}$;
   (g) $f(x) = \sqrt{x}$ with domain $[0, \infty[$;
   (h) $f(x) = x(x + 2)$ with domain $\mathbb{R}$;

Relations and Functions

(i) $f(x) = \sqrt{4-x^2}$ with domain $[-2, 2]$;
(j) $f(x) = \cos x$ with domain $[0, \pi]$;
(k) $f(x) = 1 + 2^x$ with domain $\mathbb{R}$;
(l) $f(x) = \arctan x$ with domain $\mathbb{R}$.

2. Find the 'largest' domain for which the function defined by $f(x) = 1 - x^2$ is one-to-one. For the domain chosen, find the inverse function $f^{-1}(x)$.

3. In each of the following, express $f^{-1}(x)$ in terms of x:
   (a) $f(x) = 3x$;
   (b) $f(x) = 4 - x$;
   (c) $f(x) = 3x + 2$;
   (d) $f(x) = x^3$;
   (e) $f(x) = \dfrac{4}{x}$;
   (f) $f(x) = x^3 + 1$;
   (g) $f(x) = 2 - \dfrac{1}{x}$;
   (h) $f(x) = \dfrac{3}{1-x}$.

4. For the function $f(x) = 2x + 3$, find the inverse function, $f^{-1}(x)$ and sketch the graphs of the function and its inverse on the same set of coordinates axes.

5. In each of the following, express $f^{-1}(x)$ in terms of x:
   (a) $f(x) = \dfrac{1}{x+1}$;
   (b) $f(x) = \dfrac{x+2}{x-1}$;
   (c) $f(x) = \dfrac{1-2x}{1+2x}$;
   (d) $f(x) = \dfrac{2x-3}{3x-2}$.

6. For the function $f(x) = \dfrac{2x-3}{x-2}$, $x \neq 2$, show that $f(f(x)) = x$, and write down the inverse function $f^{-1}(x)$.

7. Show that the function $f(x) = 2x^2 - 4x + 3$ with domain $[1, \infty[$ is one-to-one. Find the inverse function, $f^{-1}(x)$, and sketch graphs of both $y = f(x)$ and $y = f^{-1}(x)$ on the same set of axes.

8. Show that the function $f(x) = \log_5(2x - 1)$ with domain $]\tfrac{1}{2}, \infty[$ is one-to-one and find the inverse function $f^{-1}(x)$. Sketch graphs of both the function and its inverse on the same set of coordinate axes.

Chapter 5

Higher Level

9. Show that the function $f(x) = \dfrac{ax+b}{cx-a}$, where $c \neq 0$ and $x \neq \dfrac{a}{c}$, is self-inversing provided $a^2 + bc \neq 0$.

*10. Let $f: \mathbb{R}^+ \to \mathbb{R}$ be a function defined by $f(x) = \dfrac{x}{2} - \dfrac{2}{x}$. Show that f has an inverse function. Find $f^{-1}(x)$, and sketch the graph of both $f(x)$ and $f^{-1}(x)$ on the same set of coordinate axes.

Required Outcomes

After completing this chapter, a student should be able to:
- define the terms relation and function.
- find the (largest) range of a given function.
- combine any two functions under composition of functions.
- determine whether or not a given function is one-to-one.
- find the inverse of a function, if it exists.

6 Sequences and Series

6.1 Sequences

Definition   A sequence is a function whose domain is the set of positive integers.

The elements of a sequence f, as defined above, would be written as ordered pairs in the form $(1, f(1)), (2, f(2)), (3, f(3)), \ldots, (n, f(n)), \ldots$, but it is customary to list only the second components of the ordered pairs.

Thus the sequence is written as $f(1), f(2), f(3), \ldots, f(n), \ldots$.

The elements in the range of the sequence are called its ***terms***, and $f(n)$ is called the nth term of the sequence.

In practice we denote the nth term of the sequence by a_n, t_n, u_n, etc., and we denote the sequence by $\{a_n\}, \{t_n\}, \{u_n\}$ respectively.

Example   Write down the first 5 terms of the following sequences:

(a) $\{5n\}$;   (b) $\{3n - 2\}$;   (c) $\left\{2 - \dfrac{1}{n}\right\}$.

(a) $\{5n\} = \{5, 10, 15, 20, 25, \ldots\}$

(b) $\{3n - 2\} = \{1, 4, 7, 10, 13, \ldots\}$

(c) $\left\{2 - \dfrac{1}{n}\right\} = \{1, 1\tfrac{1}{2}, 1\tfrac{2}{3}, 1\tfrac{3}{4}, 1\tfrac{4}{5}, \ldots\}$

If we are given the first few terms of a sequence and asked for an expression for the nth term, there is an infinite number of possible answers, if indeed one exists. An example of a sequence for which a simple expression for the nth term does not exist (as of this time), is the sequence $\{2, 3, 5, 7, \ldots\}$, the sequence of consecutive prime numbers. However there is generally a *simple* pattern to the sequence, and it is normal to choose this pattern as your answer. For example the nth term of the sequence which begins $\{1, 2, 3, 4, \ldots\}$, could be of the form $u_n = n + (n-1)(n-2)(n-3)(n-4)f(n)$, for any $f(n)$. But the *simple* answer is clearly $u_n = n$.

Chapter 6

In any problem requiring a formula for the *n*th term of a sequence, it is an *obvious* formula that is requested.

A sequence is sometimes defined ***recursively***. That is the terms are not expressed as a function of *n*, but instead each term, except for possibly the first few, is found from previous terms.

Example   Sequence $\{u_n\}$ is defined as follows: $u_1 = u_2 = 1, u_{n+2} = 2u_n + u_{n+1}$, $n \geq 1$. Write down the first 5 terms of the sequence.

$u_3 = 2u_1 + u_2 = 3, u_4 = 2u_2 + u_3 = 5, u_5 = 2u_3 + u_4 = 11$, etc.
Therefore the first 5 terms are $\{1, 1, 3, 5, 11, \ldots\}$.

Exercise 6.1

1. Write down the first three terms and the 10th term of the following sequences:
   (a) $\{3n\}$;
   (b) $\{2n - 5\}$;
   (c) $\{2^{n-1}\}$;
   (d) $\left\{\dfrac{2n+1}{n+1}\right\}$;
   (e) $\{n(0.1)^n\}$;
   (f) $\{2 - (-1)^n\}$.

2. Find a suitable *n*th term for each of the following sequences:
   (a) $\{2, 4, 6, 8, \ldots\}$;
   (b) $\{5, 7, 9, 11, \ldots\}$;
   (c) $\{5, 8, 11, 14, \ldots\}$;
   (d) $\{67, 58, 49, 40, \ldots\}$;
   (e) $\{4, 16, 36, 64, \ldots\}$;
   (f) $\{3, \frac{16}{5}, \frac{25}{7}, 4, \frac{49}{11}, \ldots\}$;
   (g) $\{3, 7, 15, 31, \ldots\}$;
   (h) $\{1, 3, 6, 10, 15, 21, \ldots\}$.

3. (a) A sequence is defined recursively as follows:
   $u_1 = u_2 = 1, u_{n+2} = u_{n+1} + u_n$ for all $n \geq 1$.
   Write down the first 10 terms of this sequence.
   (b) Calculate the first 5 terms of the sequence $\{u_n\}$ where
   $$u_n = \frac{1}{\sqrt{5}}\left(\frac{1+\sqrt{5}}{2}\right)^n - \frac{1}{\sqrt{5}}\left(\frac{1-\sqrt{5}}{2}\right)^n.$$

6.2   Arithmetic Sequences (Arithmetic Progressions)

In the sequence $\{3n + 2\} = \{5, 8, 11, 14, \ldots\}$, each term can be found by adding a constant number (in this case 3) to the previous term.

142

Similarly in the sequence $\{16 - 6n\} = \{10, 4, -2, -8, \ldots \}$, each term can be found by subtracting a constant number (in this case 6) from the previous term. This is of course equivalent to adding a constant (in this case –6) to the previous term.

These sequences are examples of **arithmetic sequences** or **arithmetic progressions** (APs), in which there is a **common difference** between successive terms.

Definition   A sequence $\{u_n\}$ is said to be **arithmetic** if $u_{n+1} - u_n = d \ (n \geq 1)$, where d is a constant. The constant d is called the **common difference** of the sequence.

An arithmetic sequence $\{u_n\}$ can be written $\{u_n\} = \{u_1, u_1 + d, u_1 + 2d, u_1 + 3d, \ldots \}$.
The nth term is given by
$$u_n = u_1 + (n-1)d.$$

Example   Find a formula for the nth term of the arithmetic sequence $\{7.5, 6.6, 5.7, \ldots \}$. Which term of the sequence has the value –4.2?

$u_1 = 7.5$ and $d = -0.9$
$u_n = u_1 + (n-1)d = 7.5 - 0.9(n-1)$, giving $u_n = 8.4 - 0.9n$.

If $u_n = -4.2$, $4.2 = 8.4 - 0.9n$ $\Rightarrow$ $0.9n = 12.6$ $\Rightarrow$ $n = 14$.
Therefore the 14th term has the value –4.2.

Example   The 5th term of an arithmetic sequence is 15 and the 10th term is 45. Find the first three terms of the sequence, and find an expression for the nth term.

$u_5 = 15$ $\Rightarrow$ $15 = u_1 + 4d$ and $u_{10} = 45$ $\Rightarrow$ $45 = u_1 + 9d$.
Solving these simultaneous equations gives: $d = 6$ and $u_1 = -9$.
Thus the first 3 terms are $\{-9, -3, 3, \ldots \}$ and $u_n = -9 + 6(n-1) = 6n - 15$.

Example   Find the number of positive terms of the arithmetic sequence $\{59.2, 58.4, 57.6, \ldots \}$, and find the value of the first negative term.

$u_1 = 59.2$ and $d = -0.8$
$u_n = u_1 + (n-1)d = 59.2 - 0.8(n-1) = 60 - 0.8n$
The terms are positive provided $u_n > 0$ or $60 - 0.8n > 0$.
Solving gives $n < 75$ and so there are 74 positive terms.

Now since the 75th term is zero, the first negative term is –0.8.

Chapter 6

Example Prove that the sequence whose nth term is given by $u_n = 32 - 5n$ is arithmetic, and find the first negative term.

$u_{n+1} - u_n = 32 - 5(n+1) - (32 - 5n) = -5$ which is constant.
Therefore the sequence is arithmetic.

$u_n < 0 \Rightarrow n > 6.4$ and so the first negative term is $u_7 = -3$.

Example Given that 24, $5x + 1$ and $x^2 - 1$ are three consecutive terms of an arithmetic progression, find the values of x and the numerical value of the fourth term for each value of x found.

The common difference is $d = 5x + 1 - 24 = x^2 - 1 - (5x + 1)$.
This gives $x^2 - 10x + 21 = 0 \Rightarrow (x - 7)(x - 3) = 0 \Rightarrow x = 3$ or 7.

If $x = 3$ the terms are 24, 16, 8 and so the 4th term is 0.
If $x = 7$ the terms are 24, 36, 48 and so the 4th term is 60.

Example Three numbers are in arithmetic progression. Find the numbers if their sum is 30 and the sum of their squares is 332.

Let the numbers be $a - d$, a, $a + d$. Their sum is $3a = 30$ and so $a = 10$.
The sum of their squares is $(10 - d)^2 + 100 + (10 + d)^2 = 332$.
This gives $2d^2 = 32 \Rightarrow d^2 = 16 \Rightarrow d = \pm 4$.
Therefore the numbers are 6, 10, 14 (or 14, 10, 6).

Exercise 6.2

1. Decide which of the following sequences could be arithmetic. If the sequence is arithmetic, find the common difference.
   (a) {9, 13, 17, 21, ... } ;
   (b) {a, $4a$, $7a$, $10a$, ... } ;
   (c) {2, 4, 8, 16, ... } ;
   (d) {7, 1, –5, –11, ... } ;
   (e) {5, –8, 11, –14, ... } ;
   (f) {$3a - 2b$, $2a - b$, a, ... }.

2. The following pairs of numbers are respectively the first term and the common difference of an arithmetic sequence. Find the first 4 terms and the 10th term of each sequence.
   (a) 5, 6 ;
   (b) 43, –5 ;
   (c) –7, 4 ;
   (d) –1, –7 ;
   (e) $8\frac{1}{2}$, $1\frac{1}{4}$;
   (f) $a - 3b$, $2a + b$.

3. Find the nth term and the 10th term of each of the following APs:
   (a) $\{5, 9, 13, \ldots\}$;
   (b) $\{87, 75, 63, \ldots\}$;
   (c) $\{2.3, 2.9, 3.5, \ldots\}$;
   (d) $\{2 + 3x, 2 + x, 2 - x, \ldots\}$.

4. In each of the following find the first 3 terms of the arithmetic sequence $\{u_n\}$ where
   (a) $u_3 = 46, u_8 = 101$;
   (b) $u_5 = 35, u_9 = -1$;
   (c) $u_5 = a, u_8 = b$;
   (d) $u_4 = a + 2b, u_{11} = 8a - 5b$.

5. Find the first term and the common difference of the arithmetic sequence $\{u_n\}$, if
   (a) $u_3 + u_5 = 52, u_2 + u_8 = 66$;
   (b) $u_3 + u_7 = -14, u_5 + u_{10} = -44$;
   (c) $u_9 = 2u_2, u_3 + u_{11} = 96$;
   (d) $2u_8 - u_2 = 8, u_{19} + u_{25} = 8$.

6. (a) Find the number of positive terms of the arithmetic sequence $\{97, 96.2, 95.4, \ldots\}$.
   (b) Find the first positive term of the sequence $\{-40.3, -38.8, -37.3, \ldots\}$, given that the sequence is arithmetic.
   (c) How many negative terms are there of the arithmetic sequence $\{-16, -15\frac{5}{8}, -15\frac{1}{4}, \ldots\}$?
   (d) Write down the nth term of the arithmetic sequence $\{4, 11, 18, \ldots\}$. What is the term nearest 140? Find the least value of n for which the nth term is greater than 250.

7. (a) The sum of three numbers in arithmetic progression is 51, and the difference between the squares of the greatest and least is 408. Find the numbers.
   (b) The sum of four numbers in arithmetic progression is 38, and the sum of their squares is 406. Find the numbers.
   (c) The sum of five numbers in arithmetic progression is 10, and the product of the first, third and fifth is –64. Find the numbers.

Higher Level

8. Given that $\dfrac{1}{b-2a}, \dfrac{1}{b}, \dfrac{1}{b-2c}$ are in arithmetic progression, prove that $a(b-c)$, ac, $c(b-a)$ are also in arithmetic progression.

Chapter 6

6.3 Arithmetic Series and Sigma Notation

Sigma Notation

The symbol Σ is used to simplify expressions for the sum of many terms. For example the sum $3 + 5 + 7 + \ldots + 201$ can be written more concisely in the form $\sum_{r=1}^{100}(2r+1)$. This is read as "the sum of terms of the form '$2r + 1$' where r takes the integer values from 1 to 100 inclusive". Thus $2r + 1$ takes the values $3, 5, 7, \ldots, 201$, and these values are added together.

Example Evaluate $\sum_{r=1}^{5}(3r+2)$.

$$\sum_{r=1}^{5}(3r+2)$$
$= (3 \times 1 + 2) + (3 \times 2 + 2) + (3 \times 3 + 2) + (3 \times 4 + 2) + (3 \times 5 + 2)$
$= 5 + 8 + 11 + 14 + 17$
$= 55.$

The sum $\sum_{r=1}^{n} u_r = u_1 + u_2 + u_3 + \ldots + u_n$ of terms of an arithmetic sequence is called an *arithmetic series*.

We denote the sum of the first n terms of the sequence by S_n.

Thus $S_n = u_1 + u_2 + u_3 + \ldots + u_n = \sum_{r=1}^{n} u_r$.

To find a formula for S_n, we note that:
$S_n = u_1 + (u_1 + d) + (u_1 + 2d) + \ldots + (u_n - 2d) + (u_n - d) + u_n$, and in reverse order,
$S_n = u_n + (u_n - d) + (u_n - 2d) + \ldots + (u_1 + 2d) + (u_1 + d) + u_1$.
Adding these we obtain:
$2S_n = (u_1 + u_n) + (u_1 + u_n) + (u_1 + u_n) + \ldots + (u_1 + u_n) + (u_1 + u_n) + (u_1 + u_n)$
$= n(u_1 + u_n)$.

Hence $S_n = \dfrac{n}{2}(u_1 + u_n)$.

146

An alternative formula is found by replacing u_n with $u_1 + (n-1)d$, as follows:

$$S_n = \frac{n}{2}(u_1 + u_1 + (n-1)d), \quad \text{simplifying to}$$

$$S_n = \frac{n}{2}(2u_1 + (n-1)d).$$

Example Find the sum of 9 terms of the arithmetic series $-12 - 5 + 2 + \ldots$.

Here $u_1 = -12$ and $d = 7$.

Thus the required sum $= \frac{9}{2}(-24 + 8 \times 7) = 144$.

Example Evaluate $\sum_{n=1}^{100}\left(\frac{n-100}{2}\right)$.

Firstly the sequence for which $u_n = \frac{n-100}{2}$ is arithmetic since

$u_{n+1} - u_n = \left(\frac{n-99}{2}\right) - \left(\frac{n-100}{2}\right) = \frac{1}{2}$, which is constant.

Thus $\sum_{n=1}^{100}\left(\frac{n-100}{2}\right) = \frac{100}{2}(u_1 + u_{100}) = 50(-49.5 + 0) = -2475$.

Example Find the first term and the common difference of the arithmetic sequence in which $u_{10} = -29$ and $S_{10} = -110$.

$S_{10} = \frac{10}{2}(u_1 + u_{10}) = 5(u_1 - 29) = -110 \quad \Rightarrow \quad u_1 = 7$.

$u_{10} = u_1 + 9d = 7 + 9d = -29 \quad \Rightarrow \quad d = -4$.

Example The sum of the first 8 terms of an arithmetic series is 100, and the sum of the first 15 terms is 555. Find the first term and the common difference.

$S_8 = \frac{8}{2}(2u_1 + 7d) = 100 \quad \Rightarrow \quad 2u_1 + 7d = 25$

$S_{15} = \frac{15}{2}(2u_1 + 14d) = 555 \quad \Rightarrow \quad 2u_1 + 14d = 74$

Solving these equations gives $u_1 = -12$ and $d = 7$.

Chapter 6

Example Consider the arithmetic series for which $u_n = 72 - 6n$. If the sum of the first n terms of the series is 378, find n. Explain why it is that there are two possible values of n.

$$S_n = \frac{n}{2}(u_1 + u_n) = \frac{n}{2}(66 + 72 - 6n) = 378$$
$$\Rightarrow \quad n(23 - n) = 126$$
$$\Rightarrow \quad n^2 - 23n + 126 = 0$$
$$\Rightarrow \quad (n - 9)(n - 14) = 0.$$
Thus $n = 9$ or $n = 14$.

The sum of 9 terms is the same as the sum of 14 terms since the sum of the 10th to 14th terms is zero. ($u_{10} = 12, u_{11} = 6, u_{12} = 0, u_{13} = -6, u_{14} = -12$.)

Example Find the sum of all the multiples of 11 which are less than 1000.

The required sum is
$$11 + 22 + 33 + \ldots + 990 = \frac{90}{2}(11 + 990) = 45\,045.$$

Example Consider the series $29.8 + 29.1 + 28.4 + \ldots$. Find the sum of all the positive terms.

If $u_n > 0$ then $29.8 - 0.7(n - 1) > 0 \Rightarrow 0.7n < 30.5 \Rightarrow n \leq 43$.

Therefore the sum of the positive terms is $\frac{43}{2}(59.6 - 42 \times 0.7) = 649.3$.

Exercise 6.3

1. Find the sum of the following arithmetic series to the number of terms given in parentheses:
   (a) $2 + 5 + 8 + \ldots$ (50);
   (b) $14 + 22 + 30 + \ldots$ (20);
   (c) $2.4 + 3.6 + 4.8 + \ldots$ (25);
   (d) $34 + 31.5 + 29 + \ldots$ (12);
   (e) $a + 3a + 5a + \ldots$ (2a);
   (f) $5a + 2a - a - \ldots$ (3a).

2. Prove that each of the following series is arithmetic and find each sum:
   (a) $\sum_{n=1}^{14}(7n + 1)$;
   (b) $\sum_{n=1}^{20}(65 - 3n)$;
   (c) $\sum_{n=1}^{201}\left(\frac{101 - n}{5}\right)$;
   (d) $\sum_{n=1}^{40}\left(\frac{2n}{5} + 25\right)$.

Sequences and Series

3. Find the first three terms and the 12th term of the arithmetic sequence for which:
   (a) $u_1 = -8$ and $S_{12} = 102$;
   (b) $u_4 = 12$ and $S_{10} = 45$;
   (c) $u_{16} = -4$ and $S_{16} = -4$;
   (d) $S_8 = 192$ and $S_{12} = 432$;
   (e) $S_{10} = -160$ and $S_{14} = -392$;
   (f) $S_{15} = -255$ and $S_{30} = 165$.

4. The first three terms of an arithmetic sequence are 12, $4a^2$ and $10a$. Find the possible values of a and the sum of the first 10 terms of the sequence for each value of a found.

5. (a) If a, b are positive constants, prove that the sequence defined by $u_n = \log(ab^{n-1})$ is arithmetic.
   (b) Show that the sum of the first n terms of the series
   $$\log 4 + \log 12 + \log 36 + \ldots$$
   is $2n \log 2 + \tfrac{1}{2} n(n-1) \log 3$.

6. The sum of the first 5 terms of an arithmetic series is 5 and the sum of the next 5 terms is 80. Find the first term and the common difference.

7. Find the maximum sum of the arithmetic series $100 + 93 + 86 + \ldots$.

8. (a) Find the sum of all the positive integers less than 100 which are not multiples of 4.
   (b) Find the sum of all the integers between 500 and 1000 which are not divisible by 7.
   (c) Find the sum of all the positive integers less than 100 which do not contain the digit 4.

9. A large water tank begins to leak, and the amount of water escaping is 25 litres more each hour that the leak remains undetected. When the leak was eventually discovered, the rate of leakage was 900 litres per hour. For how long had the tank been leaking? What was the total amount of water lost between the beginning of the leakage and the time of discovery?

10. Find the values of n for which the sum of the first n terms of the arithmetic series $1 + 1.2 + 1.4 + \ldots$ exceeds 1000.

11. Two friends, A and B, begin work together but with differing pay scales. A's starting salary is $30\,000 per year with a yearly increment of $1000. B's starting salary is $6\,000 per half-year with a half-yearly increment of $500. At the end of how many years will B have received more than A in total?

149

Chapter 6

Higher Level

12. If a is a positive integer and b is any real number, prove that the sum of the series $a + (a + b) + (a + 2b) + (a + 3b) + \ldots + (a + ab)$ is $(b + 2)$ times the sum of the series $1 + 2 + 3 + \ldots + a$.

13. Prove that the series for which $S_n = 2n^2 + 9n$ is arithmetic, and find the first four terms.

*14. An arithmetic progression consists of $3p$ terms. The sum of the first p terms is a; the sum of the next p terms is b; the sum of the last p terms is c. Prove that $(a - c)^2 = 4(b^2 - ac)$.

*15. Find the sum of n terms (brackets) of the series:
  (a) $(1) + (3 + 5) + (7 + 9 + 11) + \ldots$;
  (b) $(2) + (4 + 6 + 8) + (10 + 12 + 14 + 16 + 18) + \ldots$.

*16. The first term of an arithmetic progression is $n^2 - n + 1$ and the common difference is 2. Prove that the sum of the first n terms is n^3 and hence show that $1^3 = 1, 2^3 = 3 + 5, 3^3 = 7 + 9 + 11$, and so on. Deduce the sum of the cubes of the first n positive integers.

6.4 Geometric Sequences (Geometric Progressions)

In the sequence 2, 6, 18, 54, 162, ... , the ratio of any one term to the preceding term is 3 (a constant) and each term is obtained by multiplying the previous term by 3.

Similarly in the sequence 32, –16, 8, –4, 2, –1, ... , the ratio of any one term to the preceding term is $-\frac{1}{2}$ (a constant) and each term is obtained by multiplying the previous term by $-\frac{1}{2}$.

These are examples of **geometric sequences** or **geometric progressions** (GPs) in which there is a **common ratio** between successive terms.

Definition   A sequence $\{u_n\}$ is said to be **geometric** if $\dfrac{u_{n+1}}{u_n} = r$ $(n \geq 1)$, where r is a constant. The constant r is called the **common ratio** of the sequence.

A geometric sequence $\{u_n\}$ can be written $\{u_n\} = \{u_1, u_1 r, u_1 r^2, u_1 r^3, \cdots\}$. The nth term is given by $$u_n = u_1 r^{n-1}.$$

Sequences and Series

Example Find the first three terms of the geometric sequence in which the common ratio is $-\frac{1}{3}$ and the 7th term is $-\frac{2}{81}$.

$u_7 = u_1 r^6 = -\frac{2}{81} \Rightarrow u_1\left(-\frac{1}{3}\right)^6 = -\frac{2}{81} \Rightarrow u_1 = -\frac{2}{81} \times 3^6 = -18$.

Thus the first three terms are $-18, 6, -2$.

Example Prove that the sequence defined by $u_n = 3(-2)^n$ is geometric.

$\dfrac{u_{n+1}}{u_n} = \dfrac{3(-2)^{n+1}}{3(-2)^n} = -2$ which is constant for all n.

Therefore the given sequence is geometric.

Example (a) Show that if a, b, c are three consecutive terms of a geometric sequence, then $b^2 = ac$.
(b) If $a - 4$, $a + 8$ and 54 are three consecutive terms of a geometric sequence, find the possible values of a, and the numerical value of the next term for each value of a found.

(a) Since a, b, c are terms of a geometric sequence, $r = \dfrac{b}{a} = \dfrac{c}{b}$, which gives $b^2 = ac$ as required.

(b) From part (a), $a - 4$, $a + 8$, 54 are terms of a geometric sequence provided $(a+8)^2 = 54(a-4)$.

Thus $a^2 - 38a + 280 = 0$
$\Rightarrow (a-10)(a-28) = 0$
$\Rightarrow a = 10$ or $a = 28$.

If $a = 10$ the terms are 6, 18, 54, and so the next term is 162.
If $a = 28$ the terms are 24, 36, 54, and so the next term is 81.

Example The product of three consecutive numbers in geometric progression is 27. The sum of the first two and nine times the third is -79. Find the numbers.

Let the numbers be $\dfrac{x}{a}$, x, ax.

The product of the numbers is 27, and so $x^3 = 27$, giving $x = 3$.
The sum of the first two and nine times the third is -79.

Thus $\dfrac{3}{a} + 3 + 27a = -79 \Rightarrow 27a^2 + 82a + 3 = 0 \Rightarrow (27a+1)(a+3) = 0$.

This gives $a = -3$ or $-\frac{1}{27}$, and the numbers are: $-1, 3, -9$ or $-81, 3, -\frac{1}{9}$.

151

Chapter 6

Exercise 6.4

1. Decide which of the following sequences could be geometric. If the sequence is geometric, find the common ratio.
   (a) $\{3, 12, 48, \cdots\}$;
   (b) $\{16, -8, 4, \cdots\}$;
   (c) $\{6, 12, 18, \cdots\}$;
   (d) $\{-a^2, a, -1, \cdots\}$;
   (e) $\{2\frac{1}{4}, 2, 1\frac{7}{9}, \cdots\}$;
   (f) $\{\sqrt{2}, 1, \frac{1}{2}\sqrt{2}, \cdots\}$.

2. The following pairs of numbers are respectively the first term and the common ratio of a geometric sequence. Find the first three terms of each sequence.
   (a) 5, 2 ;
   (b) 25, $\frac{1}{2}$;
   (c) $-4, -3$;
   (d) 10, 0.3 ;
   (e) $a, -b$;
   (f) $-a, -b$.

3. Write down in exponent form, the 5th, 12th and nth terms of each of the following geometric sequences:
   (a) $\{5, 15, 45, \cdots\}$;
   (b) $\{2, -6, 18, \cdots\}$;
   (c) $\{27, 18, 12, \cdots\}$;
   (d) $\{x, -x^2 y, x^3 y^2, \cdots\}$.

4. Find the common ratio and write down the first three terms of the geometric sequence, $\{u_n\}$, for which
   (a) $u_5 = 30, u_8 = -3.75$;
   (b) $u_3 = 36, u_7 = \frac{64}{9}$;
   (c) $u_6 = -3, u_8 = -\frac{3}{4}$;
   (d) $u_3 = -a^3, u_6 = b^6$.

5. Find the first term to exceed 1000 in each of the following geometric sequences: (a) $\{1, 1.3, 1.69, \cdots\}$; (b) $\{16, 24, 36, \cdots\}$.

6. (a) The sum of three numbers in geometric progression is 13 and their product is 27. Find the numbers.
   (b) The product of three numbers in geometric progression is 125 and their sum is -21. Find the numbers.
   (c) Three numbers are in geometric progression. The product of the first and third is 9, and the sum of the first and third is 7.5. Find the numbers.

Higher Level

7. If a, b, c are in geometric progression, show that $\dfrac{1}{b+a}, \dfrac{1}{2b}, \dfrac{1}{b+c}$ are in arithmetic progression provided each term exists.

8. The 1st, 5th and 12th terms of an arithmetic sequence are 3 consecutive terms of a geometric sequence. Find the common ratio of the geometric sequence.

6.5 Geometric Series

The sum $u_1 + u_2 + u_3 + \cdots$ of terms of a geometric sequence, $\{u_n\}$, is called a *geometric series*.

We denote the sum of the first n terms of the sequence, as with an arithmetic sequence, by S_n.

To find a formula for S_n we note that:

$S_n = u_1 + u_1 r + u_1 r^2 + \cdots + u_1 r^{n-2} + u_1 r^{n-1}$,   and multiplying by r,

$r S_n = \phantom{u_1 +{}} u_1 r + u_1 r^2 + \cdots + u_1 r^{n-2} + u_1 r^{n-1} + u_1 r^n$.

Subtracting the second row from the first we obtain $(1-r)S_n = u_1 - u_1 r^n$.

Hence $S_n = \dfrac{u_1(1 - r^n)}{1 - r}$ or $S_n = \dfrac{u_1(r^n - 1)}{r - 1}$, $r \neq 1$.

Note: If $r = 1$, $S_n = u_1 + u_1 + u_1 + \cdots + u_1 = nu_1$ which is a *'trivial'* result.

Example Find the sum of the first 8 terms of the geometric series
$32 - 16 + 8 - \ldots$.

$u_1 = 32, r = -\tfrac{1}{2}$, and so $S_8 = \dfrac{32\left(1 - \left(-\tfrac{1}{2}\right)^8\right)}{1 - \left(-\tfrac{1}{2}\right)}$

$\phantom{u_1 = 32, r = -\tfrac{1}{2}, \text{ and so } S_8} = \tfrac{64}{3}\left(1 - \tfrac{1}{256}\right)$

$\phantom{u_1 = 32, r = -\tfrac{1}{2}, \text{ and so } S_8} = 21.25$.

Example Evaluate $\sum_{n=1}^{50} 0.99^n$.

Firstly the sequence $\{0.99^n\}$ is geometric since $\dfrac{u_{n+1}}{u_n} = \dfrac{0.99^{n+1}}{0.99^n} = 0.99$ which is constant. Here $u_1 = 0.99$ and $r = 0.99$.

Thus $\sum_{n=1}^{50} 0.99^n = \dfrac{0.99(1 - 0.99^{50})}{1 - 0.99} = 99(1 - 0.99^{50}) = 39.1$.

Example Find the first term and the common ratio of the geometric series for which $S_n = \dfrac{5^n - 4^n}{4^{n-1}}$.

Firstly $u_1 = S_1 = 1$.

Also $u_2 = S_2 - S_1 = \tfrac{9}{4} - 1 = \tfrac{5}{4}$, and so $r = \tfrac{5}{4}$.

Chapter 6

Note: $S_n = 4\left(\dfrac{5^n - 4^n}{4^n}\right) = 4\left(\left(\dfrac{5}{4}\right)^n - 1\right) = \dfrac{\left(\dfrac{5}{4}\right)^n - 1}{\dfrac{5}{4} - 1}$ giving $u_1 = 1$ and $r = \dfrac{5}{4}$.

Exercise 6.5

1. Find the sum of the following geometric series to the number of terms given in parentheses. (Simplify the expressions, including any adjustment of signs, but do not evaluate.)
   (a) $3 + 12 + 48 + \cdots (10)$;
   (b) $2 + 2.4 + 2.88 + \cdots (12)$;
   (c) $0.3 - 0.6 + 1.2 - \cdots (20)$;
   (d) $-4.2 + 1.68 - 0.672 + \cdots (15)$.

2. Prove that each of the following series is geometric and find each sum:
   (a) $\sum\limits_{n=1}^{10} 8(0.5)^n$;
   (b) $\sum\limits_{n=1}^{20} (-0.1)^{n-1}$;
   (c) $\sum\limits_{n=1}^{15} 20(5^{-n})$;
   (d) $\sum\limits_{n=1}^{100} 100(0.99^n)$.

3. Find the sum of n terms of the geometric series $3 + 4.5 + 6.75 + \ldots$. How large must n be so that this sum is greater than 6000?

4. Nine numbers are in GP and are alternately positive and negative. If the first number is 24 and the last is $\dfrac{3}{32}$, find the sum of the numbers.

5. Evaluate $\sum\limits_{n=1}^{100} \left(0.01n + 0.99^n\right)$.

6. The growth of a certain tree during any year is 90% of its growth during the previous year. It is now 10 m high and one year ago it was 9 m high. What will be its growth this year? Next year? Find the height of the tree in 15 years time.

7. A man is appointed to a position at a salary of $30 000 per annum, and each year his salary is to be increased by 5%. Find his salary at the end of 20 years of service and his average annual salary for the twenty years.

Higher Level

8. The bob of a pendulum swings through an arc of length 25 cm in its first swing from left to right, and with each successive swing (from right to left or left to right), the distance decreases by 1% of the previous swing. Find the extent of the 100th swing and the total distance covered by the bob in the first 100 swings.

*9. (a) Show that if $\{u_n\}$ is geometric and S_n is the sum of the first n terms, then the common ratio is given by $r = \dfrac{S_n - u_1}{S_n - u_n}$ $(r \neq 1)$.

(b) A geometric progression has a first term of –0.08, a last term of 0.6075, and a sum of 0.3325. Find the common ratio.

6.6 Compound Interest

One of the most important practical applications of geometric series is in the calculation of *compound interest*.

Suppose $P is invested at r% per annum compound interest. What is the value of the investment (the *amount*) after n years?

At the beginning of the first year the investment is worth $P. During that year interest of $\$\left(P \times \dfrac{r}{100}\right)$ is added to the account so that at the end of year 1 the amount in the account is $\$\left(P + P \times \dfrac{r}{100}\right) = \$P\left(1 + \dfrac{r}{100}\right)$. Similarly at the end of year 2 the amount in the account is $\$P\left(1 + \dfrac{r}{100}\right)^2$. Continuing in this way, the amount in the account at the end of year n is $\$A_n$ where $A_n = P\left(1 + \dfrac{r}{100}\right)^n$.

Higher Level

Now consider the case where an amount of $P is deposited in the account at the beginning of *each* year. What is the amount in the account at the end of n years?

If the annual deposit of $P is made into a new account each year, then after n years, $P will have been in the first account for n years; $P will have been in a second account for $(n - 1)$ years; ...... ; $P will have been in the last (nth) account for 1 year.
Therefore the total amount after the n years is given by S_n, where
$$S_n = A_1 + A_2 + A_3 + \cdots + A_n$$
with each term of the series found from the compound interest formula above.
Thus $S_n = P(R + R^2 + R^3 + \cdots + R^n) = \dfrac{PR(R^n - 1)}{R - 1}$ where $R = 1 + \dfrac{r}{100}$.

Chapter 6

Example A woman makes an annual deposit of $1000 into an account which pays 5% interest compounded annually. How much money should be in the account at the end of 10 years?

$$\text{The required amount} = 1000(1.05 + 1.05^2 + 1.05^3 + \cdots + 1.05^{10})$$
$$= \frac{1050(1.05^{10} - 1)}{1.05 - 1}$$
$$= \$13\,206.79.$$

Often interest is paid monthly, weekly or even daily. If there are N equal time periods for each of which $r\%$ interest (per time period) is paid, the amount after these N time periods $= \$P\left(1 + \dfrac{r}{100}\right)^N$ if $\$P$ is the initial investment.

Example Calculate the amount in an account after 1 year if $1000 is invested at 6% per annum compound interest, and interest is paid
(a) annually ; (b) every 6 months ; (c) quarterly ;
(d) monthly ; (e) weekly ; (f) daily.

(a) The amount $= 1000 \times 1.06$ $= \$1060.$
(b) The amount $= 1000 \times 1.03^2$ $= \$1060.90.$ [3% per period]
(c) The amount $= 1000 \times 1.015^4$ $= \$1061.36.$ [1.5% per period]
(d) The amount $= 1000 \times 1.005^{12}$ $= \$1061.68.$ [0.5% per period]
(e) The amount $= 1000\left(1 + \dfrac{6}{5200}\right)^{52} = \$1061.80.$
(f) The amount $= 1000\left(1 + \dfrac{6}{36\,500}\right)^{365} = \$1061.83.$

How can we calculate the amount if interest is paid *continuously*?
We need a result the proof of which is beyond the requirements of this course.

That is $\lim\limits_{n \to \infty}\left(1 + \dfrac{x}{n}\right)^n = e^x$ where e is the transcendental number mentioned briefly in Chapter 1.

In the case of payment of continuous interest, $1000 invested at 6% per annum compound interest for 1 year will amount to

$$\lim_{n \to \infty} 1000\left(1 + \frac{6}{100n}\right)^n = 1000 e^{0.06} = \$1061.84.$$

Sequences and Series

Exercise 6.6

1. (a) Calculate the amount at compound interest on an investment of $5000 for 10 years at 6% per annum.
   (b) Calculate the total amount of interest payable on a loan of $2500 at 8.5% per annum compounded annually over 4 years.

2. How many years will it take for an investment to double in value at compound interest if the rate is 7.18% per annum?

3. An amount of $1000 is invested at a rate of 5.05% per annum compound interest.
   (a) Find the amount in the account after 5 years.
   (b) Find the time taken for the investment to amount to $1500.

4. (a) Land which was purchased n years ago for $\$A$ is now worth $\$A_n$ where $A_n = Ae^{0.15n}$. Find the current value of land which was purchased 30 years ago for $500.
   (b) If the original cost of the land from part (a) had been invested at compound interest, what rate of interest would have been required to keep pace with land investment?

Higher Level

5. (a) A woman makes an annual deposit of $500 into an account which pays 5.75% interest compounded annually. How much money should be in the account
    (i) immediately after the tenth deposit;
    (ii) at the end of 10 years (before the 11th deposit)?
   (b) A mother puts aside an amount of $100 every half-year for her daughter in an account which pays 4.75% interest compounded annually. How much money should be in the account immediately before the 21st payment?

6. An amount of $10 000 is deposited in an account. How much money is in the account after 5 years if compound interest of 6.5% per annum is added
   (a) annually; (b) monthly; (c) weekly; (d) daily?

7. An amount of $10 000 is invested in an account which pays 7.2% interest per annum over a period of 10 years. Calculate how much more would be in the account if interest was paid continuously instead of yearly.

Chapter 6

> 8. If $A is borrowed at r% per annum compound interest and equal repayments of $x are made at the end of each year over a period of 5 years, show that the annual repayment is given by
> $$x = \frac{AR^5(R-1)}{R^5 - 1} \text{ where } R = 1 + \frac{r}{100}.$$
>
> 9. (a) Calculate the annual repayment if $1000 is borrowed at 10% per annum compound interest for 5 years.
> (b) Calculate the annual repayment if $20 000 is borrowed at 12.5% per annum compound interest for 8 years.

6.7 Infinite Geometric Series

Consider the infinite geometric series $\sum_{n=1}^{\infty} u_1 r^{n-1}$ where the sum of the first n terms is

$$S_n = \frac{u_1(1 - r^n)}{1 - r} \quad (r \neq 1).$$

If $-1 < r < 1$, $r^n \to 0$ as $n \to \infty$.

Thus $\lim_{n \to \infty} S_n = \lim_{n \to \infty} \frac{u_1(1 - r^n)}{1 - r} = \frac{u_1}{1 - r}$ provided $-1 < r < 1$.

If $r > 1$ then $r^n \to +\infty$ as $n \to \infty$ and so $\lim_{n \to \infty} S_n$ does not exist.

If $r < -1$ then r^n takes alternating negative and positive values which increase numerically without bound as $n \to \infty$. Once again $\lim_{n \to \infty} S_n$ does not exist.

If $r = -1$ then the values of r^n alternate between -1 and $+1$, and so $S_n = u_1$ if n is odd and $S_n = 0$ if n is even. Thus $\lim_{n \to \infty} S_n$ does not exist.

If $r = 1$ the series consists of constant terms all equal to u_1. Thus $S_n = nu_1$ and $\lim_{n \to \infty} S_n$ does not exist.

We say that the infinite geometric series $\sum_{n=1}^{\infty} u_1 r^{n-1}$ has a ***sum to infinity*** which is equal to $\frac{u_1}{1 - r}$ provided $-1 < r < 1$ or that the series ***converges*** to $\frac{u_1}{1 - r}$ if $-1 < r < 1$.

158

Sequences and Series

Example Express the recurring decimal $0.\overline{32}$ as a rational number.

$0.\overline{32} = \dfrac{32}{10^2} + \dfrac{32}{10^4} + \dfrac{32}{10^6} + \cdots$ which is an infinite geometric series with first term $u_1 = 0.32$ and common ratio $r = 0.01$.

Since $-1 < r < 1$ the sum to infinity exists and is equal to $\dfrac{u_1}{1-r} = \dfrac{0.32}{0.99} = \dfrac{32}{99}$.

Therefore $0.\overline{32} = \dfrac{32}{99}$.

Example Consider the infinite geometric series $\displaystyle\sum_{n=1}^{\infty} 10\left(1 - \dfrac{3x}{2}\right)^n$.

(a) For what values of x does a sum to infinity exist?
(b) Find the sum of the series if $x = 1.3$.

(a) A sum to infinity exists provided $-1 < 1 - \dfrac{3x}{2} < 1$

$\Rightarrow \quad -2 < -\dfrac{3x}{2} < 0$

$\Rightarrow \quad -4 < -3x < 0$

$\Rightarrow \quad 0 < x < \tfrac{4}{3}$.

(b) When $x = 1.3$, $r = 1 - \dfrac{3.9}{2} = -0.95$ and $u_1 = -9.5$.

The sum to infinity $= \dfrac{u_1}{1-r} = \dfrac{-9.5}{1.95} = -\dfrac{190}{39}$.

Example A ball is dropped from a height of 10 m and after each bounce, returns to a height which is 84% of the previous height. Calculate the total distance travelled by the ball before coming to rest.

The total distance travelled by the ball
$= 10 + 2\{10 \times 0.84 + 10 \times 0.84^2 + 10 \times 0.84^3 + \cdots\}$
$= 10 + 2\left\{\dfrac{8.4}{1 - 0.84}\right\}$ [Sum to infinity of a geometric series with $r = 0.84$.]
$= 115$ m.

Chapter 6

Exercise 6.7

1. Find the sum to infinity of each of the following geometric series:
   (a) $1 + \frac{1}{2} + \frac{1}{4} + \frac{1}{8} + \cdots$;
   (b) $27 + 9 + 3 + 1 + \cdots$;
   (c) $64 - 16 + 4 - 1 + \cdots$;
   (d) $10 - 2 + 0.4 - 0.08 + \cdots$;
   (e) $1000 - 900 + 810 - 729 + \cdots$;
   (f) $\frac{125}{27} - \frac{25}{9} + \frac{5}{3} - 1 + \cdots$.

2. Find the values of x for which the sum to infinity of each of the following geometric series exists:
   (a) $3 + \frac{3x}{2} + \frac{3x^2}{4} + \cdots$;
   (b) $(2-x) + \frac{(2-x)^2}{3} + \frac{(2-x)^3}{9} + \cdots$;
   (c) $\frac{5x}{3} - \left(\frac{5x}{3}\right)^2 + \left(\frac{5x}{3}\right)^3 - \cdots$;
   (d) $\left(\frac{x-5}{2}\right) - \left(\frac{x-5}{2}\right)^2 + \left(\frac{x-5}{2}\right)^3 - \cdots$.

3. By considering the sum to infinity of an appropriate geometric sequence, express each of the following recurring decimals in rational form:
   (a) $0.\overline{45}$;
   (b) $2.1\overline{02}$;
   (c) $0.23\overline{678}$.

4. In each of the following geometric series, find the number of terms needed to give a sum which is within 10^{-6} of the sum to infinity:
   (a) $2 + 1 + \frac{1}{2} + \frac{1}{4} + \cdots$;
   (b) $\frac{2}{3} + \frac{4}{9} + \frac{8}{27} + \frac{16}{81} + \cdots$;
   (c) $0.9 + 0.9^2 + 0.9^3 + 0.9^4 + \cdots$;
   (d) $1 - \frac{5}{6} + \frac{25}{36} - \frac{125}{216} + \cdots$.

5. Find the sum to infinity of the geometric series $3 + 1.8 + 1.08 + \cdots$. How many terms of the series are needed to give a sum which is within 1% of the sum to infinity?

6. An infinite geometric series $\sum_{n=1}^{\infty} u_n$ has a common ratio r and a sum to infinity of 4. The infinite series found by adding the odd numbered terms, $\sum_{n=1}^{\infty} u_{2n-1}$, has a sum to infinity of 3. Find the value of r.

7. Find the smallest value of n for which the sum of the first n terms of the geometric series, $\sum_{n=0}^{\infty} 0.999^n$, is greater than half of its sum to infinity.

Sequences and Series

8. A ball is dropped from a height of 2 metres and every time it strikes the ground it rebounds to a height which is 75% of its previous height.
   (a) Find the height reached by the ball after the fifth bounce.
   (b) Find the total distance travelled by the ball as it comes to rest.

9. The sum to infinity of a geometric series with a common difference r is equal to twenty five times the sum of the first twenty five terms.
   (a) Find the value of r.
   (b) Find the sum to infinity, correct to three significant figures, if the first term is 3.1.

10. (a) Expand the product $(x+2)(2x^2 - 2x - 1)$.
    (b) Use the result of part (a) to find the exact solutions of the equation $2x^3 + 2x^2 - 5x - 2 = 0$.
    (c) The first term of an arithmetic sequence is 1 and the common difference is x. The first term of a geometric sequence is 2 and the common ratio is x. If the sum of the third and fourth terms of the arithmetic sequence is equal to the sum of the third and fourth terms of the geometric sequence, show that $2x^3 + 2x^2 - 5x - 2 = 0$.
    (d) For which value of x found in part (c) does a sum to infinity of the geometric series exist? For this value of x, find the sum to infinity, writing your answer in the form $a\sqrt{3} + b$ where a and b are integers.

Higher Level

11. For what positive values of k is the sum to infinity of a geometric series equal to k times the first term?

12. (a) Expand the product $(r+2)(r^2 - r - 1)$ and find all three roots of the equation $r^3 + r^2 - 3r - 2 = 0$.
    (b) The geometric series $1 + r + r^2 + r^3 + \cdots$ is such that the sum of the 1st, 3rd and 4th terms is three times the sum of the 1st and 2nd terms. Find the possible values of r.
    (c) Find the value of r for which the sum to infinity of the geometric series in part (b) exists, and write this sum to infinity in the form $a + b\sqrt{5}$ where a and b are rational numbers.

Chapter 6

13. A ball is dropped from a height of h m. Each time it strikes the ground it rebounds to a height which is $p\%$ of the previous height, $(0 < p < 100)$. Show that the total distance, D m, travelled by the ball before coming to rest is given by $D = \dfrac{h(100 + p)}{100 - p}$.

14. Consider the geometric series $1 + r + r^2 + r^3 + \cdots$, $0 < r < 1$, in which the sum of the first n terms is denoted by S_n and the sum to infinity by S. If S_n is to be used as an approximate value for S, find an expression for the percentage error and show that if this percentage error is to be less than $p\%$, then $n > \dfrac{\log p - 2}{\log r}$.

15. (a) Find the range of values of x for which the geometric series $10 + 10(2^x) + 10(2^x)^2 + 10(2^x)^3 + \cdots$ has a sum to infinity.
    (b) Find the sum to infinity of the geometric series in part (a) if $x = -0.1$, and the smallest value of n for which the sum of the first n terms exceeds 99% of the sum to infinity.

Required Outcomes

After completing this chapter, a student should be able to:
- recognise an arithmetic sequence and be able to find formulae for the nth term and the sum of the first n terms.
- recognise a geometric sequence and be able to find formulae for the nth term and the sum of the first n terms.
- calculate the compound interest on $\$P$ at $r\%$ per annum for n years.
- calculate the compound interest on $\$P$ at $r\%$ per annum for n time periods (years, months, quarters, weeks, days). **(HL)**
- determine whether or not a given geometric series has a sum to infinity, and if it has, find it.

7 Statistics 1

7.1 Frequency Tables and Frequency Histograms

In this chapter we shall be concerned with various methods of collecting, representing and analysing data. The data will refer to a given set of 'objects' called a *population*. A population could be the set of all children in a school which is finite, or the set of all prime numbers which is infinite. We often gather information concerning a large population, for example the ages of all inhabitants of Germany, by measuring that particular characteristic (variable) of a smaller *sample* from the population. Some variables are *discrete*, others *continuous*. If the variable can take only certain values, for example the number of apples on a tree, then the variable is discrete. If however, the variable can take any decimal value (in some range), for example the heights of the children in a school, then the variable is continuous.

Frequency Tables

(1) Discrete Data

The test scores out of 20 achieved by a class of 15 children are (in class-list order): 15, 20, 18, 18, 8, 16, 18, 10, 17, 14, 11, 11, 12, 16, 13. This data could be represented in tabular form as follows:

Test Score (x)	Frequency (f)	Test Score (x)	Frequency (f)
8	1	15	1
9	0	16	2
10	1	17	1
11	2	18	3
12	1	19	0
13	1	20	1
14	1		

If there is a large number of students taking the test, tally marks can be used to find the required frequencies. The following are the marks out of 20 obtained by 50 students.

```
18 15 17 17 12  9 16 10 12 12
12  5 18 13 19 15  7 18 15 16
20 11 18  9 19 16 14 18 10 11
16 18 20 15 15 10 12 17  8 16
19 17 15  8  5 17 11 16 16  7
```

163

Chapter 7

Score (x)	Tally	Frequency (f)	Score (x)	Tally	Frequency (f)
5	11	2	13	1	1
6		0	14	1	1
7	11	2	15	⊬⊤⊤ 1	6
8	11	2	16	⊬⊤⊤ 11	7
9	11	2	17	⊬⊤⊤	5
10	111	3	18	⊬⊤⊤ 1	6
11	111	3	19	111	3
12	⊬⊤⊤	5	20	11	2
					Total: 50

The test scores in the previous example could be grouped into various classes.

Score	Tally	Frequency
0-5	11	2
6-10	⊬⊤⊤ 1111	9
11-15	⊬⊤⊤ ⊬⊤⊤ ⊬⊤⊤ 1	16
16-20	⊬⊤⊤ ⊬⊤⊤ ⊬⊤⊤ ⊬⊤⊤ 111	23
		Total: 50

(2) Continuous Data

The weights in kg correct to the nearest kg of 25 students are given in the following list

55 63 67 59 62 65 54 46 57 59
58 48 56 52 56 53 55 51 57 58
52 61 63 55 53

Although the data given appear to be discrete, a measurement of 55 kg could have come from any weight w kg such that $54.5 \le w < 55.5$. This is an example of continuous data. A grouped frequency table could be constructed as follows:

Weight (kg)	Tally	Frequency
$44.5 \le w < 49.5$	11	2
$49.5 \le w < 54.5$	⊬⊤⊤ 1	6
$54.5 \le w < 59.5$	⊬⊤⊤ ⊬⊤⊤ 1	11
$59.5 \le w < 64.5$	1111	4
$64.5 \le w < 69.5$	11	2
		Total: 25

Note: The interval $44.5 \le w < 49.5$ can be written 45 – 49.

The values 44.5, 49.5, 54.5, 59.5, 64.5, 69.5 are called the *class boundaries* and the differences between the consecutive class boundaries are called the *interval widths*. Here the interval widths are all 5 kg.

The *mid-interval values* are equal to the means of the class boundaries. Here the mid-interval values are 47, 52, 57, 62 and 67 kg.

The *lower interval boundaries* are respectively 44.5, 49.5, 54.5, 59.5 and 64.5 kg. The *upper interval boundaries* are respectively 49.5, 54.5, 59.5, 64.5 and 69.5 kg.

The interval width = upper interval boundary – lower interval boundary.

Frequency Histograms

Frequency histograms are used to give a graphical description of grouped data. A frequency histogram may contain equal class widths.

Example The following frequency table gives the results of a test taken by 80 students. Draw a histogram to represent the data.

Score	Frequency
0 – 9	6
10 – 19	12
20 – 29	18
30 – 39	22
40 – 49	8
50 – 59	4
	Total: 80

A suitable histogram could be:

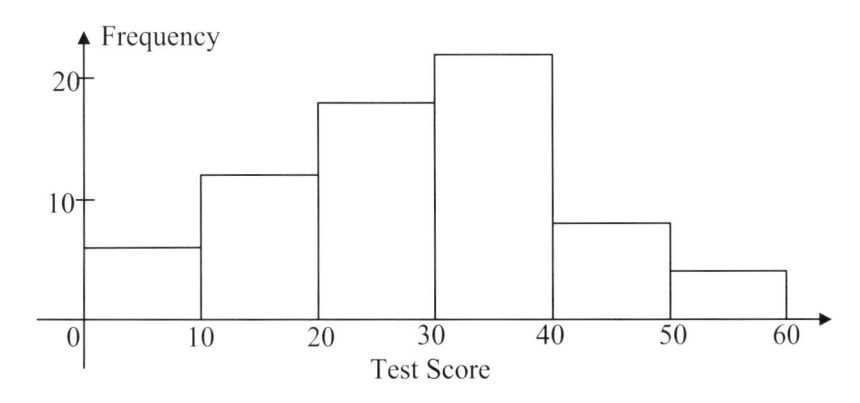

Chapter 7

A frequency histogram may also have unequal class widths.

Example   The ages in completed years of 100 students in a certain school are given by the following table:

Age (years)	Frequency
5 – 9	22
10 – 11	18
12 – 13	20
14 – 15	25
16 – 19	15
	Total: 100

Draw a histogram to represent the data.

The interval widths are 5, 2, 2, 2, 4.
Now the area of a given rectangle is proportional to the frequency of the interval and so in order to find the height of each rectangle we calculate the value of the *frequency density* for the corresponding class.

$$\text{Frequency Density} = \frac{\text{Frequency}}{\text{Interval Width}}$$

The frequency densities are $\frac{22}{5}$, $\frac{18}{2}$, $\frac{20}{2}$, $\frac{25}{2}$, $\frac{15}{4}$ or 4.4, 9, 10, 12.5, 3.75.
A suitable histogram follows:

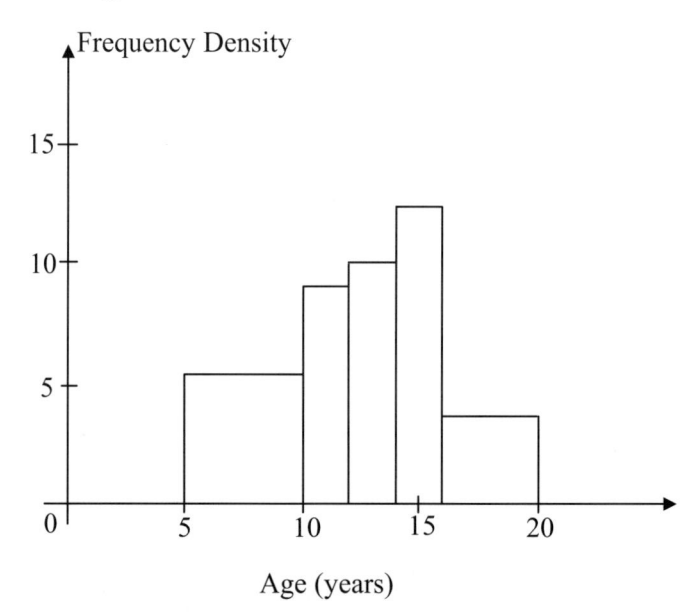

Exercise 7.1

1. From the following histogram find the number of screws with lengths in the intervals
   (a) 5–10 mm;   (b) 10–20 mm;   (c) 20–30 mm;   (d) 30–50 mm.

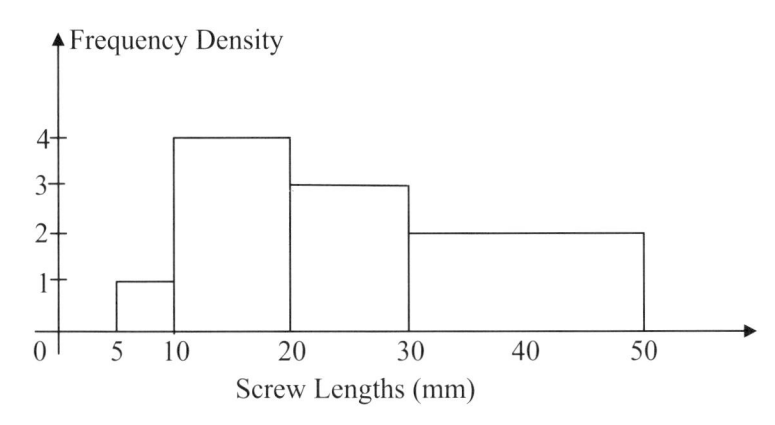

   Find the total number of screws in the population.

2. Draw a histogram to represent the following data.

Rainfall (mm)	Frequency
0–30	24
30–40	55
40–60	105
60–100	64
100–150	52

3. The heights in cm of 40 students are given in the following list.

   148 144 149 130 137 143 132 134 143 147
   130 137 141 140 132 146 145 140 135 148
   145 126 140 123 139 127 144 121 130 145
   120 131 134 142 127 147 138 136 139 126

   Draw a histogram to illustrate the data using class boundaries 120, 125, 130, 135, 140, 145, 150.

4. Draw a histogram to represent the data given at the top of the next page which concerns the scores in a test achieved by a group of year 12 students.

Chapter 7

Score	Frequency
0 – 5	6
6 – 10	11
11 – 15	24
16 – 20	30
21 – 25	21
26 – 30	8
Total:	100

5. The times in completed minutes of the telephone calls made by a group of children on a particular evening are given in the following table.

Time (min)	0 – 9	10 – 14	15 – 19	20 – 24	30 – 39
Frequency	15	24	22	12	6

Draw a histogram to represent the data.

6. Draw a histogram to represent the data in Question 5 if the times given are correct to the nearest minute.

7. The masses in kg to the nearest kg of 100 students are given in the following table.

Mass (kg)	40 – 49	50 – 54	55 – 59	60 – 64	65 – 75
Frequency	16	17	26	23	18

Draw a histogram to represent the data.

7.2 Measures of Central Tendency – Mean, Median, Mode

There are several measures of an 'average' value for a given set of data. The *mean*, *median* and *mode* are three of the most commonly used.

The Mean

Consider the set of n values $x_1, x_2, x_3, \cdots, x_n$. The mean of these values is denoted and defined by

$$\bar{x} = \frac{x_1 + x_2 + x_3 + \cdots + x_n}{n} = \sum_{i=1}^{n} \frac{x_i}{n} = \frac{1}{n} \sum_{i=1}^{n} x_i.$$

1. Finding the mean from raw data.

Statistics 1

Example Find the mean of the numbers 24, 26, 27, 29, 32, 42.

The mean is $\bar{x} = \frac{1}{6}\sum x = \frac{1}{6}(24 + 26 + 27 + 29 + 32 + 42) = \frac{1}{6}(180) = 30$.

Example The mean number of matches in nine boxes is 51. How many matches must there be in the tenth box if the mean number of matches in all ten boxes is 52?

The total number of matches in the first nine boxes = $9 \times 51 = 459$.
For the mean number in all ten boxes to be 52, there must be a total of $10 \times 52 = 520$ matches in the ten boxes.
Therefore the number of matches in the tenth box is 61.

2. Finding the mean from a frequency distribution.

The mean value from a frequency distribution is given by $\bar{x} = \frac{\sum fx}{\sum f}$.

Example Find the mean from the following frequency distribution.

x	10	11	12	13	14
Frequency (f)	5	10	11	8	6

x	f	fx
10	5	50
11	10	110
12	11	132
13	8	104
14	6	84
	$\sum f = 40$	$\sum fx = 480$

Therefore the mean is $\bar{x} = \frac{\sum fx}{\sum f} = \frac{480}{40} = 12$.

If the data has been grouped into intervals, we cannot know the values of the variable for all instances within the interval and so we must *estimate* the mean of these values. To achieve this, we assume that the interval frequency is spread 'evenly' throughout the interval by using the mid-interval value for all values of the variable.

169

Chapter 7

Example   Estimate the mean of the following frequency distribution.

Mass (kg)	32–36	37–41	42–46	47–51	52–56	57–61
Frequency (f)	6	14	24	15	12	9

The mid-interval values are 34 $[=\frac{1}{2}(31.5+36.5)]$, 39, 44, 49, 54 and 59.

Mass (kg)	Mid-interval, x	f	fx
32–36	34	6	204
37–41	39	14	546
42–46	44	24	1056
47–51	49	15	735
52–56	54	12	648
57–61	59	9	531
		$\sum f = 80$	$\sum fx = 3720$

Therefore the mean $= \dfrac{\sum fx}{\sum f} = \dfrac{3720}{80} = 46.5$

The Median

If a set of numbers is arranged in ascending order, the middle number or the mean of the two middle numbers is called the *median*. Thus there are as many numbers less than the median as there are greater than the median.

Consider the numbers $x_1, x_2, x_3, \cdots, x_n$ written in ascending order.

If n is odd, the median is x_i where $i = \frac{1}{2}(n+1)$.

If n is even, the median is $\frac{1}{2}(x_i + x_{i+1})$ where $i = \frac{1}{2}n$.

Example   Find the median of the numbers
(a)   12, 9, 17, 16, 10, 10, 13 ;
(b)   56, 46, 61, 57, 48, 50, 47, 44.

(a)   In ascending order the numbers are: 9, 10, 10, 12, 13, 16, 17.
There are 7 numbers and so the median is in position $\frac{1}{2}(7+1) = 4$.
Therefore the median is 12.

(b)   In ascending order the numbers are 44, 46, 47, 48, 50, 56, 57, 61.
There are 8 numbers and so the two middle numbers are in $\frac{1}{2}(8) = $ 4th and 5th positions.
Therefore the median is $\frac{1}{2}(48+50) = 49$.

Statistics 1

From grouped data, the median like the mean, can only be estimated. Linear interpolation is generally used.

Example   Estimate the median from the following grouped data.

Class	Frequency
0–4	4
5–9	5
10–14	5
15–19	4
20–24	2
	$\sum f = 20$

There are 20 values and so the median is the mean of the 10th and 11th values in ascending order, i.e. the '10.5th' value. There are 9 values in the first two classes and so the median must lie in the next class. This class has a lower class boundary of 9.5, a width of 5 and contains 5 values. The difference between 10.5 and 9 is 1.5.

Therefore the median is approximately $9.5 + \frac{1.5}{5} \times 5 = 11$.

The Mode

The *mode* is the value with the largest frequency. For grouped data, the *modal class* is the class with the greatest frequency. There may of course be more than one mode or modal class.

Exercise 7.2

1. Find the mean, median and mode of each of the following sets of numbers.
   (a) 6, 10, 4, 13, 11, 9, 1, 6, 12 ;
   (b) 193, 195, 202, 190, 189, 195 ;
   (c) 0.77, 0.73, 0.61, 0.73, 0.65, 0.83, 0.81, 0.65, 0.73, 0.69, 0.74, 0.76.

2. Find the mean and median of the following frequency distributions.
   (a)

x	1	2	3	4	5	6
f	32	27	26	35	33	27

   (b)

x	10	11	12	13	14
f	23	19	16	24	18

171

Chapter 7

(c)

x	2	3	5	7	11	13
f	28	32	35	17	6	2

(d)

x	3.2	3.6	4.0	4.4	4.8	5.2
f	4	9	26	35	42	34

3. Estimate the mean and median of the following grouped frequency distributions.

(a)

x	0–2	2–4	4–6	6–8
f	20	19	25	16

(b)

x	10–14	15–19	20–24	25–29	30–34	34–39
f	61	50	37	27	16	9

(c)

x	0–9	10–19	20–29	30–39	40–49	50–60
f	3	24	46	38	28	11

4. The frequency table giving the ages in completed years of the population of a small country town is shown below. Estimate the mean age and the median age of the towns-people.

Age (x years)	0–19	29–29	30–39	40–49	50–59
Frequency (f)	258	761	906	832	756

Age (x years)	60–69	70–79	80–89	90–99	100–109
Frequency (f)	484	305	148	44	6

7.3 Cumulative Frequency – Quartiles and Percentiles

To find the cumulative frequency corresponding to a given class we sum the frequencies up to the upper class boundary.

Example The frequencies of the scores of 80 students in a test are given in the following table. Complete the corresponding cumulative frequency table.

Statistics 1

Test Score	Frequency
0–9	8
10–19	18
20–29	24
30–39	20
40–49	10

A suitable table is as follows:

Test Score	Cumulative Frequency	
≤ -0.5	0	
≤ 9.5	8	
≤ 19.5	26	(8 + 18)
≤ 29.5	50	(26 + 24)
≤ 39.5	70	(50 + 20)
≤ 49.5	80	(70 + 10)

The information provided by a cumulative frequency table can be displayed in graphical form by plotting the cumulative frequencies given in the table against the upper class boundaries, and joining these points with a smooth curve.

The cumulative frequency curve corresponding to the data in the previous example is as follows:

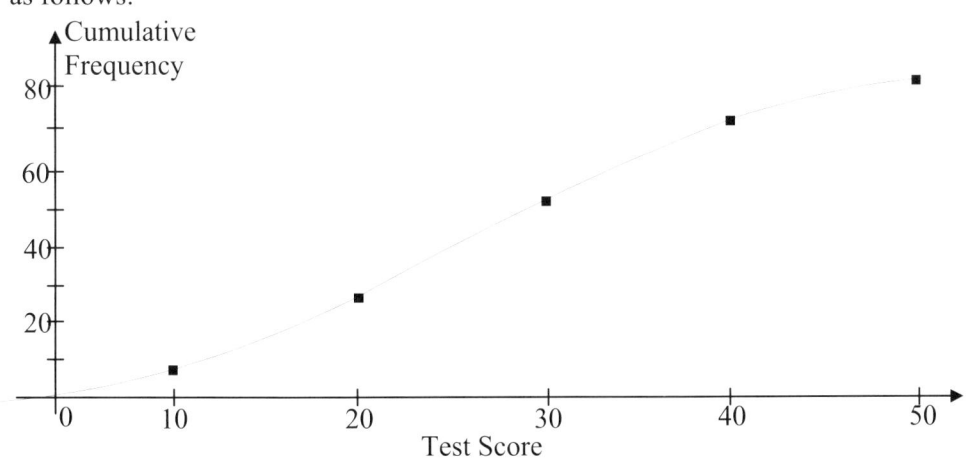

Note: If we join the points with straight lines, we form a cumulative frequency polygon.

Example The results obtained by 200 students in a mathematics test (maximum mark = 50) are given in the following table.

Chapter 7

Mark	Frequency
1–10	22
11–20	35
21–30	72
31–40	60
41–50	11

Draw a cumulative frequency curve and use it to estimate
- (a) the median mark ;
- (b) the number of students who scored less than 22 marks;
- (c) the pass mark if 120 students passed the test ;
- (d) the minimum mark required to obtain an A grade if 10% of the students received an A grade.

The required cumulative frequency curve is as follows:

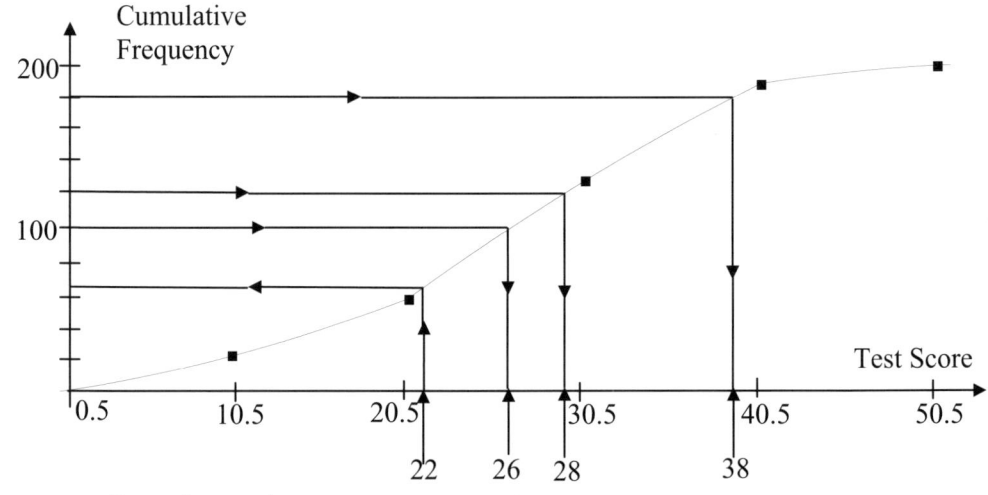

From the graph:
- (a) the median mark was 26.
- (b) approximately 69 students scored less than 22 marks.
- (c) the pass marks was 28.
- (d) the minimum mark required to obtain an A grade was 38.

Quartiles and Percentiles

Twenty five percent of the observations have values which are less than the **lower quartile**, Q_1, and twenty five percent of the observations have values which are greater than the **upper quartile**, Q_3. Thus the lower quartile, median and upper quartile divide the distribution into four equal parts.

Statistics 1

When data is grouped, we can only estimate the values of the quartiles and we use a cumulative frequency curve for this purpose.

A useful measure of the dispersion of the observations is the the ***interquartile range***, $Q_3 - Q_1$. This is not affected by any extreme values at either end of the range of values. The 'middle' 50% of the population have values between Q_1 and Q_3 and so the interquartile range provides a measure of the spread of the middle half of the population.

A similarly useful measure of dispersion is the ***semi-interquartile range***, $\frac{1}{2}(Q_3 - Q_1)$.

The ***nth percentile***, P_n, has n% of the observations with values less than P_n. Thus 20% of the observations have values less than P_{20} and 20% of the observations have values greater than P_{80}. As with quartiles, we estimate the percentiles from grouped data by reading information from the corresponding cumulative frequency curve.

Example   The heights of 500 students in a school were measured and the results were as follows.

Height (cm)	Frequency	Height (cm)	Frequency
140–144	8	165–169	92
145–149	19	170–174	88
150–154	35	175–179	53
155–159	68	180–184	42
160–164	81	185–189	14

Draw a cumulative frequency curve and use it to
(a)   estimate the median height of the students ;
(b)   determine the interquartile range ;
(c)   determine $P_{90} - P_{10}$ (the 10 to 90 percentile range).

The cumulative frequencies are given in the following table:

Height (cm)	Cumulative Frequency	Height (cm)	Cumulative Frequency
< 139.5	0	< 169.5	303
< 144.5	8	< 174.5	391
< 149.5	27	< 179.5	444
< 154.5	62	< 184.5	486
< 159.5	130	< 189.5	500
< 164.5	211		

175

Chapter 7

The cumulative frequency curve is as follows:

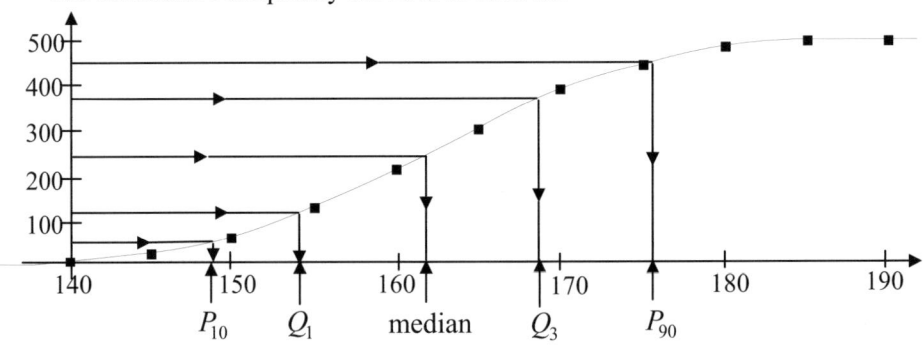

(a) Median height = 162 cm.
(b) Interquartile range = (168 − 154) cm = 14 cm.
(c) $P_{90} - P_{10} = (176 - 148)$ cm $= 28$ cm.

Exercise 7.3

1. Construct a cumulative frequency curve for the following data.

Test Mark	Frequency
1–20	4
21–40	25
41–60	71
61–80	38
81–100	12

   Use your graph to estimate
   (a) the median score ;
   (b) the interquartile range ;
   (c) the pass mark if 60% of the candidates passed ;
   (d) the smallest mark required to obtain an A grade if 10% of the candidates received an A grade.

2. The following table gives the number of people (in thousands) in a large town in the given age ranges.

Age	Number of people	Age	Number of people
0–9	24	50–59	18
10–19	27	60–69	15
20–29	31	70–79	12
30–39	28	80–89	8
40–49	23	90–	3

Statistics 1

Draw a cumulative frequency curve for these data.
Use your graph to
(a) estimate the median age ;
(b) the semi-interquartile range ;
(c) estimate the number of people in the town who are eligible to drive a car if the legal driving age is between 18 years and 75 years ;
(d) the age exceeded by 90% of the population.

3. The following graph is that of a cumulative frequency curve for the distribution of the number of marks obtained by 180 candidates in an examination. Use the graph to estimate
   (a) the median mark ;
   (b) the interquartile range ;
   (c) the number of candidates who passed if 60 were the pass mark.

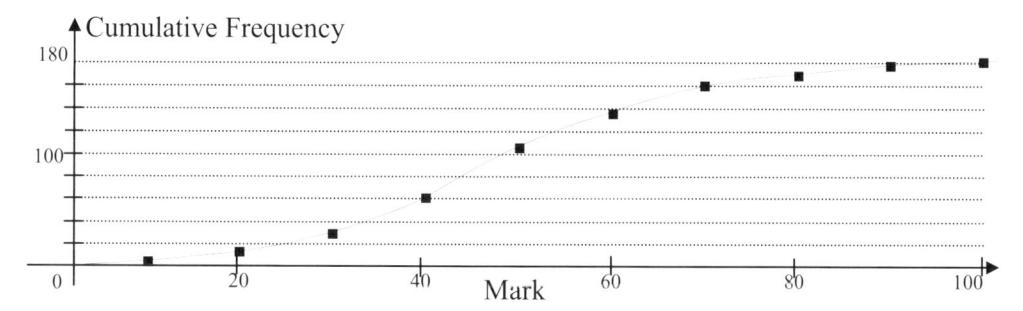

Complete the grouped frequency table corresponding to the class intervals 0–9, 10–19, 20–29, etc. and draw the corresponding histogram.

4. The life-times (in hours) of 200 light bulbs were determined and the results are given in the following table.

Life-time	Frequency	Life-time	Frequency
800–849	4	1000–1049	31
850–899	18	1050–1099	48
900–949	23	1100–1149	38
950–999	28	1150–1199	10

Draw a cumulative frequency curve and use it to estimate
(a) the median life-time of the bulbs ;
(b) the percentage of bulbs lasting between 970 hours and 1070 hours ;
(c) the life-time below which 30% of the bulbs failed ;
(d) the probability that a similar light bulb will fail before 840 hours ;
(d) the number of similar light bulbs from a batch of 3000 which could be expected to last at least 1120 hours.

177

Chapter 7

7.4 Measures of Dispersion – Discrete Data

The simplest measure of the dispersion (spread) of data is the **range**. The range is equal to the difference between the largest value and the smallest value and is completely determined by these extreme values.

Consider the following data sets:
(a)   8, 9, 10, 11, 12 ;   (b)   −45, −8, 2, 42, 59.
Each has a mean of 10, but the second set is clearly more spread out than the first. In fact the range of the first is $12 - 8 = 4$ but the range of the second is $59 - (-45) = 104$.

We have already met at least one other measure of **spread** – the **interquartile range**. Its disadvantage is that it is quite 'insensitive' to changes in the the lower and upper quarters of the data. Much more useful measures of the spread of data are the **variance** and its square root, the **standard deviation**.

Variance   Let the observed values be $x_1, x_2, x_3, \cdots, x_n$. Then the variance is denoted and defined by

$$s^2 = \frac{\sum_{i=1}^{n}(x_i - \bar{x})^2}{n}$$

where $\bar{x} = \dfrac{\sum_{i=1}^{n} x_i}{n}$ is the **mean** of the data.

Standard Deviation   The standard deviation of the observed values above is denoted and defined by

$$s = \sqrt{s^2} = \sqrt{\frac{\sum_{i=1}^{n}(x_i - \bar{x})^2}{n}}$$

which is clearly the square root of the variance.

The variance is thus the **mean of the squared deviations from the mean** and the standard deviation is simply the square root of this.

The following results follow directly from the definitions of mean and standard deviation.

(1)   When all the data values are multiplied by a constant a, the new mean and new standard deviation are equal to a times the original mean and standard deviation.
         The mean of $ax_1, ax_2, ax_3, \cdots, ax_n$ is $a\bar{x}$, and the standard deviation is as.

Statistics 1

(2) When a constant value, b, is added to all the data values, then the mean is also increased by b. However, the standard deviation does not change.
The mean of $x_1 + b, x_2 + b, \cdots, x_n + b$ is $\bar{x} + b$; the standard deviation is s.

Example The six runners in a 200 metre race clocked times (in seconds) of 24.2, 23.7, 25.0, 23.7, 24.0, 24.6.

(a) Find the mean and standard deviation of these times.
(b) These readings were found to be 10% too low due to faulty timekeeping. Write down the new mean and standard deviation.

(a) $\bar{x} = \dfrac{24.2 + 23.7 + 25.0 + 23.7 + 24.0 + 24.6}{6} = 24.2$ seconds.

$s = \sqrt{\dfrac{(24.2 - 24.2)^2 + (23.7 - 24.2)^2 + \cdots + (24.6 - 24.2)^2}{6}}$

$= \sqrt{\dfrac{0 + 0.25 + 0.64 + 0.25 + 0.04 + 0.16}{6}}$

$= 0.473$ seconds.

(b) We must divide each time by 0.9 to find the correct time.
The new mean = $24.2/0.9 = 26.9$ seconds, and the new standard deviation = $0.4726/0.9 = 0.525$ seconds.

The method which uses the formula for the standard deviation is not necessarily the most efficient. Consider the following:

$$\text{Variance} = \dfrac{\sum(x - \bar{x})^2}{n}$$

$$= \dfrac{\sum(x^2 - 2x\bar{x} + (\bar{x})^2)}{n}$$

$$= \dfrac{\sum x^2}{n} - 2\bar{x}\dfrac{\sum x}{n} + (\bar{x})^2 \dfrac{\sum 1}{n} \quad \text{(since } \bar{x} \text{ is a constant)}$$

$$= \dfrac{\sum x^2}{n} - 2\bar{x}\,\bar{x} + (\bar{x})^2$$

$$= \dfrac{\sum x^2}{n} - (\bar{x})^2.$$

Example The heights (in metres) of six children are 1.42, 1.35, 1.37, 1.50, 1.38 and 1.30. Calculate the mean height and the standard deviation of the heights.

Chapter 7

$$\text{Mean} = \tfrac{1}{6}(1.42 + 1.35 + 1.37 + 1.50 + 1.38 + 1.30) = 1.39 \text{ m}.$$
$$\text{Variance} = \tfrac{1}{6}(1.42^2 + 1.35^2 + 1.37^2 + 1.50^2 + 1.38^2 + 1.30^2) - 1.387^2$$
$$= 0.00386 \text{ m}^2.$$
$$\text{Standard Deviation} = \sqrt{0.0038556} = 0.0621 \text{ m}.$$

Exercise 7.4

1. Find the mean and standard deviation of 25.2, 22.8, 22.1, 25.3, 24.6, 25.0, 24.3 and 22.7.

2. Using the mean and standard deviation of the set of numbers {3, 5, 6, 8, 10}, find the mean and standard deviation of each of the following sets of numbers.
   (a) {6, 8, 9, 11, 13} ;
   (b) {9, 15, 18, 24, 30} ;
   (c) {2.7, 4.5, 5.4, 7.2, 9.0}.
   Find the mean and standard deviation of the set containing numbers which are 5% higher than those in the original set.

3. The mean height of a group of 5 people is $\bar{h} = 155$ cm and the standard deviation of their heights is 5 cm.
   (a) Calculate $\sum h$ and $\sum h^2$ for this data.
   (b) If an extra person of height 165 cm is added to the group, calculate the new mean and standard deviation of the heights.

4. Consider the set of data $\{x_1, x_2, x_3, \cdots, x_n\}$ with mean, $\bar{x}$, and standard deviation, s.
   (a) Prove that the set of data $\{x_1 + a, x_2 + a, x_3 + a, \cdots, x_n + a\}$, where a is a constant, has mean $\bar{x} + a$ and standard deviation s.
   (b) Prove that the set of data $\{bx_1, bx_2, bx_3, \cdots, bx_n\}$, where b is a constant, has mean $b\bar{x}$ and standard deviation bs.

5. The mean and variance of $x_1, x_2, x_3, \cdots, x_n$ are $\bar{x}$ and s^2 respectively. State the mean and variance of $2 - 3x_1, 2 - 3x_2, 2 - 3x_3, \cdots, 2 - 3x_n$.

6. Twenty values of a random variable have a mean of 15 and a variance of 1.5. Another thirty values of the same random variable have a mean of 14 and a variance of 1.4. Find the mean and variance of the combined fifty values.

7. The mean daily maximum temperature at a fixed location for the month of January was 22°C and the standard deviation of these daily maxima was 2°C. For the following February which was not a leap year, the mean was 24°C

Statistics 1

and the standard deviation was 4°C. Calculate the mean and standard deviation of the daily maximum temperatures for the two months combined.

8. A sample of 165 values of a random variable have a mean of 23.4 and a standard deviation of 1.6. When combined with another 219 values, the mean of all 384 values was 24.8 and the standard deviation was 2.2. Find the mean and standard deviation of the 219 values which were added to the original sample.

9. Twenty values of a random variable have a mean value of 12.5 and a variance of 1.35. If two more values are added to the original 20 the mean remains at 12.5 but the variance is increased by 0.082. Find the two values added.

7.5 The Mean and Variance for Grouped Data

The mean for grouped data where the mid-interval value for the *i*th group is x_i which occurs with frequency f_i (so that $\sum f_i = n$) is

$$\bar{x} = \frac{\sum f_i x_i}{\sum f_i} = \frac{\sum f_i x_i}{n}.$$

The variance is
$$s^2 = \frac{\sum f_i (x_i - \bar{x})^2}{\sum f_i} = \frac{\sum f_i (x_i - \bar{x})^2}{n} = \frac{\sum f_i x_i^2}{n} - \bar{x}^2,$$

and the standard deviation, *s*, as always, is the square root of the variance.

In the first example we will use the above formulae, but a GDC should be used in practice.

Example   The number of customers served lunch in a restaurant over a period of 60 days is as follows:

Number of customers served lunch	Number of days in the 60-day period
20–29	6
30–39	12
40–49	16
50–59	14
60–69	8
70–79	4

Find the mean and standard deviation of the number of customers served lunch using this grouped data.

181

Chapter 7

In the 40–49 group, we do not know on exactly how many of the 16 days, 44 customers say, were served lunch so we assume that on each of these days, 44.5 customers were served. We choose the **mid-interval** value ($\frac{1}{2}[40+49]$) as representative of the group as a whole and use this to **estimate** the mean and standard deviation.

Group	Mid-interval value (x_i)	Frequency (f_i)	$f_i x_i$	$f_i x_i^2$
20–29	24.5	6	147	3 601.5
30–39	34.5	12	414	14 283.0
40–49	44.5	16	712	31 684.0
50–59	54.5	14	763	41 583.5
60–69	64.5	8	516	33 282.0
70–79	74.5	4	298	22 201.0
		$\sum f_i = 60$	$\sum f_i x_i = 2\,850$	$\sum f_i x_i^2 = 146\,635$

The mean is $\bar{x} = \dfrac{\sum f_i x_i}{\sum f_i} = \dfrac{2850}{60} = 47.5$.

The standard deviation is $s = \sqrt{\dfrac{\sum f_i x_i^2}{\sum f_i} - \bar{x}^2} = \sqrt{\dfrac{146\,635}{60} - 47.5^2} = 13.7$.

Example

The age distribution of the population of a small country town on January 1, 2001 is given in the following table:

Age Group (years)	Mid-interval (x_i years)	Frequency (f_i)
0–14	7.5	856
15–29	22.5	1120
30–44	37.5	1054
45–59	52.5	792
60–74	67.5	651
75–89	82.5	288
90–104	97.5	96

Calculate the mean and standard deviation of the population age on January 1, 2001.

It should be noted that people from age 15 to almost 30 lie in the group 15–29 and so it is correct to take $\frac{1}{2}(15+30) = 22.5$ as the mid-interval value.

From the GDC, $\bar{x} = 39.1$ years and $s = 23.6$ years.

Exercise 7.5

1. Calculate the mean and standard deviation from the following data.

   (a)
Test Score	Frequency
1–10	6
11–20	17
21–30	29
31–40	23
41–50	10

   (b)
Test Mark	Frequency
1–20	4
21–40	25
41–60	71
61–80	38
81–100	12

2. The ages of the people living in a country town are grouped as follows:

Age group	Frequency	Age group	Frequency
0–9	1205	60–69	954
10–19	1528	70–79	532
20–29	2006	80–89	187
30–39	1857	90–99	56
40–49	1621	100–109	4
50–59	1483		

   Calculate the mean and standard deviation of the ages of the people in the town.

3. The heights in centimetres of a group of people are given in the following table. Calculate the mean and standard deviation of these heights.

Height (cm)	Frequency	Height (cm)	Frequency
140–144	5	170–174	173
145–149	18	175–179	130
150–154	27	180–184	72
155–159	62	185–189	35
160–164	87	190–194	12
165–169	115	195–199	2

Chapter 7

4. The life-times (in hours) of 200 light bulbs were determined and the results are given in the following table.

Life-time	Frequency	Life-time	Frequency
800–849	4	1000–1049	31
850–899	18	1050–1099	48
900–949	23	1100–1149	38
950–999	28	1150–1199	10

Calculate the mean and standard deviation of the life-times of the bulbs.

Required Outcomes

After completing this chapter, a student should be able to:
- draw frequency histograms for both discrete and continuous data.
- calculate the mean, median and mode of discrete data.
- construct cumulative frequency curves and use them to estimate the median, quartiles and percentiles of a distribution.
- calculate various measures of dispersion – the range and interquartile range, and the variance and standard deviation of a sample.

8 Counting Techniques and the Binomial Theorem

8.1 The Product Principle

The aim of this chapter is to develop some techniques for determining, without direct enumeration, the number of possible outcomes of a particular experiment or the number of elements in a given set. Such techniques are sometimes referred to as *combinatorial analysis*.

A procedure known as the ***product principle*** is very useful in this endeavour.

If some procedure can be performed in n_1 different ways, and if, following this procedure, a second procedure can be performed in n_2 different ways, and if, following this procedure, a third procedure can be performed in n_3 different ways, and so on; then the number of ways the procedures can be performed one after the other in the given order is the product $n_1 \times n_2 \times n_3 \times \cdots$.

Example A car license plate is to contain three letters of the alphabet, the first of which must be R, S, T or U, followed by three decimal digits. How many different license plates are possible?

The first letter can be chosen in 4 different ways, the second and third letters in 26 different ways each, and each of the three digits can be chosen in ten ways.

Hence there are $4 \times 26 \times 26 \times 10 \times 10 \times 10 = 2\,704\,000$ plates possible.

Example
(a) How many numbers of four different digits can be formed?
(b) How many of these are odd?
(c) How many are multiples of 5?

(a) There are nine ways to choose the first digit since 0 cannot be the first digit, and nine, eight and seven ways to choose the next three digits since no digit may be repeated.

Therefore there are $9 \times 9 \times 8 \times 7 = 4536$ numbers possible.

Chapter 8

(b) The last digit must be a 1, 3, 5, 7 or 9. There are five ways of choosing it. Then the first digit can be chosen in eight different ways since it cannot be a zero or the number chosen for the last digit. The other two digits can be chosen in eight and seven ways respectively.

Therefore the number of odd numbers is $8 \times 8 \times 7 \times 5 = 2240$.

(c) The last digit must be 0 or 5. But there is a problem when we next wish to establish the number of possibilities for the first digit which cannot be 0. This choice depends on our choice for the last digit. We therefore count the number which end in 0 and the number which end in 5 separately and add the two answers.
The number ending in 0 is $9 \times 8 \times 7 \times 1 = 504$, and the number ending in 5 is $8 \times 8 \times 7 \times 1 = 448$.

The required number = 952.

Example In how many different ways can 6 people sit in a row? In how many ways if 2 of them, A and B, must sit together?

The number of ways = $6 \times 5 \times 4 \times 3 \times 2 \times 1 = 720$.

If A and B are considered as one 'person', the number of arrangements is equal to $5 \times 4 \times 3 \times 2 \times 1 = 120$. For each of these arrangements there are 2 ways of seating A and B.
Therefore the required number of ways = $2 \times 120 = 240$.

Exercise 8.1

1. If there are three different roads joining town A to town B and four different roads joining town B to town C, in how many different ways can I travel from A to C via B and return if
   (a) there are no restrictions ;
   (b) I am not able to return on any road I used on the outward journey?

2. An ant is at one vertex of a cube. In how many different ways can it travel along three different edges to arrive at the opposite vertex?

3. (a) How many 3-digit numbers can be formed using only the digits 1, 2, 3, 5, 7 and 8?
   (b) How many of these numbers are even?
   (c) How many are less than 500?

4. Answer Question 3 if no digit may be repeated.

Counting Techniques & The Binomial Theorem

5. (a) How many numbers of 4 different digits can be formed using only the digits 0, 1, 2, 5, 7 and 8?
   (b) How many of these numbers are even?
   (c) How many of these numbers are multiples of 5?

6. (a) In how ways can three boys and two girls sit in a row?
   (b) In how many ways can they sit in a row if the boys are to sit together and the girls are to sit together?
   (c) In how many ways can they sit in a row if only the girls must sit together?

7. (a) How many four-digit numbers are there?
   (b) How many four-digit numbers contain at least one digit 3?

8. There are 20 teams in the local football competition. In how many ways can the first four places in the premiership table be filled?

Higher Level

9. Four distinct letters are to be placed in four differently addressed envelopes with one letter in each envelope.
   (a) In how many ways can this be done?
   (b) In how many ways will exactly one letter be placed in its correct envelope?
   (c) In how many ways will exactly three letters be placed in their correct envelopes?

10. In how many different ways can the digits 0 to 9 be arranged in a row if the first four must all be odd and zero is not among the last three?

11. (a) In how many different ways can 4 distinct objects be placed in 3 separate compartments?
    (b) In how many ways can this be done if no compartment is to contain more than two objects?

8.2 Factorial Notation

The product of the positive integers from 1 to n inclusive occurs often in mathematics and is therefore denoted by the special symbol $n!$ (read "n factorial").

Thus $n! = 1 \times 2 \times 3 \times \cdots \times (n-2) \times (n-1) \times n$.

It is also convenient to **define** $0! = 1$.

Chapter 8

Example   Simplify: (a) $\dfrac{6!}{4!}$; (b) $\dfrac{(n+1)!}{(n-1)!}$; (c) $\dfrac{(n+1)!-(n-1)!}{n!}$.

(a) $\dfrac{6!}{4!} = \dfrac{6 \times 5 \times 4!}{4!} = 6 \times 5 = 30$.

(b) $\dfrac{(n+1)!}{(n-1)!} = \dfrac{(n+1)n(n-1)!}{(n-1)!} = n(n+1)$.

(c) $\dfrac{(n+1)!-(n-1)!}{n!} = \dfrac{(n+1)n(n-1)!-(n-1)!}{n(n-1)!} = \dfrac{(n+1)n-1}{n} = n+1-\dfrac{1}{n}$.

Exercise 8.2

1. Find the value of:
   (a) $\dfrac{15!}{13!}$; (b) $\dfrac{9!}{11!}$; (c) $6!-5!$; (d) $\dfrac{10!-9!}{8!}$.

2. Express in factorial notation:
   (a) $5 \times 4 \times 3 \times 2$; (b) 4×3 ; (c) $10 \times 9 \times 8 \times 7$;
   (d) $\dfrac{1}{8 \times 7 \times 6 \times 5}$; (e) $\dfrac{12 \times 11 \times 10 \times 9}{4 \times 3 \times 2 \times 1}$; (f) $n(n-1)(n-2)$.

3. Simplify:
   (a) $\dfrac{(n-1)!}{(n-2)!}$; (b) $\dfrac{(n+2)!}{n!}$; (c) $\dfrac{(n+2)!-n!}{(n-1)!}$.

4. Without multiplying all the terms, show that
   (a) $10! = 6!\ 7!$; (b) $10! = 7!\ 5!\ 3!$;
   (c) $16! = 14!\ 5!\ 2!$; (d) $9! = 7!\ 3!\ 3!\ 2!$.

8.3   Permutations

An arrangement of a set of n objects *in a given order* is called *a permutation* of the objects (taken all at a time).

An arrangement of any $r \le n$ of these objects in a given order is called an *r*-permutation of the n objects or a permutation of the n objects taken r at a time.

The number of permutations of n objects taken r at a time is denoted by P_r^n, $_nP_r$ or ^nP_r.

188

Counting Techniques & The Binomial Theorem

Example Find the number of permutations of 4 objects (taken all at a time).

The first object can be chosen in 4 different ways, and following this, the second object can be chosen in 3 different ways, and following this, the third object can be chosen in 2 different ways and finally the last object can be chosen in 1 way.

Thus, by the product principle, the number of permutations of the 4 objects is $4 \times 3 \times 2 \times 1 = 24$.

The result of the previous example can be generalised:

> the number of permutations of n different objects is $n!$.

Example Find the number of permutations of 6 objects taken 3 at a time.

The first object can be chosen in 6 different ways; following this, the second object can be chosen in 5 different ways; following this, the last object can be chosen in 4 different ways. Thus, by the product principle, there are $6 \times 5 \times 4$ or 120 possible permutations of 6 objects taken 3 at a time. $\left(\text{i.e., } P_3^6 = 120\right)$

The derivation of the formula for P_r^n follows the procedure in the preceding example. The first object in an r-permutation of n objects can be chosen in n different ways; following this, the second object can be chosen in $(n - 1)$ different ways; and following this, the third object can be chosen in $(n - 2)$ different ways. Continuing in this way, we have that the rth (last) object can be chosen in $(n - r + 1)$ different ways. Thus,

$$P_r^n = n \times (n-1) \times (n-2) \times \cdots \times (n-r+1) = \frac{n!}{(n-r)!}.$$

In particular, when $r = n$ we have $P_n^n = n \times (n-1) \times (n-2) \times \cdots \times 3 \times 2 \times 1 = n!$.

Example
(a) In how many different ways can six people be arranged in a row?
(b) In how many different ways can these six people be arranged in a circle?
(c) Answer parts (a) and (b) if two particular people, A and B must sit next to each other.

(a) The number of ways is equal to the number of permutations of six objects which is $6! = 720$.

Chapter 8

(b) The six different permutations in a row, *ABCDEF*, *BCDEFA*, *CDEFAB*, *DEFABC*, *EFABCD* and *FABCDE* all give the same permutation in a circle since the circle has no "ends". Therefore, the number of permutations of six people arranged in a circle is

$$\frac{6!}{6} = 5! = 120.$$

[This is a particular case of the rule: The number of different ways of arranging n different objects in a circle is $(n-1)!$.]

(c) First we join *A* and *B* together and consider them as one 'object'. We can arrange the five objects "*AB*", *C*, *D*, *E*, *F* in 5! = 120 different ways in a row and 4! = 24 different ways in a circle. Now the pair "*AB*" can be arranged in 2! = 2 different ways for each of these ways. Therefore the number of ways of arranging 6 people in a row so that *A* and *B* sit next to each other = 2 × 120 = 240.
Also, the number of ways of arranging 6 people in a circle so that *A* and *B* sit next to each other = 2 × 24 = 48.

Permutations with Repetitions

Frequently we would like to know the number of permutations of objects some of which are alike.

Example Suppose we have 6 identical discs except for the fact that one is black, one is white, one is yellow and the other 3 are green. How many distinct permutations of the 6 discs are there?

If the 6 discs were all distinguishable, there would be 6! different permutations.
However, the 3 green discs are indistinguishable, and there are 3! (indistinguishable) ways of arranging them.
Therefore the number of (distinguishable) ways of arranging the given discs is $\frac{6!}{3!} = 120$.

This is a particular example of the following general result:

> The number of permutations of n objects of which n_1 are indistinguishable, n_2 are indistinguishable, ... , n_r are indistinguishable is $\dfrac{n!}{n_1! \times n_2! \times \cdots \times n_r!}$.

Example How many different words of 5 letters (not necessarily sensible), can be formed from the letters of the word
(a) *MATHS* ; (b) *POPPY* ?

(a)　　Number of different words = 5! = 120.

(b)　　Number of different words = $\dfrac{5!}{3!}$ (the 3 "*P*"s are indistinguishable)

$= 5 \times 4$

$= 20.$

Example　　In how many different ways can 4 identical red balls, 3 identical green balls and a yellow ball be arranged in a row?

Number of ways $= \dfrac{8!}{4! \times 3!} = \dfrac{8 \times 7 \times 6 \times 5}{3 \times 2 \times 1} = 280.$

Exercise 8.3

1. Find the number of ways of arranging 5 different books in a row in a bookshelf.

2. In how many ways can I arrange six of eight different books in a row in a bookshelf?

3. There are ten teams in the local football competition. In how many ways can the first four places in the premiership table be filled?

4. In how many ways can seven people arrange themselves
   (a)　in a row of seven chairs ;
   (b)　around a circular table?

5. In how many different ways can five identical blue balls, two identical red balls and a yellow ball be arranged in a row?

6. How many different permutations can be formed from all the letters of the word
   (a)　*NEWTON* ;　(b)　*INTERNATIONAL* ;　(c)　*BACCALAUREATE*?

7. Find the number of arrangements of four different letters chosen from the word *PROBLEM* which
   (a)　begin with a vowel ;　　(b)　end with a consonant.

8. How many different 4-digit numbers can be formed using the digits 0, 1, 2, 3, 4, 5 if
   (a)　no digit may be repeated ;　(b)　repetitions are allowed?

Chapter 8

> **Higher Level**
>
> 9. In how many ways can three books be distributed among ten people if
>   (a) each person may receive any number of books ;
>   (b) no person may be given more than one book ;
>   (c) no person may be given more than 2 books?
>
> 10. (a) How many different 6-digit numbers can be formed using the digits 1, 2, 2, 3, 3, 3?
>   (b) How many of the 6-digit numbers in part (a) are even?
>
> 11. (a) In how many ways can five men, four women and three children be arranged in a row so that the men sit together, the women sit together and the children sit together?
>   (b) Answer part (a) if they sit around a circular table.
>
> 12. Three boys and four girls sit in a row. How many different arrangements are possible if
>   (a) there are no restrictions ;
>   (b) the girls are to sit together ;
>   (c) the girls are to sit together and the boys are to sit together ;
>   (d) the sexes alternate?

8.4 Combinations and Partitions

In some problems the order in which objects are arranged is not important. For example, when three people are selected from a group to form a social committee, the order in which the three people are selected is irrelevant.

The number of *combinations* of r different objects out of a set of n is the number of different *selections*, irrespective of order, and is denoted by C_r^n or ${}_nC_r$ or ${}^nC_r$ or sometimes by $\binom{n}{r}$.

Suppose that we wish to select three people from a group of five people. The number of permutations of five people taken three at a time is $5 \times 4 \times 3 = 60$. However, since any selection of three people can be arranged in $3! = 6$ different ways, each selection of three people will appear six times in the set of possible permutations. Hence the total number of selections of three people from five people is $\dfrac{5 \times 4 \times 3}{6} = 10$. Thus the number of combinations of 5 objects taken three at a time $= \dfrac{P_3^5}{3!} = \dfrac{5!}{2!3!} = 10$.

Counting Techniques & The Binomial Theorem

This is a particular case of the general rule which tells us that the number of different selections of r objects chosen from n distinct objects is given by

$$C_r^n = \frac{P_r^n}{r!} = \frac{n!}{r!(n-r)!}.$$

For each selection of r objects from n distinct objects, there is a corresponding selection of $n - r$ objects, those not among the r objects selected. Therefore we have

$$C_r^n = C_{n-r}^n. \qquad \ldots\ldots\ldots\ldots(*)$$

Also it is clear that all n objects can be chosen in one way only, and there is only one selection of no objects possible. Thus

$$C_n^n = C_0^n = 1.$$

To evaluate C_r^n "by hand", the simplest rule to apply is

$$C_r^n = \frac{n(n-1)(n-2)\cdots(n-r+1)}{r!}$$

where the numerator is the product of exactly r consecutive integers starting from n down.

By using rule (*) on the previous page, we can simplify the working for values of r which are close to n, as follows:

$$C_{18}^{20} = \frac{20 \times 19 \times 18 \times \cdots \times 4 \times 3}{18!} \quad \text{(eighteen integers in the numerator)}$$

where sixteen of the eighteen integers in the numerator cancel with sixteen of the factors of 18!, and so a better approach is

$$C_{18}^{20} = C_2^{20} = \frac{20 \times 19}{2!} = 190.$$

Example Evaluate: (a) C_3^8; (b) C_{16}^{20}.

(a) $C_3^8 = \dfrac{8 \times 7 \times 6}{3 \times 2 \times 1} = 56$.

(b) $C_{16}^{20} = C_4^{20} = \dfrac{20 \times 19 \times 18 \times 17}{4 \times 3 \times 2 \times 1} = 4845$.

Example How many different committees of 3 people can be chosen from a group of 12 people?

193

Chapter 8

$$\text{Number of committees} = C_3^{12} = \frac{12 \times 11 \times 10}{3 \times 2 \times 1} = 220.$$

Example   A football team of 11 is to be chosen from 15 players.
- (a) How many different teams can be selected?
- (b) How many teams can be selected if the Captain must be in the team?
- (c) How many teams can be selected if the Captain must be in the team and one of the other players is injured and cannot play?

(a) The number of teams possible = $C_{11}^{15} = C_4^{15} = \dfrac{15 \times 14 \times 13 \times 12}{4 \times 3 \times 2 \times 1} = 1365$.

(b) Since the Captain must be in the team, the selectors' job is to select another 10 players from the 14 available to join the Captain.

The number of teams possible = $C_{10}^{14} = C_4^{14} = \dfrac{14 \times 13 \times 12 \times 11}{4 \times 3 \times 2 \times 1} = 1001$.

(c) Since the Captain must be in the team and one player is ruled out due to injury, the selectors' job is the select another 10 players from the 13 available to join the Captain.

The number of teams possible = $C_{10}^{13} = C_3^{13} = \dfrac{13 \times 12 \times 11}{3 \times 2 \times 1} = 286$.

Example   A committee of 5 is to be chosen from 12 men and 8 women. In how many ways can this be done if there are to be 3 men and 2 women on the committee?

The number of ways of choosing the men = $C_3^{12} = \dfrac{12 \times 11 \times 10}{3 \times 2 \times 1} = 220$.

The number of ways of choosing the women = $C_2^{8} = \dfrac{8 \times 7}{2 \times 1} = 28$.

Therefore the total number of ways of choosing the committee (given by the product rule) = $220 \times 28 = 6160$.

Note: There is no general formula for dealing with selections made from sets containing objects which are not all distinct, as there was for permutations in such cases.

Example   How many different ways are there of selecting 4 letters from the letters of the word *POPPED*?

The number of selections containing one *P* and three other letters = 1.

Counting Techniques & The Binomial Theorem

The number of selections containing two *P*s and two other letters from the remaining three = $C_2^3 = 3$.

The number of selections containing three *P*s and one other letter from the remaining three = $C_1^3 = 3$.

Therefore the total number of selections of 4 letters = 1 + 3 + 3 = 7.

Higher Level

Example Prove Pascal's identity: $C_r^{n+1} = C_r^n + C_{r-1}^n$ for $1 \leq r \leq n$.

$$C_r^n + C_{r-1}^n = \frac{n!}{r!(n-r)!} + \frac{n!}{(r-1)!(n-[r-1])!}$$

$$= \frac{n!}{r!(n-r+1)!}\{(n-r+1)+r\}$$

$$= \frac{(n+1)!}{r!([n+1]-r)!}$$

$$= C_r^{n+1}.$$

Partitions of Like Objects

Consider the problem of deciding how many different ways 6 indistinguishable balls can be placed in a box which has 4 compartments.

One such arrangement might be: | OO | | OO | OO |

We can designate this as OO||OO|OO where the Os represent the balls and the |s represent the "walls" between compartments. (For 4 cells there are 3 "walls".) All possible arrangements of the 6 balls into 4 compartments is thus equivalent to the number of ways of arranging 9 objects (6 balls and 3 "walls") of which 6 are alike and 3 are alike.

Therefore the number of ways = $\frac{9!}{6!3!} = 84$.

The general rule is as follows:

> The number of ways in which n identical objects can be placed in r cells = $\frac{(n+r-1)!}{n!(r-1)!} = C_n^{n+r-1}$ or C_{r-1}^{n+r-1}.

Chapter 8

Example In how many ways can I distribute eight $2 coins between 5 girls?

The problem is to find the number of ways of placing 8 identical objects into 5 cells.

The number of ways = $C_8^{8+5-1} = C_4^{12} = \dfrac{12 \times 11 \times 10 \times 9}{4 \times 3 \times 2 \times 1} = 495$.

Ordered Partitions

Suppose an urn A contains seven marbles numbered 1–7. We wish to find the number of ways we can draw firstly 2 marbles from the urn, then 3 marbles from the urn and lastly 2 marbles from the urn. In other words we wish to calculate the number of ordered partitions $\{A_1, A_2, A_3\}$ of the set of 7 marbles into cells A_1 containing 2 marbles, A_2 containing 3 marbles and A_3 containing 2 marbles. We call these ***ordered partitions*** since we wish to distinguish between $(\{1,2\}, \{3,4,5\}, \{6,7\})$ and $(\{6,7\}, \{3,4,5\}, \{1,2\})$ each of which yields the same partition of A.

Since we begin with 7 marbles in the urn, there are C_2^7 ways of drawing the first two marbles to determine the first cell A_1; following this, there are 5 marbles left in the urn and so there are C_3^5 ways of drawing the three marbles to determine the second cell A_2; finally there are 2 marbles left in the urn and so there are C_2^2 ways of determining the last cell A_3.

Thus there are $C_2^7 \times C_3^5 \times C_2^2 = \dfrac{7 \times 6}{2 \times 1} \times \dfrac{5 \times 4 \times 3}{3 \times 2 \times 1} \times \dfrac{2 \times 1}{2 \times 1} = 210$ different ordered partitions of A into (named) cells containing 2, 3 and 2 marbles respectively.

Note: $\quad C_2^7 \times C_3^5 \times C_2^2 = \dfrac{7!}{2! \times 5!} \times \dfrac{5!}{3! \times 2!} \times \dfrac{2!}{2! \times 0!} = \dfrac{7!}{2! \times 3! \times 2!}$.

The general rule is as follows:

> Let set A contain n elements and let $n_1, n_2, n_3, \cdots, n_r$ be positive integers such that $n_1 + n_2 + n_3 + \cdots + n_r = n$. Then there are
>
> $\dfrac{n!}{n_1! \times n_2! \times n_3! \times \cdots \times n_r!}$ different ordered partitions of A of the form
>
> $(A_1, A_2, A_3, \cdots A_r)$ where A_1 contains n_1 elements, A_2 contains n_2 elements, A_3 contains n_3 elements, ... , A_r contains n_r elements.

Example In how many ways can 9 toys be distributed between 4 children if the youngest child is to receive 3 toys and each of the others 2 toys?

This problem is really asking how many different ordered partitions of 9 objects into cells containing 3, 2, 2, 2 objects respectively are possible?

$$\text{The number} = \frac{9!}{3! \times 2! \times 2! \times 2!} = 7560.$$

Example In how many ways can 9 people be partitioned into
(a) two teams of 5 and 4 people each ;
(b) three named teams of 3 people each ;
(c) three unnamed teams of 3 people each?

(a) If we select the team of 5, the team of 4 automatically consists of those who were not selected.

$$\text{The number of ways} = C_5^9 = C_4^9 = \frac{9 \times 8 \times 7 \times 6}{4 \times 3 \times 2 \times 1} = 126.$$

(b) We require the number of ordered partitions of 9 people into cells containing 3, 3 and 3 people respectively.

$$\text{The number of ways} = \frac{9!}{3! \times 3! \times 3!} = 1680.$$

(c) There are $3! = 6$ ways of ordering three sets A_1, A_2, A_3.

$$\text{The number of ways} = \frac{1680}{6} = 280.$$

Exercise 8.4

1. Evaluate: (a) C_2^4 ; (b) C_3^9 ; (c) C_4^{16}.

2. Evaluate: (a) C_5^7 ; (b) C_{13}^{16} ; (c) C_7^{10}.

3. How many committees of 3 students can be selected from 20 students?

4. How many teams of 6 can be selected from 10 players?

5. In how many ways can a team of 2 men and 3 women be selected from 6 men and 7 women?

Chapter 8

6. A student is to answer 8 out of 10 questions in an examination.
   (a) How many choices has she?
   (b) How many choices has she if she must answer the first 4 questions?
   (c) How many choices has she if she must answer at least 4 of the first 5 questions?

7. A basketball team of 6 is to be chosen from 11 available players. In how many ways can this be done if
   (a) there are no restrictions ;
   (b) 3 of the players are automatic selections ;
   (c) 3 of the players are automatic selections and 2 other players are injured and cannot play?

8. In how many ways can 7 toys be divided between 3 children if the youngest gets 3 toys and each of the others gets 2 toys?

9. In how many ways can 9 people be placed in 3 cars which can take 2, 3 and 4 passengers respectively, assuming that the seating arrangements inside the cars are not important?

10. A committee of 5 people is to be selected from 10 men and 8 women. In how many ways can the committee be selected if
    (a) it must contains 2 men and 3 women ;
    (b) it must have at least one member of each sex?

Higher Level

11. In how many ways can 12 people be divided into
    (a) two groups each with 6 people ;
    (b) three groups each with 4 people?

12. How many numbers of 5 different digits can be constructed so that the digits are in ascending order from left to right?

13. Use Pascal's identity (page 195) to prove that if $0 \leq k \leq n-2$, then
    $$C_{k+2}^{n+2} = C_{k+2}^{n} + 2C_{k+1}^{n} + C_{k}^{n}.$$

14. Show that if n, r and k are integers with $0 \leq k \leq r \leq n$ then
    $$C_{r}^{n} \times C_{k}^{r} = C_{k}^{n} \times C_{r-k}^{n-k}.$$

15. Find the number of ways in which 6 coins can be distributed between 3 people if the coins are
    (a) all different ;   (b) indistinguishable.

16. Find the number of ways in which 10 different books can be divided between 2 girls such that no girl receives more than 8 books.

17. If 5 straight lines are drawn in a plane with no two parallel and no three concurrent, how many different triangles can be formed by joining sets of 3 of the points of intersection?

18. If m straight lines and n circles are drawn in a plane, what is the maximum number of possible points of intersection?

19. Twelve points in a plane are such that 4 of them lie on a straight line but no other set of 3 or more points are collinear. What is the maximum number of different straight lines which can be drawn through pairs of the given points?

20. In how many ways can 12 distinct objects be partitioned into
    (a) two sets of 5 and 7 each ;
    (b) two named sets of 6 objects each ;
    (c) two unnamed sets each containing 6 objects ;
    (d) three named sets of 4 objects each ;
    (e) three unnamed sets each containing 4 objects?

21. In how many ways can 5 identical rings be placed on the 4 fingers of one hand?

22. In how many ways can I place 8 identical objects into 5 compartments so that each compartment contains at least one object?

23. (a) How many distinct non-negative integer solutions to the equation
    $$x_1 + x_2 + x_3 + x_4 = 10$$
    exist?

    (b) How many positive integer solutions to the equation in part (a) exist?

8.5 The Binomial Theorem

Theorem   For every integer $n \geq 1$, $(a+b)^n = \sum_{r=0}^{n} \binom{n}{r} a^{n-r} b^r$.

Proof (Combinatorial)

199

Chapter 8

If the product $(a+b)^n = (a+b)(a+b)(a+b)\cdots(a+b)$ is multiplied out, each term of the answer will be of the form $c_1 c_2 c_3 \cdots c_k \cdots c_n$ where, for all k, c_k is either a or b.

Thus if $c_k = a$ for all k we obtain the term a^k. If $c_k = b$ for one of the terms and $c_k = a$ for the rest, we obtain terms such as $b \times a \times a \times \cdots \times a \times a$, $a \times b \times a \times \cdots \times a \times a$, $a \times a \times b \times a \times \cdots \times a \times a$, $\cdots$, $a \times a \times a \times \cdots \times b \times a$, $a \times a \times a \times \cdots \times a \times b$, and their sum is $na^{n-1}b$.

If $c_k = b$ for r of the terms and $c_k = a$ for the rest we obtain a number of terms of the form $a^{n-r}b^r$.

The number of such terms is the number of ways in which r of the $c_1, c_2, \cdots, c_n$ can be selected as equal to b. This number is $\binom{n}{r}$.

Thus $\binom{n}{r}$ is the coefficient of $a^{n-r}b^r$ in the expansion of $(a+b)^n$.

Hence $(a+b)^n = \sum_{r=0}^{n} \binom{n}{r} a^{n-r} b^r$.

The following properties of the expansion of $(a+b)^n$ should be observed:
- There are $n+1$ terms.
- The sum of the exponents of a and b in each term is n.
- The exponents of a decrease term by term from n to 0; the exponents of b increase term by term from 0 to n.
- The coefficient of any term is $\binom{n}{k}$ where k is the exponent of either a or b.
- The coefficients of terms equidistant from the ends are equal.

Example Write down the expansions of each of the following:
(a) $(a+b)^4$;   (b) $(a-b)^3$;   (c) $(x+2y)^5$.

(a) $(a+b)^4 = \binom{4}{0}a^4 b^0 + \binom{4}{1}a^3 b + \binom{4}{2}a^2 b^2 + \binom{4}{3}ab^3 + \binom{4}{4}a^0 b^4$

$= a^4 + 4a^3 b + 6a^2 b^2 + 4ab^3 + b^4$.

(b) $(a-b)^3 = \binom{3}{0}a^3(-b)^0 + \binom{3}{1}a^2(-b) + \binom{3}{2}a(-b)^2 + \binom{3}{3}a^0(-b)^3$
$= a^3 - 3a^2b + 3ab^2 - b^3$.

(c) $(x+2y)^5 = \binom{5}{0}x^5(2y)^0 + \binom{5}{1}x^4(2y) + \binom{5}{2}x^3(2y)^2 + \cdots + \binom{5}{5}a^0(2y)^5$
$= x^5 + 10x^4y + 40x^3y^2 + 80x^2y^3 + 80xy^4 + 32y^5$.

Example Find the coefficient of x^3 in the expansion of each of the following:

(a) $(2x-1)^5$;

(b) $\left(x^2 - \dfrac{1}{x}\right)^6$;

(c) $(1-x+2x^2)(3-x)^9$.

(a) The term in x^r is $\binom{5}{r}(2x)^r(-1)^{5-r}$ and so the term in x^3 has $r = 3$.

The coefficient of this term $= \binom{5}{3}2^3(-1)^2 = 80$.

(b) The $(r+1)$st term in descending powers of x^2 is $\binom{6}{r}(x^2)^{6-r}\left(-\dfrac{1}{x}\right)^r$

which can be written $\binom{6}{r}x^{12-3r}(-1)^r$; the term in x^3 has $r = 3$.

The required coefficient $= \binom{6}{3}(-1)^3 = -20$.

(c) The $(r+1)$st term in descending powers of x in the expansion of $(3-x)^9$ is $\binom{9}{r}(-x)^r 3^{9-r}$. Therefore the coefficients of the terms containing x, x^2, x^3 are $\binom{9}{1}(-1)3^8, \binom{9}{2}(-1)^2 3^7, \binom{9}{3}(-1)^3 3^6$ respectively ; that is $-9 \times 3^8, 36 \times 3^7, -84 \times 3^6$.

To obtain the term in x^3 we must multiply the first by 2, the second by -1, the third by 1, and then add the results.
The required coefficient is $3^6\{-162 - 108 - 84\} = -258\,066$.

Chapter 8

Note: If only one term of the binomial contains the variable (x), then the exponent of x is known and so is the exponent of the other term. The required binomial coefficient is then also known. Observe the following example.

Example Find the term independent of x in the expansion of $\left(x - \dfrac{2}{x^2}\right)^{12}$.

$\left(x - \dfrac{2}{x^2}\right)^{12} = x^{12}\left(1 - \dfrac{2}{x^3}\right)^{12}$ and $x^{12} \times \dfrac{k}{x^{12}}$ is independent of x.

The coefficient of x^{-12} in the expansion of $\left(1 - \dfrac{2}{x^3}\right)^{12}$ is $\binom{12}{4} \times 1^8 \times (-2)^4$.

Therefore the term independent of $x = \binom{12}{4} 1^8 (-2)^4 = 7920$.

Example Find the coefficient of the term in x^{-3} in the expansion of $\left(\dfrac{2}{x} + \dfrac{x}{2}\right)^7$.

$\left(\dfrac{2}{x} + \dfrac{x}{2}\right)^7 = \dfrac{1}{x^7}\left(2 + \dfrac{x^2}{2}\right)^7$ so the coefficient of x^{-3} is $\binom{7}{2} \times 2^5 \times \dfrac{1}{2^2} = 168$.

Exercise 8.5

1. Use the binomial theorem to expand each of the following:
   (a) $(x+y)^4$;
   (b) $(a-b)^7$;
   (c) $(2+p^2)^6$;
   (d) $(2h-k)^5$;
   (e) $\left(x + \dfrac{1}{x}\right)^3$;
   (f) $\left(z - \dfrac{1}{2z}\right)^8$.

2. If n is a positive integer, use the binomial expansion of $(a+b)^n$ with suitable values of a and b to prove that
   (a) $2^n = \binom{n}{0} + \binom{n}{1} + \binom{n}{2} + \cdots + \binom{n}{n}$;
   (b) $0 = \binom{n}{0} - \binom{n}{1} + \binom{n}{2} - \cdots + (-1)^n \binom{n}{n}$.

3. Expand and simplify: $\left(2x + \dfrac{1}{x^2}\right)^5 + \left(2x - \dfrac{1}{x^2}\right)^5$.

Counting Techniques & The Binomial Theorem

4. Find the coefficients of x^3 and x^4 in the expansion of $\left(x+\dfrac{1}{x}\right)^5$.

5. Find the value of n if the coefficient of x^3 in the expansion of $(2+3x)^n$ is twice the coefficient of x^2.

6. In the binomial expansion of $\left(1+\dfrac{1}{3}\right)^n$ in ascending powers of $\dfrac{1}{3}$, the fourth and fifth terms are equal. Find the value of n.

7. The coefficient of x^5 in the expansion of $(1+5x)^8$ is equal to the coefficient of x^4 in the expansion of $(a+5x)^7$. Find the value of a.

8. Use the expansion of $(a+b)^4$ to evaluate 1.03^4 correct to 2 decimal places.

9. Use the expansion of $(2-x)^5$ to evaluate 1.98^5 correct to 5 decimal places.

10. Find the first 4 terms in ascending powers of x in the expansion of $(1+2x)^{15}$. Hence evaluate 1.002^{15} correct to 5 decimal places.

11. In the expansion of each of the following, find the coefficient of the specified power of x:
    (a) $(1+2x)^7$, x^4;
    (b) $(5x-3)^7$, x^4;
    (c) $(3x-2x^3)^5$, x^{11};
    (d) $\left(x-\dfrac{2}{x}\right)^8$, x^2;
    (e) $\left(x^2+\dfrac{4}{x}\right)^{10}$, x^{-1};
    (f) $\left(\dfrac{x}{6}-\dfrac{3}{x}\right)^9$, x^3.

12. Find the term independent of x in the expansion of $\left(\dfrac{2}{x^8}-x^{2}\right)^{12}$.

Higher Level

13. Find the possible values of the constant k if the coefficient of x^{-5} in the expansion of $\left(kx^2+\dfrac{1}{x}\right)^9$ is 16.

Chapter 8

14. (a) Determine the coefficient of x in the expansion of $(2+x)(2-x)^7$.
    (b) Find the integer p, $0 \le p \le 8$, for which the coefficient of x^p in the expansion of $(2+x)(2-x)^7$ is zero.

15. Find the coefficient of
    (a) x^6 in the expansion of $(2+x)(1+x)^8$;
    (b) x^5 in the expansion of $(1+x)(2-x)^7$;
    (c) x^8 in the expansion of $(2-x^2)(3+x)^9$;
    (d) x^4 in the expansion of $(1-x+x^2)(2+x)^6$.

16. The first 3 terms of the expansion of $(1+ax)^n$ in ascending powers of x are given below. In each case find a and b and the next term in the expansion.
    (a) $1+8x+28x^2+\cdots$; (b) $1+4x+7.5x^2+\cdots$;
    (c) $1-\dfrac{x}{4}+\dfrac{3x^2}{100}+\cdots$; (d) $1-6x+\dfrac{63}{4}x^2+\cdots$.

17. Show that the coefficient of x^2 in the expansion of $(2+x)(1+x)^n$ is n^2 and the coefficient of x^3 is $\frac{1}{6}n(n-1)(2n-1)$.

18. Find the coefficient of x^r, $1 \le r \le 8$, in the expansion of $(2+x)(1+x)^8$ and hence show that one of the coefficients of the expansion is zero.

*19. Find the coefficient of the given power of x in the expansion of
    (a) $(1-x)^2(1+x)^8$, x^4 ; (b) $\left(x+\dfrac{1}{x}\right)^2(1-x)^5$, x^2 ;
    (c) $(1+x+x^2)^{12}$, x^3 ; (d) $(1-x^2+x^3)^5$, x^4 .

Required Outcomes

After completing this chapter, a student should be able to:
- use the product principle to count the number of possible outcomes of an 'experiment'.
- use permutations and combinations to ennumerate possible outcomes.
- count possible arrangements of objects some of which are alike.
- use ordered and unordered partitions to count possible outcomes. **(HL)**
- apply the binomial theorem in a variety of contexts.

9 Matrices and Linear Equations

9.1 Matrix Addition

Definition   A ***matrix*** is a rectangular array of mn numbers arranged in m horizontal rows and n vertical columns.

We write

$$A = \begin{pmatrix} a_{11} & a_{12} & a_{13} & \cdots & a_{1n} \\ a_{21} & a_{22} & a_{23} & \cdots & a_{2n} \\ a_{31} & a_{32} & a_{33} & \cdots & a_{3n} \\ \vdots & \vdots & \vdots & \vdots & \vdots \\ a_{m1} & a_{m2} & a_{m3} & \cdots & a_{mn} \end{pmatrix},$$

where a_{ij} is the element in the ith row and jth column of A.

[A short-hand notation which is often used is $A = (a_{ij})$.]

Dimension   A matrix which has m rows and n columns is said to have ***dimension*** $m \times n$.

The dimension of a matrix is sometimes called its ***shape*** or its ***size***.

If $m = n$ the matrix is called ***square***.

For example the matrix

$$A = \begin{pmatrix} 2 & 3 \\ -1 & 0 \end{pmatrix}$$

is a 2×2 matrix or a square matrix of order 2.

Example   Consider the matrix $A = \begin{pmatrix} 2 & 3 & 1 \\ 3 & -1 & 0 \end{pmatrix}$.

A has dimension 2×3.

$a_{12} = 3$, $a_{23} = 0$, but a_{31} does not exist since A does not have 3 rows.

A matrix of dimension $m \times 1$ is called a ***column vector***, or simply a ***vector***.
A matrix of dimension $1 \times n$ is called a ***row vector***.
We sometimes denote a column vector by $\boldsymbol{a}$ and a row vector by $\boldsymbol{a}^{\mathrm{T}}$.

Chapter 9

Equality Two matrices A and B are *equal* iff (if and only if) they have the same dimension and $a_{ij} = b_{ij}$ for all i and j.

Example Consider the matrices

$$A = \begin{pmatrix} 3 & 2 & 1 \\ 5 & 0 & -4 \end{pmatrix}, B = \begin{pmatrix} 3 & 5 \\ 2 & 0 \\ 1 & -4 \end{pmatrix}, C = \begin{pmatrix} 3 & 2 & 1 \\ 5 & 0 & -4 \\ 0 & 0 & 0 \end{pmatrix}, D = \begin{pmatrix} 3 & 2 & 1 \\ 5 & 0 & -4 \end{pmatrix}.$$

A and B are not equal since they do not have the same dimension.
A and C are not equal since C has an extra row.
A and D are equal.

Addition If two matrices are of the same dimension we can define their sum, $C = A + B$, as that matrix with the same shape as A and B and such that $c_{ij} = a_{ij} + b_{ij}$. Thus each entry in C is equal to the sum of the corresponding entries in A and B.

Example Consider the matrices

$$A = \begin{pmatrix} 2 & -1 \\ 4 & 0 \\ -2 & 3 \end{pmatrix}, B = \begin{pmatrix} 1 & 2 \\ -1 & 2 \\ 2 & -4 \end{pmatrix}, C = \begin{pmatrix} 3 & 2 & 1 \\ 5 & -1 & 2 \end{pmatrix}.$$

Then $A + B = \begin{pmatrix} 2+1 & -1+2 \\ 4+(-1) & 0+2 \\ -2+2 & 3+(-4) \end{pmatrix} = \begin{pmatrix} 3 & 1 \\ 3 & 2 \\ 0 & -1 \end{pmatrix}.$

However $A + C$ and $B + C$ do not exist since C does not have the same dimension as A or B.

The Zero Matrix The *zero matrix* has all its entries equal to zero.

We denote the zero matrix of dimension $m \times n$ by $O_{m \times n}$, or if the dimension is obvious, by O or more simply by 0. The zero matrix of dimension 2×2 is

$$O_{2 \times 2} = \begin{pmatrix} 0 & 0 \\ 0 & 0 \end{pmatrix}.$$

If A has dimension $m \times n$ and O is the zero matrix of dimension $m \times n$ then $A + O = O + A = A$. Thus O is the *additive identity matrix*.

Matrices & Linear Equations

Multiplication by a Scalar If $A = (a_{ij})$ and s is a scalar, $sA = (sa_{ij})$. The entries of sA are found by multiplying each of the entries of A by s.

Example Given $A = \begin{pmatrix} 2 & -1 \\ 4 & 6 \end{pmatrix}$, find $2A$, $-A$ and $\frac{1}{2}A$.

$$2A = \begin{pmatrix} 4 & -2 \\ 8 & 12 \end{pmatrix}, \quad -A = \begin{pmatrix} -2 & 1 \\ -4 & -6 \end{pmatrix} \text{ and } \frac{1}{2}A = \begin{pmatrix} 1 & -\frac{1}{2} \\ 2 & 3 \end{pmatrix}.$$

Transpose The *transpose* of the $m \times n$ matrix $A = (a_{ij})$ is defined to be the $n \times m$ matrix $A^T = (a_{ji})$. Thus A^T is the matrix whose rows are the same as the columns of A and whose columns are the same as the rows of A.

Example If $A = \begin{pmatrix} 3 & 2 & -1 \\ 4 & 5 & 2 \end{pmatrix}$ then $A^T = \begin{pmatrix} 3 & 4 \\ 2 & 5 \\ -1 & 2 \end{pmatrix}$.

Exercise 9.1

1. If $A = \begin{pmatrix} 2 & 1 \\ 3 & -2 \end{pmatrix}$ and $B = \begin{pmatrix} 3 & -1 \\ 2 & -3 \end{pmatrix}$, find

    (a) $A + B$; (b) $2A - B$; (c) $(A+B)^T$; (d) $A^T + B^T$.

2. If $A = \begin{pmatrix} 2 & 1 & 3 \\ -2 & 4 & 0 \end{pmatrix}$, $B = \begin{pmatrix} -4 & 2 & 5 \\ 0 & -2 & 0 \end{pmatrix}$, $C = \begin{pmatrix} 5 & -1 \\ 2 & 4 \\ -3 & -2 \end{pmatrix}$, find if possible

    (a) $A + B$; (b) $A + C^T$; (c) $3A - 2B$; (d) $3B - C$; (e) $B - 2C^T$.

3. Verify the associative law of addition for 2×2 matrices. That is, show that for any 2×2 matrices A, B and C, $A + (B + C) = (A + B) + C$.

4. Where possible, find the unknown matrix X in each of the following:

    (a) $X + 3\begin{pmatrix} 1 & -2 \\ 4 & 5 \end{pmatrix} = \begin{pmatrix} 4 & -2 \\ 14 & 16 \end{pmatrix}$;

207

Chapter 9

(b) $\quad 2\begin{pmatrix} 3 & 4 \\ -1 & 2 \end{pmatrix} = 3X - \begin{pmatrix} 6 & 1 \\ -1 & 5 \end{pmatrix};$

(c) $\quad 2X - 3\begin{pmatrix} 1 & 0 \\ 2 & -2 \\ 4 & -1 \end{pmatrix}^T = \begin{pmatrix} 1 & 4 & -2 \\ 4 & 0 & -1 \end{pmatrix};$

(d) $\quad \begin{pmatrix} 1 & -1 & 2 \\ 4 & 3 & -2 \end{pmatrix} - 3X^T = \begin{pmatrix} 4 & 2 & 8 \\ -2 & 9 & 1 \end{pmatrix}.$

5. In each of the following, A, B, C and X are matrices of the same dimension. Express X in terms of A, B and C.
(a) $\quad X - (2A + 3B - C) = 2(X - A) - (X + B) + 3(X - 2C)$;
(b) $\quad 3(X - 2A) + 2(B - 2X) - 3(2X + C) = 0.$

9.2 Matrix Multiplication

Let A be an $m \times n$ matrix and B be an $n \times p$ matrix. Then the product $C = AB$ is a matrix of dimension $m \times p$ such that $c_{ij} = \sum_{k=1}^{n} a_{ik} b_{kj} = a_{i1} b_{1j} + a_{i2} b_{2j} + \cdots + a_{in} b_{nj}$ for $i = 1, 2, 3, \ldots, m$ and $j = 1, 2, 3, \ldots, p.$

Thus the element in the ith row and jth column of the product AB is found by multiplying the corresponding entries in the ith row of A and the jth column of B and then adding the n products.

Matrix multiplication is not always possible. Matrices are said to be ***conformable for multiplication*** if the number of columns of the first matrix is equal to the number of rows of the second matrix. The product matrix has the same number of rows as the first matrix and the same number of columns as the second matrix.

In the product AB, we say that A ***pre-multiplies*** B or B ***post-multiplies*** A.

As the next example shows, matrix multiplication is ***not commutative***, i.e., $AB \neq BA$ in general.

Example $\quad$ If $A = \begin{pmatrix} 2 & 3 & -1 \\ 4 & 0 & 2 \end{pmatrix}$ and $B = \begin{pmatrix} 3 & 1 \\ 1 & 2 \\ 0 & 4 \end{pmatrix}$, find AB and BA.

A is 2×3 and B is 3×2. Therefore AB exists and has dimension 2×2.

Now $AB = \begin{pmatrix} 2\times 3 + 3\times 1 + (-1)\times 0 & 2\times 1 + 3\times 2 + (-1)\times 4 \\ 4\times 3 + 0\times 1 + 2\times 0 & 4\times 1 + 0\times 2 + 2\times 4 \end{pmatrix} = \begin{pmatrix} 9 & 4 \\ 12 & 12 \end{pmatrix}.$

B is 3×2 and A is 2×3. Therefore BA exists and has dimension 3×3.

Now $BA = \begin{pmatrix} 3\times 2 + 1\times 4 & 3\times 3 + 1\times 0 & 3\times (-1) + 1\times 2 \\ 1\times 2 + 2\times 4 & 1\times 3 + 2\times 0 & 1\times (-1) + 2\times 2 \\ 0\times 2 + 4\times 4 & 0\times 3 + 4\times 0 & 0\times (-1) + 4\times 2 \end{pmatrix} = \begin{pmatrix} 10 & 9 & -1 \\ 10 & 3 & 3 \\ 16 & 0 & 8 \end{pmatrix}.$

It is obvious from this that $AB \neq BA$.

Example If $A = \begin{pmatrix} 2 & 4 & 1 \\ 3 & 0 & 2 \end{pmatrix}$ and $B = \begin{pmatrix} 3 & 5 \\ 1 & 4 \end{pmatrix}$, find, if possible

(a) AB; (b) $A^\mathrm{T}B$; (c) BA.

(a) AB does not exist since the number of columns of A (3) is not equal to the number of rows of B (2).

(b) A^T is 3×2 and B is 2×2. Thus $A^\mathrm{T}B$ exists with dimension 3×2.

Now $A^\mathrm{T}B = \begin{pmatrix} 2 & 3 \\ 4 & 0 \\ 1 & 2 \end{pmatrix}\begin{pmatrix} 3 & 5 \\ 1 & 4 \end{pmatrix} = \begin{pmatrix} 6+3 & 10+12 \\ 12+0 & 20+0 \\ 3+2 & 5+8 \end{pmatrix} = \begin{pmatrix} 9 & 22 \\ 12 & 20 \\ 5 & 13 \end{pmatrix}.$

(c) B is 2×2 and $A = 2 \times 3$. Thus BA exists with dimension 2×3.

Now $BA = \begin{pmatrix} 3 & 5 \\ 1 & 4 \end{pmatrix}\begin{pmatrix} 2 & 4 & 1 \\ 3 & 0 & 2 \end{pmatrix} = \begin{pmatrix} 6+15 & 12+0 & 3+10 \\ 2+12 & 4+0 & 1+8 \end{pmatrix}$

$= \begin{pmatrix} 21 & 12 & 13 \\ 14 & 4 & 9 \end{pmatrix}.$

Example The school shop sells two sorts of cold drinks, milk and orange juice. Milk costs $1.20 per carton and orange juice is $1.50 per carton. On a given week 100 cartons of milk and 75 cartons of orange juice were sold. In the succeeding week 120 cartons of milk and 70 cartons of orange juice were sold. In the third week 110 cartons of milk and 80 cartons of orange juice were sold. Find, using matrices, the total number of cartons of each sort of drink sold in the three weeks and the total cost.

Chapter 9

Let $X = \begin{pmatrix} 1.2 & 1.5 \end{pmatrix}$, $Y = \begin{pmatrix} 100 & 120 & 110 \\ 75 & 70 & 80 \end{pmatrix}$ and $Z = \begin{pmatrix} 1 \\ 1 \\ 1 \end{pmatrix}$.

Now $YZ = \begin{pmatrix} 100 & 120 & 110 \\ 75 & 70 & 80 \end{pmatrix} \begin{pmatrix} 1 \\ 1 \\ 1 \end{pmatrix} = \begin{pmatrix} 330 \\ 225 \end{pmatrix}$ and

$X(YZ) = \begin{pmatrix} 1.2 & 1.5 \end{pmatrix} \begin{pmatrix} 330 \\ 225 \end{pmatrix} = \begin{pmatrix} 733.5 \end{pmatrix}$.

Therefore 330 cartons of milk and 225 cartons of orange juice were sold at a total cost of $733.50.

Note: Positive integer powers of a matrix have a similar meaning to positive integer powers of a real number.

For example, $A^2 = A \times A$, $A^3 = A \times A \times A = A \times A^2 = A^2 \times A$, etc.

Exercise 9.2

1. The matrix A has 4 rows and 5 columns while the matrix B has 5 rows and 4 columns. Find the dimensions of AB and BA.

2. Given matrices A, B, C and D such that $D = A(B - C)$ where A, B have dimensions $m \times n$ and $n \times p$ respectively, find the dimensions of C and D.

3. Find the following products:

    (a) $\begin{pmatrix} 2 & -1 \\ 3 & 2 \end{pmatrix} \begin{pmatrix} -1 & 3 \\ 4 & 0 \end{pmatrix}$;

    (b) $\begin{pmatrix} 3 & 1 \\ 5 & 2 \end{pmatrix} \begin{pmatrix} 2 & -1 \\ -5 & 3 \end{pmatrix}$;

    (c) $\begin{pmatrix} 2 & 5 \\ -2 & 3 \\ 1 & 0 \end{pmatrix} \begin{pmatrix} 1 & 2 & 3 \\ -2 & -3 & -4 \end{pmatrix}$;

    (d) $\begin{pmatrix} 1 & 4 & 2 \\ 3 & -2 & 1 \end{pmatrix} \begin{pmatrix} 4 & -1 \\ -2 & 3 \\ 0 & -3 \end{pmatrix}$;

    (e) $\begin{pmatrix} 1 & 3 & 5 \end{pmatrix} \begin{pmatrix} -2 \\ -1 \\ 1 \end{pmatrix}$;

    (f) $\begin{pmatrix} 2 \\ 3 \\ 1 \end{pmatrix} \begin{pmatrix} 3 & -1 & -2 \end{pmatrix}$;

    (g) $\begin{pmatrix} 2 & 4 \\ -1 & 2 \end{pmatrix} \begin{pmatrix} 3 & 2 & 4 & 1 \\ -1 & 0 & 2 & 3 \end{pmatrix}$;

    (h) $\begin{pmatrix} 5 & -2 \end{pmatrix} \begin{pmatrix} -1 & 2 \\ 4 & 3 \end{pmatrix}$;

(i) $\begin{pmatrix} 3 & -2 & -2 \\ 5 & 1 & 4 \\ -3 & 0 & 2 \end{pmatrix} \begin{pmatrix} 2 & 4 \\ 5 & 0 \\ -3 & -2 \end{pmatrix}$; (j) $\begin{pmatrix} 6 & 2 & -5 \\ 3 & 1 & 2 \\ 5 & 4 & -6 \end{pmatrix} \begin{pmatrix} 2 & 7 & 8 \\ -2 & 0 & 2 \\ 6 & 2 & -1 \end{pmatrix}$;

(k) $\begin{pmatrix} 2 & 3 & -1 \\ 1 & 2 & 1 \\ -1 & -1 & 3 \end{pmatrix} \begin{pmatrix} 7 & -8 & 5 \\ -4 & 5 & -3 \\ 1 & -1 & 1 \end{pmatrix}$; (l) $\begin{pmatrix} 3 & -1 & 2 & 5 \\ 4 & 6 & -5 & 7 \end{pmatrix} \begin{pmatrix} 2 \\ 1 \\ 5 \\ 3 \end{pmatrix}$.

4. Show that multiplication of 2×2 matrices distributes over addition, i.e., show that for any 2×2 matrices A, B and C, $A(B + C) = AB + AC$.

5. If $A = \begin{pmatrix} 2 & -4 \\ 1 & -2 \end{pmatrix}$ and $B = \begin{pmatrix} 4 & 2 \\ 2 & 1 \end{pmatrix}$, show that AB is the zero matrix and find the matrix BA.

6. Show that $AB = AC$ does not imply that $B = C$ by showing that $AB = AC$ for the matrices $A = \begin{pmatrix} 4 & 2 \\ 2 & 1 \end{pmatrix}$, $B = \begin{pmatrix} 3 & 1 \\ -5 & 2 \end{pmatrix}$, $C = \begin{pmatrix} 2 & 1 \\ -3 & 2 \end{pmatrix}$.

7. Find the numbers a and b for which $\begin{pmatrix} -1 & a \\ 3 & 3 \end{pmatrix} \begin{pmatrix} b & 4 \\ 6 & 6 \end{pmatrix} = \begin{pmatrix} b & 4 \\ 6 & 6 \end{pmatrix} \begin{pmatrix} -1 & a \\ 3 & 3 \end{pmatrix}$.

8. Consider the matrices $A = \begin{pmatrix} 1 & 2 \\ 3 & 4 \end{pmatrix}$, $B = \begin{pmatrix} 2 & -3 \\ -1 & 5 \end{pmatrix}$ and $C = \begin{pmatrix} 1 & 4 \\ 5 & 1 \end{pmatrix}$.

Evaluate both $(A + B)^2$ and $A^2 + 2AB + B^2$. Explain why they are not equal. Can you suggest the correct expansion for $(A + B)^2$? Show that your suggestion is correct for these matrices.

9. Find real numbers a and b for which

(a) $(a \ 3)\begin{pmatrix} 2 & 1 \\ -3 & 2 \end{pmatrix} = (b \ 11)$; (b) $(4 \ 1)\begin{pmatrix} a & -b \\ b & a \end{pmatrix} = (1 \ -21)$;

(c) $\begin{pmatrix} 6 & 5 \\ -2 & -3 \end{pmatrix}\begin{pmatrix} a \\ b \end{pmatrix} = \begin{pmatrix} 14 \\ -10 \end{pmatrix}$; (d) $\begin{pmatrix} a & 0 \\ b & -2 \end{pmatrix}\begin{pmatrix} 1 & 3 \\ -2 & 4 \end{pmatrix} = \begin{pmatrix} -2 & -6 \\ 4 & -8 \end{pmatrix}$.

10. If $A = \begin{pmatrix} a & b \\ c & d \end{pmatrix}$ and $I = \begin{pmatrix} 1 & 0 \\ 0 & 1 \end{pmatrix}$, show that $A^2 - (a + d)A + (ad - bc)I = 0$.

Chapter 9

11. The percentages by mass of silver, lead and zinc in four samples of ore are given by the matrices $A = \begin{pmatrix} 0.5 \\ 2.2 \\ 1.5 \end{pmatrix}$, $B = \begin{pmatrix} 0.2 \\ 2.5 \\ 1.0 \end{pmatrix}$, $C = \begin{pmatrix} 0.1 \\ 3.1 \\ 0.9 \end{pmatrix}$ and $D = \begin{pmatrix} 0.3 \\ 2.6 \\ 1.4 \end{pmatrix}$. If these samples have masses of 12kg, 8kg, 10kg and 6kg respectively, find the matrix $\frac{1}{36}(12A + 8B + 10C + 6D)$ and interpret the result.

12. A company manufactures three different TV sets – Types A, B and C. Each set requires one or more of each of three components – X, Y and Z.
Type A requires 5 of X, 3 of Y and 1 of Z ;
Type B requires 4 of X, 4 of Y and 2 of Z ;
Type C requires 6 of X, 2 of Y and 3 of Z.
Each component X costs $2.50, each component Y costs $4.20 and each component Z costs $3.00, and in a given week the company manufactures 10 sets of type A, 8 sets of type B and 12 sets of type C.

Let L, M, N be the matrices $L = \begin{pmatrix} 2.5 & 4.2 & 3 \end{pmatrix}$, $M = \begin{pmatrix} 5 & 4 & 6 \\ 3 & 4 & 2 \\ 1 & 2 & 3 \end{pmatrix}$, $N = \begin{pmatrix} 10 \\ 8 \\ 12 \end{pmatrix}$.

(a) Find the matrix MN and interpret the result.
(b) Find the matrix LMN and interpret the result.

13. If $A = \begin{pmatrix} 1 & 3 & -2 \\ 2 & 1 & 4 \\ 3 & 2 & 1 \end{pmatrix}$ and $B = \begin{pmatrix} -7 & -7 & 14 \\ 10 & 7 & -8 \\ 1 & 7 & -5 \end{pmatrix}$, show that $AB = BA$.

14. If $A = \frac{1}{\sqrt{2}}\begin{pmatrix} 1 & -1 \\ 1 & 1 \end{pmatrix}$, show that $A^4 = \begin{pmatrix} -1 & 0 \\ 0 & -1 \end{pmatrix}$.

15. (a) If $A = \frac{1}{2}\begin{pmatrix} -1 & -\sqrt{3} \\ \sqrt{3} & -1 \end{pmatrix}$, evaluate A^3.

    (b) If $B = \frac{1}{2}\begin{pmatrix} 1 & -\sqrt{3} \\ \sqrt{3} & 1 \end{pmatrix}$, evaluate B^6.

16. If $A = \begin{pmatrix} 2 & 1 \\ 3 & 1 \end{pmatrix}$ and $I = \begin{pmatrix} 1 & 0 \\ 0 & 1 \end{pmatrix}$, show that $A^2 - 3A = I$. Use the result that $AI = IA = A$ to find an expression for A^4 in terms of A and I.

Matrices & Linear Equations

Higher Level

17. Find the matrix B for which $AB = I$ where $A = \begin{pmatrix} 5 & 4 \\ 4 & 3 \end{pmatrix}$ and $I = \begin{pmatrix} 1 & 0 \\ 0 & 1 \end{pmatrix}$.

18. Show that it is not possible to find a matrix B for which $AB = I$ in the case where $A = \begin{pmatrix} 6 & 4 \\ 3 & 2 \end{pmatrix}$ and $I = \begin{pmatrix} 1 & 0 \\ 0 & 1 \end{pmatrix}$.

19. Let $A = \begin{pmatrix} a & 1 \\ -1 & b \end{pmatrix}$ where a and b are scalars. Find a and b if

    (a) $A^2 = O$; (b) $A^2 = \begin{pmatrix} 1 & 0 \\ 0 & 1 \end{pmatrix}$; (c) $A^2 = A$.

20. If $A = \begin{pmatrix} 3 & -4 \\ 1 & -2 \end{pmatrix}$ and $X = \begin{pmatrix} x \\ y \end{pmatrix}$ where x and y are not both zero, find the values of k for which $AX = kX$ and for each value of k find a suitable matrix X.

9.3 The Determinant of a (Square) Matrix

The Determinant of a 2 × 2 Matrix

Definition The *determinant* of the 2 × 2 matrix $A = \begin{pmatrix} a & b \\ c & d \end{pmatrix}$ is denoted and defined by $\det A = \begin{vmatrix} a & b \\ c & d \end{vmatrix} = ad - bc$.

Clearly the determinant is a scalar (number).

Example Find the determinant of the matrix $A = \begin{pmatrix} 3 & 5 \\ -2 & -1 \end{pmatrix}$.

$\det A = 3 \times (-1) - (-2) \times 5 = -3 + 10 = 7$.

Definition A matrix A for which $\det A = 0$ is called a *singular* matrix.

Example For what values of a is the matrix $A = \begin{pmatrix} 3-a & 8 \\ 1 & -4-a \end{pmatrix}$ singular?

Chapter 9

$\det A = (3-a)(-4-a) - 8 = a^2 + a - 20 = (a+5)(a-4)$
Thus A is singular when $a = -5$ or $a = 4$.

The Determinant of a 3 × 3 Matrix

Consider the matrix $A = \begin{pmatrix} a_{11} & a_{12} & a_{13} \\ a_{21} & a_{22} & a_{23} \\ a_{31} & a_{32} & a_{33} \end{pmatrix}$.

Before we look at the general method, it is helpful to define what is meant by the *minor* of an element of a matrix.

Definition The *minor* of the element a_{ij} of matrix A, denoted by m_{ij}, is the determinant of the matrix left when the ith row and jth column of A are removed. Thus $m_{11} = \begin{vmatrix} a_{22} & a_{23} \\ a_{32} & a_{33} \end{vmatrix}$, $m_{12} = \begin{vmatrix} a_{21} & a_{23} \\ a_{31} & a_{33} \end{vmatrix}$, etc.

The determinant of A can be evaluated in 6 different ways (a graphic display calculator may also be used).

$$\det A = \sum_{k=1}^{3} (-1)^{i+k} a_{ik} m_{ik} \text{ for } i = 1, 2 \text{ or } 3 \qquad \text{[by rows]}$$

or $$\det A = \sum_{k=1}^{3} (-1)^{k+j} a_{kj} m_{kj} \text{ for } j = 1, 2 \text{ or } 3 \qquad \text{[by columns]}.$$

Thus $\det A = a_{11} m_{11} - a_{12} m_{12} + a_{13} m_{13}$ evaluating by the first row, or
$\det A = -a_{21} m_{21} + a_{22} m_{22} - a_{23} m_{23}$ evaluating by the second row, or
$\det A = a_{13} m_{13} - a_{23} m_{23} + a_{33} m_{33}$ evaluating by the third column, etc.

Note: If the sum of the row and column number of the element and its minor is *odd*, a negative sign is attached to the product of the matrix element and its minor. If the sum is *even*, a positive sign is used.

Example Evaluate the determinant of the matrix $A = \begin{pmatrix} 1 & 3 & -2 \\ 2 & 1 & 4 \\ 3 & 4 & 2 \end{pmatrix}$.

$\det A = \begin{vmatrix} 1 & 4 \\ 4 & 2 \end{vmatrix} - 3\begin{vmatrix} 2 & 4 \\ 3 & 2 \end{vmatrix} + (-2)\begin{vmatrix} 2 & 1 \\ 3 & 4 \end{vmatrix}$
$= (2 - 16) - 3(4 - 12) - 2(8 - 3)$
$= -14 + 24 - 10$
$= 0$.

Matrices & Linear Equations

The work required to evaluate a 3 × 3 determinant can be reduced considerably if a row or column contains one or more zeros among its entries.

Example   Find the value of the determinant $\begin{vmatrix} 2 & 5 & 0 \\ -3 & 2 & 3 \\ 4 & 1 & 0 \end{vmatrix}$.

Since the third column has two zero entries, we evaluate the determinant by the third column.

This gives $\begin{vmatrix} 2 & 5 & 0 \\ -3 & 2 & 3 \\ 4 & 1 & 0 \end{vmatrix} = 0\begin{vmatrix} -3 & 2 \\ 4 & 1 \end{vmatrix} - 3\begin{vmatrix} 2 & 5 \\ 4 & 1 \end{vmatrix} + 0\begin{vmatrix} 2 & 5 \\ -3 & 2 \end{vmatrix}$

$= -3\begin{vmatrix} 2 & 5 \\ 4 & 1 \end{vmatrix}$

$= 54.$

Theorem   If A and B are any two square matrices of order n, then
$$\det(AB) = (\det A)(\det B).$$

Proof   ($n = 2$ only)

Let $A = \begin{pmatrix} a & b \\ c & d \end{pmatrix}$ and $B = \begin{pmatrix} e & f \\ g & h \end{pmatrix}$.

Then $\det A = ad - bc$, $\det B = eh - fg$ and $AB = \begin{pmatrix} ae + bg & af + bh \\ ce + dg & cf + dh \end{pmatrix}$.

$\det(AB) = (ae + bg)(cf + dh) - (af + bh)(ce + dg)$
$= acef + adeh + bcfg + bdgh - acef - adfg - bceh - bdgh$
$= adeh - adfg - bceh + bcfg$
$= ad(eh - fg) - bc(eh - fg)$
$= (ad - bc)(eh - fg)$
$= (\det A)(\det B).$

Theorem   If A is any square matrix of order n, then $\det(A^T) = \det A$.

Proof

If $n = 2$, let $A = \begin{pmatrix} a & b \\ c & d \end{pmatrix}$.

Then $A^T = \begin{pmatrix} a & c \\ b & d \end{pmatrix}$ and $\det(A^T) = ad - bc = \det A$.

215

Chapter 9

If $n = 3$, let $A = \begin{pmatrix} a_{11} & a_{12} & a_{13} \\ a_{21} & a_{22} & a_{23} \\ a_{31} & a_{32} & a_{33} \end{pmatrix}$ then $A^T = \begin{pmatrix} a_{11} & a_{21} & a_{31} \\ a_{12} & a_{22} & a_{32} \\ a_{13} & a_{23} & a_{33} \end{pmatrix}$.

Then $\det A^T = a_{11}\begin{vmatrix} a_{22} & a_{32} \\ a_{23} & a_{33} \end{vmatrix} - a_{21}\begin{vmatrix} a_{12} & a_{32} \\ a_{13} & a_{33} \end{vmatrix} + a_{31}\begin{vmatrix} a_{12} & a_{22} \\ a_{13} & a_{23} \end{vmatrix}$

$= a_{11}\begin{vmatrix} a_{22} & a_{23} \\ a_{32} & a_{33} \end{vmatrix} - a_{21}\begin{vmatrix} a_{12} & a_{13} \\ a_{32} & a_{33} \end{vmatrix} + a_{31}\begin{vmatrix} a_{12} & a_{13} \\ a_{22} & a_{23} \end{vmatrix}$

$= \det A.$

Note: $\det A = \sum_{k=1}^{n}(-1)^{i+k} a_{ik} m_{ik} = \sum_{k=1}^{n}(-1)^{k+i} a_{ki}^T m_{ki}^T = \det A^T$, $n \in \mathbb{Z}^+$.

The expression for $\det A$ is evaluated by the ith row and the expression for $\det A^T$ is evaluated by the ith column.

Exercise 9.3

1. Evaluate the determinant of each of the following matrices:

   (a) $\begin{pmatrix} 5 & -1 \\ 4 & 1 \end{pmatrix}$; (b) $\begin{pmatrix} -1 & 3 \\ 1 & -4 \end{pmatrix}$;

   (c) $\begin{pmatrix} \cos\alpha & -\sin\alpha \\ \sin\alpha & \cos\alpha \end{pmatrix}$; (d) $\begin{pmatrix} \cos\alpha & \sin\alpha \\ \sin\alpha & -\cos\alpha \end{pmatrix}$;

   (e) $\begin{pmatrix} 3 & 2 & -1 \\ 2 & 0 & 0 \\ -1 & 3 & -1 \end{pmatrix}$; (f) $\begin{pmatrix} -1 & 2 & 1 \\ 3 & 1 & -1 \\ 1 & 4 & 1 \end{pmatrix}$;

   (g) $\begin{pmatrix} 2 & -2 & -1 \\ 1 & 1 & -2 \\ 3 & 1 & -3 \end{pmatrix}$; (h) $\begin{pmatrix} 1 & 2 & 2 \\ -1 & 1 & -2 \\ 2 & 7 & 4 \end{pmatrix}$.

2. Show that the matrix $A = \begin{pmatrix} 1 & 2 & 3 \\ 4 & 5 & 6 \\ 7 & 8 & 9 \end{pmatrix}$ is singular.

3. If $A = \begin{pmatrix} k & -4 \\ 1 & 2 \end{pmatrix}$ and $B = \begin{pmatrix} k & 3 \\ 1 & k-2 \end{pmatrix}$, find the values of k for which the matrix product AB is singular.

Matrices & Linear Equations

4. Find the values of k for which $\begin{vmatrix} 1 & -2 & k \\ 2 & 1 & 2 \\ k & 4 & 1 \end{vmatrix} = 0$.

5. Find, in terms of λ, the determinant of the matrix
$$A = \begin{pmatrix} 2-\lambda & 1 & 3 \\ 1 & 1-\lambda & 1 \\ -1 & -1 & -2-\lambda \end{pmatrix}.$$

6. For what real value of k is the matrix $A = \begin{pmatrix} k-1 & -2 & 1 \\ -2 & k-1 & -1 \\ -1 & 6 & k+2 \end{pmatrix}$ singular?

Higher Level

7. Solve the equation $\begin{vmatrix} 2-x & 1 & -1 \\ 1 & -1-x & 2 \\ 8 & 2 & -1-x \end{vmatrix} = 0$ for $x \in \mathbb{R}$.

8. (a) If $A = \begin{pmatrix} 3 & 2 \\ -1 & 1 \end{pmatrix}$, find $\det(A^2)$ and $\det(2A)$.

   (b) If $A = \begin{pmatrix} 1 & 2 & 1 \\ 2 & -1 & 2 \\ 3 & 4 & 1 \end{pmatrix}$, find $\det(A^2)$ and $\det(2A)$.

9. If A is a square matrix of order n such that $\det A = x$, what is the value of
   (a) $\det(A^2)$;
   (b) $\det(A^m)$, $m \in \mathbb{Z}^+$?

10. If A is a square matrix of order n such that $\det A = x$, what is the value of
    (a) $\det(2A)$;
    (b) $\det(mA)$, $m \in \mathbb{R}$?

11. (a) Expand $(a-b)(b-c)(c-a)$.

    (b) If a, b and c are distinct real numbers show that the matrix
    $\begin{pmatrix} 1 & 1 & 1 \\ a & b & c \\ a^2 & b^2 & c^2 \end{pmatrix}$ is not singular.

217

Chapter 9

9.4 Matrix Algebra

Addition

Consider the set S of all square matrices of order n.

(a1) S is **closed** under addition. Thus the sum of any two $n \times n$ matrices is an $n \times n$ matrix.

(a2) S is **commutative** under addition. That is if A and B are any two $n \times n$ matrices, then $A + B = B + A$.

(a3) S is **associative** under addition. That is if A, B and C are any $n \times n$ matrices, then $A + (B + C) = (A + B) + C$.

(a4) There exists an additive **identity**, $O \in S$, such that for any $n \times n$ matrix A, $A + O = O + A = A$.

(a5) Each $n \times n$ matrix A has an **additive inverse** $n \times n$ matrix $(-A)$ such that $A + (-A) = (-A) + A = O$.

Multiplication

Consider the set S of all square matrices of order n.

(m1) S is **closed** under multiplication. That is the product of any two $n \times n$ matrices is an $n \times n$ matrix.

(m2) S is **associative** under multiplication. That is if A, B and C are any $n \times n$ matrices, $A(BC) = (AB)C$.

(m3) There exists a multiplicative **identity**, $I \in S$, such that for any $n \times n$ matrix A, $AI = IA = A$.

(m4) Matrix multiplication **distributes over addition**. That is if A, B and C are any $n \times n$ matrices, then $A(B + C) = AB + AC$.

Note: I from (m3) is called **the** identity matrix and has all elements on its "leading diagonal" equal to 1 and all other elements equal to zero. The 2×2 identity matrix is $I_{2 \times 2} = \begin{pmatrix} 1 & 0 \\ 0 & 1 \end{pmatrix}$ and the 3×3 identity matrix is $I_{3 \times 3} = \begin{pmatrix} 1 & 0 & 0 \\ 0 & 1 & 0 \\ 0 & 0 & 1 \end{pmatrix}$.

Matrices & Linear Equations

The following 'field axioms' are not generally satisfied by matrices in *S*:

(a) Commutativity of multiplication.
It is **not** generally true that if *A* and *B* are any two $n \times n$ matrices then $AB = BA$.

(b) Inverse under multiplication.
Not all $n \times n$ matrices *A* have multiplicative inverses and so it may not be possible to find an $n \times n$ matrix A^{-1} such that $AA^{-1} = A^{-1}A = I$.

Example For matrices $A = \begin{pmatrix} 2 & 1 \\ 1 & 2 \end{pmatrix}$ and $B = \begin{pmatrix} 3 & 2 \\ -2 & 1 \end{pmatrix}$, show that $AB \neq BA$.

$$AB = \begin{pmatrix} 2 & 1 \\ 1 & 2 \end{pmatrix}\begin{pmatrix} 3 & 2 \\ -2 & 1 \end{pmatrix} = \begin{pmatrix} 4 & 5 \\ -1 & 4 \end{pmatrix}$$

$$BA = \begin{pmatrix} 3 & 2 \\ -2 & 1 \end{pmatrix}\begin{pmatrix} 2 & 1 \\ 1 & 2 \end{pmatrix} = \begin{pmatrix} 8 & 7 \\ -3 & 0 \end{pmatrix}$$

Clearly $AB \neq BA$.

Example Show that the matrix $A = \begin{pmatrix} 2 & 3 \\ 4 & 6 \end{pmatrix}$ does not have an inverse but the matrix $B = \begin{pmatrix} 2 & 1 \\ 5 & 3 \end{pmatrix}$ does. Find the matrix B^{-1}.

Let the inverse of *A* be $A^{-1} = \begin{pmatrix} a & b \\ c & d \end{pmatrix}$, if it exists.

Then $AA^{-1} = I$ or $\begin{pmatrix} 2 & 3 \\ 4 & 6 \end{pmatrix}\begin{pmatrix} a & b \\ c & d \end{pmatrix} = \begin{pmatrix} 1 & 0 \\ 0 & 1 \end{pmatrix}$.

$\Rightarrow \begin{pmatrix} 2a+3c & 2b+3d \\ 4a+6c & 4b+6d \end{pmatrix} = \begin{pmatrix} 1 & 0 \\ 0 & 1 \end{pmatrix}$

$\Rightarrow \begin{cases} 2a+3c = 1 \\ 4a+6c = 0 \end{cases}$ and $\begin{cases} 2b+3d = 0 \\ 4b+6d = 1 \end{cases}$

$\Rightarrow \begin{cases} 2a+3c = 1 \\ 2a+3c = 0 \end{cases}$ and $\begin{cases} 2b+3d = 0 \\ 2b+3d = \frac{1}{2} \end{cases}$.

Since $2a + 3c$ cannot be both 1 and 0, and $2b + 3c$ cannot be both 0 and $\frac{1}{2}$, no values of *a*, *b*, *c* and *d* can be found. Thus *A* does not have an inverse.

Chapter 9

Let the inverse of B be $B^{-1} = \begin{pmatrix} e & f \\ g & h \end{pmatrix}$, if it exists.

Then $BB^{-1} = I$ or $\begin{pmatrix} 2 & 1 \\ 5 & 3 \end{pmatrix}\begin{pmatrix} e & f \\ g & h \end{pmatrix} = \begin{pmatrix} 1 & 0 \\ 0 & 1 \end{pmatrix}$.

$\Rightarrow \begin{pmatrix} 2e+g & 2f+h \\ 5e+3g & 5f+3h \end{pmatrix} = \begin{pmatrix} 1 & 0 \\ 0 & 1 \end{pmatrix}$

$\Rightarrow \begin{cases} 2e+g=1 \\ 5e+3g=0 \end{cases}$ and $\begin{cases} 2f+h=0 \\ 5f+3h=1 \end{cases}$

Solving these simultaneous equations gives $e = 3, f = -1, g = -5, h = 2$.

Therefore B has an inverse and $B^{-1} = \begin{pmatrix} 3 & -1 \\ -5 & 2 \end{pmatrix}$.

Example For matrices $A = \begin{pmatrix} 2 & 3 \\ 4 & 6 \end{pmatrix}$ and $B = \begin{pmatrix} 3 & 6 \\ -2 & -4 \end{pmatrix}$, find AB.

$AB = \begin{pmatrix} 2 & 3 \\ 4 & 6 \end{pmatrix}\begin{pmatrix} 3 & 6 \\ -2 & -4 \end{pmatrix} = \begin{pmatrix} 0 & 0 \\ 0 & 0 \end{pmatrix} = O$

Note: From this example we see that if $AB = O$, we cannot say that $A = O$ or $B = O$. However, the following 'cancellation' theorem should be known.

Theorem If $AB = O$ and the inverse of A exists, then $B = O$.

Proof $AB = O$

$\Rightarrow \quad A^{-1}(AB) = (A^{-1})O \quad$ since A^{-1} exists and S is closed under multiplication

$\Rightarrow \quad (A^{-1}A)B = O \quad$ since matrix multiplication is associative

$\Rightarrow \quad IB = O$

$\Rightarrow \quad B = O$.

The Existence of a (Multiplicative) Inverse of a 2 × 2 Matrix

Consider the 2 × 2 matrix $A = \begin{pmatrix} a & b \\ c & d \end{pmatrix}$ and its inverse, if it exists, $A^{-1} = \begin{pmatrix} e & f \\ g & h \end{pmatrix}$.

Then $AA^{-1} = \begin{pmatrix} a & b \\ c & d \end{pmatrix}\begin{pmatrix} e & f \\ g & h \end{pmatrix} = \begin{pmatrix} ae+bg & af+bh \\ ce+dg & cf+dh \end{pmatrix} = \begin{pmatrix} 1 & 0 \\ 0 & 1 \end{pmatrix}$.

Thus $ae + bg = 1$ .... (i) $\qquad af + bh = 0$ .... (iii)
$\quad\quad ce + dg = 0$ .... (ii) $\qquad cf + dh = 1$ .... (iv).

Multiply (i) by c: $ace + bcg = c$ .... (v).
Multiply (ii) by a: $ace + adg = 0$ .... (vi).

Subtract (v) from (vi): $(ad - bc)g = -c$ so $g = \dfrac{-c}{ad - bc}$ provided $ad - bc \neq 0$.

Multiply (i) by d: $ade + bdg = d$ .... (vii).
Multiply (ii) by b: $bce + bdg = 0$ .... (viii).

Subtract (viii) from (vii): $(ad - bc)e = d$ so $e = \dfrac{d}{ad - bc}$ provided $ad - bc \neq 0$.

Since equations (iii) and (iv) are essentially the same as equations (i) and (ii) with a replaced by c, b replaced by d, e replaced by f and g replaced by h, the solutions for f and h are $f = \dfrac{b}{cb - da} = \dfrac{-b}{ad - bc}$ and $h = \dfrac{-a}{cb - da} = \dfrac{a}{ad - bc}$ provided $ad - bc \neq 0$.

Thus $A^{-1} = \begin{pmatrix} \dfrac{d}{ad - bc} & \dfrac{-b}{ad - bc} \\ \dfrac{-c}{ad - bc} & \dfrac{a}{ad - bc} \end{pmatrix} = \dfrac{1}{ad - bc} \begin{pmatrix} d & -b \\ -c & a \end{pmatrix}$ provided $ad - bc \neq 0$.

But $ad - bc = \det A$ so $A^{-1} = \dfrac{1}{\det A} \begin{pmatrix} d & -b \\ -c & a \end{pmatrix}$ provided $\det A \neq 0$.

Thus the inverse of A exists provided A is not singular.

If $\det A \neq 0$, then the inverse of the 2×2 matrix A is found by multiplying by $1/\det A$ that matrix formed by interchanging the elements on the 'leading diagonal' and changing the signs of the elements on the 'off diagonal'.

Example   Find the inverse of the matrix $A = \begin{pmatrix} 8 & 3 \\ 6 & 2 \end{pmatrix}$.

$$\det A = 16 - 18 = -2 \quad \Rightarrow \quad A^{-1} = \dfrac{1}{-2} \begin{pmatrix} 2 & -3 \\ -6 & 8 \end{pmatrix} = \begin{pmatrix} -1 & 1\tfrac{1}{2} \\ 3 & -4 \end{pmatrix}.$$

Higher Level

The Inverse of a 3 × 3 Matrix (Optional)

Finding the inverse of a 3×3 matrix is most easily done using a graphic display calculator, however we should know how to accomplish the task 'by hand'.

Consider the matrix $A = \begin{pmatrix} a_{11} & a_{12} & a_{13} \\ a_{21} & a_{22} & a_{23} \\ a_{31} & a_{32} & a_{33} \end{pmatrix}$.

Chapter 9

Definition The *cofactor* of the element a_{ij} is defined by $c_{ij} = (-1)^{i+j} m_{ij}$ where m_{ij} is the minor defined in section 9.3.

Thus cofactors are simply 'signed minors'.

Definition The *adjoint* of matrix A is the transpose of the matrix of its cofactors. Thus $\text{adj } A = (c_{ij})^T$.

Definition The *inverse* of matrix A is $A^{-1} = \dfrac{1}{\det A} \text{adj } A$ provided $\det A \neq 0$.

Example Find the inverse of the matrix $A = \begin{pmatrix} 1 & 2 & 3 \\ 2 & 0 & 1 \\ -1 & -3 & -4 \end{pmatrix}$.

$$\det A = -2 \begin{vmatrix} 2 & 3 \\ -3 & -4 \end{vmatrix} - \begin{vmatrix} 1 & 2 \\ -1 & -3 \end{vmatrix} = -2 + 1 = -1$$

$c_{11} = \begin{vmatrix} 0 & 1 \\ -3 & -4 \end{vmatrix} = 3 \qquad c_{12} = -\begin{vmatrix} 2 & 1 \\ -1 & -4 \end{vmatrix} = 7 \qquad c_{13} = \begin{vmatrix} 2 & 0 \\ -1 & -3 \end{vmatrix} = -6$

$c_{21} = -\begin{vmatrix} 2 & 3 \\ -3 & -4 \end{vmatrix} = -1 \qquad c_{22} = \begin{vmatrix} 1 & 3 \\ -1 & -4 \end{vmatrix} = -1 \qquad c_{23} = -\begin{vmatrix} 1 & 2 \\ -1 & -3 \end{vmatrix} = 1$

$c_{31} = \begin{vmatrix} 2 & 3 \\ 0 & 1 \end{vmatrix} = 2 \qquad c_{32} = -\begin{vmatrix} 1 & 3 \\ 2 & 1 \end{vmatrix} = 5 \qquad c_{33} = \begin{vmatrix} 1 & 2 \\ 2 & 0 \end{vmatrix} = -4$

Therefore $A^{-1} = -\begin{pmatrix} 3 & -1 & 2 \\ 7 & -1 & 5 \\ -6 & 1 & -4 \end{pmatrix} = \begin{pmatrix} -3 & 1 & -2 \\ -7 & 1 & -5 \\ 6 & -1 & 4 \end{pmatrix}$.

Sometimes the inverse of a 3×3 matrix can easily be found from information given in the question.

Example If $A = \begin{pmatrix} 1 & 2 & -2 \\ 1 & 1 & 1 \\ 2 & -1 & -2 \end{pmatrix}$, evaluate A^3 and **hence** find A^{-1}.

$A^2 = \begin{pmatrix} 1 & 2 & -2 \\ 1 & 1 & 1 \\ 2 & -1 & -2 \end{pmatrix} \begin{pmatrix} 1 & 2 & -2 \\ 1 & 1 & 1 \\ 2 & -1 & -2 \end{pmatrix} = \begin{pmatrix} -1 & 6 & 4 \\ 4 & 2 & -3 \\ -3 & 5 & -1 \end{pmatrix}$

$$A^3 = \begin{pmatrix} 1 & 2 & -2 \\ 1 & 1 & 1 \\ 2 & -1 & -2 \end{pmatrix} \begin{pmatrix} -1 & 6 & 4 \\ 4 & 2 & -3 \\ -3 & 5 & -1 \end{pmatrix} = \begin{pmatrix} 13 & 0 & 0 \\ 0 & 13 & 0 \\ 0 & 0 & 13 \end{pmatrix}$$

Therefore $A^3 = 13I$ which gives $A(\frac{1}{13}A^2) = I$

$$\Rightarrow \quad A^{-1} = \frac{1}{13}A^2 = \frac{1}{13}\begin{pmatrix} -1 & 6 & 4 \\ 4 & 2 & -3 \\ -3 & 5 & -1 \end{pmatrix}.$$

Theorem If $AB = I$ then $B = A^{-1}$. (There is no need to show that $BA = I$.)

Proof $AB = I \Rightarrow \det(AB) = \det I \Rightarrow (\det A)(\det B) = 1 \Rightarrow \det A \neq 0$.
Therefore A^{-1} exists.
Hence $A^{-1}(AB) = A^{-1}I$
$(A^{-1}A)B = A^{-1}$
$IB = A^{-1}$
$B = A^{-1}$.

Theorem $(AB)^{-1} = B^{-1}A^{-1}$

Proof $(AB)(B^{-1}A^{-1}) = A\{(BB^{-1})A^{-1}\} = A\{IA^{-1}\} = AA^{-1} = I$
Thus from the previous theorem $(AB)^{-1} = B^{-1}A^{-1}$.

Example Find the matrix A such that

(a) $\begin{pmatrix} 3 & 5 \\ 4 & 7 \end{pmatrix} A = \begin{pmatrix} 2 & -1 \\ 2 & -5 \end{pmatrix}$;   (b) $\begin{pmatrix} 7 & 2 \\ 9 & -4 \end{pmatrix} A = \begin{pmatrix} 11 \\ 47 \end{pmatrix}$.

(a) Let $B = \begin{pmatrix} 3 & 5 \\ 4 & 7 \end{pmatrix}$ then $B^{-1} = \begin{pmatrix} 7 & -5 \\ -4 & 3 \end{pmatrix}$ and $BA = \begin{pmatrix} 2 & -1 \\ 2 & -5 \end{pmatrix}$.

Therefore $B^{-1}(BA) = \begin{pmatrix} 7 & -5 \\ -4 & 3 \end{pmatrix}\begin{pmatrix} 2 & -1 \\ 2 & -5 \end{pmatrix} = \begin{pmatrix} 4 & 18 \\ -2 & -11 \end{pmatrix}$.

$\Rightarrow \quad (B^{-1}B)A = IA = A = \begin{pmatrix} 4 & 18 \\ -2 & -11 \end{pmatrix}$.

Chapter 9

(b) Let $B = \begin{pmatrix} 7 & 2 \\ 9 & -4 \end{pmatrix}$ then $B^{-1} = -\dfrac{1}{46}\begin{pmatrix} -4 & -2 \\ -9 & 7 \end{pmatrix}$ and $BA = \begin{pmatrix} 11 \\ 47 \end{pmatrix}$.

Therefore, $B^{-1}(BA) = -\dfrac{1}{46}\begin{pmatrix} -4 & -2 \\ -9 & 7 \end{pmatrix}\begin{pmatrix} 11 \\ 47 \end{pmatrix} = \begin{pmatrix} 3 \\ -5 \end{pmatrix}$.

$\Rightarrow \qquad (B^{-1}B)A = IA = A = \begin{pmatrix} 3 \\ -5 \end{pmatrix}$.

Example Given matrices $A = \begin{pmatrix} 2 & -2 \\ 1 & 4 \end{pmatrix}$ and $B = \begin{pmatrix} 3 & 2 \\ 5 & 4 \end{pmatrix}$, find $\det\{(AB)^{-1}\}$.

$\det\{(AB)^{-1}\} = \dfrac{1}{\det(AB)} = \dfrac{1}{(\det A)(\det B)} = \dfrac{1}{10 \times 2} = \dfrac{1}{20}$.

Exercise 9.4

1. Find the inverse of each of the following matrices:

   (a) $\begin{pmatrix} 5 & 2 \\ 7 & 3 \end{pmatrix}$;

   (b) $\begin{pmatrix} 4 & 3 \\ 11 & 8 \end{pmatrix}$;

   (c) $\begin{pmatrix} -5 & 3 \\ -3 & 2 \end{pmatrix}$;

   (d) $\begin{pmatrix} 2 & 5 \\ 2 & 4 \end{pmatrix}$;

   (e) $\begin{pmatrix} 3 & 0 \\ 6 & -1 \end{pmatrix}$;

   (f) $\begin{pmatrix} \cos\alpha & -\sin\alpha \\ \sin\alpha & \cos\alpha \end{pmatrix}$;

   (g) $\begin{pmatrix} \cos\alpha & \sin\alpha \\ \sin\alpha & -\cos\alpha \end{pmatrix}$;

   (h) $\begin{pmatrix} \dfrac{1-m^2}{1+m^2} & \dfrac{2m}{1+m^2} \\ \dfrac{2m}{1+m^2} & -\dfrac{1-m^2}{1+m^2} \end{pmatrix}$.

2. If $A = \begin{pmatrix} 7 & 2 \\ 10 & 3 \end{pmatrix}$ and $B = \begin{pmatrix} -4 & 5 \\ 4 & -6 \end{pmatrix}$, find

   (a) A^{-1};   (b) B^{-1};   (c) $A^{-1}B^{-1}$;
   (d) $B^{-1}A^{-1}$;   (e) $(AB)^{-1}$;   (f) $(BA)^{-1}$.

3. If $A = \begin{pmatrix} 2 & 4 \\ 6 & 8 \end{pmatrix}$ and $B = \begin{pmatrix} 3 & -2 \\ -5 & -2 \end{pmatrix}$, find the value of $\det\{(AB)^{-1}\}$.

Matrices & Linear Equations

4. Show that the matrix $A = \dfrac{1}{13}\begin{pmatrix} 5 & 12 \\ 12 & -5 \end{pmatrix}$ is its own inverse.

5. If A is the matrix $\dfrac{1}{2}\begin{pmatrix} -1 & -\sqrt{3} \\ \sqrt{3} & -1 \end{pmatrix}$, find A^3. Hence, or otherwise, find the inverse of A.

6. Given $A = \begin{pmatrix} 6 & -5 \\ 5 & -4 \end{pmatrix}$, show that $A^2 - 2A + I = 0$. Hence deduce A^{-1}.

7. In each of the following, find the matrix A:

   (a) $\begin{pmatrix} 2 & 5 \\ 1 & 3 \end{pmatrix} A = \begin{pmatrix} 1 & -2 \\ 2 & 1 \end{pmatrix}$; (b) $\begin{pmatrix} 9 & -5 \\ 6 & 2 \end{pmatrix} A = \begin{pmatrix} -1 \\ 10 \end{pmatrix}$;

   (c) $A\begin{pmatrix} 6 & -7 \\ -3 & 4 \end{pmatrix} = \begin{pmatrix} 0 & 2 \\ 3 & -5 \end{pmatrix}$; (d) $A\begin{pmatrix} 2 & -7 \\ 5 & 4 \end{pmatrix} = \begin{pmatrix} 11 & -17 \end{pmatrix}$;

   (e) $\begin{pmatrix} 7 & 2 \\ 4 & 1 \end{pmatrix} A \begin{pmatrix} 4 & -7 \\ -3 & 5 \end{pmatrix} = \begin{pmatrix} 2 & 1 \\ 0 & -2 \end{pmatrix}$.

8. Given $A = \begin{pmatrix} 2 & 1 & 2 \\ 2 & -1 & 5 \\ 0 & -1 & -1 \end{pmatrix}$, evaluate A^3 and deduce A^{-1}.

9. Given $A = \begin{pmatrix} 1 & 2 & -1 \\ 0 & 1 & 2 \\ -1 & 0 & -2 \end{pmatrix}$,

   (a) show that $A^3 - 4A + 7I = 0$ and deduce A^{-1};

   (b) find the matrix X such that $AX = \begin{pmatrix} 2 \\ 1 \\ -7 \end{pmatrix}$.

10. If $A = \begin{pmatrix} 2 & -1 & 1 \\ 1 & 0 & -2 \\ -1 & 1 & -2 \end{pmatrix}$, find A^3 and A^{-1}, and hence find a matrix X such that

    $\begin{pmatrix} 2 & -1 & 2 \\ 4 & -3 & 5 \\ 1 & -1 & 1 \end{pmatrix} X = \begin{pmatrix} 3 \\ 5 \\ 2 \end{pmatrix}$.

225

Chapter 9

11. Let $A = \begin{pmatrix} 1 & 1 & 1 \\ 2 & 0 & 1 \\ -2 & -1 & -2 \end{pmatrix}$. Find A^2 and write down the inverse of A.

12. Show that the inverse of the matrix $A = \begin{pmatrix} 3 & 2 & 2 \\ 1 & -2 & 1 \\ -1 & 1 & -1 \end{pmatrix}$ is the matrix $B = A^2 - 8I$.

13. Let $A = \begin{pmatrix} 1 & 3 & -3 \\ 2 & 1 & 1 \\ 4 & 5 & -2 \end{pmatrix}$ and let $B = 2I - A^2$. Evaluate AB and find A^{-1}.

14. Given $A = \begin{pmatrix} 2 & 1 & -1 \\ 1 & -1 & 2 \\ 8 & 2 & -1 \end{pmatrix}$, find A^3 and deduce the inverse of A.

15. If $A = \begin{pmatrix} 2 & 3 & -4 \\ 3 & 2 & -4 \\ 3 & 3 & -5 \end{pmatrix}$, evaluate A^2 and find the inverse of A.

Higher Level (Optional)

16. Find the inverses of the following matrices:

(a) $\begin{pmatrix} 1 & -1 & 1 \\ -3 & 1 & 2 \\ 1 & 2 & -6 \end{pmatrix}$; (b) $\begin{pmatrix} 2 & 1 & 3 \\ 3 & 1 & 2 \\ 1 & 2 & 8 \end{pmatrix}$;

(c) $\begin{pmatrix} 3 & -2 & -4 \\ 2 & 0 & -1 \\ -4 & 1 & 4 \end{pmatrix}$; (d) $\begin{pmatrix} -2 & 0 & 2 \\ 1 & 2 & 2 \\ -2 & 1 & 4 \end{pmatrix}$.

17. Solve the equation $Ax = b$ where $A = \begin{pmatrix} 2 & -1 & 2 \\ 1 & -1 & 1 \\ 3 & 2 & 4 \end{pmatrix}$ and $b = \begin{pmatrix} 16 \\ 9 \\ 21 \end{pmatrix}$.

9.5 Systems of Linear Equations

Consider the following system of linear equations:

$$a_{11}x_1 + a_{12}x_2 + a_{13}x_3 + \cdots + a_{1n}x_n = b_1$$
$$a_{21}x_1 + a_{22}x_2 + a_{23}x_3 + \cdots + a_{2n}x_n = b_2$$
$$a_{31}x_1 + a_{32}x_2 + a_{33}x_3 + \cdots + a_{3n}x_n = b_3$$
$$\vdots$$
$$a_{m1}x_1 + a_{m2}x_2 + a_{m3}x_3 + \cdots + a_{mn}x_n = b_m$$

There are m equations in n unknowns $(x_1, x_2, x_3, \cdots, x_n)$. The a's are called the **coefficients** and the b's are called the **constants**.

A **solution** of the system is any set of values for $x_1, x_2, x_3, \cdots, x_n$ which satisfies each of the m equations.

In solving such a system, any one of three possiblities may be encountered.

1. A unique solution exists. That is there is one set, and only one set, of values for $x_1, x_2, x_3, \cdots, x_n$ which satisfies all m equations.
2. An infinite number of solutions exist. That is there is an infinite number of sets of values for $x_1, x_2, x_3, \cdots, x_n$ which satisfy all m equations.
3. No solution exists. That is no set of values for $x_1, x_2, x_3, \cdots, x_n$ satisfies all m equations.

Definition   A system of linear equations in which all the constants are zero is said to be **homogeneous**.

In the case of a homogeneous system of equations, possibility 3 cannot occur since $x_1 = x_2 = x_3 = \cdots = x_n = 0$ is a trivial solution.

Note that the given equations can be presented in matrix form:

$$\begin{pmatrix} a_{11} & a_{12} & a_{13} & \cdots & a_{1n} \\ a_{21} & a_{22} & a_{23} & \cdots & a_{2n} \\ a_{31} & a_{32} & a_{33} & \cdots & a_{3n} \\ \vdots & \vdots & \vdots & \vdots & \vdots \\ a_{m1} & a_{m2} & a_{m3} & \cdots & a_{mn} \end{pmatrix} \begin{pmatrix} x_1 \\ x_2 \\ x_3 \\ \vdots \\ x_n \end{pmatrix} = \begin{pmatrix} b_1 \\ b_2 \\ b_3 \\ \vdots \\ b_m \end{pmatrix}$$

which is in the form $AX = B$ or $Ax = b$ where A is called **the matrix of the system**.

Chapter 9

The matrix
$$\begin{pmatrix} a_{11} & a_{12} & a_{13} & \cdots & a_{1n} & b_1 \\ a_{21} & a_{22} & a_{23} & \cdots & a_{2n} & b_2 \\ a_{31} & a_{32} & a_{33} & \cdots & a_{3n} & b_3 \\ \vdots & \vdots & \vdots & \vdots & \vdots & \vdots \\ a_{m1} & a_{m2} & a_{m3} & \cdots & a_{mn} & b_m \end{pmatrix}$$

is called *the augmented matrix* of the system.

Example   Consider the system
$$\begin{aligned} x + y + z &= 3 \\ 3x - y + z &= -1 \\ 4x + 2y + 3z &= 7. \end{aligned}$$

There are 3 equations in 3 unknowns. The unknowns are x, y, z. The matrix of the system is $\begin{pmatrix} 1 & 1 & 1 \\ 3 & -1 & 1 \\ 4 & 2 & 3 \end{pmatrix}$ and the augmented matrix is $\begin{pmatrix} 1 & 1 & 1 & 3 \\ 3 & -1 & 1 & -1 \\ 4 & 2 & 3 & 7 \end{pmatrix}$.

The constants are 3, –1, 7.

It can be shown that this system has an infinite number of solutions which are given by $x = t$, $y = t + 2$, $z = 1 - 2t$ for all real values of t.

Methods of Solution

Method 1   *Gaussian Elimination (Row Operations)*

Given a system of linear equations, S, we obtain systems which are equivalent to S (i.e. they have the same set of solutions as S), but which are progressively simpler, until we reach a system from which all solutions (if any) may be read off. To do this, we can use any of the following **basic row operations**.

(i)   Interchange any two rows.
(ii) Multiply any row by a non-zero constant.
(iii) Add to any row any multiple of any other row.

Example   Solve the following system of equations:
$$\begin{aligned} R_1: & \quad x + 2y + 5z = 4 \\ R_2: & \quad 3x + 5y + 9z = 7 \\ R_3: & \quad 2x - y - 3z = 0. \end{aligned}$$

Matrices & Linear Equations

Procedure: (i) Replace R_2 by $R_2 - 3R_1$ and R_3 by $R_3 - 2R_1$.
This gives the equivalent system
$R_1:$ $\quad x + 2y + 5z = 4$
$R_2:$ $\quad -y - 6z = -5$
$R_3:$ $\quad -5y - 13z = -8$

(ii) Replace R_3 by $R_3 - 5R_2$, This gives
$R_1:$ $\quad x + 2y + 5z = 4$
$R_2:$ $\quad -y - 6z = -5$
$R_3:$ $\quad 17z = 17.$

The solution can now be read off by the method of *back substitution*.
From R_3, $z = 1$. Then substituting this in R_2 gives $y = -1$.
Finally, substituting these values for y and z in R_1 gives $x = 1$.
Therefore the solution is $x = 1, y = -1, z = 1$.

Example Solve the system
$R_1:$ $\quad x + 2y + 5z = 4$
$R_2:$ $\quad 3x - 2y - z = 4$
$R_3:$ $\quad 2x + 5y + 12z = 9.$

Replace R_2 by $R_2 - 3R_1$ and R_3 by $R_3 - 2R_1$.
This gives
$R_1:$ $\quad x + 2y + 5z = 4$
$R_2:$ $\quad -8y - 16z = -8$
$R_3:$ $\quad y + 2z = 1$

Replace R_2 by $-\frac{1}{8}R_2$.
This gives
$R_1:$ $\quad x + 2y + 5z = 4$
$R_2:$ $\quad y + 2z = 1$
$R_3:$ $\quad y + 2z = 1$

Replace R_3 by $R_3 - R_2$.
This gives
$R_1:$ $\quad x + 2y + 5z = 4$
$R_2:$ $\quad y + 2z = 1$
$R_3:$ $\quad 0 = 0$

Putting $z = t$, we obtain $y = 1 - 2t$ from R_2 and then $x = 2 - t$ from R_1.
Therefore the solution is $x = 2 - t, y = 1 - 2t, z = t$ for all real t.

229

Chapter 9

Since the process of reduction depends only on the coefficients on the left hand side and not on the names of the variables or on the right hand side, we may set the calculation out in ***detached coefficient form***.

Example   Solve the system
$$x + 2y - z = 9$$
$$2x + 3y + 2z = 19$$
$$3x - 2y + z = 7.$$

The system is

1	2	−1	9
2	3	2	19
3	−2	1	7

$R_2 \leftarrow R_2 - 2R_1$:
$R_3 \leftarrow R_3 - 3R_1$:

1	2	−1	9
0	−1	4	1
0	−8	4	−20

$R_3 \leftarrow R_3 - 8R_2$:

1	2	−1	9
0	−1	4	1
0	0	−28	−28

The third equation reads $-28z = -28$ and so $z = 1$.
The second equation reads $-y + 4z = 1$ and so $y = 3$ since $z = 1$.
The first equation reads $x + 2y - z = 9$ and so $x = 4$ since $y = 3$ and $z = 1$.
Hence the solution is $x = 4, y = 3, z = 1$.

Example   Solve the system
$$x + 3y - z = 4$$
$$2x + 5y + 2z = 9$$
$$3x + 8y + z = 14.$$

The system is

1	3	−1	4
2	5	2	9
3	8	1	14

$R_2 \leftarrow R_2 - 2R_1$:
$R_3 \leftarrow R_3 - 3R_1$:

1	3	−1	4
0	−1	4	1
0	−1	4	2

$R_3 \leftarrow R_3 - R_2$:

1	3	−1	4
0	−1	4	1
0	0	0	1

The last equation reads $0x + 0y + 0z = 1$ which is impossible and so no solution exists.

Method 2   Inverse Matrix Method

Consider the system $4x - 3y = 1$
$\quad 3x - 2y = 2.$

This system can be written in matrix form as $\begin{pmatrix} 4 & -3 \\ 3 & -2 \end{pmatrix}\begin{pmatrix} x \\ y \end{pmatrix} = \begin{pmatrix} 1 \\ 2 \end{pmatrix}$ which is in the form $AX = B$ where A is the 2×2 matrix of coefficients and X and B are each 2×1 column matrices. Now, since $AX = B \Rightarrow A^{-1}(AX) = A^{-1}B \Rightarrow (A^{-1}A)X = A^{-1}B \Rightarrow IX = A^{-1}B$ provided A^{-1} exists, the solution of the system is given by $X = A^{-1}B$.

We have $A = \begin{pmatrix} 4 & -3 \\ 3 & -2 \end{pmatrix}$ where $\det A = -8 + 9 = 1$ and $A^{-1} = \begin{pmatrix} -2 & 3 \\ -3 & 4 \end{pmatrix}$.

Therefore $X = \begin{pmatrix} x \\ y \end{pmatrix} = \begin{pmatrix} -2 & 3 \\ -3 & 4 \end{pmatrix}\begin{pmatrix} 1 \\ 2 \end{pmatrix} = \begin{pmatrix} 4 \\ 5 \end{pmatrix}$ or $x = 4$ and $y = 5$.

Note: If $\det A = 0$ (A^{-1} does not exist), there is no unique solution to the given system and this method cannot be used.

Example   Use this method to solve the system $5x - 7y = 37$
$\ 9x + 4y = 50.$

The solution is $\begin{pmatrix} x \\ y \end{pmatrix} = \begin{pmatrix} 5 & -7 \\ 9 & 4 \end{pmatrix}^{-1}\begin{pmatrix} 37 \\ 50 \end{pmatrix} = \frac{1}{83}\begin{pmatrix} 4 & 7 \\ -9 & 5 \end{pmatrix}\begin{pmatrix} 37 \\ 50 \end{pmatrix} = \begin{pmatrix} 6 \\ -1 \end{pmatrix}$.

This is $x = 6$ and $y = -1$.

Example   Solve the system $x + 2y + 4z = 0$
$\ 2x - y + 3z = 5$
$\ 4x - 3y + 5z = 11.$

$\begin{vmatrix} 1 & 2 & 4 \\ 2 & -1 & 3 \\ 4 & -3 & 5 \end{vmatrix} = -5 + 9 - 2(10 - 12) + 4(-6 + 4) = 0$ and so the inverse of the matrix of the system does not exist and the solution by matrix methods cannot be found. Gaussian elimination must be used.

The system is

1	2	4	0
2	-1	3	5
4	-3	5	11

Chapter 9

$$R_2 \leftarrow R_2 - 2R_1:$$
$$R_3 \leftarrow R_3 - 4R_1:$$

$$\begin{array}{cccc} 1 & 2 & 4 & 0 \\ 0 & -5 & -5 & 5 \\ 0 & -11 & -11 & 11 \\ \hline 1 & 2 & 4 & 0 \\ 0 & -5 & -5 & 5 \\ 0 & 0 & 0 & 0 \end{array}$$

$$R_3 \leftarrow R_3 - \tfrac{11}{5} R_2:$$

Put $z = t$ and from the second equation we get $y = -t - 1$. Then from the first equation we get $x = 2 - 2t$.

Therefore the solution is $x = 2 - 2t$, $y = -t - 1$ and $z = t$ for all real values of t.

Example Given $A = \begin{pmatrix} 2 & -2 & -1 \\ 1 & 1 & -2 \\ 3 & 1 & -3 \end{pmatrix}$, find A^3 and hence solve the equations:

(a) $2x - 2y - z = -18$
$x + y - 2z = -2$
$3x + y - 3z = -10$;

(b) $x + 7y - 5z = 9$
$x + y - z = 3$
$x + 4y - 2z = 6$.

$$A^2 = \begin{pmatrix} -1 & -7 & 5 \\ -3 & -3 & 3 \\ -2 & -8 & 4 \end{pmatrix} \quad \text{and} \quad A^3 = \begin{pmatrix} 6 & 0 & 0 \\ 0 & 6 & 0 \\ 0 & 0 & 6 \end{pmatrix} = 6I$$

Therefore $A^{-1} = \tfrac{1}{6} A^2$ and $(A^2)^{-1} = \tfrac{1}{6} A$.

(a) The system is $A \begin{pmatrix} x \\ y \\ z \end{pmatrix} = \begin{pmatrix} -18 \\ -2 \\ -10 \end{pmatrix}$ so

$$\begin{pmatrix} x \\ y \\ z \end{pmatrix} = A^{-1} \begin{pmatrix} -18 \\ -2 \\ -10 \end{pmatrix} = \tfrac{1}{6} \begin{pmatrix} -1 & -7 & 5 \\ -3 & -3 & 3 \\ -2 & -8 & 4 \end{pmatrix} \begin{pmatrix} -18 \\ -2 \\ -10 \end{pmatrix} = \begin{pmatrix} -3 \\ 5 \\ 2 \end{pmatrix}.$$

The solution is therefore $x = -3$, $y = 5$ and $z = 2$.

(b) The system is
$-x - 7y + 5z = -9$
$-3x - 3y + 3z = -9$
$-2x - 8y + 4z = -12$

Matrices & Linear Equations

which is $A^2 \begin{pmatrix} x \\ y \\ z \end{pmatrix} = \begin{pmatrix} -9 \\ -9 \\ -12 \end{pmatrix}$ and so

$$\begin{pmatrix} x \\ y \\ z \end{pmatrix} = (A^2)^{-1} \begin{pmatrix} -9 \\ -9 \\ -12 \end{pmatrix} = \frac{1}{6} \begin{pmatrix} 2 & -2 & -1 \\ 1 & 1 & -2 \\ 3 & 1 & -3 \end{pmatrix} \begin{pmatrix} -9 \\ -9 \\ -12 \end{pmatrix} = \begin{pmatrix} 2 \\ 1 \\ 0 \end{pmatrix}.$$

The solution is therefore $x = 2$, $y = 1$ and $z = 0$.

Exercise 9.5

1. Solve each of the following systems of linear equations by each of the two methods described:
   (a) $x + 2y = 13$
   $3x + y = 14$;
   (b) $3x + 4y = 10$
   $4x + y = 9$;
   (c) $6x - 5y = 28$
   $5x + 3y = 9$;
   (d) $14x - 3y = 39$
   $6x + 17y = 35$.

2. Find the values of k for which the system $\quad kx + 2y = 1$
   $3x + (k-1)y = 1$
   does not have a unique solution. If k does not have these values, find the unique solution. For each value of k for which no unique solution exists, determine whether or not any solution of the system exists.

3. Find the values of k for which the system $\quad x + 2y + 2z = 1$
   $2x + y + 2z = 4$
   $3x + 3y + kz = 5$
   has a unique solution. Find this unique solution and solve the system for any values of k for which the unique solution does not exist.

4. Solve each of the following systems of equations:

   (a) $x - y + 3z = 3$
   $2x - 3y + 2z = 1$
   $4x - 5y + 7z = 6$;
   (b) $x + y + z = 3$
   $2x + y - 2z = 0$
   $3x - 2y + 5z = 23$;

   (c) $x + 2y - 2z = 5$
   $3x + 5y + z = 2$
   $4x + 7y - 3z = 11$;
   (d) $x - y + 3z = 1$
   $x + 2y + 6z = 7$
   $2x + y + 9z = 8$;

   (e) $2x + 3y + z = 6$
   $5x + y - 4z = -4$
   $x - y - 2z = -2$;
   (f) $2x + y - 2z = 10$
   $x - 2y + z = 3$
   $x - y + 2z = 20$;

233

Chapter 9

(g) $3x + y - 2z = 6$
       $4x - 3y - 4z = 0$
       $5x + 2y + 6z = 6$;

(h) $6x + 5y - 2z = 7$
       $8x - 16y + 3z = 3$
       $2x - 5y + z = 1.$

5.   Show that the inverse of the matrix $\begin{pmatrix} 3 & 1 & -2 \\ 2 & -2 & 1 \\ 1 & 2 & -2 \end{pmatrix}$ is $\begin{pmatrix} -2 & 2 & 3 \\ -5 & 4 & 7 \\ -6 & 5 & 8 \end{pmatrix}$ and hence solve the following systems of equations:

    (a) $3x + y - 2z = 7$
           $2x - 2y + z = 7$
           $x + 2y - 2z = 1$;

    (b) $2x - 2y - 3z = 2$
           $5x - 4y - 7z = 1$
           $6x - 5y - 8z = 10.$

6.   Solve the system $x + 3y - 2z = 0$
                                 $x - 2y + 3z = 0$
                                    $- 2y + z = 1.$

7.   For the matrices $A = \begin{pmatrix} 1 & 2 & 2 \\ 2 & -1 & -4 \\ 2 & 2 & 1 \end{pmatrix}$ and $B = \begin{pmatrix} -7 & -2 & 6 \\ 10 & 3 & -8 \\ -6 & -2 & 5 \end{pmatrix}$ find AB. Use this result to solve the system $7x + 2y - 6z = 3$
                                                       $10x + 3y - 8z = 2$
                                                       $6x + 2y - 5z = k.$

8.   For the equations $x + 2y - z = b_1$
                                   $2x + 3y + 2z = b_2$
                                   $3x + 5y + z = b_3,$
find the condition on b_1, b_2, b_3 for which a solution exists.
Find all solutions of the system when $b_1 = 1, b_2 = 2$ and $b_3 = 3$.

9.   (a)   Find the values of x, y and z in terms of a, b and c for which
$$\begin{pmatrix} 1 & 2 & 3 \\ 2 & 5 & 8 \\ 4 & 7 & 11 \end{pmatrix} \begin{pmatrix} x \\ y \\ z \end{pmatrix} = \begin{pmatrix} a \\ b \\ c \end{pmatrix}.$$

    (b)   Use the result of part (a) to write down the inverse of $\begin{pmatrix} 1 & 2 & 3 \\ 2 & 5 & 8 \\ 4 & 7 & 11 \end{pmatrix}$.

Matrices & Linear Equations

10. (a) Solve the system of equations $2x + y - z = 0$
    $2x + 3y + 5z = 0$
    $5x + 4y + 2z = 0.$

   (b) Use the result of part (a) to find a set of values for x, y and z (not all zero) for which $AX = X$ where $A = \begin{pmatrix} 3 & 1 & -1 \\ 2 & 4 & 5 \\ 5 & 4 & 3 \end{pmatrix}$ and $X = \begin{pmatrix} x \\ y \\ z \end{pmatrix}$.

Higher Level

9.6 Geometrical Interpretation of Solutions (Optional)

Two Equations in two Unknowns

In general, the system $ax + by = e$
$cx + dy = f$
represents two straight lines in a plane and the solution of the system represents the coordinates of the point(s) where the two lines meet.

There are three possibilities:
(i)   The two lines meet in a single point – a unique solution.
(ii) The two lines coincide – an infinite number of solutions.
(iii) The two lines are parallel and distinct – no solution.

Since $A = \begin{pmatrix} a & b \\ c & d \end{pmatrix}$ is the matrix of the system, there is a unique solution iff $\det A \neq 0$. If $\det A = 0$, there will be either no solution or an infinite number of solutions.

Two Equations in three Unknowns

In general, the system $a_1 x + b_1 y + c_1 z = d_1$
$a_2 x + b_2 y + c_2 z = d_2$
represents two planes in 3-space and the solution of the system represents the coordinates of the points where the two planes meet.

There are three possibilities:
(i)   The two planes meet in a line – an infinite number of solutions.
(ii) The two planes coincide – an infinite number of solutions.
(iii) The two planes are parallel and distinct – no solution.

A unique solution is not possible in this case.

Chapter 9

Three Equations in three Unknowns

In general, the system
$$a_{11}x_1 + a_{12}x_2 + a_{13}x_3 = b_1$$
$$a_{21}x_1 + a_{22}x_2 + a_{23}x_3 = b_2$$
$$a_{31}x_1 + a_{32}x_2 + a_{33}x_3 = b_3$$

represents three planes in 3-space and the solution of the system represents the coordinates of all points which lie on each of the three planes.

There are eight possibilities:

(i) The three planes coincide – an infinite number of solutions.
(ii) Two of the planes coincide and the third is parallel to and distinct from the others – no solution.
(iii) Two of the planes coincide and the third is not parallel to them – an infinite number of solutions.
(iv) All three planes are parallel and distinct – no solution.
(v) Two of the planes are parallel and distinct and the third is not parallel to them – no solution.
(vi) The line of intersection of any two of the planes is parallel to and distinct from the third plane – no solution.

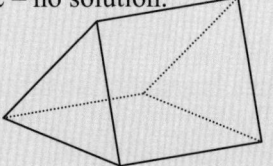

(vii) All three planes meet in a line – an infinite number of solutions.

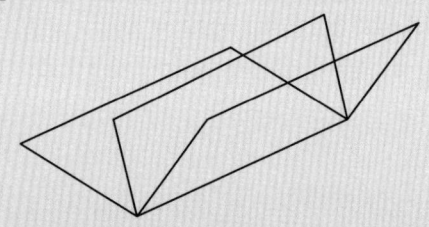

(viii) The three planes meet in a single point – a unique solution.

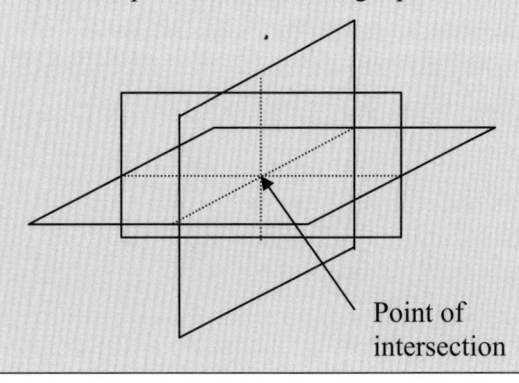

Point of intersection

236

Since $A = \begin{pmatrix} a_{11} & a_{12} & a_{13} \\ a_{21} & a_{22} & a_{23} \\ a_{31} & a_{32} & a_{33} \end{pmatrix}$ is the matrix of the system, there will be a unique solution [possibility (viii)] iff $\det A \neq 0$, and either an infinite number of solutions or no solution at all if $\det A = 0$.

In practice, any of possibilities (i) to (v) can be detected at sight. There is no need to solve the system of equations using the techniques described. This is because at least two of the given equations represent the same plane or parallel planes.

Example   Solve the equations
$$x - 2y + 4z = 5$$
$$2x + y + 3z = 7$$
$$2x - 4y + 8z = 3.$$

The first and last equations represent parallel and distinct planes. These planes never meet and so the system has no solution. [Possibility (v).]

Example   Solve the equations
$$x + 2y - z = 3$$
$$-2x - 4y + 2z = -6$$
$$3x + 5y - 2z = 9.$$

The first two equations represent the same plane and the third plane is not parallel to them. Therefore there are an infinite number of solutions. [Possibility (iii).]

The system is

1	2	−1	3
3	5	−2	9
1	2	−1	3

$R_2 \leftarrow R_2 - 3R_1$: $\quad$ 0 $\quad$ −1 $\quad$ 1 $\quad$ 0

Put $z = t$. Then $y = t$ (from R_2) and $x = 3 - t$ (from R_1).
Therefore the solution is $x = 3 - t, y = t, z = t$ for all real t.

Example   Solve the following systems of linear equations and give a geometrical interpretation of each.

(a) $x + 3y - z = 5$ (b) $x - y - 2z = 0$
$2x + 7y + 2z = 2$ $2x - 3y - 5z = -1$
$3x + 7y - 11z = 8$; $5x + 4y - z = 9$;

(c) $x - 4y + 3z = 2$
$3x - 8y + z = 2$
$4x - 5y + z = -3.$

237

Chapter 9

(a) The system is

$$\begin{array}{cccc|c} 1 & 3 & -1 & 5 \\ 2 & 7 & 2 & 2 \\ 3 & 7 & -11 & 8 \end{array}$$

$R_2 \leftarrow R_2 - 2R_1$:
$R_3 \leftarrow R_3 - 3R_1$:

$$\begin{array}{cccc} 1 & 3 & -1 & 5 \\ 0 & 1 & 4 & -8 \\ 0 & -2 & -8 & -7 \end{array}$$

$R_3 \leftarrow R_3 + 2R_2$:

$$\begin{array}{cccc} 1 & 3 & -1 & 5 \\ 0 & 1 & 4 & -8 \\ 0 & 0 & 0 & -23 \end{array}$$

Therefore the system has no solution.

Geometrically the line of intersection of any two of the planes is parallel to and distinct from the third plane.

(b) The system is

$$\begin{array}{cccc} 1 & -1 & -2 & 0 \\ 2 & -3 & -5 & -1 \\ 5 & 4 & -1 & 9 \end{array}$$

$R_2 \leftarrow R_2 - 2R_1$:
$R_3 \leftarrow R_3 - 5R_1$:

$$\begin{array}{cccc} 1 & -1 & -2 & 0 \\ 0 & -1 & -1 & -1 \\ 0 & 9 & 9 & 9 \end{array}$$

$R_3 \leftarrow R_3 + 9R_2$:

$$\begin{array}{cccc} 1 & -1 & -2 & 0 \\ 0 & -1 & -1 & -1 \\ 0 & 0 & 0 & 0 \end{array}$$

Put $z = t$, then $y = 1 - t$ (from R_2) and $x = 1 + t$ (from R_1). Therefore the solution is $x = 1 + t$, $y = 1 - t$, $z = t$ for all real t.

Geometrically the three planes meet in the line whose parametric equations are $x = 1 + t$, $y = 1 - t$, $z = t$.

(c) The system is

$$\begin{array}{cccc} 1 & -4 & 3 & 2 \\ 3 & -8 & 1 & 2 \\ 4 & -5 & 1 & -3 \end{array}$$

$R_2 \leftarrow R_2 - 3R_1$:
$R_3 \leftarrow R_3 - 4R_1$:

$$\begin{array}{cccc} 1 & -4 & 3 & 2 \\ 0 & 4 & -8 & -4 \\ 0 & 11 & -11 & -11 \end{array}$$

$R_2 \leftarrow \frac{1}{4}R_2$:
$R_3 \leftarrow \frac{1}{11}R_3$:

$$\begin{array}{cccc} 1 & -4 & 3 & 2 \\ 0 & 1 & -2 & -1 \\ 0 & 1 & -1 & -1 \end{array}$$

238

Matrices & Linear Equations

$$R_3 \leftarrow R_3 - R_2 : \quad \begin{array}{cccc} 1 & -4 & 3 & 2 \\ 0 & 1 & -2 & -1 \\ \hline 0 & 0 & 1 & 0 \end{array}$$

Therefore $z = 0$ (from R_3), $y = -1$ (from R_2), $x = -2$ (from R_1).
The solution is $x = -2, y = -1, z = 0$.

Geometrically the three planes meet in the point $(-2, -1, 0)$.

Exercise 9.6

1. Solve the following systems of linear equations and give a geometrical interpretation of the result:

   (a) $x + 2y = 5$
          $3x + y = 10$;

   (b) $3x - y = 4$
          $5x + 3y = 9$;

   (c) $x - 5y = 2$
          $-2x + 10y = 4$;

   (d) $2x - 3y = 1$
          $9y - 6x = -3$;

   (e) $x + 2y + 3z = 2$
          $2x + 5y + 5z = 3$;

   (f) $2x - 2y + 4z = 7$
          $3x - 3y + 6z = 2$;

   (g) $x + 2y + z = 6$
          $3x + 5y - z = 13$
          $4x + 7y - 2z = 17$;

   (h) $x + 3y - z = 4$
          $2x - y + 3z = 3$
          $3x + 9y - 3z = 1$;

   (i) $2x - y - 3z = 11$
          $x - y - 2z = 5$
          $3x - 2y - 5z = 16$;

   (j) $x + 4y - 2z = 1$
          $2x + 7y - 5z = 0$
          $3x + 8y - 10z = -1$;

   (k) $2x - y + 3z = 7$
          $x - 2y - 3z = 2$
          $-4x + 2y - 6z = -14$;

   (l) $x + y - 3z = 0$
          $2x + 3y - 7z = -1$
          $5x - 6y - 4z = 16$;

   (m) $2x - y - z = 3$
          $x + 2y - z = 3$
          $3x - 4y + 2z = -1$;

   (n) $x + 2y + 3z = 0$
          $2x - y - z = 2$
          $5x - 5y - 6z = 0$.

2. Solve for x, y and z, the system of equations

   $x + 2y - z = 5$
   $3x - y + 4z = 1$
   $5x + 3y + 2z = 11$.

   What is the geometrical significance of your result?

239

Chapter 9

3. Show that the system
$$x + 3y + 2z = 5$$
$$2x + 5y - z = 1$$
$$3x + 8y + z = 5$$
does not possess any solution. State the geometrical significance of this.

4. Show that the system of linear equations
$$3x - 5y + 2z = 2$$
$$x - 2y + 3z = 6$$
$$5x - 9y + 8z = a$$
can be solved iff $a = 14$.
Interpret this geometrically.

5. For what values of k does the system of equations
$$x + ky = 2$$
$$(k-2)x + 3y = 2$$
not have a unique solution? If k does not have those values, find the unique solution. For each value of k for which no unique solution exists, determine whether or not any solution to the system of equations exists. Interpret your results geometrically.

6. Prove that if $a \neq 3$, the equations
$$x + 2y - 2z = 1$$
$$2x + ay - 3z = 4$$
$$3x + 4y - 4z = b$$
have a unique solution.

If $a = 3$, classify the values of b such that
(a) the equations have no solution ;
(b) the equations have an infinite number of solutions.
Interpret each result geometrically.

Required Outcomes

After completing this chapter, a student should be able to:
- add and multiply matrices.
- evaluate the determinants of 2 × 2 and 3 × 3 matrices.
- determine whether a 2 × 2 or a 3 × 3 matrix has an inverse and find the inverse if it exists.
- solve systems of linear equations with up to 3 equations in 3 unknowns by either Gaussian elimination or by matrix inverse methods.
- discuss the geometrical significance of a system of linear equations and its solution. **(HL – Optional)**

10 Differential Calculus 1

10.1 Limits

The concept of a limit is a very important one in mathematics and an understanding of it is crucial to the study of both the differential calculus and the integral calculus.

In the work of this chapter we shall be discussing the behaviour of a given function $f(x)$ as x takes values closer and closer to a given value a, i.e., as x approaches a. Sometimes $f(x)$ will be defined at $x = a$ and sometimes it will not be defined there. We shall use the notation $\lim_{x \to a} f(x)$ for the limit of $f(x)$ as x approaches a.

Example   Consider the function $f : x \mapsto 2x + 1$. Find $\lim_{x \to 2} f(x)$.

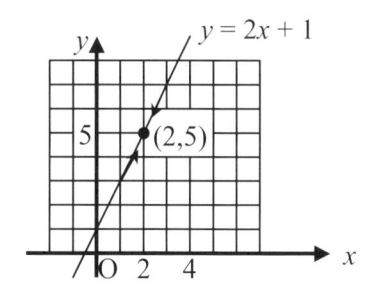

It is immediately obvious that as x approaches 2, the values of $f(x)$ approach $2(2) + 1 = 5$ and so $\lim_{x \to 2} f(x) = 5$.

Note: In this example $\lim_{x \to 2} f(x) = f(2)$.

Example   For the function $f : x \mapsto \dfrac{2x^2 - 3x - 2}{x - 2}$, find $\lim_{x \to 2} f(x)$.

Here $f(x) = \dfrac{2x^2 - 3x - 2}{x - 2} = \dfrac{(2x + 1)(x - 2)}{x - 2} = 2x + 1$ provided $x \neq 2$, and since $f(x)$ does not exist at $x = 2$, the graph of $y = f(x)$ will be the same as that of $y = 2x + 1$ except that the point (2, 5) is missing.

Chapter 10

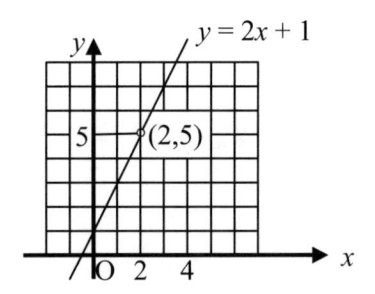

It is obvious that we can make $f(x)$ as close to 5 as we please by choosing x sufficiently close to 2. That is $\lim_{x \to 2} f(x) = 5$. However in this case $f(2)$ does not exist and so the statement $\lim_{x \to 2} f(x) = f(2)$ is obviously **not** true.

In practice we write:
$$\lim_{x \to 2} \frac{2x^2 - 3x - 2}{x - 2} = \lim_{x \to 2} \frac{(2x+1)(x-2)}{x-2}$$
$$= \lim_{x \to 2} (2x+1)$$
$$= 2(2) + 1$$
$$= 5.$$

Note: When we are asked to find $\lim_{x \to a} f(x)$ we are not at all interested in what happens *at x = a*. As we have seen, $\lim_{x \to a} f(x)$ may exist even though $f(a)$ does not.

Example Evaluate $\lim_{t \to 0} \frac{t^3 - 2t^2 + 4t}{2t}$.

$$\lim_{t \to 0} \frac{t^3 - 2t^2 + 4t}{2t} = \lim_{t \to 0} \frac{t^2 - 2t + 4}{2} = \frac{4}{2} = 2.$$

Example Find, if possible, $\lim_{x \to 2} \frac{2x^2 - 3x - 2}{(x-2)^2}$.

$\lim_{x \to 2} \frac{2x^2 - 3x - 2}{(x-2)^2} = \lim_{x \to 2} \frac{(x-2)(2x+1)}{(x-2)^2} = \lim_{x \to 2} \frac{2x+1}{x-2}$ which ***does not exist*** since the denominator, $x - 2$ becomes smaller and smaller numerically as $x \to 2$ and so $\frac{2x+1}{x-2}$ becomes larger and larger numerically.

Example Discuss the behaviour of the function $f(x) = \dfrac{1}{x^2}$ as $x \to 0$.

As we select values of x closer and closer to 0, $\dfrac{1}{x^2}$ takes increasingly larger values. Thus $f(x) = \dfrac{1}{x^2}$ has no limit as x tends to 0.

We may describe the behaviour of $f(x) = \dfrac{1}{x^2}$ by writing: as $x \to 0$, $f(x) \to \infty$ where "$\to \infty$" is read "tends to infinity".

Example Discuss the behaviour of $f(x) = \dfrac{1}{x}$ as $x \to \infty$ and as $x \to -\infty$.

As x takes increasingly larger positive values, $f(x)$ takes smaller and smaller positive values. Thus as $x \to \infty$, $f(x) \to 0$ from above $\{f(x) \to 0^+\}$.

As x takes numerically larger negative values, $f(x)$ takes numerically smaller and smaller negative values. Thus as $x \to -\infty$, $f(x) \to 0$ from below $\{f(x) \to 0^-\}$.

Higher Level

Example Evaluate $\lim\limits_{x \to 4} \dfrac{\sqrt{x} - 2}{x - 4}$.

$$\lim_{x \to 4} \frac{\sqrt{x} - 2}{x - 4} = \lim_{x \to 4} \frac{\sqrt{x} - 2}{(\sqrt{x} - 2)(\sqrt{x} + 2)} = \lim_{x \to 4} \frac{1}{\sqrt{x} + 2} = \frac{1}{4}.$$

Alternative Method:

$$\lim_{x \to 4} \frac{\sqrt{x} - 2}{x - 4} = \lim_{x \to 4} \frac{(\sqrt{x} - 2)(\sqrt{x} + 2)}{(x - 4)(\sqrt{x} + 2)}$$

$$= \lim_{x \to 4} \frac{x - 4}{(x - 4)(\sqrt{x} + 2)}$$

$$= \lim_{x \to 4} \frac{1}{\sqrt{x} + 2}$$

$$= 1/4.$$

Chapter 10

Limit Theorems

If $\lim_{x \to a} f(x) = m$ and $\lim_{x \to a} g(x) = n$, then

1. $\lim_{x \to a} \{f(x) + g(x)\} = m + n$
2. $\lim_{x \to a} k\{f(x)\} = km$ for any number k
3. $\lim_{x \to a} \{f(x)g(x)\} = mn$
4. $\lim_{x \to a} \dfrac{f(x)}{g(x)} = \dfrac{m}{n}$ provided $n \neq 0$

   (a) If $n = 0$ and $m \neq 0$, *no limit exists*.
   (b) If $n = 0$ and $m = 0$ there *may* be a limit.

   (i) $\lim_{x \to 1} \dfrac{(x-1)^2}{x-1} = \lim_{x \to 1}(x-1) = 0$

   (ii) $\lim_{x \to 1} \dfrac{x-1}{(x-1)^2} = \lim_{x \to 1} \dfrac{1}{x-1}$ which *does not exist*.

Example Evaluate $\lim_{x \to \infty} \dfrac{3x^2 - 4x + 5}{3 - 4x - 5x^2}$.

$$\lim_{x \to \infty} \frac{3x^2 - 4x + 5}{3 - 4x - 5x^2} = \lim_{x \to \infty} \frac{3 - \frac{4}{x} + \frac{5}{x^2}}{\frac{3}{x^2} - \frac{4}{x} - 5} = -\frac{3}{5} \quad \text{\{Theorems 1 and 4\}}$$

Exercise 10.1

1. Discuss the behaviour as $x \to 0$ of each of the following:

   (a) $3x$;
   (b) $\dfrac{x}{4}$;
   (c) $\dfrac{x}{2x}$;
   (d) $\dfrac{1}{2x}$;
   (e) $\dfrac{x^2 - 2x}{x}$;
   (f) $\dfrac{3x}{x+3}$;
   (g) $\dfrac{x+3}{3x}$;
   (h) $\dfrac{2x}{x^2 + 2x}$;
   (i) $\dfrac{5x^2 - 2x}{x^2}$;
   (j) $\dfrac{x^2 - 2x}{x - 2}$;
   (k) $\dfrac{(3-x)^2 - 9}{x}$;
   (l) $\dfrac{8x}{(4+x)^2 - 16}$.

Differential Calculus 1

2. Discuss the behaviour as $x \to 1$ of each of the following:

(a) $\dfrac{2x+1}{x}$; (b) $\dfrac{x^2-x}{x-1}$; (c) $\dfrac{x^2+x}{x-1}$;

(d) $\dfrac{x^2+x-2}{x^2-x}$; (e) $\dfrac{x^2-2x+1}{x^2-1}$; (f) $\dfrac{2x^2-5x+3}{2x^2-2x}$;

(g) $\dfrac{x^3-2x^2+1}{3-4x+x^2}$; (h) $\dfrac{x^3-1}{x^2-1}$; (i) $\dfrac{\sqrt{x}-1}{x-1}$;

(j) $\dfrac{(3-x)^2-4}{x-1}$; (k) $\dfrac{x^2-1}{x^3-2x^2+x}$; (l) $\dfrac{x-1}{\sqrt{x+3}-2}$.

3. Evaluate the limit, if it exists, of each of the following as $t \to 2$:

(a) $\dfrac{t^4-16}{t^3-2t^2}$; (b) $\dfrac{3t^3-t-22}{4-t^2}$; (c) $\dfrac{t^2-t-2}{t^3-2t^2}$;

(d) $\dfrac{2t^3-5t^2+4}{t^2-3t+2}$; (e) $\dfrac{t^{3/2}-2\sqrt{2}}{t^3-8}$; (f) $\dfrac{\sqrt{t+7}-3}{t-2}$.

4. Find the limit as $x \to \infty$ of each of the following:

(a) $\dfrac{1}{x-1}$; (b) $\dfrac{1}{x^2}$; (c) $\dfrac{2x+3}{x-1}$;

(d) $\dfrac{3x-5}{4-3x}$; (e) $\dfrac{2x^2+2x+1}{3-x-4x^2}$; (f) $\dfrac{2x^2-3x+4}{3x^3-3x+2}$.

5. Evaluate each of the following:

(a) $\lim\limits_{x \to -2} \dfrac{2x^2-3x-14}{3x^2+7x+2}$; (b) $\lim\limits_{x \to \infty} \dfrac{2x^2-3x-14}{3x^2+7x+2}$;

(c) $\lim\limits_{x \to 3} \dfrac{x^3-4x^2+x+6}{2x^3-5x^2-2x-3}$; (d) $\lim\limits_{x \to \infty} \dfrac{x^3-4x^2+x+6}{2x^3-5x^2-2x-3}$.

10.2 Gradient

The Gradient of a Straight Line

Consider the straight line joining the points $P(x_1, y_1)$ and $Q(x_2, y_2)$. We define the **gradient** of the line (PQ) to be the value of the quotient $\dfrac{y_2-y_1}{x_2-x_1}$ or $\dfrac{y_1-y_2}{x_1-x_2}$.

245

Chapter 10

Therefore the gradient of the line (PQ) is defined to be

$$\frac{\text{the change in } y}{\text{the corresponding change in } x} = \frac{\delta y}{\delta x},$$

and its value is independent of the positions of P and Q. Thus the gradient of a straight line is *constant*.

The Gradient of a Curve

The gradient of a curve at a point depends on the position of the point on the curve and is defined to be *the gradient of the tangent to the curve at that point*.

In the diagram below, $P(x, f(x))$ is any point on the graph of $y = f(x)$ and Q is a neighbouring point $(x+h, f(x+h))$.

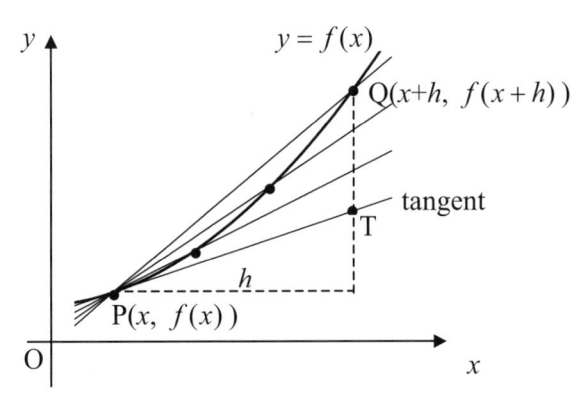

As Q approaches P along the curve, the gradient of the secant PQ approaches the gradient of the tangent PT at P. The gradient of the tangent at P is thus defined to be *the limit of the gradient of the secant PQ as Q approaches P along the curve*, i.e., as $h \to 0$.

Now the gradient of PQ is $\dfrac{f(x+h) - f(x)}{(x+h) - x} = \dfrac{f(x+h) - f(x)}{h}$.

Thus we define the gradient of the tangent at P and hence the *gradient of the curve at P* to be

$$\boxed{\lim_{h \to 0} \frac{f(x+h) - f(x)}{h}.}$$

Example   Find the gradient of the curve $f(x) = x^2$ at the point P(1, 1).

Differential Calculus 1

Let Q be the point $(1+h, (1+h)^2)$.

Then the gradient of the secant $PQ = \dfrac{(1+h)^2 - 1}{(1+h) - 1} = \dfrac{h^2 + 2h}{h}$.

Therefore the gradient of the curve at $P = \lim\limits_{h \to 0} \dfrac{h^2 + 2h}{h} = \lim\limits_{h \to 0}(h+2) = 2$.

Example Find the equation of the tangent to the curve $f(x) = x^2 - 5x + 6$ at the point P(2, 0).

Let Q be the point on the curve with coordinates $(2+h, f(2+h))$.

Then the gradient of $PQ = \dfrac{f(2+h) - 0}{h} = \dfrac{(2+h)^2 - 5(2+h) + 6}{h} = \dfrac{h^2 - h}{h}$.

Thus the gradient of the tangent at P is $\lim\limits_{h \to 0} \dfrac{h^2 - h}{h} = \lim\limits_{h \to 0}(h-1) = -1$.

The equation of the tangent is therefore $x + y = 2$.

Example Show that the gradient of the curve $f(x) = 3x - 1$ is constant and equal to 3 (as we expect!).

Let P(a, 3a − 1) be a general point on the curve and let Q(a + h, 3(a + h) − 1) be a neighbouring point. The gradient of the curve at P is defined to be

$$\lim\limits_{h \to 0}(\text{gradient of PQ}) = \lim\limits_{h \to 0} \dfrac{3(a+h) - 1 - (3a - 1)}{h}$$
$$= \lim\limits_{h \to 0} \dfrac{3h}{h}$$
$$= \lim\limits_{h \to 0} 3$$
$$= 3 \text{ regardless of the position of P.}$$

Therefore the gradient of the curve is constant and equal to 3.

Exercise 10.2

1. P is the point (1, 10) on the curve $y = 12 - 2x^2$ and Q is the point on the curve whose x-coordinate is 1 + h. Write down the gradient of the secant PQ and determine its limit as $h \to 0$. Interpret the result.

2. Find the gradient of the tangent to the curve $f(x) = x^2 - 2x$ at the point on the curve where $x = 1$. Can you explain this result?

Chapter 10

3. Find the gradient of each of the following curves at the given point:
   (a) $f(x) = x^2 - 1$ at $(1, 0)$;
   (b) $f(x) = x - x^2$ at $(2, -2)$;
   (c) $f(x) = x^3$ at $(2, 8)$;
   (d) $f(x) = \dfrac{1}{x}$ at $(-1, -1)$;
   (e) $f(x) = x + x^2$ at $(-3, 6)$;
   (f) $f(x) = 3x - 7$ at $(5, 8)$.

Higher Level

4. Find the equation of the tangent to each of the following curves at the given point:
   (a) $f(x) = 3x^2$ at $(1, 3)$;
   (b) $f(x) = 1 - x^2$ at $(-2, -3)$;
   (c) $f(x) = x^3 - x$ at $(-1, 0)$;
   (d) $f(x) = 2x^2 - 3x^3$ at $(0, 0)$;
   (e) $f(x) = \dfrac{x+1}{x}$ at $(1, 2)$;
   (f) $f(x) = \dfrac{x+3}{x+1}$ at $(1, 2)$;
   (g) $f(x) = \dfrac{x^2}{x+1}$ at $(1, \tfrac{1}{2})$;
   (h) $f(x) = \sqrt{x}$ at $(4, 2)$.

10.3 The Derived Function

If $P(x, f(x))$ and $Q(x+h, f(x+h))$ are any two points on the curve $y = f(x)$, we have shown that the gradient of the curve at P is given by

$$\lim_{h \to 0}(\text{gradient of PQ}) = \lim_{h \to 0} \frac{f(x+h) - f(x)}{h}.$$

The expression on the right above defines another function of x called the **derived function of $f(x)$**, or the **derivative of $f(x)$**. We denote the derived function of $f(x)$ [the derivative of $f(x)$] by $f'(x)$.

Thus,
$$f'(x) = \lim_{h \to 0} \frac{f(x+h) - f(x)}{h}.$$

When we evaluate $f'(x)$ for a particular value of x, say $x = a$, we obtain the number $f'(a)$ which is called the **differential coefficient of $f(x)$ at $x = a$**.

Thus,
$$f'(a) = \lim_{h \to 0} \frac{f(a+h) - f(a)}{h}.$$

Differential Calculus 1

Put $x = a + h$ so that $h = x - a$. Then as $h \to 0$, $(x - a) \to 0$ or $x \to a$. Thus from the above definition of the differential coefficient of $f(x)$ at $x = a$ we have

$$f'(a) = \lim_{x \to a} \frac{f(x) - f(a)}{x - a}.$$

The two definitions of $f'(a)$ are equivalent.

The process used to obtain $f'(x)$ is called **differentiation**.

Higher Level

Example Find the differential coefficient of $f(x) = x^2 + 2x$ at $x = 1$.

The differential coefficient of $f(x) = x^2 + 2x$ at $x = 1$ is denoted and defined by

either

$$f'(1) = \lim_{x \to 1} \frac{f(x) - f(1)}{x - 1}$$
$$= \lim_{x \to 1} \frac{x^2 + 2x - 3}{x - 1}$$
$$= \lim_{x \to 1} \frac{(x - 1)(x + 3)}{x - 1}$$
$$= \lim_{x \to 1} (x + 3)$$
$$= 4.$$

or

$$f'(1) = \lim_{h \to 0} \frac{f(1 + h) - f(1)}{h}$$
$$= \lim_{h \to 0} \frac{(1 + h)^2 + 2(1 + h) - 3}{h}$$
$$= \lim_{h \to 0} \frac{h^2 + 4h}{h}$$
$$= \lim_{h \to 0} (h + 4)$$
$$= 4.$$

Example If $f(x) = 3x^2 - 1$, find $f'(a)$.

Either

$$f'(a) = \lim_{x \to a} \frac{f(x) - f(a)}{x - a}$$
$$= \lim_{x \to a} \frac{3x^2 - 1 - (3a^2 - 1)}{x - a}$$
$$= \lim_{x \to a} \frac{3(x - a)(x + a)}{x - a}$$
$$= \lim_{x \to a} 3(x + a)$$
$$= 6a.$$

or

$$f'(a) = \lim_{h \to 0} \frac{f(a + h) - f(a)}{h}$$
$$= \lim_{h \to 0} \frac{3(a + h)^2 - 1 - (3a^2 - 1)}{h}$$
$$= \lim_{h \to 0} \frac{6ah + 3h^2}{h}$$
$$= \lim_{h \to 0} (6a + 3h)$$
$$= 6a.$$

Chapter 10

Example Find the derivative of $f(x) = \dfrac{2}{x}$.

Either

$$f'(a) = \lim_{x \to a} \frac{f(x) - f(a)}{x - a}$$

$$= \lim_{x \to a} \frac{\dfrac{2}{x} - \dfrac{2}{a}}{x - a}$$

$$= \lim_{x \to a} \frac{2a - 2x}{ax(x - a)}$$

$$= \lim_{x \to a} \frac{-2(x - a)}{ax(x - a)}$$

$$= \lim_{x \to a} \frac{-2}{ax}$$

$$= \frac{-2}{a^2} \Rightarrow f'(x) = \frac{-2}{x^2}.$$

or

$$f'(x) = \lim_{h \to 0} \frac{f(x + h) - f(x)}{h}$$

$$= \lim_{h \to 0} \frac{\dfrac{2}{x+h} - \dfrac{2}{x}}{h}$$

$$= \lim_{h \to 0} \frac{2x - 2(x + h)}{x(x + h)h}$$

$$= \lim_{h \to 0} \frac{-2h}{x(x + h)h}$$

$$= \lim_{h \to 0} \frac{-2}{x(x + h)}$$

$$= \frac{-2}{x^2}.$$

Using the definitions to obtain the derivative or differential coefficient is said to be *differentiating from first principles*.

Exercise 10.3

1. Find from first principles the differential coefficient of $f(x) = x^2$ at $x = -3$.

2. Find from the definition the differential coefficient at $x = 1$ of each of the following functions:
   (a) $3x^2 - 1$;
   (b) $2x^3 - 4x$;
   (c) $2x + 5$;
   (d) $\dfrac{3}{x}$;
   (e) $\dfrac{x}{x+1}$;
   (f) $\dfrac{2x - 3}{x^2}$.

3. Find from first principles the derivative of each of the following functions:
   (a) $2x^2 + 3x - 1$;
   (b) $2x^3$;
   (c) $\dfrac{2x + 1}{x + 1}$ $(x \neq -1)$;
   (d) $\dfrac{2}{x^2}$ $(x \neq 0)$;
   (e) $\dfrac{x^2 + 1}{2x - 1}$ $(x \neq \tfrac{1}{2})$;
   (f) $\sqrt{x}$ $(x > 0)$.

250

Differential Calculus 1

4. For the function $f(x) = 2x^2 - 3x$, find $f(1)$, and the value of $f'(1)$ from first principles. Hence write down the equation of the tangent to the curve $y = f(x)$ at the point where $x = 1$.

5. Find the equation of the tangent at $x = 1$ on each of the following curves:
   (a) $f(x) = x^2 + x$;
   (b) $f(x) = 1 - 3x^2$;
   (c) $f(x) = \dfrac{4}{x}$;
   (d) $f(x) = \dfrac{x}{x+2}$;
   (e) $f(x) = \dfrac{2x+3}{x+1}$;
   (f) $\dfrac{x^2}{x+3}$.

6. Find from first principles the derivative of each of the following functions:
   (a) $f(x) = 1$;
   (b) $f(x) = x$;
   (c) $f(x) = x^2$;
   (d) $f(x) = x^3$;
   (e) $f(x) = x^4$;
   (f) $f(x) = x^{1/2}$.

   For the function $f(x) = x^n$ where n is a rational number, guess a formula for $f'(x)$.

10.4 The Derivative of x^n

Theorem If $f(x) = x^n$ where n is rational, then $f'(x) = nx^{n-1}$.

Proof (For positive integers n only.)
Let $f(x) = x^n$ where n is a positive integer.
Then from the definition we have

$$f'(x) = \lim_{h \to 0} \frac{f(x+h) - f(x)}{h}$$

$$= \lim_{h \to 0} \frac{(x+h)^n - x^n}{h}$$

$$= \lim_{h \to 0} \frac{x^n + \binom{n}{1}x^{n-1}h + \binom{n}{2}x^{n-2}h^2 + \cdots + \binom{n}{n-1}xh^{n-1} + h^n - x^n}{h}$$

$$= \lim_{h \to 0} \left\{ \binom{n}{1}x^{n-1} + \binom{n}{2}x^{n-2}h + \cdots + \binom{n}{n-1}xh^{n-2} + h^{n-1} \right\}$$

$$= \binom{n}{1}x^{n-1}$$

$$= nx^{n-1}.$$

251

Chapter 10

In general, if $f(x) = ax^n$ where a and n are real numbers, then $f'(x) = anx^{n-1}$.

Example   Find $f'(x)$ for each of the following:
(a) $f(x) = 3x^4$;   (b) $f(x) = 5x$;   (c) $f(x) = x^{1/3}$;
(d) $f(x) = x^{5/4}$;   (e) $f(x) = \dfrac{4}{x}$;   (f) $f(x) = \sqrt{5x}$.

(a) $f(x) = 3x^4$ $\Rightarrow$ $f'(x) = (3)4x^3 = 12x^3$
(b) $f(x) = 5x$ $\Rightarrow$ $f'(x) = (5)1x^0 = 5$
(c) $f(x) = x^{1/3}$ $\Rightarrow$ $f'(x) = \frac{1}{3}x^{-2/3}$
(d) $f(x) = x^{5/4}$ $\Rightarrow$ $f'(x) = \frac{5}{4}x^{1/4}$
(e) $f(x) = 4x^{-1}$ $\Rightarrow$ $f'(x) = (4)(-1)x^{-2} = -4x^{-2}$
(f) $f(x) = \sqrt{5}x^{1/2}$ $\Rightarrow$ $f'(x) = \sqrt{5}(\frac{1}{2}x^{-1/2}) = \frac{1}{2}\sqrt{5}x^{-1/2}$

The Derivative of a Constant

If $f(x) = k$ where k is a constant, then $f(x) = kx^0$ and so $f'(x) = k \times 0 \times x^{-1}$, i.e., $f'(x) = 0$ for all x. This is consistent with the fact that geometrically $f(x) = k$ represents a straight line parallel to the x-axis and hence its gradient is 0 at all points on it.

The Derivative of a Polynomial

The derivative of any polynomial may be found using the limit theorems together with the rule for the derivative of x^n.

Example   Differentiate each of the following with respect to x:
(a) $x^2 - 5x + 6$;   (b) $4 + 3x - 5x^3$;
(c) $(3x-4)^2$;   (d) $\dfrac{2x^2 - 3x + 4}{\sqrt{x}}$.

(a) Let $f(x) = x^2 - 5x + 6$, then $f'(x) = 2x - 5$.

(b) Let $f(x) = 4 + 3x - 5x^3$, then $f'(x) = 3 - 15x^2$.

252

(c) Let $f(x) = (3x-4)^2 = 9x^2 - 24x + 16$, then $f'(x) = 18x - 24$.

(d) Let $f(x) = \dfrac{2x^2 - 3x + 4}{\sqrt{x}} = 2x^{3/2} - 3x^{1/2} + 4x^{-1/2}$, then

$f'(x) = 3x^{1/2} - \tfrac{3}{2}x^{-1/2} - 2x^{-3/2}$.

An Alternative Notation for the Derivative of a Function

A small change in the value of x is sometimes denoted by δx ("delta x"). The corresponding change in y is therefore denoted by δy.

Consider the point P(x, y) on the curve $y = f(x)$ and let Q($x + \delta x$, $y + \delta y$) be a nearby point.

The derivative of $y = f(x)$ is defined to be

$\lim\limits_{\delta x \to 0} \dfrac{f(x + \delta x) - f(x)}{\delta x}$

$= \lim\limits_{\delta x \to 0} \dfrac{y + \delta y - y}{\delta x}$

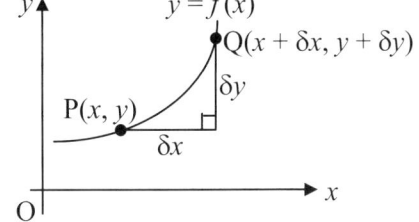

$= \lim\limits_{\delta x \to 0} \dfrac{\delta y}{\delta x}$ which is written $\dfrac{dy}{dx}$ and read as "dee y by dee x".

Note: In this context, d, dx and dy have no separate meanings.

Both $f'(x)$ and $\dfrac{dy}{dx}$ are widely used and so differentiation results may be written using either notation.

Thus if $f(x) = x^2$, then $f'(x) = 2x$; if $y = x^2$, then $\dfrac{dy}{dx} = 2x$.

We can think of $\dfrac{d}{dx}$ as a symbol indicating the operation of differentiation, writing

$\dfrac{d}{dx}(x^2) = 2x$.

Exercise 10.4

1. Write down the derivative of each of the following functions:
   (a) x^4 ; (b) $4x^3$; (c) $8x^2$;
   (d) $x^2 - 4x + 5$; (e) $4 - 3x - 2x^3$; (f) $3x^4 - 4x^3 + 5x$;
   (g) $3x^2 - 2x^3$; (h) $2x^5 - 3x^3 - 6$; (i) $5 - 4x - 2x^2 - 5x^3$.

253

Chapter 10

2. Differentiate each of the following with respect to x:

(a) $\dfrac{5}{x^2}$; (b) $\sqrt[3]{x}$; (c) $\dfrac{3x+2}{5x}$;

(d) $(2x-5)^2$; (e) $\dfrac{2}{\sqrt{x}}$; (f) $\dfrac{x^2-3x-5}{x\sqrt{x}}$;

(g) $\sqrt[3]{\dfrac{6}{x}}$; (h) $(2x-1)^3$; (i) $\dfrac{(4x+1)^2}{2x}$.

3. (a) Find the equation of the tangent to the curve $y = x^2 - 4x - 5$ at the point $(1, -8)$.

(b) Find the equation of the tangent to the curve $y = 3x^2 - 4x$ at the point where $x = \tfrac{2}{3}$.

4. (a) The gradient of the curve $y = ax^2 + bx$ at the point $(2, 8)$ is 10. Find the values of the real numbers a and b.

(b) The gradient of the curve $y = ax^2 + \dfrac{b}{x}$ at the point $(1, -5)$ is -1. Find the values of the real numbers a and b.

(c) Find the values of the real numbers a and b if $7x - 2y = 4$ is a tangent to the curve $y = \dfrac{ax^2 + b}{2x}$ at the point $\left(\tfrac{1}{2}, -\tfrac{1}{4}\right)$.

5. Differentiate each of the following with respect to t:

(a) $\dfrac{3}{t} - \dfrac{4}{t^2} + \dfrac{5}{t^3}$; (b) $3 - 4t - \dfrac{6}{5t^3}$;

(c) $\dfrac{3}{(2t)^2} - \dfrac{4}{(3t)^3}$; (d) $(3-2t)(2+5t)$;

(e) $2\pi t^3 + 2\pi t(5-t)$; (f) $\dfrac{2}{\pi t^2} - \dfrac{25}{t} + 3$.

6. For the function $f(x) = 2x^3 - 3x^2 - 12x + 20$,

(a) show that $f(2) = 0$ and factorise $f(x)$;

(b) find the values of x for which (i) $f(x) = 0$; (ii) $f'(x) = 0$.

7. For the function $f(x) = 3x^2 - 10x + 3$,

(a) find the values of $f'(x)$ when $f(x) = 0$;

(b) find the value of $f(x)$ when $f'(x) = 0$.

Differential Calculus 1

8. (a) If $P = \dfrac{3-2t}{\sqrt{t}}$, find the value of $\dfrac{dP}{dt}$ when

   (i) $t = 4$;   (ii) $P = 1$.

   (b) Find $\dfrac{dm}{dt}$ if $mt^2 = 6t^2 - 3t + 4$.

   (c) If $y = \dfrac{4-3x^4}{x}$, show that $x\dfrac{dy}{dx} + \dfrac{16}{x} = 3y$.

10.5 The Second Derivative – Motion of a Body on a Straight Line

Consider the function $y = f(x)$. If we differentiate the derivative of y with respect to x, we obtain the **second derivative of y** with respect to x. We denote this second derivative by $f''(x)$ or $\dfrac{d^2y}{dx^2}$ (pronounced "dee two y by dee x squared").

Thus $\dfrac{d}{dx}\left(\dfrac{dy}{dx}\right) = \dfrac{d^2y}{dx^2}$ or $\dfrac{d}{dx}(f'(x)) = f''(x)$.

Similarly, higher order derivatives may be found:

$\dfrac{d}{dx}\left(\dfrac{d^2y}{dx^2}\right) = \dfrac{d^3y}{dx^3} = \dfrac{d}{dx}(f''(x)) = f'''(x)$, the third derivative;

$\dfrac{d}{dx}\left(\dfrac{d^3y}{dx^3}\right) = \dfrac{d^4y}{dx^4} = \dfrac{d}{dx}(f'''(x)) = f^{(iv)}(x)$, the fourth derivative, and so on.

Motion of a Body on a Straight Line

Consider a body moving along the x-axis such that its displacement, x metres to the right of the origin O after a time t seconds ($t \geq 0$), is given by $x = f(t)$.

The **average velocity** of the body in the time interval $[t, t + h]$ is given by

$$\bar{v} = \dfrac{\text{total displacement}}{\text{total time taken}} = \dfrac{f(t+h) - f(t)}{h} \quad (h \neq 0).$$

In order to find the **instantaneous velocity** of the body at time t seconds, we find the average velocity in the time interval $[t, t + h]$ and let h take smaller and smaller values. In fact, the **instantaneous velocity** of the body at time t seconds is defined by

$$v(t) = \lim_{h \to 0} \dfrac{f(t+h) - f(t)}{h}.$$

This is clearly the first derivative of the function $x = f(t)$ and so $v(t) = \dfrac{dx}{dt} = f'(t)$.

255

Chapter 10

Note: When differentiating with respect to time, the "dot" notation is often used.

Thus if $x = f(t)$, then $v(t) = \dfrac{dx}{dt} = \dot{x} = f'(t)$.

We define velocity to be *the time rate of change of displacement*.

Acceleration is defined to be *the time rate of change of velocity*. Thus the acceleration of the body moving along a straight line with displacement $x(t)$ at time t seconds is given by $a(t) = \dfrac{d}{dt}(v(t)) = v'(t) = \dot{v}(t) = \dfrac{d^2}{dt^2}(x(t)) = x''(t) = \ddot{x}(t)$.

Velocity is a vector quantity and so the direction is critical. If the body is moving towards the right (the positive direction of the *x*-axis), its velocity is positive and if it is moving towards the left, its velocity is negative. Therefore, the body changes direction when the velocity changes sign. A sign diagram of the velocity provides a great deal of information regarding the motion of the body.

Example   A body moves along the *x*-axis so that at time t seconds its displacement x metres to the right of the origin is given by $x = 2t^3 - 9t^2 + 12t$. Draw a sign diagram of the velocity and describe the motion of the body for the first three seconds.

$x = 2t^3 - 9t^2 + 12t \;\;\Rightarrow\;\; v = 6t^2 - 18t + 12 = 6(t^2 - 3t + 2) = 6(t-1)(t-2)$.

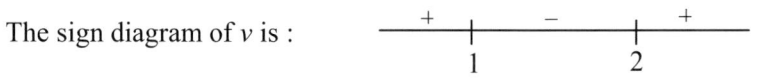

The sign diagram of v is : $\begin{array}{c} + \quad\; - \quad\; + \\ \overline{\underset{1}{|}\underset{2}{|}} \end{array}$

When $t = 0$, $x = 0$ and $v = 12$; when $t = 1$, $x = 5$ and $v = 0$; when $t = 2$, $x = 4$ and $v = 0$; when $t = 3$, $x = 9$ and $v = 12$.

Therefore the body starts at the origin with velocity 12 m s^{-1}. After 1 second, it comes to rest at 5 metres to the right of the origin and changes direction. After 2 seconds it comes to rest again 4 metres to the right of the origin and changes direction once more. For the third second, the body moves off to the right and reaches a point 9 metres to the right of the origin with a speed of 12 m s^{-1}.

On the other hand, *speed* is a scalar quantity and direction is irrelevant. In fact speed = |velocity| or the magnitude of the velocity.

We define the average speed of the body over the time interval $[t, t + h]$ to be
$$\overline{|v|} = \dfrac{\text{total distance travelled}}{\text{total time taken}}.$$

Differential Calculus 1

Example A body moves along the *x*-axis such that its displacement *x* metres to the right of the origin at time *t* seconds is given by $x(t) = t^3 - 3t^2$. Find the acceleration of the body when its velocity is zero and the velocity of the body when its acceleration is zero.

$$x(t) = t^3 - 3t^2 \Rightarrow v(t) = 3t^2 - 6t = 3t(t-2)$$
$$\Rightarrow a(t) = 6t - 6 = 6(t-1).$$

Thus the velocity is zero when $t = 0$ or $t = 2$, and the acceleration is zero when $t = 1$.

Therefore when the velocity is zero, $a(0) = -6 \text{ m s}^{-2}$ and $a(2) = 6 \text{ m s}^{-2}$.

When the acceleration is zero, $v(1) = -3 \text{ m s}^{-1}$.

Example A body moves along the *x*-axis so that at time *t* seconds $x(t) = t^3 + 3t^2 - 9t$.

Find: (a) the position and velocity of the body at $t = 0, 1, 2$;
(b) where and when the body comes to rest ;
(c) the maximum speed of the body in the first 1 second of motion ;
(d) the maximum velocity of the body in the first 1 second of motion ;
(e) the total distance travelled by the body in the first 2 seconds of motion.

(a) $x(t) = t^3 + 3t^2 - 9t \Rightarrow v(t) = 3t^2 + 6t - 9 = 3(t-1)(t+3)$
When $t = 0$, $x = 0$ and $v = -9$; when $t = 1$, $x = -5$ and $v = 0$; when $t = 2$, $x = 2$ and $v = 15$.
At $t = 0$ the body is at the origin with a velocity of -9 m s^{-1}.
At $t = 1$ the body is 5 m to the left of O with velocity 0.
At $t = 2$, the body is 2 m to the right of O with velocity 15 m s^{-1}.

(b) The body is at rest when $v = 0$. This occurs when $t = 1$ ($t \geq 0$).
At this time the body is 5 m to the left of the origin.

(c) The velocity is increasing in the interval [0, 1] since $v'(t) = 6t + 6 > 0$.
$v(0) = -9$ and $v(1) = 0$.
Therefore the maximum speed in the first 1 second is 9 m s^{-1}.

(d) From part (c), the maximum velocity is 0 m s^{-1}.

Chapter 10

(e) The following diagram illustrates the position of the body from $t = 0$ to $t = 2$.

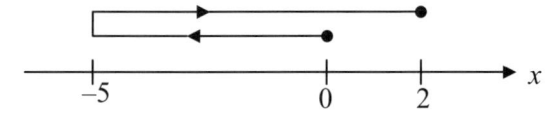

From the diagram the total distance travelled is 12 m.

Exercise 10.5

1. Write down the first and second derivatives of each of the following:
   (a) $5x - 4$;
   (b) $3x^2 - 6x - 5$;
   (c) $2x^3 - 5x^2 + 4x + 2$;
   (d) $x^3 - \dfrac{2}{x}$;
   (e) $\dfrac{3x + 2}{\sqrt{x}}$;
   (f) $\dfrac{6}{x} - \dfrac{3}{x^2} + \dfrac{4}{x^3}$.

2. (a) If $f(x) = \dfrac{4 - x^2}{\sqrt{x}}$ show that $4x^2 f''(x) = 3f(x)$.

   (b) If $f(x) = x^3 - 3x^2 + 3x - 1$ show that $f''(x)f'(x) = 18f(x)$.

   (c) If $f(x) = 3x^3 - 5x^2 + x + 1$, find
   (i) $f'(x)$ when $f''(x) = 0$;
   (ii) $f''(x)$ when $f'(x) = 0$.

3. A body moves along the x-axis so that its position is $x(t)$ metres to the right of the origin at time t seconds.
   (a) If $x(t) = t^3 - 3t^2$ explain why the total distance travelled in the first three seconds of motion is not equal to the displacement in that time.
   (b) If $x(t) = t^3 - 3t^2 + 3t$ explain why the distance travelled in the first three seconds of motion is now equal to the displacement in that time.

4. A particle moves along the x-axis such that its displacement is $x(t)$ metres to the right of the origin at time t seconds.
   (a) If $x(t) = 3t^3 - 4t^2 + 5t$ find the initial velocity and displacement.
   (b) If $x(t) = 3t^3 - 7t^2 + 2t$ find the velocity and acceleration of the particle at the times when it is at the origin.
   (c) If $x(t) = 2t^3 - 21t^2 + 60t$ find the position and acceleration of the particle when it is momentarily at rest.
   (d) If $x(t) = 2t^3 - 15t^2 + 36t + 1$ find the times at which the particle is at rest and calculate the distance travelled by the particle in the first 4 seconds of motion.

5. A particle is moving along the x-axis such that its position, x(t) metres to the right of the origin at time t seconds, is given by $x(t) = t^3 - 9t^2 + 24t - 18$. Describe the particle's motion during the first five seconds and calculate the distance travelled in that time.

6. A particle moves along the x-axis so that at time t seconds its position is given by $x(t) = (t^3 - 6t^2 - 36t)$ m. Find the maximum speed and the maximum velocity in the first seven seconds of motion and also the average speed and average velocity in that time.

7. A ball is thrown vertically into the air so that it reaches a height of $y = 19.6t - 4.9t^2$ metres in t seconds.
    (a) Find the velocity and acceleration of the ball at time t seconds.
    (b) Find the time taken for the ball to reach its highest point.
    (c) How high did the ball rise?
    (d) At what time(s) would the ball be at half its maximum height?

10.6 Points of Increase, Points of Decrease and Stationary Points

If at a point where $x = a$ on the curve $y = f(x)$, $f'(a) > 0$, i.e., the tangent at $x = a$ has a positive slope, then on passing through this point in the direction of increasing x, the value of $f(x)$ is increasing. We call such a point $x = a$ a *point of increase of f*.

Definition   If $f'(a) > 0$, then $x = a$ is a point of increase of f.

For such a point, $f(x) > f(a)$ if x is just greater than a, and
                    $f(x) < f(a)$ if x is just less than a.
Thus whenever $f'(x) > 0$, the function $f(x)$ is said to be *increasing*.

Definition   If $f'(a) < 0$, then $x = a$ is a point of decrease of f.

For such a point, $f(x) < f(a)$ if x is just greater than a, and
                    $f(x) > f(a)$ if x is just less than a.
Thus whenever $f'(x) < 0$, the function $f(x)$ is said to be *decreasing*.

If, however, $f'(a) = 0$, the point $x = a$ may be
    (1) a point of increase (point D in the following diagram);
    (2) a point of decrease (point C in the following diagram);
or (3) neither a point of increase nor a point of decrease (points A and B in the following diagram).

Chapter 10

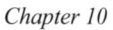

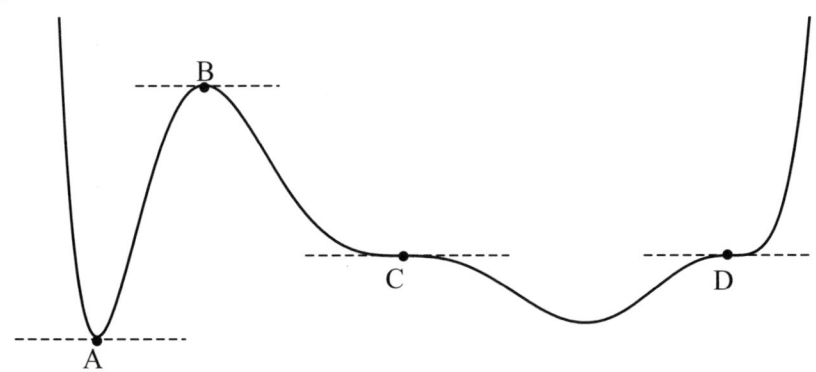

Concavity

Whenever a curve bends at a given point so that the tangent at the point lies above it, we say that the curve is *concave downwards* (or *concave*), and as we pass through the point in the direction of increasing x, **the slope of the curve is continually decreasing**. That is $f'(x)$ is decreasing and so its derivative $f''(x)$ must be negative.

Definition A curve is said to be *concave downwards* (or *concave*) in an interval $]a, b[$ if $f''(x) < 0$ for all $x \in\,]a, b[$.

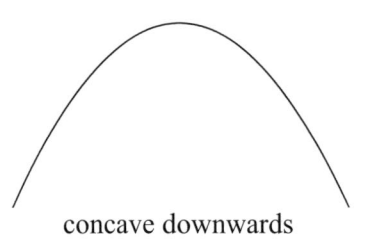

concave downwards

Similarly, if the curve bends at a given point so that the tangent at the point lies below it, we say that the curve is *concave upwards* (or *convex*), and as we pass through the point in the direction of increasing x, **the slope of the curve is continually increasing**. That is, $f'(x)$ is increasing and so its derivative, $f''(x)$, must be positive.

Definition A curve is said to be *concave upwards* (or *convex*) in an interval $]a, b[$ if $f''(x) > 0$ for all $x \in\,]a, b[$.

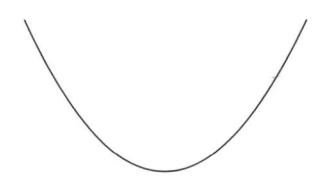

concave upwards

Differential Calculus 1

Points of Inflexion

Definition The point $x = c$ is a ***point of inflexion*** of the continuous function $y = f(x)$ if $f''(x) > 0$ on one side of $x = c$ and $f''(x) < 0$ on the other.

If $x = c$ is a point of inflexion of f, then the concavity of the curve $y = f(x)$ changes at $x = c$.

For a continuous curve with a continuous first derivative, $x = c$ is a point of inflexion of $y = f(x)$ if $f''(c) = 0$ ***and*** $f''(x)$ changes sign at $x = c$.

Example Find the point of inflexion on the curve $f(x) = x^3 + 2x - 3$.

$f(x) = x^3 + 2x - 3 \Rightarrow f'(x) = 3x^2 + 2 \Rightarrow f''(x) = 6x$
Therefore $x = 0$ is a point of inflexion since $f''(0) = 0$ and $f''(x)$ clearly changes sign at $x = 0$.

Although $f''(c) = 0$ is a ***necessary*** condition for $x = c$ to be a point of inflexion on a continuous curve with a continuous first derivative, it is not a ***sufficient*** condition.

Consider the function $f(x) = x^4$ for which $f'(x) = 4x^3$ and $f''(x) = 12x^2$. The point $x = 0$ is not a point of inflexion (it is a minimum) even though $f''(0) = 0$

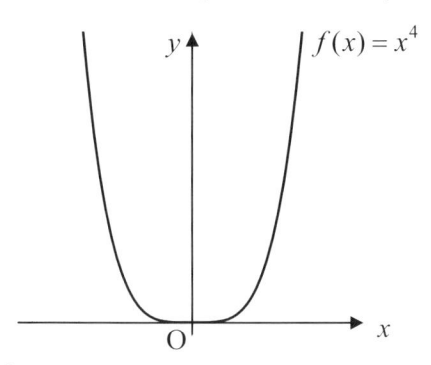

Stationary Points

Definition The point $x = a$ for which $f'(a) = 0$ is defined to be a ***stationary point*** of the function f.

A stationary point may be either
- a **local maximum**, so-called because it is the highest point in the immediate neighbourhood ;
- a **local minimum** ;   or
- a **horizontal (stationary) inflexion**.

The tangent to a curve at any stationary point must be horizontal.

261

Chapter 10

Determining the Natures of Stationary Points

The *position* of any stationary point is found by solving the equation $f'(x) = 0$. There are two common methods for determining the *nature* of these points.

The first method uses the sign of the first derivative. The second method considers the sign of the second derivative. Unfortunately the second method fails when both $f'(c) = 0$ and $f''(c) = 0$ since then the point $x = c$ may be a point of increase of $f'(x)$, a point of decrease of $f'(x)$, or neither a point of increase nor a point of decrease of $f'(x)$.

These methods are summarised in the following tables.

Method 1   The First Derivative Test

Maximum	Minimum	Horizontal Inflexion
If $x = c$ is a local maximum, $f'(c) = 0$. Just before $x = c$, $f'(x)$ is positive and just after $x = c$, $f'(x)$ is negative. The sign of $f'(x)$ is:	If $x = c$ is a local minimum, $f'(c) = 0$. Just before $x = c$, $f'(x)$ is negative and just after $x = c$, $f'(x)$ is positive. The sign of $f'(x)$ is:	At such points, $f'(c) = 0$, but the sign of $f'(x)$ is the same just before $x = c$ as just after $x = c$. The sign of $f'(x)$ is:

Method 2   The Second Derivative Test

Maximum	Minimum	Not Determined
If $f'(c) = 0$ and $f''(c) < 0$ then $f'(x)$ is decreasing at $x = c$. Hence $x = c$ is a local maximum.	If $f'(c) = 0$ and $f''(c) > 0$ then $f'(x)$ is increasing at $x = c$. Hence $x = c$ is a local minimum.	If $f'(c) = 0$ and $f''(c) = 0$ this method fails since $x = c$ may then be a point of increase of $f'(x)$, a point of decrease of $f'(x)$ or neither.

As an example of the failure of the second derivative test to establish the type of stationary point, consider the functions $f(x) = x^3$, $f(x) = x^4$ and $f(x) = -x^4$. Here

$f'(0) = f''(0) = 0$ in all three cases, but $x = 0$ is a point of inflexion, a local minimum and a local maximum respectively.

Example Determine the position and nature of the stationary points and points of inflexion of the function $f(x) = x^3 - 6x^2 + 9x$ and sketch its graph.

Method 1 $f'(x) = 3x^2 - 12x + 9$
$= 3(x^2 - 4x + 3)$
$= 3(x - 3)(x - 1)$
$= 0$ when $x = 3, 1$.

sign of $f'(x)$: + on $(-\infty, 1)$, − on $(1, 3)$, + on $(3, \infty)$

Method 2 $f'(x) = 3x^2 - 12x + 9$
$= 3(x^2 - 4x + 3)$
$= 3(x - 3)(x - 1)$
$= 0$ when $x = 3, 1$.

$f''(x) = 6x - 12$
$\begin{cases} < 0 \text{ for } x = 1 \\ > 0 \text{ for } x = 3. \end{cases}$

$f''(x) = 6x - 12 = 0$ when $x = 2$ and $f''(x)$ changes sign at $x = 2$.

Therefore (1, 4) is a local maximum and (3, 0) is a local minimum and (2, 2) is a (non-stationary) point of inflexion.

The graph of $y = f(x) = x^3 - 6x^2 + 9x$ follows.

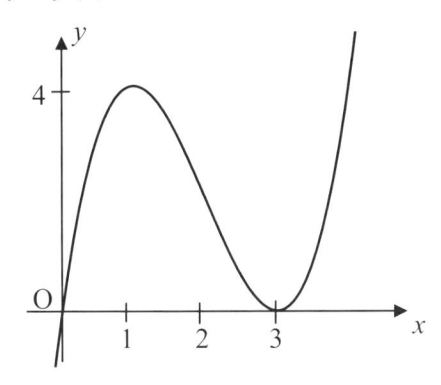

Example Find the stationary points and points of inflexion of $f(x) = 3x^5 - 20x^3$ and sketch its graph.

$f(x) = 3x^5 - 20x^3$
$f'(x) = 15x^4 - 60x^2$
$= 15x^2(x^2 - 4)$
$= 15x^2(x - 2)(x + 2)$
$= 0$ when $x = 0, 0, \pm 2$.

263

Chapter 10

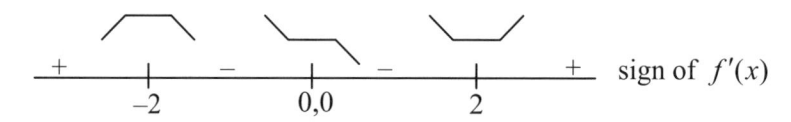

```
    +    |    -    |    -    |    +     sign of f'(x)
        -2        0,0       2
```

Therefore (−2, 64) is a local maximum,
(0, 0) is a stationary inflexion (on a falling curve),
and (2, −64) is a local minimum.

$$f''(x) = 60x^3 - 120x$$
$$= 60x(x^2 - 2)$$
$$= 60x(x - \sqrt{2})(x + \sqrt{2})$$

Clearly $f''(x) = 0$ and changes sign at $x = 0, \pm\sqrt{2}$.
Thus points $(\sqrt{2}, -28\sqrt{2}) \approx (1.41, -39.6)$ and $(-\sqrt{2}, 28\sqrt{2}) \approx (-1.41, 39.6)$ are (non-stationary) inflexions.
[The stationary inflexion at $x = 0$ has already been found from the first derivative.]

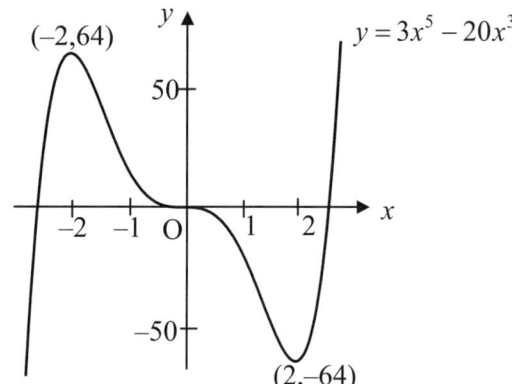

Exercise 10.6

1. Find the position and nature of any stationary points that the following functions may have and sketch the graph of each:
   (a) $y = x^3 - 3x^2 + 3$;
   (b) $y = x^4 + 4x^3$;
   (c) $y = x^3 - 6x^2 + 12x - 8$;
   (d) $y = 2x^2 - x^4$;
   (e) $y = x^4 - 3x^3$;
   (f) $y = x^3(x - 5)^2$.

2. Find the position and nature of each stationary point of the following functions. Find also any non-stationary points of inflexion. In each case sketch the graph of the function.

Differential Calculus 1

(a) $f(x) = x^3 - 3x^2 + 2$;   (b) $f(x) = x^3 + 9x$;

(c) $f(x) = 6x - 2x^3$;   (d) $f(x) = x^3 - 3x + 2$;

(e) $f(x) = 3x^2 - x^3$;   (f) $f(x) = \frac{1}{4}x^4 + x$.

3.   The function $f(x) = 2x^3 + ax^2 + bx$ has stationary points at $x = -1$, $x = 2$. Find the values of the real numbers a and b and sketch the graph of the function.

4.   Find the values of x for which the values of $f(x) = x^4 - 4x^3 + 4x^2$ exceed the y-coordinate of the local maximum.

5.   (a)   The curve $y = x^3 + ax^2 + bx$ has a stationary point at (1, 4). Find the values of the real numbers a, b and the coordinates of the second stationary point.

     (b)   The curve $y = x^3 + ax^2 + bx + c$ has a stationary point at (2, 1) and passes through the point (−1, 1). Find the values of the real numbers a, b, c. Find also the equation of the tangent to the curve at (−1, 1).

     (c)   The curve $y = ax^3 + bx^2 + cx$ has a stationary point at (2, d), and the equation of the tangent at the point $x = 1$ on the curve is $12x + y = -1$. Find the values of the real numbers a, b, c, d and find the coordinates of the point where the tangent at $x = 1$ *cuts* the curve.

Higher Level

6.   Find the position and nature of each stationary point of $y = 8x^3 - 6x$ and hence show that $8x^3 - 6x + 1 = 0$ has three real roots all numerically less than 1. By putting $x = \cos\theta$ and using the identity $\cos 3\theta = 4\cos^3\theta - 3\cos\theta$, find these roots.

10.7   Extreme Points

If $y = f(x)$ is a continuous function on the interval [a, b] and $c \in $ [a, b], then $f(c)$ is

- a *local maximum* value of f if $f(x) \leq f(c)$ for all values of x in the 'immediate neighbourhood' of $x = c$.
- a *local minimum* value of f if $f(x) \geq f(c)$ for all values of x in the 'immediate neighbourhood' of $x = c$.

Chapter 10

- the ***absolute maximum*** value of f if $f(x) \leq f(c)$ for all $x \in [a, b]$.
- the ***absolute minimum*** value of f if $f(x) \geq f(c)$ for all $c \in [a, b]$.

We call any maximum or minimum value of the function, an ***extreme value*** of the function.

[The plurals of maximum, minimum and extreme are maxima, minima and extrema respectively.]

Extreme values of a function can occur not only at interior points but also at the end-points of the domain of definition.
In the following diagram we find a number of extreme values of the function f. Two extreme values occur at the end-points of the domain (points A and E), two occur at interior points (points B and C) where $f'(x) = 0$, and one occurs at an interior point (point D) where $f'(x)$ does not exist.

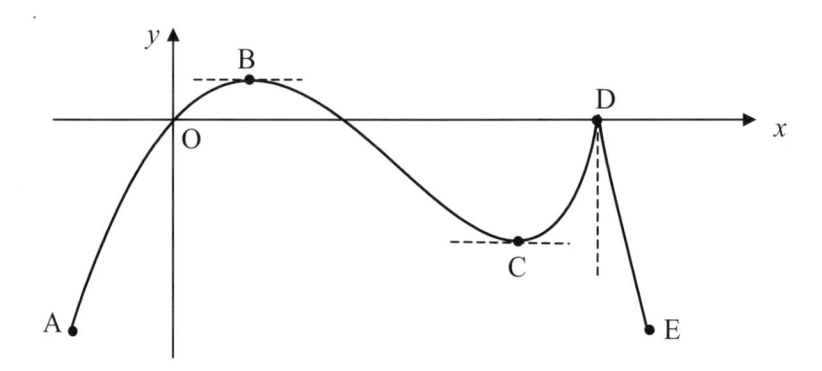

Thus for a continuous curve $y = f(x)$ defined on $[a, b]$, the only points where f can have an extreme point are
- at $x = c$ where $a < c < b$ and $f'(c) = 0$.
- at $x = c$ where $a < c < b$ and $f'(c)$ does not exist.
- at $x = a$ or at $x = b$, the endpoints of the domain.

Definition The point $x = c$ is called a ***critical point*** of the continuous function f defined on $[a, b]$ if $c \in \,]a, b[$ and $f'(c) = 0$ or $f'(c)$ does not exist.

Theorem The ***absolute maximum*** and ***absolute minimum*** values of a continuous function f defined on $[a, b]$ will be found at critical points within the domain or at the end points of the domain.

Therefore, to find the absolute maximum and minimum values of a continuous function f defined on $[a, b]$

Differential Calculus 1

1. find the critical points $x = c_1, c_2, \cdots$;

2. evaluate $f(c_1), f(c_2), \cdots$;

3. evaluate $f(a)$ and $f(b)$;

4. take the largest and smallest values of $f(a), f(c_1), f(c_2), \cdots, f(b)$.

Example Find the largest and smallest values of $f(x) = 2x^3 - 3x^2 - 12x + 24$ for $0 \leq x \leq 3$.

$$f(x) = 2x^3 - 3x^2 - 12x + 24$$
$$f'(x) = 6x^2 - 6x - 12$$
$$= 6(x^2 - x - 2)$$
$$= 6(x + 1)(x - 2)$$
$$= 0 \text{ when } x = -1, 2.$$

sign of $f'(x)$: $+$ at -1, $-$ between, $+$ at 2.

Hence $(-1, 31)$ is a local maximum and $(2, 4)$ is a local minimum – the critical points.

Now $f(0) = 24$, $f(2) = 4$ and $f(3) = 15$ and so the largest value of $f(x)$ in the interval $0 \leq x \leq 3$ is 24 and the smallest value is 4. [See the following diagram.]

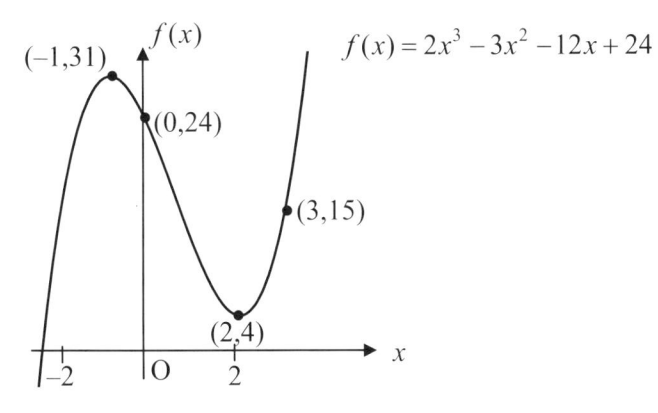

Graph showing $f(x) = 2x^3 - 3x^2 - 12x + 24$ with points $(-1, 31)$, $(0, 24)$, $(3, 15)$, $(2, 4)$.

Example Find the absolute maximum and minimum values of $f(x) = x^{1/3}$ on the interval $[-1, 8]$.

$$f(x) = x^{1/3} \implies f'(x) = \tfrac{1}{3} x^{-2/3} = \frac{1}{3x^{2/3}}$$

Therefore $f'(x)$ is never zero but is undefined at $x = 0$.

Chapter 10

Thus there is only one extreme point at $x = 0$, and $f(0) = 0$.
Also $f(-1) = -1$ and $f(8) = 2$.
Therefore the maximum value of f is 2 and the minimum value is -1.

The graph of $y = f(x) = x^{1/3}$, $-1 \leq x \leq 8$, follows.

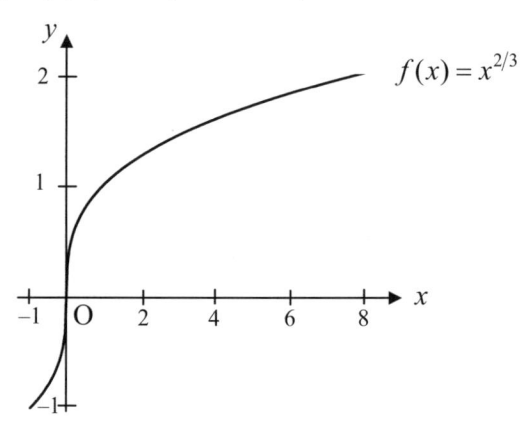

Example Find the coordinates of any critical points that the graph of the function $f(x) = |x| - 2|x-1|$, $-2.5 \leq x \leq 3$, may have.

The graph of $y = f(x)$ is as follows:

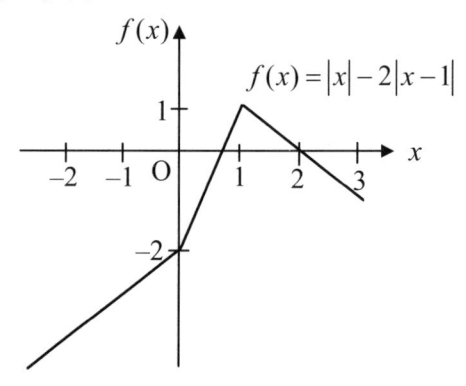

The critical points are $(0, -2)$ and $(1, 1)$ where $f'(x)$ does not exist.

Higher Level

Example Find the values of x at the extreme points of the graph of $f(x) = x^{2/3}(\frac{5}{2} - x)$, $-0.7 \leq x \leq 2$, and sketch its graph. Calculate the absolute maximum and minimum values of $f(x)$ in the given domain. Find also the x-coordinates of any points of inflexion.

Differential Calculus 1

$f(x) = x^{2/3}(\frac{5}{2} - x) = \frac{5}{2}x^{2/3} - x^{5/3} \Rightarrow f'(x) = \frac{5}{3}x^{-1/3} - \frac{5}{3}x^{2/3} = \frac{5}{3}x^{-1/3}(1 - x)$

Therefore extreme points occur at $x = 0$ (where $f'(x)$ is undefined) and at $x = 1$ (where $f'(x) = 0$).

Now $f(-0.7) = 2.52$, $f(0) = 0$, $f(1) = 1.5$ and $f(2) = 0.794$.
Thus the absolute minimum of $f(x)$ is 0; the absolute maximum is 2.52.

$f''(x) = -\frac{5}{9}x^{-4/3} - \frac{10}{9}x^{-1/3} = -\frac{5}{9}x^{-4/3}(2x + 1)$

The only point of inflexion occurs at $x = -\frac{1}{2}$ (where the second derivative changes sign). $\{x^{-4/3} = (x^{-1/3})^4$ which is never negative.$\}$

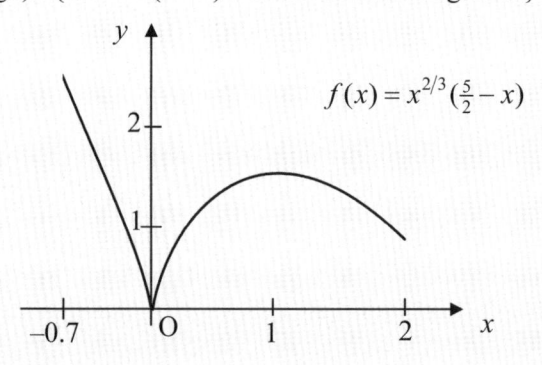

Example The first derivative of the function $y = f(x)$ is $f'(x) = x(x + 1)$.
(i) State the values of x for which f is increasing.
(ii) Find the x-coordinate of each extreme point of f.
(iii) State the values of x for which the curve of f is concave upwards.
(iv) Find the x-coordinate of each point of inflexion.
(v) Sketch the general shape of the graph of f indicating the extreme points and points of inflexion.

(i) $f'(x) = x(x+1)$ sign of $f'(x)$: $+$ at -1, $-$ between, $+$ at 0

Therefore f is increasing for $x < -1$ or $x > 0$.

(ii) The extreme points are at $x = -1$ and $x = 0$.

(iii) $f'(x) = x^2 + x \Rightarrow f''(x) = 2x + 1$ sign of $f''(x)$: $-$ then $+$ at $-\frac{1}{2}$

The curve is concave upwards for $x > -\frac{1}{2}$.

269

Chapter 10

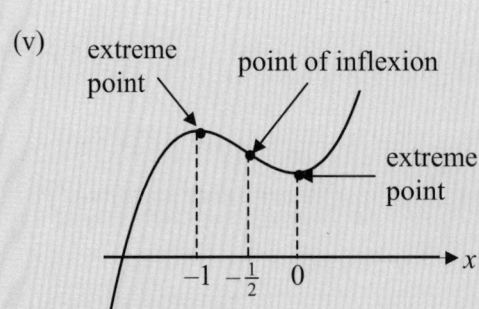

(iv) $x = -\frac{1}{2}$ is a point of inflexion.

(v)   extreme point, point of inflexion, extreme point

Exercise 10.7

1. Find the largest and smallest values of $f(x) = x(x-2)^2$ for
   (a) $-1 \leq x \leq 3$;   (b) $0 \leq x \leq 2$;   (c) $-\frac{1}{3} \leq x \leq \frac{2}{3}$.

2. Find the largest and smallest values of each of the following functions defined on the domain, and sketch the graph of each function for this domain:
   (a) $f(x) = x^3 - 3x$, $0 \leq x \leq 2$;
   (b) $f(x) = 3x^4 - 4x^3 + 1$, $-1 \leq x \leq 1.5$;
   (c) $f(x) = x^{3/5}$, $-1 \leq x \leq 1$;
   (d) $f(x) = x^{2/3}(1 - x^2)$, $-1 \leq x \leq 1$.

3. For each of the following functions
   (i)   find all of the zeros;
   (ii) find the extreme points;
   (iii) find the points of inflexion;
   (iv) sketch the graph of f.
   (a) $f(x) = x^3 + 3x^2$;   (b) $f(x) = x^3 + 3x^2 - 9x - 2$;
   (c) $f(x) = 3x^4 + 8x^3$;   (d) $f(x) = x^2(x-2)^2$;
   (e) $f(x) = 3x^{2/3}(x^2 - 16)$;   (f) $f(x) = x^{1/3}(x+3)$.

Higher Level

4. For each of the following functions where the first derivative is given
   (i) find the values of x for which f is increasing ;
   (ii) find the extreme points ;
   (iii) state the values of x for which the curve of f is concave upwards ;
   (iv) find any points of inflexion ;
   (v) sketch the general shape of the graph of f indicating the extreme points and points of inflexion.

   (a) $f'(x) = x(x-2)$;   (b) $f'(x) = (1-x)(2+x)$;
   (c) $f'(x) = (x-1)(x-2)(x+3)$;   (d) $f'(x) = (x-1)^2(x+2)$;
   (e) $f'(x) = x^{-1/3}(x+2)$;   (f) $f'(x) = x^{-2/3}(x-2)$.

10.8 Optimisation

We would often like to know the greatest or least value of some variable. Least cost, greatest profit, maximum volume for a given surface area, etc,. The techniques developed in the early part of this chapter are extremely useful in the solution of such problems. We seek the absolute maximum or minimum of a given function defined on some finite interval.

Example   The sum of the lengths of the radius and height of a cylinder is 15 cm. Find the largest volume of the cylinder.

$r + h = 15$ and $0 \leq r \leq 15$

The volume of the cylinder is
$$\begin{aligned} V &= \pi r^2 h \\ &= \pi r^2 (15 - r) \\ &= \pi(15r^2 - r^3). \end{aligned}$$

$$\begin{aligned} \frac{dV}{dr} &= \pi(30r - 3r^2) \\ &= 3\pi r(10 - r) \\ &= 0 \text{ when } r = 0, 10. \end{aligned}$$

sign of $\dfrac{dV}{dr}$: $-$ on $(-\infty, 0)$, $+$ on $(0, 10)$, $-$ on $(10, \infty)$

Therefore V is a maximum when $r = 10$.
The maximum volume $= \pi(10^2)(15 - 10) = 500\pi$ cm^3.

Chapter 10

Example An open-topped baking tin with a volume of 864 cm^3 is to be constructed with a square base and vertical sides. Find the least amount of tin plate required.

$x^2 y = 864$

$\Rightarrow y = \dfrac{864}{x^2}$.

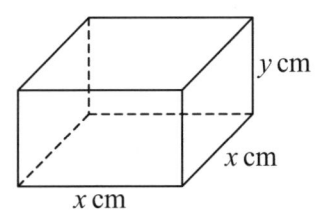

The area of tin plate required is given by

$$A = x^2 + 4xy = x^2 + 4x\left(\dfrac{864}{x^2}\right) = x^2 + \dfrac{3456}{x}$$

$$\begin{aligned}\dfrac{dA}{dx} &= 2x - \dfrac{3456}{x^2} \\ &= \dfrac{2(x^3 - 1728)}{x^2} \\ &= 0 \text{ when } x^3 = 1728 \text{ or } x = 12.\end{aligned}$$

sign of $\dfrac{dA}{dx}$: − then + at $x = 12$

Then A is least when $x = 12$.

Therefore the least amount of tin plate required is $12^2 + \dfrac{3456}{12} = 432 \text{ cm}^2$.

Summary of Method

1. Draw a large, neat, labelled diagram.
2. Find an expression for the variable to be maximised (or minimised).
3. Eliminate all but one independent variable from this expression.
4. Differentiate with respect to this remaining independent variable.
5. Equate this derivative to zero.
6. Draw a sign diagram of this derivative (or use the second derivative method).
7. Answer the question required taking care to check the end-points of the domain of the independent variable.

[Clearly, after step 3 one can use a GDC.]

Differential Calculus 1

Example   Triangle ABC is isosceles with AB = AC = 5 cm and BC = 6 cm. Another isosceles triangle is formed with one vertex at M, the mid-point of BC, and the other vertices P and Q being any two points on [AB] and [AC] respectively such that MP = MQ. Find the greatest area of the triangle MPQ.

1.

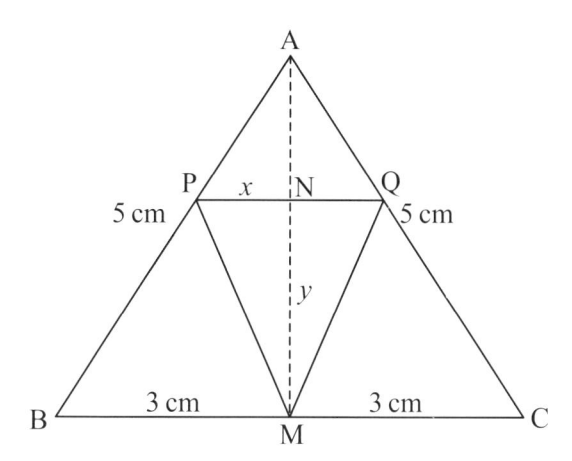

2. Let N be the mid-point of [PQ], let PN = x cm and let NM = y cm. The area A of the triangle MPQ is given by $A = xy$.

3. BM = 3 cm and AM = 4 cm (Pythagoras)

    Triangles APN, ABM are similar and so $\dfrac{AN}{AM} = \dfrac{PN}{BM}$.

    Thus $\dfrac{4-y}{4} = \dfrac{x}{3}$ ⇒ $x = \tfrac{3}{4}(4-y)$.

    Hence $A = \tfrac{3}{4}(4-y)y = 3y - \tfrac{3}{4}y^2$.

4. $\dfrac{dA}{dy} = 3 - \tfrac{3}{2}y$

5. $\dfrac{dA}{dy} = 0$ when $y = 2$.

6.

    + /‾‾\ −   sign of $\dfrac{dA}{dy}$
    ────┼──────
        2

7. A is a maximum when $y = 2$. [$0 \le y \le 4$ and $A = 0$ at both end-points.]
    The greatest area of triangle MPQ = 3 cm².

273

Chapter 10

Exercise 10.8

1. The radius of the base r cm and the perpendicular height h cm of a right-circular cone together equal 12 cm. Find the ratio $r : h$ when the volume of the cone is a maximum.

2. A rectangular block, the length of whose base is twice the width, has a total surface area of 108 cm^2. Find the depth of the block if it has the maximum possible volume.

3. A closed rectangular box has a square base and the sum of the lengths of its twelve edges is 8 m. Find the largest volume of the box.

4. What are the inside dimensions of a closed cylindrical can of capacity 1 litre if the amount of material required to make the can is the least possible?

5. A closed cylinder has a total inside surface area of 800 cm^2. Find the radius of the cylinder when its volume is a maximum.

6. The sum of two non-negative numbers is 50. Find the greatest and least values of the sum of their squares.

7. Show that of all rectangles of a given area, a square has the largest perimeter.

8. Show that of all rectangles with a given perimeter, the one with the largest area is a square.

9. An existing fence is to form one side of a rectangular enclosure. What minimum length of fencing is required to form the other three sides if the area enclosed is to be 1.28 hectare?

10. A sheet of cardboard 15 cm by 8 cm has four equal squares cut out of its corners and the sides turned up to form an open rectangular box. Find the length of the edge of the squares cut out so that the box may have the maximum possible volume.

11. A wire of length 10 m has a piece of length x m cut from one end and formed into a circle. The other piece is then formed into a square. Find the value of x if the sum of the areas enclosed by the two pieces of wire is
    (a)   a minimum ;
    (b)   a maximum.

Higher Level

12. A rectangle has an area of 128 m². Find its dimensions if the distance from one vertex to the mid-point of a non-adjacent side is a minimum, and find this minimum distance.

13. A swimming pool is in the shape of a rectangle with semi-circles at each end as in the following diagram. The surface area of the pool is a constant A m².

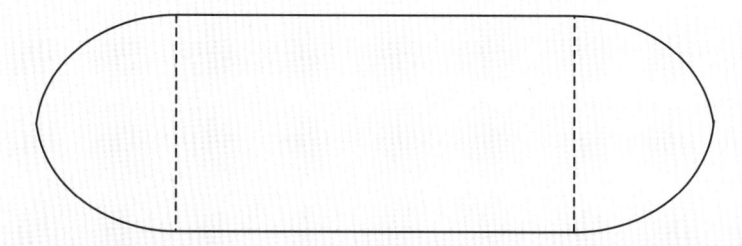

    The cost of tiling the curved walls of the pool is 25% more expensive than that for tiling the straight walls. Show that the total cost of tiling the walls is a minimum when the area of the the semi-circular ends is $\frac{2}{3}A$.

14. A right-circular cone is to be cut from a sphere. Show that the volume of the the cone cannot exceed $\frac{8}{27}$ of the volume of the sphere.

    [*Hint:* Let the distance from the centre of the sphere to the centre of the base of the cone be x and express the volume of the cone in terms of x.]

*15. A cylindrical can has a radius of r, a height of h and a slip-on lid of fixed length a. If the can and its lid are to be made from A cm² of metal, show that the tin has a maximum volume when $h = 2r + a$.

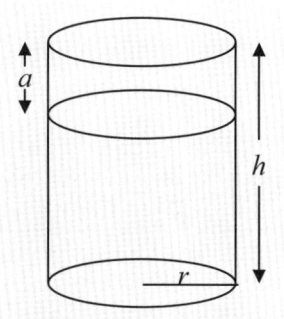

275

Chapter 10

Required Outcomes

After completing this chapter, a student should be able to:
- evaluate simple limits.
- differentiate simple functions from first principles. **(HL)**
- know and apply the limit theorems.
- differentiate functions of the form x^n.
- find the gradient of a curve at any point on it.
- find the gradient and equation of the tangent to a curve at a given point.
- find higher order derivatives.
- discuss the motion of a particle along a straight line given the displacement as a function of time.
- find the extreme and critical points of a continuous curve.
- find the position of any point of inflexion on a continuous curve.
- determine the general shape of a continuous curve from the first and second derivatives of the function. **(HL)**
- optimise simple functions.

11 Vector Geometry

11.1 Addition of Vectors – The Zero Vector – Scalar Multiples

Definition A *vector* is a directed line segment. That is to say, a vector has a given length and a given direction.

We denote the vector which joins the point A to the point B by $\overrightarrow{AB}$ or by *v*.

A is called the *initial point* or *tail* of $\overrightarrow{AB}$ and B is called the *terminal point* or *tip* of $\overrightarrow{AB}$.

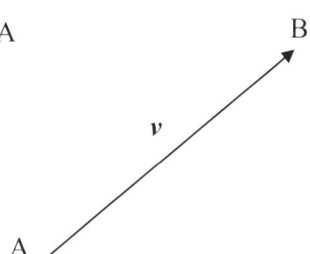

If the initial point of a vector is fixed, the vector is called a *bound* or *localised* vector. All other vectors are called *free* vectors.

The *length* of a vector $\overrightarrow{AB}$ is denoted by $|\overrightarrow{AB}|$ or simply AB.

The length of the vector *v* is denoted by $|v|$ or *v*.

The length of a vector is a *scalar* (real number).

The Zero Vector

The zero vector is a vector of length zero. We denote the zero vector by **0** or, because it is completely defined by its length, 0.

Multiplication by a Scalar

If *v* is a vector and *s* is a scalar, then *sv* is a vector.
(1)   If $s > 0$, *sv* has the same direction as *v* and a length equal to *s* times the length of *v*.
(2)   If $s < 0$, *sv* has a direction opposite to that of *v* and a length equal to $|s|$ times the length of *v*.
(3)   If $s = 0$, *sv* is the zero vector.

Chapter 11

Parallel Vectors

Two vectors *a* and *b* are parallel iff
(1)   they have the same direction, or
(2)   they have opposite directions.

Thus *a* and *b* are parallel iff one can be expressed as a scalar multiple of the other. That is, *a* and *b* are parallel iff *a* = *sb* where *s* is a scalar, or *b* = *ta* where *t* is a scalar.

Equal Vectors

Two vectors *a* and *b* are equal, *a* = *b*, iff they have the same length and the same direction.

Addition of Vectors

Vectors may be added in two different ways:
(1)   If the tails of *a* and *b* are together, complete the parallelogram which has *a* and *b* as two adjacent sides. Then *a* + *b* is that vector which joins the point of intersection of *a* and *b* to the opposite vertex.

ABCD is a parallelogram.

$\overrightarrow{AC}$ = *a* + *b*

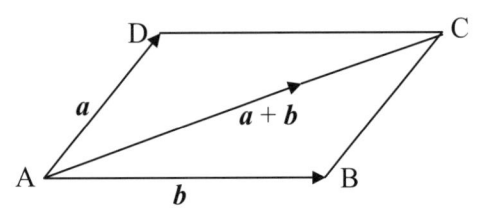

(2)   If the tail of *b* is joined to the tip of *a*, then *a* + *b* is that vector which joins the tail of *a* to the tip of *b*.

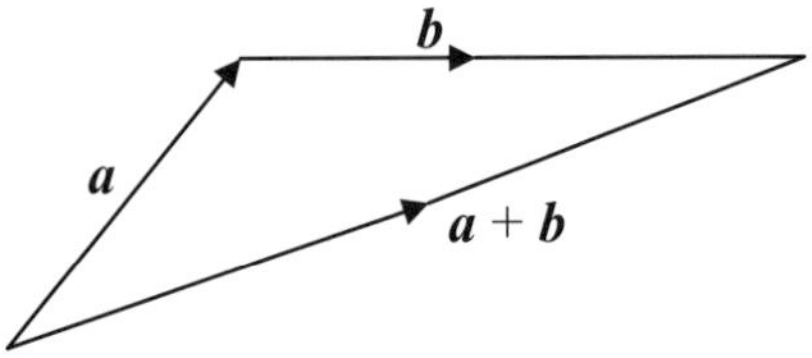

Subtraction of Vectors

The vector *a* − *b* is equal to *a* + (−*b*). Thus *a* − *b* can be found by adding vector −*b* to vector *a*.

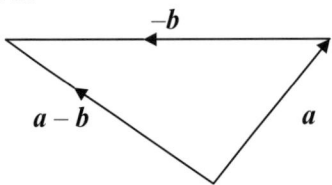

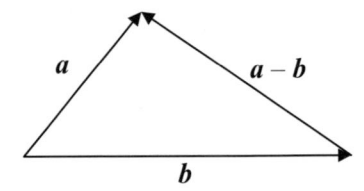

Vector Geometry

Example Simplify $\overrightarrow{AB} + \overrightarrow{BC} - \overrightarrow{DC}$.

$$\overrightarrow{AB} + \overrightarrow{BC} - \overrightarrow{DC} = \overrightarrow{AB} + \overrightarrow{BC} + \overrightarrow{CD}$$
$$= \overrightarrow{AC} + \overrightarrow{CD}$$
$$= \overrightarrow{AD}.$$

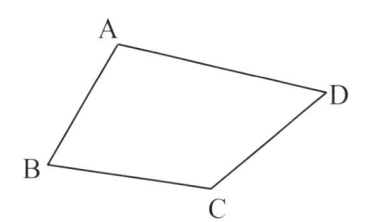

Example In the diagram, ABCD is a quadrilateral and E is an interior point. $\overrightarrow{AB} = a$, $\overrightarrow{BE} = b$, $\overrightarrow{DC} = c$ and $\overrightarrow{DE} = d$.

(a) Find expressions for $\overrightarrow{AE}$, $\overrightarrow{EC}$, $\overrightarrow{BC}$ and $\overrightarrow{AC}$ in terms of a, b, c and/or d.

(b) What can you say about the position of E if $a + b = c - d$?

(a) In triangle ABE, $\overrightarrow{AE} = a + b$.
In triangle CDE, $\overrightarrow{EC} = c - d$.
In triangle BCE, $\overrightarrow{BC} = b + c - d$.
In triangle ACE, $\overrightarrow{AC} = \overrightarrow{AE} + \overrightarrow{EC}$
$= a + b + c - d$.

(b) If $a + b = c - d$, then $\overrightarrow{AE} = \overrightarrow{EC}$ and so E is the mid-point of AC.

Example Vector a has a length of 6 and a direction east.
Vector b has a length of 8 and a direction north.
Find the length and direction of the vector
(a) $a + b$; (b) $a - b$.

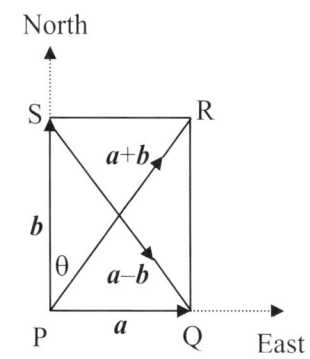

PR = SQ = 10 and $\tan \theta = \frac{3}{4} \Rightarrow \theta = 36.9°$.
$\overrightarrow{PR} = a + b$ and $\overrightarrow{SQ} = a - b$.

Therefore $a + b$ has length 10 and direction 036.9°.

Also $a - b$ has length 10 and direction 143.1°.

279

Chapter 11

Example Prove that the line joining the mid-points of any two sides of a triangle is parallel to the third side and equal in length to half the length of the third side.

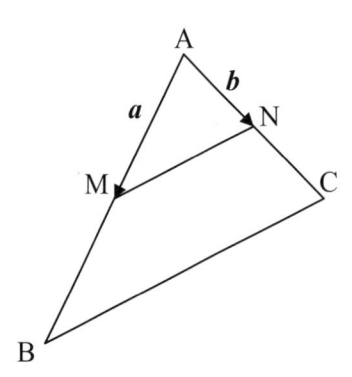

Let M and N be the mid-points of the sides AB and AC of the triangle ABC.

Let $\overrightarrow{AM} = a$ and $\overrightarrow{AN} = b$.

Then $\overrightarrow{AB} = 2a$ and $\overrightarrow{AC} = 2b$.

Now $\overrightarrow{MN} = b - a$ and

$\overrightarrow{BC} = 2b - 2a = 2(b - a) = 2\overrightarrow{MN}$.

Therefore MN is parallel to BC and $MN = \tfrac{1}{2} BC$, as required.

Exercise 11.1

1. With the aid of diagrams, show that vectors are both associative and commutative under addition.

2. Simplify each of the following:
   (a) $\overrightarrow{AB} + \overrightarrow{BC}$;
   (b) $\overrightarrow{AB} + \overrightarrow{BA}$;
   (c) $\overrightarrow{AB} - \overrightarrow{CB}$;
   (d) $\overrightarrow{AB} + \overrightarrow{BC} + \overrightarrow{CD}$;
   (e) $\overrightarrow{AB} - \overrightarrow{DC} - \overrightarrow{CB}$;
   (f) $\overrightarrow{AB} - \overrightarrow{DC} - \overrightarrow{DA} - \overrightarrow{CB}$.

3. In the pentagon ABCDE, $\overrightarrow{AB} = a$, $\overrightarrow{BC} = b$, $\overrightarrow{CD} = c$ and $\overrightarrow{BE} = d$. Find each of the following vectors in terms of *a*, *b*, *c* and *d*.
   (a) $\overrightarrow{AC}$;
   (b) $\overrightarrow{AD}$;
   (c) $\overrightarrow{AE}$;
   (d) $\overrightarrow{DE}$.

4. In the regular hexagon ABCDEF, $\overrightarrow{AB} = a$ and $\overrightarrow{BC} = b$. Find expressions in terms of *a* and/or *b* for each of the following vectors:
   (a) $\overrightarrow{DE}$;
   (b) $\overrightarrow{AD}$;
   (c) $\overrightarrow{CD}$;
   (d) $\overrightarrow{DF}$.

5. In the triangle PQR, M and N are points on PQ and PR such that $\overrightarrow{PM} = \tfrac{1}{3}\overrightarrow{PQ}$ and $\overrightarrow{PN} = \tfrac{1}{3}\overrightarrow{PR}$. Prove that MN is parallel to QR and equal in length to one-third of the length of QR.

6. In the triangle ABC, $\overrightarrow{AB} = a$, $\overrightarrow{AC} = b$ and $B\hat{A}C = 120°$. Given $|a| = 3$ and $|b| = 5$, find $|a - b|$.

Vector Geometry

7. The points P, Q, R, S are the mid-points of the sides AB, BC, CD, DA of the quadrilateral ABCD. Prove that PQRS is a parallelogram.

8. In the parallelogram ABCD, E is the mid-point of the diagonal AC. If $\overrightarrow{AB} = a$ and $\overrightarrow{AD} = b$, find in terms of *a* and/or *b* the vectors $\overrightarrow{BD}$ and $\overrightarrow{BE}$. Hence prove that the diagonals of a parallelogram bisect each other.

9. Vector *a* has a length 10 and a direction west, vector *b* has a length 20 and a direction south and vector *c* has a length 10 and a direction 030°. Find the length and direction of each of the following vectors:
   (a) $a + b$; (b) $a - b$; (c) $a + c$; (d) $b - c$.

*10. In parallelogram ABCD, M is the mid-point of BC and BD meets AM at E. Use vectors to prove that E is a point of trisection of both BD and AM.

*11. Prove that for any vectors *a* and *b*,
   (a) $|a - b| \geq |a| - |b|$; (b) $|a - b| \geq |b| - |a|$.

   Deduce that $|a - b| \geq ||a| - |b||$.
   Under what conditions is $|a - b| = ||a| - |b||$?

11.2 Position Vectors and the Ratio Formula (Optional)

Definition   The vector which joins the origin, O, to the point P is called the *position vector* of P with respect to O, or simply the position vector of P.

We sometimes denote the position vector of P by $\overrightarrow{P}$. That is, $\overrightarrow{OP} = \overrightarrow{P}$.

Consider the vector $\overrightarrow{AB}$, the position vector of B with respect to A.

From the diagram, $\overrightarrow{AB} = \overrightarrow{OB} - \overrightarrow{OA}$
Thus $\overrightarrow{AB} = \overrightarrow{B} - \overrightarrow{A}$.

(The rule is "*second minus first*".)

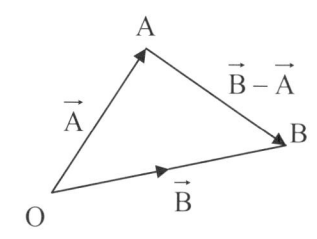

Chapter 11

Higher Level
(Optional)

The Ratio Formula

If P is a point on the line (AB) such that $\vec{AP} = \rho \vec{AB}$ where ρ is a scalar, then

$\vec{P} - \vec{A} = \rho(\vec{B} - \vec{A})$

$\Rightarrow \vec{P} = \vec{A} + \rho\vec{B} - \rho\vec{A}$

$\Rightarrow \vec{P} = (1-\rho)\vec{A} + \rho\vec{B}$.

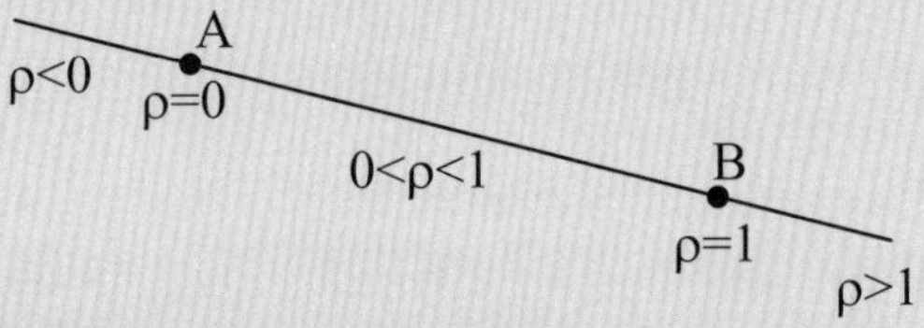

Thus:
(1) If $\rho = 0$, $\vec{AP} = 0$ and so P coincides with A.
(2) If $\rho = 1$, $\vec{AP} = \vec{AB}$ and so P coincides with B.
(3) If $\rho < 0$, $\vec{AP}$ and $\vec{AB}$ have opposite directions and so P is external to the line segment [AB] and closer to A than B.
(4) If $0 < \rho < 1$, $\vec{AP}$ and $\vec{AB}$ have the same direction and AP < AB. Thus P lies between A and B.
(5) If $\rho > 1$, $\vec{AP}$ and $\vec{AB}$ have the same direction and AP > AB. Thus P is external to the line segment [AB] and closer to B than to A.

In particular, if P is the mid-point of the line segment [AB], $\rho = \frac{1}{2}$ and $\vec{P} = \frac{1}{2}\vec{A} + \frac{1}{2}\vec{B}$.

Example Find the position vector of the point P which divides [AB]
(a) internally ; (b) externally,
in the ratio 1 : 3.

(a) Here $\vec{AP} = \frac{1}{4}\vec{AB}$.

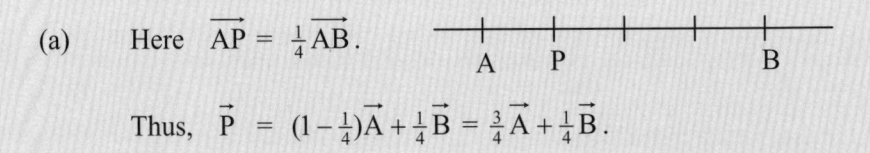

Thus, $\vec{P} = (1-\frac{1}{4})\vec{A} + \frac{1}{4}\vec{B} = \frac{3}{4}\vec{A} + \frac{1}{4}\vec{B}$.

(b) Here $\vec{AP} = -\frac{1}{2}\vec{AB}$.

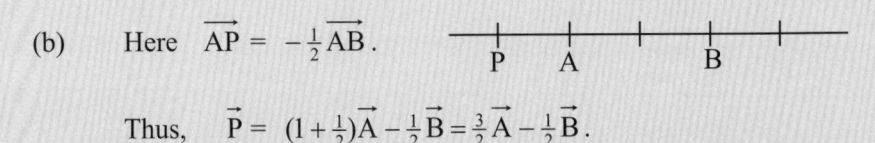

Thus, $\vec{P} = (1+\frac{1}{2})\vec{A} - \frac{1}{2}\vec{B} = \frac{3}{2}\vec{A} - \frac{1}{2}\vec{B}$.

Example Prove that the medians of a triangle are concurrent in a point which divides each median in the ratio 2 : 1.

Let AM be a median of the triangle ABC and let G be a point on [AM] such that AG : GM = 2 : 1.

Then $\vec{M} = \frac{1}{2}\vec{B} + \frac{1}{2}\vec{C}$ and $\vec{AG} = \frac{2}{3}\vec{AM}$.

Thus $\vec{G} = (1 - \frac{2}{3})\vec{A} + \frac{2}{3}\vec{M}$

$= \frac{1}{3}\vec{A} + \frac{2}{3}(\frac{1}{2}\vec{B} + \frac{1}{2}\vec{C})$

$= \frac{1}{3}(\vec{A} + \vec{B} + \vec{C})$.

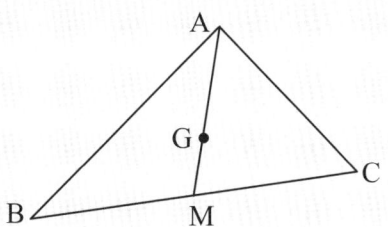

Similarly if BN is a median and H lies on BN such that BH : HN = 2 : 1, then $\vec{H} = \frac{1}{3}(\vec{A} + \vec{B} + \vec{C})$; and if CP is a median and K lies on CP such that CK : KP = 2 : 1, then $\vec{K} = \frac{1}{3}(\vec{A} + \vec{B} + \vec{C})$.

Therefore G, H and K coincide (same position vector) and the result is proved.

Exercise 11.2

1. Prove that if ABCD is a parallelogram then $\vec{A} + \vec{C} = \vec{B} + \vec{D}$ and hence prove that the diagonals of a parallelogram bisect each other.

2. The position vectors of A, B, C and D are $3a + b$, $a - b + 2c$, $a - 2b + c$ and $3a - c$ respectively. Show that ABCD is a parallelogram.

3. Prove that the points A, B, C with position vectors $a - 2b$, $2a + b$, $4a + 7b$ respectively are collinear.

4. In a regular hexagon OABCDE the position vectors of A and B relative to O are a and b respectively. Find expressions in terms of a and b for the vectors $\vec{AB}$ and $\vec{BC}$. Find also the position vectors of C, D and E.

Higher Level (Optional)

5. Find in terms of $\vec{A}$ and $\vec{B}$ the position vector of the point P which divides the line segment [AB]
   (a) internally in the ratio 2 : 1 ;
   (b) internally in the ratio 4 : 3 ;
   (c) internally in the ratio 2 : 3 ;
   (d) externally in the ratio 1 : 3 ;
   (e) externally in the ratio 5 : 2 ;
   (f) externally in the ratio 3 : 4.

Chapter 11

6. Prove that the points A, B and P are collinear iff $\vec{P}$ can be expressed in the form $\vec{P} = s\vec{A} + t\vec{B}$ where $s + t = 1$.

7. (a) If P is a point on the line (AB) such that P divides AB internally in the ratio $m : n$, show that $\vec{P} = \dfrac{m\vec{B} + n\vec{A}}{m + n}$.

   (b) If P is a point on the line (AB) such that P divides AB externally in the ratio $m : n$, show that $\vec{P} = \dfrac{m\vec{B} - n\vec{A}}{m - n}$.

8. In trapezium ABCD where AB is parallel to DC, AB has length 4 and CD has length 9. If P and Q are the mid-points of the diagonals BD and AC respectively, prove that PQ is parallel to AB and find the length of PQ.

*9. The vertices A, B and C of the parallelogram ABCD have position vectors *a*, *b* and *c* respectively. If M is the mid-point of BC and BD meets AM at N, find in terms of *a*, *b* and/or *c* the position vectors of M and N.

*10. A median of a tetrahedron is the line joining a vertex to the point of intersection of the medians of the opposite face. Prove that the medians of a tetrahedron are concurrent in a point which divides each median in the ratio 3 : 1.

*11. Prove that the three lines joining pairs of mid-points of opposite edges of a tetrahedron are concurrent.

11.3   Vectors in Cartesian 2-Space

We denote the position vector of the the point P(x, y) by $\overrightarrow{OP} = \vec{P} = \begin{pmatrix} x \\ y \end{pmatrix}$.

x is called the ***first component*** of the vector $\begin{pmatrix} x \\ y \end{pmatrix}$ and y is called the ***second component***. Thus the components of the position vector of a given point are simply the coordinates of that point.

Unit Vectors

A ***unit vector*** has a length of one unit.

Vector Geometry

The unit vector in the direction of the vector a is the vector $\hat{a} = \left(\dfrac{1}{|a|}\right) a$.

The unit vectors in the directions of the positive x and y axes are denoted by i and j respectively. Thus $i = \begin{pmatrix} 1 \\ 0 \end{pmatrix}$ and $j = \begin{pmatrix} 0 \\ 1 \end{pmatrix}$.

The Vector $s \begin{pmatrix} x \\ y \end{pmatrix}$

If s is a scalar and $\begin{pmatrix} x \\ y \end{pmatrix}$ is a vector then $s \begin{pmatrix} x \\ y \end{pmatrix} = \begin{pmatrix} sx \\ sy \end{pmatrix}$.

Addition

If $\begin{pmatrix} x_1 \\ y_1 \end{pmatrix}$ and $\begin{pmatrix} x_2 \\ y_2 \end{pmatrix}$ are vectors their sum is defined by $\begin{pmatrix} x_1 \\ y_1 \end{pmatrix} + \begin{pmatrix} x_2 \\ y_2 \end{pmatrix} = \begin{pmatrix} x_1 + x_2 \\ y_1 + y_2 \end{pmatrix}$.

This definition is consistent with the original definition of addition for if A, B are points with coordinates (x_1, y_1), (x_2, y_2) respectively and OACB is a parallelogram, then C has coordinates $(x_1 + x_2, y_1 + y_2)$.

Equality

Two vectors $\begin{pmatrix} x_1 \\ y_1 \end{pmatrix}$ and $\begin{pmatrix} x_2 \\ y_2 \end{pmatrix}$ are equal iff $x_1 = x_2$ and $y_1 = y_2$, that is their respective components are equal.

Unit Vector Representation

Since $\begin{pmatrix} x \\ y \end{pmatrix} = x \begin{pmatrix} 1 \\ 0 \end{pmatrix} + y \begin{pmatrix} 0 \\ 1 \end{pmatrix}$, every vector $\begin{pmatrix} x \\ y \end{pmatrix}$ can be expressed as a linear combination of i and j as follows: $\begin{pmatrix} x \\ y \end{pmatrix} = xi + yj$.

Chapter 11

The Vector $\overrightarrow{AB}$

Let $A(x_1, y_1)$ and $B(x_2, y_2)$ be any two points in Cartesian 2-space. Then the vector $\overrightarrow{AB}$ is given by $\overrightarrow{AB} = \begin{pmatrix} x_2 - x_1 \\ y_2 - y_1 \end{pmatrix}$.

Note: The rule is "*second minus first*".

The Length of Vector $\overrightarrow{AB}$

From the distance formula we have $|\overrightarrow{AB}| = \sqrt{(x_2 - x_1)^2 + (y_2 - y_1)^2}$.

Example   If $A = (3, 1)$ and $B = (6, -3)$ find
- (a) $\overrightarrow{AB}$;
- (b) $|\overrightarrow{AB}|$;
- (c)   the components of the unit vector in the direction of $\overrightarrow{AB}$.

(a) $\overrightarrow{AB} = \begin{pmatrix} 6-3 \\ -3-1 \end{pmatrix} = \begin{pmatrix} 3 \\ -4 \end{pmatrix}$.

(b) $|\overrightarrow{AB}| = \sqrt{9 + 16} = 5$.

(c)   The required unit vector is $\frac{1}{5}\begin{pmatrix} 3 \\ -4 \end{pmatrix} = \begin{pmatrix} \frac{3}{5} \\ -\frac{4}{5} \end{pmatrix}$.

Higher Level (Optional)

Example   Points A and B have coordinates $(1, 5)$ and $(-2, 2)$ respectively. Find the coordinates of the point
- (a)   P which divides [AB] internally in the ratio $2 : 1$;
- (b)   Q which divides [AB] externally in the ratio $2 : 1$.

(a)   A   P   B $\overrightarrow{AP} = \frac{2}{3}\overrightarrow{AB}$

$\Rightarrow \overrightarrow{P} = \frac{1}{3}\overrightarrow{A} + \frac{2}{3}\overrightarrow{B} = \frac{1}{3}\begin{pmatrix} 1 \\ 5 \end{pmatrix} + \frac{2}{3}\begin{pmatrix} -2 \\ 2 \end{pmatrix} = \begin{pmatrix} -1 \\ 3 \end{pmatrix}$ and $P = (-1, 3)$.

(b)   A   B   Q $\overrightarrow{AQ} = 2\overrightarrow{AB}$

$\Rightarrow \overrightarrow{Q} = -\overrightarrow{A} + 2\overrightarrow{B} = -\begin{pmatrix} 1 \\ 5 \end{pmatrix} + 2\begin{pmatrix} -2 \\ 2 \end{pmatrix} = \begin{pmatrix} -5 \\ -1 \end{pmatrix}$ and $Q = (-5, -1)$.

Vector Geometry

Example Show that the points A(1, –1), B(5, 1) and C(–1, –2) are collinear and find the ratio in which C divides [AB].

$$\overrightarrow{CB} = \begin{pmatrix} 6 \\ 3 \end{pmatrix} \text{ and } \overrightarrow{AC} = \begin{pmatrix} -2 \\ -1 \end{pmatrix} = -\frac{1}{3}\overrightarrow{CB}$$

Therefore $\overrightarrow{AC}$ and $\overrightarrow{CB}$ are parallel vectors and C is common to both.
Thus A, B and C are collinear and C divides [AB] externally in the ratio 1 : 3.

Exercise 11.3

1. Find $\overrightarrow{OA}$, $|\overrightarrow{OA}|$ and the components of the unit vector in the direction of $\overrightarrow{OA}$ when A has coordinates (a) (2, –1); (b) (2.5, 6); (c) (–5, –2).

2. Find $\overrightarrow{AB}$, $|\overrightarrow{AB}|$ and the components of the unit vector in the direction of $\overrightarrow{AB}$ when (a) A = (–2, 3) and B = (1, –1); (b) A = (6, 7) and B = (4, 9).

3. Find the unit vector in the direction of $\overrightarrow{AB}$ in each of the following:
   (a) A = (2, 3), B = (5, –1); (b) A = (3, –4), B = (–2, 8);
   (c) A = (3, 5), B = (–3, –1); (d) A = (5, 4), B = (8, 2).

4. Show that the points with position vectors $2i + 3j$, $5i - j$ and $-4i + 11j$ are collinear.

5. Prove that the points with position vectors $3i - j$, $6i + 3j$, $2i + 6j$ and $-i + 2j$ are vertices of a square.

6. The position vectors of the points A, B and D are $2i - 3j$, $6i + j$, $-3i + 7j$ respectively. Find the position vector of the point C if ABCD is a parallelogram.

7. Find the shape of the triangle with the following vertices.
   (a) (4, 1), (6, 4), (1, 3);
   (b) (–1, –2), (–6, 3), (6, –1);
   (c) (1, 1), (1, 4), $(1+\frac{3}{2}\sqrt{3}, \frac{5}{2})$.

8. Find the shape of the quadrilateral with the following vertices.
   (a) (3, –1), (5, 4), (6, 8), (4, 3);
   (b) (9, 2), (7, –2), (3, 0), (5, 4);
   (c) (–1, –2), (7, –1), (11, 6), (3, 5);
   (d) (2, 3), (10, 9), (5, –1), (13, 5).

Chapter 11

Higher Level (Optional)

9. Find the coordinates of the point P which divides the line segment [AB] internally such that
   (a) A = (3, 1), B = (−2, 11) and AP : PB = 1 : 4 ;
   (b) A = (−5, 2), B = (4, −1) and AP : PB = 2 : 1 ;
   (c) A = (−3, −2), B = (7, 13) and AP : PB = 3 : 2 ;
   (d) A = (4, −1), B = (8, 5) and AP : PB = 1 : 3.

10. Find the coordinates of the point Q which divides the line segment [AB] externally such that
    (a) A = (5, −1), B = (3, 4) and AQ : QB = 3 : 2 ;
    (b) A = (−1, 4), B = (8, 7) and AQ : QB = 1 : 4 ;
    (c) A = (6, −5), B = (−2, −9) and AQ : QB = 1 : 3 ;
    (d) A = (10, −1), B = (8, 0) and AQ : QB = 3 : 4.

11. Show that the centroid of triangle ABC where A = (x_1, y_1), B = (x_2, y_2) and C = (x_3, y_3) has coordinates $\left(\frac{1}{3}[x_1 + x_2 + x_3], \frac{1}{3}[y_1 + y_2 + y_3]\right)$.

12. The points A, B and C have coordinates (2, −1), (6, 3) and (4, 7) respectively. Point P is the mid-point of [AC], point Q divides [AB] internally in the ratio 3 : 1 and point R divides [BC] externally in the ratio 1 : 3. Find the coord-inates of P, Q and R and show that these points are collinear. Find also the ratio in which R divides [PQ].

13. The point P divides the line segment [AB] where A = (3, 1) and B = (−2, 6) internally in the ratio k : 1 (k > 0). Find the value of k if P also lies on the line 5x − y = 2.

14. In the triangle ABC, P divides AB internally in the ratio 2 : 1 and Q divides BC internally in the ratio 3 : 2. The side AC is produced to meet the line PQ produced in R. Find the position vector of R in terms of the position vectors of A and C.

11.4 The Scalar Product of Two Vectors – Orthogonal Projection

Definition   The *scalar product* of the vectors $a = a_1 i + a_2 j$ and $b = b_1 i + b_2 j$ is denoted and defined by

$$a \cdot b = |a||b|\cos\theta$$

where θ is the angle between *a* and *b*.

This product is called the *scalar* product since the result $|a||b|\cos\theta$ is a scalar.

The scalar product is often called the *dot* product since a 'dot' is used to denote it.

Alternative definition $\quad a \bullet b = a_1 b_1 + a_2 b_2$

Thus $\quad |a||b|\cos\theta = a_1 b_1 + a_2 b_2$.

Proof Let A, B have coordinates $(a_1, a_2), (b_1, b_2)$ respectively.
Then $\overrightarrow{OA} = a$ and $\overrightarrow{OB} = b$.

Let θ be the angle between vectors *a* and *b* as shown in the diagram on the right.

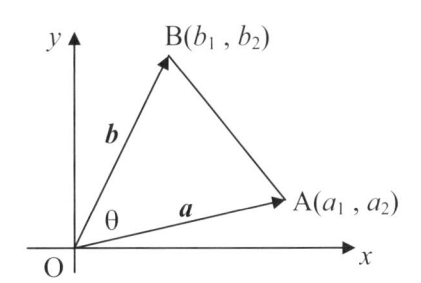

Now in triangle OAB, $AB^2 = OA^2 + OB^2 - 2(OA)(OB)\cos\theta$ {cosine rule}
and so $(a_1 - b_1)^2 + (a_2 - b_2)^2 = a_1^2 + a_2^2 + b_1^2 + b_2^2 - 2(OA)(OB)\cos\theta$ which gives
$a_1^2 - 2a_1 b_1 + b_1^2 + a_2^2 - 2a_2 b_2 + b_2^2 = a_1^2 + a_2^2 + b_1^2 + b_2^2 - 2(OA)(OB)\cos\theta$
and finally gives $a_1 b_1 + a_2 b_2 = |\overrightarrow{OA}||\overrightarrow{OB}|\cos\theta = |a||b|\cos\theta$ which completes the proof.

Perpendicular Vectors

Non-zero vectors *a* and *b* are perpendicular iff $a \bullet b = 0$.

Proof (a) If *a* and *b* are perpendicular the angle between them is 90° and so $\cos\theta = 0$.
Therefore $a \bullet b = |a||b|(0) = 0$.

(b) If $a \bullet b = 0$ then $|a||b|\cos\theta = 0$.
But *a* and *b* are non-zero vectors and so $|a| \neq 0$ and $|b| \neq 0$.
Therefore $\cos\theta = 0$ and so θ = 90°.
Thus *a* and *b* are perpendicular.

Note: Since $\begin{pmatrix} a \\ b \end{pmatrix} \bullet \begin{pmatrix} b \\ -a \end{pmatrix} = ab - ba = 0$, we can always find a vector which is perpendicular to a given vector in 2-space by simply interchanging the components and then changing the sign of either one of them.

289

Chapter 11

Some Properties of the Scalar Product

1. $a \cdot b = b \cdot a$
2. $a \cdot a = |a|^2$
3. $a \cdot (b + c) = (a \cdot b) + (a \cdot c)$
4. $(a + b) \cdot (a - b) = |a|^2 - |b|^2$

The Angle between Two Vectors

Since $a \cdot b = |a||b|\cos\theta$, then $\cos\theta = \dfrac{a \cdot b}{|a||b|}$.

Thus given any two non-zero vectors, the cosine of the angle between them (and hence the angle itself) can be found by dividing the scalar product of the vectors by the product of their lengths.

Example Given vectors $a = 3i - 2j$ and $b = -2i - 4j$, find $a \cdot b$.

$a \cdot b = 3(-2) + (-2)(-4) = 2$.

Example Show that the triangle ABC in which A = (3, −1), B = (5, 4) and C = (0, 6) is right-angled.

$\overrightarrow{AB} = 2i + 5j$, $\overrightarrow{BC} = -5i + 2j$ and so $\overrightarrow{AB} \cdot \overrightarrow{BC} = 2(-5) + 5(2) = 0$.
Thus AB is perpendicular to BC and so the triangle ABC is right-angled.

Example Find the size of $A\hat{B}C$ of the triangle ABC where A = (3, 10), B = (1, 4) and C = (7, 7). [**Note:** $A\hat{B}C$ is the angle between vectors $\overrightarrow{BA}$ and $\overrightarrow{BC}$ (B is the tail of both vectors).]

$$\cos A\hat{B}C = \dfrac{\overrightarrow{BA} \cdot \overrightarrow{BC}}{|\overrightarrow{BA}||\overrightarrow{BC}|} = \dfrac{(2i+6j) \cdot (6i+3j)}{\sqrt{40} \times \sqrt{45}} = \dfrac{12+18}{2\sqrt{10} \times 3\sqrt{5}} = \dfrac{1}{\sqrt{2}}.$$

Therefore $A\hat{B}C = 45°$.

Orthogonal Projection

Definition The **orthogonal projection** of a on b is defined to be

$$|a|\cos\theta = \dfrac{a \cdot b}{|b|} = a \cdot \hat{b}$$

where $\hat{b}$ is the unit vector in the direction of b.

Vector Geometry

When the angle θ between *a* and *b* is acute, the orthogonal projection of *a* on *b* is equal to the length of the vector $\overrightarrow{OM}$.

(See diagram on the right.)

When the angle θ between *a* and *b* is obtuse, the orthogonal projection of *a* on *b* is equal to the negative of the length of $\overrightarrow{OM}$. (See diagram on the right.)

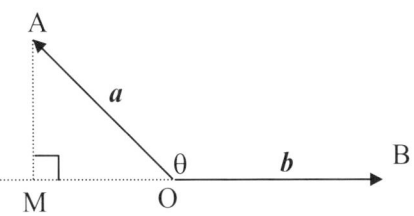

When *a* and *b* are perpendicular, the orthogonal projection of *a* on *b* is zero.

The orthogonal projection of *a* on *b* is sometimes described as the projection of *a* in the direction of *b*.

Example   Find the (orthogonal) projection of $a = 3i + 10j$ on $b = 4i - 3j$.

$$\text{The required projection} = \frac{a \cdot b}{|b|} = -\frac{18}{5} = -3.6.$$

Example   The vertices of the triangle ABC are A(2, −1), B(7, 5) and C(4, 13).
  (a)   Write down vectors $\overrightarrow{AB}$ and $\overrightarrow{AC}$.
  (b)   Find a vector *n* which is perpendicular to $\overrightarrow{AB}$.
  (c)   Find the projection of $\overrightarrow{AC}$ in the direction of *n*.
  (d)   Calculate the area of the triangle ABC.

(a) $\overrightarrow{AB} = \begin{pmatrix} 5 \\ 6 \end{pmatrix}$ and $\overrightarrow{AC} = \begin{pmatrix} 2 \\ 14 \end{pmatrix}$

(b) $n = \begin{pmatrix} -6 \\ 5 \end{pmatrix}$

(c)   The required projection $h = \dfrac{|\overrightarrow{AC} \cdot n|}{|n|} = \dfrac{\left|\begin{pmatrix} 2 \\ 14 \end{pmatrix} \cdot \begin{pmatrix} -6 \\ 5 \end{pmatrix}\right|}{\sqrt{6^2 + 5^2}} = \dfrac{58}{\sqrt{61}}.$

(d)   The area of ABC $= \dfrac{1}{2}(AB)\left(\dfrac{58}{\sqrt{61}}\right) = \dfrac{1}{2}\left(\sqrt{61}\right)\left(\dfrac{58}{\sqrt{61}}\right) = 29.$

291

Chapter 11

Exercise 11.4

1. Evaluate the following scalar products:

    (a) $\begin{pmatrix}5\\3\end{pmatrix} \cdot \begin{pmatrix}2\\-3\end{pmatrix}$; (b) $\begin{pmatrix}-3\\1\end{pmatrix} \cdot \begin{pmatrix}-2\\-4\end{pmatrix}$; (c) $\begin{pmatrix}7\\2\end{pmatrix} \cdot \begin{pmatrix}3\\5\end{pmatrix}$;

    (d) $\begin{pmatrix}1\\4\end{pmatrix} \cdot \begin{pmatrix}-2\\3\end{pmatrix}$; (e) $\begin{pmatrix}12\\5\end{pmatrix} \cdot \begin{pmatrix}-3\\4\end{pmatrix}$; (f) $\begin{pmatrix}-2\\5\end{pmatrix} \cdot \begin{pmatrix}10\\4\end{pmatrix}$;

    (g) $\begin{pmatrix}15\\-2\end{pmatrix} \cdot \begin{pmatrix}3\\20\end{pmatrix}$; (h) $\begin{pmatrix}-8\\0\end{pmatrix} \cdot \begin{pmatrix}-3\\12\end{pmatrix}$; (i) $\begin{pmatrix}-7\\2\end{pmatrix} \cdot \begin{pmatrix}-5\\6\end{pmatrix}$.

2. Evaluate each of the following scalar products:
    (a) $i \cdot i$; (b) $i \cdot j$;
    (c) $j \cdot j$; (d) $(2i - 3j) \cdot (4i + j)$;
    (e) $(5i + 4j) \cdot (2i - 3j)$; (f) $i \cdot (4i - 5j)$;
    (g) $(j - 2i) \cdot (3i + 5j)$; (h) $2j \cdot (4i + 3j)$;
    (i) $(4i + 3j) \cdot (3j - 4i)$; (j) $(12i - 5j) \cdot (-8i - 20j)$.

3. Use the scalar product to show that in each of the following the three given points are the vertices of a right-angled triangle:
    (a) (5, −1), (−2, 4), (3, 11) ; (b) (1, 8), (−3, 4), (2, 7) ;
    (c) (−2, −1), (2, 5), (3, 0) ; (d) (−1, −1), (5, 7), (−2, 0).

4. Find the cosine of the angle between each of the following pairs of vectors:
    (a) $a = -i + 2j$, $b = -2i + j$; (b) $a = 4i + 3j$, $b = -5i + 12j$;
    (c) $a = 8i - j$, $b = 4i + 7j$; (d) $a = 3i - 2j$, $b = 6i + 9j$;
    (e) $a = 3j - 2i$, $b = -3i + 4j$; (f) $a = -4i + j$, $b = 5i - 12j$.

5. Find the projection of a on b in each of the following:
    (a) $a = 4i - j$, $b = 3i + 4j$; (b) $a = 2j - 3i$, $b = 5i + 12j$;
    (c) $a = 4i - 3j$; $b = i + j$; (d) $a = 6j - 2i$, $b = 24i + 7j$;
    (e) $a = i - 3j$, $b = 9i + j$; (f) $a = -5i - 2j$, $b = 3i + j$.

6. Write down a vector which is perpendicular to each of the following vectors:
    (a) $\begin{pmatrix}-2\\9\end{pmatrix}$; (b) $\begin{pmatrix}6\\5\end{pmatrix}$; (c) $\begin{pmatrix}-7\\-11\end{pmatrix}$; (d) $3i + 5j$;
    (e) $2i + 7j$; (f) $4j - i$; (g) $-4i - 5j$; (h) $j - 2i$.

7. Use the technique of the last example on the previous page to calculate the area of each of the following triangles whose vertices are given:
    (a) (2, −3), (9, 2), (−1, 12) ; (b) (−3, −3), (3, 5), (−5, 11) ;
    (c) (1, 7), (−3, 5), (2, 8) ; (d) (10, 9), (25, 17), (13, 5).

Vector Geometry

8. The vertices of the quadrilateral ABCD are A(1, 3), B(10, 15), C(7, 18) and D(1, 10).
   (a) Show that (AB) is parallel to (CD).
   (b) Find the lengths of [AB] and [CD].
   (c) What can you say about the quadrilateral?
   (d) Write down a vector **n** which is perpendicular to vector $\overrightarrow{AB}$.
   (e) Find the projection of $\overrightarrow{AD}$ in the direction of **n**.
   (f) Calculate the area of ABCD.

9. Show that the points A(2, 5), B(6, –2) and D(–5, 1) are three vertices of a square ABCD and find the coordinates of C.

10. The origin O and the point A(4, –3) are two vertices of a square OABC. Find the coordinates of B and C.

11. Find the angles of the triangle whose vertices are:
    (a) (2, 1), (5, 3), (0, 4);   (b) (–5, 2), (–3, 3), (–1, –1);
    (c) (1, 3), (5, 5), (4, 12);   (d) (–1, 2), (7, 1), (0, 8).

12. If $a = i - j$ and $b = 7i + j$, find
    (a) $|a|$;   (b) the projection of **b** on **a**;
    (c) $|a - b|$;   (d) the projection of **b** on **a** – **b**.

13. Prove that the diagonals of a rhombus are perpendicular.

14. Prove that the angle in a semi-circle is a right-angle.

Higher Level

15. Prove that the three altitudes of a triangle are concurrent.

16. Triangle ABC is right-angled at A and P, Q are points of trisection of [BC]. Prove that $AP^2 + AQ^2 = \frac{5}{9} BC^2$.

17. In the trapezium ABCD, AB is parallel to DC. If AB = x, AD = y, DC = 2x and E divides [BC] in the ratio 1 : 2, prove that $\overrightarrow{AC} \cdot \overrightarrow{DE} = \frac{2}{3}(4x^2 - y^2)$.

18. Prove that if $|a - b| = |a + b|$ then **a** and **b** are perpendicular.

19. Find the value of t if the angle between the vectors $a = 8i + j$ and $b = 4i + tj$ is $\arccos \frac{5}{13}$.

Chapter 11

> 20. Use the result $\mathbf{a} \cdot \mathbf{a} = |\mathbf{a}|^2$ to prove Apollonius' theorem: The sum of the squares of the lengths of two sides of a triangle is equal to twice the sum of the squares of the lengths of half the third side and the median to the third side.
>
> i.e. If M is the mid-point of BC, prove that $AB^2 + AC^2 = 2(BM^2 + AM^2)$.
>
> [**Hint:** Let $\overrightarrow{AB} = \mathbf{b}$ and $\overrightarrow{AC} = \mathbf{c}$.]

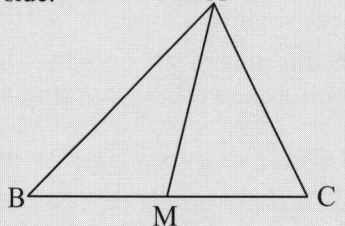

11.5 The Vector Equation of a Line in Cartesian 2-Space

Consider the line ℓ which passes through the point A with position vector $\mathbf{a} = \begin{pmatrix} a_1 \\ a_2 \end{pmatrix}$ and which is parallel to the vector $\mathbf{v} = \begin{pmatrix} v_1 \\ v_2 \end{pmatrix}$. Let $\mathbf{r} = \begin{pmatrix} x \\ y \end{pmatrix}$ be the position vector of a general point P on ℓ.

Since AP is parallel to $\mathbf{v}$, $\overrightarrow{AP} = t\mathbf{v}$ where t is a scalar.

Thus $\overrightarrow{OP} - \overrightarrow{OA} = t\mathbf{v}$ or $\mathbf{r} = \mathbf{a} + t\mathbf{v}$.

The equation $\mathbf{r} = \mathbf{a} + t\mathbf{v}$ or $\begin{pmatrix} x \\ y \end{pmatrix} = \begin{pmatrix} a_1 \\ a_2 \end{pmatrix} + t \begin{pmatrix} v_1 \\ v_2 \end{pmatrix}$ is the vector equation of the line ℓ.

The conversion of the vector equation to the normal Cartesian equation is illustrated in the following example.

Example Write the vector equation $\begin{pmatrix} x \\ y \end{pmatrix} = \begin{pmatrix} 2 \\ -1 \end{pmatrix} + t \begin{pmatrix} 3 \\ -2 \end{pmatrix}$ in the form $ax + by = c$.

$\begin{pmatrix} x \\ y \end{pmatrix} = \begin{pmatrix} 2 \\ -1 \end{pmatrix} + t \begin{pmatrix} 3 \\ -2 \end{pmatrix} = \begin{pmatrix} 2 + 3t \\ -1 - 2t \end{pmatrix}$ which gives $x = 2 + 3t$ and $y = -1 - 2t$.

Making t the subject of each equation gives $t = \dfrac{x-2}{3}$ and $t = \dfrac{y+1}{-2}$. Hence $\dfrac{x-2}{3} = \dfrac{y+1}{-2}$ and so $-2x + 4 = 3y + 3$ or $2x + 3y = 1$.

Motion of a Body Moving in a Straight Line in Cartesian 2-Space

In the following work, the vector components each represent a displacement of 1 unit of distance either in the direction of the x-axis or in the direction of the y-axis.

A body moves in a straight line in the direction of the vector $v = \begin{pmatrix} v_1 \\ v_2 \end{pmatrix}$. If the body starts from point A with position vector a at time $t = 0$, the position vector of the body at any subsequent time t is given by $r = a + tv$.

The vector v is called the **velocity vector** of the body and the length of this vector, $|v|$, denotes the (constant) **speed** of the body.

Example   A body initially at the point $(2, 1)$ has a velocity vector $3i - 4j$. If the distance unit is a metre and the time unit is a second, find:
(a)   the position vector of the body after 8 seconds ;
(b)   the speed of the body.

The position vector of the body after t seconds is given by
$$r = 2i + j + t(3i - 4j).$$
(a)   When $t = 8$, $r = 2i + j + 8(3i - 4j) = 26i - 31j$.
(b)   The speed of the body $= |3i - 4j| = 5 \text{ m s}^{-1}$.

Example   If a body moving in the direction of the vector $7i + 24j$ where the component unit is a kilometre and the speed of the body is 200 m s^{-1}, find the body's velocity vector.

Let $v = k(7i + 24j)$. Then $|v| = k\sqrt{7^2 + 24^2} = 25k$. But $|v| = 200$, so $k = 8$. Therefore the velocity vector is $v = 8(7i + 24j) = 56i + 192j$.

The Intersection of two Lines with Equations given in Vector Form

In order to find the point of intersection of two lines whose equations are given in vector form, each equation must have a separate parameter. The method is illustrated in the following example.

Chapter 11

Example Find the point of intersection of the lines $r = 3i + j + t_1(2i - j)$ and $r = 5i - 12j + t_2(i + j)$.

[These lines must meet since their direction vectors, $2i - j$ and $i + j$ are not parallel.]

By equating the components in the first line we have $x = 3 + 2t_1$, $y = 1 - t_1$, and in the second line we have $x = 5 + t_2$, $y = -12 + t_2$.

For intersection
$$3 + 2t_1 = 5 + t_2 \quad \text{and}$$
$$1 - t_1 = -12 + t_2.$$

Solving these equations simultaneously we find that $t_1 = 5$ and $t_2 = 8$. Hence these lines meet at the point with position vector $13i - 4j$.

Exercise Car A has position vector $r = \begin{pmatrix} 5 \\ 1 \end{pmatrix} + t_1 \begin{pmatrix} 2 \\ 3 \end{pmatrix}$ t_1 minutes after noon on a given day. Car B has position vector $r = \begin{pmatrix} -5 \\ 11 \end{pmatrix} + t_2 \begin{pmatrix} 3 \\ 2 \end{pmatrix}$ t_2 minutes after noon the same day. Show that the cars collide and find the time of collision.

For intersection of the two directions we have $5 + 2t_1 = -5 + 3t_2$ and
$$1 + 3t_1 = 11 + 2t_2.$$

Solving gives $t_1 = t_2 = 10$.

Since the cars start at the same time and each travels for 10 minutes to reach the point of intersection, the cars collide. The time of collision is 12:10.

Exercise 11.5

1. Find a vector equation for the straight line passing through the point A and parallel to the vector v in each of the following:
   (a) $A = (3, 2)$, $v = 3i + j$;
   (b) $A = (2, -1)$, $v = -i - j$;
   (c) $A = (-4, 0)$, $v = i$;
   (d) $A = (2, -5)$, $v = -7i + 3j$.

2. Find a vector equation for the line joining the following pairs of points:
   (a) $(4, 1)$ and $(6, 2)$;
   (b) $(3, -2)$ and $(-1, 1)$;
   (c) $(-2, -3)$ and $(-2, 5)$;
   (d) $(4, -2)$ and $(9, 8)$.

3. Find a Cartesian equation in the form $ax + by = c$ for each line in Question 1.

4. (a) Show that if the gradient of a straight line is $\dfrac{p}{q}$, then the vector $v = qi + pj$ is parallel to the line and find a vector which is perpendicular to the line.

Vector Geometry

(b) Show that vector $v = ai + bj$ is perpendicular to the line $ax + by = c$ and find a vector which is parallel to the line.

5. Find a vector which is parallel to each of the following lines:
   (a) $y = 2x - 1$;
   (b) $y = 3 - \frac{2}{3}x$;
   (c) $x + 2y = 5$;
   (d) $3x - 4y = 5$;
   (e) $3x + 2 = 0$;
   (f) $5y = 2$.

6. Find a Cartesian equation for a line which passes through the point A and which is perpendicular to the vector v, in each of the following:
   (a) $A = (2, 3), v = 3i + j$;
   (b) $A = (2, -1), v = 2i - 5j$;
   (c) $A = (-1, 0), v = -6i + 5j$;
   (d) $A = (4, -2), v = 5i + j$;
   (e) $A = (5, -3), v = 5i$;
   (f) $A = (-2, -5), v = j$.

7. Find a Cartesian equation for a line which passes through the point A and which is parallel to the vector v, in each of the following:
   (a) $A = (-1, 2), v = 2i + 3j$;
   (b) $A = (-2, -3), v = 5i - 4j$;
   (c) $A = (0, 3), v = i - 2j$;
   (d) $A = (5, 3), v = i + 3j$;
   (e) $A = (-2, -4), v = i$;
   (f) $A = (3, 4), v = -2j$.

8. A car is moving in the direction of vector a with a speed s km hr^{-1}. Find the velocity vector in each of the following.
   (a) $a = 3i - 4j, s = 65$;
   (b) $a = i + j, s = 70.7$;
   (c) $a = 0.28i + 0.96j, s = 75$;
   (d) $a = 2i + 3j, s = 78$.

9. At time t seconds, the position vector r of a moving particle is given by
   $r = \begin{pmatrix} 3 \\ -1 \end{pmatrix} + t \begin{pmatrix} -5 \\ 12 \end{pmatrix}$. (Displacement unit = 1 metre.)
   (a) What is the initial position vector of the particle?
   (b) Find the (constant) speed of the particle.
   (c) Show that the particle passes through the point with position vector $-49.5i + 125j$. At what time is this?

10. Particle A starts from the point with position vector $46i + 8j$ and travels with velocity vector $3i + 4j$. Three minutes later, particle B starts from the point with position vector $10i + 15j$ and travels with velocity vector $12i + 5j$.
    (a) If a displacement unit is 1 metre, find the speed of each particle.
    (b) Show that these cars collide.
    (c) Find the position vector of the point of collision.
    (d) When do they collide?

Chapter 11

11.6 Vectors in Cartesian 3-Space

Almost all of the work with vectors in 2-space can be extended to 3-space by adding another coordinate or another component.

We denote the *position vector* of the point $P(x, y, z)$ by $\overrightarrow{OP} = \vec{P} = x\mathbf{i} + y\mathbf{j} + z\mathbf{k}$.
z is called the *third component* of the vector $x\mathbf{i} + y\mathbf{j} + z\mathbf{k}$.

The *unit vectors* in 3-space are $\mathbf{i} = \begin{pmatrix} 1 \\ 0 \\ 0 \end{pmatrix}$, $\mathbf{j} = \begin{pmatrix} 0 \\ 1 \\ 0 \end{pmatrix}$ and $\mathbf{k} = \begin{pmatrix} 0 \\ 0 \\ 1 \end{pmatrix}$.

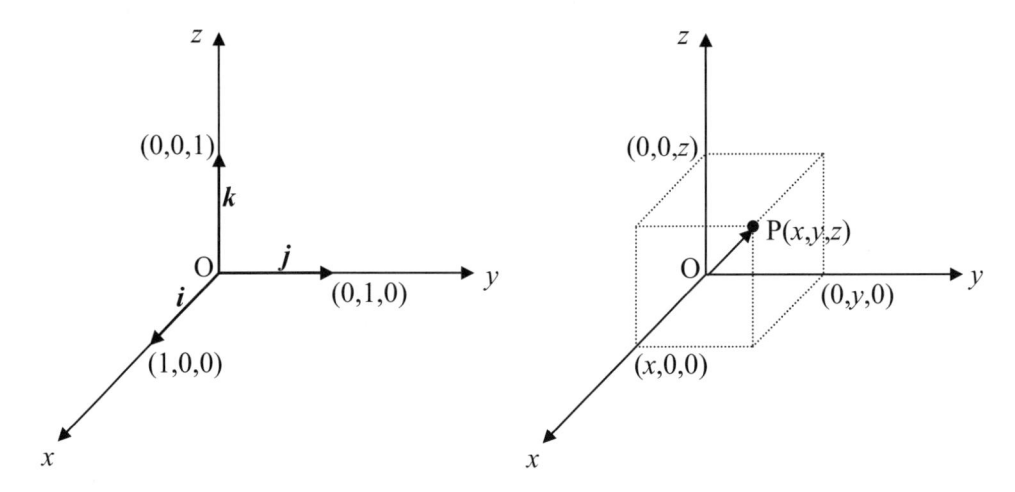

A *scalar multiple* of the vector $x\mathbf{i} + y\mathbf{j} + z\mathbf{k}$ is defined as follows:
$$s(x\mathbf{i} + y\mathbf{j} + z\mathbf{k}) = (sx)\mathbf{i} + (sy)\mathbf{j} + (sz)\mathbf{k}.$$

Vectors $x\mathbf{i} + y\mathbf{j} + z\mathbf{k}$ and $u\mathbf{i} + v\mathbf{j} + w\mathbf{k}$ are *equal* iff $x = u$, $y = v$ and $z = w$.

The *sum* of two vectors $x\mathbf{i} + y\mathbf{j} + z\mathbf{k}$ and $u\mathbf{i} + v\mathbf{j} + w\mathbf{k}$ is defined as follows:
$$(x\mathbf{i} + y\mathbf{j} + z\mathbf{k}) + (u\mathbf{i} + v\mathbf{j} + w\mathbf{k}) = (x + u)\mathbf{i} + (y + v)\mathbf{j} + (z + w)\mathbf{k}.$$

If $A = (x_1, y_1, z_1)$ and $B = (x_2, y_2, z_2)$, then $\overrightarrow{AB} = (x_2 - x_1)\mathbf{i} + (y_2 - y_1)\mathbf{j} + (z_2 - z_1)\mathbf{k}$ and $|\overrightarrow{AB}| = \sqrt{(x_2 - x_1)^2 + (y_2 - y_1)^2 + (z_2 - z_1)^2}$.

The *scalar product* of the vectors $\mathbf{a} = a_1\mathbf{i} + a_2\mathbf{j} + a_3\mathbf{k}$ and $\mathbf{b} = b_1\mathbf{i} + b_2\mathbf{j} + b_3\mathbf{k}$ is denoted and defined by $\mathbf{a} \cdot \mathbf{b} = |\mathbf{a}||\mathbf{b}|\cos\theta$ where θ is the angle between $\mathbf{a}$ and $\mathbf{b}$. As before we can show that $\mathbf{a} \cdot \mathbf{b} = a_1b_1 + a_2b_2 + a_3b_3$.

Example Quadrilateral ABCD has A = (2, 1, –2), B = (4, 2, –4), C = (5, –1, 2) and D = (3, –2, 4). Show that ABCD is a parallelogram.

$\overrightarrow{AB} = 2i + j - 2k$ and $\overrightarrow{DC} = 2i + j - 2k$.

Therefore $\overrightarrow{AB} = \overrightarrow{DC}$ and so AB = DC and (AB) is parallel to (DC). Thus ABCD is a parallelogram.

Higher Level

Example Show that the components of the unit vector in the direction of vector *a* are equal to the cosines of the angles that *a* makes with the coordinate axes.

If α, β, γ are the angles that $a = a_1 i + a_2 j + a_3 k$ makes with the x-axis, y-axis, z-axis respectively, then $\cos\alpha = \dfrac{a \cdot i}{|a||i|} = \dfrac{a_1}{|a|}$, $\cos\beta = \dfrac{a \cdot j}{|a||j|} = \dfrac{a_2}{|a|}$ and $\cos\gamma = \dfrac{a \cdot k}{|a||k|} = \dfrac{a_3}{|a|}$. [These are called the *direction cosines* of *a*.]

The unit vector in the direction of *a* is $\dfrac{1}{|a|} a = \begin{pmatrix} a_1/|a| \\ a_2/|a| \\ a_3/|a| \end{pmatrix} = \begin{pmatrix} \cos\alpha \\ \cos\beta \\ \cos\gamma \end{pmatrix}$ which is as required.

Example Find the angles between the vector $a = 2i - 2j + k$ and the coordinate axes.

$|a| = \sqrt{2^2 + (-2)^2 + 1^2} = 3$ and so the unit vector in the direction of *a* is $\hat{a} = \begin{pmatrix} 2/3 \\ -2/3 \\ 1/3 \end{pmatrix}$. Thus $\cos\alpha = \tfrac{2}{3} \Rightarrow \alpha = 48.2°$, $\cos\beta = -\tfrac{2}{3} \Rightarrow \beta = 131.8°$, and $\cos\gamma = \tfrac{1}{3} \Rightarrow \gamma = 70.5°$.

Exercise 11.6

1. Given $a = 2i + 3j - 6k$ and $b = 4j + 3k$ find
   (a) $a + b$;
   (b) $a - b$;
   (c) $2a - 3b$;
   (d) $|a|$;
   (e) $|b|$;
   (f) $|a + b|$.

Chapter 11

2. For each of the following points P, find the length of the vector $\overrightarrow{OP}$ and the angles which $\overrightarrow{OP}$ makes with the coordinate axes:
   (a) (2, 2, 1); (b) (1, 0, –1); (c) (3, –2, 6); (d) (0, –4, 3).

3. Points A and B have coordinates (3, 2, –1) and (6, –4, 8) respectively. Find the coordinates of the point which divides [AB] internally in the ratio
   (a) 2 : 1; (b) 1 : 3; (c) 4 : 5; (d) 5 : 2.

4. Points A and B have coordinates (–1, 4, 3) and (5, –8, –3) respectively. Find the coordinates of the point which divides [AB] externally in the ratio
   (a) 1 : 2; (b) 3 : 2; (c) 4 : 1; (d) 2 : 5.

5. Find the shape of the quadrilateral whose vertices are:
   (a) (3, 4, 1), (5, 0, –1), (1, –1, 6), (–1, 3, 8) ;
   (b) (2, 3, –1), (12, 14, 1), (2, 4, –4), (–8, –7, –6) ;
   (c) (–1, 2, –2), (3, –2, 1), (5, 3, 5), (1, 7, 2) ;
   (d) (2, –1, –1), (4, 1, 0), (3, 3, 2), (1, 1, 1) ;
   (e) (5, 1, 2), (3, –1, 5), (9, 0, 4), (7, –2, 7).

6. Find the angles of the triangle whose vertices are
   (a) (3, 1, 1), (4, –1, 3), (5, 3, 2) ;
   (b) (2, –1, 3), (3, 2, 4), (6, –4, –1) ;
   (c) (1, 2, –2), (3, 3, –1), (–1, 1, –1) ;
   (d) (2, 4, 1), (0, 2, 2), (3, 5, 1).

7. Prove that the points (1, 1, 0), (5, –5, 2), (–1, 4, –1) and (13, –17, 6) are collinear.

8. Find the scalar t if the vectors $ti - 2j + (t - 2)k$ and $3i + j + tk$ are perpendicular.

Higher Level

9. Find the coordinates of the point in which the line (AB) meets the XOY plane if A = (4, –2, 1) and B = (–5, 4, –2).

10. Given vectors $a = 2i + 3j - k$ and $b = 4i - j - 2k$ find vectors c and d such that c is parallel to a, d is perpendicular to b and $a + b = c + d$.

11. Points P and Q divide the sides AB and AC of triangle ABC in the ratios 1 : 2 and 3 : 1 respectively. Show that $\overrightarrow{PQ}$ is parallel to $5\overrightarrow{AC} + 4\overrightarrow{BC}$.

12. (a) Prove that the vector $|b|a + |a|b$ bisects the angle between a and b.
    (b) Find a unit vector in the plane of $a = -2i - j - 2k$ and $b = 6i + 2j + 3k$ which bisects the angle between a and b.

Vector Geometry

13. Find the scalar p if the angle between the vectors $a = 2i + 3j - k$ and $b = 6i + pj + 4k$ is 60°.

14. If A, B, C and D are the vertices of a tetrahedron with coordinates (1, 2, 3), (3, 3, 5), (2, 4, 1) and (−1, 4, 4) respectively, show that (AB), (AC) and (AD) are mutually perpendicular and find the volume of the tetrahedron.

11.7 The Vector Product of Two Vectors

Definition The ***vector product*** of $a = a_1 i + a_2 j + a_3 k$ and $b = b_1 i + b_2 j + b_3 k$ is denoted and defined by

$$a \times b = (a_2 b_3 - a_3 b_2)i + (a_3 b_1 - a_1 b_3)j + (a_1 b_2 - a_2 b_1)k.$$

This product is called the ***vector*** product since the result is a vector. The vector product is sometimes called the ***cross product***.

The vector product of any two vectors is perpendicular to each of these vectors.

Proof With vectors a and b as given in the above definition,
$a \cdot (a \times b) = a_1(a_2 b_3 - a_3 b_2) + a_2(a_3 b_1 - a_1 b_3) + a_3(a_1 b_2 - a_2 b_1) = 0$ and
$b \cdot (a \times b) = b_1(a_2 b_3 - a_3 b_2) + b_2(a_3 b_1 - a_1 b_3) + b_3(a_1 b_2 - a_2 b_1) = 0$.
Therefore $a \times b$ is perpendicular to both a and b.

Although the proof is beyond the scope of this book, it can be shown that

$$a \times b = |a||b|\sin\theta \,\hat{u}$$

where θ is the angle between a and b, and $\hat{u}$ is a unit vector which is perpendicular to both a and b and with a direction given by the following "right hand" rule: If the index finger of the right hand gives the direction of a and the second finger the direction of b, then the thumb (held perpendicular to both the first two fingers) gives the direction of $\hat{u}$.

Clearly $b \times a$ must have a direction ***opposite*** to that of $a \times b$.
That is $b \times a = -a \times b$.

Evaluation of a Vector Product

$$a \times b = (a_2 b_3 - a_3 b_2)i + (a_3 b_1 - a_1 b_3)j + (a_1 b_2 - a_2 b_1)k$$

Chapter 11

$$= \begin{vmatrix} a_2 & a_3 \\ b_2 & b_3 \end{vmatrix} i - \begin{vmatrix} a_1 & a_3 \\ b_1 & b_3 \end{vmatrix} j + \begin{vmatrix} a_1 & a_2 \\ b_1 & b_2 \end{vmatrix} k$$

$$= \begin{vmatrix} i & j & k \\ a_1 & a_2 & a_3 \\ b_1 & b_2 & b_3 \end{vmatrix}.$$

Thus, the evaluation of the 3 × 3 determinant which has vectors i, j, k in the first row, the components of a in the second row and the components of b in the third row, provides us with an easy-to-remember process that we can use to find the components of $a \times b$.

Example Given $a = 2i + 2j - k$ and $b = i + 2j + 2k$, find $a \times b$ and show that it is perpendicular to both a and b.

$$a \times b = \begin{vmatrix} i & j & k \\ 2 & 2 & -1 \\ 1 & 2 & 2 \end{vmatrix} = \begin{vmatrix} 2 & -1 \\ 2 & 2 \end{vmatrix} i - \begin{vmatrix} 2 & -1 \\ 1 & 2 \end{vmatrix} j + \begin{vmatrix} 2 & 2 \\ 1 & 2 \end{vmatrix} k = 6i - 5j + 2k.$$

$a \cdot (a \times b) = 2 \times 6 + 2 \times (-5) - 2 = 0$ and $b \cdot (a \times b) = 6 + 2 \times (-5) + 2 \times 2 = 0$.
Therefore $a \times b$ is perpendicular to both a and b.

The Length of the Vector Product

From the second definition, $|a \times b| = |a||b||\sin\theta||\hat{u}| = |a||b|\sin\theta$ since $0 \leq \theta \leq \pi$ and $|\hat{u}| = 1$.

Example Given $a = i - 2j + 2k$ and $b = 3i - 4k$, justify the equivalence of the two definitions of $a \times b$.

From the first definition:

$$a \times b = \begin{vmatrix} i & j & k \\ 1 & -2 & 2 \\ 3 & 0 & -4 \end{vmatrix} = \begin{vmatrix} -2 & 2 \\ 0 & -4 \end{vmatrix} i - \begin{vmatrix} 1 & 2 \\ 3 & -4 \end{vmatrix} j + \begin{vmatrix} 1 & -2 \\ 3 & 0 \end{vmatrix} k = 8i + 10j + 6k.$$

Now $|a \times b| = 2|4i + 5j + 3k| = 2\sqrt{50} = 10\sqrt{2}$ (from the first definition).
The lengths of the vectors are $|a| = 3$, $|b| = 5$ and if θ is the angle between a and b then $\cos\theta = \dfrac{a \cdot b}{|a||b|} = \dfrac{-5}{3 \times 5} = -\dfrac{1}{3}$ and so $\sin\theta = \dfrac{2\sqrt{2}}{3}$.

Thus from the second definition we have:

$|a \times b| = |a||b|\sin\theta = 3 \times 5 \times \dfrac{2\sqrt{2}}{3} = 10\sqrt{2}$ which agrees with the result obtained from the first definition.

Example Prove that $|a \times b|^2 = |a|^2|b|^2 - (a \cdot b)^2$

$$\begin{aligned}|a \times b|^2 &= |a|^2|b|^2 \sin^2\theta \text{ where } \theta \text{ is the angle between } a \text{ and } b, \\ &= |a|^2|b|^2\{1-\cos^2\theta\} \\ &= |a|^2|b|^2 - |a|^2|b|^2\cos^2\theta \\ &= |a|^2|b|^2 - (a \cdot b)^2. \quad \text{Q.E.D.}\end{aligned}$$

Properties of the Vector Product

(a) $b \times a = -a \times b$
(b) $s(a \times b) = (sa) \times b = a \times (sb)$ for any scalar s.
(c) $a \cdot (a \times b) = b \cdot (a \times b) = 0$
(d) If a and b are non-zero vectors, then $a \times b = 0$ iff a is parallel to b.
(e) $a \times a = 0$
(f) $a \times (b + c) = (a \times b) + (a \times c)$
(g) $(a + b) \times (a - b) = 2b \times a$

It is worth noting the evaluation of the various cross products involving the unit vectors i, j and k. Clearly $i \times i = j \times j = k \times k = 0$.
Now consider the following diagram:

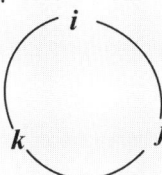

Using a clockwise movement we obtain $i \times j = k$, $j \times k = i$ and $k \times i = j$ while using an anticlockwise movement we obtain $i \times k = -j$, $k \times j = -i$ and $j \times i = -k$.

Example Evaluate the cross product $(2i + 3j) \times (2j - 5k)$.

$$\begin{aligned}(2i + 3j) \times (2j - 5k) &= 4i \times j - 10i \times k + 6j \times j - 15j \times k \\ &= 4k + 10j + 0 - 15i \\ &= -15i + 10j + 4k.\end{aligned}$$

Chapter 11

Areas using Vector Products

Theorem The area of a parallelogram with vectors a and b as two adjacent sides is given by $A = |a \times b|$.

Proof

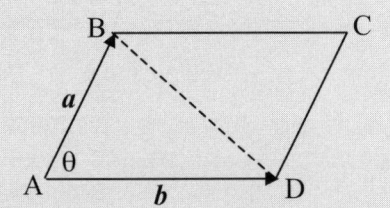

The area of parallelogram ABCD
= 2 × area of triangle ABD
= 2 × $\frac{1}{2}$(AB)(AD) sin θ
= $|a \times b|$.

Note: Since the area of the parallelogram is also twice the area of the triangle ABC, we may use the cross product of any side with a diagonal. Thus the area of the parallelogram in the diagram above is also given by $A = |\overrightarrow{AB} \times \overrightarrow{AC}|$.

It is obvious from the above proof that the area of the triangle with vectors a and b as two sides is given by $A = \frac{1}{2}|a \times b|$.

Example Find the coordinates of the fourth vertex, D, and the area of the parallelogram ABCD in which A = (2, −1, −3), B = (5, 2, 0) and C = (−3, 7, 2).

In any parallelogram ABCD, $\overrightarrow{OA} + \overrightarrow{OC} = \overrightarrow{OB} + \overrightarrow{OD}$ and so D = (−6, 4, −1).

The area of the parallelogram = $|\overrightarrow{AB} \times \overrightarrow{AC}|$

$$= \begin{Vmatrix} i & j & k \\ 3 & 3 & 3 \\ -5 & 8 & 5 \end{Vmatrix}$$

$$= |-9i - 30j + 39k|$$

$$= 3\sqrt{278}.$$

Since we cannot define the cross product of vectors in 2-space, we need to modify our approach if we wish to find the area of a triangle or parallelogram.

Example Calculate the area of the triangle ABC with A = (1, 2), B = (7, 4) and C = (−2, 10).

The equation of the XOY-plane is $z = 0$ so we simply give each point a third coordinate of zero and proceed as before.

The area of triangle ABC $= \frac{1}{2}\left|\overrightarrow{AB} \times \overrightarrow{AC}\right|$

$$= \frac{1}{2}\begin{Vmatrix} i & j & k \\ 6 & 2 & 0 \\ -3 & 8 & 0 \end{Vmatrix}$$

$$= \tfrac{1}{2}|54k|$$

$$= 27.$$

A second method, which requires the evaluation of a determinant, does not require any vector techniques. However, it is a useful method for finding the area of a triangle or parallelogram in 2-space.

Consider triangle ABC in which $A = (x_1, y_1)$, $B = (x_2, y_2)$ and $C = (x_3, y_3)$. Using the method described above, we find that the area of the triangle ABC is given by $A = \frac{1}{2}\left|\overrightarrow{AB} \times \overrightarrow{AC}\right|$

$$= \frac{1}{2}\begin{Vmatrix} i & j & k \\ x_2 - x_1 & y_2 - y_1 & 0 \\ x_3 - x_1 & y_3 - y_1 & 0 \end{Vmatrix}$$

$$= \tfrac{1}{2}\left|((x_2 - x_1)(y_3 - y_1) - (x_3 - x_1)(y_2 - y_1))k\right|$$

$$= \tfrac{1}{2}\left|(x_2 - x_1)(y_3 - y_1) - (x_3 - x_1)(y_2 - y_1)\right|.$$

Note $\frac{1}{2}\begin{Vmatrix} 1 & x_1 & y_1 \\ 1 & x_2 & y_2 \\ 1 & x_3 & y_3 \end{Vmatrix} = \tfrac{1}{2}\left|x_2 y_3 - x_3 y_2 - x_1(y_3 - y_2) + y_1(x_3 - x_2)\right|$

$$= \tfrac{1}{2}\left|x_2 y_3 - x_2 y_1 - x_1 y_3 - x_3 y_2 + x_3 y_1 + x_1 y_2\right|$$

$$= \tfrac{1}{2}\left|x_2 y_3 - x_2 y_1 - x_1 y_3 + x_1 y_1 - x_3 y_2 + x_3 y_1 + x_1 y_2 - x_1 y_1\right|$$

$$= \tfrac{1}{2}\left|x_2(y_3 - y_1) - x_1(y_3 - y_2) - x_3(y_2 - y_1) + x_1(y_2 - y_1)\right|$$

$$= \tfrac{1}{2}\left|(x_2 - x_1)(y_3 - y_1) - (x_3 - x_1)(y_2 - y_1)\right| \quad \text{which is the}$$

result already obtained.

Example Use the method just outlined to calculate the area of the triangle ABC in which $A = (-1, -2)$, $B = (5, 9)$ and $C = (8, 1)$.

Chapter 11

The area $= \dfrac{1}{2}\begin{Vmatrix} 1 & -1 & -2 \\ 1 & 5 & 9 \\ 1 & 8 & 1 \end{Vmatrix}$

$= \tfrac{1}{2}|5 - 72 + (1-9) - 2(8-5)|$

$= 40.5.$

Exercise 11.7

1. Find each of the following vector products:
   $i \times i$; $\quad i \times j$; $\quad i \times k$; $\quad j \times k$; $\quad i \times (j \times k)$; $\quad (i \times j) \times k$.

2. Find $a \times b$ in each of the following:
   (a) $a = i + 2j - k$, $\quad b = 4i - j - k$;
   (b) $a = i + 2j + 2k$, $\quad b = 3i - 4j + k$;
   (c) $a = i + j - k$, $\quad b = 2i - j - 3k$;
   (d) $a = 3i - 2j - 2k$, $\quad b = 6i + 6j$;
   (e) $a = 3i + j - 2k$, $\quad b = 4i - 4j - 7k$.

3. Verify each of the following using $a = 3i - 2j - 2k$, $b = i + 2j + 4k$ and $c = 2i + j - 2k$:
   (a) $a \times a = 0$;
   (b) $b \times a = -a \times b$
   (c) $a \times (b + c) = (a \times b) + (a \times c)$
   (d) $(a \times b) \times c = (a \cdot c)b - (b \cdot c)a$.

4. In each of the following the vertices of a triangle are given. Find the area of each triangle.
   (a) $(1, 2, 3), (2, 0, 1), (3, 2, 1)$;
   (b) $(0, -1, -2), (4, 1, 0), (1, 0, -1)$;
   (c) $(-3, 9), (5, -2), (2, 11)$;
   (d) $(2, -3), (7, 3), (3, 9)$.

5. The vectors a and b are two sides of a parallelogram in each of the following. Calculate the area of each parallelogram.
   (a) $a = 3i + j, b = -3i - 2j + 2k$;
   (b) $a = 4i - j + 3k, b = 8i + 3j + k$;
   (c) $a = 2i - 2j + k, b = i - 5k$;
   (d) $a = 2i + 3j - 5k, b = i + 5j - 6k$.

6. Find the fourth vertex of the parallelogram ABCD and the area of the parallelogram in each of the following:

   (a) $A = (2, 1, 0)$, $B = (3, 2, -1)$, $C = (0, -1, 1)$;
   (b) $A = (3, 2, -1)$, $B = (2, 4, 5)$, $D = (4, 6, 2)$;
   (c) $A = (0, 3, 2)$, $C = (2, 5, 1)$, $D = (1, 4, 0)$;
   (d) $B = (5, 2, 3)$, $C = (1, 2, 2)$, $D = (-1, -1, -1)$.

7. Verify the formula $|a \times b| = |a||b|\sin\theta$ for each of the following:

   (a) $a = -i + 2j + 2k$, $b = 2i - 2j + k$;
   (b) $a = 4i + 3k$, $b = 6i - 2j - 3k$;
   (c) $a = i + j + k$, $b = 5i - j + 7k$;
   (d) $a = 3i - 2j - 2k$, $b = 5i + 2j - 6k$;
   (e) $a = 3i + 2j - 5k$, $b = -i + 2j - k$.

8. (a) Use the scalar product of the vectors $a = (\cos\alpha)i + (\sin\alpha)j$ and $b = (\cos\beta)i + (\sin\beta)j$ to derive the formula
   $\cos(\alpha - \beta) = \cos\alpha\cos\beta + \sin\alpha\sin\beta$.

   (b) Use the vector product of the vectors $a = (\cos\alpha)i + (\sin\alpha)j + 0k$ and $b = (\cos\beta)i + (\sin\beta)j + 0k$ to derive the formula
   $\sin(\alpha - \beta) = \sin\alpha\cos\beta - \cos\alpha\sin\beta$.

9. The vectors a, b and c form three edges of a cuboid ABCDEFGH.

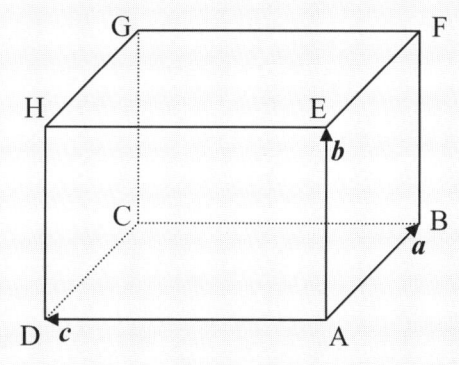

   If the lengths of a, b and c are a, b and c respectively, find the value of each of the following:
   (a) $|a \times (b \times c)|$; (b) $a \cdot (b \times c)$.

10. Prove that the area of a parallelogram ABCD is given by $A = \frac{1}{2}|\overrightarrow{AC} \times \overrightarrow{BD}|$.

307

11.8 Volumes of Parallelepipeds and Tetrahedra (Optional)

Definition A *parallelepiped* is a solid with six faces with opposite faces pairs of congruent parallelograms.

Theorem The volume of a parallelepiped which has vectors ***a***, ***b*** and ***c*** are three concurrent edges is given by $V = |a \bullet b \times c|$.

Proof

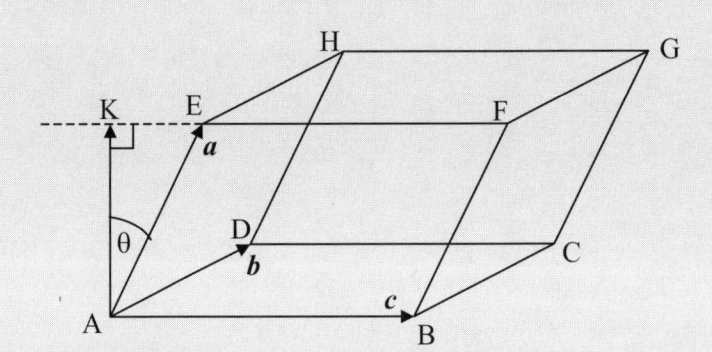

A parallelepiped is clearly a prism, and the volume of a prism is equal to the area of the base multiplied by the 'height'.

The volume of the prism in the above diagram
= Area of parallelogram ABCD × AK.

Now the area of the base = $|b \times c|$ and the length of AK is given by $|a| \cos \theta$.
Thus the volume of the parallelepiped is given by
$$V = |a||b \times c|\cos \theta.$$

But θ is the angle between ***a*** and the normal to the plane ABCD, and since $b \times c$ is normal to the plane ABCD, $\cos \theta = \dfrac{|a \bullet b \times c|}{|a||b \times c|}$. This gives $|a \bullet b \times c| = |a||b \times c|\cos \theta$, and the area of the parallel-epiped is $V = |a \bullet b \times c|$, as required.

Note: There is no need for parentheses around $b \times c$ for $(a \bullet b) \times c$ has no meaning since $a \bullet b$ is a scalar.
Also, the volume can be found using any order of ***a***, ***b*** and ***c*** in the formula. Thus $V = |b \bullet a \times c| = |c \bullet b \times a|$ and so on.

Example In the parallelepiped ABCDEFGH, AE, BF, CG and DH are parallel edges. Given that A, B, D and E have coordinates (1, 2, 3), (3, 1, 1), (2, 4, 0) and (−1, 4, 4) respectively, find the coordinates of the other vertices and the volume of the parallelepiped.

The coordinates of C, F, G and H are shown in the following diagram.

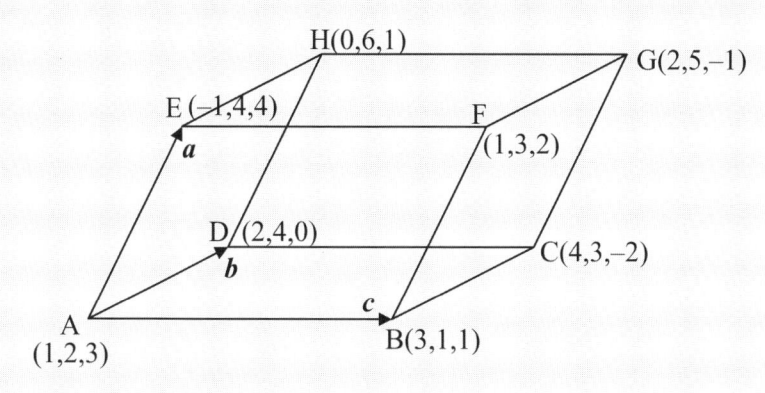

$\overrightarrow{AB} = 2i - j - 2k$, $\overrightarrow{AD} = i + 2j - 3k$ and $\overrightarrow{AE} = -2i + 2j + k$

$$\overrightarrow{AD} \times \overrightarrow{AE} = \begin{vmatrix} i & j & k \\ 1 & 2 & -3 \\ -2 & 2 & 1 \end{vmatrix} = 8i + 5j + 6k$$

Required volume = $|(2i - j - 2k) \cdot (8i + 5j + 6k)| = 1$.

Theorem The volume of a tetrahedron with vectors *a*, *b* and *c* as three concurrent edges is given by $V = \frac{1}{6}|a \cdot b \times c|$.

Proof

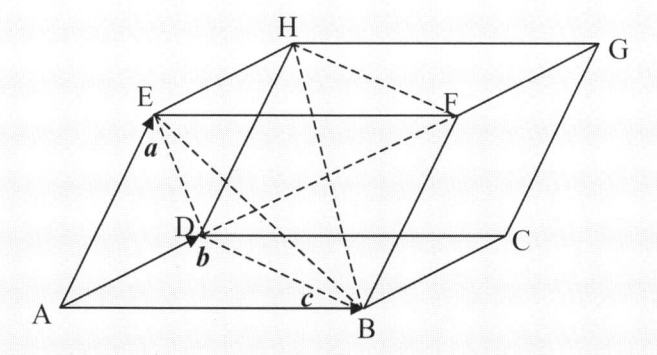

The parallelepiped can be split into 6 tetrahedra of equal volumes. In the diagram above one can easily see that the tetrahedra ABDE, BDEH and BEFH together provide exactly half the volume of the parallelepiped.

Chapter 11

Each of the first two have BDE as a base and, by symmetry, equal heights since A and H are equal distances from this base. Therefore the tetrahedra ABDE and BDEH have equal volumes. The second and third tetrahedra have equal bases BEH and equal heights since D and F are equal distances from this base. Therefore the tetrahedra BDEH and BDFH have equal volumes. Thus all three tetrahedra have the same volume.

Thus the volume of the tertahedron ABDE is given by

$V = \frac{1}{6}$(volume of the parallelepiped) $= \frac{1}{6}|a \bullet b \times c|$, as required.

Example   The points A(1, 2, 3), B(2, 4, 1), C(−2, 3, −1) and D(0, −2, 4) are the vertices of a tetrahedron. Calculate the volume of this tetrahedron.

$\overrightarrow{AB} = i + 2j - 2k$ and $\overrightarrow{AC} \times \overrightarrow{AD} = \begin{vmatrix} i & j & k \\ -3 & 1 & -4 \\ -1 & -4 & 1 \end{vmatrix} = -15i + 7j + 13k$.

The required volume $= \frac{1}{6}|\overrightarrow{AB} \bullet \overrightarrow{AC} \times \overrightarrow{AD}|$

$= \frac{1}{6}|(i + 2j - 2k) \bullet (-15i + 7j + 13k)|$

$= \frac{1}{6} \times 27$

$= 4\frac{1}{2}$.

Exercise 11.8

1. In each of the following, the coordinates of the vertices of a tetrahedron are given. Calculate the volume of each tetrahedron.
   (a)   (1, 2, 3), (3, 6, 4), (4, 0, 5), (1, −4, −3) ;
   (b)   (−3, −5, 1), (3, −4, −1), (6, −1, −1), (0, −6, 5) ;
   (c)   (2, 4, 5), (−3, 0, 4), (3, 3, 4), (5, 3, 8) ;
   (d)   (−4, −3, 1), (1, −6, 5), (−3, −1, 0), (−6, −5, 0).

2. In each of the following, *a*, *b* and *c* are three concurrent edges of a parallelepiped. Calculate the volume of the parallelepiped in each case.
   (a) $a = i + 2j + 2k$, $b = 2i + j$, $c = -i + 3k$;
   (b) $a = i + j + k$, $b = 2i - j - k$, $c = 3i - 2k$;
   (c) $a = 3i + 2j - k$, $b = -2i - j + 2k$, $c = 4i + 3j + k$;
   (d) $a = 2j - 3k$, $b = 3i + j$, $c = 2i - 5k$.

3. (a)   Prove that if $\overrightarrow{AB} \bullet \overrightarrow{AC} \times \overrightarrow{AD} = 0$, points A, B, C, D are coplanar.
   (b)   Determine which of the following sets of points are coplanar:
   (i)   (2, 1, −2), (8, 2, 6), (−4, 5, 0), (0, 5, 4) ;
   (ii)   (1, 1, 3), (2, 3, 4), (−1, 0, 5), (0, −3, −1) ;
   (iii)   (3, −1, 2), (−1, 3, −3), ($\frac{1}{2}$, 2, $-\frac{3}{2}$), ($\frac{3}{4}$, $\frac{1}{2}$, $-\frac{1}{4}$).

4. Consider the points (4, 2, 1), (6, x, 0), (x, 4, 2) and (11, 5, 3). Find the values of x for which
   (a)   the points are collinear ;
   (b)   the points are vertices of a tetrahedron which has a volume of 2.

11.9   The Equations of a Line

The Line passing through a Given Point with a Given Direction

Consider the line passing through the point A with position vector *a* and parallel to the vector *v*. Let *r* be the position vector of a general point P on the line.

Since $\overrightarrow{AP}$ is parallel to *v*, $\overrightarrow{AP} = \lambda v$ for some scalar λ.

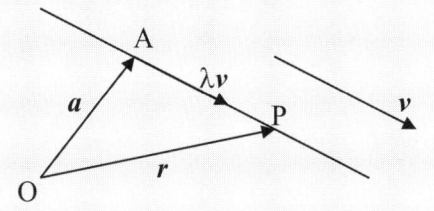

Thus $r - a = \lambda v$ or $r = a + \lambda v$.

This form of the equation of a line is known as the ***vector-parametric*** form.

The vector *v* is called the ***direction vector*** of the line.

Example   Find the vector-parametric equation of the line which passes through the point A(2, −1, 3) and which is parallel to the vector $2j - k$.

The required equation is $r = 2i - j + 3k + \lambda(2j - k)$.

This can be written in the form $r = 2i + (2\lambda - 1)j + (3 - \lambda)k$.

The Equation of a Line passing through Two Points

Consider the line passing through the points A, B with position vectors *a*, *b* respectively.

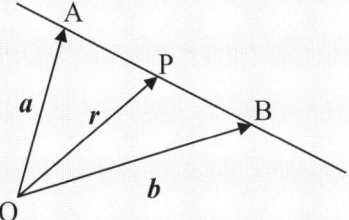

311

Chapter 11

If a point P with position vector r is any point on this line then $\overrightarrow{AP} = \lambda \overrightarrow{AB}$ for some scalar λ. Thus $r - a = \lambda(b - a)$ or $r = a + \lambda(b - a)$.

Hence the equation of the line passing through two points is $r = (1 - \lambda)a + \lambda b$.

In the form $r = a + \lambda(b - a)$ this is essentially the same as the vector-parametric form.

Example Find an equation for the line passing through the points A(2, 3, 1) and B(3, −1, 4).

A suitable equation is $r = (1 - \lambda)(2i + 3j + k) + \lambda(3i - j + 4k)$ or
$r = (2 + \lambda)i + (3 - 4\lambda)j + (1 + 3\lambda)k$.

The Equation of a Line in Parametric Form

The equation $r = a + \lambda v$ can be written in the form $\begin{pmatrix} x \\ y \\ z \end{pmatrix} = \begin{pmatrix} a_1 \\ a_2 \\ a_3 \end{pmatrix} + \lambda \begin{pmatrix} v_1 \\ v_2 \\ v_3 \end{pmatrix}$.

Equating components gives
$$x = a_1 + \lambda v_1$$
$$y = a_2 + \lambda v_2$$
$$z = a_3 + \lambda v_3.$$

These three equations make up what are known as the parametric equations of the line.

Example Write the equation of the line passing through the points A(2, 3, −1) and B(1, 2, 3) in parametric form.

$\overrightarrow{AB} = -i - j + 4k$ and so the parametric equations of the line are
$$x = 2 - \lambda$$
$$y = 3 - \lambda$$
$$z = 4\lambda - 1.$$

The Cartesian Form of the Equation of a Straight Line

From the parametric equations we find $\lambda = \dfrac{x - a_1}{v_1}$, $\lambda = \dfrac{y - a_2}{v_2}$, $\lambda = \dfrac{z - a_3}{v_3}$

provided v_1, v_2, v_3 are non-zero.

Eliminating λ gives $\dfrac{x - a_1}{v_1} = \dfrac{y - a_2}{v_2} = \dfrac{z - a_3}{v_3}$

which is known as the *Cartesian form* of the equation of a straight line.

Vector Geometry

By convention, the Cartesian form may be used for all values of v_1, v_2, v_3 including zero. For example the equations $\dfrac{x+4}{2} = \dfrac{y-2}{0} = \dfrac{z}{-1}$ represent the straight line through the point (–4, 2, 0) and parallel to the vector $2i - k$.

Exercise 11.9

1. Find the vector-parametric equation of the line with direction vector **v** and passing through the point with position vector **a** given that
   (a) $a = 2i - 3j + k$ and $v = i + 2j - 3k$;
   (b) $a = i - j$ and $v = i + j + k$.

2. Write the equation of each of the following straight lines in
   (i) vector-parametric form ; (ii) parametric form ; (iii) Cartesian form.

   (a)   through (2, 1, –1) and parallel to $i + 2j + 3k$;
   (b)   through the points (3, 2, 1) and (5, 0, 2) ;
   (c)   through the points with position vectors $3i - 2j$ and $j + 4k$;
   (d)   through (3, 2, 4) and parallel to the *x*-axis ;
   (e)   through the origin and the point (–2, –1, 5).

3. The position vectors of four points A, B, C and D are $a = i + k$, $b = j - k$, $c = 7j - 7k$ and $d = -i + 5j - 6k$ respectively.
   (a)   Find a vector-parametric equation for the line (AB).
   (b)   Find a vector-parametric equation for the line (CD). [Do *not* use the parameter already used in part (a).]
   (c)   Assuming that the lines (AB) and (CD) meet, find the position vector of their point of intersection.

4. The position vectors of the four coplanar points A, B, C and D are $a = i + 2j$, $b = 3i + 4j + 2k$, $c = 4i - j$ and $d = 7j + 2k$ respectively. Find the position vector of the point of intersection of (AB) and (CD).

5. In the parallelogram ABCD, M is the mid-point of [BC]. The lines (AM) and (BD) meet at E. If the position vectors of A, B, C are **a**, **b** and **c** respectively, find the equations of the lines (AM) and (BD) in terms of vectors **a**, **b** and/or **c** and hence show that E is a point of trisection of both [AM] and [BD].

*6. The vertices A, B, C of a triangle have position vectors **a**, **b**, **c** respectively. The point P divides [BC] internally in the ratio 2 : 1 and the point Q divides [AC] externally in the ratio 2 : 1. Find vector-parametric equations for the lines (AP) and (BQ). Hence find the position vector of the point of intersection of the lines (AP) and (BQ) in terms of **a**, **b** and/or **c**.

313

11.10 The Equations of a Plane

The Vector-Parametric Form

There is only one plane which can contain any two intersecting lines, and if vectors *a* and *b* are parallel to these lines, every vector in the plane can be expressed uniquely as a linear combination of *a* and *b*.

That is, if vector *c* lies in a plane which contains the non-parallel vectors *a* and *b*, then *c* can be written (uniquely) in the form $c = \lambda a + \mu b$.

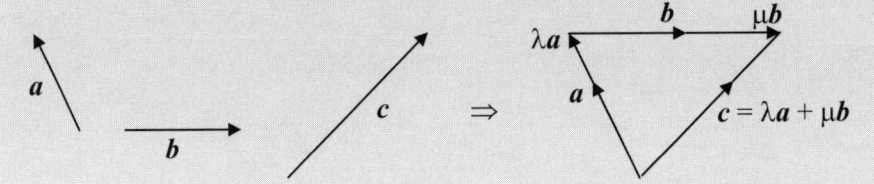

Consider a plane which passes through a point A with position vector *a* and which contains two non-parallel vectors *p* and *q*. Let *r* be the position vector of a general point P in the plane.

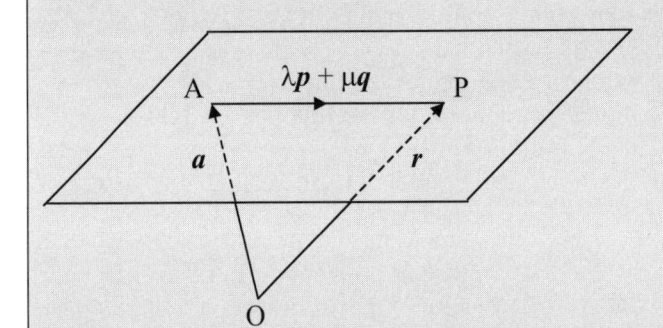

Since $\overrightarrow{AP}$ lies in the plane, there are scalars λ and μ such that

$$\overrightarrow{AP} = \lambda p + \mu q$$
$$r - a = \lambda p + \mu q$$
$$r = a + \lambda p + \mu q.$$

The equation $r = a + \lambda p + \mu q$ is called the **vector-parametric** equation of the plane.

Example Write down the vector-parametric form of the equation of the plane passing through the point A(1, 0 –2) and containing the vectors $i + j$ and k.

The vector-parametric form of the equation is
$$r = i - 2k + \lambda(i + j) + \mu k$$
or $\quad r = (\lambda + 1)i + \lambda j + (\mu - 2)k.$

Vector Geometry

The Vector Parametric Equation of a Plane passing through three Non-Collinear Points

Let r be the position vector of a general point P on a plane containing three points A, B, C whose position vectors are a, b and c respectively.

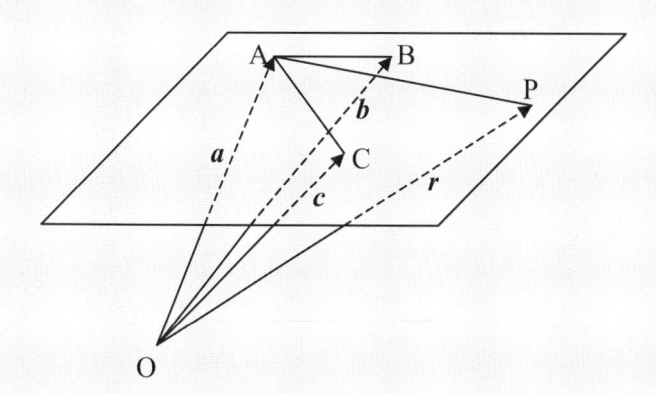

Since $\overrightarrow{AP}$ lies in the plane containing $\overrightarrow{AB}$ and $\overrightarrow{AC}$, there are scalars λ and μ such that

$$\overrightarrow{AP} = \lambda \overrightarrow{AB} + \mu \overrightarrow{AC}$$
$$r - a = \lambda(b - a) + \mu(c - a)$$
$$r = (1 - \lambda - \mu)a + \lambda b + \mu c.$$

The form $\quad r = (1 - \lambda - \mu)a + \lambda b + \mu c$

is an alternative vector-parametric form of the equation of the plane.

Example Find in vector parametric form the equation of the plane passing through the points A(0, 0, −3), B(1, 4, 0) and C(3, 0, −2).

The equation is $r = (1 - \lambda - \mu)(-3k) + \lambda(i + 4j) + \mu(3i - 2k)$
or $\quad r = (\lambda + 3\mu)i + 4\lambda j + (3\lambda + \mu - 3)k.$

The Normal (Scalar Product) Form of the Equation of a Plane

Definition A vector is said to be ***normal*** to a plane if it is perpendicular to every non-zero vector in the plane.

A vector will be perpendicular to every non-zero vector in the plane if it is perpendicular to any two intersecting non-zero vectors in that plane.

Consider the plane which passes through a point A with position vector a and which has vector n as a normal.

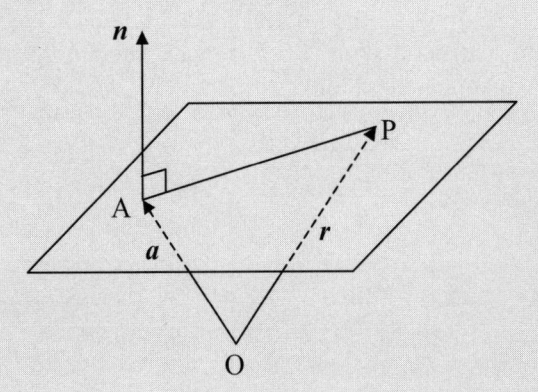

If r is the position vector of a general point P in the plane, then n is perpendicular to $\overrightarrow{AP}$ and so $\overrightarrow{AP} \cdot n = 0$

$\Rightarrow \quad (r - a) \cdot n = 0$
$\Rightarrow \quad r \cdot n - a \cdot n = 0$
$\Rightarrow \quad r \cdot n = a \cdot n.$

This is a vector equation of the plane in **normal** or **scalar product** form.

Now since vectors a and n are constant vectors, $a \cdot n$ is a constant, p say. Therefore the normal form of the equation of the plane can be written in the form
$$r \cdot n = p.$$

If the point D with position vector d is the foot of the normal from the origin O to the plane $r \cdot n = p$, then the perpendicular distance from the origin to the plane is $d = |d|$.

But D lies in the plane so $d \cdot n = p$ and d is parallel to n giving $d \cdot n = \pm d|n|$.

Thus $|d \cdot n| = |p| = d|n|$ which gives $d = \dfrac{|p|}{|n|}$.

Thus the plane $r \cdot n = p$ is perpendicular to a vector n and its shortest distance from the origin is $\dfrac{|p|}{|n|}$.

Vector Geometry

Example Find in scalar product form the equation of the plane passing through the point (2, 1, 3) with normal vector $2i - j + 4k$ and find the shortest distance from the origin to the plane.

The equation is $r \cdot (2i - j + 4k) = (2i + j + 3k) \cdot (2i - j + 4k)$ which can be written as $r \cdot (2i - j + 4k) = 15$.

The shortest distance from the origin to the plane is $\dfrac{15}{\sqrt{4+1+16}} = \dfrac{15}{\sqrt{21}}$.

Example Find in scalar product form the equation of the plane passing through the points A(0, 0 –3), B(1, 4, 0), C(3, 0, –2).

Vectors $\overrightarrow{AB}$ and $\overrightarrow{AC}$ are two intersecting vectors in the plane and their cross product $\overrightarrow{AB} \times \overrightarrow{AC} = \begin{vmatrix} i & j & k \\ 1 & 4 & 3 \\ 3 & 0 & 1 \end{vmatrix} = 4i + 8j - 12k = 4(i + 2j - 3k)$ is a normal.

Therefore, the equation of the plane is
$r \cdot (i + 2j - 3k) = (-3k) \cdot (i + 2j - 3k)$ or $r \cdot (i + 2j - 3k) = 9$.

The Cartesian Equation of a Plane

Consider the plane which has $n = ai + bj + ck$ as a normal and which passes through the point $Q(x_0, y_0, z_0)$. Let P(x, y, z) be a general point in the plane. Then n is perpendicular to $\overrightarrow{QP}$ and so $n \cdot \overrightarrow{QP} = 0$.
Thus $(ai + bj + ck) \cdot ([x - x_0]i + [y - y_0]j + [z - z_0]k) = 0$
$\Rightarrow a(x - x_0) + b(y - y_0) + c(z - z_0) = 0$
$\Rightarrow ax + by + cz = ax_0 + by_0 + cz_0$ and since Q is a fixed point and n is fixed
$ax + by + cz = d$ where d is the constant $ax_0 + by_0 + cz_0$.

This form is called the ***Cartesian*** form of the equation of a plane.

Example Find in Cartesian form the equation of the plane passing through the point (2, 1, –3) with normal vector $2i - 3j + 5k$.

The required equation is $2x - 3y + 5z = 2(2) - 3(1) + 5(-3) = -14$
or $2x - 3y + 5z + 14 = 0$.

317

Chapter 11

Example Find an equation of the plane $r = i + 3k + \lambda(2i - k) + \mu(i + 3j + k)$ in Cartesian form.

The plane passes through point A(1, 0, 3) and contains vectors $u = 2i - k$ and $v = i + 3j + k$.

The vector $u \times v = \begin{vmatrix} i & j & k \\ 2 & 0 & -1 \\ 1 & 3 & 1 \end{vmatrix} = 3i - 3j + 6k = 3(i - j + 2k)$ is normal to the plane and so the Cartesian equation is $x - y + 2z = 1 - 0 + 2(3) = 7$ which is $x - y + 2z = 7$.

Exercise 11.10

1. Write down in vector-parametric form the equation of the plane passing through the point A and parallel to the vectors p and q in each of the following:
    (a) A = (2, 3, 4), $p = 2i - 3j + 2k$, $q = j + 2k$;
    (b) A = (0, 0, –2), $p = 3i + 3j - k$, $q = i - j + k$;
    (c) A = (–2, –1, –3), $p = i + k$, $q = 2i + j + k$;
    (d) A = (5, 1, –4), $p = i - j + k$, $q = 3i - j - k$.

2. Write the equations of each of the planes given in Question 1 in Cartesian form.

3. Write the equation of the plane containing the points with position vectors $i + j, j + 4k$ and $i + 5k$ in
    (a) vector-parametric form ;
    (b) scalar product form ;
    (c) Cartesian form.

4. Find in scalar product form the equation of the plane passing through the point with position vector a and with normal vector n in each of the following:
    (a) $a = 3i - j + 2k$, $n = i + j - 2k$;
    (b) $a = 2i - 3k$, $n = 4i + 7j + 2k$;
    (c) $a = 3i$, $n = j$.

5. Find in the form $r \cdot n = p$ the equation of the plane passing through the given point and perpendicular to the given vector in each of the following:
    (a) (2, 1, –1), $3i - j - 2k$; (b) (5, 1, 1), $2i + 3j - 4k$.

318

6. Find the shortest distance from the origin to each of the following planes:
   (a) $r \cdot (2i - j + 2k) = 12$;
   (b) $r \cdot (6i + 2j - 3k) = 35$;
   (c) $r \cdot (i + j) = 3$;
   (d) $r \cdot (3i + 4j - 5k) = 100$.

7. Find in scalar product form the equation of the plane parallel to the vectors $i + 3k$, $2i - j + 2k$ and passing through the point with position vector $i + j$.

8. Find in scalar product form the equation of the plane
   (a) $r = (1 + \lambda + 3\mu)i + (1 + 4\lambda - 2\mu)j + (7\mu - \lambda)k$;
   (b) containing the lines $r = (1 + \lambda)i + (1 + \lambda)j + (1 - 2\lambda)k$ and $r = (11 + 3\mu)i - j + (1 - \mu)k$.

9. Write down direction vectors for the lines $r = (1 + \lambda)i + (2 - \lambda)j + (4 + \lambda)k$ and $r = (4 + 5\mu)i + j + \mu k$ and find an equation for the plane which contains the first line and is parallel to the second.

10. Find an equation for the plane determined by the points
    (a) $(3, -1, 6), (1, 2, 2), (-2, 4, 1)$;
    (b) $(1, -1, 3), (-1, 1, -4), (3, 3, 7)$;
    (c) $(1, 4, -5), (-2, -1, -6), (3, 5, -9)$;
    (d) $(-5, 2, 6), (5, -3, 2), (0, -8, -5)$;
    (e) $(1, -1, -1), (2, 1, 1), (0, -7, 1)$.

11. In each of the following find an equation of the plane determined by the data:
    (a) through the point with position vector $a = 2i + 3j - 4k$ and perpendicular to a ;
    (b) through the points with position vectors $6i, -3k$ and $3i + 6j$;
    (c) through the points $(5, 2, -7), (-2, 4, -2)$ and the origin ;
    (d) through the point $(1, 1, -1)$ and containing the vectors $2i + j + 2k$ and $5j + 4k$;
    (e) through the points $(3, 2, -1), (4, 4, 0)$ and perpendicular to the plane $2x + 4y - 4z = 3$;
    (f) through the points $(2, -1, -3), (4, -3, 2)$ and parallel to the x-axis ;
    (g) through the point $(3, 4, 2)$ and perpendicular to the x-axis.

12. Find an equation for the plane passing through the point $(3, 2, 1)$ and perpendicular to each of the planes $2x + 3y - z = 5$ and $3x + 3z = 2$.

13. Find an equation for the plane containing the line of intersection of the planes $x + y + 5z = 0$, $2x + 3y + 12z = 0$ and passing through $(3, 1, 1)$.

Chapter 11

11.11 The Angles between Lines and Planes

The Angle between Two Lines

Consider two lines whose direction vectors are a and b. If θ is the angle between these lines then we already know that $\cos\theta = \dfrac{a \cdot b}{|a||b|}$, $0° \le \theta \le 180°$.

Example   Find the angle between the lines
$$\frac{x-1}{2} = \frac{y+1}{2} = -z \quad \text{and} \quad \frac{x+2}{3} = \frac{y}{6} = \frac{z-1}{2}.$$

Vectors $a = 2i + 2j - k$ and $b = 3i + 6j + 2k$ are parallel to the given lines. If θ is the acute angle between these lines then $\cos\theta = \dfrac{|a \cdot b|}{|a||b|} = \dfrac{16}{3 \times 7} = \dfrac{16}{21}$.

Therefore the required angle is $40.4°$.

The Angle between Two Planes

Definition   The angle between two planes is defined to be the angle between their normals.

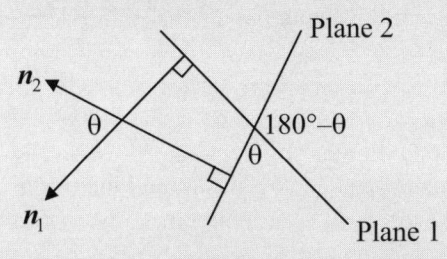

Clearly, if θ is an angle between the planes, then $180° - \theta$ is also an angle between them. Thus we are free to find the acute angle θ where $\cos\theta = \dfrac{|n_1 \cdot n_2|}{|n_1||n_2|}$.

Example   Find the angle between planes $2x + y - z = 6$ and $3x + y + 2z = 2$.

The vectors $a = 2i + j - k$ and $b = 3i + j + 2k$ are the normals to the planes. If θ is the (acute) angle between the planes then $\cos\theta = \dfrac{|a \cdot b|}{|a||b|} = \dfrac{5}{\sqrt{6}\sqrt{14}}$

and so the required angle is $56.9°$.

Vector Geometry

The Angle between a Line and a Plane

Definition The angle between a line and a plane is defined to be the angle between the line and its orthogonal projection in the plane.

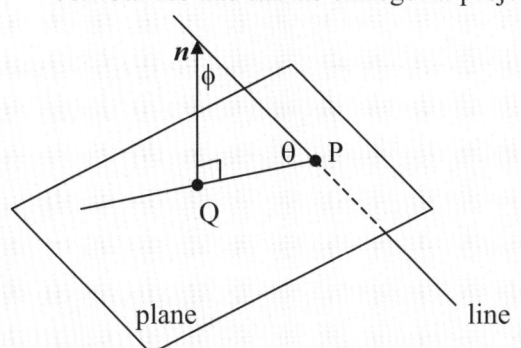

In the diagram above, P is the point of intersection of the line and the plane, (PQ) is the orthogonal projection of the line in the plane, vector n is a normal to the plane, ϕ is the angle between n and the line, and θ is the angle between the line and the plane.

Since ϕ is the acute angle between the line (with direction vector v) and n, then $\cos\phi = \dfrac{|n \cdot v|}{|n||v|}$. But $\phi = 90° - \theta$, and so $\cos\phi = \sin\theta$.

Therefore $\sin\theta = \dfrac{|n \cdot v|}{|n||v|}$.

Example Find the angle between the line $r = i + j - k + \lambda(2i - 2j - k)$ and the plane $3x - 6y + 2z = 1$.

$n = 3i - 6j + 2k$ is normal to the plane and $v = 2i - 2j - k$ is parallel to the line.

If θ is the angle between the line and the plane then $\sin\theta = \dfrac{|n \cdot v|}{|n||v|} = \dfrac{16}{21}$.

The required angle is 49.6°.

Exercise 11.11

1. Find the (acute) angle between the following pairs of lines:
   (a) $r = i + j + \lambda(2i - j - k)$, $r = k + \mu(i + j + 2k)$;
   (b) $r = 2i - 2j + k + \lambda(3i - k)$, $r = i - j + \mu(j - 3k)$;
   (c) $r = i + j + k + \lambda(i + j - k)$, $r = 3k + \mu(i + 5j - k)$.

Chapter 11

2. Find the (acute) angle between the following pairs of planes:
   (a) $2x + 3y - z = 12$ and $3x + y + 2z = 16$;
   (b) $4x - y + z = 14$ and $x + y - z = 0$;
   (c) $r = i + \lambda(i + j + k) + \mu(2i - j + 2k)$ and $r = 2j + \lambda i + \mu(2j + 3k)$;
   (d) $r = (1 + \lambda)i + (2 + \mu)j + (\lambda + \mu)k$ and $r = 2i + (\lambda + \mu)j + 3\mu k$.

3. Find the (acute) angle between the following pairs of line and plane:
   (a) $\dfrac{x-3}{2} = \dfrac{y+1}{2} = \dfrac{z+3}{-1}$, $x - 2y + 2z = 9$;
   (b) $4x = 3y = 2z$, $4x + 3y + 2z = 12$;
   (c) $r = i + \lambda(2i - 2j - k)$, $r = \lambda i + \mu j + (\lambda + \mu + 1)k$;
   (d) $r = (1 - \lambda)i + (2 - 3\lambda)j + k$, $r = (1 - \lambda)i + (\lambda + 2\mu)j + (1 - \lambda + 2\mu)k$.

4. In each of the following find the angle between the given line and plane:
   (a) $r = i - 3j + \lambda(2i - j - k)$, $r \cdot (i - 2j - 7k) = 10$;
   (b) $r = (2 + \lambda)i - 3j + (1 - \lambda)k$, $r \cdot (4i + j - k) = 6$;
   (c) $r = 3\lambda i + 2\lambda j - 6\lambda k$, $r \cdot (4i - 3k) = 20$.

11.12 The Intersections of Lines and Planes

The Intersection of Two Lines

Definition Lines in space which are not parallel and do not meet are called *skew* lines.

In three dimensional space, the lines $r = a + \lambda u$ and $r = b + \mu v$ may represent
(i) the same line ;
(ii) parallel and distinct lines ;
(iii) intersecting lines ;
(iv) skew lines.

Example Show that the lines $r = 5i + 4j + 5k + \lambda(2i + j + k)$ and $x - 1 = 2(y - 2) = 2(z - 3)$ are the same line.

The second line has parametric equations $x = 2t + 1$, $y = t + 2$, $z = t + 3$ (equate each expression in the equation with $2t$).
The lines meet when $2t + 1 = 5 + 2\lambda$, $t + 2 = 4 + \lambda$ and $t + 3 = 5 + \lambda$.
Each of these equations can be reduced to the same equation $t - \lambda = 2$, so there are an infinite number of points of intersection. Thus the lines coincide.

Vector Geometry

Example Show that the lines $r = (2 - \lambda)i + 2(1 + \lambda)j + (1 + 3\lambda)k$ and $6(1 - x) = 3(y - 1) = 2(z - 1)$ are parallel and distinct.

L1: $r = 2i + 2j + k + \lambda(-i + 2j + 3k)$ has direction vector $-i + 2j + 3k$.

L2: $\dfrac{x-1}{-1} = \dfrac{y-1}{2} = \dfrac{z-1}{3}$ has direction vector $-i + 2j + 3k$.

Clearly the lines are parallel.

Now A(2, 2, 1) lies on the first line, B(1, 1, 1) lies on the second line and $\overrightarrow{AB} = -i - j$ which is not parallel to the direction vector of either line. Therefore the lines are also distinct.

Example Show that the lines $\dfrac{x-5}{3} = \dfrac{y-3}{2} = z$ and $x - 4 = y - 3 = z - 1$ intersect and find the coordinates of the point of intersection.

The parametric equations of the lines are
L1: $x = 3t + 5,\quad y = 2t + 3,\quad z = t.$
L2: $x = u + 4,\quad y = u + 3,\quad z = u + 1.$

For intersection:
$3t + 5 = u + 4$
$2t + 3 = u + 3$
$t = u + 1.$

Solving the first two equations gives $t = -1$ and $u = -2$. These values also satisfy the third equation ($t = u + 1$) and so the lines meet at $(2, 1, -1)$.

Example Prove that the lines $x - 1 = 2 - y = \dfrac{z+5}{2}$ and $r = 2\mu i - 3j + (\mu - 2)k$ are skew.

The vector $u = i - j + 2k$ is parallel to the first line and the vector $v = 2i + k$ is parallel to the second line. Since u cannot be expressed as a scalar multiple of v, the lines are not parallel.

The parametric equations of the first line are $x = 1 + t, y = 2 - t, z = 2t - 5$.
The parametric equations of the second line are $x = 2\mu, y = -3, z = \mu - 2$.
These lines meet when $1 + t = 2\mu$
$2 - t = -3$
$2t - 5 = \mu - 2.$

From the first two equations $t = 5$ and $\mu = 3$, but these do not satisfy the third equation since $2(5) - 5 = 5$ and $3 - 2 = 1$.

Therefore the lines are skew.

323

Chapter 11

The Intersection of a Line and a Plane

The intersection of a line and a plane has three possibilities.

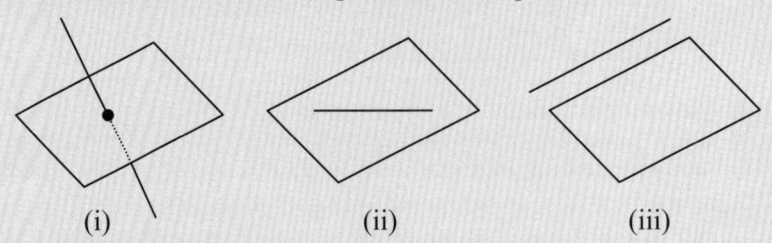

(i)   (ii)   (iii)

(i) The line meets the plane in exactly one point. (A unique solution.)
(ii) The line lies in the plane. (An infinite number of solutions.)
(iii) The line is parallel to and distinct from the plane. (No solution.)

Example   Show that the line $r = (6 + 2\lambda)i - \lambda j + (3 + 2\lambda)k$ meets the plane $3x + y + 4z = 4$ in exactly one point and find the position vector of this point.

The parametric equations of the line are $x = 6 + 2\lambda$, $y = -\lambda$, $z = 3 + 2\lambda$.
The line and plane meet when $3(6 + 2\lambda) - \lambda + 4(3 + 2\lambda) = 4 \Rightarrow \lambda = -2$.
Therefore the line and plane meet in exactly one point which has position vector $2i + 2j - k$.

Example   Show that the line $r = i + j + k + \lambda(i - j - k)$ lies in the plane with equation $r \cdot (5i + 2j + 3k) = 10$.

For intersection we have $5(1 + \lambda) + 2(1 - \lambda) + 3(1 - \lambda) = 10$
$\Rightarrow$ $5 + 5\lambda + 2 - 2\lambda + 3 - 3\lambda = 10$
$\Rightarrow$ $10 = 10$, which is true for all values of λ.

Therefore the line lies in the plane.

Example   Show that the line $x = -y = z + 2$ does not meet the plane with equation $x - 2y - 3z = 8$.

For intersection $t - 2(-t) - 3(t - 2) = 8$
$\Rightarrow$ $t + 2t - 3t + 6 = 8$
$\Rightarrow$ $6 = 8$, which is never true.

Therefore the line and plane do not meet (the line is parallel to the plane).

Vector Geometry

The Intersection of Two Planes

As with two lines, there are three possible intersection types.

(i)   (ii)   (iii)

(i) The planes meet in a line (an infinite number of solutions).
(ii) The planes coincide (an infinite number of solutions).
(iii) The planes are parallel and distinct (no solution).

Example Find parametric equations for the line of intersection of the planes $x + y - z = 6$ and $2x + y - 3z = 5$.

We use Gaussian elimination:

$$\begin{array}{cccc} 1 & 1 & -1 & 6 \\ 2 & 1 & -3 & 5 \\ \hline 1 & 1 & -1 & 6 \\ 0 & -1 & -1 & -7 \end{array}$$

$R_2 \leftarrow R_2 - 2R_1$:

Therefore, the planes meet in the line whose parametric equations are
$x = 2t - 1, \ y = 7 - t, \ z = t$.

Example Show that the planes $r = (2 + \lambda + 3\mu)i + (1 - \lambda - 2\mu)j + (1 - \lambda)k$ and $2x + 3y - z = 6$ are the same plane.

For intersection, $2(2 + \lambda + 3\mu) + 3(1 - \lambda - 2\mu) - (1 - \lambda) = 6$
$\Rightarrow \quad 4 + 2\lambda + 6\mu + 3 - 3\lambda - 6\mu - 1 + \lambda = 6$
$\Rightarrow \quad 6 = 6$, which is true for all λ, μ and so the planes coincide.

Alternative solution
The first plane is $r = 2i + j + k + \lambda(i - j - k) + \mu(3i - 2j)$ and contains the vectors $i - j - k$ and $3i - 2j$.

325

Chapter 11

Therefore the vector $\begin{vmatrix} i & j & k \\ 1 & -1 & -1 \\ 3 & -2 & 0 \end{vmatrix} = -2i - 3j + k$ is normal to this plane and the plane passes through the point (2, 1, 1). The second plane has normal vector $2i + 3j - k = -(-2i - 3j + k)$ and also contains the point (2, 1, 1). Therefore the planes coincide.

Exercise 11.12

1. Decide whether the following pairs of lines are coincident, intersecting, parallel or skew. If they intersect (but are not coincident), find the position vector of the point of intersection:

   (a) $x = y - 3 = \dfrac{z}{2}$, $\dfrac{x-2}{3} = \dfrac{y}{2} = z$;

   (b) $x - 3 = \dfrac{y+1}{2} = z + 3$, $x - 1 = y + 1 = 3 - z$;

   (c) $x = t - 2, y = 6 - t, z = 4 - t$, $x = 1 - 2u, y = 3 + 2u, z = 1 + 2u$;

   (d) $x = 1 - 3t, y = 2(t + 1), z = t + 2$, $\dfrac{x-1}{2} = y - 9 = \dfrac{z-10}{2}$;

   (e) $r = i + \lambda(2i - 2j - k)$, $r = -2\mu i + 2(1 + \mu)j + \mu k$.

2. Show that the lines $\dfrac{x}{2} = -(y+1) = \dfrac{z+1}{-2}$ and $x - 6 = y = \tfrac{1}{2}(z - 1)$ meet and find the coordinates of the point of intersection. Find also an equation for the plane determined by the lines.

3. Show that the lines $r = (1 + 2\lambda)i + (2 + \lambda)j + (3 + \lambda)k$ and $x = 2y = 2z$ are parallel and distinct. Find the equation of the plane which contains both of these lines.

4. The sides of the triangle ABC have equations as follows:
   AB: $x - 2 = y - 4 = 1 - z$;
   AC: $x + 2 = \dfrac{y+2}{2} = \dfrac{z+1}{2}$;
   BC: $\dfrac{x-7}{5} = \dfrac{y-12}{8} = \dfrac{z-5}{4}$.
   Find the coordinates of A, B and C, and the area of the triangle.

5. Find the position vector of the point where the line $\dfrac{x}{2} = y + 3 = \dfrac{z+6}{2}$ meets the plane $3x + 2y - z = 12$.

6. Find the position vector of the point of intersection of the line $r = ti - 2j + (2t - 1)k$ and the plane
   (a) $2x + y - 3z = 5$;
   (b) $r \cdot (i + 5j + 2k) = -2$;
   (c) $r = i - 2j - 2k + \lambda(i + 4j + k) + \mu(j - k)$;
   (d) $r = (1 + \lambda + 3\mu)i + (3\mu - \lambda - 2)j + (1 - \lambda - \mu)k$.

7. In each of the following determine whether or not the given line meets the given plane. If the point of intersection is unique, find the position vector of the common point. In all other cases describe the situation geometrically.
   (a) $r = -3i + \lambda(i + j)$, $r \cdot (2i + j + k) = 3$;
   (b) $r = (\lambda - 1)i + (\lambda + 2)j + (2\lambda + 3)k$, $x - y + 4z = 9$;
   (c) $r = \begin{pmatrix} 2 \\ 3 \\ 1 \end{pmatrix} + \lambda \begin{pmatrix} 1 \\ -1 \\ -1 \end{pmatrix}$, $r \cdot (3i + 4j - k) = 6$;
   (d) $x - 3 = \dfrac{y}{2} = \dfrac{z + 7}{3}$, $2x - y + 2z = 4$;
   (e) $x - 1 = 2 - y = \dfrac{3 - z}{4}$, $r \cdot (2i - 2j + k) = 1$;
   (f) $r = 4i + j + 2k + \lambda(3i + j + 2k)$, $r \cdot (i + j + k) = 4$.

8. (a) Show that the length of the orthogonal projection of the line segment [AB] on the plane $r \cdot n = d$ is $\dfrac{|\mathbf{AB} \times \mathbf{n}|}{|\mathbf{n}|}$.
   (b) If A = (4, 9, 5) and B = (8, 11, 10), calculate the length of the projection of the line segment [AB] on the plane $x + 2y + 2z = 14$.
   (c) Find the position vector of the point of intersection of the line (AB) and its projection on the plane in part (b).

9. Find in parametric form the equation of the line of intersection of the following pairs of planes:
   (a) $x + 3y + 2z = 1$, $2x + 5y + 3z = 2$;
   (b) $x - y + 5z = 4$, $3x - 4y + 11z = 11$;
   (c) $r = i + j + 2k + \lambda(i + j - k) + \mu(3i - k)$, $r = (\lambda + 2\mu)i + \lambda j - \mu k$;
   (d) $r \cdot (3i + 2j + k) = 4$, $r \cdot (3i + j + 2k) = 5$.

Chapter 11

10. Show that in each of the following the planes are identical:
    (a) $8x - y + 2z = 16$, $\boldsymbol{r} = (2 + \lambda)\boldsymbol{i} + 2\mu\boldsymbol{j} + (\mu - 4\lambda)\boldsymbol{k}$;
    (b) $\boldsymbol{r} = (1 + 2\lambda + \mu)\boldsymbol{i} + (2 + \lambda + 2\mu)\boldsymbol{j} + (3 - 2\lambda - \mu)\boldsymbol{k}$, $\boldsymbol{r} \cdot (\boldsymbol{i} + \boldsymbol{k}) = 4$.

11. Find an equation for the line of intersection of the planes $x + y + 2z = 1$ and $3x + 2y - z = 10$. Hence determine the point of intersection of the given planes and the plane $x + 2y + 12z = 0$.

12. Consider the planes $P_1 : 3x - y - z = 7$,
    $P_2 : 5x + 2y - 9z = 8$,
    $P_3 : x - y + z = 12$.

    Show that the line of intersection of P_1 and P_2 is parallel to P_3 and hence find an equation for the plane which is parallel to P_3 and contains the line of intersection of P_1 and P_2.

11.13 Shortest Distance between Points, Lines and Planes

Shortest Distance from a Point to a Line

Consider the line which is parallel to vector $\boldsymbol{v}$ and passes through the point A with position vector $\boldsymbol{a}$ and a point P which does not lie on the line. Let Q be the point on the line closest to P. Let d be the shortest distance from P to the line, i.e. $d = |\overrightarrow{PQ}|$. Let θ be the angle between $\overrightarrow{AP}$ and $\boldsymbol{v}$.

Now, $d = |\overrightarrow{AP}| \sin \theta$

$= \dfrac{|\overrightarrow{AP}||\boldsymbol{v}| \sin \theta}{|\boldsymbol{v}|}$

$= \dfrac{|\overrightarrow{AP} \times \boldsymbol{v}|}{|\boldsymbol{v}|}$ since θ is the angle between $\overrightarrow{AP}$ and $\boldsymbol{v}$.

Alternative Method This method is best described by an example.

Find the shortest distance from P(5, 6, 2) to the line $\boldsymbol{r} = \begin{pmatrix} 2 \\ 1 \\ 4 \end{pmatrix} + \lambda \begin{pmatrix} 1 \\ 1 \\ -2 \end{pmatrix}$.

With a diagram similar to that above, the line (AQ) has parametric equations $x = 2 + \lambda$, $y = 1 + \lambda$, $z = 4 - 2\lambda$, so we can let $Q = (2 + \lambda, 1 + \lambda, 4 - 2\lambda)$.

Then $\overrightarrow{QP} = \begin{pmatrix} 3-\lambda \\ 5-\lambda \\ 2\lambda-2 \end{pmatrix}$ and this is perpendicular to $v = \begin{pmatrix} 1 \\ 1 \\ -2 \end{pmatrix}$ and so $\overrightarrow{QP} \cdot v = 0$.

Hence $3 - \lambda + 5 - \lambda - 2(2\lambda - 2) = 0$ or $\lambda = 2$, and the shortest distance from P to the line $= \left\| \begin{pmatrix} 1 \\ 3 \\ 2 \end{pmatrix} \right\| = \sqrt{14}$.

Note: The advantage of the alternative method is that not only did we find the required distance, but the coordinates of Q, the closest point to P, have also been determined. {Q = (4, 3, 0) in our example.}

We can check this result using the first method:

$$d = \frac{|\overrightarrow{AP} \times v|}{|v|} \text{ where } A = (2, 1, 4) \text{ lies on the line,}$$

$$= \frac{|(3i + 5j - 2k) \times (i + j - 2k)|}{|i + j - 2k|}$$

$$= \frac{|-8i + 4j - 2k|}{\sqrt{6}}$$

$$= \frac{\sqrt{84}}{\sqrt{6}}$$

$$= \sqrt{14}, \text{ which checks our previous solution.}$$

The Distance between a Point and a Plane

Consider the plane $ax + by + cz = d$ and the point $P(x_0, y_0, z_0)$ not on the plane.

Let D be the distance from P to the plane, and let Q be the foot of the normal from P to the plane.

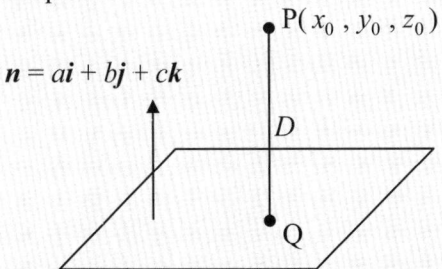

329

Chapter 11

Since $\overrightarrow{PQ}$ is parallel to $\mathbf{n}$, the parametric equations of (PQ) are:
$$x = x_0 + ta, \ y = y_0 + tb, \ z = z_0 + tc.$$

For intersection with the plane we have
$$a(x_0 + at) + b(y_0 + bt) + c(z_0 + ct) = d$$
$$\Rightarrow a^2 t + b^2 t + c^2 t = d - (ax_0 + by_0 + cz_0)$$
$$\Rightarrow t = \frac{d - (ax_0 + by_0 + cz_0)}{a^2 + b^2 + c^2}.$$

Since Q lies on the plane, the coordinates of Q can be written in the form
$$(x_0 + at, \ y_0 + bt, \ z_0 + ct), \text{ where } t = \frac{d - (ax_0 + by_0 + cz_0)}{a^2 + b^2 + c^2}$$
and the vector $\overrightarrow{PQ} = at\mathbf{i} + bt\mathbf{j} + ct\mathbf{k}$.

Now $D = |\overrightarrow{PQ}| = \sqrt{(at)^2 + (bt)^2 + (ct)^2} = |t|\sqrt{a^2 + b^2 + c^2}$.

Thus $D = \left| \frac{d - (ax_0 + by_0 + cz_0)}{a^2 + b^2 + c^2} \right| \sqrt{a^2 + b^2 + c^2} = \frac{|d - (ax_0 + by_0 + cz_0)|}{\sqrt{a^2 + b^2 + c^2}}$ which

is usually written $D = \dfrac{|ax_0 + by_0 + cz_0 - d|}{\sqrt{a^2 + b^2 + c^2}}$.

Example Find the shortest distance from the point P(2, 3, −1) to the plane $2x + 2y - z = 20$.

The required distance is $\dfrac{|2(2) + 2(3) - (-1) - 20|}{\sqrt{2^2 + 2^2 + (-1)^2}} = \dfrac{|-9|}{3} = 3.$

If the coordinates of the point on the plane closest to P are required, the following method could be used.

Example Find the point on the plane $x + y + 2z = 6$ which is closest the point P(5, 5, 7), and the shortest distance from P to the plane.

Let Q be the point on the plane closest to P.
Since (PQ) is parallel to the normal to the plane, $\mathbf{n} = \mathbf{i} + \mathbf{j} + 2\mathbf{k}$, then the equations of (PQ) are $x = t + 5, \ y = t + 5, \ z = 2t + 7$.
This line meets the plane when $(t + 5) + (t + 5) + 2(2t + 7) = 6$, i.e. $t = -3$.

Vector Geometry

Therefore the point on the plane closest to P is Q(2, 2, 1), and the shortest distance = PQ = $\sqrt{3^2 + 3^2 + 6^2} = 3\sqrt{6}$.

The Shortest Distance between Two Skew Lines

Consider two skew lines. One is parallel to vector *u* and contains the point A with position vector *a*; the second is parallel to vector *v* and contains the point B with position vector *b*.

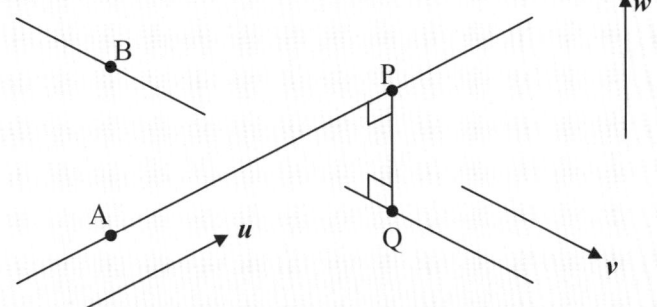

If P and Q are the points, one on each line, which are closest together then (PQ) is perpendicular to both lines and hence parallel to $w = u \times v$.

The required distance is then $|\overrightarrow{PQ}|$ which is the projection of $\overrightarrow{AB}$ on *w*.

Thus the shortest distance between two points, one on each line, is given by

$$|\overrightarrow{PQ}| = \frac{|\overrightarrow{AB} \cdot w|}{|w|} = \frac{|(b-a) \cdot (u \times v)|}{|u \times v|}.$$

Example Find the shortest distance between the skew lines
$r = 5i + 3j + \lambda(2i - j)$ and $r = 2i + 9k + \mu(j - k)$.

$u = 2i - j$ is parallel to the first line and $v = j - k$ is parallel to the second.

$w = u \times v = \begin{vmatrix} i & j & k \\ 2 & -1 & 0 \\ 0 & 1 & -1 \end{vmatrix} = i + 2j + 2k$; A(5, 3, 0) lies on the first line and B(2, 0, 9) lies on the second.

Now $\overrightarrow{AB} = -3i - 3j + 9k$ so the distance is $\frac{|\overrightarrow{AB} \cdot w|}{|w|} = \frac{|-3 - 6 + 18|}{3} = 3$.

Chapter 11

Note: This method does not find the two points, one on each line, which are closest together. If the coordinates of these points, P and Q in the previous diagram, are required, the method shown in the following example could be used.

Example Find the shortest distance between the lines in the previous example, and find the coordinates of the two points, one on each line, which are closest together.

The parametric equations of the lines are:
$x = 2\lambda + 5, y = 3 - \lambda, z = 0$ and
$x = 2, y = \mu, z = 9 - \mu$.

Let $P(2\lambda + 5, 3 - \lambda, 0)$ and $Q(2, \mu, 9 - \mu)$ be points, one on each line.

Now $\overrightarrow{PQ} = \begin{pmatrix} -3 - 2\lambda \\ -3 + \mu + \lambda \\ 9 - \mu \end{pmatrix}$ is parallel to $\mathbf{u} \times \mathbf{v} = \begin{pmatrix} 1 \\ 2 \\ 2 \end{pmatrix}$.

Thus $-3 + \mu + \lambda = 9 - \mu$
and $-3 + \mu + \lambda = 2(-3 - 2\lambda)$.

Solving these equations gives $\lambda = -2, \mu = 7$.

Therefore, $P = (1, 5, 0)$ and $Q = (2, 7, 2)$ are the required points and the shortest distance is $PQ = |\mathbf{i} + 2\mathbf{j} + 2\mathbf{k}| = 3$, as before.

A third method may also be used. The following example illustrates the method.

Example Show that the lines $\dfrac{x+6}{6} = \dfrac{y+3}{2} = \dfrac{z-7}{3}$ and $\dfrac{x+8}{6} = \dfrac{y-9}{-2} = z + 4$ are skew, and find the shortest distance between any two points, one on each line.

$\mathbf{u} = 6\mathbf{i} + 2\mathbf{j} + 3\mathbf{k}$ is parallel to the first line and $\mathbf{v} = 6\mathbf{i} - 2\mathbf{j} + \mathbf{k}$ is parallel to the second. Since these vectors are not scalar multiples of each other, the lines are not parallel.
The parametric equations of the lines are:
$x = 6t - 6$
$y = 2t - 3$ and
$z = 3t + 7$

$x = 6u - 8$
$y = 9 - 2u$
$z = u - 4$.

For intersection:
$6t - 6 = 6u - 8$ ....(1)
$2t - 3 = 9 - 2u$ ....(2)
and $3t + 7 = u - 4$. ....(3)

From equations (1) and (3) we obtain $t = -\frac{16}{3}$, $u = -5$ but these values do not satisfy equation (2) since LHS = $-\frac{32}{3} - 3 = -\frac{41}{3}$ and RHS = $9 + 10 = 19$. Therefore the lines do not meet and since the lines are not parallel they must be skew.

Now $u \times v = 8i + 12j - 24k = 4(2i + 3j - 6k)$ and $w = 2i + 3j - 6k$ is perpendicular to both lines.

Also the plane $2x + 3y - 6z = 2(-8) + 3(9) - 6(-4) = 35$ is parallel to the first line and contains the second line since it contains point $B(-8, 9, -4)$ on the second line.

Now $A(-6, -3, 7)$ lies on the first line and its distance from the plane just found is given by $\dfrac{|2(-6) + 3(-3) - 6(7) - 35|}{\sqrt{2^2 + 3^2 + (-6)^2}} = \dfrac{98}{7} = 14$ which is the required distance.

Check: Using the first method the distance is
$$\dfrac{|\overrightarrow{AB} \cdot w|}{|w|} = \dfrac{|(-2i + 12j - 11k) \cdot (2i + 3j - 6k)|}{7} = \dfrac{98}{7} = 14.$$

Exercise 11.13

1. Find the shortest distance from the origin to each of the following lines:
   (a) $x = 1 + t, y = 2 + t, z = 3 + t$;
   (b) $x = 2t, y = 3 - t, z = 3 - 2t$;
   (c) $x = t + 4, y = 2t + 2, z = 3t + 2$;
   (d) $x = 3t - 1, y = 2t + 1, z = t - 6$.

2. Find the shortest distance between the given point and the given line in each of the following:
   (a) $(6, 4, -3)$, $r = i - k + \lambda(2i + j - 2k)$;
   (b) $(2, -1, 4)$, $r = i + 3j + \lambda(i + 2j + k)$;
   (c) $(7, 4, 6)$, $r = (1 + \lambda)i + \lambda j - \lambda k$;
   (d) $(\frac{1}{2}, 1, -1)$, $r = (2 + 3\lambda)i + (1 + \lambda)j + \lambda k$.

Chapter 11

3. Find the shortest distance from the point P to the given line and the coordinates of the point on the line closest to P in each of the following:
   (a) $P = (3, 5, 9)$, $r = i + (6 + 2\lambda)j + (1 - \lambda)k$;
   (b) $P = (6, 1, 1)$, $r = \lambda i + (2\lambda - 5)j + (7 - 4\lambda)k$;
   (c) $P = (8, -2, 4)$, $r = (8 + 2\lambda)i + (4 + 2\lambda)j - (2 + \lambda)k$;
   (d) $P = (3, 1, 2)$, $r = (1 + \lambda)i + (2 - \lambda)j + (3 + \lambda)k$.

4. (a) Show that the shortest distance d from point P to the line through point A and in the direction of vector v is given by
   $$d = \sqrt{\overrightarrow{AP} \cdot \overrightarrow{AP} - \frac{(\overrightarrow{AP} \cdot v)^2}{v \cdot v}}.$$
   (b) Use the result in part (a) to find the shortest distance from the point P to the given line in each part of Question 3.

5. Find the shortest distance from the origin to each of the following planes:
   (a) $x + 2y + 3z = 14$;   (b) $2x - y - 2z = 12$;
   (c) $3x - 4y + 5z = 10$;   (d) $r \cdot (2i + j + k) = 10$;
   (e) $r \cdot (i - 3j - 5k) = 18$; (f) $r \cdot (3i + 2j - 2k) = 15$.

6. Calculate the shortest distance from the given point to the given plane in each of the following:
   (a) $(2, 3, 4)$, $2x - 2y + z = 11$;
   (b) $(-1, -4, 5)$, $3x + 4y - 5z = 6$;
   (c) $(3, -1, -2)$, $r \cdot (i - 2j - 2k) = 15$;
   (d) $(-2, 3, 1)$, $r \cdot (i - j + 2k) = 15$.

7. Calculate the distance between each of the following pairs of parallel planes:
   (a) $3x + y - z = 5$ and $6x + 2y - 2z = 43$;
   (b) $2x + 3y - 6z = 12$ and $2x + 3y - 6z = 40$.

8. The vertices of a tetrahedron ABCD are $(4, -2, -3)$, $(5, 3, 5)$, $(3, -1, 1)$ and $(8, -3, -4)$ respectively. Find:
   (a) the area of the triangle ABC;
   (b) the equation of the plane ABC;
   (c) the shortest distance from D to the plane ABC;
   (d) the volume of the tetrahedron given that the volume is one-third of the area of the base × perpendicular height.

9. Find the volume of the tetrahedra with vertices:
   (a) $(0, 0, 0)$, $(6, 0, 0)$, $(0, 12, 0)$, $(1, 6, 4)$;
   (b) $(1, 2, 3)$, $(7, 2, -3)$, $(3, 4, 0)$, $(-5, 1, 4)$.

Vector Geometry

10. Show that the lines
$$r = (4 + 2\lambda)i + (2 + \lambda)j - (2 + 3\lambda)k \quad \text{and}$$
$$r = (\mu + 1)i + 2j + (4 - \mu)k$$
are skew and find the shortest distance between them.

11. In each of the following decide whether the given lines are skew or they intersect. If they intersect find the coordinates of their common point; if they are skew, find the shortest distance between them.

   (a) $4x = 4y = z + 3$ and $\dfrac{x-7}{2} = y - 5 = \dfrac{z-12}{6}$;

   (b) $r = -2i - 3j - 13k + \lambda(2i + 2j + 3k)$ and
   $r = -i + 3j - 5k + \mu(3i - 2j - 2k)$;

   (c) $\dfrac{x-1}{2} = y - 1 = z - 2$ and $x - 4 = \dfrac{y+2}{3} = \dfrac{z+1}{-2}$;

   (d) $\dfrac{x-1}{2} = \dfrac{2y+1}{2} = \dfrac{1-z}{2}$ and $\dfrac{x}{2} = \dfrac{1-y}{3} = z$.

12. In each of the following show that the lines are skew and find the coordinates of the two points, one on each line, which are closest together.

   (a) $\dfrac{x-11}{3} = \dfrac{y+5}{-2} = \dfrac{z-8}{2}$ and $x = -2t - 4, y = 5t + 9, z = 6t + 15$;

   (b) $r = 5i + 5j + 7k + \lambda(2i + 2j + 3k)$ and $r = 2i - 4k + \mu(j + k)$;

   (c) $\dfrac{x-9}{-6} = \dfrac{y}{2} = z - 4$ and $\dfrac{x-7}{-3} = \dfrac{y-3}{-2} = \dfrac{z+5}{2}$;

   (d) $r = 4i + 4j + \lambda(3i + 2j - k)$ and $r = 14i + j + \mu(6i + j + k)$.

13. The equations of two lines are: L_1: $r = 4i + j - k + \lambda(i + 2k)$;
   L_2: $r = 3i - 2j + \mu(2i - j + 3k)$.

   Find the equation of the plane which contains L_1 and is parallel to L_2. Hence find the shortest distance between the two points, one on L_1 and the other on L_2, which are closest together.

335

Chapter 11

Required Outcomes

After completing this chapter, a student should be able to:
- use vectors in a variety of situations.
- solve geometric problems using vectors.
- define and use the scalar product of 2 vectors.
- define and use the vector product of 2 vectors. **(HL)**
- calculate the (orthogonal) projection of one vector on another.
- find the size of the angle between two directions in 2-space.
- define and use velocity vectors in 2-space.
- find the size of the angle between two lines, between a line and a plane and between two planes. **(HL)**
- calculate the area of a triangle and a parallelogram in 2- and 3-space. **(HL)**
- find the equation of a line in 2-space in both Cartesian and vector forms.
- find the equation of a line in any of several forms. **(HL)**
- find the equation of a plane in any of several forms. **(HL)**
- find the intersection of two lines, a line and a plane, and two planes. **(HL)**
- determine whether two lines intersect, are parallel or are skew. **(HL)**
- calculate the shortest distance between a point and a line, a point and a plane, and two skew lines. **(HL)**

12 Differential Calculus 2

12.1 Derivatives of Composite Functions – The Chain Rule

The functions that we have been able to differentiate to this point have been restricted to those which can be written as a sum of terms of the form kx^p where k and p are real constants. Many functions are not readily expressed in this way and we need techniques which enable us to differentiate them. Fortunately such techniques are available and are simple to apply.

Higher Level

Derivatives of Composite Functions

Consider the composite function $(f \circ g)(x) = f(u)$ where $u = g(x)$. If function g is differentiable at x and the function f is differentiable at $g(x)$, then the composite function $f \circ g$ is differentiable at x and $(f \circ g)'(x) = f'(g(x))g'(x)$ where $f'(g(x))$ is the derivative of $f(u)$ with respect to u.

Using the alternative notation, we can express this rule in a more easily remembered form.

If y is a function of u defined by $y = f(u)$ and $\dfrac{dy}{du}$ exists, and if u is a function of x defined by $u = g(x)$ and $\dfrac{du}{dx}$ exists, then y is a function of x and $\dfrac{dy}{dx}$ exists and is given by $\dfrac{dy}{dx} = \dfrac{dy}{du} \times \dfrac{du}{dx}$.

This is a convenient form for remembering the chain rule. It simply *appears* that du cancels but we must keep in mind that neither dy nor dx has been given independent meaning in the notation $\dfrac{dy}{dx}$.

Chapter 12
The Chain Rule

If $y = F(x) = (f \circ g)(x) = f(u)$ where $u = g(x)$, then

$$F'(x) = f'(u) \times g'(x) = (f'(g(x)))(g'(x))$$

or

$$\frac{dy}{dx} = \frac{dy}{du} \times \frac{du}{dx}.$$

Note: The proof of the Chain Rule is beyond the scope of this course.

Example If $y = F(x) = (2x-1)^3 = f(u) = u^3$ where $f(x) = x^3$ and $u = g(x) = 2x-1$, then

$$F'(x) = (f'(u))(g'(x)) \qquad \text{or} \qquad \frac{dy}{dx} = \left(\frac{dy}{du}\right)\left(\frac{du}{dx}\right)$$

$$= (3u^2)(2) \qquad\qquad\qquad\qquad = (3u^2)(2)$$

$$= 6u^2 \qquad\qquad\qquad\qquad\qquad = 6u^2$$

$$= 6(2x-1)^2. \qquad\qquad\qquad\qquad = 6(2x-1)^2.$$

Note: The process can be accomplished using the following steps:
(1) Differentiate first as if the expression in parentheses were a single variable.
  e.g. $3(2x-1)^2$ in the above example.

(2) Multiply by the derivative of the expression in parentheses.
  e.g. Multiply by the derivative of $2x-1$ in our example.
  This gives $3(2x-1)^2(2) = 6(2x-1)^2$.

Example Differentiate $y = \sqrt{1-x^2}$ with respect to x.

$$y = \sqrt{1-x^2} = (1-x^2)^{1/2}$$

$$\frac{dy}{dx} = \tfrac{1}{2}(1-x^2)^{-1/2}(-2x) = \frac{-x}{\sqrt{1-x^2}}.$$

Differential Calculus 2

Exercise 12.1

1. Differentiate each of the following functions with respect to x:
   (a) $(3x+1)^3$;
   (b) $(5x-2)^4$;
   (c) $(1-3x)^5$;
   (d) $(x^2+2)^3$;
   (e) $(1+x-x^2)^5$;
   (f) $\dfrac{1}{2x+5}$;
   (g) $\dfrac{4}{3-x^2}$;
   (h) $\dfrac{2}{(x^3+1)^2}$;
   (i) $\dfrac{5}{(1-3x^2)^3}$;
   (j) $\sqrt{5x-4}$;
   (k) $\sqrt[3]{2x^2+5}$;
   (l) $\dfrac{3}{\sqrt{12-5x^2}}$.

2. (a) Find the equation of the tangent to the curve $y=(2x-1)^3$ at the point (1, 1).
   (b) Find the equations of the tangent and normal to the curve $y\sqrt{2x+3}=6$ at the point (3, 2).
   (c) Find the values of a and b if $y = ax + b$ is a tangent to the curve $y = 2x+(3x-2)^3$ at the point (1, 3).

3. (a) Find the maximum perimeter of a right-angled triangle which has a hypotenuse of length 20 cm.
   (b) A right-angled triangle has a hypotenuse of length $\sqrt{5}$. If the lengths of the other sides are x and y, find the maximum value of $2x + y$.

4. (a) A particle moves along the x-axis so that its position, x m to the right of the origin at time t seconds, is given by $x = t - \sqrt{t^2+4t}$. Find the position and velocity of the particle at $t = 2$ and show that the particle never comes to rest.
   (b) A particle moves along the x-axis so that its position, x m to the right of the origin at time t seconds, is given by $x = 3t^2 - 18\sqrt{2t+7}$. Find where and when the particle is stationary, and find the acceleration at this time.
   (c) A particle moves along the x-axis so that its position, x m to the right of the origin at time t seconds, is given by $x = 5t + \dfrac{24}{t^2+3}$. Find the initial velocity, the average velocity for the first three seconds, and the maximum velocity.

339

Chapter 12

5. Find the position and nature of the stationary points of each of the following curves:

(a) $y = x + 2 + \dfrac{1}{x+2}$;

(b) $y = 3x - \sqrt{6x - 5}$;

(c) $y = 3x + \dfrac{4}{3x - 2}$;

(d) $y = (3x - 1)^3 + 3(3 - x)^3$.

6. A man is in a boat at A which is 3 km from the nearest point O on a straight beach. His destination is C which is 6 km along the beach from O. If he can row at 4 km h^{-1} and walk at 5 km h^{-1}, towards which point B on the beach should he row to reach his destination in the least possible time? What is the least possible time required?

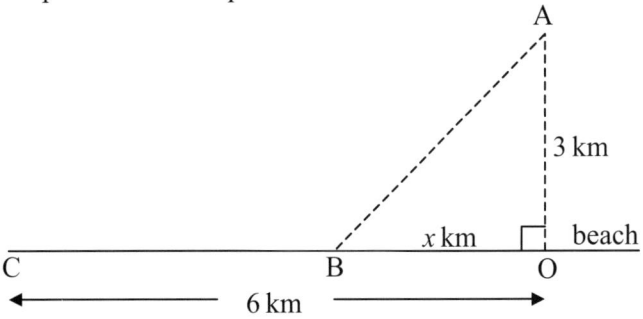

Answer the question above if his destination is only 3 km from O, giving the time required correct to the nearest minute.

Higher Level

7. (a) Find the equation of the tangent to the curve $y = \dfrac{3}{(ax - 1)^2}$ at the point where $x = 1$ and find the value of a for which this tangent passes through the origin.

(b) Find the values of the real numbers a and b if $4x + 3y = 7$ is a tangent to the curve $y = \dfrac{a}{3x + 1} + \dfrac{b}{x + 5}$ at the point where $x = 1$.

(c) Find the equation of the tangent to the curve $y = (2x + 1)^4$ at the point where $x = a$. Deduce the equations of the tangents to the curve from the origin.

8. Find the equation of the tangent to the curve $y = (3x - 1)^3$ at the point (1,8). By solving for points of intersection of the tangent and the curve, find the coordinates of the point where the tangent at (1, 8) *cuts* the curve.

12.2 The Product Rule

If $f(x)$ and $g(x)$ are differentiable functions of x and $F(x) = f(x)g(x)$, then the derivative of $F(x)$, the product of the functions $f(x)$ and $g(x)$, is given by

$$F'(x) = g(x)f'(x) + f(x)g'(x).$$

Proof (Higher Level only.)

From the definition,
$$F'(x) = \lim_{h \to 0} \frac{F(x+h) - F(x)}{h}$$
$$= \lim_{h \to 0} \frac{f(x+h)g(x+h) - f(x)g(x)}{h}$$
$$= \lim_{h \to 0} \frac{f(x+h)g(x+h) - f(x)g(x+h) + f(x)g(x+h) - f(x)g(x)}{h}$$
$$= \lim_{h \to 0} \left(g(x+h) \left(\frac{f(x+h) - f(x)}{h} \right) \right) + \lim_{h \to 0} \left(f(x) \left(\frac{g(x+h) - g(x)}{h} \right) \right)$$
$$= g(x)f'(x) + f(x)g'(x) \quad \{\text{limit theorems}\}.$$

Example   Find $\dfrac{dy}{dx}$ in each of the following:

(a) $y = (x^3 + 2)(5x - 4)$;   (b) $y = x^2\sqrt{3x + 2}$.

(a) $y = (x^3 + 2)(5x - 4)$

$\Rightarrow \dfrac{dy}{dx} = (5x - 4)(3x^2) + (x^3 + 2)(5) = 20x^3 - 12x^2 + 10$.

[*Note:* The product rule is not necessary here since the given expression for y may be expanded before differentiating:
$y = 5x^4 - 4x^3 + 10x - 8 \;\Rightarrow\; \dfrac{dy}{dx} = 20x^3 - 12x^2 + 10$.]

(b) $y = x^2\sqrt{3x + 2}$

$\Rightarrow \dfrac{dy}{dx} = \sqrt{3x+2}\,(2x) + x^2\{\tfrac{1}{2}(3x+2)^{-1/2}(3)\}$

$= 2x\sqrt{3x+2} + \dfrac{3x^2}{2\sqrt{3x+2}}$.

341

Chapter 12

Exercise 12.2

1. Differentiate each of the following functions using the product rule. [There is no need to simplify your answers.]
   (a) $(3x+2)(2x+3)$;
   (b) $(x^2+1)(2x-1)$;
   (c) $(2x^2-1)(1-3x)$;
   (d) $(3x^2-2)(2x^3+5)$;
   (e) $2x\sqrt{2x^2+3}$;
   (f) $(3x^2-1)\sqrt{3x^2+1}$;
   (g) $(3x+1)(2x-1)^3$;
   (h) $(1-x^2)^2(1+x^2)^3$;
   (i) $3x(2x+1)^{3/2}$;
   (j) $(2x-3)^3(3x-2)^2$;
   (k) $(2x-1)^{1/2}(3x+1)^{3/2}$;
   (l) $(1-x)\sqrt{2-3x^2}$.

2. (a) What are the dimensions of an isosceles triangle which has a perimeter of 18 cm and whose area is as large as possible?
   (b) The diagonal of a rectangle is 2 m. Find the maximum area that the rectangle may have.

3. Find the positions and natures of the stationary points of each of the following curves:
   (a) $y = x\sqrt{2x+3}$;
   (b) $y = x\sqrt[3]{x+4}$;
   (c) $y = x(x-14)^{2/3}$.

4. A triangular prism has a square base and an isosceles triangle at each end. If the perimeter of each end is 25 m, find the length of the prism with maximum possible volume.

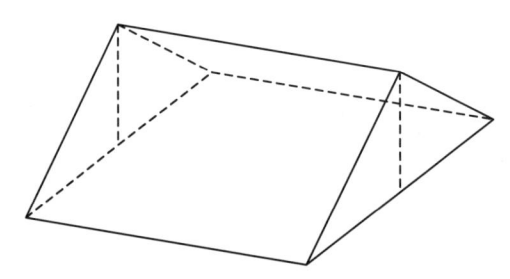

5. A particle moves along the x-axis such that is position, $x(t)$ m to the right of the origin at time t s, is given by $x(t) = (t-1)(t-4)^2$.
   (a) When and where is the particle stationary?
   (b) During which time interval is the particle moving towards the left? What is the maximum speed of the particle during this time?
   (c) Sketch the graph of $x(t)$ versus t for $0 \le t \le 6$, and the graph of the velocity of the particle for the same values of t.

(d) For the first six seconds of motion find:
  (i) the total distance travelled by the particle ;
  (ii) the average speed of the particle ;
  (iii) the average velocity of the particle.

Higher Level

6. (a) Find the equation of the tangent to the curve $y = x\sqrt{3x-2}$ at the point (2, 4).
   (b) Find the equation of the tangent to the curve $y = (2x-3)^3(5x-8)^2$ at the point where $x = 2$.
   (c) Find the values of the real numbers a and b if $y = ax + 6$ is a tangent to the curve $y = (x^2 + b)\sqrt{3-2x}$ at the point where $x = -3$.

7. The stiffness of a rectangular beam is proportional to the product of the width of the beam and the cube of its depth. Find the dimensions of the stiffest beam that may be cut from a log of diameter 24 cm.

12.3 The Quotient Rule

If $f(x)$ and $g(x)$ are differentiable functions of x and $F(x) = \dfrac{f(x)}{g(x)}$, then the derivative of $F(x)$, the quotient of the functions $f(x)$ and $g(x)$, is given by

$$F'(x) = \frac{g(x)f'(x) - f(x)g'(x)}{[g(x)]^2} \quad \text{provided } g(x) \neq 0.$$

Proof (Higher Level only.)

From the definition,

$$F'(x) = \lim_{h \to 0} \frac{F(x+h) - F(x)}{h}$$

$$= \lim_{h \to 0} \frac{\dfrac{f(x+h)}{g(x+h)} - \dfrac{f(x)}{g(x)}}{h}$$

$$= \lim_{h \to 0} \frac{f(x+h)g(x) - f(x)g(x+h)}{hg(x)g(x+h)}$$

343

Chapter 12

$$= \lim_{h \to 0} \frac{f(x+h)g(x) - f(x)g(x) - f(x)g(x+h) + f(x)g(x)}{hg(x)g(x+h)}$$

$$= \lim_{h \to 0} \frac{1}{g(x)g(x+h)} \left(g(x) \frac{f(x+h) - f(x)}{h} - f(x) \frac{g(x+h) - g(x)}{h} \right)$$

$$= \frac{g(x)f'(x) - f(x)g'(x)}{[g(x)]^2}.$$

Example Find $\frac{dy}{dx}$ in each of the following:

(a) $y = \frac{x+3}{x-2}$; (b) $y = \frac{2x^2}{3x+4}$; (c) $y = \frac{2x}{(2x+3)^2}$.

(a) $y = \frac{x+3}{x-2} \Rightarrow \frac{dy}{dx} = \frac{(x-2)(1) - (x+3)(1)}{(x-2)^2} = \frac{-5}{(x-2)^2}$

(b) $y = \frac{2x^2}{3x+4} \Rightarrow \frac{dy}{dx} = \frac{(3x+4)(4x) - 2x^2(3)}{(3x+4)^2} = \frac{2x(3x+8)}{(3x+4)^2}$

(c) $y = \frac{2x}{(2x+3)^2} \Rightarrow \frac{dy}{dx} = \frac{(2x+3)^2(2) - 2x\{2(2x+3)(2)\}}{(2x+3)^4} = \frac{2(3-2x)}{(2x+3)^3}$

Example Find the equations of the tangents to the curve $y = \frac{4x}{x^2+1}$ at the origin and at the point (1,2).

$\frac{dy}{dx} = \frac{(x^2+1)(4) - 4x(2x)}{(x^2+1)^2} = 4, 0$ at the origin and at the point (1,2).

The tangents are $y = 4x$ and $y = 2$.

The Derivative of x^n when n is a Negative Integer

We stated in Section 10.4 that the derivative of x^n for rational n is nx^{n-1} but were only able to prove this for positive integer values of n. However, we are now in a position to prove that the rule holds when n is a negative integer.

[We shall prove later that the rule holds for rational values of n and later still for all real values of n.]

Theorem If $y = x^n$ where n is a negative integer, then $\frac{dy}{dx} = nx^{n-1}$.

Proof   Let $m = -n$ where n is a negative integer.
Then m is a positive integer.

Now using the quotient rule with the function $y = x^n = x^{-m} = \dfrac{1}{x^m}$ we

obtain $\dfrac{dy}{dx} = \dfrac{x^m \dfrac{d}{dx}(1) - (1)\dfrac{d}{dx}(x^m)}{x^{2m}}$

$= \dfrac{-mx^{m-1}}{x^{2m}}$ [since m is a positive integer]

$= -mx^{-m-1}$

$= nx^{n-1}$ [since $-m = n$].

Exercise 12.3

1. Differentiate the following with respect to x and simplify each answer:

   (a) $\dfrac{x}{x+2}$;   (b) $\dfrac{x-4}{x+5}$;   (c) $\dfrac{3x^2}{2x+5}$;

   (d) $\dfrac{3x+1}{(2x-1)^2}$;   (e) $\dfrac{3x+4}{x^2+1}$;   (f) $\dfrac{x}{\sqrt{x^2+2}}$;

   (g) $\dfrac{x^2-4}{x^2+4}$;   (h) $\dfrac{2x}{(2x+3)^3}$;   (i) $\dfrac{(x+1)(2x+1)}{x-2}$;

   (j) $\dfrac{x}{\sqrt{2x+1}}$;   (k) $\sqrt{\dfrac{2x+1}{2x-1}}$;   (l) $\dfrac{3x}{\sqrt{2x^2-x+7}}$.

2. (a) Find the equation of the tangent to the curve $y = \dfrac{3x+1}{x+1}$ at (1, 2).

   (b) Find the equation of the tangent to the curve $y = x^2 - \dfrac{x}{2x+1}$ at $x = -1$.

   (c) Find the equations of the tangents to the curve $y = \dfrac{2x-3}{x+1}$ which are parallel to $5x - y = 0$.

3. Determine the position and nature of the stationary points of each of the following functions:

   (a) $y = \dfrac{4x}{(x-1)^2}$;   (b) $y = \dfrac{x^2-5x+4}{x^2+5x+4}$;   (c) $y = \dfrac{x^2-4x+1}{x-4}$.

345

Chapter 12

4. If the tangent and normal to the curve $y = \dfrac{25 - 2x}{15 - x}$ at the point P(10, 1) meet the y-axis in Q and R, calculate the area of the triangle PQR.

5. A certain epidemic spreads so that after x months $P\%$ of the population will be infected where $P = \dfrac{30x^2}{(1+x^2)^2}$. In which month will the largest number of people be infected? What percentage of the population will be infected during this month?

6. A particle moves along the x-axis so that at time t s after passing through O, its displacement, $x(t)$ m to the right of O, is given by $x(t) = \dfrac{2t}{1+t^2}$. Find when and where the particle comes to rest. With what velocities does it pass through the point 60 cm to the right of O?

Higher Level

7. (a) Find the values of the real numbers a and b if $8x - 4y = 7$ is a tangent to the curve $y = \dfrac{ax+b}{(x+1)^2}$ at $x = 1$.

   (b) Find the values of the real numbers a and b if the point $(4, \frac{1}{4})$ is a stationary point of $y = \dfrac{ax+b}{x^2-4}$.

   (c) The line $y = ax + b$ is a tangent to the curve $y = \dfrac{13x+1}{x+1}$ at $x = 1$. Find the values of the real numbers a and b, and find the equation of the tangent parallel to $y = ax + b$.

8. If $y = \dfrac{x-2}{1+2x}$, show that $\dfrac{dy}{dx} = \dfrac{1+y^2}{1+x^2}$.

9. Find the position and nature of the stationary points of the curve $y = \dfrac{2x^2 + 5x}{x^2 + 6x - 7}$.

10. Find the position and nature of the stationary point of the graph of the function $f(x) = \dfrac{x^2 + 4x}{(x+3)(x+1)}$.

11. A rectangular sheet of paper of dimensions 21 cm × 28 cm is folded so that one of its corners is placed on the opposite edge [see diagram].

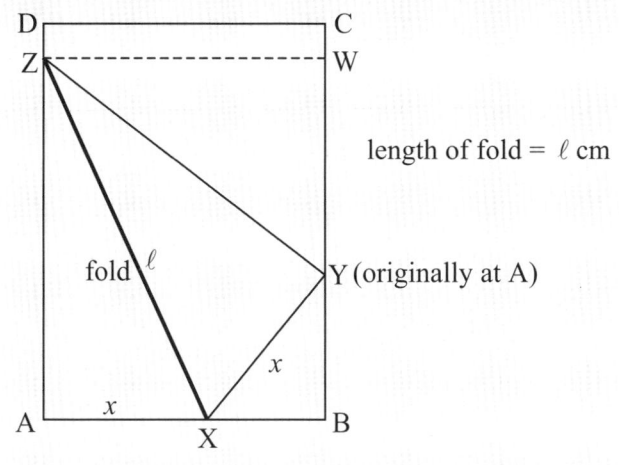

length of fold = ℓ cm

Y (originally at A)

(a) Prove that triangles XBY and YWZ are similar.

(b) Show that $\ell^2 = \dfrac{2x^3}{2x-21}$.

(c) Find the value of x for which ℓ^2 is a minimum.

(d) Find the minimum length of the fold.

12.4 Implicit Differentiation

If we are given a relation of the form $x^2 + y^2 = 9$ and we attempt to express y in terms of x, we obtain two possible results; namely $y = \sqrt{9-x^2}$ and $y = -\sqrt{9-x^2}$.

Each of these functions defines y **explicitly** and uniquely in terms of x. Hence we can work with two functions, f and g, where $f(x) = \sqrt{9-x^2}$ and $g(x) = -\sqrt{9-x^2}$.

Now $f'(x) = \frac{1}{2}\left(9-x^2\right)^{-1/2}(-2x) = \dfrac{-x}{\sqrt{9-x^2}} = \dfrac{-x}{y}$ where $y = f(x)$.

Also $g'(x) = -\frac{1}{2}\left(9-x^2\right)^{-1/2}(-2x) = \dfrac{x}{\sqrt{9-x^2}} = \dfrac{-x}{y}$ where $y = g(x)$.

Note that the derivative, $\dfrac{-x}{y}$, is independent of the function chosen.

Chapter 12

However, it is not always easy (or possible) to transpose the expressions of certain *implicit* relations in order to express y explicitly in terms of x, e.g. $x^5 + 2x^2 y^3 + y^5 = 6$. In such cases $\frac{dy}{dx}$ is most easily found by a method known as *implicit differentiation*.

To illustrate the method, consider the equation introduced at the beginning of this section: $x^2 + y^2 = 9$.

We differentiate both sides with respect to x: $\frac{d}{dx}(x^2) + \frac{d}{dx}(y^2) = \frac{d}{dx}(9)$

$$2x + \frac{d}{dy}(y^2) \times \frac{dy}{dx} = 0 \text{ [chain rule]}$$

$$\Rightarrow \quad 2x + 2y\frac{dy}{dx} = 0$$

$$\Rightarrow \quad \frac{dy}{dx} = \frac{-x}{y}, \text{ as before.}$$

Example Find $\frac{dy}{dx}$ if $2x^2 + 3xy + 5y^2 = 10$.

We need to differentiate the product xy with respect to x.

Now $\frac{d}{dx}(xy) = y\frac{dx}{dx} + x\frac{dy}{dx} = y + x\frac{dy}{dx}$ using the product rule.

Also $\frac{d}{dx}(5y^2) = \frac{d}{dy}(5y^2) \times \frac{dy}{dx} = 10y\frac{dy}{dx}$ using the chain rule.

Differentiating the given relation with respect to x gives:

$$4x + 3\left(y + x\frac{dy}{dx}\right) + 10y\frac{dy}{dx} = 0 \quad \Rightarrow \quad \frac{dy}{dx} = \frac{-(4x+3y)}{3x+10y}.$$

Example Find the equation of the normal to the curve $2x^2 - 6xy + y^2 = 9$ at the point (4, 1).

Differentiating the given equation with respect to x gives:

$$4x - 6\left(y + x\frac{dy}{dx}\right) + 2y\frac{dy}{dx} = 0.$$

At the point (4, 1), $16 - 6\left(1 + 4\frac{dy}{dx}\right) + 2\frac{dy}{dx} = 0$ or $\frac{dy}{dx} = \frac{10}{22} = \frac{5}{11}$.

348

Therefore the gradient of the normal is $\dfrac{-11}{5}$ and the required equation is $11x + 5y = 49$.

The Derivative of x^n when n is Rational

Theorem If $y = x^n$ where n is rational, then $\dfrac{dy}{dx} = nx^{n-1}$.

Proof Let $n = \dfrac{p}{q}$ where p and q are integers.

Then $y = x^n = x^{p/q} \;\Rightarrow\; y^q = x^p$.

Differentiating with respect to x gives: $qy^{q-1}\dfrac{dy}{dx} = px^{p-1}$.

Thus $\dfrac{dy}{dx} = \dfrac{px^{p-1}}{qy^{q-1}}$

$= \left(\dfrac{p}{q}\right)\left(\dfrac{x^{p-1}}{\left(x^{p/q}\right)^{q-1}}\right)$ [since $y = x^{p/q}$]

$= \left(\dfrac{p}{q}\right)\left(\dfrac{x^{p-1}}{x^{p-p/q}}\right)$

$= \left(\dfrac{p}{q}\right) x^{(p-1)-(p-p/q)}$

$= \left(\dfrac{p}{q}\right) x^{(p/q)-1}$

$= nx^{n-1}$.

Exercise 12.4

1. If y is a function of x, write down the derivative with respect to x of each of the following:

   (a) $5y$; $3y^2$; $\sqrt{y}$; $\dfrac{1}{y}$; $\dfrac{2}{y^3}$.

   (b) xy ; $x^2 y$; $3xy^2$; $\dfrac{x^2}{y}$; $6x^2 y^3$.

   (c) $(5x - 2y)^2$; $(2x + 3y)^3$; $\dfrac{2x}{x+y}$; $\sqrt{x^2 + y^2}$; $\dfrac{x + 2y}{x - 2y}$.

349

Chapter 12

2. Find $\dfrac{dy}{dx}$ in each of the following:

   (a) $4x^2 + y^2 = 5$;   (b) $y^3 = x^2$;

   (c) $3x^2 - 4y^2 = 12$;   (d) $x^3 + x^2y^2 + 2y^3 = 20$;

   (e) $\sqrt{x} + 2\sqrt{y} = 5$;   (f) $x^3 - 2x^2y + 4y^2 = 12$.

3. Find the equations of the tangent and normal to each of the following curves at the indicated point:

   (a) $2x^3 - 3xy + y^3 = 11$ at (2, 1) ;   (b) $x^3 + xy + y^2 = 7$ at (−2, −3) ;

   (c) $(2x + 3y)(x - 2y) = 4$ at (−2, 1) ;   (d) $\dfrac{1}{y} = x^2 + xy + y^2$ at (0, 1).

4. If the tangent and normal to the curve $x^2 + xy - y^2 = 5$ at P(3, −1) meet the y-axis at Q and R, find the area of the triangle PQR.

5. Find where the following curves are parallel to the coordinate axes:

   (a) $4x^2 + 9y^2 = 36$;   (b) $(x + 2y)^2 + 4(x - 2y) = 0$.

6. Prove that the curves $y^2 = 2x - 2$ and $2(x - 1)^2 + y^2 = 12$ cut at right angles at the point (3, 2).

7. Prove that the curves $x^2 + 3y^2 = 24$ and $3x^2 - y^2 = 12$ intersect at right angles at the point $\left(\sqrt{6}, \sqrt{6}\right)$.

8. Find the equations of the tangents to the curve $xy^2 = 2x^2 + 8$ which are parallel to the x-axis.

9. The total surface area of a right-circular cone of height h and base-radius r is given by $S = \pi\left(r^2 + r\sqrt{r^2 + h^2}\right)$. If S is constant, find $\dfrac{dr}{dh}$ when $r = 3$ and $h = 4$.

10. A cone has a fixed volume of 144π cm^3. Calculate the length of its least possible slant edge.

12.5 Related Rates

In this section we will learn how to calculate a rate of change we cannot measure from a rate of change that we can measure. For example, while a balloon is being inflated, we may know the rate at which air is being pumped in is 1 litre per second. However we would really like to know the rate at which the radius of the balloon is increasing. The procedure which provides us with a suitable method to achieve our aim, is best illustrated by an example.

Example   Air is pumped into a spherical balloon at the rate of 1 litre per second. Find the rate at which the radius of the balloon is changing when the radius is 10 cm.

The first task is to select appropriate units. In this problem centimetres and seconds have been chosen.

Next, it is helpful to express the rate of change we are given in terms of a derivative with respect to time. In our problem, we are given $\frac{dV}{dt} = 1\,000 \text{ cm}^3 \text{ s}^{-1}$.

We require the rate of change of the radius of the balloon, i.e., $\frac{dr}{dt}$, when $r = 10$ cm.

The equation connecting the volume of the balloon with its radius is $V = \frac{4}{3}\pi r^3$.

Differentiating this equation with respect to time gives: $\frac{dV}{dt} = 4\pi r^2 \frac{dr}{dt}$.

Thus $\frac{dr}{dt} = \frac{1\,000}{4\pi r^2}$ and clearly the rate at which the radius changes is dependent upon the radius at the time.

In our case $r = 10$ cm.

Finally we obtain $\frac{dr}{dt} = \frac{1\,000}{400\pi} = \frac{5}{2\pi} \approx 0.796$ and so when $r = 10$ cm, the radius is increasing at the rate of 0.796 cm s^{-1}.

Chapter 12

The procedure is as follows:
(1) Draw a large, neat, labelled diagram.
(2) Choose suitable units for length, angle, time, etc.
(3) Write down the rate(s) of change given.
(4) Write down the rate of change required by the question.
(5) Find a relation between the variables the rates of change of which are given and required.
(6) Differentiate this relation with respect to time.
(7) Substitute the rate given in (3) and the values of any variables at the time required.
(8) Answer the question in sentence form.

Example A ladder 5 m long is placed against a vertical wall. If the upper end slides down the wall at a rate of 6 cm s^{-1}, find the rate at which the lower end is moving when it is 3 m from the base of the wall.

Units: m, s

Given: $\dfrac{dy}{dt} = -0.06$

[This is negative since y is decreasing with time.]

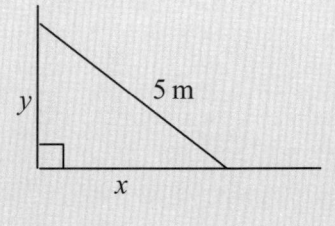

To Find: $\dfrac{dx}{dt}$ when $x = 3$

Calculation: $x^2 + y^2 = 25$

Differentiate with respect to t: $2x\dfrac{dx}{dt} + 2y\dfrac{dy}{dt} = 0$ which

gives $\dfrac{dx}{dt} = \dfrac{-y}{x}\dfrac{dy}{dt} = \dfrac{0.06y}{x}$.

When $x = 3$, $y = 4$ and $\dfrac{dx}{dt} = 0.08$. Therefore the lower end is moving at the rate of 8 cm s^{-1}.

Example Two ships leave a port P at noon and sail in directions which make an angle of 120° with each other. Ship X sails at 24 km h^{-1} and ship Y at 40 km h^{-1}. Find the rate at which the distance between the two ships is changing at 4 pm.

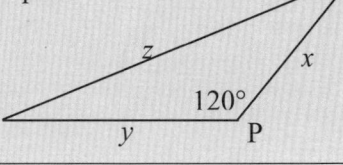

Units: km, h

Given: $\dfrac{dx}{dt} = 24$, $\dfrac{dy}{dt} = 40$

To Find: $\dfrac{dz}{dt}$ when $t = 4$

Calculation: $z^2 = x^2 + y^2 - 2xy\cos 120° = x^2 + y^2 + xy$

Differentiate with respect to t:

$$2z\frac{dz}{dt} = 2x\frac{dx}{dt} + 2y\frac{dy}{dt} + y\frac{dx}{dt} + x\frac{dy}{dt}$$

$$= 48x + 80y + 24y + 40x$$

$$= 88x + 104y.$$

When $t = 4$, $x = 96$, $y = 160$ and $z^2 = 96^2 + 160^2 + 96 \times 160$,

and so $z = 224$ which gives $\dfrac{dz}{dt} = \dfrac{88 \times 96 + 104 \times 160}{448} = 56.$

At 4 pm the distance between the two ships is increasing at the rate of 56 km h^{-1}.

Consider the following method which may be used in related rate problems in which all variables can be expressed as functions of time.

After t hours, $x = 24t$, $y = 40t$ and $z^2 = (24t)^2 + (40t)^2 + (24t)(40t)$ which gives $z = 56t$.

Now clearly $\dfrac{dz}{dt} = 56$ and so the distance between the ships is *always* changing at the rate of 56 km h^{-1}.

Exercise 12.5

1. The length of the side of a square is increasing at the rate of 3 cm s^{-1}. At what rate is the area of the square increasing when the length of the side is 10 cm?

2. The length of the radius of a circle is increasing at the rate of 20 cm s^{-1}. Find the rate at which the area is changing when the radius is 1 m.

3. The volume of a sphere is increasing at the rate of $80\pi \text{ cm}^3 \text{ s}^{-1}$. Find the rate at which the surface area is increasing when the radius is 5 cm.

4. A triangle ABC is right-angled at B and $AC = 10$ cm. If AB is increasing at the rate of 0.5 cm s^{-1}, find the rate of change of the length of BC when $BC = 6$ cm.

Chapter 12

5. A spherical balloon is being inflated at the rate of 36π cm^3 s^{-1}. Find the rate at which the radius of the balloon is increasing
   (a) when the volume is 288π cm^3 ;
   (b) 19 seconds after the above instant.

6. Grain is pouring out of a hole in the bottom of a large inverted cone with a vertical angle of 90° at the rate of 2 m^3 min^{-1}. Find the rate, in cm s^{-1}, at which the depth of grain in the cone is changing when the depth is 0.8 m.

7. A hemispherical bowl has a radius of 1.5 m. Water is added to the bowl at the rate of $\frac{1}{3}\pi$ m^3 min^{-1}. When the depth of water in the bowl is h m, the volume of water is given by $V = \frac{1}{6}\pi h^2(9-2h)$ m^3. How deep is the water after 1 minute? At what rate is the depth increasing at this time?

8. Two trains, X and Y, leave a station A simultaneously and travel in straight paths at an angle of 60° to each other, and with speeds of 40 km h^{-1} and 50 km h^{-1} respectively. Find the distance between the trains at the end of 30 minutes and the rate of change of this distance at that time.

9. Three towns, A, B and C, are at the vertices of an equilateral triangle with sides 100 km. One train leaves A and travels towards B at a rate of 80 km h^{-1} and a second train leaves B at the same time and travels towards C at a speed of 60 km h^{-1}. Find the rate at which the distance between the trains is changing after 45 minutes.

10. The volume of a cylinder is increasing at the rate of 20 cm^3 s^{-1}. The radius of the base is increasing at the rate of 3 cm s^{-1}. How fast is the height of the cylinder changing when the volume is 120 cm^3 and the radius is 5 cm.

12.6 The Graphs of Rational Functions

Although a graphic display calculator can be used to draw a graph of any rational function, the resolution of various critical parts of the curve may not be sufficiently clear enough. If a window is not well set-up, asymptotes may sometimes appear like cusps (sharp points), and indeed asymptotes may not be obvious at all. Before using a GDC, a student should have already determined the equations of all asymptotes, and should have a reasonable idea of the shape of the graph to be drawn.

Differential Calculus 2

Asymptotes

If the graph of a function approaches a fixed line (or curve) as the graph moves further from the origin, we say that the line is an asymptote of the graph.

Asymptotes arise in two distinct ways.

(i) If $\lim_{x \to \infty} f(x) = a$ or $\lim_{x \to -\infty} f(x) = a$, then $y = a$ is a ***horizontal asymptote*** of the graph of $y = f(x)$.

(ii) If $\lim_{x \to a^+} f(x) = \pm\infty$ or $\lim_{x \to a^-} f(x) = \pm\infty$, then $x = a$ is a ***vertical asymptote*** of the graph of $y = f(x)$.

Note: If the degree of the numerator of the rational function is greater than the degree of the denominator, an oblique line or a curve will replace the horizontal line as an asymptote.

Example Find the equations of the the horizontal and vertical asymptotes of the curve $y = \dfrac{2x+3}{x+1}$.

The first step is to divide $(x + 1)$ into $(2x + 3)$ giving $y = 2 + \dfrac{1}{x+1}$.

The vertical asymptote occurs when the denominator is zero, i.e., $x = -1$ is the vertical asymptote.

To establish the horizontal asymptote we must determine the behaviour of the function for large positive and negative values of x.

As $x \to +\infty$, $y \to 2^+$ since $\dfrac{1}{x+1}$ is very small and positive for large positive values of x.

As $x \to -\infty$, $y \to 2^-$ since $\dfrac{1}{x+1}$ is very small and negative for large negative values of x.

Therefore $y = 2$ is the horizontal asymptote.

From the information gathered in the determination of the behaviour near the asymptotes, together with any axes intercepts, we are able to make a very reasonable sketch of the graph. The only information which would assist with the sketch that we have not established at this point, is the position and nature of any stationary points.

In the previous example, the axes intercepts are $(0, 3)$ and $(-1.5, 0)$.

355

Chapter 12

The graph of $y = \dfrac{2x+3}{x+1}$ is as follows:

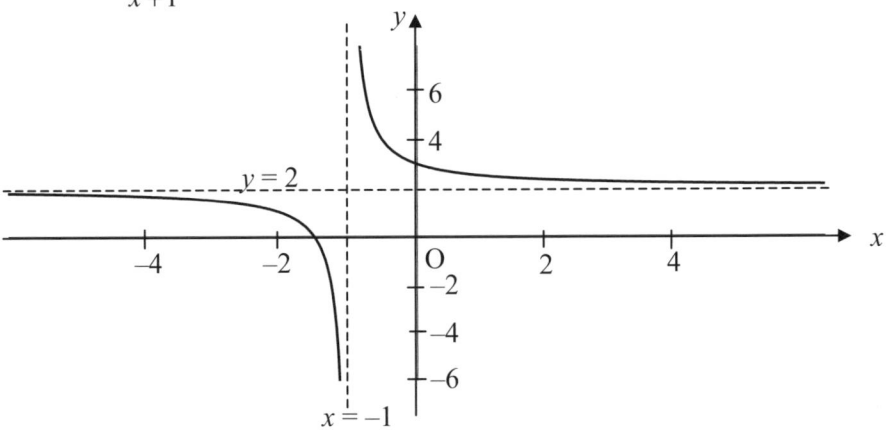

[Graph showing the function with horizontal asymptote $y = 2$ and vertical asymptote $x = -1$]

Exercise 12.6

1. Find the equations of the asymptotes of each of the following functions:

   (a) $y = \dfrac{2x-1}{x-1}$;
   (b) $y = \dfrac{-3}{x}$;

   (c) $y = \dfrac{4-3x}{2x+5}$;
   (d) $y = \dfrac{2}{(x-1)(x-2)}$;

   (e) $y = \dfrac{2x}{(x-1)(x-2)}$;
   (f) $y = \dfrac{x^2-2x+1}{x^2-2x-3}$;

   (g) $y = \dfrac{x^2+5x+4}{x}$;
   (h) $y = \dfrac{(x+1)^2}{2-2x}$.

2. Discuss the behaviour near the asymptotes and find the axes intercepts of the graph of each of the following functions. Sketch each graph.

   (a) $y = \dfrac{2x-3}{x+1}$;
   (b) $y = \dfrac{x+3}{(x-1)(x+2)}$;

   (c) $y = \dfrac{x^2-4x}{x^2-4x-5}$;
   (d) $y = \dfrac{2x^2-x-1}{x+2}$.

   Check each sketch using the facilities of your GDC.

12.7 Graphs of Rational Functions with and without the Use of a GDC

Although knowledge of the exact shape of the graph is not necessary when a GDC is available, students should know how to establish the details "by hand".

Differential Calculus 2

The following general procedure can be used to enable a graph of a rational function $y = f(x) = \dfrac{g(x)}{h(x)}$ to be sketched:

(i) Factorise both $g(x)$ and $h(x)$, if possible.
(ii) If $\deg(g) \geq \deg(h)$, use the division process (section 19.1) to express $f(x)$ in the form $f(x) = f_1(x) + \dfrac{g_1(x)}{h(x)}$ where $\deg(g_1) < \deg(h)$.
(iii) Differentiate $f(x)$ with respect to x.
(iv) Equate the derivative with zero.
(v) Determine the nature and position of any stationary points.
(vi) Find the zeros of $g(x)$ and $h(x)$ and draw a sign diagram of $f(x)$.
(vii) Discuss the behaviour of $f(x)$ as x approaches any zero of $h(x)$ to determine the nature of the graph near the vertical asymptotes.
(viii) Discuss the behaviour of $f(x)$ as x approaches $\pm\infty$ to determine the nature of the graph near any horizontal, oblique or curved asymptote.
(ix) Find the coordinate axes intercepts, if any.
(x) Sketch the graph using any extra points that may help determine the shape more accurately.

Example   Discuss and sketch the graph of $y = \dfrac{x^2 - 1}{x^2 - 4}$.

(i) $y = \dfrac{(x-1)(x+1)}{(x-2)(x+2)}$

(ii) $y = 1 + \dfrac{3}{x^2 - 4}$

(iii) $\dfrac{dy}{dx} = \dfrac{-6x}{(x^2-4)^2} = \dfrac{-6x}{(x-2)^2(x+2)^2}$

(iv) $\dfrac{dy}{dx} = 0$ when $x = 0$

(v) The sign diagram of $\dfrac{dy}{dx}$ is as follows:

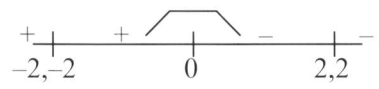

```
    +    +        −      −
  ──┼────┼────┬────┼──
   −2,−2      0   2,2
```

Therefore, $(0, \tfrac{1}{4})$ is a local maximum.

(vi) The critical values of $f(x)$ are ± 1 and ± 2.
The sign diagram of $f(x)$ is as follows:

357

Chapter 12

```
      +    -    +    -    +
    ──┼────┼────┼────┼──
     -2   -1    1    2
```

(vii) As $x \to 2^+$, $y \to +\infty$, and

as $x \to 2^-$, $y \to -\infty$.

As $x \to -2^+$, $y \to -\infty$, and

as $x \to -2^-$, $y \to +\infty$.

$x = \pm 2$ are vertical asymptotes.

(viii) As $x \to +\infty$, $y \to 1^+$, and

as $x \to -\infty$, $y \to 1^+$.

$y = 1$ is a horizontal asymptote.

(ix) When $x = 0$, $y = \frac{1}{4}$ (the stationary point) and when $x = \pm 1$, $y = 0$.

(x) When $x = \pm 3$, $y = 1.6$.

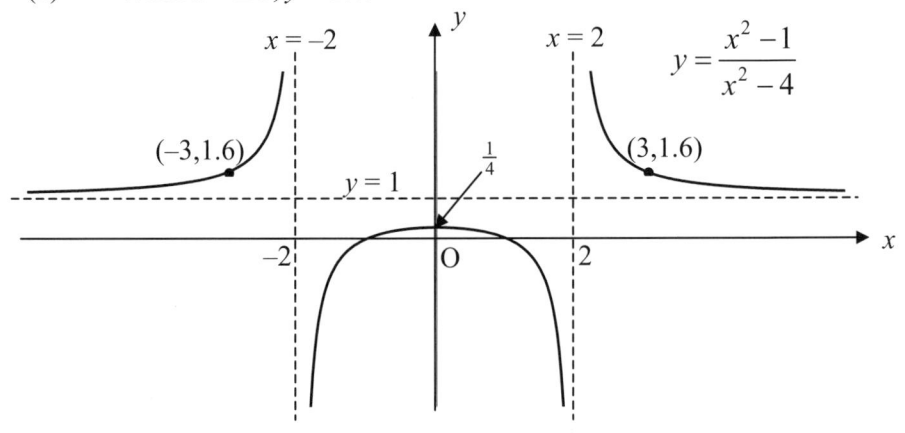

Example Discuss and sketch the graph of $y = \dfrac{6x+3}{(x-1)^2}$.

$$\frac{dy}{dx} = \frac{(x-1)^2(6) - (6x+3)(2)(x-1)(1)}{(x-1)^4}$$

$$= \frac{6x - 6 - 12x - 6}{(x-1)^3}$$

$$= \frac{-6(x+2)}{(x-1)^3}$$

$= 0$ when $x = -2$.

```
       -      +      -
    ───┼─────┼──────
      -2    1,1,1
```
sign of $\dfrac{dy}{dx}$

Therefore $(-2, -1)$ is a local minimum.

Differential Calculus 2

y is undefined when $x = 1$ and so $x = 1$ is a vertical asymptote.

$$\begin{array}{c|c|c|c} - & + & + & \\ \hline & -\frac{1}{2} & 1,1 & \end{array} \quad \text{sign of } y$$

As $x \to 1^+$, $y \to +\infty$, and

as $x \to 1^-$, $y \to +\infty$.

As $x \to +\infty$, $y \to 0^+$, and

as $x \to -\infty$, $y \to 0^-$.

Therefore $y = 0$ is a horizontal asymptote.

The graph meets the coordinate axes at $(-\frac{1}{2}, 0)$ and $(0, 3)$.

Extra points: $(4, 3)$ and $(3, \frac{21}{4})$.

The graph is as follows:

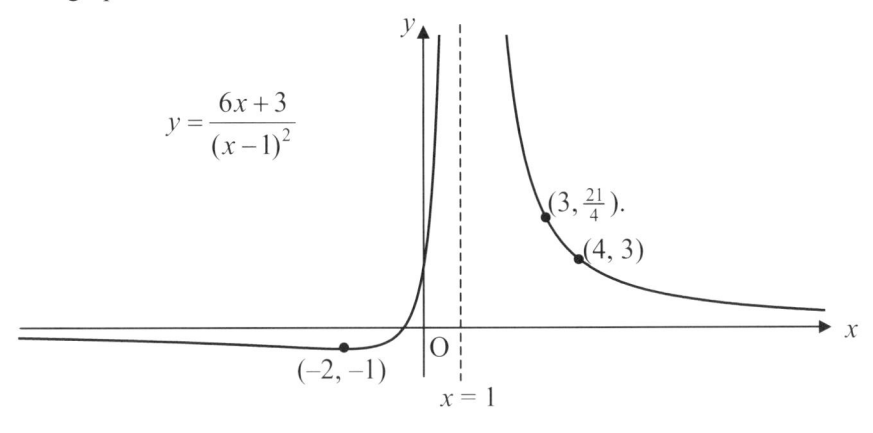

$$y = \frac{6x+3}{(x-1)^2}$$

$(3, \frac{21}{4})$.
$(4, 3)$
$(-2, -1)$
$x = 1$

Exercise 12.7

1. Discuss and sketch the graphs of the following functions. Check all graphs using the facilities of your GDC.

   (a) $y = \dfrac{2x}{x+1}$;

   (b) $y = \dfrac{2x}{x^2 - 1}$;

   (c) $y = \dfrac{7 - 2x}{x^2 - 8x}$;

   (d) $y = \dfrac{x^2 + x + 1}{x^2 + 1}$;

   (e) $y = \dfrac{2x^2 + 5x}{x^2 + 6x - 7}$;

   (f) $y = \dfrac{x^2}{x - 2}$;

   (g) $y = \dfrac{x^2 + x}{(3x + 2)^2}$;

   (h) $y = \dfrac{x^2 - 8}{x - 3}$;

Chapter 12

(i) $y = \dfrac{1-x}{(x+1)^2}$; (j) $y = \dfrac{2}{(x+1)^2}$;

(k) $y = \dfrac{2x+3}{x^2+2x+3}$; (l) $y = \dfrac{4x^2+16x+15}{x^2+4x+3}$.

2. Sketch the graph of $y = \dfrac{x(x-1)}{x^2-x-6}$ and determine, correct to 4 significant figures, the smallest positive value of x for which $x^2 - x - 1 = \dfrac{x(x-1)}{x^2-x-6}$.

3. Sketch the graph of $y = \dfrac{x^2+3x-4}{x^2+3x+2}$ and determine, correct to 4 significant figures, the largest value of x for which $x^2 + 3x - 10 = \dfrac{x^2+3x-4}{x^2+3x+2}$.

4. Sketch the curve whose equation is $y = \dfrac{8x^2+7}{4x^2-4x-3}$ and find the values of k for which the line $y = k$ does not meet the graph.

5. Find the smallest value of $y = x^2 + \dfrac{16}{x^2}$ and sketch its graph.

6. Show that the function $y = \dfrac{2x-1}{(x-1)^2}$ has a minimum value of -1 and sketch its graph. Determine from your graph the number of real roots of the equation $x(x-1)^2 = k(2x-1)$ if k is (a) positive ; (b) negative.

Required Outcomes

After completing this chapter, a student should be able to:
- use the chain, product and quotient rules to find derivatives.
- apply the rules in related rates problems. **(HL)**
- differentiate implicitly. **(HL)**
- sketch the graph of a rational function with and without the use of a GDC.

13 Probability

13.1 Sets

The concept of a set is basic to all of mathematics and mathematical applications. This chapter deals with the language of sets and its application to the theory of probability.

A *set* is simply any collection of objects. The objects in the set are called its *elements* or *members*. We denote elements of a set by small alphabetic characters, a, b, c etc., and the sets themselves by block capitals A, B, C etc. We therefore write $p \in A$ if p is an element in the set A.

If every element of the set A is also an element of the set B, then A is called a *subset* of B. We denote this by $A \subseteq B$. If A is a subset of B and there is at least one element of B which is not an element of A, then A is a *proper subset* of B and we write $A \subset B$.

Two sets are *equal* if each is a subset of the other. Thus sets A and B are equal, $A = B$, if

$$p \in A \iff p \in B.$$

The negations of $p \in A$, $A \subset B$ and $A = B$ are $p \notin A$, $A \not\subset B$ and $A \neq B$ respectively.

We specify a particular set by either listing its elements or by stating any property which may characterise the elements of the set. For example
$$A = \{1, 3, 5, 7, 9\}$$
is the set of all positive odd numbers less than 10 and $B = \{x \mid x \text{ is prime and } x < 15\}$ means that B is the set of all prime numbers less than 15.

We sometimes deal with sets all of which are subsets of a set U called the *universal set* or *universe*. The set U is either given or it can be inferred from the context.
The set with no elements is called the *empty* (*null* or *void*) *set* and is denoted by $\emptyset$. The empty set is a subset of every other set. Thus for any set A we have $\emptyset \subseteq A \subseteq U$.

Set Operations

Union   The *union* of two sets, A and B, is denoted by $A \cup B$ and consists of all the elements which are members of either A or B or both A and B. i.e. $A \cup B = \{x \mid x \in A \text{ or } x \in B\}$.

Chapter 13

Intersection The ***intersection*** of sets A and B is denoted by $A \cap B$ and consists of all those elements which belong to both A and B.
i.e. $A \cap B = \{ x \mid x \in A \text{ and } x \in B \}$.

Difference The ***difference*** of A and B is denoted by $A - B$ and consists of those elements which belong to A but not to B, i.e. belong to A only.
i.e. $A - B = \{ x \mid x \in A, x \notin B \}$.

Complement The complement of A is denoted by A' and consists of all those elements in the universe which do not belong to A.
i.e. $A' = \{ x \mid x \in U, x \notin A \}$.

Note that $A - B = A \cap B'$.

Definition Two sets, A and B, are said to be ***disjoint*** if they do not have any common elements. Thus A and B are disjoint if $A \cap B = \varnothing$.

Venn Diagrams

Venn diagrams are used to illustrate the relationships between sets. We use a rectangle to represent the universal set and often circles within this rectangle to represent other sets.

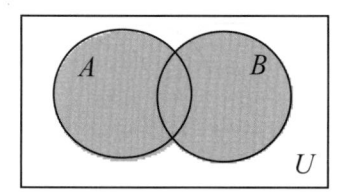

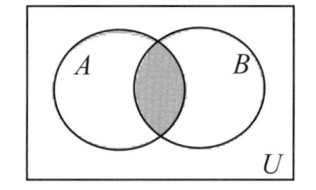

$A \cup B$ is shaded

$A \cap B$ is shaded

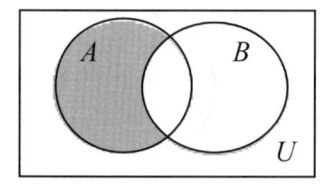

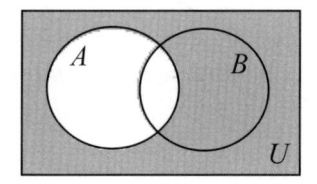

$A - B$ is shaded

A' is shaded

Probability

Laws of the Algebra of Sets

Associative Laws $(A \cup B) \cup C = A \cup (B \cup C)$
$(A \cap B) \cap C = A \cap (B \cap C)$

Commutative Laws $A \cup B = B \cup A$
$A \cap B = B \cap A$

Distributive Laws $A \cup (B \cap C) = (A \cup B) \cap (A \cup C)$
$A \cap (B \cup C) = (A \cap B) \cup (A \cap C)$

De Morgan's Laws $(A \cup B)' = A' \cap B'$
$(A \cap B)' = A' \cup B'$

These laws can be verified using Venn diagrams, but such verifications are not considered to be rigorous proofs.

Example Illustrate the first of de Morgan's laws using a Venn diagram.

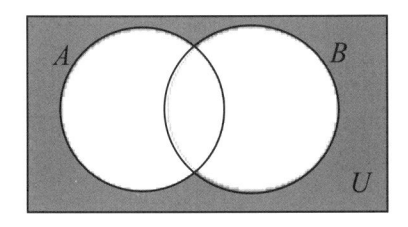

$(A \cup B)'$ is shaded

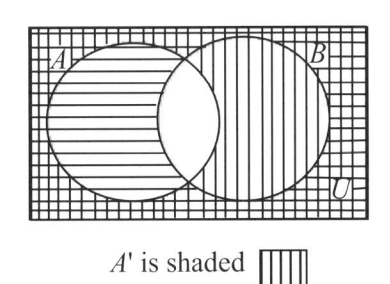

A' is shaded

B' is shaded

$A' \cap B'$ is shaded

From the above diagrams, it can be seen that $(A \cup B)' = A' \cap B'$.

Higher Level

A rigorous proof of the previous law could be something like:

Let $a \in (A \cup B)'$.	Let $a \in A' \cap B'$.
Then $a \notin A \cup B$ and so $a \notin A$ and $a \notin B$.	Then $a \in A'$ and $a \in B'$.
Therefore $a \in A'$ and $a \in B'$	Therefore $a \notin A$ and $a \notin B$
$\Rightarrow a \in A' \cap B'$.	$\Rightarrow a \notin A \cup B$ and so $a \in (A \cup B)'$.
Hence $(A \cup B)' \subseteq A' \cap B'$.	Hence $A' \cap B' \subseteq (A \cup B)'$.

Therefore $(A \cup B)' = A' \cap B'$ which is the required result.

Chapter 13

The Order of a Set

The *order* of a finite set A is denoted by $|A|$ and is equal to the number of elements in it.

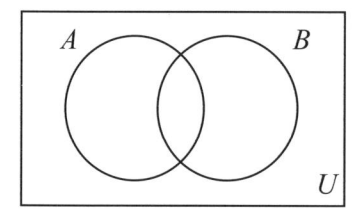

We can see from the diagram on the left that the sum $|A| + |B|$ includes the number of elements in $A \cap B$ twice.

Thus $|A \cup B| = |A| + |B| - |A \cap B|$.

Note that union and intersection may be interchanged in this equation. Thus we also have

$$|A \cap B| = |A| + |B| - |A \cup B|.$$

Example In a class of 25 students, 19 study geography and 14 study history and all students study at least one of these subjects. How many students study both geography and history?

Let G and H represent the set of all students who study geography and history respectively. Then $|G| = 19$, $|H| = 14$, $|G \cup H| = 25$.

Then $|G \cap H| = |G| + |H| - |G \cup H|$
$= 19 + 14 - 25$
$= 8$.

Thus 8 students study both geography and history.

Example In a survey of 100 households, 59 read newspaper X and 71 read newspaper Y. What can be said about the number of households who read both papers?

Let X and Y represent the sets of households reading those newspapers.
Then $|X \cap Y| = |X| + |Y| - |X \cup Y|$
$= 59 + 71 - |X \cup Y|$
$= 130 - |X \cup Y|$.

But $|X \cup Y|$ cannot exceed 100, i.e., $|X \cup Y| \le 100$ and so $|X \cap Y| \ge 30$.

Also, the number of elements in $X \cap Y$ cannot exceed the number of elements in X (or the number of elements in Y). Thus $|X \cap Y| \le 59$.

Thus at least 30, but no more than 59, households read both newspaper X and newspaper Y.

Exercise 13.1

1. Let $U = \{1, 2, 3, 4, 5, 6, 7\}$, $A = \{1, 3, 5, 7\}$, $B = \{4, 5, 6, 7\}$ and $C = \{2, 3, 5, 7\}$. List the elements in each of the following sets:
   (a) $A \cup B$;
   (b) $A \cap C$;
   (c) $B - C$;
   (d) $A \cup B'$;
   (e) $C' - A'$;
   (f) $A' \cap C$.

2. If $U = \{a, b, c, d, e, f, g\}$, $A = \{a, b, c, d\}$, $B = \{c, d, e, f, g\}$ and $C = \{b, d, f\}$, list the elements of each of the following sets:
   (a) $A \cap (B \cup C)$;
   (b) $(A \cap B) \cup (A \cup C)$;
   (c) $(A \cap C) - B$;
   (d) $(A - B)'$;
   (e) $B' - C'$;
   (f) $(A' \cup B)'$.

3. If $n \in \mathbb{Z}^+$, list the elements of each of the following sets:
   (a) $\{ n \mid 2 < n < 7 \}$;
   (b) $\{ n^2 \mid 1 \leq n \leq 5 \}$;
   (c) $\{ n \mid 1 < n^2 \leq 40 \}$;
   (d) $\{ n \mid n \text{ divides } 143 \}$;
   (e) $\{ n \mid n^2 = 5 - 4n \}$;
   (f) $\{ n \mid 2n^2 = 5n - 3 \}$.

4. If $x \in \mathbb{R}$, find $A \cup B$ and $A \cap B$ in each of the following:
   (a) $A = \{ x \mid -3 < x < 2 \}$ and $B = \{ x \mid -1 < x < 8 \}$;
   (b) $A = \{ x \mid 2 \leq x \leq 11 \}$ and $B = \{ x \mid x \geq -2 \}$;
   (c) $A = \{ x \mid |x| < 4 \}$ and $B = \{ x \mid -5 < x \leq 1 \}$;
   (d) $A = \{ x \mid -5 < x < -1 \}$ and $B = \{ x \mid -1 \leq x \leq 10 \}$.

5. In each of the following, draw a Venn diagram showing the universal set U and the sets A, B and C, then shade the region representing:
   (a) $A \cup B \cup C$;
   (b) $A \cap B \cap C$;
   (c) $A' \cap B$;
   (d) $(A \cup B) \cap C'$;
   (e) $B' \cup C'$;
   (f) $A' \cap B' \cap C'$.

6. If $U = \{1, 2, 3, 4, 5, 6, 7, 8, 9, 10\}$, $A = \{1, 2, 3, 4, 5, 6\}$ and $B = \{2, 4, 6, 8\}$, find expressions for the following sets in terms of A, B, A', B', $\cup$, $\cap$ and/or $-$:
   (a) $\{2, 4, 6\}$;
   (b) $\{8\}$;
   (c) $\{7, 8, 9, 10\}$;
   (d) $\{1, 3, 5, 8\}$;
   (e) $\{1, 3, 5, 7, 9, 10\}$.

7. Use Venn diagrams to verify the following:
   (a) $A - B = A \cap B'$;
   (b) $(A \cup B)' = A' \cap B'$;
   (c) $A \cap (B \cup C) = (A \cap B) \cup (A \cap C)$.

8. In a group of 100 children it was found that 65 play football and 58 play basketball. If 15 of these children play neither football nor basketball, how many play (a) both football and basketball; (b) football but not basketball; (c) exactly one of the two sports.

Chapter 13

9. How many of the positive integers less than or equal to 1000 are
   (a) divisible by 4 ;
   (b) divisible by 6 ;
   (c) divisible by 4 and 6 ;
   (d) divisible by neither 4 nor 6?

10. In a survey of 100 students, it was found that
    77 students were studying Mathematics;
    47 students were studying Physics;
    44 students were studying Chemistry;
    43 students were studying both Mathematics and Physics;
    37 students were studying both Mathematics and Chemistry;
    12 students were studying both Physics and Chemistry;
    12 students were studying all three sciences.
    (a) Find the number of students from among the 100 who were not studying any one of the three subjects.
    (b) Find the number of students from among the 100 who were studying both Physics and Chemistry but not Mathematics.

13.2 Elementary Probability Theory

Rolling a die (die is the singular of dice) is an example of a random experiment and probability is the study of such random experiments. When we roll a die, we know that the set of possible outcomes is $S = \{1, 2, 3, 4, 5, 6\}$, called the ***sample space***. We have no idea exactly which of the elements of S will appear in any toss but we know *intuitively* that each of these ***outcomes*** is ***equally likely***. That is, a '6' is no more likely to appear than a '1', which is no more likely to appear than a '2', and so on. If this experiment is performed on n separate occasions and s is the number of times a '6' appears, we know from observation that the ratio s/n becomes close to 1/6 as n increases.

Historically, probability theory began when the mathematician Blaise Pascal was asked to decide just how a given stake must be divided when a game of chance was interrupted before it could be completed, but its modern applications are evident in many facets of our everyday life.

Sample Spaces and Events

The set S of all possible outcomes of a given experiment is called the ***sample space***. A particular outcome, i.e., an element of S, is called a ***sample point***. Any subset of the sample space is called an ***event***. The event $\{a\}$ consisting of a single element of S is called a ***simple event***.

Probability

Definition Suppose that in an experiment there are n different possible outcomes and that these outcomes are all equally likely. Suppose also that an event E occurs in s of these outcomes. Then P(E), the probability of the event E, is given by

$$P(E) = \frac{s}{n} = \frac{\text{number of favourable outcomes}}{\text{number of possible outcomes}} = \frac{|E|}{|S|}.$$

Theorem Suppose that an experiment has only a finite number of equally likely outcomes. If E is an event, then $0 \leq P(E) \leq 1$.

Proof Since E is a subset of S, the set of equally likely outcomes, then $0 \leq |E| \leq |S|$. Hence

$$0 \leq \frac{|E|}{|S|} \leq 1 \quad \text{or} \quad 0 \leq P(E) \leq 1.$$

Note that if $E = S$, then clearly $|E| = |S|$ and P(E) = 1 (the event is certain to occur), and if $E = \emptyset$, then $|E| = 0$ and P(E) = 0 (the event cannot occur).

Example A letter is chosen from the letters of the word "MATHEMATICS". What is the probability that the letter chosen is an "A"?

Since two of the eleven letters are "As", the probability of choosing a letter "A" is 2/11.

Definition If E is an event, then E' is the event which occurs when E does not occur. Events E and E' are said to be ***complementary events***.

Theorem P(E') = 1 − P(E) or P(E) = 1 − P(E')

Proof $$P(E') = \frac{|E'|}{|S|} = \frac{|S|-|E|}{|S|} = 1 - \frac{|E|}{|S|}$$

Thus P(E') = 1 − P(E) and clearly P(E) = 1 − P(E').

Consider two different events, A and B, which may occur when an experiment is performed.

The event $A \cup B$ is the event which occurs if A or B or both A and B occur, i.e., at least one of A or B occurs.

The event $A \cap B$ is the event which occurs when both A and B occur.

The event $A − B$ is the event which occurs when A occurs and B does not occur.

The event A' is the event which occurs when A does not occur.

Chapter 13

Example An integer is chosen at random from the set
$$S = \{x \mid x \in \mathbb{Z}^+, x < 14\}.$$
Let A be the event of choosing a multiple of 2 and let B be the event of choosing a multiple of 3. Find
(a) $P(A \cup B)$;   (b) $P(A \cap B)$;   (c) $P(A - B)$.

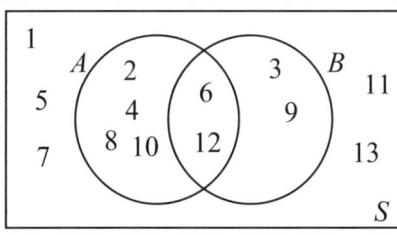

From the diagram,
(a) $P(A \cup B) = 8/13$;
(b) $P(A \cap B) = 2/13$;
(c) $P(A - B) = 4/13$.

Example If 4 people A, B, C, D sit in a row on a bench, what is the probability that A and B sit next to each other?

The number of ways of arranging 4 people in a row is $|S| = 4! = 24$.

The number of ways of arranging the 4 people so that A and B are next to each other = $|X| = 2 \times 3! = 12$.

Thus P(A and B sit next to each other) = $|X|/|S| = \frac{1}{2}$.

Example If 5 cards are selected at random from an ordinary deck of 52 cards, find the probability that exactly 2 of them are aces.

The number of ways of selecting 2 aces from the 4 aces is $C_2^4 = 6$.

The number of ways of selecting 3 non-aces from the 48 non-aces is C_3^{48}.
Therefore the number of ways of selecting 5 cards of which exactly 2 are aces is $|A| = 6 \times C_3^{48}$.

The number of ways of selecting 5 cards from 52 is $|S| = C_5^{52}$.

Thus the required probability
$$= \frac{|A|}{|S|}$$
$$= \frac{6 \times C_3^{48}}{C_5^{52}}$$
$$= 6 \times \frac{48 \times 47 \times 46}{3 \times 2 \times 1} \times \frac{5 \times 4 \times 3 \times 2 \times 1}{52 \times 51 \times 50 \times 49 \times 48}$$
$$= 0.0399.$$

Probability

Tree Diagrams

A tree diagram is a means which can be used to find the number of possible outcomes of experiments where each experiment occurs in a finite number of ways.

Example　Two dice are tossed. Find the set of possible outcomes.

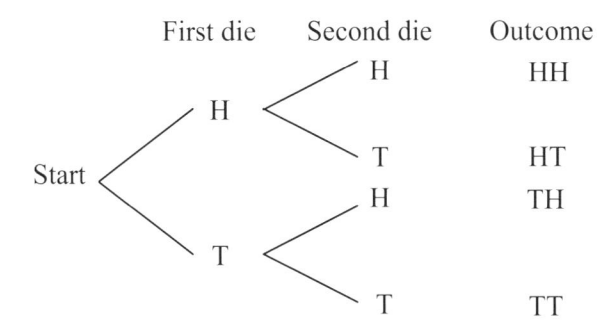

```
         First die    Second die    Outcome
                         H            HH
              H
                         T            HT
    Start
                         H            TH
              T
                         T            TT
```

Thus the set of possible outcomes = {HH, HT, TH, TT}.

Example　Alan and Bob play a game of tennis in which the winner is the first to win 2 sets. In how many different ways can this be done? (The individual game scores in each set are immaterial.)

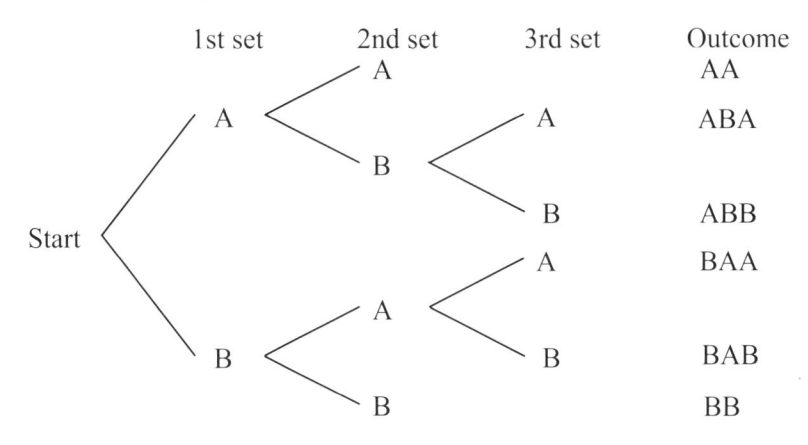

```
        1st set    2nd set    3rd set    Outcome
                      A                    AA
             A
                                 A         ABA
                      B
                                 B         ABB
    Start
                                 A         BAA
                      A
             B                   B         BAB
                      B                    BB
```

Thus the number of possible outcomes of the game = 6.

Finite Probability Spaces

Let $S = \{a_1, a_2, a_3, \cdots, a_n\}$ be a finite sample space. A finite probability space is obtained by assigning to each point $a_r \in S$ a real number p_r called the probability of a_r, satisfying the following:　(a)　$p_r \geq 0$ for all integers r, $1 \leq r \leq n$;

(b)　$\sum_{r=1}^{n} p_r = 1$.

Chapter 13

If A is any event, then the probability $P(A)$ is defined to be the sum of the probabilities of the sample points in A.

Example   A coin is weighted so that heads is three times as likely to appear as tails. Find $P(T)$ and $P(H)$.

Let $P(T) = p$, then $P(H) = 3p$.
But $P(T) + P(H) = 1$.
Therefore $4p = 1$ or $p = \frac{1}{4}$.
Thus $P(T) = \frac{1}{4}$ and $P(H) = \frac{3}{4}$.

Exercise 13.2

1. An unbiased cubic die is thrown. Find the probability that the number showing is
   (a)   even ;   (b)   prime ;   (c)   less than 4.

2. Three coins are tossed. Find the probability of obtaining
   (a)   3 heads ;   (b)   at least 2 tails ;
   (c)   at least 1 head and 1 tail.

3. A die is loaded in such a way that $P(1) = P(3) = P(5) = \frac{1}{12}$, $P(2) = P(6) = \frac{1}{8}$ and $P(4) = \frac{1}{2}$. Find the probability that the number appearing is
   (a)   odd ;   (b)   even ;   (c)   prime ;   (d)   not 3.

4. A die is thrown. Let A be the event: "an odd number appears", let B be the event: "a number greater than 2 appears", and let C be the event: "a prime number appears". Find:
   (a) $P(A \cup B)$;   (b) $P(A \cap B)$;   (c) $P(B \cup C)$;
   (d) $P(B \cap C)$.

5. Let A, B and C be events. Illustrate with a Venn diagram and find an expression for the event:
   (a) A and B but not C occurs ;   (b)   only B occurs.

6. A die is weighted so that the probability of a number x appearing is proportional to x. Thus the probability of scoring a 6 is twice as likely as that of scoring a 3 and three times as likely as that of scoring a 2. Find the probability of scoring
   (a)   each of the numbers from 1 to 6 ;
   (b)   an even number ;
   (c)   a prime number less than 4 ;
   (d)   a number which is prime but not greater than 3.

Probability

7. An integer is chosen at random from the first 200 positive integers. Find the probability that the number is
   (a) divisible by 2 ;
   (b) divisible by 7 ;
   (c) divisible by 2 and 7 ;
   (d) divisible by neither 2 nor 7.

8. The probability that a man is colourblind is 0.05 and the probability that a woman is colourblind is 0.0025. In a school of 1200 students, 400 are male. A student is selected at random. Find the probability that this student is
   (a) male and colourblind ;
   (b) neither male nor colourblind.

9. Two cards are drawn at random from an ordinary deck of 52 cards. Find the probability that
   (a) both cards are spades ;
   (b) at least one card is a spade.

10. Three light bulbs are selected at random from 15 bulbs of which 3 are defective. Find the probability that
    (a) none is defective ;
    (b) exactly one is defective ;
    (c) at least one is defective.

11. The letters of the word *FACETIOUS* are arranged in a row. Find the probability that
    (a) the first 2 letters are consonants ;
    (b) all the vowels are together.

12. A hand of three cards is dealt from a well-shuffled pack of 52. Find the probability that the hand contains
    (a) exactly one ace ;
    (b) three cards of the same suit ;
    (c) no two cards of the same suit.

13.3 Sum and Product Laws

Theorem   If A and B are events from a sample space S, then
$$P(A \cup B) = P(A) + P(B) - P(A \cap B).$$

Proof
$$\begin{aligned}
P(A \cup B) &= \frac{|A \cup B|}{|S|} \\
&= \frac{|A| + |B| - |A \cap B|}{|S|} \quad \text{(result from Section 13.1)} \\
&= \frac{|A|}{|S|} + \frac{|B|}{|S|} - \frac{|A \cap B|}{|S|} \\
&= P(A) + P(B) - P(A \cap B).
\end{aligned}$$

This is known as the **addition law** of probability.

371

Chapter 13

Definition Events A and B are said to be ***mutually exclusive*** if the events A and B are disjoint, i.e., A and B cannot occur at the same time.

For mutually exclusive events, $A \cap B = \emptyset \Rightarrow P(A \cap B) = P(\emptyset) = 0$, and so the addition law reduces to
$$P(A \cup B) = P(A) + P(B).$$

Example A card is drawn from a pack of 52. A is the event of drawing an ace and B is the event of drawing a spade. Find $P(A)$, $P(B)$, $P(A \cap B)$ and $P(A \cup B)$.

$P(A) = P(\text{an ace}) = \frac{4}{52} = \frac{1}{13}$
$P(B) = P(\text{a spade}) = \frac{13}{52} = \frac{1}{4}$
$P(A \cap B) = P(\text{the ace of spades}) = \frac{1}{52}$
$P(A \cup B) = P(A) + P(B) - P(A \cap B) = \frac{4}{52} + \frac{13}{52} - \frac{1}{52} = \frac{16}{52} = \frac{4}{13}$.

Example A marble is drawn from an urn containing 10 marbles of which 5 are red and 3 are blue. Let A be the event: the marble is red, and let B be the event: the marble is blue. Find $P(A)$, $P(B)$ and $P(A \cup B)$.

$P(A) = \frac{5}{10} = \frac{1}{2}$, $P(B) = \frac{3}{10}$, and since the marble cannot be both red and blue, A and B are mutually exclusive so $P(A \cup B) = P(A) + P(B) = \frac{4}{5}$.

Definition Two events are ***independent*** iff $P(A \cap B) = P(A) \times P(B)$.

This definition of independence seems to bear no relationship to the meaning of the English word *independent*. Sometimes it is obvious that events A and B are independent. For example, if we are selecting 2 discs, one after the other with replacement, from a box containing red and blue discs, then the events "the first disc is red" and "the second disc is red" are clearly independent – the result of the first selection has no bearing on the result of the second selection. However, it is sometimes impossible to tell from the description of the events whether or not they are independent. The following example illustrates this.

Example A factory runs two machines, A and B. Machine A operates for 80% of the time while machine B operates for 60% of the time and at least one machine operates for 92% of the time. Do these machines operate independently?

The data does not give any clues. However we are given $P(A) = 0.8$, $P(B) = 0.6$ and $P(A \cup B) = 0.92$.

Now $P(A \cap B) = P(A) + P(B) - P(A \cup B)$
$= 0.8 + 0.6 - 0.92$
$= 0.48$
$= 0.8 \times 0.6$
$= P(A) \times P(B)$
and so these machines do operate independently.

Example   Find the probability that in 3 throws of a fair die, the 3 numbers are all even.

The probability that the first number is even $= \frac{1}{2}$.

The probability that the second number is even $= \frac{1}{2}$.

The probability that the third number is even $= \frac{1}{2}$.

Since these events are independent, the probability that the 3 numbers are all even $= \frac{1}{2} \times \frac{1}{2} \times \frac{1}{2} = \frac{1}{8}$.

This rule is the simplest form of the **multiplication law** of probability. The extension to events which are not independent will be considered in the next section.

Exercise 13.3

1. If A and B are any two events with $P(A) = \frac{3}{8}$, $P(B) = \frac{5}{12}$ and $P(A \cap B) = \frac{1}{4}$, find $P(A \cup B)$.

2. In a certain school, the probability that a student takes mathematics is 0.65, the probability that a student takes physics is 0.4 and the probability that a student takes both mathematics and physics is 0.3. Find the probability that a student takes at least one of mathematics and physics.

3. Events A and B are such that $P(A - B) = 0.3$, $P(B - A) = 0.4$ and $P(A' \cap B') = 0.1$. Find
   (a) $P(A \cap B)$;   (b) $P(A)$;   (c) $P(B)$.

4. Events A and B are independent with $P(A - B) = 0.3$ and $P(B - A) = 0.2$. Find $P(A)$ and $P(B)$

5. If A and B are independent events such that $P(A) = \frac{3}{4}$ and $P(B) = \frac{5}{6}$, find
   (a) $P(A \cap B)$;   (b) $P(A \cup B)$;   (c) $P(A \cap B')$;
   (d) $P(A' \cup B')$.

6. Two independent events A and B are such that $P(A \cap B) = \frac{1}{6}$, $P(A \cup B) = \frac{3}{4}$ and $P(A) > P(B)$. Find $P(A)$ and $P(B)$.

Chapter 13

7. A bag contains 10 red marbles numbered 1 to 10 and 10 blue marbles numbered 1 to 10. One marble is drawn at random from the bag. Let A be the event that a marble numbered 1 is drawn, let B be the event that a red marble is drawn, and let C be the event that an odd-numbered marble is drawn. Show that the events A and B are independent but that the events A and C are not independent.

8. Three unbiased dice are thrown. Find the probability that
   (a) the number on each die is the same ;
   (b) the sum of the three numbers is 5 ;
   (c) the sum of the three numbers is even.

9. A die is biased so that the probability of throwing a six is $\frac{1}{3}$. If the die is thrown twice, find the probability of obtaining
   (a) two sixes ; (b) at least one six.

10. A bag contains 3 red discs, 4 blue discs and 5 green discs. Three discs are withdrawn from the bag one after the other with each being replaced before the next disc is withdrawn. Find the probability of obtaining
    (a) three red discs ;
    (b) exactly two blue discs ;
    (c) one green and two blue discs.

11. An unbiased die is thrown three times. Find the probability of obtaining
    (a) 3 sixes ; (b) exactly 2 sixes ;
    (c) at least one six.

12. Cards are drawn at random, with replacement, from a pack of 52. Find the probability that
    (a) the first two cards drawn are both spades ;
    (b) exactly two of the first three cards drawn are spades ;
    (c) the third card drawn is a club ;
    (d) the fourth card is the first black card to be drawn.

13. Three marksmen fire simultaneously at a target. The probabilities of scoring a 'bull' with any given shot are respectively $\frac{1}{2}, \frac{1}{3}$ and $\frac{1}{4}$. Find the probability that at least one of the marksmen scores a bull with their first shot.

14. Prove that if A and B are independent events, then
    (a) A' and B are independent ;
    (b) A' and B' are independent.

15. A and B are events with non-zero probabilities. Prove that if
    (a) A and B are mutually exclusive they cannot be independent ;
    (b) A and B are independent they cannot be mutually exclusive.

13.4 Conditional Probability and Bayes' Theorem

The probability of an event B given that event A has occurred is called the *conditional probability* of B given A and is written $P(B|A)$.

Here $P(B|A)$ is the probability that B occurs considering A as the sample space, and since the subset of A in which B occurs is $A \cap B$, then $P(B|A) = \dfrac{P(B \cap A)}{P(A)}$.

The general statement of the *multiplication law* is obtained by rearranging this result:
$$P(A \cap B) = P(A) \times P(B|A).$$

Thus the probability that two events will both occur is the product of the probability that one will occur and the conditional probability that the other will occur given that the first has occurred. As A and B are clearly interchangeable, we can also write
$$P(A \cap B) = P(B) \times P(A|B).$$

If A and B are independent, then the probability of B is not affected by the occurrence of A and so $P(B|A) = P(B)$ giving $P(A \cap B) = P(A) \times P(B)$ which is our definition of independence given in the previous section.

Example   A die is tossed. Find the probability that the number obtained is a 5 given that the number is greater than 3.

Let A be the event that the number is a 5 and B the event that the number is greater than 3. Then $A \cap B = \{5\}$ and $B = \{4, 5, 6\}$.

We require $P(A|B) = \dfrac{P(A \cap B)}{P(B)}$

$= \dfrac{1/6}{3/6}$

$= \tfrac{1}{3}.$

Example   A coin is tossed twice in succession. Let A be the event that the first toss is heads and let B be the event that the second toss is heads. Find:
(a) $P(A)$;   (b) $P(B)$;   (c) $P(B \cap A)$;   (d) $P(B|A)$.

(a) $P(A) = \tfrac{1}{2}$

(b) $P(B) = \tfrac{1}{2}$

(c) $P(B \cap A) = P(HH) = \tfrac{1}{2} \times \tfrac{1}{2} = \tfrac{1}{4}$

(d) $P(B|A) = \dfrac{P(B \cap A)}{P(A)} = \dfrac{1/4}{1/2} = \tfrac{1}{2}$

Chapter 13

A tree diagram can be quite useful in the calculation of certain probabilities. The following example illustrates the method.

Example Two discs are selected one at a time without replacement from a box containing 5 red and 3 blue discs. Find the probability that
(a) the discs are of the same colour ;
(b) if the discs are of the same colour, both are red.

1st disc	2nd disc	Outcome	Probability
red $\tfrac{5}{8}$	red $\tfrac{4}{7}$	RR	$\tfrac{5}{8} \times \tfrac{4}{7} = \tfrac{20}{56}$
	blue $\tfrac{3}{7}$	RB	$\tfrac{5}{8} \times \tfrac{3}{7} = \tfrac{15}{56}$
blue $\tfrac{3}{8}$	red $\tfrac{5}{7}$	BR	$\tfrac{3}{8} \times \tfrac{5}{7} = \tfrac{15}{56}$
	blue $\tfrac{2}{7}$	BB	$\tfrac{3}{8} \times \tfrac{2}{7} = \tfrac{6}{56}$

(a) Required probability = P(RR) + P(BB) = $\tfrac{20}{56} + \tfrac{6}{56} = \tfrac{13}{28}$.

(b) Required probability = P(RR | BB) = $\dfrac{20/56}{20/56 + 6/56} = \dfrac{10}{13}$.

Higher Level

Bayes' Theorem

The symbols $P(A|B)$ and $P(B|A)$ look alike but there is a great deal of difference between the probabilities which they represent. For example, suppose that A represents the event that a person has been innoculated against a disease and let B be the event that this person has the disease. Then $P(A|B)$ is the probability that the person has been innoculated given that they have the disease and $P(B|A)$ is the probability that the person has the disease given that they were innoculated.

Since there are many instances where both probabilities are involved, we need a general formula which expresses one in terms of the other.

From the fact that $P(A \cap B) = P(A) \times P(B|A) = P(B) \times P(A|B)$, we have
$$P(A) \times P(B|A) = P(B) \times P(A|B)$$
and hence

$$\boxed{P(B|A) = \frac{P(B) \times P(A|B)}{P(A)}}$$ **Bayes' Theorem**

Example Two machines A and B produce 60% and 40% respectively of the total output of a factory. Of the parts produced by machine A, 3% are defective and of the parts produced by machine B, 5% are defective. A part is selected at random from a day's production and found to be defective. What is the probability that it came from machine A?

Let A be the event that the part came from machine A and let D be the event that the part is defective. We require $P(A|D)$.
Now $P(A) \times P(D|A) = 0.6 \times 0.03 = 0.018$ and
$P(D) = P(A \cap D) + P(B \cap D)$
$= 0.018 + 0.4 \times 0.05$
$= 0.038$.

Therefore the required probability $= \dfrac{0.018}{0.038}$

$= \dfrac{9}{19}$.

Exercise 13.4

1. A pair of fair dice is tossed. If the sum is 6, find the probability that one of the dice is a 2.

2. A pair of fair dice is thrown. Find the probability that the sum is greater than 9 if
   (a) a 5 appears on the first die ;
   (b) a 5 appears on at least one of the dice.

3. A bag contains 5 red marbles and 3 blue marbles. If marbles are withdrawn from the bag one at a time, find the probability that 3 red marbles are followed by 1 blue marble if
   (a) each marble is replaced before another is withdrawn ;
   (b) the marbles are not replaced after they are withdrawn.

Chapter 13

4. About 5% of males are colourblind while only 0.25% of females are colourblind. A person is chosen from a group of 50 people, 20 of which are male and 30 female. If the person chosen is colourblind, what is the probability that this person is a male?

5. The probabilities of the weather being fine, raining or snowing are respectively $\frac{1}{2}$, $\frac{1}{3}$ and $\frac{1}{6}$. The probabilities that a student arrives on time for school under each of these conditions is $\frac{3}{4}$, $\frac{2}{5}$ and $\frac{3}{10}$ respectively. What is the probability that
    (a) the student arrives at school on time on any given day ;
    (b) if the student is late, it was raining?

6. A pair of fair dice is thrown. If the two numbers showing are different, find the probability that
    (a) the sum is 6 ;
    (b) a 1 appears ;
    (c) the sum is 4 or less.

7. An urn contains 4 red and 6 blue marbles. Three marbles are withdrawn from the urn one after the other (without replacement). Find the probability that
    (a) the first two are red and the third is blue ;
    (b) two of the marbles are red and the other is blue.

8. A team is ranked fourth best in a 10-team competition. The probability that this team beats a higher ranked team is 0.4 and the probability this team beats a lower ranked team is 0.6. In its last match, the team won. What is the probability it played a lower ranked team on that day? (Ignore tied games.)

9. A man is dealt 4 cards which include 3 aces (exactly) from an ordinary deck of 52. What is the probability that the fifth card dealt to him is the fourth ace?

10. Let A and B be events with $P(A) = \frac{1}{2}$, $P(B) = \frac{1}{3}$ and $P(A \cap B) = \frac{1}{4}$. Find:
    (a) $P(A|B)$;   (b) $P(B|A)$;   (c) $P(A \cup B)$;
    (d) $P(A'|B')$;   (e) $P(B'|A')$.

11. In a certain college, 5% of the men and 1% of the women are taller than 180cm. Also, 60% of the students are women. If a student is selected at random and found to be taller than 180cm, what is the probability that this student is a woman?

12. The probabilities that two men hit a target from a certain distance are respectively $\frac{1}{2}$ and $\frac{1}{3}$. Each man shoots once at the target from that distance.

Probability

    (a)   Find the probability that exactly one of them hits the target.
    (b)   If only one hits the target, what is the probability that it was the first man?

13.   A certain type of missile hits its target with probability 0.3. How many missiles should be fired so that there is at least a 95% chance of hitting the target?

14.   The probability that a man will live 10 more years is $\frac{1}{4}$ and the probability that his wife will live 10 more years is $\frac{1}{3}$. Find the probabilities that
    (a)   at least one will be alive in 10 years ;
    (b)   exactly one will be alive in 10 years ;
    (c)   if exactly one is alive in ten years, it will be the wife.

Higher Level

13.5   Binomial Probabilities

A **Bernoulli trial** is one in which there are two possible outcomes usually referred to as "success" and "failure". If the probability of success is p, then the probability of failure is $q = 1 - p$.

For the **binomial probability distribution** we are interested in the probabilities of obtaining "r **successes** in n **trials**", or in other words, "r successes and $n - r$ failures in n attempts".

In this section we shall always make the following assumptions:
    (1)   There is a fixed number (n) of trials.
    (2)   The probability of success (p) is the same for each trial.
    (3)   Each trial is independent of all the other trials.

If p and $q = 1 - p$ are the probabilities of a success and failure on any given trial, then the probability of exactly r successes and $(n - r)$ failures *in some specific order* is $p^r q^{n-r} = p^r (1-p)^{n-r}$. The number of ways of arranging r 'ps' and $(n - r)$ 'qs' in a row is $\dfrac{n!}{r!(n-r)!} = \binom{n}{r}$. Thus the number of ways to choose the r trials in which the successes are to occur is $\binom{n}{r}$. This leads to the following result:

Chapter 13

> The probability of obtaining r successes in n independent trials is
> $$P(n, r, p) = \binom{n}{r} p^r (1-p)^{n-r} \quad \text{for } 0 \leq r \leq n$$
> where p is the probability of a success in each trial.

This is the *binomial distribution*.

It is called "binomial" since the values of the probabilities are successive terms of the binomial expansion of $[(1-p) + p]^n$.

Example A die is tossed 7 times in succession. Find the probability that
(a) exactly 2 tosses resulted in a 6;
(b) at least 2 tosses resulted in a 6.

(a) Let $p = P(\text{a six}) = \frac{1}{6}$.

The required probability $= P(7, 2, \frac{1}{6})$

$$= \binom{7}{2}\left(\frac{1}{6}\right)^2 \left(\frac{5}{6}\right)^5$$

$$= 0.234.$$

(b) P(at least 2 successes in 7 trials)
$= 1 - P(\text{less than 2 successes in 7 trials})$
$= 1 - \{P(7, 0, \frac{1}{6}) + P(7, 1, \frac{1}{6})\}$

$$= 1 - \left\{\binom{7}{0} p^0 q^7 + \binom{7}{1} pq^6\right\}$$

$= 1 - q^6 \{q + 7p\}$
$= 0.330.$

Example A doctor estimates that his treatment of a particular illness is successful 75% of the time. Find the probability that he will successfully treat exactly 5 of 6 patients who seek his help.

Let $p = 0.75$.
The required probability $= P(6, 5, 0.75)$

$$= \binom{6}{5}(0.75)^5 (0.25)$$

$$= 0.356.$$

Example How many throws of two dice are required to ensure that the probability of obtaining at least one 'double 6' is greater than 0.95?

Let p = P(a 'double 6' with two dice) = $\frac{1}{36}$ and let the pair of dice be thrown n times. Then the probability of at least one 'double 6' in n throws = 1 − P(no 'double 6' in n throws)

$$= 1 - \left(\frac{35}{36}\right)^n.$$

We require the least value of n for which $1 - \left(\frac{35}{36}\right)^n \geq 0.95$.

$\Rightarrow \left(\frac{35}{36}\right)^n \leq 0.05$

$\Rightarrow n \log \frac{35}{36} \leq \log 0.05$

$\Rightarrow n \geq \dfrac{\log 0.05}{\log(35/36)}$ (The inequality changes direction since $\log \frac{35}{36} < 0$!)

$\Rightarrow n \geq 106.3$ and so at least 107 throws are needed.

Example Five percent of a large consignment of fruit is inedible. Find the probability that in a random selection of 10 pieces of fruit from this consignment, exactly two pieces are inedible.

Although this is strictly *not* a binomial distribution since the probability that a second piece of fruit is inedible after the selection of the first piece, is not constant (i.e., the trials are not independent), the change in p after each trial will be very small for a *large* consignment. Hence the binomial distribution provides us with an excellent *approximation* to the given situation.

Hence the required probability $\approx$ P(10, 2, 0.05)

$$= \binom{10}{2}(0.05)^2(0.95)^8$$

$$= 0.0746.$$

Note: If, for example, the consignment contained 10 000 pieces of fruit, the actual probability that exactly 2 of 10 pieces selected at random are inedible is $\binom{10}{2} \times \dfrac{500}{10000} \times \dfrac{499}{9999} \times \dfrac{9500}{9998} \times \dfrac{9499}{9997} \times \cdots \times \dfrac{9493}{9991} = 0.0746$

which is the same answer (correct to 3 significant figures) as that obtained from the binomial distribution.

Exercise 13.5

1. A fair coin is tossed 4 times. Find the probability of obtaining
   - (a) exactly 2 heads ;
   - (b) exactly 3 tails ;
   - (c) no tails ;
   - (d) at least 2 heads.

2. A family has 5 children. Find the probability that the family contains
   - (a) 2 boys and 3 girls ;
   - (b) more boys than girls.

3. What is the probability of throwing a sum of 9 once only in 4 throws of a pair of dice?

4. When Peter plays Paul at tennis, Peter wins on average 2 games out of every 3 played. If Peter and Paul play 6 games, find the probability that
   - (a) Peter wins exactly 4 of these games ;
   - (b) Paul wins more games than Peter.

5. Of 1000 families with four children, how many would you expect to have
   - (a) 2 boys ;
   - (b) 4 girls ;
   - (c) either 2 or 3 boys?

6. How many throws of a single die are required so that there is a better than 90% chance of obtaining a six?

7. Two percent of the total daily output of transistors from a factory are defective. If 5 transistors are selected at random, estimate the probability that exactly 2 of them are defective.

8. The probability that a man hits a target is 1/3.

   (a) If he fires 10 times, what is the probability that he hits the target at least twice?

   (b) How many times does he need to fire to ensure that the probability he hits the target at least once is greater than 2/3?

9. Peter and Bill play a game of tennis. Peter is the more skilful, having a probability of 2/3 of winning any game against Bill. Find the probability that Peter wins the first set 6 games to 4.

10. A bag contains 6 red and 3 blue marbles. One marble is withdrawn at random, its colour noted, and then replaced. This procedure is performed 6 times. What is the probability of drawing
    (a) one blue marble only ;
    (b) no blue marble ;
    (c) no more than one blue marble ;
    (d) at least 5 red marbles ;
    (e) no more than 4 red marbles?

11. Each trial of a binomial experiment has a constant probability of p of yielding a success.

    (a) Given that 7 successes occurred in 15 independent trials, find the probability that the first and last trials were successes.
    (b) Find the value of p given that in 20 independent trials the probability of exactly 4 successes is twice that of exactly 6 successes.

*12. An unbiased die is tossed until a 6 appears for the second time. Find the most likely number of throws required.

*13. If p is the probability of a success in any single performance of n Bernoulli trials, then for fixed n and p, the value of $P(n, r, p)$ initially increases as r, the number of successes increases, and finally decreases as r increases. Thus $P(n, r, p)$ is a maximum for one particular value of r or perhaps for two consecutive values of r.

    (a) If $P(n, r, p) \geq P(n, r-1, p)$, show that $r \leq p(n + 1)$.

    (b) If $P(n, r, p) \geq P(n, r+1, p)$, show that $r \geq p(n + 1) - 1$.

    (c) Find the most likely number of successes in

        (i) 16 Bernoulli trials where $p = 0.3$;

        (ii) 19 Bernoulli trials where $p = 0.4$.

*14. In a multiple choice test there are $2n + 1$ questions, $(n \in \mathbb{Z}^+)$, each with a choice of two answers. In order to pass the test, a student must answer correctly more than half of the questions. Show that the probability the student passes is independent of the value of n.

Chapter 13

Required Outcomes

After completing this chapter, a student should be able to:
- establish and justify set relations using Venn diagrams.
- use de Morgan's laws in a variety of contexts.
- understand the terms 'sample space', 'sample point', 'simple event', 'event', 'complementary events', 'independent events', 'mutually exclusive events' and 'conditional probability'.
- use Bayes' theorem and the Binomial probability distribution. **(HL)**
- apply any of the above ideas to establish the probability of a variety of events.

14 Statistics 2

14.1 Discrete Probability Distributions

In a proper study of statistics we are interested in numbers that are associated with situations resulting from elements of chance, i.e., in the values of *random variables*. In the study of random variables we are interested in the probabilities with which they take in the range of their possible values, called their *probability distributions*. In this section we will introduce some important random variables and their probability distributions. Two in particular will be discussed in some detail. The *binomial probability distribution* (which is discrete) and the *normal distribution* (which is continuous).

Higher Level

Definition   A *discrete random variable* is a variable quantity which occurs randomly in a given experiment and which can assume only certain, well-defined values, usually integral.

Definition   A *discrete probability distribution* describes a discrete random variable in terms of the probabilities associated with each individual value that the variable may take.

For a probability distribution concerning a discrete random variable X where X may assume any of the values $x_1, x_2, x_3, \cdots, x_n$, then

$$\sum_{i=1}^{n} P(X = x_i) = 1.$$

Example   Three coins are tossed. Let the random variable X represent the number of heads obtained. Construct a table to represent this probability distribution.

The probability of exactly x heads in 3 tosses
$= P(X = x)$
$= \binom{3}{x}\left(\frac{1}{2}\right)^x \left(\frac{1}{2}\right)^{3-x}, x = 0, 1, 2, 3.$

Therefore, the probability distribution table is as follows:

x	$P(X=x)$
0	0.125
1	0.375
2	0.375
3	0.125

Example Of the 15 light bulbs in a box, 5 are defective. Four bulbs are selected at random from the box. Let the random variable X represent the number of defective bulbs selected. Construct a table to represent this distribution and show that the sum of the probabilities is 1.

The probability that there are x defective bulbs in the selection of 4 must mean that there are $(4-x)$ non-defective bulbs in the selection..

The number of ways of selecting x defective bulbs = $\binom{5}{x}$ and the number of ways of selecting $(4-x)$ non-defective bulbs = $\binom{10}{4-x}$.

The number of ways of selecting 4 bulbs from 15 = $\binom{15}{4}$.

Therefore $P(X=x) = \dfrac{\binom{5}{x}\binom{10}{4-x}}{\binom{15}{4}}$.

x	$P(X=x)$
0	0.154
1	0.440
2	0.330
3	0.073
4	0.004

The sum of the probabilities = 1.00.

Exercise 14.1

1. Which of the following represent discrete variables:
   (a) the number of rivers in the world ;
   (b) the heights of children in a school ;
   (c) the winning margin in a football match ;

(d) the time taken to run 100m;
(e) a human lifetime;
(f) the number of children born in the world per year?

2. A tetrahedral die, with a number from 1 to 4 on each of 4 faces, is thrown twice. Let X represent the sum of the two scores. Find $P(X = x)$ for $x = 2, 3, 4, \ldots, 8$ and show that the sum of these probabilities is 1.

3. A box contains 6 marbles of which 4 are red. Three marbles are withdrawn one after the other. Let X represent the number of red marbles drawn. Construct a table to represent this probability distribution if
(a) each marble is replaced before another is withdrawn;
(b) the marbles are not replaced before the next is withdrawn.
Show that in each case the sum of the probabilities is 1.

4. A box contains 8 marbles of which 3 are red. A random sample of 4 marbles is taken without replacement. Write down, in tabular form, the probability distribution for the number of red marbles in the sample.

5. A discrete random variable X has the following probability distribution:

x	$P(X = x)$
0	p
1	p
2	$3p$
3	$3p$
4	$2p$

Find: (a) the value of p;
       (b) $P(X > 1)$;
       (c) $P(X = 3 \mid X \geq 2)$.

6. Urn A contains 5 red and 3 blue marbles; urn B contains 2 red and 4 blue marbles. A marble is selected from each urn and the colour noted. Let X represent the number of red marbles selected. Tabulate the probability distribution for X.

7. From a pack of 52 cards, a poker-hand of 5 cards is dealt. Let the random variable X represent the number of aces in this hand. Construct a table to represent this probability distribution and show that the sum of the probabilities is 1.

Chapter 14

8. A coin is tossed until a head appears. Find a formula for the probability distribution of X, the number of times the coin must be tossed, and show that $\sum_{i=1}^{\infty} P(X = x_i) = 1$.

9. Find a formula for the probability distribution of the random variable X which represents the number of times that a die must be thrown until a six appears.

10. A discrete random variable X has the following probability distribution:

x	$P(X = x)$
0	p
1	$2p$
2	$1 - 2p^2$
3	$2p - 3p^2$

    Find: (a) the value of p;
    (b) $P(X = 2)$;
    (c) $P(X \geq 2)$;
    (d) $P(X = 2 \mid X \geq 2)$.

11. Three men and three women enter a room one after the other. If X represents the number of women entering the room before any man, describe the probability distribution of X, i.e., construct a table representing the probability distribution.

12. The probability of "success" in a single Bernoulli trial is p. Let X represent the number of successes in three trials. Describe the probability distribution of X and show that

$$\sum_{x=0}^{3} P(X = x) = 1.$$

13. In a multiple choice test with 5 questions, there is a choice of 4 answers for each question, only one of which is correct, and a student guesses each answer. Tabulate the probability distribution of the random variable X which represents the number of correct answers given by this student.

*14. For a discrete random variable X,

$$P(X = x) = \frac{e^{-2.5}(2.5)^x}{x!} \text{ for } x = 0, 1, 2, \ldots .$$

(a) Find: (i) $P(X = 3)$; (ii) $P(X > 2)$;
(iii) $P(X = 3 \mid X > 2)$; (iv) $\{x \mid P(X = x) > P(X = x + 1)\}$.

(b) Calculate that value of x for which $P(X = x)$ is a maximum.

(c) Prove that $\sum_{x=0}^{\infty} P(X = x) = 1$.

[*Hint:* You may find the result $\sum_{x=0}^{\infty} \frac{a^x}{x!} = e^a$ useful.]

14.2 The Mean (or Expected Value) of a Distribution

For a discrete random variable X which can assume the values $x_1, x_2, x_3, \cdots, x_n$, the *mean* or *expected value* of X is denoted and defined by

$$\mu = E(X) = \sum_{i=1}^{n} x_i P(X = x_i).$$

Just like the mean of a population, it is denoted by the Greek letter μ (mu).

The expected value represents the long-term average of the variable X. In effect, multiplying each value of x by its probability gives it a "weight" which depends on the likelihood of its occurrence. Summing these weighted values gives an overall expectation for X.

Example Consider the probability distribution of the number of heads that appear in 3 tosses of a coin.

x	$P(X = x)$	$xP(X = x)$
0	0.125	0.000
1	0.375	0.375
2	0.375	0.750
3	0.125	0.375
		$\sum_{x=0}^{3} xP(X = x) = 1.5$

The expected number of heads obtained when 3 coins are tossed is 1.5.

Example The probability distribution X is given on the next page.

x	1	2	3
$P(X = x)$	0.2	0.5	0.3

(a) Find the expected value of X, $E(X)$.
(b) Find the expected value of $X + a$, $E(X + a)$.
(c) Find the expected value of aX, $E(aX)$.
(d) Find the expected value of X^2, $E(X^2)$.

Chapter 14

x	$P(X=x)$	$xP(x=x)$	$(x+a)P(x=x)$	$axP(X=x)$	$x^2P(X=x)$
1	0.2	0.2	$0.2(1+a)$	$0.2a$	0.2
2	0.5	1.0	$0.5(2+a)$	$1.0a$	2.0
3	0.3	0.9	$0.3(3+a)$	$0.9a$	2.7
Sum	1.0	2.1	$2.1+a$	$2.1a$	4.9

(a) $E(X) = 2.1$
(b) $E(X+a) = 2.1 + a$
(c) $E(aX) = 2.1a$
(d) $E(X^2) = 4.9$

Note: $E(X^2) \neq [E(X)]^2$ $(2.1^2 = 4.41 \neq 4.9)$.

However the above example does illustrate the following theorem.

Theorem If X is a discrete random variable and a, k are constants, then
(a) $E(X+a) = E(X) + a$;
(b) $E(kX) = kE(X)$;
(c) $E(kX+a) = kE(X) + a$.

Proof (a) $E(X+a) = \sum_{i=1}^{n}(x_i + a)P(X = x_i)$

$= \sum_{i=1}^{n} x_i P(X = x_i) + a\sum_{i=1}^{n} P(X = x_i)$

$= E(X) + a$.

(b) $E(kX) = \sum_{i=1}^{n} kx_i P(X = x_i)$

$= k\sum_{i=1}^{n} x_i P(X = x_i)$

$= kE(X)$.

(c) $E(kX+a) = \sum_{i=1}^{n}(kx_i + a)P(X = x_i)$

$= k\sum_{i=1}^{n} x_i P(X = x_i) + a\sum_{i=1}^{n} P(X = x_i)$

$= kE(X) + a$.

Exercise 14.2

1. Find the mean of each of the following discrete probability distributions:

   (a)

x	$P(X=x)$
0	0.1
1	0.2
2	0.4
3	0.3

   (b)

x	$P(X=x)$
1	0.1
2	0.4
3	0.5

   (c)

x	$P(X=x)$
-1	0.2
0	0.3
1	0.4
2	0.1

   (d)

x	$P(X=x)$
4	0.002
6	0.040
8	0.299
10	0.659

2. Find the expected value of X, $E(X)$, for each of the following probability distributions:

   (a) $P(X=x) = \binom{4}{x}\left(\dfrac{2}{3}\right)^{4-x}\left(\dfrac{1}{3}\right)^{x}$, $x = 0, 1, 2, 3, 4$;

   (b) $P(X=x) = \dfrac{\binom{4}{x}\binom{3}{3-x}}{\binom{7}{3}}$, $x = 0, 1, 2, 3$;

   (c) $P(X=x) = p^{x}(1-p)^{x}$, $x = 0, 1$;

   *(d) $P(X=x) = \left(\dfrac{1}{2}\right)^{x}$, $x = 1, 2, 3, 4, \ldots$.

3. A committee of 3 people is to be selected from 4 men and 2 women. Let X represent the number of women chosen. Find the expected value of X.

4. Among 12 transistors, 4 are defective. If 6 transistors are randomly selected, how many of them can be expected to be defective?

5. Two people, A and B, play a game in which A tosses 3 coins and B tosses 2 coins. If A obtains more heads than B, B pays A $5; if B obtains more heads than A, A pays B $10; if they obtain the same number of heads, A pays B $2. Let X represent A's profit from any one game. Find the expected value of X.

Chapter 14

6. A manufacturer makes TV sets to order. The cost of manufacture depends on the number of sets made. If X sets are ordered in any week, the the cost of manufacture, $C, is given by $C = 400X + 850$. From past experience it has been determined that the probability distribution for X is given by

x	$P(X=x)$
0	0.02
1	0.12
2	0.29
3	0.39
4	0.15
5	0.03

Calculate the expected number of orders to be filled in any one week and the expected weekly cost of manufacture.

7. Two events, A and B are independent with $P(A) = 0.4$, $P(B) = 0.6$, and X is a discrete random variable for which $X = 0$ when both events A and B occur, $X = 1$ when only one of the events A and B occurs, and $X = 2$ when neither event A nor event B occurs. Calculate the expected value of X.

8. A company manufactures metal washers of 4 differing diameters. The cost of manufacturing each washer is directly proportional to the square of its diameter. The unit cost is 6 cents when the diameter is 2 cm, and the proportion of the output for each size is given in the following table:

Diameter (cm)	Proportion of Output
2	0.2
3	0.4
4	0.3
6	0.1

Calculate the average manufacturing cost per washer.

14.3 Variance and Standard Deviation

For a discrete random variable X which can assume the values $x_1, x_2, x_3, \cdots, x_n$, the *variance* and *standard deviation* of X are denoted and defined by

$$\sigma^2 = \text{Var}(X) = E\left((X-\mu)^2\right) = \sum_{i=1}^{n}(x_i - \mu)^2 P(X = x_i), \text{ and}$$

$$\sigma = \text{SD}(X) = \sqrt{\text{Var}(X)} = \sqrt{\sum_{i=1}^{n}(x_i - \mu)^2 P(X = x_i)} \text{ where } \mu = E(X).$$

Statistics 2

Just like the variance and standard deviation of a population, we use σ^2 and σ.

The standard deviation gives a measure of the way in which the values are spread about the mean.

Example Consider the following the probability distribution.

x	$P(X=x)$
2	0.3
3	0.5
4	0.2

Find: (a) Var(X);   (b) SD(X).

x	$P(X=x)$	$xP(X=x)$	$x-\mu$	$(x-\mu)^2$	$(x-\mu)^2 P(X=x)$
2	0.3	0.6	-0.9	0.81	0.243
3	0.5	1.5	0.1	0.01	0.005
4	0.2	0.8	1.1	1.21	0.242
		$\mu = \Sigma = 2.9$			$\Sigma = 0.490$

$$\text{Var}(X) = E(x-\mu)^2 = 0.490; \quad \text{SD}(X) = \sqrt{0.49} = 0.7$$

An alternative formula for the variance, involving less arithmetic, can be found by manipulation of the basic definition. [Compare with the alternative formula for the variance of a sample distribution.]

$$\begin{aligned}
\text{Var}(X) &= E(x-\mu)^2 \\
&= \sum_{i=1}^{n}(x_i - \mu)^2 P(X = x_i) \\
&= \sum_{i=1}^{n}(x_i^2 - 2\mu x_i + \mu^2) P(X = x_i) \\
&= \sum_{i=1}^{n} x_i^2 P(X = x_i) - 2\mu \sum_{i=1}^{n} x_i P(X = x_i) + \mu^2 \sum_{i=1}^{n} P(X = x_i) \\
&= E(X^2) - 2\mu \times \mu + \mu^2 \times 1 \\
&= E(X^2) - \mu^2 \\
&= E(X^2) - [E(X)]^2.
\end{aligned}$$

Let us see how the alternative formula works for the variable X in the previous example.

393

Chapter 14

x	x^2	$P(X=x)$	$xP(X=x)$	$x^2P(X=x)$
2	4	0.3	0.6	1.2
3	9	0.5	1.5	4.5
4	16	0.2	0.8	3.2
			$\Sigma = 2.9$	$\Sigma = 8.9$

$$\text{Var}(X) = E(X^2) - [E(X)]^2$$
$$= 8.9 - (2.9)^2$$
$$= 0.49.$$

Theorem If X is a discrete random variable and a, k are constants, then
(a) $\quad$ SD$(X + a) = $ SD(X); (b) $\quad$ SD$(kX) = k$SD(X);
(c) $\quad$ SD$(kX + a) = k$SD(X).

Proof (a) $\quad$ Var$(X + a)$

$$= \sum_{i=1}^{n}(x_i + a)^2 P(X = x_i) - [E(X + a)]^2$$

$$= \sum_{i=1}^{n} x_i^2 P(X = x_i) + 2a\sum_{i=1}^{n} x_i P(X = x_i) + a^2 \sum_{i=1}^{n} P(X = x_i) -$$
$$[E(X)]^2 - 2aE(X) - a^2$$

$$= E(X^2) - [E(X)]^2 + 2aE(X) - 2aE(X) + a^2 - a^2$$
$$= \text{Var}(X)$$

$\Rightarrow \quad$ SD$(X + a) = $ SD(X).

(b) $\quad$ Var(kX)

$$= \sum_{i=1}^{n} k^2 x_i^2 P(X = x_i) - [E(kX)]^2$$

$$= k^2 \sum_{i=1}^{n} x_i^2 P(X = x_i) - k^2 [E(X)]^2$$

$$= k^2 \left(E(X^2) - [E(X)]^2\right)$$

$$= k^2 \text{Var}(X)$$

$\Rightarrow \quad$ SD$(kX) = k$SD(X).

(c) $\text{Var}(kX + a)$
$$= \sum_{i=1}^{n}(kx_i + a)^2 P(X = x_i) - [E(kX + a)]^2$$
$$= k^2\sum_{i=1}^{n}x_i^2 P(X = x_i) + 2ka\sum_{i=1}^{n}x_i P(X = x_i) + a^2\sum_{i=1}^{n}P(X = x_i)$$
$$\quad - [kE(X) + a]^2$$
$$= k^2 E(X^2) + 2kaE(X) + a^2 - k^2[E(X)]^2 - 2kaE(X) - a^2$$
$$= k^2\left(E(X^2) - [E(X)]^2\right)$$
$$= k^2 \text{Var}(X)$$

$\Rightarrow \quad \text{SD}(kX + a) = k\text{SD}(X).$

Theorem For independent random variables X and Y,
(a) $E(X \pm Y) = E(X) \pm E(Y)$;
(b) $\text{Var}(X \pm Y) = \text{Var}(X) + \text{Var}(Y)$.

The proof is left as an exercise for the reader.

Exercise 14.3

1. Find the mean and standard deviation of each of the following random variables X with the given probability distributions:

(a)

x	$P(X = x)$
0	$\frac{1}{16}$
1	$\frac{1}{4}$
2	$\frac{3}{8}$
3	$\frac{1}{4}$
4	$\frac{1}{16}$

(b)

x	$P(X = x)$
0	0.264
1	0.494
2	0.220
3	0.022

2. The random variable X has the following probability distribution:

x	$P(X = x)$
2	a
4	$2a^2 - a$
6	$a^2 + a - 1$

Chapter 14

Find:
(a) the value of a; (b) $E(X)$; (c) $Var(X)$; (d) $SD(X)$.

3. Find the mean and standard deviation of the random variable X whose probability distribution is defined by:

(a) $P(X=x) = \binom{4}{x}(0.2)^{4-x}(0.8)^x$, for $x = 0, 1, 2, 3, 4$;

(b) $P(X=x) = \dfrac{\binom{6}{x}\binom{4}{4-x}}{\binom{10}{4}}$, for $x = 0, 1, 2, 3, 4$.

4. A debating team of 4 is to be chosen from 6 girls and 3 boys. Let X be the number of boys chosen. Find:
(a) $E(X)$; (b) $Var(X)$; (c) $SD(X)$.

5. An urn contains 5 red and 3 blue marbles. Three marbles are drawn at random from the urn. Find the mean and standard deviation of the number of red marbles chosen.

6. From an urn containing 4 red and 6 blue marbles, marbles are drawn (without replacement) until a blue marble is drawn. If X represents the number of marbles drawn, including the final blue one, find the mean and standard deviation of X.

7. Three men and three women enter a room one at a time. Let X represent the number of women who enter before the first man enters. Find $E(X)$ and $Var(X)$. [The probability distribution of X has already been found in Question 11 of Exercise 14.1.]

8. Given that the mean and variance of $x_1, x_2, x_3, \cdots, x_n$ are μ and σ^2 respectively, state the mean and variance of
(a) $3x_1, 3x_2, 3x_3, \cdots, 3x_n$;
(b) $2-3x_1, 2-3x_2, 2-3x_3, \cdots, 2-3x_n$.

9. If X and Y are independent random single digit numbers from 1 to 9 inclusive, calculate the variance of
(a) X; (b) $3X$; (c) $X+Y$; (d) $3X-Y$.

10. Four different letters are to be put one each into four differently addressed envelopes. Only one envelope is the correct one for any given letter. Let X represent the number of letters which are placed in the correct envelope. Find the mean and standard deviation of X. [It can be shown that as the number of letters increases, the probability that not one single letter is placed in its correct envelope, i.e., $P(X=0)$, approaches $e^{-1} \approx 0.36788!$]

14.4 The Binomial Random Variable

The binomial random variable X represents the number of successes in n independent trials where the probability of a success in any one trial is p.

We have already seen that X has the following probability distribution:

$$P(X = x) = \binom{n}{x} p^x (1-p)^{n-x}, \text{ for } x = 0, 1, 2, 3, \ldots, n.$$

The Mean of a Binomial Random Variable

In order to establish the formula for the mean of a binomial random variable, we need the following theorem:

Theorem $\quad r\binom{n}{r} = n\binom{n-1}{r-1}$, for $1 \le r \le n$.

Proof $\quad n\binom{n-1}{r-1} = \dfrac{n(n-1)!}{(r-1)!(n-r)!}$

$\qquad\qquad\qquad = \dfrac{rn!}{r!(n-r)!}$

$\qquad\qquad\qquad = r\binom{n}{r}.$

Theorem The mean of a binomial random variable X in which n independent trials are performed and the probability of success in any trial is p is given by
$$E(X) = np.$$

397

Chapter 14

Proof

$$E(X) = \sum_{x=0}^{n} x P(X = x)$$

$$= \sum_{x=0}^{n} x \binom{n}{x} p^x (1-p)^{n-x}$$

$$= \sum_{x=1}^{n} x \binom{n}{x} p^x (1-p)^{n-x} \quad \text{(since the first term is zero)}$$

$$= \sum_{x=1}^{n} n \binom{n-1}{x-1} p^x (1-p)^{n-x} \quad \text{(from the previous theorem)}$$

$$= \sum_{x=1}^{n} np \binom{n-1}{x-1} p^{x-1} (1-p)^{n-x}$$

$$= np \sum_{x=1}^{n} \binom{n-1}{x-1} p^{x-1} (1-p)^{(n-1)-(x-1)}$$

$$= np \sum_{x=0}^{n-1} \binom{n-1}{x} p^x (1-p)^{(n-1)-x}$$

$$= np (p + [1-p])^{n-1} \quad \text{(binomial theorem)}$$

$$= np.$$

The Standard Deviation of a Binomial Random Variable

Theorem The standard deviation of a binomial random variable X in which n independent trials are performed and the probability of success in any trial is p is given by

$$SD(X) = \sqrt{np(1-p)}.$$

Proof

$$\text{Var}(X)$$
$$= E(X^2) - [E(X)]^2$$
$$= \sum_{x=0}^{n} x^2 P(X = x) - (np)^2$$
$$= \sum_{x=1}^{n} x^2 \binom{n}{x} p^x (1-p)^{n-x} - (np)^2 \quad \text{(the first term is zero)}$$
$$= \sum_{x=1}^{n} xn \binom{n-1}{x-1} p^x (1-p)^{n-x} - (np)^2 \quad \text{(previous theorem)}$$

398

$$= np\left(\sum_{x=1}^{n} x\binom{n-1}{x-1}p^{x-1}(1-p)^{n-x} - np\right)$$

$$= np\left(\sum_{x=1}^{n} x\binom{n-1}{x-1}p^{x-1}(1-p)^{n-x} - \sum_{x=1}^{n}\binom{n-1}{x-1}p^{x-1}(1-p)^{n-x} + \right.$$
$$\left. \sum_{x=1}^{n}\binom{n-1}{x-1}p^{x-1}(1-p)^{n-x} - np\right)$$

$$= np\left(\sum_{x=1}^{n}(x-1)\binom{n-1}{x-1}p^{x-1}(1-p)^{n-x} + \sum_{x=1}^{n}\binom{n-1}{x-1}p^{x-1}(1-p)^{n-x} - np\right)$$

$$= np\left(\sum_{x=0}^{n-1} x\binom{n-1}{x}p^{x}(1-p)^{n-1-x} + \sum_{x=0}^{n-1}\binom{n-1}{x}p^{x}(1-p)^{n-1-x} - np\right).$$

Now $\sum_{x=0}^{n-1} x\binom{n-1}{x}p^{x}(1-p)^{n-1-x} = (n-1)p$, i.e., the expected number of successes in $(n-1)$ trials, and

$$\sum_{x=0}^{n-1}\binom{n-1}{x}p^{x}(1-p)^{n-1-x} = (p + [1-p])^{n-1} = 1.$$

Thus $\text{Var}(X) = np\big((n-1)p + 1 - np\big) = np(1-p)$.

Therefore $\text{SD}(X) = \sqrt{np(1-p)}$.

Example A die is tossed 180 times. Find the mean and standard deviation of the random variable representing the total number of sixes obtained.

Let X represent the number of sixes obtained. Here $n = 180$ and $p = \frac{1}{6}$.
Therefore $E(X) = 180 \times \frac{1}{6} = 30$, and $\text{SD}(X) = \sqrt{180 \times \frac{1}{6} \times \frac{5}{6}} = 5$.

Exercise 14.4

1. A binomial experiment consists of $n = 5$ independent trials, each with a probability of success of $p = 0.2$. If X represents the number of successes in the experiment, show that

    (a) $\sum_{x=0}^{5} x P(X = x) = np$; (b) $\sum_{x=0}^{5}(x-\mu)^2 P(X = x) = np(1-p)$.

Chapter 14

> 2. Calculate the mean and variance of the number of successes resulting from a binomial experiment in which
> (a) the number of trials is 10 and the probability of success is 0.5 ;
> (b) the number of trials is 100 and the probability of success is 0.2 ;
> (c) the number of trials is 48 and the probability of success is 0.25 ;
> (d) the number of trials is 150 and the probability of success is 0.6.
>
> 3. If in a binomial experiment of n trials, the probability of success is p and the mean and variance are 3 and 2 respectively, find the probability of
> (a) exactly 1 success ;
> (b) at least 1 success.
>
> 4. If an unbiased die is tossed 12 times and X represents the number of sixes obtained, find the probability that the value of X lies within one standard deviation of the mean, i.e., find $P(\mu - \sigma < X < \mu + \sigma)$.

14.5 The Normal Probability Distribution

One of the most common continuous probability distributions is the ***normal distribution***. Measured quantities such as age, height, mass and life-times of batteries etc., have a normal distribution.

These distributions begin with low frequencies, rising to a maximum near the centre and falling away to low frequencies again. The shape of the probability density function of such a random variable is 'bell-shaped'.

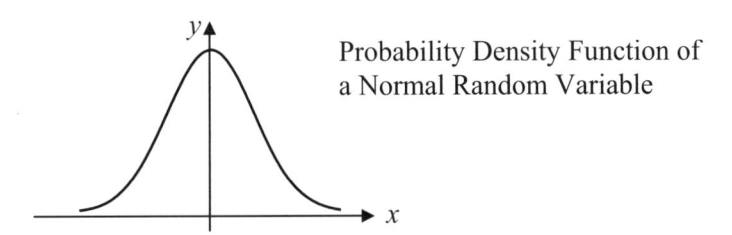

Probability Density Function of a Normal Random Variable

The equation of this probability density function is $y = \dfrac{1}{\sigma\sqrt{2\pi}} \exp\left(-\dfrac{1}{2}\left(\dfrac{x-\mu}{\sigma}\right)^2\right)$,

$[\exp(x) = e^x]$. How this equation is derived is beyond the scope of this book.

However, the following properties should be known:
(1) The function is completely defined by the mean μ and the variance σ^2 of the distribution. For a normal variable X we write $X \sim N(\mu, \sigma^2)$.

Statistics 2

(2) The values of the function are positive for all values of x.
(3) The total area under the curve is 1.
(4) The curve is symmetrical about the line $x = \mu$ and so the mean, mode and median of the distribution all coincide.
(5) Almost all of the population, 99.7%, lies in the interval $[\mu - 3\sigma, \mu + 3\sigma]$; about 95% lies in the interval $[\mu - 2\sigma, \mu + 2\sigma]$; about 67% lies in the interval $[\mu - \sigma, \mu + \sigma]$.
(6) The probability that X takes a value between a and b is the area between the curve and the x-axis between the lines $x = a$ and $x = b$.

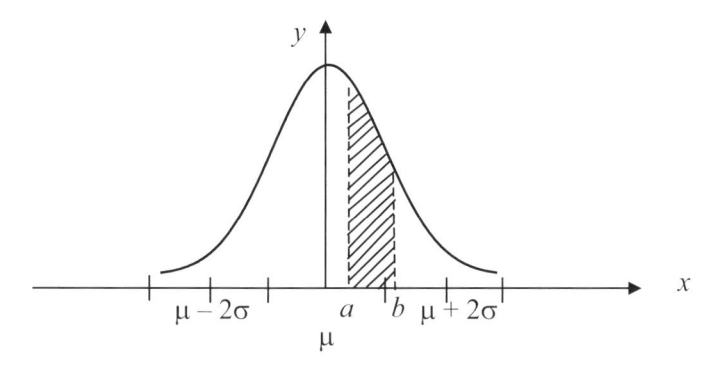

Thus $P(a < X < b)$ is equal to the shaded area.

Normal distribution curves will differ in location and degree of spread according to the values of the parameters μ and σ respectively.

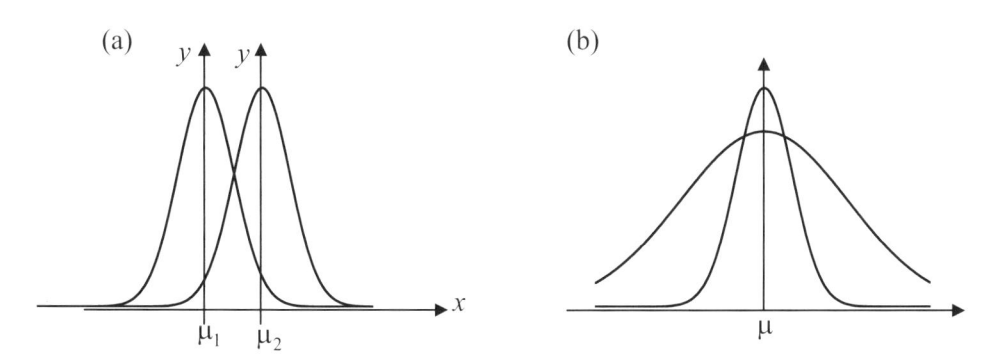

(a)

(b)

In diagram (a), the curves differ in location but have the same degree of spread. That is, σ is the same for both but μ is different. In diagram (b), the curves have the same location but different degrees of spread. That is, μ is the same for both but σ is different.

Chapter 14

To calculate a given area under this curve is outside the scope of this course. Added to this, there are an infinte number of normal distributions each with its own probability density function. Fortunately a set of tables from which any area can be found is readily available. But in order to use these tables, the normal variable X must be transformed into what we call the **standard normal** variable denoted by $Z \sim N[0,1]$. That is Z is a normal variable with mean 0 and variance 1. It is areas under this curve which are given in the tables.

To transform X to the standard normal Z, we use the transformation formula

$$z = \frac{x - \mu}{\sigma}.$$

Thus
$$y = f(x) = \frac{1}{\sigma\sqrt{2\pi}} \exp\left(-\frac{1}{2}\left(\frac{x-\mu}{\sigma}\right)^2\right)$$

becomes
$$f(z) = \sigma y = \frac{1}{\sqrt{2\pi}} \exp\left(-\frac{1}{2}z^2\right).$$

The tables provide us with values of the probability $P(Z < z)$ for positive values of z.

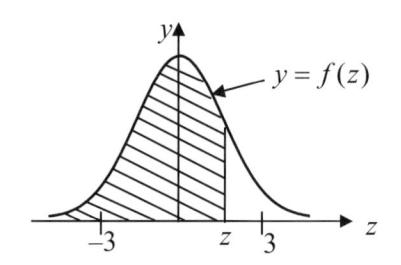

This cumulative probability is denoted by $\Phi(z)$, i.e., $\Phi(z) = P(Z < z)$.

Note: Many GDCs can be used to find the value of $\Phi(z)$ for *any* value of z.

In using standard normal tables (or a GDC), it is sensible to draw a diagram indicating the area to be found.

Example If $Z \sim N(0,1)$, find
(a) $P(Z < 0.6)$; (b) $P(Z > 1.5)$;
(c) $P(Z > -0.3)$; (d) $P(-1.2 < Z < 0.2)$.

(a) $P(Z < 0.6) = \Phi(0.6)$ which is the shaded area in the following diagram.

Statistics 2

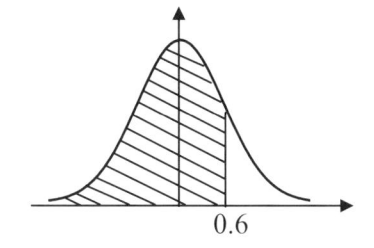

$P(Z < 0.6) = \Phi(0.6) = 0.726$ (tables or GDC)

(b) $P(Z > 1.5)$ is the area shaded in the following diagram.

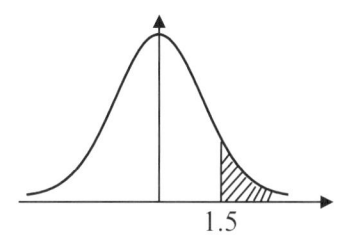

$P(Z > 1.5) = 1 - P(Z < 1.5) = 1 - \Phi(1.5) = 1 - 0.9332 = 0.0668$.

(c) $P(Z > -0.3)$ is the shaded area in the following diagram.

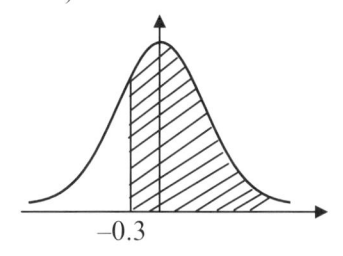

$P(Z > -0.3) = P(Z < 0.3) = \Phi(0.3) = 0.618$ (by symmetry).

(d) $P(-1.2 < Z < 0.2)$ is the shaded area in the following diagram.

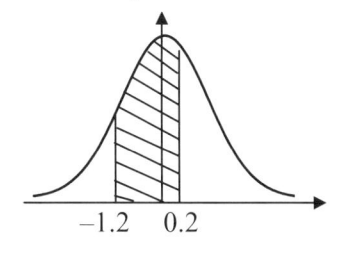

$$P(-1.2 < Z < 0.2) = P(Z < 0.2) - [1 - P(Z < 1.2)]$$
$$= 0.5793 - [1 - 0.8849]$$
$$= 0.464.$$

Chapter 14

Example If X is a normal variable with mean 100 and standard deviation 5, find $P(X > 108)$ and $P(97 < X < 102)$.

$$P(X > 108) = P\left(Z > \frac{108 - \mu}{\sigma}\right) = P(Z > 1.6) \text{ which is the shaded area in the following diagram.}$$

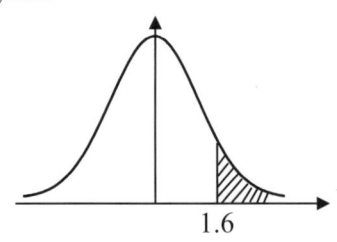

The required probability $= 1 - \Phi(1.6) = 0.0548$.

$$P(97 < X < 102) = P\left(\frac{97 - 100}{5} < Z < \frac{102 - 100}{5}\right) = P(-0.6 < Z < 0.4) \text{ which is the shaded area in the following diagram.}$$

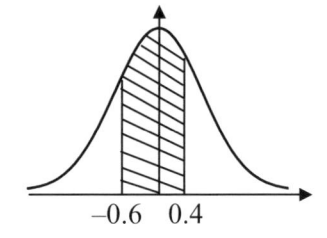

That is $P(97 < X < 102) = \Phi(0.4) - [1 - \Phi(0.6)] = 0.6554 - 0.2743 = 0.381$.

Example Five percent of the values of a normal variable X are greater than 32.3 and three percent of the values are less than 27.1. Find the mean and standard deviation of X.

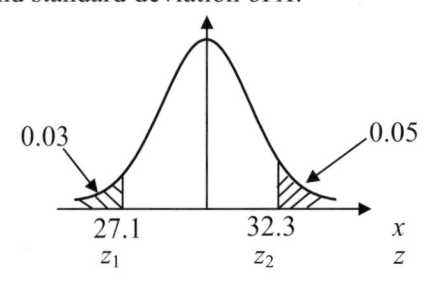

From tables, or a GDC, we find $P(Z < 1.645) = 0.95$ and $P(Z < 1.881) = 0.97$.

Thus $z_1 = -1.881$ and $z_2 = 1.645$.

This gives $\dfrac{27.1 - \mu}{\sigma} = -1.881$ and $\dfrac{32.3 - \mu}{\sigma} = 1.645$.

Hence $27.1 - \mu = -1.881\sigma$
and $32.3 - \mu = 1.645\sigma$.
Subtracting these equations gives $5.2 = 3.526\sigma$ and so $\sigma = 1.47$.
Finally we find $\mu = 29.9$.

Therefore X has mean 29.9 and standard deviation 1.47.

Example A lathe turns out cylinders with a mean diameter of 2.16 cm and a standard deviation of 0.08 cm. Assuming that the distribution of diameters is normal, find the limits to the acceptable diameters if it is found that 5% in the long run are rejected because they are oversize and 5% are rejected because they are undersize.

Each of the shaded areas in the following figure is 5% of the total area. Using the given tables for the standard normal distribution, we find that the required values of z are $z = \pm 1.645$.

Now $z = \dfrac{x - 2.16}{0.08}$ which gives $x = 2.16 \pm 0.08 \times 1.645 = 2.29$ or 2.03
which are the acceptable limits.

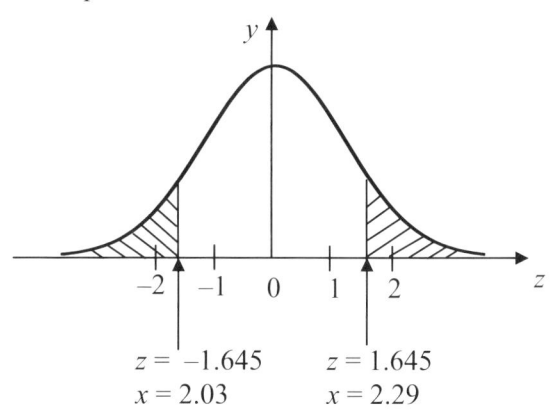

$z = -1.645$ $z = 1.645$
$x = 2.03$ $x = 2.29$

Exercise 14.5

1. For the standard normal variable Z, find
   (a) $P(Z < 1.8)$; (b) $P(Z > 0.568)$; (c) $P(0.345 < Z < 1.212)$.

2. For the normal variable X with mean 50 and standard deviation 2, find
   (a) $P(X > 45)$; (b) $P(X < 48)$; (c) $P(45 < X < 49)$.

Chapter 14

3. If $Z \sim N(0, 1)$, find the value of a if
   (a) $P(Z < a) = 0.8159$;
   (b) $P(Z < a) = 0.3446$;
   (c) $P(Z > a) = 0.409$;
   (d) $P(Z > a) = 0.9505$.

4. If $Z \sim N(0, 1)$, find the value of a if
   (a) $P(|Z| > a) = 0.05$;
   (b) $P(|Z| > a) = 0.025$;
   (c) $P(|Z| < a) = 0.9$;
   (d) $P(|Z| < a) = 0.98$.

5. If $X \sim N(20, 4)$, find
   (a) $P(X > 18)$;
   (b) $P(X < 21)$;
   (c) $P(|X - \mu| < 1.5)$;
   (d) $P(|X - \mu| > 2\sigma)$.

6. A normal variable X has a mean of 38.6 and a variance of 10. Find the probability that a value of X selected at random
   (a) is less than 35.3 ;
   (b) differs from 40 by at least 3.

7. The masses of flour packed by a machine are normally distributed with a mean of 1 kilogram and a standard deviation of 8 gram. Find the probability that a packet of flour from the machine has a mass which is
   (a) less than 990 gram ;
   (b) more than 1015 gram.

8. The random variable X is normally distributed with a standard deviation of 3. If $P(X < 25) = 0.06$, find the mean of X.

9. Steel rods are manufactured and their diameters have a mean of 5 cm. However, the rods are only considered acceptable if their diameters are between 4.98 cm and 5.02 cm. In the long run, 5% of the diameters are found to be too large and 5% are found to be too small. If the diameters are normally distributed, find the distribution's standard deviation.

10. For a normally distributed random variable X, 5% of the values are less than 65.4 and 10% of the values are greater than 69.8. Find the mean and standard deviation of X.

11. If X is a normally distributed random variable with mean 15, and the probability that X is greater than 17 is 0.245, find
    (a) the standard deviation of X ;
    (b) $P(X < 14)$;
    (c) $P(X < 14 \mid X < 17)$;
    (d) the value of x for which $P(X > x) = 0.75$.

12. The continuous random variable X is normally distributed. If $P(X < 65) = 0.02$ and $P(X < 85) = 0.98$, find the interquartile range of the distribution.

13. A manufacturer of light bulbs finds that the life-times of his products are normally distributed with a mean of 1500 burning hours and a standard deviation of 250 hours.
    (a) What is the probability that a bulb selected at random has a life-time of between 1300 hours and 1600 hours?
    (b) In a batch of 100 bulbs, how many would be expected to last for more than 1800 hours?
    (c) Find the probability that in a randomly selected batch of 5 globes, none of them will last longer than 1350 hours.

14.6 An Introduction to Sampling

In the lead-up to the election of politicians to sit in the next parliament, the main parties try to keep track of their 'popularity'. They need to know just how likely they are to win the forthcoming election so that beneficial strategies can be put in place. Since it is not viable, on economic grounds, to ask every elector in the population just how they intend to vote, a smaller sample of the population is approached and the results obtained used to estimate the population's overall intention.

Providing that the sample is fully representative of the population as a whole, the parties can draw conclusions regarding the voting trends of the entire population. From the data collected in a sample, we can calculate sample parameters such as the mean and standard deviation. These can then be used to estimate the population parameters.

Such sampling procedures should avoid the introduction of bias. Also, it is important that care is taken with the method of selection of the sample – the sample should be taken *at random*.

The size of the sample is also important. Information gained by selecting a single member of the population is not of much use. The more information gathered, the more reliable are the parameter estimates.

From sample data we can calculate the sample mean and the sample deviation known as *point estimates* of the corresponding population parameters. But what about the mean and standard deviation of the population as a whole? How can these be estimated without introducing a bias?

Definition   A *statistic* is a function of a random sample $X_1, X_2, X_3, \cdots, X_n$ taken from a population.

The sample mean is $\overline{X} = \dfrac{1}{n}\sum_{i=1}^{n} X_i$ and the sample variance is $S^2 = \dfrac{1}{n}\sum_{i=1}^{n} (X_i - \overline{X})^2$.

These are *estimators* of the population mean, μ, and the population variance, σ^2.

Chapter 14
Unbiased Estimators

The standard deviation of a sampling distribution of a statistic is called the *standard error*.

Consider the following proofs which give the mean and variance of the distribution $\overline{X} = \frac{1}{n}\sum_{i=1}^{n} X_i$ where each sample is taken from a population with mean μ and variance σ^2.

$$\begin{aligned}
\mathrm{E}(\overline{X}) &= \mathrm{E}\left(\frac{1}{n}\sum_{i=1}^{n} X_i\right) \\
&= \frac{1}{n}\mathrm{E}(X_1 + X_2 + X_3 + \cdots + X_n) \\
&= \frac{1}{n}\{\mathrm{E}(X_1) + \mathrm{E}(X_2) + \mathrm{E}(X_3) + \cdots + \mathrm{E}(X_n)\} \\
&= \frac{1}{n}\{n\mu\} \\
&= \mu.
\end{aligned}$$

$$\begin{aligned}
\mathrm{Var}(\overline{X}) &= \mathrm{Var}\left(\frac{1}{n}\sum_{i=1}^{n} X_i\right) \\
&= \frac{1}{n^2}\mathrm{Var}(X_1 + X_2 + X_3 + \cdots + X_n) \\
&= \frac{1}{n^2}\{\mathrm{Var}(X_1) + \mathrm{Var}(X_2) + \mathrm{Var}(X_3) + \cdots + \mathrm{Var}(X_n)\} \\
&= \frac{1}{n^2}\{n\sigma^2\} \\
&= \frac{\sigma^2}{n}.
\end{aligned}$$

Therefore the mean of the sampling distribution is equal to the population mean and the standard error of the mean is equal to $\sigma/\sqrt{n}$.

Clearly, the standard error decreases as the number of samples taken increases. Also, the smaller the standard error, the more reliable is the sample mean as an estimate of the population mean.

An estimator is said to be *unbiased* if the mean of the estimator is equal to the corresponding population parameter.

We have shown that $E(\overline{X}) = \mu$ and so $\overline{X}$ is an unbiased estimator of μ.

Note: Although the unbiased estimate of the population standard deviation is not required by the Standard Level course, it is presented here with the Standard Level material. The calculation of this unbiased estimate using a GDC is quite a simple task.

Now $E(S^2) \neq \sigma^2$ and so S^2 is not an unbiased estimator of σ^2.

Consider the following:

$$\begin{aligned}
E(S^2) &= E\left(\frac{1}{n}\sum_{i=1}^{n}(X_i - \overline{X})^2\right) \\
&= \frac{1}{n}E\left\{\sum_{i=1}^{n}\left(X_i^2 - 2X_i\overline{X} + \overline{X}^2\right)\right\} \\
&= \frac{1}{n}E\left(\sum_{i=1}^{n}X_i^2 - 2\overline{X}\sum_{i=1}^{n}X_i + n\overline{X}^2\right) \\
&= \frac{1}{n}E\left(\sum_{i=1}^{n}X_i^2 - 2n\overline{X}^2 + n\overline{X}^2\right) \\
&= \frac{1}{n}E\left(\sum_{i=1}^{n}X_i^2 - n\overline{X}^2\right) \\
&= \frac{1}{n}\left\{\sum_{i=1}^{n}E(X_i^2) - nE(\overline{X}^2)\right\} \\
&= \frac{1}{n}\left\{n(\sigma^2 + \mu^2) - n\left(\frac{\sigma^2}{n} + \mu^2\right)\right\} \quad \left[E(Y^2) = \text{Var}(Y) + \{E(Y)\}^2\right] \\
&= \sigma^2 + \mu^2 - \frac{\sigma^2}{n} - \mu^2 \\
&= \frac{n-1}{n}\sigma^2.
\end{aligned}$$

If we let $s_{n-1}^2 = E\left(\frac{n}{n-1}S^2\right)$, then $s_{n-1} = \sqrt{\frac{n}{n-1}}S = \sqrt{\frac{n}{n-1}}s_n$ is an unbiased estimate of σ and $S = s_n$ is the standard deviation of the sample.

The ***most efficient*** estimator of a population parameter is one which is unbiased and has the smallest possible variance.

Chapter 14

Example A distribution has a known mean, μ, and variance, σ^2. X_1 and X_2 is a random sample of two independent observations taken from the population. Given that $Y_1 = \dfrac{X_1 + X_2}{2}$ and $Y_2 = \dfrac{X_1 + 2X_2}{3}$ are two estimators for μ, show that both Y_1 and Y_2 are unbiased and determine which of them is the more efficient.

$$\begin{aligned}
\text{E}(Y_1) &= \text{E}\left(\frac{X_1 + X_2}{2}\right) \\
&= \frac{1}{2}\{\text{E}(X_1) + \text{E}(X_2)\} \\
&= \frac{1}{2}(2\mu) \\
&= \mu.
\end{aligned}$$

$$\begin{aligned}
\text{E}(Y_2) &= \text{E}\left(\frac{X_1 + 2X_2}{3}\right) \\
&= \frac{1}{3}\text{E}(X_1) + \frac{2}{3}\text{E}(X_2) \\
&= \frac{1}{3}\mu + \frac{2}{3}\mu \\
&= \mu.
\end{aligned}$$

Therefore both Y_1 and Y_2 are unbiased estimators for μ.

$$\begin{aligned}
\text{Var}(Y_1) &= \text{Var}\left(\frac{X_1 + X_2}{2}\right) \\
&= \frac{1}{4}\text{Var}(X_1 + X_2) \\
&= \frac{1}{4}\{\text{Var}(X_1) + \text{Var}(X_2)\} \\
&= \frac{1}{4}(2\sigma^2) \\
&= \frac{1}{2}\sigma^2.
\end{aligned}$$

$$\begin{aligned}
\text{Var}(Y_2) &= \text{Var}\left(\frac{X_1 + 2X_2}{3}\right) \\
&= \frac{1}{9}\text{Var}(X_1 + 2X_2)
\end{aligned}$$

Statistics 2

$$= \frac{1}{9}\{\text{Var}(X_1) + 4\text{Var}(X_2)\}$$

$$= \frac{1}{9}(\sigma^2 + 4\sigma^2)$$

$$= \frac{5}{9}\sigma^2.$$

Clearly $\text{Var}(Y_1) < \text{Var}(Y_2)$ and so Y_1 is the more efficient.

Exercise 14.6

1. In each of the following, find the best unbiased estimates of the mean and variance of the population from which the sample was taken.
   (a) 18, 20, 21, 18, 22, 19, 19, 23 ;
   (b) 49, 59, 54, 57, 50, 55 ;
   (c) 110, 105, 98, 112, 102, 104, 99, 96, 109, 101.

2. In each of the following, find the best unbiased estimates of the mean and variance of the population from which the sample was taken.
   (a)
x	5	6	7	8	9	10
f	6	16	20	27	19	12

   (b)
x	51	52	53	54	55	56	57	58
f	3	9	16	28	29	24	17	7

3. The following grouped frequency tables are samples from larger populations. Find unbiased estimates of the mean and standard deviation of each population from which the sample was taken.

   (a)
Interval	Frequency
1 – 10	5
11 – 20	8
21 – 30	12
31 – 40	16
41 – 50	15
51 – 60	4

   (b)
Interval	Frequency
0 –	43
10 –	58
20 –	129
30 –	218
40 –	325
50 –	528
60 –	456
70 –	235
80 –	102
90 –	38

Chapter 14

4. Bags of sugar are filled by machine. A sample of 10 bags selected at random have masses (in kilograms) of

  1.02 0.99 1.03 1.01 1.02
  0.98 1.01 1.04 0.98 1.01

Calculate unbiased estimates of the mean and variance of the population from which this sample was taken.

5. A distribution has a known mean, μ, and variance, σ^2. X_1, X_2, X_3 is a random sample of three independent observations taken from the population.

Given that $Y_1 = \dfrac{X_1 + X_2 + X_3}{3}$, $Y_2 = \dfrac{X_1 + 2X_2 + 3X_3}{6}$, $Y_3 = \dfrac{X_1 + 2X_2 + X_3}{3}$

are three estimators for μ, determine which of them are unbiased and which of any unbiased estimators is the most efficient.

14.7 Interval Estimates and Confidence

Point estimates give only a single numerical approximation to a population parameter. Interval estimates give a much more useful estimate of the accuracy of any such approximation.

An *interval estimate* of an unknown population parameter is an interval which has a given probability of including the parameter.

If we can find an interval (a, b) such that $P(a < \lambda < b) = 0.95$ for an unknown parameter λ, we can say that (a, b) is a 95% confidence interval for the parameter λ. Here we have the probability that the interval includes λ is 0.95, which is **not** the probability that λ lies in the interval.

The Central Limit Theorem

If X is a random variable with mean μ and standard deviation σ which is not normally distributed, then $\overline{X} \sim N\left(\mu, \dfrac{\sigma^2}{n}\right)$ for large values of n ($n \geq 30$).

If X is normally distributed with mean μ and standard deviation σ, i.e., $X \sim N(\mu, \sigma^2)$, then $\overline{X} \sim N\left(\mu, \dfrac{\sigma^2}{n}\right)$ for any n.

[The proof of this theorem is beyond the scope of this text.]

Standardising $\overline{X}$ we have $Z = \dfrac{\overline{X} - \mu}{\sigma/\sqrt{n}}$ where $Z \sim N(0, 1)$.

From tables (or a GDC) we know that $P(-1.96 \leq Z \leq 1.96) = 0.95$.

Thus $P\left(-1.96 \leq \dfrac{\overline{X} - \mu}{\sigma/\sqrt{n}} \leq 1.96\right) = 0.95$.

$\Rightarrow \quad P\left(-1.96\dfrac{\sigma}{\sqrt{n}} \leq \overline{X} - \mu \leq 1.96\dfrac{\sigma}{\sqrt{n}}\right) = 0.95$

$\Rightarrow \quad P\left(-\overline{X} - 1.96\dfrac{\sigma}{\sqrt{n}} \leq -\mu \leq -\overline{X} + 1.96\dfrac{\sigma}{\sqrt{n}}\right) = 0.95$

$\Rightarrow \quad P\left(\overline{X} - 1.96\dfrac{\sigma}{\sqrt{n}} \leq \mu \leq \overline{X} + 1.96\dfrac{\sigma}{\sqrt{n}}\right) = 0.95$.

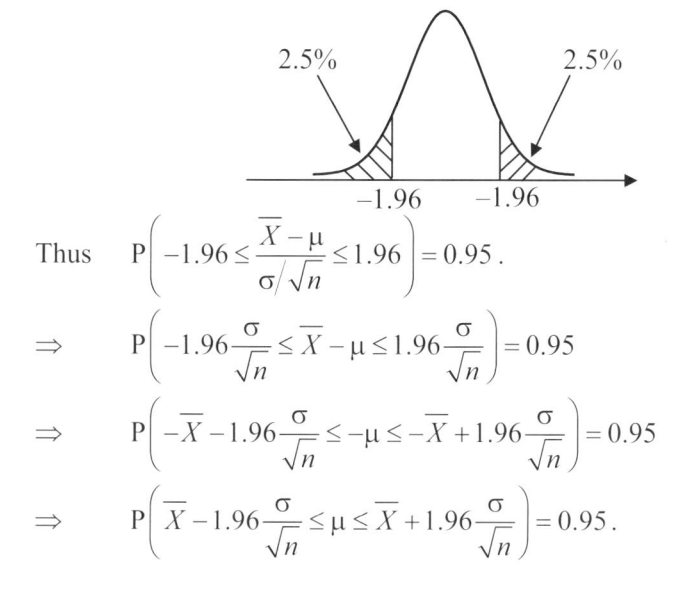

Thus the 95% confidence interval for μ is $\left(\overline{X} - 1.96\dfrac{\sigma}{\sqrt{n}},\ \overline{X} + 1.96\dfrac{\sigma}{\sqrt{n}}\right)$ which is often written as $\overline{X} \pm 1.96\dfrac{\sigma}{\sqrt{n}}$.

If $\overline{x}$ is the mean of a random sample of size n taken from either
(a) a normal population with known standard deviation σ, or
(b) any population with known standard deviation σ where $n \geq 30$,
then a 95% confidence interval for μ is given by

$$\overline{x} \pm 1.96\dfrac{\sigma}{\sqrt{n}} = \left(\overline{x} - 1.96\dfrac{\sigma}{\sqrt{n}},\ \overline{x} + 1.96\dfrac{\sigma}{\sqrt{n}}\right).$$

Example A sample of 16 values of a normal random variable with standard deviation 2 are:

8.6   10.2   9.5   9.8   10.0   8.8   9.2   10.8
10.5   9.9   10.2   8.6   10.4   9.2   9.7   9.5

Calculate a 95% confidence interval for the mean value μ of the variable.

Let X represent the variable.
$n = 16$, $\overline{x} = 9.68$ (from a GDC) and $X \sim N(\mu, 4)$.

Chapter 14

A 95% confidence interval for the mean value μ of the variable is

$$\bar{x} \pm 1.96 \frac{\sigma}{\sqrt{n}} = 9.68 \pm 1.96 \frac{2}{\sqrt{16}} = 9.68 \pm 0.98 = (8.70, 10.66).$$

Example The lengths of a random sample of 100 steel rods produced in a manufacturing process are measured and the mean length is found to be 1.24 metres. If the lengths of the rods are normally distributed with a standard deviation of 0.04 metres, find a 99% confidence interval for the population mean.

Note that if $P(Z > 0.005)$ and $P(Z < -0.005)$, then $Z = 2.5758$.
Let X represent the lengths of the rods, then $X \sim N(\mu, 0.04^2)$.
For a random variable with $n = 100$ and $\bar{x} = 1.24$, a 99% confidence interval for the population mean μ is

$$\bar{x} \pm 2.5758 \frac{\sigma}{\sqrt{n}} = 1.24 \pm 2.5758 \times \frac{0.04}{10} = (1.23, 1.25).$$

Example The 95% confidence interval for the mean of a normally distributed random variable X on the basis of a random sample of 64 measurements was found to be (146.54, 149.48). Find the mean $\bar{x}$ of the sample and the standard deviation σ of the population from which this sample was taken.

The 95% confidence interval is $\bar{x} \pm 1.96 \frac{\sigma}{\sqrt{n}} = (146.54, 149.48)$.

Thus $\bar{x} + 1.96 \frac{\sigma}{8} = 149.48$ ....(i)

and $\bar{x} - 1.96 \frac{\sigma}{8} = 146.54$ ....(ii).

Adding (i) and (ii) gives $2\bar{x} = 296.02$ and so $\bar{x} = 148.0$.

Subtracting (ii) from (i) gives $2(1.96) \frac{\sigma}{8} = 2.94$ and so $\sigma = \frac{4 \times 2.94}{1.96} = 6$.

Example The random variable X is normally distributed with mean μ and standard deviation 4.0. A symmetrical 95% confidence interval for μ with a width which is less than 1 is required. Find the size of the smallest sample needed to achieve this.

$X \sim N(\mu, 4.0^2)$

The 95% confidence interval is $\left(\bar{x} - 1.96 \frac{\sigma}{\sqrt{n}}, \bar{x} + 1.96 \frac{\sigma}{\sqrt{n}} \right)$.

414

The width of this interval is $2(1.96)\dfrac{\sigma}{\sqrt{n}} = \dfrac{15.68}{\sqrt{n}}$.

We require the smallest n for which $\dfrac{15.68}{\sqrt{n}} < 1$, so $\sqrt{n} > 15.68$ or $n > 245.8$.

The size of the smallest sample is 246.

If the standard deviation σ of a population is not known and n is large, we must use an estimator for σ if we wish to find a confidence interval for the mean μ of the population.

Now the estimator used for σ is the unbiased estimator calculated from the standard deviation of the sample. This is $s_{n-1} = \sqrt{\dfrac{n}{n-1}}\, s_n$ where s_n is the sample standard deviation.

Therefore if $\bar{x}$ and s_n are the mean and standard deviation of a random sample of size n (where n is large) taken from a normal population with unknown mean μ and unknown standard deviation σ, then a 95% confidence interval for μ is

$$\bar{x} \pm 1.96 \dfrac{s_{n-1}}{\sqrt{n}} = \left(\bar{x} - 1.96 \dfrac{s_{n-1}}{\sqrt{n}},\, \bar{x} + 1.96 \dfrac{s_{n-1}}{\sqrt{n}}\right).$$

Example The following is a random sample of 30 measurements taken from a normal population.

8.7 8.9 8.4 9.2 9.0 8.7 8.6 9.5 8.6 8.5
8.6 9.5 8.9 8.7 9.1 9.4 8.5 8.6 9.2 9.1
9.3 8.4 8.8 9.6 8.5 9.5 9.3 9.0 8.6 9.4.

Find a 90% confidence interval for the population mean μ.

$P(Z > -1.64) = P(Z < 1.64) = 0.95$ (from tables or a GDC).
Let X represent the random variable, so $\bar{x} = 8.94$ and $s_{n-1} = 0.380$ (GDC).
A 90% confidence interval for the population mean is

$$\bar{x} \pm 1.64 \dfrac{s_{n-1}}{\sqrt{n}} = 8.94 \pm 1.64 \dfrac{0.380}{\sqrt{30}} = (8.83,\, 9.05).$$

Exercise 14.7

1. A sample of size 16 is drawn from a Normal population with standard deviation 5. The sample mean is 30. Calculate 90%, 95% and 98% confidence intervals for the population mean.

Chapter 14

2. A sample of size 36 is drawn from the population given in Question 1. Do you expect the confidence intervals found to be larger or smaller? Calculate the new confidence intervals for the population mean to verify your answer.

3. Find a 95% confidence interval for a population mean μ given a population standard deviation σ, sample size n and sample mean $\bar{x}$ in each of the following:
   (a) $n = 25$, $\bar{x} = 25.2$, $\sigma = 1.85$;
   (b) $n = 64$, $\bar{x} = 3.42$, $\sigma = 0.214$;
   (c) $n = 120$, $\bar{x} = 155$, $\sigma = 70.2$.

4. Find a 99% confidence interval for a population mean μ given a population variance of σ^2, a sample of size n and a sample mean $\bar{x}$ in each of the following:
   (a) $n = 35$, $\bar{x} = 0.23$, $\sigma^2 = 0.052$;
   (b) $n = 50$, $\bar{x} = 35.2$, $\sigma^2 = 20.4$;
   (c) $n = 120$, $\bar{x} = 345$, $\sigma^2 = 210$.

5. The mean waiting time for 100 people at a bus stop is 16 minutes. If the population standard deviation is 3 minutes, find a 95% confidence interval for the population mean waiting time.

6. The masses of a sample of 200 packets of cereal are measured and the mean mass is found to be 502 g with a standard deviation of 2.4 g. Find a 98% confidence interval for the mean mass of packets of cereal.

7. A random sample of n measurements is taken from a Normal population with unknown mean μ and standard deviation $\sigma = 20$. Find the size of the sample required to give
   (a) a 90%; (b) a 95%; (c) a 99%
   confidence interval of width 2.

8. The following is a random sample of 30 measurements taken from a normal population.
   22 25 19 21 24 27 20 19 22 22
   24 26 21 18 23 23 21 27 21 25
   18 23 24 21 19 22 28 23 21 26
   Find
   (a) a 90% confidence interval;
   (b) a 95% confidence interval;
   (c) a 99% confidence interval
   for the population mean μ.

14.8 Significance Testing

Any point estimate of a population parameter almost always differs from its expected value. If the difference is too large it may be that the true value of the population parameter differs from its expected value. We use significance testing to decide whether or not an assumption about a population parameter is true or not.

The first step requires that we make a statistical hypothesis about the population parameter or parameters. This may take the form of : "The mean of the population is 20", or "Two population means are equal", etc. This hypothesis is called the *null hypothesis* and is denoted by H_0.

In order to test this hypothesis, we perform a statistical test based on observations made from random samples taken from a population. The results of this test may suggest that we should reject H_0 in favour of an *alternative hypothesis* denoted by H_1. This may take the form of : "The mean of the population is not 20", or "The two population means are not equal", etc.

Suppose we wish to test whether an observed sample mean taken from a normal population of known standard deviation is significantly different from the assumed population mean. We must first decide upon the *level of significance* of our test. Suppose we choose a 5% level of significance. If H_0 is true, 95% of sample means will lie in the interval

$$\left(\mu - 1.96\frac{\sigma}{\sqrt{n}}, \mu + 1.96\frac{\sigma}{\sqrt{n}}\right).$$

The interval for the corresponding standard normal distribution is (−1.96, 1.96).

This leaves a 5% chance that a sample mean lies outside this interval, in which case we should reject H_0 in favour of H_1.

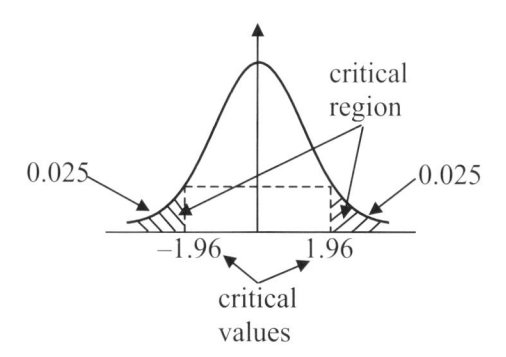

To reject H_0 at the 5% level of significance, the sample mean lies in one of the two tails of the normal distribution curve called the *critical region*. (See diagram above.)

Chapter 14

The values at the boundaries of the critical region are called the ***critical values***. This test is called a ***two-tailed test***.

A ***one-tailed test*** is required if H_1 suggests a definite increase or a definite decrease in the population mean. For example, we could test

(a) $H_0:$ $\mu = 20$
  $H_1:$ $\mu < 0$

for a definite decrease in μ, or

(b) $H_0:$ $\mu = 20$
  $H_1:$ $\mu > 0$

for a definite increase in μ.

For a 5% level of significance, the rejected region is a 5% tail on one side only. For test (a) we have:

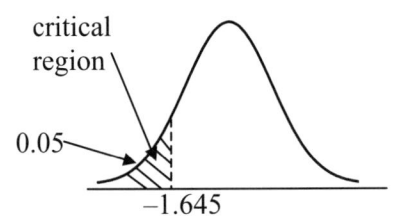

and for test (b) we have:

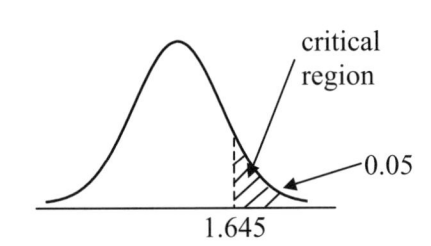

The most common levels of significance are 5% which is deemed to be ***significant***, 1% which is deemed to be ***very significant*** and 0.1% which is deemed to be ***highly significant***. The test statistic Z is from a standard normal distribution. For these levels of significance, the critical values are as follows:

Level of significance	Two-tailed Test	One-tailed Test
Significant (5%)	±1.960	1.645 or −1.645
Very Significant (1%)	±2.576	2.326 or −2.326
Highly Significant (0.1%)	±3.291	3.090 or −3.090

Statistics 2

Example   Consider a normal distribution $X \sim N(\mu, 25)$. A value is selected at random from the population and is found to be 55. Test at the 5% level of significance whether the population mean μ could be 64.

We must first state the null and alternative hypotheses.
$$H_0 : \mu = 64$$
$$H_1 : \mu \neq 64.$$

We test at the 5% level using the test statistic $Z = \dfrac{X - \mu}{\sigma}$ and H_0 is rejected if $|z| > 1.96$.

Now $z = \dfrac{x - \mu}{\sigma} = \dfrac{55 - 64}{5} = -1.8$ and since $|z| < 1.96$, we do not reject H_0 and conclude that at the 5% level of significance, the population mean could be 64.

Example   A normal distribution has a standard deviation of 5 but an unknown mean, μ. A random sample of 16 items from this distribution gives a sample mean of 32.8. Test at the 5% level of significance whether the population mean could be 30.

$$H_0 : \mu = 30$$
$$H_1 : \mu \neq 30$$

The test statistic is $Z = \dfrac{\bar{X} - \mu}{\sigma/\sqrt{n}}$ and we reject H_0 if $|z| > 1.96$.

Now $z = \dfrac{\bar{x} - \mu}{\sigma/\sqrt{n}} = \dfrac{32.8 - 30}{5/4} = 2.24$ and so $|z| > 1.96$.

Hence we reject H_0 and conclude that there is significant evidence at the 5% level to suggest that the population mean is not 30.

Example   A gardener knows from experience that the masses of cabbages grown under standard conditions are normally distributed with mean 0.78 kg and standard deviation 0.2 kg. He decides to try a new fertiliser, and when he weighs a random sample of 90 cabbages from the treated crop, he finds that their mean mass is 0.83 kg. Is there sufficient evidence at the 1% level that the fertiliser did provide a general increase in the masses of the cabbages?

$$H_0 : \mu = 0.78$$
$$H_1 : \mu > 0.78 \text{ (a one-tailed test)}.$$

Chapter 14

Let X be the random variable representing the masses of cabbages. Then $X = N(0.78, 0.2^2)$.

The test statistic is $Z = \dfrac{\bar{X} - \mu}{\sigma/\sqrt{n}}$ and so $z = \dfrac{0.83 - 0.78}{0.2/\sqrt{90}} = 2.372$ which lies in the critical region since $z > 2.326$.

Therefore we reject H_0 and conclude that at the 1% level of significance, the fertiliser did seem to increase the masses of the cabbages.

Exercise 14.8

1. A random variable X is normally distributed with unknown mean and known standard deviation 4. A random sample of 36 values of X had a mean of 22.5. Test, at the 5% level of significance, whether the population mean μ could be 25.

2. A machine produces washers whose internal diameters are normally distributed with mean 8.0 mm and standard deviation 0.05 mm. A random sample of 100 washers is found to have a mean internal diameter of 7.9 mm. Does the machine need adjusting? (Test at the 5% level of significance.)

3. Experience has shown that a standardised test administered to all Year 12 students produces results which are normally distributed with a mean of 60 and a standard deviation of 6. When the test was administered to a random group of 100 students in Year 12, the mean score was 58.5. Is there sufficient evidence, at the 1% level, that these students did not perform as well as expected?

4. A variable, X, with known variance of 28 is thought to have a mean of 65. A random sample of 64 independent observations of X have a mean of 63.8. Is there evidence that the mean is not 65 at
    (a)   the 10% level of significance ;
    (b)   the 5% level of significance ;
    (c)   the 1% level of significance?

5. The masses of mass-produced components in a manufacturing process are normally distributed with variance 4.5 g^2. In order to check on the accuracy of this process, a random sample of 36 components is weighed and the mean mass is calculated. Find the interval in which the value of the sample mean must lie so that the hypothesis H_0 : "The production mean is 20.5 g", will not be rejected at the
    (a)   5% level ;   (b)   1% level ;   (c)   0.1% level.

Statistics 2

14.9 Contingency Tables

It is sometimes desirable to classify individuals into two or more mutually exclusive categories. For example:
  (1)   IB grades achieved by students in two different schools.
  (2)   Voting intentions of males and females.

We would like to investigate whether there is any significant difference between the observed frequencies and the expected frequencies of the distribution. Are the categories independent or is there some evidence of a link between them. The test statistic used is $\sum \dfrac{(O-E)^2}{E}$ where O is the observed frequency and E the corresponding expected frequency of a particular event. The χ^2 (chi-squared) distribution can often be used in such cases.

The χ^2 distribution has a complicated probability density function with one parameter, v (nu), known as the number of degrees of freedom. It is equal to the number of *independent* variables used to calculate χ^2.

The example which follows illustrates the method required to determine whether or not two attributes are independent of each other.

Example   A survey of results in an examination administered in two different schools is given in the following table:

	\multicolumn{4}{c}{Grade}
	7	6	5	4
School A	12	20	18	10
School B	10	12	10	8

Test at the 5% level whether or not there is a significant difference between the two schools in the proportions of candidates awarded the four grades.

The results of the examination are presented in a 2 × 4 *contingency table* as follows:

	7	6	5	4	Total
School A	12	20	18	10	60
School B	10	12	10	8	40
Total	22	32	28	18	100

421

Chapter 14

We must state the null and alternative hypotheses.

H_0 : There is no difference between the two schools in the proportions of candidates awarded the four grades.

H_1 : There is a difference and the proportions are not independent of the school.

We now calculate the expected frequencies:
$$P(\text{candidate is from school A}) = \frac{60}{100}, \text{ and}$$
$$P(\text{candidate receives a grade 7}) = \frac{22}{100}.$$

These two events are independent if H_0 is true and so
$$P(\text{candidate is from school A and receives a grade 7}) = \frac{60}{100} \times \frac{22}{100}.$$

Therefore, of the total of 100 candidates, the expected number from school A receiving a grade 7 is $100 \times \frac{60}{100} \times \frac{22}{100} = 13.2$.

Although this seems to be a somewhat lengthy calculation, it can be shortened by using the formula:
$$\text{expected frequency} = \frac{(\text{row total}) \times (\text{column total})}{\text{grand total}}.$$

All other **expected frequencies** can be calculated and then presented in tabular form:

	7	6	5	4	Total
School A	$\frac{60 \times 22}{100} = 13.2$	$\frac{60 \times 32}{100} = 19.2$	$\frac{60 \times 28}{100} = 16.8$	10.8	60
School B	8.8	12.8	11.2	7.2	40
Total	22	32	28	18	100

Only three expected frequencies have been calculated using the formula. The rest are automatic since the row and column totals must agree with those in the table of observed frequencies.

Since all the entries are known once three are known, the number of degrees of freedom is given by $\nu = 3$.

[The number of degrees of freedom for an $r \times c$ contingency table is given by $\nu = (r-1)(c-1)$.]

From χ^2 tables we find $\chi^2_{0.05}(3) = 7.815$ and we reject H_0 at the 5% level of significance if $\chi^2 > 7.815$.

The values of $\dfrac{(O-E)^2}{E}$ are calculated and presented in tabular form:

O	E	$(O-E)^2/E$
12	13.2	0.109
20	19.2	0.033
18	16.8	0.086
10	10.8	0.059
10	8.8	0.163
12	12.8	0.050
10	11.2	0.129
8	7.2	0.089
100	100	0.718

Since $\chi^2 = 0.718$ is much smaller than the critical value $\chi^2_{0.05}(3) = 7.815$, we do not reject H_0 and conclude that there is not a significant difference between the results of the two schools at the 5% level.

Note: In any χ^2 test for independence of two attributes where the contingency table is 2 × 2, the number of degrees of freedom is 1. In these cases, it is advisable to use **Yates' correction**. The chi-squared statistic is then

$$\chi^2 = \sum \frac{(|O-E|-0.5)^2}{E}.$$

Example Trials are conducted to determine the effectiveness of a new drug in the treatment of a certain disease. Three hundred patients with the disease were selected with 150 of them given doses of the drug and the rest given a completely non-active placebo. After the trial, patients were asked whether or not their condition had improved. The results are given in the following table:

	Drug	Placebo
Improved	91	64
Not Improved	59	86

Use a chi-squared test at the 1% level of significance to determine whether or not there is evidence that the drug is an effective aid in the treatment of the disease.

Chapter 14

The 2 × 2 contingency table is:

	Drug	Placebo	Total
Improved	91	64	155
Not Improved	59	86	145
Total	150	150	300

The expected frequencies are:

	Drug	Placebo	Total
Improved	77.5	77.5	155
Not Improved	72.5	72.5	145
Total	150	150	300

H_0 : The drug is not effective.

H_1 : The drug is effective.

The number of degrees of freedom is 1 and at the 1% level $\chi^2_{0.01}(1) = 6.635$.
We will reject H_0 if $\chi^2 > 6.635$.

O	E	$(|O-E|-0.5)^2/E$
91	77.5	2.1806
64	77.5	2.1806
59	72.5	2.3310
86	72.5	2.3310
300	300	9.0232

Since $\chi^2 > 6.635$, there is sufficient evidence at the 1% level to disprove the assumption that the drug is not effective and conclude that the drug is effective in the treatment of the disease.

Exercise 14.9

1. Consider the following contingency tables and for each one, test at the 5% level whether A and B are independent.

(a)

	B_1	B_2
A_1	52	48
A_2	15	25
A_3	33	27

(b)

	B_1	B_2	B_3
A_1	35	32	33
A_2	38	22	10
A_3	8	5	7
A_4	29	11	20

(c)

	B_1	B_2	B_3
A_1	16	20	14
A_2	24	20	26

(d)

	B_1	B_2
A_1	25	45
A_2	12	43

2. Two football teams in the Premier Division scored the following number of goals in a given season:

	Home	Away
Team A	41	24
Team B	39	36

Test at the 1% level of significance whether there is any difference in the proportions of goals scored home and away by the two teams.

3. In a survey of teachers and administrators in a certain school, staff were asked whether they were happy in their job. The results were as follows:

	Satisfied	Not satisfied
Administrators	15	5
Upper School teachers	28	22
Lower School teachers	32	23

Is there a significant difference in job satisfaction between the three sections of the school?

4. A survey of the colourblindness of a large number of adults provided the following information:

	Colourblind	Not Colourblind
Male	45	855
Female	15	795

At the 5% level, determine whether or not the sex of a person plays a part in whether or not they are colourblind.

5. The following data regarding the voting intentions of a random group of eligible voters who intend to exercise their right at the next election, gave the following results:

	18–30	31–40	41–50	50+
Labour	26	48	32	28
Conservative	14	46	48	56

Age (years)

Chapter 14

Is there any evidence at the 5% level of an association between political affiliation and age.

6. The number of deaths and injuries suffered as a result of car accidents in a given year are given in the following table:

	Wearing a seat belt	Not wearing a seat belt
Deaths	42	38
Injuries	328	52

Is there any evidence to suggest that motorists are safer if they wear a seat belt?

Required Outcomes

After completing this chapter, a student should be able to:
- draw up a probability table for a number of discrete distributions. **(HL)**
- find the mean, variance and standard deviation of a discrete distribution whose probability table is known. **(HL)**
- find the mean and standard deviation of a given binomial distribution. **(HL)**
- convert a given normal probability to standard normal and use tables or a GDC to calculate probabilities of a normal distribution.
- calculate the mean, variance and standard deviation of a sample taken from from a large population.
- calculate unbiased estimates of population parameters.
- calculate confidence intervals for the mean of a population which is normally distributed with known standard deviation.
- perform a significance test for the the mean of a normal population where σ is known.
- determine whether there is sufficient evidence that two attributes of a distribution are independent using a contingency table and the χ^2 statistic.

15 Integral Calculus

15.1 Antiderivatives and the Integral Notation

In our work in the differential calculus chapters we considered methods for finding the derivative $f'(x)$ corresponding to a given function $f(x)$. We now consider the reverse problem of finding $f(x)$ when $f'(x)$ is known. This process is known as *integration*.

If $F'(x) = f(x)$, then $f(x)$ is **the** derivative of $F(x)$, but $F(x)$ is called **an *anti-derivative*** or ***primitive*** of $f(x)$.

We say **an** antiderivative since a given function can have infinitely many anti-derivatives. For example, $\frac{d}{dx}(x^3) = 3x^2$, $\frac{d}{dx}(x^3 - 5) = 3x^2$, $\frac{d}{dx}(x^3 + \pi) = 3x^2$ and so $x^3, x^3 - 5$ and $x^3 + \pi$ are all antiderivatives of $3x^2$.

Theorem Any two antiderivatives of $f(x)$ differ by a constant.

Proof Let $F_1(x)$ and $F_2(x)$ be any two antiderivatives of $f(x)$.

$$\begin{aligned}\text{Then } \frac{d}{dx}(F_1(x) - F_2(x)) &= \frac{d}{dx}(F_1(x)) - \frac{d}{dx}(F_2(x)) \\ &= f(x) - f(x) \\ &= 0.\end{aligned}$$

But only the derivative of a constant is zero.
Therefore $F_1(x) - F_2(x) = $ constant.
Thus any two antiderivatives of $f(x)$ differ by a constant.

Since any two antiderivatives of $3x^2$ differ by a constant, we denote **all** antiderivatives of $3x^2$ by $x^3 + c$ where c is an arbitrary constant.

We write $\int 3x^2 \, dx = x^3 + c$ where the symbol "$\int \; dx$" means "*the integral with respect to x*". We call this ***an indefinite integral***.
The function between "$\int$" and "dx" is called the ***integrand***. [The function to be integrated.]

Chapter 15

For example $\int 5x^3 \, dx = \frac{5}{4}x^4 + c$ and $\int \frac{1}{\sqrt{x}} \, dx = 2\sqrt{x} + c$.

The rule is: $\boxed{\int x^n \, dx = \frac{1}{n+1}x^{n+1} + c, \text{ provided } n \neq -1.}$

Example Write down three differing antiderivatives of $3x^2 - 6x + 2$.

One antiderivative is $x^3 - 3x^2 + 2x$.
Two others are $x^3 - 3x^2 + 2x + 1$ and $x^3 - 3x^2 + 2x - \pi$.

Example Integrate each of the following functions with respect to x:
(a) $(3x-1)^2$;   (b) $\dfrac{2x^2 - 3x + 4}{\sqrt{x}}$.

(a) $\int (3x-1)^2 \, dx = \int (9x^2 - 6x + 1) \, dx = 3x^3 - 3x^2 + x + c$

(b) $\int \dfrac{2x^2 - 3x + 4}{\sqrt{x}} \, dx = \int \left(2x^{3/2} - 3x^{1/2} + 4x^{-1/2}\right) dx$

$\qquad = \frac{4}{5}x^{5/2} - 2x^{3/2} + 8x^{1/2} + c$.

Example Find an antiderivative of $(3x + 1)(x + 3)$ which is zero when $x = -1$.

All antiderivatives of $(3x + 1)(x + 3) = 3x^2 + 10x + 3$ are of the form $F(x) = x^3 + 5x^2 + 3x + c$ where c is an arbitrary constant.
Now $F(-1) = -1 + 5 - 3 + c = 0$ when $c = -1$.
The required antiderivative is $x^3 + 5x^2 + 3x - 1$.

Example A car travelling at 30 m s^{-1} has its brakes applied and comes to rest after 2 s. Assuming that the acceleration was constant, find the acceleration and the distance travelled by the car as it came to rest.

$a = \dfrac{dv}{dt} = -k$ where k is a positive constant

Therefore $v = -kt + c_1$ [v is the integral of a with respect to time].
When $t = 0$, $v = 30$ and so $c_1 = 30$; when $t = 2$, $v = 0$ and so $2k = 30$ which gives $k = 15$.
The acceleration is -15 m s^{-2}.

428

Integral Calculus

Now $v = 30 - 15t$ and the distance travelled is given by $s = 30t - \frac{15}{2}t^2 + c_2$
[where s is the integral of v with respect to time].

At $t = 0$, $s = c_2$ and when $t = 2$, $s = 30 + c_2$.
The distance travelled is 30 m.

Example   A train starts from rest and for the first minute its acceleration after t s is given by $a = \left(c - \frac{1}{18}t\right)$ m s^{-2} where c is a constant. At the end of the first minute the speed is 20 m s^{-1}. Find the distance travelled in the first minute.

$a = c - \frac{1}{18}t \;\Rightarrow\; v = ct - \frac{1}{36}t^2 + c_1$

But $v = 0$ when $t = 0$ and so $c_1 = 0$.

Also $v = 20$ when $t = 60$ and so $60c - \frac{1}{36}(3600) = 20 \;\Rightarrow\; c = 2$.

Thus $v = 2t - \frac{1}{36}t^2 \;\Rightarrow\; s = t^2 - \frac{1}{108}t^3 + c_2$.

Since $v = \frac{1}{36}t(72 - t)$, v does not change sign in the first minute (in the first 72 s in fact).

When $t = 0$, $s = c_2$: when $t = 60$, $s = (60^2) - \frac{1}{108}(60^3) + c_2 = 1600 + c_2$.

Therefore the distance travelled in that first minute is 1600 m.

Exercise 15.1

1. Find an antiderivative of each of the following functions:
   (a) $4x - 5$;
   (b) $6x^2 + 4x + 3$;
   (c) $x^3 + x^2 + x$;
   (d) $(3x - 4)^2$;
   (e) 5;
   (f) $(2x - 3)(3 - 4x)$;
   (g) $(x^2 + 1)(2x^2 - 5)$
   (h) $(2x^2 - 1)^2$;
   (i) $(x - 2)^3$.

2. Integrate each of the following functions with respect to x:
   (a) $\frac{4}{x^3} - \frac{3}{x^2}$;
   (b) $\frac{6}{x^2} - 3 + 4x$;
   (c) $\frac{4x^3 - 4x + 8}{x^3}$;
   (d) $\frac{(x^2 - 2)^2}{x^2}$;
   (e) $\left(x - \frac{4}{x}\right)^2$;
   (f) $5 - \frac{3}{\sqrt{x}}$;
   (g) $\frac{2x^2 - 3x - 2}{\sqrt{x}}$; (h)
   $\frac{(4\sqrt{x} - 3)^2}{x^3}$; (i) $\sqrt{\frac{5}{x}}$.

429

Chapter 15

3. The curve for which $\dfrac{dy}{dx} = 6x^2 - 6x - 2$ passes through the point $(-1, 0)$. Find the equation of the curve and the equation of the tangent at the point $x = 1$.

4. A body moving along a straight line with constant acceleration a m s^{-2} begins at the origin ($s = 0$) with an initial velocity u ms^{-1}. Derive the equation $s = ut + \tfrac{1}{2}at^2$ which gives the distance travelled in t seconds.

5. Express y in terms of x if

   (a) $\dfrac{dy}{dx} = 4x - 1$ and $y = 4$ when $x = 1$;

   (b) $\dfrac{dy}{dx} = \dfrac{3x^2 - 2}{x^2}$ and $y = 1$ when $x = -2$;

   (c) $\dfrac{dy}{dx} = (3\sqrt{x} - 2)^2$ and $y = 4$ when $x = 4$;

   (d) $\dfrac{dy}{dx} = (\sqrt[3]{x} - 1)^3$ and $y = 4$ when $x = 8$.

6. Find each of the following integrals:

   (a) $\displaystyle\int (1 - 3x)^2 \, dx$;

   (b) $\displaystyle\int \left(2t^2 - 4 - \dfrac{3}{t^2}\right) dt$;

   (c) $\displaystyle\int 2(\sqrt{t} - 1)^2 \, dt$;

   (d) $\displaystyle\int \left(x\sqrt{x} - \sqrt{2x} + 3\right) dx$.

7. (a) The gradient of a curve at any point $P(x, y)$ is given by $6x^2 - 2x + 1$. Find the equation of the curve if it passes through the point $(2, 10)$.

   (b) The gradient of a curve at any point $P(x, y)$ is given by $3x^2 + 2x - 2$. If the curve cuts the x-axis at the point $(1, 0)$, find the equation of the curve and the coordinates of the other points where it cuts the x-axis.

8. Find the distance travelled in the first 2 seconds by a particle moving in a straight line such that

   (a) its velocity at time t seconds is given by $v = \tfrac{3}{4}(2t + t^2)$ m s^{-1};

   (b) it starts from rest and its acceleration at time t seconds is given by $a = \left(1 - \tfrac{3}{4}t^2\right)$ m s^{-2}.

9. The curve for which $\dfrac{d^2y}{dx^2} = 6x - 2$ has a turning point at $(1, 1)$. Find the equation of the curve and the position and nature of the other turning point.

430

Integral Calculus

10. The line $3x - 2y = 8$ is a tangent to the curve for which $\dfrac{dy}{dx} = \sqrt{x} - \dfrac{a}{\sqrt{x}}$ at the point (4, 2). Find the value of a and the equation of the curve.

Higher Level

11. A particle moves along the x-axis so that at time t s its acceleration is given by $a(t) = \left(15\sqrt{t} - \dfrac{3}{\sqrt{t}}\right)$ m s^{-2} ($t > 0$). If the particle is at the origin with velocity 4 m s^{-1} when $t = 1$, find when and where the particle is stationary.

*12. What constant acceleration is required to bring a car travelling at 72 km h^{-1} to a complete stop in a distance of 40 m?

15.2 The Use of the Chain Rule to Find $\int (ax+b)^n \, dx$ $(n \neq -1)$

We know that the derivative of the composite function $f(g(x))$ is $f'(g(x))g'(x)$. When $g(x)$ is a linear function of x, $g(x) = ax + b$, then $g'(x) = a$ which is a constant.

Thus $\dfrac{d}{dx}(f(ax+b)) = af'(ax+b)$.

For example $\dfrac{d}{dx}(3x+2)^2 = 3 \times 2(3x+2) = 6(3x+2)$.

Hence an antiderivative of $f'(ax+b)$ is $\dfrac{1}{a} f(ax+b)$.

The rule is: $\quad \int (ax+b)^n \, dx = \dfrac{1}{a(n+1)} (ax+b)^{n+1} + c$, provided $n \neq -1$.

Example Find the following integrals in their most general form:

(a) $\int (2x-3)^4 \, dx$; (b) $\int \dfrac{1}{\sqrt{2t-1}} \, dt$ $(t > \tfrac{1}{2})$.

(a) $\int (2x-3)^4 \, dx = \dfrac{1}{(2)(5)}(2x-3)^5 + c = \dfrac{1}{10}(2x-3)^5 + c$

(b) $\int \dfrac{1}{\sqrt{2t-1}} \, dt = \int (2t-1)^{-1/2} \, dt = \dfrac{1}{(2)(\tfrac{1}{2})}(2t-1)^{1/2} + c = \sqrt{2t-1} + c$

Chapter 15

Example   A curve for which $\frac{dy}{dx} = (3x+1)^3 + 8$ has a turning point with a y-coordinate of $\frac{10}{3}$. Find the position and nature of this turning point and the equation of the curve.

The turning point occurs when $\frac{dy}{dx} = 0$, i.e., when

$(3x+1)^3 = -8 \Rightarrow 3x+1 = -2 \Rightarrow x = -1$.

$$\begin{array}{c c c} - & + & \text{sign of } \frac{dy}{dx} \\ \hline & -1 & \end{array}$$

Therefore the point $\left(-1, \frac{10}{3}\right)$ is a local minimum.

Now $\frac{dy}{dx} = (3x+1)^3 + 8 \Rightarrow y = \frac{1}{12}(3x+1)^4 + 8x + c$ and $y = \frac{10}{3}$ when $x = -1$

giving $\frac{10}{3} = \frac{1}{12}(-2)^4 - 8 + c$ or $c = 10$.

Therefore $y = \frac{1}{12}(3x+1)^4 + 8x + 10$.

Example   A body moves in a straight line with velocity $v = \left(\sqrt{3t+1} - 4\right)$ m s^{-1} at time t seconds. Find the acceleration when it comes to rest and the distance travelled before it comes to rest.

Body comes to rest when $v = 0$ or $\sqrt{3t+1} = 4 \Rightarrow 3t+1 = 16 \Rightarrow t = 5$.

$a = \frac{d}{dt}\left(\sqrt{3t+1} - 4\right) = \frac{3}{2}(3t+1)^{-1/2}$ which gives $a = \frac{3}{8}$ m s^{-2}.

Also, the position at time t is given by $s = \int v\, dt = \frac{2}{9}(3t+1)^{3/2} - 4t + c$.

Since the body does not change direction until $t = 5$, the distance travelled before coming to rest = | position at $t = 5$ − position at $t = 0$ |

$= \left| \frac{2}{9}(16)^{3/2} - 20 + c - \left(\frac{2}{9} + c\right) \right|$

$= 6$ m.

Exercise 15.2

1. Integrate each of the following with respect to x:
    (a) $(2x+1)^2$;   (b) $(3-5x)^3$;
    (c) $(5x+2)^3 - 2$;   (d) $6x - (1-3x)^4$;

432

Integral Calculus

- (e) $3x^2 - 2(5-2x)^3$;
- (f) $(2x^2-3)^2$;
- (g) $4x+1-12(2-3x)^3$;
- (h) $(2x+3)^2 + (3x^2+2)^2$.

2. Integrate each of the following with respect to x:
   - (a) $4(2x-3)^{-2}$;
   - (b) $6(5-2x)^{-4}$;
   - (c) $\dfrac{3}{(2x+3)^2}$;
   - (d) $\dfrac{4}{(2-3x)^3}$;
   - (e) $6 - \dfrac{8}{(2x+3)^3}$;
   - (f) $(2x+3)^3 - \dfrac{1}{(2x+3)^3}$.

3. Find an antiderivative of each of the following functions:
   - (a) $(6x+3)^{1/2}$;
   - (b) $(5x-2)^{3/2}$;
   - (c) $(8x+1)^{-1/3}$;
   - (d) $(2-3x)^{-3/2}$;
   - (e) $\sqrt{2x+5}$;
   - (f) $\sqrt[3]{3x-8}$;
   - (g) $\dfrac{4}{\sqrt{2x+3}}$;
   - (h) $\dfrac{2}{(1-3x)^{3/2}}$.

4. A curve for which $\dfrac{dy}{dx} = 2(2x+1)^2$ meets the x-axis at $(1, 0)$.
   Find:
   - (a) the equation of the curve ;
   - (b) the equation of the tangent to the curve at $x = -2$;
   - (c) the coordinates of the point where the tangent in part (b) meets the curve again.

5. A curve for which $\dfrac{dy}{dx} = \dfrac{a}{\sqrt{2x+1}} - b$, where a, b are constants, passes through the points P(4, 7) and Q(0, 7). The equation of the tangent to the curve at P is $x + y = 11$.
   - (a) Find the values of a and b, and the equation of the curve.
   - (b) Find the equation of the tangent to the curve at Q.

6. A body moves along the x-axis with velocity $v = \left(\dfrac{6}{5} + \dfrac{4}{\sqrt{4t+9}}\right)$ m s^{-1} at time t seconds after it passes through the origin.
   Find:
   - (a) the position of the body at $t = 18$ seconds ;
   - (b) the average speed of the body during the first 4 seconds of motion ;
   - (c) the acceleration of the body at time $t = 4$ seconds.

Chapter 15

Higher Level

15.3 Integration by Substitution

Sometimes a change of variable helps to turn an unfamiliar integral into one that is easily recognisable. The method of integration by substitution can often accomplish this task. We will see why and how this method works in the following pages, but a thorough study of the subject will not be made until Chapter 23 dealing with more advanced integration techniques.

Let $F(x)$ be an antiderivative of $f(x)$ so that

$$F'(x) = f(x) \text{ and } F(x) = \int f(x)\,dx.$$

If u is a differentiable function of x then by the chain rule

$$\frac{d}{dx}(F(u)) = F'(u)\frac{du}{dx} = f(u)\frac{du}{dx}.$$

Thus $F(u)$ is an antiderivative of $f(u)\dfrac{du}{dx}$.

Then $\int \left(f(u)\dfrac{du}{dx}\right) dx = F(u) + c$.

But $\int f(u)\,du = F(u) + c$.

Hence
$$\boxed{\int \left(f(u)\frac{du}{dx}\right) dx = \int f(u)\,du.}$$

This rule is called *integration by substitution*.

Example Find $\int 2x(x^2+1)\,dx$.

Let $u = x^2 + 1$ and so $\dfrac{du}{dx} = 2x$.

Then $\int 2x(x^2+1)\,dx = \int \left(u\dfrac{du}{dx}\right)dx = \int u\,du = \tfrac{1}{2}u^2 + c = \tfrac{1}{2}(x^2+1)^2 + c$.

Example Integrate $\dfrac{x+1}{(x^2+2x+3)^2}$ with respect to x.

The fact that $x^2 + 2x + 3$ has derivative $2x + 2 = 2(x+1)$ suggests that we try putting $u = x^2 + 2x + 3$ which gives $\dfrac{du}{dx} = 2x + 2 = 2(x+1)$.

434

Integral Calculus

Therefore $\int \dfrac{x+1}{(x^2+2x+3)^2}\,dx = \int \left(\dfrac{1}{2}u^{-2}\dfrac{du}{dx}\right)dx$

$$= \int \tfrac{1}{2}u^{-2}\,du$$

$$= -\tfrac{1}{2}u^{-1} + c$$

$$= c - \dfrac{1}{2(x^2+2x+3)}.$$

Note: In a large number of integrals, the substitution method is easy to apply by writing the integrand in the form $k(f(x))^n f'(x)$ for some constant k to be determined. Then $\int k(f(x))^n f'(x)\,dx = \dfrac{k(f(x))^{n+1}}{n+1} + c$ provided $n \neq -1$.

Example Find $\int \dfrac{x}{(1-x^2)^2}\,dx$.

$\int \dfrac{x}{(1-x^2)^2}\,dx = \int -\tfrac{1}{2}(1-x^2)^{-2}(-2x)\,dx$

$$= \int -\tfrac{1}{2}(f(x))^{-2}f'(x)\,dx \quad \text{where } f(x)=1-x^2$$

$$= \dfrac{-\tfrac{1}{2}(1-x^2)^{-1}}{-1} + c$$

$$= \dfrac{1}{2(1-x^2)} + c.$$

Example Find $\int \dfrac{4x-3}{\sqrt{4x^2-6x+3}}\,dx$.

$\int \dfrac{4x-3}{\sqrt{4x^2-6x+3}}\,dx = \int \tfrac{1}{2}(4x^2-6x+3)^{-1/2}(8x-6)\,dx$

$$= \dfrac{\tfrac{1}{2}(4x^2-6x+3)^{1/2}}{\tfrac{1}{2}} + c$$

$$= \sqrt{4x^2-6x+3} + c.$$

Note: For some integrals the method just outlined will not be applicable.

435

Chapter 15

The following example illustrates an integration using substitution which cannot be solved by the method just outlined. However problems of this type will be not be presented until a later chapter.

Example Find $\int \dfrac{x}{\sqrt{x-3}}\, dx$ $(x > 3)$.

Put $u = x - 3$ which gives $\dfrac{du}{dx} = 1$ and $x = u + 3$.

Then
$$\int \dfrac{x}{\sqrt{x-3}}\, dx = \int \left(\dfrac{u+3}{\sqrt{u}} \dfrac{du}{dx}\right) dx$$
$$= \int \dfrac{u+3}{\sqrt{u}}\, du$$
$$= \int (u^{1/2} + 3u^{-1/2})\, du$$
$$= \tfrac{2}{3} u^{3/2} + 6u^{1/2} + c$$
$$= \tfrac{2}{3}(x-3)^{3/2} + 6(x-3)^{1/2} + c.$$

Exercise 15.3

1. Integrate each of the following with respect to x:
   (a) $6x(3x^2 - 1)^3$; (b) $2x(1 - x^2)^4$; (c) $x^2(x^3 + 2)^2$;
   (d) $x^3(x^4 - 3)^3$; (e) $3x^4(2x^5 + 3)^3$; (f) $8x(1 - 3x^2)^5$;
   (g) $3x^6(2 - 3x^7)^2$; (h) $9x^3(4x^4 - 2)^5$; (i) $5x^2(3 - x^3)^3$;
   (j) $x^{-2}(2x^{-1} - 3)^2$; (k) $x^3(3x^4 + 1)^{-2}$; (l) $x^{-3}(x^{-2} + 1)^{-2}$;
   (m) $x(x^3 + 2)^2$; (n) $x^2(x^3 + 2)^2$; (o) $x^3(x^3 + 2)^2$.

2. Find the following indefinite integrals:
   (a) $\int x\sqrt{x^2 + 3}\, dx$; (b) $\int \dfrac{x^2}{(2x^3 + 5)^2}\, dx$;
   (c) $\int \dfrac{x^2}{\sqrt{x^3 + 1}}\, dx$; (d) $\int \dfrac{x+1}{(x^2 + 2x + 5)^3}\, dx$;
   (e) $\int \dfrac{(1+\sqrt{x})^3}{\sqrt{x}}\, dx$; (f) $\int \dfrac{(2\sqrt{x} - 3)^2}{\sqrt{x}}\, dx$;
   (g) $\int \dfrac{(2\sqrt{x} - 3)^2}{x^3}\, dx$; (h) $\int \dfrac{3x - 6}{\sqrt{x^2 - 4x + 11}}\, dx$;

(i) $\int \dfrac{2-x}{(x^2-4x+7)^{3/2}}\,dx$; (j) $\int x^{-1/2}(2\sqrt{x}-3)^{-2}\,dx$.

3. Find the numbers a, b and c such that
$$\dfrac{x^5-2x^4+2x^3-4x^2+4x-2}{(x^2+1)^2} = ax+b+\dfrac{cx}{(x^2+1)^2}$$ for all x, and hence find
$$\int \dfrac{x^5-2x^4+2x^3-4x^2+4x-2}{(x^2+1)^2}\,dx.$$

4. A particle moves in a straight line such that its velocity at time t s is given by $v = \dfrac{100t}{(t^2+1)^3}$ m s^{-1}. Find the distance travelled by the particle in the first 2 seconds of motion.

15.4 The Definite Integral and Area

The definite integral $\int_a^b f(x)\,dx$

(read as "*the integral from a to b of f (x) with respect to x*"), where a and b are numbers, is defined by the rule:
$$\int_a^b f(x)\,dx = F(b)-F(a)$$

where $F(x)$ is **any** antiderivative of $f(x)$ defined on the interval $[a, b]$.

For example, suppose that $f(x) = 2x$. Then $F(x) = x^2$ is an antiderivative of $f(x)$. Thus $\int_1^2 2x\,dx = 2^2-1^2 = 3$ and $\int_{-1}^5 2x\,dx = 5^2-(-1)^2 = 24$.

Note: It does not matter which antiderivative we choose since $G(x) = F(x) + c$ is an antiderivative of $f(x)$ whenever $F(x)$ is, and
$G(b) - G(a) = [F(b) + c] - [F(a) + c] = F(b) - F(a)$ as before.

A convenient and commonly used notation for $F(b) - F(a)$ is $\big[F(x)\big]_a^b$ or $F(x)\big|_a^b$.
Thus $\big[x^2\big]_{-1}^5 = 5^2-(-1)^2 = 24$ and $\big[x^3-2x^2\big]_1^3 = (3^3-(2)3^2)-(1^3-(2)1^2) = 10$.

Chapter 15

With this notation we may express our definition of the definite integral as:

$$\int_a^b f(x)\, dx = \left[F(x)\right]_a^b$$

where $F(x)$ is **any** antiderivative of $f(x)$ which is defined on $[a, b]$.

Note: We require that $F'(x) = f(x)$ for **all** $x \in [a, b]$.

For example, it is incorrect to write $\int_{-1}^{1} \dfrac{dx}{x^2} = \left[-\dfrac{1}{x}\right]_{-1}^{1} = -1 - 1 = -2$ since the relation

$\dfrac{d}{dx}\left(-\dfrac{1}{x}\right) = \dfrac{1}{x^2}$ does not hold when $x = 0$ which is in the domain $[-1, 1]$.

Example Evaluate the following definite integrals:

(a) $\displaystyle\int_1^4 \left(3x^2 - \dfrac{2}{\sqrt{x}} + 1\right) dx$; (b) $\displaystyle\int_2^3 \dfrac{2x-1}{\sqrt{2x^2 - 2x - 3}}\, dx$.

(a) $\displaystyle\int_1^4 \left(3x^2 - \dfrac{2}{\sqrt{x}} + 1\right) dx = \left[x^3 - 4\sqrt{x} + x\right]_1^4$

$\qquad = (64 - 8 + 4) - (1 - 4 + 1)$
$\qquad = 62.$

(b) $\displaystyle\int_2^3 \dfrac{2x-1}{\sqrt{2x^2 - 2x - 3}}\, dx = \left[\sqrt{2x^2 - 2x - 3}\right]_2^3$

$\qquad = \sqrt{18 - 6 - 3} - \sqrt{8 - 4 - 3}$
$\qquad = 2.$

Areas

One of the most important uses for the definite integral is in the calculation of areas whose boundaries may not be straight lines.

Consider the function $y = f(x)$ which is continuous, positive and strictly increasing for all x in the interval $[a, b]$. Our aim is to find the area between the curve, the x-axis and the lines $x = a$ and $x = b$. The required area is shaded in the diagram at the top of the next page.

Denote by $A(c)$, the area bounded by the graph of f, the x-axis and the lines $x = a$ and $x = c$, where $a \le c \le b$.

Then $A(a) = 0$ and the required area is $A(b)$.

Integral Calculus

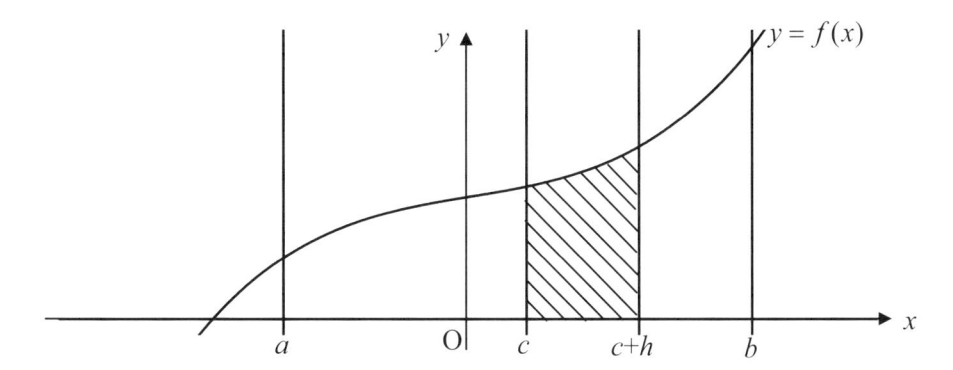

This defines a function $A(x)$ which is continuous for $a \leq x \leq b$.

Consider the ratio $\dfrac{A(c+h) - A(c)}{h}$.

Now $A(c)$ is the area between $x = a$ and $x = c$, and $A(c + h)$ is the area between $x = a$ and $x = c + h$. If $h > 0$, then $A(c + h) - A(c)$ is the area between $x = c$ and $x = c + h$. This area is shaded in the following diagram.

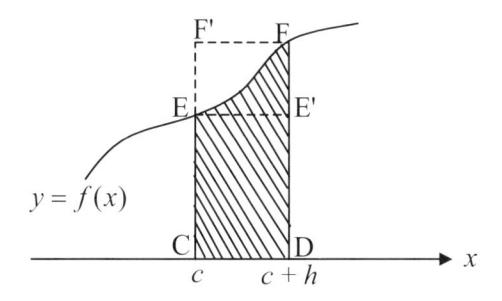

It can be clearly seen that this area lies between the areas of two rectangles with base h and with heights equal to $f(c)$ and $f(c + h)$. These rectangles are CDE'E (which has the smaller area) and CDFF' (which has the larger area).

If h is sufficiently small these rectangles have areas which are as close as we please to $A(c + h) - A(c)$. In fact, for all positive h and for all $a \leq c \leq b$,
$$hf(c) < A(c+h) - A(c) < hf(c+h).$$

Thus $f(c) < \dfrac{A(c+h) - A(c)}{h} < f(c+h)$.

Hence $\lim\limits_{h \to 0} \dfrac{A(c+h) - A(c)}{h} = f(c)$ or $A'(c) = f(c)$.

This result means that $A(x)$ is an antiderivative of $f(x)$.

Hence $A(b) = A(b) - A(a) = \int_a^b f(x)\,dx$.

439

Chapter 15

Thus the area between the graph of $y = f(x)$, the x-axis and the lines $x = a$ and $x = b$, is given by

$$A = \int_a^b y \, dx.$$

Since $f(x)$ is continuous and strictly increasing over the interval $[a, b]$, then the inverse function $x = f^{-1}(y)$ is strictly increasing over the interval $[f(a), f(b)]$. Hence the area between the curve $y = f(x)$, the y-axis and the lines $y = f(a)$ and $y = f(b)$ [the area shaded in the following diagram], is given by

$$A = \int_{f(a)}^{f(b)} x \, dy.$$

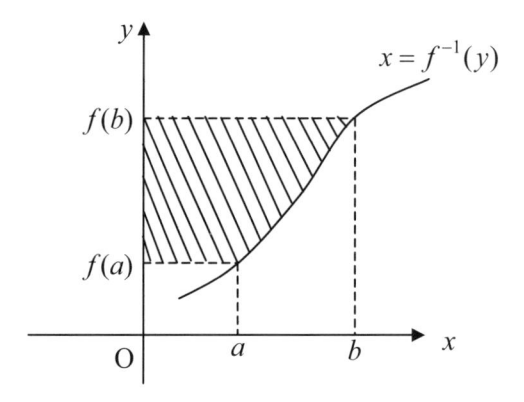

Although we have confined the above discussion to that of a function $y = f(x)$ which is positive and strictly increasing over the interval $[a, b]$, these restrictions are not essential. If for example $y = f(x)$ is positive and decreasing over $[a, b]$, the argument is similar but with $f(c+h) < \dfrac{A(c+h) - A(c)}{h} < f(c)$.

For $f(x)$ negative and decreasing on $[a, b]$, we observe that the reflection of $f(x)$ in the x-axis, $-f(x)$, which is positive and increasing, produces the same area.
Thus the area under the x-axis but above the curve $y = f(x)$ and between the lines $x = a$ and $x = b$ is given by

$$A = -\int_a^b y \, dx.$$

(See graph at the top of the next page.)

440

Integral Calculus

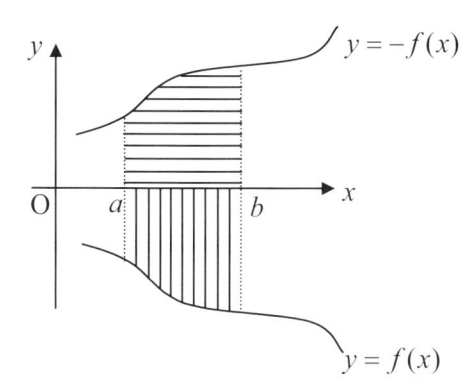

If over the interval [a, b], $f(x)$ is not monotonic (i.e., strictly increasing or strictly decreasing) and takes opposite signs, we partition the interval into sub-intervals such that in each sub-interval $f(x)$ is monotonic and of the same sign. To each of these sub-intervals we apply the arguments given previously.

Example Find the area of the region between the curve $y = x^2$, the x-axis and the lines $x = 0$ and $x = 2$.

The required area $= \int_0^2 x^2 \, dx$

$= \left[\frac{1}{3}x^3\right]_0^2$

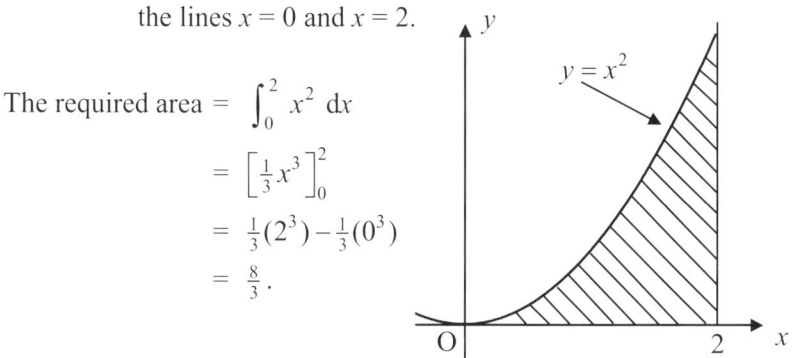

$= \frac{1}{3}(2^3) - \frac{1}{3}(0^3)$

$= \frac{8}{3}.$

Example Find the area enclosed by curve $y = x^2 - 4x + 3$ and the x-axis.

We must first draw a neat sketch and shade the area to be found. The x-intercepts are calculated first of all.

$y = x^2 - 4x + 3 = (x-1)(x-3)$ giving x-intercepts 1 and 3.

The required area $= -\int_1^3 \left(x^2 - 4x + 3\right) dx$ [the area is below the x-axis]

$= -\left[\frac{1}{3}x^3 - 2x^2 + 3x\right]_1^3$

$= -\left(\frac{1}{3}(3^3) - 2(3^2) + 3(3)\right) + \left(\frac{1}{3}(1^3) - 2(1^2) + 3(1)\right)$

$= -9 + 18 - 9 + \frac{1}{3} - 2 + 3$

$= 4/3.$

441

Chapter 15

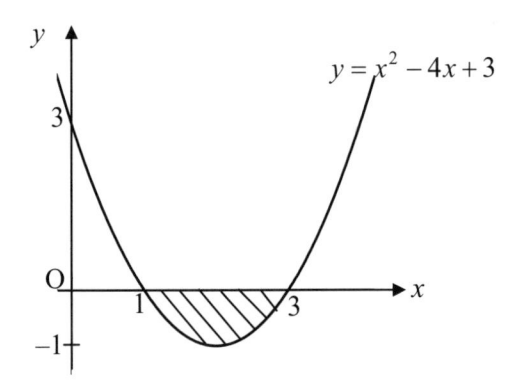

y ↑ $y = x^2 - 4x + 3$

Example   Find the area enclosed by the curve $y = x^4$ and the line $y = x$.

The required area is shaded in the diagram which follows. This area is clearly the difference between A_1, the area of the triangle OCD and A_2, the area between the curve and the *x*-axis.

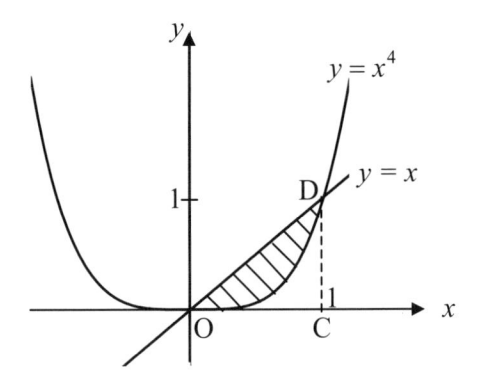

To find the coordinates of the points of intersection of $y = x^4$ and $y = x$, we solve the equation $x^4 = x \Rightarrow x(x^3 - 1) = 0 \Rightarrow x = 0, 1$. The graphs meet at (0, 0) and (1, 1).

Thus A_1 = area of triangle OCD = $\frac{1}{2}$ and $A_2 = \int_0^1 x^4 \, dx = \left[\frac{1}{5}x^5\right]_0^1 = \frac{1}{5}$.

Hence the required area = $A_1 - A_2 = \frac{1}{2} - \frac{1}{5} = \frac{3}{10}$.

Note: The area A_1 could have been found by integration:

$$A_1 = \int_0^1 x \, dx = \left[\frac{1}{2}x^2\right]_0^1 = \frac{1}{2},$$

Integral Calculus

and indeed the required area could have been found by integrating the difference between the functions:

$$A = \int_0^1 \left(x - x^4\right) dx = \left[\tfrac{1}{2}x^2 - \tfrac{1}{5}x^5\right]_0^1 = \tfrac{1}{2} - \tfrac{1}{5} = \tfrac{3}{10}.$$

If $y_1 = f_1(x)$ and $y_2 = f_2(x)$ are *any* two curves continuous for $a \le x \le b$ and such that $f_1(c) \ge f_2(c)$ for all $c \in [a, b]$, then the area enclosed by the curves and the lines $x = a$ and $x = b$ is given by

$$A = \int_a^b (y_1 - y_2) \, dx.$$

Example Find in terms of the positive constant p, the area enclosed by the curves $y^2 = px$ and $x^2 = py$.

The curves meet when $x^4 = p^2 y^2 = p^3 x \Rightarrow x(x^3 - p^3) = 0 \Rightarrow x = 0, p$.

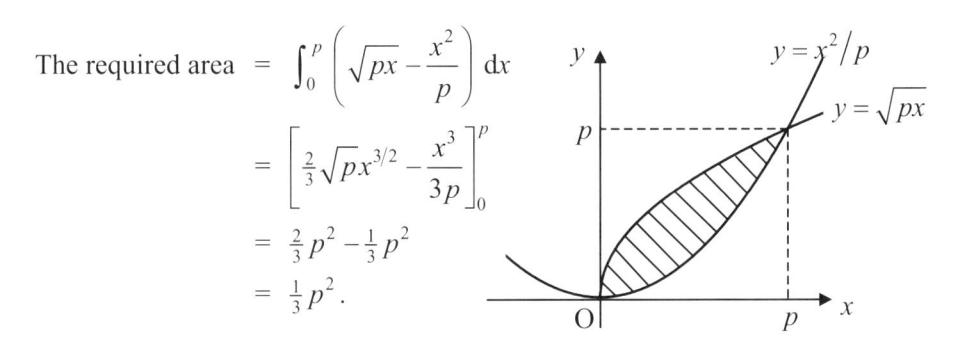

The required area $= \int_0^p \left(\sqrt{px} - \dfrac{x^2}{p}\right) dx$

$$= \left[\tfrac{2}{3}\sqrt{p}\, x^{3/2} - \dfrac{x^3}{3p}\right]_0^p$$

$$= \tfrac{2}{3}p^2 - \tfrac{1}{3}p^2$$

$$= \tfrac{1}{3}p^2.$$

The following property of the definite interval allows us to split integrals into parts and treat each separately:

$$\int_a^b f(x)\,dx = \int_a^c f(x)\,dx + \int_c^b f(x)\,dx.$$

Example Evaluate $\int_{-2}^{2} |x - 1|\,dx$.

Since $|x - 1| = x - 1$ for $x \ge 1$ and $|x - 1| = 1 - x$ for $x \le 1$, then

$$\int_{-2}^{2} |x-1|\,dx = \int_{-2}^{1} (1-x)\,dx + \int_{1}^{2} (x-1)\,dx$$

443

Chapter 15

$$= \left[x - \tfrac{1}{2}x^2\right]_{-2}^{1} + \left[\tfrac{1}{2}x^2 - x\right]_{1}^{2}$$
$$= \left(1 - \tfrac{1}{2}\right) - \left(-2 - 2\right) + \left(2 - 2\right) - \left(\tfrac{1}{2} - 1\right)$$
$$= 5.$$

Exercise 15.4

1. Evaluate each of the following:

   (a) $\int_0^2 (2x+1)\,dx$;
   (b) $\int_0^4 (x^2 - 4x)\,dx$;
   (c) $\int_1^4 \left(2\sqrt{x} + 1 - \dfrac{1}{\sqrt{x}}\right)dx$;
   (d) $\int_0^{\frac{1}{2}} (2x-1)^3\,dx$;
   (e) $\int_1^2 \dfrac{4}{(2x-1)^2}\,dx$;
   (f) $\int_{\frac{1}{2}}^{3} \sqrt{2x+3}\,dx$;
   (g) $\int_{-1}^{1} \left(x^2+1\right)^2 dx$;
   (h) $\int_0^{\frac{1}{2}} \dfrac{4}{(2x+1)^3}\,dx$.

2. Evaluate each of the following:

   (a) $\int_0^1 x(x^2+1)^3\,dx$;
   (b) $\int_{-1}^{2} \dfrac{x}{(x^2+1)^2}\,dx$;
   (c) $\int_0^3 t\sqrt{16+t^2}\,dt$;
   (d) $\int_2^3 \dfrac{6y+1}{\sqrt{3y^2+y-5}}\,dy$;
   (e) $\int_0^4 \dfrac{P}{\sqrt{9+P^2}}\,dP$;
   (f) $\int_{\frac{1}{2}}^{1} \dfrac{x+1}{(x^2+2x+2)^2}\,dx$.

3. Calculate the area between the x-axis and the curve
   (a) $y = x^3$ from $x = 1$ to $x = 2$;
   (b) $y = \dfrac{1}{\sqrt{x}}$ from $x = 1$ to $x = 4$;
   (c) $y = (x+1)^2$ from $x = -1$ to $x = 2$;
   (d) $y = \dfrac{1}{x^2}$ from $x = \tfrac{1}{4}$ to $x = \tfrac{3}{4}$.

4. Calculate the area enclosed by the curve and the coordinate axes in each of the following:
   (a) $y = x^3 + 8$;
   (b) $y = \sqrt{4-x}$;
   (c) $y = (x-3)^2$;
   (d) $y = 2x^3 - 3x^2 + 5$.

444

Integral Calculus

5. Find the total area enclosed by the x-axis and the curve
   (a) $y = (x+1)(2-x)$;
   (b) $y = x(x-1)(x-3)$;
   (c) $y = 6x - x^2 - 5$;
   (d) $y = x^3 - 2x^2 - 8x$.

6. Find the area enclosed by the curve and the straight line in each of the following:
   (a) $y = 4x^2$, $y = 2x$;
   (b) $y = x^3$, $y = 3x + 2$;
   (c) $y = 2x^3$, $y = 8x$;
   (d) $y^2 = x$, $y = \frac{1}{2}x$;
   (e) $y = x^3 - x$, $y = 3x$;
   (f) $y^2 = 4x$, $y = 2x - 4$.

7. Use your graphic display calculator to find the area between the x-axis and the curve
   (a) $y = \dfrac{4}{1+x^2}$ from $x = 0$ to $x = 1$;
   (b) $y = \dfrac{1}{x}$ from $x = -3$ to $x = -1$;
   (c) $y = \dfrac{8}{\sqrt{4-x^2}}$ from $x = -1$ to $x = \sqrt{2}$;
   (d) $y = 2 - \sqrt{7 + 2x - x^2}$ from $x = -1$ to $x = 3$.

Higher Level

8. Find the area enclosed by the graphs of $y = 7 - x^2$ and $y = 3|x-1|$.

9. Find the area between the curve $y = 2a^2x^2 - x^4$, $a > 0$, and the line joining its local maxima.

10. Find, in terms of the positive number a, the total area between the x-axis and the curve:
    (a) $y = 4a^2 - x^2$;
    (b) $y = (x-a)(x-2a)$;
    (c) $y = x^3 - 4a^2x$;
    (d) $y = a^2x^2 - x^4$.

11. Find, in terms of the positive number a, the area enclosed by
    (a) $y = 4a^2 - x^2$ and $y = 3ax$;
    (b) $y = a^2 - x^2$ and $y = x^2 - ax$;
    (c) $y = x^3 - ax^2$, $y = 2ax(x-a)$.

Chapter 15

12. Find in terms of the number $a > 1$, the area enclosed by the curves $y^2 = a^2(1-x)$ and $y^2 = a(a-x)$.

13. Prove that the area of the region enclosed between the parabolas $y^2 = mx$ and $x^2 = ny$ is $\frac{1}{3}mn$ where m, n are positive numbers.

*14. Find the area bounded by $y = \dfrac{1}{x^2}$, $y = -27x$ and $y = -\frac{1}{8}x$.

*15. Find, in terms of the positive number a, the area enclosed by the coordinate axes and the curve $y = \sqrt{x+a} - 1$.

Required Outcomes

After completing this chapter, a student should be able to:
- find a primitive of any function of the form $(ax+b)^n$, $n \neq -1$.
- solve first or second order differential equations of the forms $f'(x) = P(x)$ and $f''(x) = P(x)$ where $P(x)$ can be expressed as a sum of terms of the form $(ax+b)^n$, $n \neq -1$.
- use the method of substitution to find primitives of suitable functions. **(HL)**
- evaluate the definite integral $\int_a^b f(x)\,dx$ whenever $f(x)$ can be expressed as a sum of terms of the form $(ax+b)^n$, $n \neq -1$.
- use a definite integral to find the area between a curve and the x-axis or between two curves.

16 Trigonometric Calculus

16.1 Limits

As we have already seen in Chapter 10, a study of limits is important to the understanding of the calculus.

In the work of this chapter we need a result which students of Mathematics SL should commit to memory. The method used to prove this result is well beyond the scope of the Mathematics SL course. Naturally, students of Mathematics HL need to know both the theorem and its proof.

Higher Level

Theorem $\lim_{\theta \to 0} \dfrac{\sin \theta}{\theta} = 1.$ [θ must be in radians]

Proof In the diagram we see that if $0 < \theta < \tfrac{1}{2}\pi$,

the area of $\triangle OAB$ < the area of sector OAB < the area of $\triangle OBT$,

and so $\tfrac{1}{2}r^2 \sin\theta < \tfrac{1}{2}r^2 \theta < \tfrac{1}{2}r^2 \tan\theta$ or $\sin\theta < \theta < \tan\theta$.

Now since $\sin\theta > 0$ we have

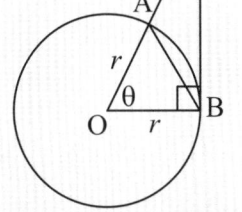

$$\dfrac{\sin\theta}{\sin\theta} < \dfrac{\theta}{\sin\theta} < \dfrac{\tan\theta}{\sin\theta} \quad \text{or} \quad 1 < \dfrac{\theta}{\sin\theta} < \dfrac{1}{\cos\theta}.$$

Every term here is positive and so we can invert to give $\cos\theta < \dfrac{\sin\theta}{\theta} < 1$.

Since every term in this expression is unaltered by replacing θ with $-\theta$, the result $\cos\theta < \dfrac{\sin\theta}{\theta} < 1$ holds for all non-zero θ for which $-\tfrac{1}{2}\pi < \theta < \tfrac{1}{2}\pi$.

Therefore as $\theta \to 0$, $\cos\theta \to 1$ and so $\lim_{\theta \to 0} \dfrac{\sin\theta}{\theta} = 1$ as required.

Chapter 16

Example Evaluate each of the following limits:

(a) $\lim\limits_{\theta \to 0} \dfrac{\sin 5\theta}{\theta}$; (b) $\lim\limits_{\theta \to 0} \dfrac{1-\cos\theta}{\theta^2}$.

(a) $\lim\limits_{\theta \to 0} \dfrac{\sin 5\theta}{\theta} = \lim\limits_{\theta \to 0} \dfrac{5\sin 5\theta}{5\theta}$

$= 5\lim\limits_{5\theta \to 0} \dfrac{\sin 5\theta}{5\theta}$

$= 5.$

(b) $\lim\limits_{\theta \to 0} \dfrac{1-\cos\theta}{\theta^2} = \lim\limits_{\theta \to 0} \dfrac{1-\left(1-2\sin^2 \frac{1}{2}\theta\right)}{\theta^2}$

$= \lim\limits_{\theta \to 0} \dfrac{2\sin^2 \frac{1}{2}\theta}{\theta^2}$

$= \dfrac{1}{2}\lim\limits_{\theta \to 0} \left(\dfrac{\sin \frac{1}{2}\theta}{\frac{1}{2}\theta}\right)^2$

$= \dfrac{1}{2}.$

Example Evaluate $\lim\limits_{n \to \infty} n\sin\dfrac{2}{n}$.

$\lim\limits_{n \to \infty} n\sin\dfrac{2}{n} = \lim\limits_{m \to 0} \dfrac{1}{m}\sin 2m$ where $m = \dfrac{1}{n}$

$= 2\lim\limits_{m \to 0} \dfrac{\sin 2m}{2m}$

$= 2.$

Exercise 16.1

1. Discuss the behaviour of the following functions as $\theta \to 0$:

   (a) $\dfrac{\sin 3\theta}{\theta}$; (b) $\dfrac{\cos 3\theta}{\theta}$; (c) $\dfrac{\tan 3\theta}{\theta}$;

   (d) $\dfrac{\sin^2 \theta}{3\theta^2}$; (e) $\dfrac{\sin 3\theta}{\theta^2}$; (f) $\dfrac{\sin^2 3\theta}{\theta^2}$.

2. Find the limit as $\theta \to 0$ of each of the following:

   (a) $\dfrac{\sin 5\theta}{\theta}$; (b) $\dfrac{\theta}{\sin 5\theta}$; (c) $\dfrac{\sin 2\theta}{\sin 4\theta}$;

448

(d) $\dfrac{\sin^2 \theta}{\sin 2\theta \sin 5\theta}$; (e) $\dfrac{\sin \theta \sin 3\theta}{\theta^2}$; (f) $\dfrac{\sin \theta \sin 5\theta}{\sin^2 4\theta}$.

3. Find the limit as $\theta \to 0$ of each of the following:
[*Note:* You may find the identities $\sin A + \sin B = 2\sin\tfrac{1}{2}(A+B)\cos\tfrac{1}{2}(A-B)$ and $\cos A - \cos B = -2\sin\tfrac{1}{2}(A+B)\sin\tfrac{1}{2}(A-B)$ useful in parts (e) and (f).]

(a) $\dfrac{1-\cos 2\theta}{\theta^2}$; (b) $\dfrac{\sqrt{1-\cos 2\theta}}{\sin 5\theta}$; (c) $\dfrac{\cos 2\theta - 1}{\tan^2 2\theta}$;

(d) $\dfrac{\sin^2 \tfrac{1}{2}\theta}{1-\cos 4\theta}$; (e) $\dfrac{\sin 4\theta}{\sin 7\theta + \sin 3\theta}$; (f) $\dfrac{\theta^2}{\cos 3\theta - \cos 7\theta}$.

4. Find the limit as $n \to \infty$ of each of the following:

(a) $n \sin \dfrac{\pi}{2n}$; (b) $n \sin^2 \dfrac{\pi}{n}$; (c) $n^2 \sin^2 \dfrac{2\pi}{n}$;

(d) $\sin \dfrac{\pi}{n} \cot \dfrac{2\pi}{n}$; (e) $n^2 \left(1 - \cos \dfrac{2\pi}{n}\right)$; (f) $n^2 \sin \dfrac{2}{n} \sin \dfrac{3}{n}$.

5. Find expressions for the circumference and area of both the regular inscribed and circumscribed n-gons of the circle of radius r. Hence show that
   (a) the circumference of the circle is $2\pi r$;
   (b) the area of the circle is πr^2.

16.2 The Derivatives of $\sin x$, $\cos x$ and $\tan x$

Students of Mathematics SL must be familiar with the derivatives of the functions $\sin x$, $\cos x$ and $\tan x$, but not with their proofs. The three derivatives are to be found in a table at the end of this section.

Higher Level

Along with the theorem from the beginning of this section, you may find the following result quite useful when seeking the derivatives of the trigonometric functions from first principles:

$$\lim_{h \to 0} \frac{1 - \cos h}{h} = 0.$$

Proof
$$\lim_{h\to 0}\frac{1-\cos h}{h} = \lim_{h\to 0}\frac{1-(1-2\sin^2\frac{1}{2}h)}{h}$$
$$= \lim_{h\to 0}\frac{2\sin^2\frac{1}{2}h}{h}$$
$$= \lim_{h\to 0}\left(\sin\tfrac{1}{2}h\right)\left(\frac{\sin\frac{1}{2}h}{\frac{1}{2}h}\right)$$
$$= (0)(1)$$
$$= 0.$$

1. The derivative of $f(x) = \sin x$ is denoted and defined by
$$f'(x) = \lim_{h\to 0}\frac{f(x+h)-f(x)}{h}$$
$$= \lim_{h\to 0}\frac{\sin(x+h)-\sin x}{h}$$
$$= \lim_{h\to 0}\frac{\sin x\cos h + \cos x\sin h - \sin x}{h}$$
$$= (\sin x)\lim_{h\to 0}\left(\frac{\cos h - 1}{h}\right) + (\cos x)\lim_{h\to 0}\left(\frac{\sin h}{h}\right)$$
$$= (\sin x)(0) + (\cos x)(1)$$
$$= \cos x.$$

2. The derivative of $f(x) = \cos x$ is denoted and defined by
$$f'(x) = \lim_{h\to 0}\frac{f(x+h)-f(x)}{h}$$
$$= \lim_{h\to 0}\frac{\cos(x+h)-\cos x}{h}$$
$$= \lim_{h\to 0}\frac{\cos x\cos h - \sin x\sin h - \cos x}{h}$$
$$= (\cos x)\lim_{h\to 0}\left(\frac{\cos h - 1}{h}\right) - (\sin x)\lim_{h\to 0}\left(\frac{\sin h}{h}\right)$$
$$= (\cos x)(0) - (\sin x)(1)$$
$$= -\sin x.$$

3. Let $f(x) = \tan x = \dfrac{\sin x}{\cos x}$.

Then by the quotient rule $f'(x) = \dfrac{\cos x\cos x - \sin x(-\sin x)}{\cos^2 x}$

$$= \frac{\cos^2 x + \sin^2 x}{\cos^2 x}$$

$$= \frac{1}{\cos^2 x}.$$

The work so far can be summarised in the following table.

These results must be memorised by students of Mathematics SL as well as by students of Mathematics HL.

$f(x)$	$f'(x)$
$\sin x$	$\cos x$
$\cos x$	$-\sin x$
$\tan x$	$1/\cos^2 x$

The correct application of the chain rule is crucial in the differentiation of trigonometric functions. For example, consider $y = \sin 3x$. If we let $u = 3x$, then $y = \sin u$ and $\frac{dy}{dx} = \frac{dy}{du} \times \frac{du}{dx} = \cos u \times 3 = 3\cos 3x$.

The method used to establish the derivative of $\sin(f(x))$ is as follows:

1. The derivative of $\sin(f(x)) = \sin u$ (with respect to u) is $\cos u$ or $\cos(f(x))$.
2. This is then multiplied by the derivative of $f(x)$ (with respect to x), i.e., $f'(x)$.
3. The final result is $\frac{d}{dx}(\sin(f(x))) = \cos(f(x)) \times f'(x)$.

To differentiate a more complicated trigonometric function such as $\sin^3 2x$, we proceed as follows:

1. Differentiate $u^3 = (\sin 2x)^3$ with respect to u giving $3u^2 = 3(\sin 2x)^2$.
2. Multiply by the derivative of $\sin v = \sin 2x$ with respect to v. This gives $3(\sin 2x)^2 \times \cos v = 3(\sin 2x)^2 \cos 2x$.
3. Finally, multiply by the derivative of $v = 2x$ with respect to x giving $\frac{dy}{dx} = \frac{dy}{du} \times \frac{du}{dv} \times \frac{dv}{dx} = (3\sin^2 2x)(\cos 2x)(2) = 6\sin^2 2x \cos 2x$.

Chapter 16

With practice this procedure should become automatic.

Example Differentiate each of the following functions with respect to x:
- (a) $f(x) = \cos 2x$;
- (b) $f(x) = \sin(x + \frac{1}{4}\pi)$;
- (c) $f(x) = \tan 3x$;
- (d) $f(x) = \sin(\frac{1}{2}\pi - 2x)$;
- (e) $f(x) = \cos^2 x$;
- (f) $f(x) = \tan^2 3x$.

(a) $f(x) = \cos 2x$ and so $f'(x) = (-\sin 2x)(2) = -2\sin 2x$.

(b) $f(x) = \sin(x + \frac{1}{4}\pi) \Rightarrow f'(x) = \cos(x + \frac{1}{4}\pi)(1) = \cos(x + \frac{1}{4}\pi)$.

(c) $f(x) = \tan 3x \Rightarrow f'(x) = \dfrac{1}{\cos^2 3x} \times 3 = \dfrac{3}{\cos^2 3x}$.

(d) $f(x) = \sin(\frac{1}{2}\pi - 2x) \Rightarrow f'(x) = \cos(\frac{1}{2}\pi - 2x)(-2) = -2\cos(\frac{1}{2}\pi - 2x)$.

(e) $f(x) = \cos^2 x \Rightarrow f'(x) = 2\cos x(-\sin x) = -2\cos x \sin x = -\sin 2x$.

(f) $f(x) = \tan^2 3x \Rightarrow f'(x) = (2\tan 3x)(1/\cos^2 3x)(3) = \dfrac{6\tan 3x}{\cos^2 3x}$.

Example Differentiate each of the following with respect to x:
- (a) $f(x) = \dfrac{\sin^2 3x}{x}$;
- (b) $f(x) = x^3 \sin \frac{1}{2} x$.

(a) $f(x) = \dfrac{\sin^2 3x}{x} \Rightarrow f'(x) = \dfrac{x\{2\sin 3x \cos 3x(3)\} - \sin^2 3x(1)}{x^2}$.

(b) $f(x) = x^3 \sin \frac{1}{2} x \Rightarrow f'(x) = \sin \frac{1}{2} x (3x^2) + x^3 \cos \frac{1}{2} x (\frac{1}{2})$
$= \frac{1}{2} x^2 (6 \sin \frac{1}{2} x + x \cos \frac{1}{2} x)$.

Example Find the equation of the tangent to the curve $y = \tan 2x$ at $x = \frac{1}{8}\pi$.

$y = \tan 2x \Rightarrow \dfrac{dy}{dx} = \dfrac{2}{\cos^2 2x} = \dfrac{2}{(1/\sqrt{2})^2} = 4$ at $x = \frac{1}{8}\pi$.

Also $y = \tan \frac{1}{4}\pi = 1$ at $x = \frac{1}{8}\pi$.

Therefore the equation of the tangent at $x = \frac{1}{8}\pi$ is $4x - y = \frac{1}{2}\pi - 1$.

Trigonometric Calculus

Example   Find the maximum value of $3\sin x + 4\cos x$ and the corresponding value of x for $0 \leq x \leq 2\pi$.

Let $f(x) = 3\sin x + 4\cos x$.
Then $f'(x) = 3\cos x - 4\sin x = 0$ when $\tan x = \frac{3}{4}$.
Thus $x = 0.6435, 3.785$ ($0 \leq x \leq 2\pi$).
Now $f''(x) = -3\sin x - 4\cos x$ which is negative at $x = 0.6435$ and positive at $x = 3.785$.
Thus $f(x)$ is a maximum when $x = \arctan\frac{3}{4} = 0.644$ and the maximum value of $f(x)$ is $3\left(\frac{3}{5}\right) + 4\left(\frac{4}{5}\right) = 5$.

Higher Level

Example   Triangle ABC is right-angled at A and AB = 6 cm long. If the length of BC is increasing at the rate of 0.5 cm s^{-1}, find the rate of change of
(a)   the length of AC ;
(b)   the angle ACB,
when AC = 8 cm.

(a)   **Units:**   cm, rad, s

Given: $\dfrac{dx}{dt} = 0.5$ cm s^{-1}

To Find: (a) $\dfrac{dy}{dt}$ (b) $\dfrac{d\theta}{dt}$
when $y = 8$.

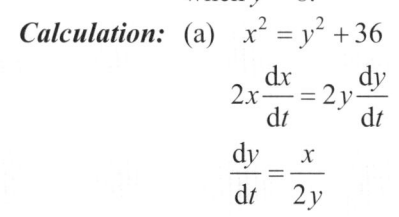

Calculation: (a) $x^2 = y^2 + 36$

$$2x\frac{dx}{dt} = 2y\frac{dy}{dt}$$

$$\frac{dy}{dt} = \frac{x}{2y}$$

When $y = 8$, $x = 10$ and $\dfrac{dy}{dt} = \dfrac{10}{16} = 0.625$ cm s^{-1}.

Therefore the length of AC is increasing at the rate of 0.625 cm s^{-1} when AC is 8 cm long.

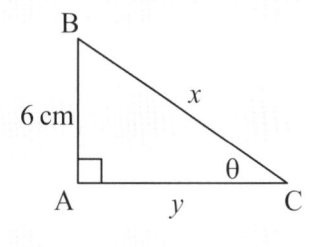

Chapter 16

(b) $\sin\theta = \dfrac{6}{x}$

$\cos\theta \dfrac{d\theta}{dt} = \dfrac{-6}{x^2}\dfrac{dx}{dt}$

$\dfrac{d\theta}{dt} = \dfrac{-3}{x^2\cos\theta}$

When $y = 8$, $x = 10$ and $\cos\theta = 0.8$.

Then $\dfrac{d\theta}{dt} = \dfrac{-3}{80} = -0.0375$.

Therefore the angle ACB is decreasing at the rate of 0.0375 rad s^{-1} = $2.15°$ s^{-1} when AC = 8 cm.

Exercise 16.2

1. Differentiate each of the following functions with respect to x:
   (a) $\sin 2x + \cos 2x$;
   (b) $2\cos x - 3\cos 2x$;
   (c) $\tan x - \tan\frac{1}{4}\pi$;
   (d) $x\cos 3x$;
   (e) $\cos 2x(2 + \sin 2x)$;
   (f) $\tan 2x - 3\cos 2x$;
   (g) $x^2 \cos 3x$;
   (h) $\dfrac{\sin x}{1 + \cos x}$;
   (i) $\dfrac{\cos 2x}{1 - \sin 2x}$;
   (j) $\dfrac{x^2}{3 - \tan 2x}$.

2. Differentiate each of the following functions with respect to x:
   (a) $2\sin^2 x$;
   (b) $\cos^2 4x$;
   (c) $\sin^2 x + \cos^2 2x$;
   (d) $x - 3\tan^2 x$;
   (e) $\cos^2 2x + \sin^3 3x$;
   (f) $\dfrac{4}{\sin^2 5x}$;
   (g) $\cos^2 x - \cos^2 2x$;
   (h) $\sin^2 \frac{1}{4}(x-\pi) + \tan^3 \frac{1}{2}(x-\pi)$.

3. (a) Find the equation of the tangent to the curve $y = 3\sin 2x + 4\cos 2x$ at the point where $x = 0$.
   (b) Find the equation of the normal to the curve $y\sin x = x$ at the point where $x = \frac{1}{2}\pi$.
   (c) Find the values of the real numbers a and b if the line $2x - y = 2$ is a tangent to the curve $y = a\sin 3x + b\cos x$ at $x = 0$.

4. A particle moves along the x-axis such that at time t seconds its position is x metres from O where $x = \frac{1}{2} + t + \frac{1}{2}\cos 4t$. Find:
   (a) the number of times the particle comes to rest in the first 6 seconds;
   (b) the distance travelled between the first two positions of rest;
   (c) the position of the particle at the first instant of zero acceleration.

5. Find the position and nature of the stationary points of each of the following functions and sketch their graphs:
   (a) $f(x) = 5\sin x + 12\cos x$, $0 \le x < 2\pi$;
   (b) $f(x) = \sin 2x + 2\cos x$, $0 \le x < 2\pi$;
   (c) $f(x) = 2 + 3\sin 2x - 4\cos 2x$, $-\pi < x \le \pi$;
   (d) $f(x) = x - \sin 2x$, $0 \le x < \pi$;
   (e) $f(x) = x\cos x - \sin x$, $0 \le x < 2\pi$;
   (f) $f(x) = \dfrac{2 - \sin x}{\cos x}$, $0 \le x < 2\pi$;
   (g) $f(x) = 8\sin x - \tan x$, $-\frac{1}{2}\pi < x < \frac{1}{2}\pi$.

Higher Level

6. Find the derivative of each of the following functions from first principles:
   (a) $\sin 2x$; (b) $\cos 3x$; (c) $\tan x$;
   (d) $\cos \frac{1}{2}x$; (e) $\tan 3x$; (f) $\sin 4x$.

7. Find, from first principles, the differential coefficient at $x = a$ of each of the following functions:
   (a) $\sin \frac{1}{2}x$; (b) $\cos 2x$; (c) $\tan 3x$;
   (d) $\tan \frac{1}{3}x$; (e) $\sin 2x$; (f) $\cos \pi x$.

8. Find the position and nature of the stationary points of the graph of $y = \cos 4x + \cos 2x$ for $0 \le x \le \pi$, and sketch the graph. Find the values of the real number k for which the equation $\cos 4x + \cos 2x = k$ has four distinct roots in the interval $[0, \pi]$.

9. A triangle ABC is right-angled at B and side BC has a fixed length of 15 cm. If the length of AC is decreasing at the rate of 2 cm s^{-1}, find the rate of change of
   (a) the angle BAC (in degrees per second) when AC = 25 cm;
   (b) the area of the triangle when AC = 17 cm.

455

Chapter 16

10. The angle ABC of the triangle ABC is increasing at the rate of 20° per minute. If the lengths of AB and BC are 6 cm and 10 cm respectively, find the rate of change of
   (a) the length of AC when AC = 14 cm;
   (b) the area of the triangle ABC when AC = 12 cm.

11. A ladder 6 m long rests in a vertical plane with one end A on a horizontal floor and the other end B against a vertical wall. The end A is pulled away from the wall at a constant rate of 50 cm s^{-1}. If θ is the inclination of the ladder to the horizontal, find the rate at which θ is changing when A is 4 m from the wall, and determine the rate at which B is descending at that time.

12. A circle with centre O and radius 8 cm has AB as a diameter. A point P moves around the circumference of the circle at the rate of 1 revolution per minute. Find the rate of change of the area of the triangle ABP and the rate of change of the area of the minor segment cut off by BP when the angle PAB is 30°.

16.3 Integration of sin x, cos x and $1/\cos^2 x$

From the derivatives of the trigonometric functions sin x, cos x and tan x, the following integrals follow automatically:

$$\int \sin x \, dx = -\cos x + c$$

$$\int \cos x \, dx = \sin x + c$$

$$\int 1/\cos^2 x \, dx = \tan x + c$$

$$\int \sin ax \, dx = \frac{-\cos ax}{a} + c$$

$$\int \cos ax \, dx = \frac{\sin ax}{a} + c$$

$$\int \frac{dx}{\cos^2 ax} = \frac{\tan ax}{a} + c$$

Example   Integrate each of the following functions with respect to x:
   (a)   sin 2x ;   (b) $\cos \tfrac{1}{3} x$;   (c) $1/\cos^2 2x$.

(a) $\int \sin 2x \, dx = \dfrac{-\cos 2x}{2} + c = -\tfrac{1}{2}\cos 2x + c$

456

(b) $\int \cos\tfrac{1}{3}x \, dx = \dfrac{\sin\tfrac{1}{3}x}{\tfrac{1}{3}} + c = 3\sin\tfrac{1}{3}x + c$

(c) $\int \dfrac{dx}{\cos^2 2x} = \dfrac{\tan 2x}{2} + c = \tfrac{1}{2}\tan 2x + c$

Example Find the area enclosed by the curves $y = \sin x$ and $y = \sin 2x$ between $x = \pi/3$ and $x = \pi$.

The graphs of $y = \sin x$ and $y = \sin 2x$ between $x = 0$ and $x = \pi$ are:

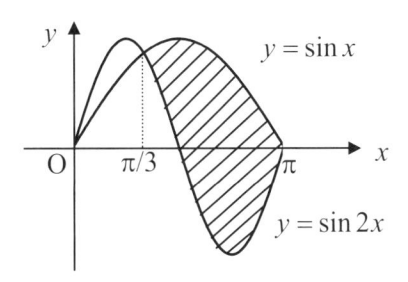

The required area $= \int_{\pi/3}^{\pi} (\sin x - \sin 2x) \, dx$

$= \left[-\cos x + \tfrac{1}{2}\cos 2x \right]_{\pi/3}^{\pi}$

$= -\cos \pi + \tfrac{1}{2}\cos 2\pi + \cos \pi/3 - \tfrac{1}{2}\cos 2\pi/3$

$= 1 + \tfrac{1}{2} + \tfrac{1}{2} + \tfrac{1}{4}$

$= 2.25$.

Example (a) Show that $\cos 2x = 1 - 2\sin^2 x$ and that $\sin^2 x = \tfrac{1}{2} - \tfrac{1}{2}\cos 2x$.
(b) Hence calculate the area enclosed by the x-axis and one arch of the curve $y = \sin^2 x$.

(a) $\cos 2x = \cos^2 x - \sin^2 x$
$= (1 - \sin^2 x) - \sin^2 x$
$= 1 - 2\sin^2 x$.

Therefore $2\sin^2 x = 1 - \cos 2x$
and so $\sin^2 x = \tfrac{1}{2} - \tfrac{1}{2}\cos 2x$.

Chapter 16

(b) The graph of the curve $y = \sin^2 x$ for $0 \leq x \leq \pi$ follows and the required area is given by

$$\int_0^\pi \sin^2 x \, dx = \int_0^\pi \left(\tfrac{1}{2} - \tfrac{1}{2}\cos 2x\right) dx$$

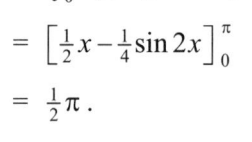

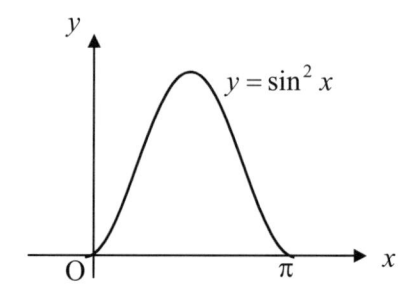

$$= \left[\tfrac{1}{2}x - \tfrac{1}{4}\sin 2x\right]_0^\pi$$

$$= \tfrac{1}{2}\pi.$$

Example The velocity, $v(t)$ m s^{-1}, of a particle moving in a straight line is given by $v(t) = 2\sin 2t + 2\sin t$. Find the distance travelled in the first π seconds of motion.

$v(t) = 4\sin t \cos t + 2\sin t = 2\sin t(2\cos t + 1)$ which is zero when $t = 0$, $\tfrac{2}{3}\pi$, π.

For $0 < t < \tfrac{2}{3}\pi$, $v > 0$ and for $\tfrac{2}{3}\pi < t < \pi$, $v < 0$.

Therefore the distance travelled in the first π seconds

$$= \int_0^{2\pi/3} v(t) \, dt - \int_{2\pi/3}^\pi v(t) \, dt$$

$$= \int_0^{2\pi/3} (2\sin 2t + 2\sin t) \, dt - \int_{2\pi/3}^\pi (2\sin 2t + 2\sin t) \, dt$$

$$= \left[-\cos 2t - 2\cos t\right]_0^{2\pi/3} - \left[-\cos 2t - 2\cos t\right]_{2\pi/3}^\pi$$

$$= \tfrac{1}{2} + 1 - (-1 - 2) - \{-1 + 2 - (\tfrac{1}{2} + 1)\}$$

$$= 5 \text{ m}.$$

[*Note:* If a graphic display calculator is used, the distance travelled in the first π seconds is given by $\int_0^\pi |2\sin 2x + 2\sin x| \, dx$.]

Exercise 16.3

1. Integrate each of the following with respect to x:
   (a) $\sin 3x$;
   (b) $\sin\left(\tfrac{1}{2}\pi - \tfrac{1}{4}x\right)$;
   (c) $\cos\tfrac{1}{2}x$;
   (d) $2\cos 5x$;
   (e) $1/\cos^2\left(\tfrac{2}{3}x\right)$;
   (f) $1/\cos^2\tfrac{1}{2}(x-2)$;
   (g) $x - 2\sin 3x$;
   (h) $3\cos 2(x-1) - 3x^2$;
   (i) $3/\cos^2(2x-3)$.

Trigonometric Calculus

2. Integrate each of the following functions with respect to x:
   - (a) $\cos 2x + \sin \frac{1}{2}x$;
   - (b) $1 - 2\sin 3x + \cos 2x$;
   - (c) $\cos 2x - 4\cos \frac{1}{2}x$;
   - (d) $2\sin(3x - 1) - 3\cos(2x + 1)$.

3. (a) Show that $\cos 2x = 2\cos^2 x - 1 = 1 - 2\sin^2 x$.
   (b) Use the results from part (a) to show that $\cos^2 x = \frac{1}{2} + \frac{1}{2}\cos 2x$ and that $\sin^2 x = \frac{1}{2} - \frac{1}{2}\cos 2x$.
   (c) Integrate each of the following with respect to x:
   - (i) $\cos^2 3x$;
   - (ii) $\sin^2 2x$;
   - (iii) $x^2 - \cos^2 \frac{1}{2}x$;
   - (iv) $\sin^2 5x - 2x + 3$.

4. (a) Differentiate $\sin^2 2x$ with respect to x.
   (b) Use the result of part (a) to integrate $\sin 2x \cos 2x$ with respect to x.

5. (a) Differentiate $\cos^2 3x$ with respect to x.
   (b) Use the result of part (a) to integrate $\sin 3x \cos 3x$ with respect to x.

6. (a) Differentiate $\sin^3 x$ with respect to x.
   (b) Use the result of part (a) to integrate $\sin^2 x \cos x$ with respect to x.

7. Solve the following differential equations:
   - (a) $\frac{dy}{dx} = 2\sin 2x - \cos x$ where $y = 1$ when $x = 0$;
   - (b) $\frac{d^2y}{dx^2} = 2\sin 4x$ where $\frac{dy}{dx} = 2$ and $y = 4$ when $x = 0$;
   - (c) $\frac{d^2y}{dx^2} = 2\cos 2x + \sin x$ where $\frac{dy}{dx} = 3$ and $y = 2\frac{1}{2}$ when $x = 0$.

Higher Level

16.4 The Derivatives of $\csc x$, $\sec x$ and $\cot x$

1. Let $f(x) = \csc x = \dfrac{1}{\sin x} = (\sin x)^{-1}$.

   Then $f'(x) = -(\sin x)^{-2}(\cos x)$

   $= -\left(\dfrac{1}{\sin x}\right)\left(\dfrac{\cos x}{\sin x}\right)$

   $= -\csc x \cot x$.

459

Chapter 16

2. Let $f(x) = \sec x = \dfrac{1}{\cos x} = (\cos x)^{-1}$.

   Then $f'(x) = -(\cos x)^{-2}(-\sin x)$

   $ = \left(\dfrac{1}{\cos x}\right)\left(\dfrac{\sin x}{\cos x}\right)$

   $ = \sec x \tan x$.

3. Let $f(x) = \cot x = \dfrac{\cos x}{\sin x}$.

   Then $f'(x) = \dfrac{\sin x(-\sin x) - \cos x \cos x}{\sin^2 x}$

   $ = \dfrac{-(\sin^2 x + \cos^2 x)}{\sin^2 x}$

   $ = \dfrac{-1}{\sin^2 x}$

   $ = -\csc^2 x$.

The complete table of of derivatives of the six trigonometric functions is as follows:

$f(x)$	$f'(x)$
$\sin x$	$\cos x$
$\cos x$	$-\sin x$
$\tan x$	$1/\cos^2 x = \sec^2 x$
$\csc x$	$-\csc x \cot x$
$\sec x$	$\sec x \tan x$
$\cot x$	$-\csc^2 x$

Exercise 16.4

1. Find, from first principles, the derivative of each of the following functions:
   (a) $\csc 2x$;   (b) $\sec \tfrac{1}{2} x$;   (c) $\cot 4x$.

2. Find, from first principles, the differential coefficient at $x = a$ of each of the following functions:
   (a) $\csc x$;   (b) $\sec 4x$;   (c) $\cot \tfrac{1}{2} x$.

3. Differentiate each of the following functions with respect to x:
   (a) $\tan 2x + \sec 2x$;
   (b) $\tan^2 x - \sec^2 2x$;
   (c) $\csc 2x - \csc^2 3x$;
   (d) $\cot^2 x - \cot^2 2x$;
   (e) $\sec x \tan 2x$;
   (f) $\sec^2 \frac{1}{4}(x-\pi) + \tan^3 \frac{1}{2}(x-\pi)$.

4. Find the position and nature of any stationary point that the following functions may have and sketch their graphs:
   (a) $f(x) = 27\csc x + 8\sec x$, $0 < x < \frac{1}{2}\pi$;
   (b) $f(x) = \tan x + \cot x$, $0 < x < \pi$, $x \neq \frac{1}{2}\pi$.

16.5 Further Integration of Trigonometric Functions

From the derivatives of the six trigonometric functions $\sin x$, $\cos x$, $\tan x$, $\csc x$, $\sec x$ and $\cot x$, the following integrals follow automatically:

$$\int \sin x \, dx = -\cos x + c$$

$$\int \cos x \, dx = \sin x + c$$

$$\int \sec^2 x \, dx = \tan x + c$$

$$\int \csc^2 x \, dx = -\cot x + c$$

$$\int \sec x \tan x \, dx = \sec x + c$$

$$\int \csc x \cot x \, dx = -\csc x + c.$$

For other functions, identities are used to express the integrand in a form which can be readily integrated using any of the above "standard" integrals.

Example Integrate $\tan^2 x$, $\sin^2 x$, $\cos^2 x$ and $\cot^2 x$ with respect to x.

From the identity $1 + \tan^2 x = \sec^2 x$ we obtain $\tan^2 x = \sec^2 x - 1$.
Thus $\int \tan^2 x \, dx = \int (\sec^2 x - 1) \, dx = \tan x - x + c$.

Rearranging the identities $\cos 2x = 1 - 2\sin^2 x$ and $\cos 2x = 2\cos^2 x - 1$ gives $\sin^2 x = \frac{1}{2} - \frac{1}{2}\cos 2x$ and $\cos^2 x = \frac{1}{2} + \frac{1}{2}\cos 2x$.

Thus $\int \sin^2 x \, dx = \int \left(\frac{1}{2} - \frac{1}{2}\cos 2x\right) dx = \frac{1}{2}x - \frac{1}{4}\sin 2x + c$,

and $\int \cos^2 x \, dx = \int \left(\frac{1}{2} + \frac{1}{2}\cos 2x\right) dx = \frac{1}{2}x + \frac{1}{4}\sin 2x + c$.

Chapter 16

From the identity $1 + \cot^2 x = \csc^2 x$ we obtain $\cot^2 x = \csc^2 x - 1$.
Thus $\int \cot^2 x \, dx = \int \left(\csc^2 x - 1 \right) dx = -\cot x - x + c$.

Note: The above-mentioned identities should be known in the forms required for integration, $\tan^2 x = \sec^2 x - 1$, $\cot^2 x = \csc^2 x - 1$, $\sin^2 x = \frac{1}{2} - \frac{1}{2}\cos 2x$ and $\cos^2 x = \frac{1}{2} + \frac{1}{2}\cos 2x$, as well as in their standard forms.

The "inverse chain rule" may be used for any integral of the form
$$\int \cos^n x \sin x \, dx \quad \text{or} \quad \int \sin^n x \cos x \, dx$$
as follows:
$$\int \cos^n x \sin x \, dx = \frac{-1}{n+1} \cos^{n+1} x + c, \quad n \neq -1,$$
and $\int \sin^n x \cos x \, dx = \frac{1}{n+1} \sin^{n+1} x + c, \quad n \neq -1$.

Integrals of the form $\int \cos^n x \, dx$ or $\int \sin^n x \, dx$ are in general a little more difficult, and there is no general rule which covers all cases.

Example Find $\int \cos^3 x \, dx$.

$$\begin{aligned}
\int \cos^3 x \, dx &= \int \cos^2 x \cos x \, dx \\
&= \int \left(1 - \sin^2 x \right) \cos x \, dx \\
&= \int \left(\cos x - \sin^2 x \cos x \right) dx \\
&= \sin x - \tfrac{1}{3} \sin^3 x + c.
\end{aligned}$$

Example Find $\int \sin^4 x \, dx$.

$$\begin{aligned}
\int \sin^4 x \, dx &= \int \sin^2 x \sin^2 x \, dx \\
&= \int \left(1 - \cos^2 x \right) \sin^2 x \, dx \\
&= \int \left(\sin^2 x - \sin^2 x \cos^2 x \right) dx \\
&= \int \left(\tfrac{1}{2} - \tfrac{1}{2}\cos 2x - \tfrac{1}{4}\sin^2 2x \right) dx \\
& \left[\sin 2x = 2\sin x \cos x \text{ and so } \sin^2 x \cos^2 x = \tfrac{1}{4}\sin^2 2x \right]
\end{aligned}$$

462

$$= \int \left(\tfrac{1}{2} - \tfrac{1}{2}\cos 2x - \tfrac{1}{4}\left[\tfrac{1}{2} - \tfrac{1}{2}\cos 4x\right]\right) dx$$
$$= \int \left(\tfrac{3}{8} - \tfrac{1}{2}\cos 2x + \tfrac{1}{8}\cos 4x\right) dx$$
$$= \tfrac{3}{8}x - \tfrac{1}{4}\sin 2x + \tfrac{1}{32}\sin 4x + c.$$

Exercise 16.5

1. Integrate each of the following with respect to x:
   (a) $\sec^2 \tfrac{2}{3}x$;
   (b) $\csc^2 \tfrac{1}{2}(x-2)$;
   (c) $\sin^2 \tfrac{1}{2}x$;
   (d) $\sin^2(1-3x)$;
   (e) $\cos^2 \tfrac{3}{4}x$;
   (f) $\cos^2 \tfrac{1}{2}\pi x$;
   (g) $\tan^2 3x$;
   (h) $\tan^2 \tfrac{2}{3}x$;
   (i) $\cot^2 5x$;
   (j) $\cot^2 \left(\tfrac{2}{3}\pi - 4x\right)$;
   (k) $\cos^2 x \sin x$;
   (l) $\sin 3x \cos^2 3x$;
   (m) $\dfrac{\cos x}{\sqrt{\sin x}}$;
   (n) $\dfrac{\cos 2x}{(\sin x \cos x)^2}$.

2. Integrate $\sin x \cos x$ in three different ways:
   (1) by considering $\sin x \cos x$ in the form $\sin'' x \cos x$;
   (2) by considering $\sin x \cos x$ in the form $\sin x \cos'' x$;
   (3) by expressing $\sin x \cos x$ as a trigonometric function of $2x$.

   Explain how it is possible to obtain three apparently differing answers here.

3. Integrate each of the following with respect to x: [Do part (a) in two ways.]
   (a) $\cos x(1 - 2\sin x)$;
   (b) $\sin 2x(\cos 2x + 1)$;
   (c) $(2 - \sin 3x)^2$;
   (d) $(\sin 2x - \cos 2x)^2$;
   (e) $2\cos x \csc^2 x$;
   (f) $\dfrac{\cos^3 3x - 1}{\cos^2 3x}$;
   (g) $\dfrac{\cos x}{\sqrt{3 - 2\sin x}}$;
   (h) $\dfrac{\sin 3x}{(1 + 2\cos 3x)^3}$;
   (i) $\dfrac{1 - \cos 2x}{1 + \cos 2x}$;
   (j) $\dfrac{4}{1 + \cos 4x}$;
   (k) $x \sin(x^2)$;
   (l) $x^2 \tan^2(x^3)$.

4. A curve for which $\dfrac{dy}{dx} = \cos x(1 + \cos x)$ passes through the origin. Find the equation of the curve and the equation of the tangent at the origin.

Chapter 16

5. (a) Show that $8\cos^4 x = 3 + 4\cos 2x + \cos 4x$, and deduce $\int \cos^4 x \, dx$.

   (b) Differentiate $x \sin x$ with respect to x and deduce $\int x \cos x \, dx$.

   (c) Find $\int x \sin x \, dx$.

6. A particle P is moving along the x-axis and at time t seconds its acceleration is given by $a(t) = 2\sin 2t$ m s^{-2}. The particle is initially at the origin and moving with velocity 1 m s^{-1}. Show that P does not return to the origin. A second particle Q is moving with a constant velocity of 2 m s^{-1}. If Q meets P when $t = \frac{1}{4}\pi$, show that Q does not pass P and find the greatest distance between P and Q.

7. Sketch the curve $y = \sin x (1 - 2\cos x)$ for $0 \leq x \leq \pi$. Find the area enclosed by the x-axis and the curve.

8. Find, in terms of the positive integer a, the value of $\int_0^{\pi a/2} |\sin 2x| \, dx$.

9. Calculate the area enclosed by the line $y = 1$ and one arch the curve $y = 2 + 2\sin\frac{1}{2}x$ lying above the line.

10. Find the area of the region bounded by the curve $y = \dfrac{2}{1 + \cos 2x}$, the x-axis, the y-axis and the line $x = \frac{1}{4}\pi$.

11. Find the area between the curve $y = 2\sin x \sqrt{\cos x}$ and the x-axis from $x = 0$ to the first positive zero.

Required Outcomes

After completing this chapter, a student should be able to:

- use the results $\lim_{\theta \to 0} \dfrac{\sin \theta}{\theta} = 1$ and/or $\lim_{\theta \to 0} \dfrac{1 - \cos \theta}{\theta} = 0$ to differentiate trigonometric functions from first principles. **(HL)**
- find the derivative of any trigonometric function involving $\sin x$, $\cos x$ and $\tan x$ and integrate any of the functions $\sin x$, $\cos x$ and $1/\cos^2 x$.
- find the derivative of any of the six trigonometric functions. **(HL)**
- integrate simple trigonometric functions using appropriate identities.
- integrate more complicated trigonometric functions by first expressing them in a suitable form. **(HL)**

17 Exponential and Logarithmic Functions

17.1 The Exponential Function a^x

Exponential functions, $f(x) = a^x$, $a > 0$, $a \neq 1$, and their inverse functions, $f^{-1}(x) = \log_a x$, have already been introduced briefly in Chapter 1. In the current section we will learn how to differentiate and integrate such functions. This will enable us to solve many problems which occur quite naturally in "growth and decay" situations such as money invested at compound interest, the decay of radio-active material, and the growth of bacteria cultures in medicine. Along the way we will learn what is meant by a^u when u is an irrational number for up to now we have defined it for rational u only.

The derivative of the function $f(x) = a^x$ with respect to x is denoted and defined by

$$\begin{aligned} f'(x) &= \lim_{h \to 0} \frac{f(x+h) - f(x)}{h} \\ &= \lim_{h \to 0} \frac{a^{x+h} - a^x}{h} \\ &= \lim_{h \to 0} \frac{a^x \left(a^h - 1 \right)}{h} \\ &= a^x \left(\lim_{h \to 0} \frac{a^h - 1}{h} \right) \\ &= ma^x \text{ where } m = \lim_{h \to 0} \frac{a^h - 1}{h} \text{ (a fixed number).} \end{aligned}$$

Note: It can be proved that m does exist for any positive value of a.

Exercise Use your graphic display calculator to sketch the graph of $\dfrac{a^h - 1}{h}$ for $a = 2, 2.2, 2.5, 2.8$ and 3, and use your graph to estimate the intercept on the vertical axis in each case.

The y-intercepts are respectively 0.69, 0.79, 0.92, 1.03 and 1.10 (see graphs at the top of the next page).

465

Chapter 17

The results should show that

$$\frac{d}{dx}(2^x) \approx 0.69(2^x),$$

$$\frac{d}{dx}(2.2^x) \approx 0.79(2.2^x),$$

$$\frac{d}{dx}(2.5^x) \approx 0.92(2.5^x),$$

$$\frac{d}{dx}(2.8^x) \approx 1.03(2.8^x) \text{ and}$$

$$\frac{d}{dx}(3^x) \approx 1.10(3^x).$$

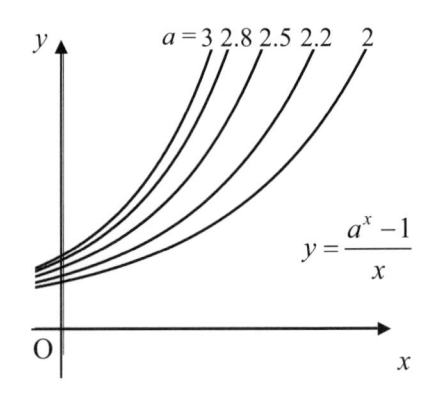

This enables us to estimate the derivatives of such functions, but is not very practical especially when we need higher order derivatives. For example,

$$\frac{d^2}{dx^2}(3^x) \approx (1.10)^2(3^x), \quad \frac{d^3}{dx^3}(3^x) \approx (1.10)^3(3^x), \quad \ldots, \quad \frac{d^n}{dx^n}(3^x) \approx (1.10)^n(3^x).$$

The Exponential Function e^x

It may have already occurred to you that there is a value of a between 2.5 and 2.8 for which $m = \lim\limits_{h \to 0} \frac{a^h - 1}{h} = 1$. In this case $\frac{d}{dx}(a^x) = a^x$ and all higher order derivatives are also given by $\frac{d^n}{dx^n}(a^x) = a^x$.

Definition We denote by e that value of a for which $\lim\limits_{h \to 0} \frac{a^h - 1}{h} = 1$ so that $\frac{d}{dx}(e^x) = e^x$.

The value of e is 2.718 282 ….. correct to 7 significant figures.

The fact that the function e^x is its own derivative makes it the most useful exponential function in calculus situations, in particular the "growth and decay" problems that often occur in nature.

The chain rule must be applied when we differentiate $e^{f(x)}$ where $f(x)$ is differentiable.

Exponential and Logarithmic Functions

This gives:

$$\frac{d}{dx}\left(e^{f(x)}\right) = e^{f(x)} f'(x).$$

Note: The notation $\exp(x)$ for e^x is sometimes used particularly when the expression for $f(x)$ in $e^{f(x)}$ has exponents in it, or is a quotient or product of other functions. For example $\exp\left(\dfrac{2+x}{2-x}\right)$ seems preferable to $e^{\left(\frac{2+x}{2-x}\right)}$.

Example Differentiate each of the following with respect to x:
(a) e^{2x} ;
(b) e^{1-3x} ;
(c) $\exp(x^2+1)$;
(d) $\exp\left(\dfrac{x-1}{x}\right)$.

(a) $\dfrac{d}{dx}\left(e^{2x}\right) = \left(e^{2x}\right)(2) = 2e^{2x}$

(b) $\dfrac{d}{dx}\left(e^{1-3x}\right) = \left(e^{1-3x}\right)(-3) = -3e^{1-3x}$

(c) $\dfrac{d}{dx}\left(\exp(x^2+1)\right) = \left(\exp(x^2+1)\right)(2x) = 2x\exp(x^2+1)$

(d) $\dfrac{d}{dx}\left(\exp\left(\dfrac{x-1}{x}\right)\right) = \exp\left(\dfrac{x-1}{x}\right)\left\{\dfrac{x-(x-1)}{x^2}\right\} = \dfrac{1}{x^2}\exp\left(\dfrac{x-1}{x}\right)$

Higher Level

The General Exponential Function a^x, $a > 0$, $a \neq 1$

If the base of the exponential function is not e, we must first change the base to e and then use the rule $\dfrac{d}{dx}\left(e^{f(x)}\right) = e^{f(x)} f'(x)$.

Firstly, we denote $\log_e x$ by $\ln x$ (the 'natural' logarithm of x).
Now if $y = a^x$, we may write $\ln y = x \ln a$ and so $y = e^{x \ln a}$ (See Chapter 1).

Chapter 17

Thus $\boxed{a^x = e^{x\ln a}}$ $(a > 0, a \neq 1)$, and so $\boxed{\dfrac{d}{dx}(a^x) = (\ln a)e^{x\ln a} = (\ln a)a^x}$.

When we first attempted to differentiate a^x we found $\dfrac{d}{dx}(a^x) = m(a^x)$ where $m = \lim\limits_{h \to 0} \dfrac{a^h - 1}{h}$. We are now able to find the value of m for each value of a, and in fact $m = \lim\limits_{h \to 0} \dfrac{a^h - 1}{h} = \ln a$.

Also, we are now able to give meaning to a^n for irrational values of n, since $a^n = e^{n\ln a}$ for all real values of n.

Example Differentiate 2^x with respect to x.

$$\dfrac{d}{dx}(2^x) = \dfrac{d}{dx}\left(e^{x\ln 2}\right) = \left(e^{x\ln 2}\right)(\ln 2) = (\ln 2)2^x$$

Example For what positive value of x is the function $f(x) = x^2 10^{-x}$ a relative maximum? Find this maximum value.

$f(x) = x^2 10^{-x}$ $\Rightarrow$ $f'(x) = 10^{-x}(2x) + x^2 10^{-x}(-\ln 10)$
$\phantom{f(x) = x^2 10^{-x} \Rightarrow f'(x)} = x10^{-x}(2 - x\ln 10)$
$\phantom{f(x) = x^2 10^{-x} \Rightarrow f'(x)} = 0$ when $x = \dfrac{2}{\ln 10}$ $(x > 0)$.

$\begin{array}{c} + \quad\quad\; - \\ \overline{\diagup\diagdown} \\ 1 \end{array}$ sign of $f'(x)$

Therefore $f(x)$ is a maximum when $x = \dfrac{2}{\ln 10} \approx 0.869$, and the maximum value of $f(x)$ is $\left(\dfrac{2}{\ln 10}\right)^2 \left(10^{-2/\ln 10}\right) \approx 0.102$.

Exercise 17.1

1. It is known that $e = \sum\limits_{n=0}^{\infty} \dfrac{1}{n!}$. By summing sufficient terms of this series, find the value of e correct to six decimal places.

Exponential and Logarithmic Functions

2. Write down the derivative with respect to x of each of the following:
   (a) $e^{3x+1} + e^{-2x}$;   (b) $3 - 2e^{2-x}$;   (c) $e - e^{-x}$;
   (d) $\left(e^x - e^{-x}\right)^2$;   (e) $e^{2x} \sin 2x$;   (f) $e^{\sin 2x}$;
   (g) $\dfrac{x^2}{e^{3x}}$;   (h) $\dfrac{e^{3x} - e^{-3x}}{e^{3x} + e^{-3x}}$;   (i) $x^2 \exp(\cos 2x)$.

3. Find the position and nature of any stationary points that the graphs of the following functions may have:
   (a) $f(x) = xe^x$;   (b) $f(x) = \left(x^2 - x - 1\right)e^x$;
   (c) $f(x) = e^{2x} + e^{-2x}$;   (d) $f(x) = 2e^{2x} + 4e^{-x}$;
   (e) $f(x) = x^3 e^{-x}$;   (f) $f(x) = e^x \sin x$, $0 \le x \le 2\pi$.

4. (a) If $y = e^{2x} + e^{-3x}$, show that $\dfrac{d^2y}{dx^2} + \dfrac{dy}{dx} - 6y = 0$.

   (b) If $y = (2x - 1)e^{-2x}$, show that $\dfrac{d^2y}{dx^2} + 4\dfrac{dy}{dx} + 4y = 0$.

5. (a) If A and B are constants and $y = Ae^{2x} + Be^{4x}$, show that
   $$\dfrac{d^2y}{dx^2} - 6\dfrac{dy}{dx} + 8y = 0.$$

   (b) If A and B are constants and $y = (A + Bx)e^{-3x}$, show that
   $$\dfrac{d^2y}{dx^2} + 6\dfrac{dy}{dx} + 9y = 0.$$

   (c) If $y = e^{mx}$ and $\dfrac{d^2y}{dx^2} - 7\dfrac{dy}{dx} + 12y = 0$, find the possible values of m.

6. (a) Show that the tangent to the curve $y = x^2 e^{-x}$ at $x = 1$ passes through the origin.
   (b) Find the values of A and B if the tangent to the curve $y = Ae^{2x} + Be^{-2x}$ at $x = 0$ has the equation $y = 8x + 2$.

Higher Level

7. (a) Find the angle between the curves $y = e^{-x}$ and $y = 2e^{-2x}$ at their point of intersection.

469

(b) The tangent at P(h, k) on the curve $y = \frac{1}{2}(e^x + e^{-x})$, $x \neq 0$, makes an angle θ with the positive x-axis. Prove that $k = |\sec \theta|$.

8. (a) The position of a particle moving along the x-axis at time t seconds is given by $x = 32(e^{-t} - 2e^{-2t})$ m. Find the acceleration of P when it is stationary and find its maximum speed in the subsequent motion.

(b) A particle moves along the x-axis so that at time t seconds its position, x metres from O, is given by $x = 20 - 10e^{-t/10}$.
   (i) Find the velocity and acceleration of the particle at the end of the tenth second.
   (ii) Find the average velocity of the particle during the first ten seconds.
   (iii) Find the acceleration of the particle when
       (1) $x = 17.5$ m;
       (2) its velocity is half its initial velocity.

9. (a) Find the coordinates of the turning point and the points of inflexion of the graph of $y = \exp(-x^2)$. Sketch the graph.

(b) A rectangle has two vertices on the x-axis and two on the curve $y = \exp(-x^2)$. If the point $P(x, \exp(-x^2))$, where $x > 0$, is one of the vertices on the curve, show that the area of the rectangle is given by $A = 2x \exp(-x^2)$. Show that the maximum area of this rectangle occurs when P is one of the points of inflexion found in part (a).

17.2 Differentiation of Logarithmic Functions

Since the function $f : \mathbb{R} \to \mathbb{R}^+$ such that $f(x) = e^x$ is differentiable and its derivative is never zero, its inverse function $f^{-1} : \mathbb{R}^+ \to \mathbb{R}$ for which $f^{-1}(x) = \ln x$ is also differentiable.

Let $y = \ln x$, then $e^y = x$.

When we differentiate each side of this last expression with respect to x we obtain

Exponential and Logarithmic Functions

$$e^y \frac{dy}{dx} = 1$$

$$\frac{dy}{dx} = \frac{1}{e^y}$$

$$\frac{dy}{dx} = \frac{1}{x}.$$

Thus
$$\boxed{\frac{d}{dx}(\ln x) = \frac{1}{x} \quad (x > 0).}$$

In the more general form, if $f(x)$ is any differentiable function such that $f(x) > 0$, then

$$\boxed{\frac{d}{dx}(\ln f(x)) = \frac{1}{f(x)} \times f'(x) = \frac{f'(x)}{f(x)}.}$$

Example Differentiate each of the following functions with respect to x:
    (a) $\ln(x^2 + 1)$;
    (b) $x^2 \ln x$;
    (c) $\ln(kx)$, where k is a constant.

(a) $\dfrac{d}{dx}\left(\ln\left(x^2+1\right)\right) = \dfrac{1}{x^2+1} \times 2x = \dfrac{2x}{x^2+1}$

(b) $\dfrac{d}{dx}\left(x^2 \ln x\right) = \ln x \times 2x + x^2 \times \dfrac{1}{x} = 2x \ln x + x$

(c) $\dfrac{d}{dx}\left(\ln(kx)\right) = \dfrac{1}{kx} \times k = \dfrac{1}{x}$

Note: The derivative of $\ln(kx)$ is the same as that of $\ln x$. This is easily explained by the fact that $\ln(kx) = \ln k + \ln x$ and the derivative of the constant $\ln k$ is zero.

Higher Level

The Derivative of $\log_a x$

To differentiate a logarithmic function where the base is not e, we use the change of base rule we first met in Chapter 1 to convert the logarithm in base a to one in base e.

Chapter 17

Let $y = \log_a x = \dfrac{\ln x}{\ln a}$ (change of base).

Then $\dfrac{d}{dx}(\log_a x) = \dfrac{d}{dx}\left(\dfrac{\ln x}{\ln a}\right) = \dfrac{1}{\ln a}\dfrac{d}{dx}(\ln x) = \dfrac{1}{x \ln a}.$

Therefore $\boxed{\dfrac{d}{dx}(\log_a x) = \dfrac{1}{x \ln a}.}$

In the more general form, if $f(x)$ is differentiable, then

$$\boxed{\dfrac{d}{dx}(\log_a f(x)) = \dfrac{f'(x)}{f(x)\ln a}.}$$

A number of scientific formulae are expressed in terms of base 10 logarithms. The **Richter scale** for measuring the intensity of an earthquake, the **decibel scale** for measuring the loudness of sound and the **pH-scale** for measuring the acidity of a solution are all constructed using base 10 logarithms.

Example If the intensity I of a sound is measured in watts per square metre, the sound level is $10\log(I \times 10^{12})$ decibels.

(a) What change in the sound level is created by a doubling of the intensity?
(b) By what factor must the intensity be multiplied to increase the sound level by 5 decibels?

(a) Let the original sound level be $10\log(I_0 \times 10^{12})$ decibels.
If we double the intensity, the sound level becomes
$10\log(2I_0 \times 10^{12}) = 10\log 2 + 10\log(I_0 \times 10^{12})$
$= 3.01 +$ original sound level.
Therefore the sound level is increased by approximately 3 decibels.

(b) Let the original sound level be $L_0 = 10\log(I_0 \times 10^{12})$ decibels.
If this is increased by 5 decibels, we have
$L = 10\log(I_0 \times 10^{12}) + 5$
$= 10\log(I_0 \times 10^{12}) + 10\log 3.162$
$= 10\log(3.162 I_0 \times 10^{12}).$
Therefore, the intensity is multiplied by a factor of 3.16.

Exponential and Logarithmic Functions

Example Differentiate the function $\log(2x)$, $x > 0$ with respect to x:

$$\frac{d}{dx}(\log(2x)) = \frac{d}{dx}\left(\frac{\ln(2x)}{\ln 10}\right) = \frac{2}{2x \ln 10} = \frac{1}{x \ln 10}$$

The rules of logarithms can often be used to simplify working.

Example Differentiate $y = \ln\sqrt{\dfrac{1+x}{1-x}}$ with respect to x.

Here $y = \ln\sqrt{1+x} - \ln\sqrt{1-x} = \tfrac{1}{2}\ln(1+x) - \tfrac{1}{2}\ln(1-x)$.

Therefore $\dfrac{dy}{dx} = \dfrac{1}{2}\left(\dfrac{1}{1+x} + \dfrac{1}{1-x}\right) = \dfrac{1}{1-x^2}$.

Example Differentiate $\dfrac{e^{3x-1}\sqrt{\sin x}}{x^2+1}$ with respect to x.

Let $y = \dfrac{e^{3x-1}\sqrt{\sin x}}{x^2+1}$. Then

$\ln y = \ln(e^{3x-1}) + \ln\sqrt{\sin x} - \ln(x^2+1) = 3x - 1 + \tfrac{1}{2}\ln \sin x - \ln(x^2+1)$.

Differentiating with respect to x gives:

$$\frac{1}{y}\frac{dy}{dx} = 3 + \frac{\cos x}{2\sin x} - \frac{2x}{x^2+1} \quad\Rightarrow\quad \frac{dy}{dx} = \frac{e^{3x-1}\sqrt{\sin x}}{x^2+1}\left(3 + \tfrac{1}{2}\cot x - \frac{2x}{x^2+1}\right).$$

At this point we have proved that $\dfrac{d}{dx}(x^n) = nx^{n-1}$ for

(a) all positive integer values of n (Section 10.4);
(b) all negative integer values of n (Section 12.3);
(c) all rational values of n (Section 12.4).

We will now prove that the rule holds when n is any real number.

Theorem If $y = x^n$ where n is any real number, then $\dfrac{dy}{dx} = nx^{n-1}$.

Proof $\dfrac{d}{dx}(x^n) = \dfrac{d}{dx}(e^{n\ln x})$

Chapter 17

$$= e^{n\ln x} \frac{d}{dx}(n\ln x)$$

$$= x^n \left(\frac{n}{x}\right)$$

$$= nx^{n-1}.$$

Exercise 17.2

1. Differentiate each of the following functions with respect to x:
   (a) $\ln 3x$;
   (b) $\ln(x+2)$;
   (c) $\ln(3x+5)$;
   (d) $\ln(2x^3+3)$;
   (e) $\ln(\sin 3x)$;
   (f) $\ln(\tan x)$;
   (g) $x^3 \ln 2x$;
   (h) $\ln(\ln x)$;
   (i) $\ln\sqrt{4x+3}$;
   (j) $\dfrac{\ln x}{x^2}$;
   (k) $x\ln\left(\dfrac{1}{x}\right)$;
   (l) $\ln(\cos^2 x)$.

2. Find the equation of the tangent to the curve $y = \ln(4x-11)$ at the point where $x = 3$.

3. The tangent to the curve $y = \ln(2x^2)$ at $x = e$ meets the x-axis at P and the y-axis at Q. Find the area of the triangle OPQ where O is the origin.

4. Given that $\ln 2 = p$ and $\ln 3 = q$, express each of the following in terms of p and/or q:
   (a) $\ln 18$;
   (b) $\ln 24e$;
   (c) $\log_2 3$;
   (d) $\log_6 e$;
   (e) 6 ;
   (f) 2.25.

5. Solve each of the following equations:
   (a) $e^{2x} - 5e^x + 6 = 0$;
   (b) $e^{3x} - 7e^{2x} + 6e^x = 0$;
   (c) $3e^{-x} - 2e^x = 5$;
   (d) $e^{2x} + e^{-2x} = 4$.

6. Find the position and nature of any turning point and sketch the graph of each of the following curves:
   (a) $y = \dfrac{x}{\ln x}$;
   (b) $y = (\ln x)^2$;
   (c) $y = x^2 - \ln(x^2)$;
   (d) $y = x - \ln x$;
   (e) $y = x^2 \ln x$;
   (f) $y = x \ln x$.

Higher Level

7. Find the derivatives of each of the following functions with respect to x:
   (a) $y = \log_6 x$;
   (b) $y = x^2 \log x$;
   (c) $y = \log_2 2x - \log_2(x^2)$.

8. Find the change in the sound level created by an increase in the intensity by a factor of 5.

9. By what factor must the intensity by multiplied to increase the sound level by 8 decibels?

10. Find $\dfrac{dy}{dx}$ in each of the following by differentiating $\ln y$:
    (a) $y = \dfrac{x^2 + 1}{e^x (2x+1)^2}$;
    (b) $y = \dfrac{\exp(x^2)}{\sqrt{2x^2 + 1}}$;
    (c) $y = \dfrac{e^{4x} \sin^2 3x}{\sqrt{5x+1}}$.

11. (a) Prove that $x^{\ln a} = a^{\ln x}$ for $x > 0$, $a > 0$.
    (b) Differentiate $x^{\ln 2}$ and $2^{\ln x}$ with respect to x and show that the derivatives are identical.

12. (a) Show that $y = \dfrac{1}{x} \log_a x$ has its maximum value at $x = e$ for each choice of a, provided $a > 1$.
    (b) Sketch the graph of $y = \dfrac{1}{x} \log_a x$ for $a > 1$ and show that the point of inflexion is at $x = e^{3/2}$.

17.3 Integration of Exponential Functions

Since the function e^x is its own derivative, we have $\boxed{\int e^x \, dx = e^x + c}$.

In a more general form we have $\boxed{\int e^{ax+b} \, dx = \dfrac{1}{a} e^{ax+b} + c}$.

Example Integrate each of the following functions with respect to x:

(a) $6e^x - 2e^{2x}$;
(b) $(e^{-x} - 1)^2$;
(c) $2 - 6e^{1-2x}$.

Chapter 17

(a) $\int \left(6e^x - 2e^{2x}\right) dx = 6e^x - \frac{2}{2}e^{2x} + c = 6e^x - e^{2x} + c$

(b) $\int \left(e^{-x} - 1\right)^2 dx = \int \left(e^{-2x} - 2e^{-x} + 1\right) dx = -\frac{1}{2}e^{-2x} + 2e^{-x} + x + c$.

(c) $\int \left(2 - 6e^{1-2x}\right) dx = 2x + 3e^{1-2x} + c$.

Example The gradient of a curve at the point (x, y) on it is $8x - 2e^{1-2x}$, and the curve passes through the point $(\frac{1}{2}, 5)$. Find the equation of the curve and the equation of the tangent at the point $(\frac{1}{2}, 5)$.

$\frac{dy}{dx} = 8x - 2e^{1-2x} \Rightarrow y = 4x^2 + e^{1-2x} + c$

Point $(\frac{1}{2}, 5)$ lies on the curve and so $5 = 1 + e^0 + c$ which gives $c = 3$.
The equation of the curve is $y = 4x^2 + e^{1-2x} + 3$.

The gradient of the tangent at $x = \frac{1}{2}$ is $8(\frac{1}{2}) - 2e^0 = 2$ and so the equation of this tangent is $2x - y = -4$.

Example The velocity at time t seconds of a particle moving in a straight line is given by $v(t) = (6 - 3e^t) \text{ m s}^{-1}$. Find the distance travelled by the particle in the first two seconds.

The particle comes to rest when $v(t) = 0$ or $e^t = 2$, i.e., when $t = \ln 2 \approx 0.693$.
The sign of $v(t)$ is: $+$ | $-$
                              $\ln 2$

Therefore the distance travelled in the first second is given by

$\int_0^{\ln 2} (6 - 3e^t) dt - \int_{\ln 2}^1 (6 - 3e^t) dt = \left[6t - 3e^t\right]_0^{\ln 2} - \left[6t - 3e^t\right]_{\ln 2}^1$

$= 6\ln 2 - 6 - 0 + 3 - 6 + 3e + 6\ln 2 - 6$
$= 3e + 12\ln 2 - 15$
$= 1.47 \text{ m}$.

Exercise 17.3

1. Integrate each of the following with respect to x:
   (a) $2e^x + 3e^{-x}$;
   (b) $4e^{2x} - 6e^{-x/2}$;
   (c) $4e^{0.2x} - e^{0.2}$;
   (d) $2e^{1-2x} - 3e^{1-3x}$;

(e) $\dfrac{2e^{2x}-1}{e^{2x}}$; (f) $\left(e^{-x}+1\right)^{2}$.

2. The gradient of the curve at the point (x, y) on it is $e^{x/2} - 2e^{-x/2}$. Find the equation of the curve given that it passes through the point $(0, 2)$.

3. The function $y = f(x)$ has a minimum value of 2.25. If the gradient of the curve at any point $P(x, y)$ on it is $2e^{2x} - e^{x}$, find the equation of the curve.

4. A particle moving along the x-axis has velocity $v(t)$ cm s^{-1} at time t seconds. Find the distance travelled by the particle in the first two seconds of motion for each of the following:
   (a) $v(t) = 5e^{-t/2}$; (b) $v(t) = 3\left(1 - e^{-t/20}\right)$; (c) $v(t) = 4 - 2e - 3e^{t/2}$.

5. Find the area enclosed by the curves $y = e^{2x}$, $y = e^{-x}$ and the line $x = 1$.

6. Find the area between the curve $y = 2^{-x}$ and the x-axis from $x = -2$ to $x = 0$.

7. A raindrop falls from rest and its acceleration at time t seconds is given by $a(t) = ke^{-t/2}$ cm s^{-2} where k is a positive constant.
   (a) Find the limiting velocity of the raindrop.
   (b) Calculate the time taken for the raindrop to reach half its limiting velocity.
   (c) Show that for the first 10 seconds the average velocity of the raindrop is approximately $1.6k$ cm s^{-1}.

Higher Level

17.4 Further Integration of Exponential Functions

We have seen in the previous section that $\int e^{ax+b}\,dx = \dfrac{1}{a}e^{ax+b} + c$.

In the more general form we have $\boxed{\int e^{f(x)} f'(x)\,dx = e^{f(x)} + c\,.}$

Also, if the base of the exponential function is not e, we must change the base to e and use the rules just stated.

Chapter 17

This gives $\int a^x \, dx = \int e^{x \ln a} \, dx = \dfrac{1}{\ln a} e^{x \ln a} + c = \dfrac{a^x}{\ln a} + c,$

or in the more general form

$\int a^{f(x)} f'(x) \, dx = \int e^{f(x) \ln a} f'(x) \, dx = \dfrac{1}{\ln a} e^{f(x) \ln a} + c = \dfrac{a^{f(x)}}{\ln a} + c.$

Example Integrate each of the following functions with respect to x:

(a) $6x e^{x^2}$;   (b) 10^{1-2x} ;   (c) $\dfrac{e^x - e^{-x}}{(e^x + e^{-x})^2}.$

(a) $\int 6x e^{x^2} \, dx = 3 \int (2x) e^{x^2} \, dx$ which is of the form $3 \int f'(x) e^{f(x)} \, dx$.

Therefore $\int 6x e^{x^2} \, dx = 3 e^{x^2} + c.$

(b) $\int 10^{1-2x} \, dx = 10 \int 0.01^x \, dx$

$= 10 \int e^{x \ln 0.01} \, dx$

$= \dfrac{10}{\ln 0.01} e^{x \ln 0.01} + c$ or $\dfrac{10^{1-2x}}{\ln 0.01} + c.$

(c) $\int \dfrac{e^x - e^{-x}}{(e^x + e^{-x})^2} \, dx = \int f'(x)(f(x))^{-2} \, dx$ where $f(x) = e^x + e^{-x}.$

Therefore $\int \dfrac{e^x - e^{-x}}{(e^x + e^{-x})^2} \, dx = c - \dfrac{1}{e^x + e^{-x}}.$

Exercise 17.4

1. Integrate each of the following with respect to x:
   (a) $e^x(e^x + 2)^3$;   (b) $(\cos x)\exp(\sin x)$;
   (c) $3x \exp(x^2 + 1)$;   (d) $\dfrac{\exp(\tan x)}{\cos^2 x}$;
   (e) $5^{2x} + 3^x$;   (f) $x 10^{x^2}$;

(g) $\dfrac{e^x}{\sqrt{e^x+1}}$; (h) $(e^x - e^{-x})(e^x + e^{-x})$.

2. The gradient of the curve $y = f(x)$ at any point (x, y) on the curve is given by $2^{0.4x}$. Find the equation of the curve if it passes through the origin.

3. Calculate the area enclosed by the y-axis and the curves $y = 2^x - 2$ and $y = 3(2^{-x})$.

4. Evaluate each of the following definite integrals:

   (a) $\displaystyle\int_{\ln 2}^{\ln 3} e^x \cos(e^x - 2)\, dx$;   (b) $\displaystyle\int_0^{\ln 4} \dfrac{2e^x}{(1+2e^x)^2}\, dx$;

   (c) $\displaystyle\int_0^{\pi/2} \sin x \exp(\cos x)\, dx$;   (d) $\displaystyle\int_0^1 (x^2+1)\exp(x^3+3x+1)\, dx$.

17.5 Integration of Functions of the Form $\dfrac{1}{ax+b}$

If $y = \ln x$, $x > 0$, we have $\dfrac{dy}{dx} = \dfrac{1}{x}$, and if $y = \ln(-x)$, $x < 0$, we have $\dfrac{dy}{dx} = \dfrac{-1}{-x} = \dfrac{1}{x}$.

Therefore $\displaystyle\int \dfrac{1}{x}\, dx = \begin{cases} \ln x + c, & x > 0 \\ \ln(-x) + c, & x < 0 \end{cases} = \ln|x| + c,\ x \neq 0.$

Example Integrate each of the following with respect to x:

(a) $\dfrac{1}{x-5}$ for $x > 5$;   (b) $\dfrac{2}{3-2x}$ for $x > \tfrac{3}{2}$.

(a) $\displaystyle\int \dfrac{1}{x-5}\, dx = \ln(x-5) + c$ since $x > 5$.

(b) $\displaystyle\int \dfrac{2}{3-2x}\, dx = -\displaystyle\int \dfrac{1}{x - \tfrac{3}{2}}\, dx = -\ln\left(x - \tfrac{3}{2}\right) + c$ since $x > \tfrac{3}{2}$.

or $\displaystyle\int \dfrac{2}{3-2x}\, dx = \dfrac{2}{-2}\ln|3-2x| + c = -\ln(2x-3) + c$ since $x > \tfrac{3}{2}$.

479

Chapter 17

In a more general form we have
$$\int \frac{k\,dx}{ax+b} = \frac{k}{a}\ln|ax+b|+c, \quad x \ne -\frac{b}{a}.$$

Example Integrate each of the following with respect to x:

(a) $\dfrac{3}{2x+3}$;   (b) $\dfrac{4}{2-3x}$;   (c) $\dfrac{2x+1}{x-1}$.

(a) $\displaystyle\int \frac{3}{2x+3}\,dx = \frac{3}{2}\ln|2x+3|+c$

(b) $\displaystyle\int \frac{4}{2-3x}\,dx = -\frac{4}{3}\ln|2-3x|+c$

(c) We must first express $\dfrac{2x+1}{x-1}$ in the form $a + \dfrac{b}{x-1}$ as follows:

$$\frac{2x+1}{x-1} = \frac{2(x-1)+3}{x-1} = 2 + \frac{3}{x-1}.$$

Therefore $\displaystyle\int \frac{2x+1}{x-1}\,dx = \int\left(2+\frac{3}{x-1}\right)dx = 2x + 3\ln|x-1|+c$.

Example Evaluate each of the following:

(a) $\displaystyle\int_1^2 \frac{3}{2+3x}\,dx$;   (b) $\displaystyle\int_{-2}^{-1} \frac{2x-1}{2x+1}\,dx$.

(a) $\displaystyle\int_1^2 \frac{3}{2+3x}\,dx = \Big[\ln|2+3x|\Big]_1^2 = \ln 8 - \ln 5 = \ln\left(\frac{8}{5}\right) = \ln 1.6 = 0.470.$

(b) $\displaystyle\int_{-2}^{-1} \frac{2x-1}{2x+1}\,dx = \int_{-2}^{-1}\left(1-\frac{2}{2x+1}\right)dx = \Big[x-\ln|2x+1|\Big]_{-2}^{-1}$

$= -1 - \ln 1 - (-2 - \ln 3)$
$= \ln 3 + 1 \quad (2.10).$

Exercise 17.5

1. Integrate each of the following with respect to x:

(a) $\dfrac{1}{x+3}$ for $x > -3$;   (b) $\dfrac{1}{3x-2}$ for $x > \frac{2}{3}$;

(c) $\dfrac{2}{1-x}$ for $x > 1$;   (d) $\dfrac{2}{3x-2}$ for $x < \frac{2}{3}$.

Exponential and Logarithmic Functions

2. Integrate each of the following with respect to x:

   (a) $\dfrac{2}{x+4}$;   (b) $\dfrac{4}{1-2x}$;   (c) $\dfrac{2}{4x+3}$;

   (d) $\dfrac{2x}{x-1}$;   (e) $\dfrac{2x}{x+1}$;   (f) $\dfrac{x^2+2x+3}{x+2}$.

3. Evaluate each of the following:

   (a) $\displaystyle\int_{2}^{4} \dfrac{2}{x}\,dx$;   (b) $\displaystyle\int_{2}^{5} \dfrac{1}{2x-1}\,dx$;   (c) $\displaystyle\int_{-5}^{-1} \dfrac{1}{1-3x}\,dx$;

   (d) $\displaystyle\int_{1}^{3} \dfrac{1}{x-5}\,dx$;   (e) $\displaystyle\int_{3}^{5} \dfrac{x}{x+1}\,dx$;   (f) $\displaystyle\int_{-1}^{1} \dfrac{3x-2}{x+2}\,dx$.

4. Find the area enclosed by the curve $y = \dfrac{1}{x-2}$ and the line $y = 7 - 2x$.

5. Express $\dfrac{1}{2x-1} - \dfrac{2}{3-x}$ as a single fraction and hence evaluate $\displaystyle\int_{1}^{2} \dfrac{1-x}{(2x-1)(3-x)}\,dx$.

Higher Level

17.6 Integration of Functions which can be Written in the Form $\dfrac{f'(x)}{f(x)}$

In the general form we have:

$$\int \dfrac{f'(x)}{f(x)}\,dx = \ln|f(x)| + c.$$

Example Integrate each of the following functions with respect to x:

   (a) $\dfrac{2x}{x^2+1}$;   (b) $\dfrac{\cos x}{\sin x - 3}$;   (c) $\dfrac{e^{2x}}{e^{2x}+1}$.

(a) $\displaystyle\int \dfrac{2x}{x^2+1}\,dx = \ln|x^2+1| + c = \ln(x^2+1) + c$ (as $x^2+1 > 0$ for all x).

(b) $\displaystyle\int \dfrac{\cos x}{\sin x - 3}\,dx = \ln|\sin x - 3| + c = \ln(3-\sin x) + c$.

(c) $\displaystyle\int \dfrac{e^{2x}}{e^{2x}+1}\,dx = \dfrac{1}{2}\ln|e^{2x}+1| + c = \ln\sqrt{e^{2x}+1} + c$.

481

Chapter 17

Example Evaluate each of the following:

(a) $\int_1^2 \frac{2x+1}{x^2+x+2} dx$; (b) $\int_{\pi/3}^{\pi/2} \frac{\sin x}{\cos x + 2} dx$;

(c) $\int_{\ln 2}^{\ln 3} \frac{e^x - e^{-x}}{e^x + e^{-x}} dx$.

(a) $\int_1^4 \frac{2x+1}{x^2+x+2} dx = \left[\ln|x^2 + x + 2| \right]_1^2$
$= \ln 8 - \ln 4$
$= \ln 2$.

(b) $\int_{\pi/3}^{\pi/2} \frac{\sin x}{\cos x + 2} dx = -\left[\ln|\cos x + 2| \right]_{\pi/3}^{\pi/2}$
$= -\ln 2 + \ln 2\tfrac{1}{2}$
$= \ln \tfrac{5}{4}$.

(c) $\int_{\ln 2}^{\ln 3} \frac{e^x - e^{-x}}{e^x + e^{-x}} dx = \left[\ln\left(e^x + e^{-x}\right) \right]_{\ln 2}^{\ln 3}$ $\left\{ \begin{array}{l} \frac{d}{dx}\left(e^x + e^{-x}\right) = e^x - e^{-x} \\ \text{and } e^x + e^{-x} > 0 \end{array} \right\}$

$= \ln \tfrac{10}{3} - \ln \tfrac{5}{2}$
$= \ln \tfrac{4}{3}$ $\left\{ \ln\left(\tfrac{10}{3} \div \tfrac{5}{2}\right) = \ln\left(\tfrac{10}{3} \times \tfrac{2}{5}\right) \right\}$.

We are now in a position to integrate the functions $\tan x$ and $\cot x$.

Example Integrate $\tan x$ with respect to x.

$\int \tan x \, dx = \int \frac{\sin x}{\cos x} dx \quad \left\{ -\frac{\sin x}{\cos x} = \frac{f'(x)}{f(x)} \text{ where } f(x) = \cos x \right\}$
$= -\ln|\cos x| + c$.

[The integral of $\cot x$ is found in a similar fashion with $f(x) = \sin x$ and $f'(x) = \cos x$ giving $\int \cot x \, dx = \int \frac{\cos x}{\sin x} dx = \ln|\sin x| + c$.]

482

Exercise 17.6

1. Integrate each of the following with respect to x:

   (a) $\dfrac{x+1}{2x^2+4x+1}$; (b) $\dfrac{\cos x}{2+\sin x}$; (c) $\cot 3x$;

   (d) $\dfrac{e^x}{1+e^x}$; (e) $\dfrac{1}{x\ln x}$; *(f) $\dfrac{1}{x+\sqrt{x}}$.

2. Evaluate each of the following:

   (a) $\displaystyle\int_0^{2\pi/3} \tan\tfrac{1}{2}x\, dx$; (b) $\displaystyle\int_{\pi/6}^{\pi/3} \dfrac{\sec^2 x}{\tan x}\, dx$;

   (c) $\displaystyle\int_0^{\ln 2} \dfrac{e^{-x}}{e^{-x}+1}\, dx$; (d) $\displaystyle\int_2^4 \dfrac{x+2}{x^2+4x-7}\, dx$;

   (e) $\displaystyle\int_0^{\pi/3} \dfrac{\sin x}{3-\cos x}\, dx$; (f) $\displaystyle\int_{-2}^{3} \dfrac{x+1}{x^2+2x+5}\, dx$.

3. Differentiate $y = \ln(\sec x + \tan x)$ and hence find $\displaystyle\int \sec x\, dx$.

4. Find the area between the graph of $y = \cot 2x$ and the x-axis from $x = \tfrac{1}{6}\pi$ to $x = \tfrac{1}{3}\pi$.

5. Evaluate $\displaystyle\int_e^{e^2} \dfrac{dx}{x\log x}$.

6. A function $y = f(x)$ for which $\dfrac{dy}{dx} = \dfrac{2x}{x^2+3}$ is such that $f(1) = \ln 2$. Find $f(x)$.

7. Solve the differential equation $\dfrac{dy}{dx} = \tan 2x \; \left(0 < x < \tfrac{1}{4}\pi\right)$ given that $y = 0$ when $x = \tfrac{1}{6}\pi$.

8. A particle moving in a straight line starts from rest and has an acceleration at time t seconds which is given by $a(t) = \left(\tan 2t + 3t^2 + 2t\right)$ m s^{-2}. Find the total distance travelled by the particle in the first second.

483

Chapter 17

Required Outcomes

After completing this chapter, a student should be able to:
- differentiate exponential functions expressed with base e.
- differentiate exponential functions of any positive base. **(HL)**
- integrate exponential functions expressed with base e.
- integrate exponential functions of any positive base. **(HL)**
- differentiate logarithmic functions expressed in base e.
- differentiate logarithmic functions of any positive base other than base 1. **(HL)**
- integrate any function of the form $\dfrac{k}{ax+b}$.
- integrate any function of the form $\dfrac{f'(x)}{f(x)}$. **(HL)**

PART 2
HIGHER LEVEL CORE

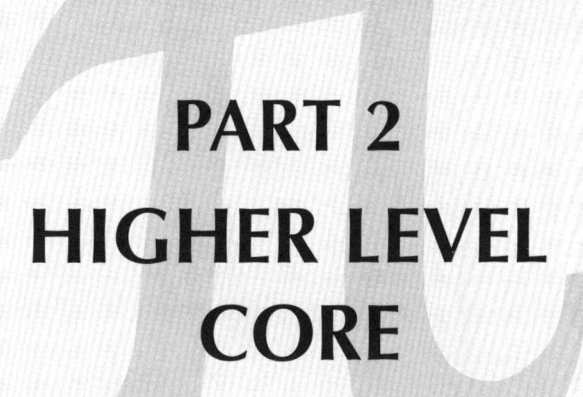

18 Mathematical Induction

18.1 The Principle of Mathematical Induction

We will be introducing several new or perhaps unfamiliar words in this chapter which are worthwhile defining, namely: proposition, deduction, induction, hypothesis and conjecture.

Proposition A formal statement which may or may not be true.

Deduction A statement arrived at after logical reasoning.

Induction The inference of a general law from particular instances.

Thus after the results of several cases are known, we may *induce* a general law which holds in all cases.

Hypothesis A supposition or proposed explanation made on the basis of limited evidence as a starting point for further investigation.

Conjecture The formation of an opinion on incomplete information – a guess.

We are often required to formulate a general rule concerning a number n after having established that the rule holds for several specific values of n. The process we employ to arrive at such a rule is called '*induction*'. However, no matter how many times we check this rule, we cannot be absolutely certain that the rule holds for *all* values of n.

For example, if we count the maximum number of regions inside a circle formed by joining each of n points on the circumference of the circle to each of the other $n-1$ points on the circumference [see diagrams at the top of the next page], we find that for $n = 1, 2, 3, 4$ and 5, the maximum number of regions is 1, 2, 4, 8 and 16 respectively. It seems that with the addition of a further point, the number of regions doubles and we may be led to induce that with n points this number is 2^{n-1}. However, if we take the trouble to check our result for 6 points we find that the number of regions is not $2^5 = 32$ but only 31! Thus our induced 'rule' is not a rule at all.

Chapter 18

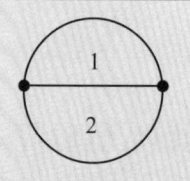

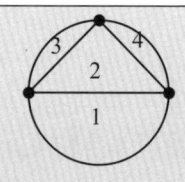

2 points – 2 regions   3 points – 4 regions   4 points – 8 regions

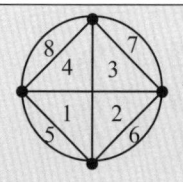

As a further more striking example, we may calculate terms of the sequence $\{n(0.999\,999)^n\}$ and find that for the first 999 999 terms, the sequence is increasing and then induce that the sequence is indeed an increasing sequence. But careful analysis shows that the 999 999th term is equal to the one-millionth term and after this the terms decrease in size, eventually approaching zero.

Consider a set of three pegs and n discs to be placed on these pegs. The discs are of differing sizes and in the starting position (see diagram below), all n discs are on one peg in increasing order of size with the largest disc on the bottom and the smallest disc at the top. The object of our exercise is to move the discs one at a time from one peg to another without placing a large disc on top of a smaller one and ending with all the discs on a different peg arranged as at the start from the largest at the bottom to the smallest at the top. How many moves are required?

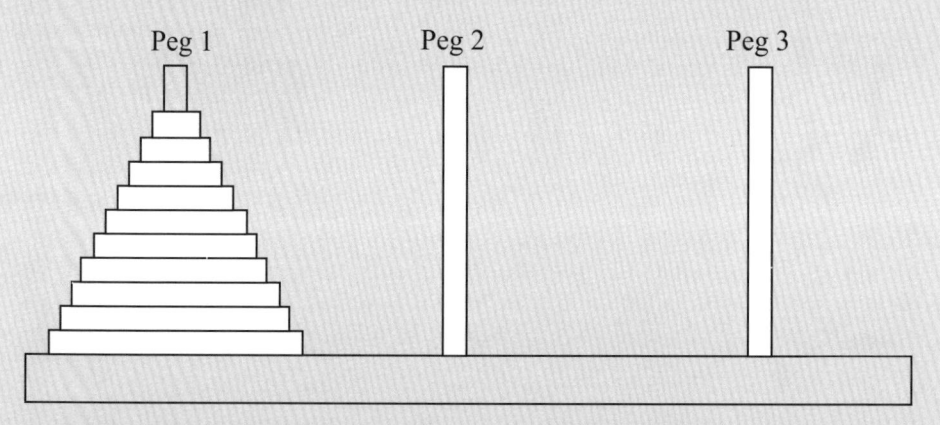

Peg 1   Peg 2   Peg 3

If we try to induce a rule by establishing the number of moves for $n = 1, 2, 3, 4$, etc., discs, we find that in these cases the number required is 1, 3, 7, 15, …. and we might be led to induce that the number of moves required is $2^n - 1$. But we have seen that this may not be true for all values of n.

Obviously it requires 1 move to move a single disc from one peg to another.

Say we start with 2 discs on the peg 1. We can move the top disc to peg 2, move the bottom disc to peg 3, and then move the small disc on peg 2 to peg 3 and we have accomplished our task in 3 moves.

Mathematical Induction

Instead of starting from the beginning with 3 discs, consider the following approach to our counting procedure. We know that we can move the two top discs to peg 2 in 3 moves. We now move the largest disc to peg 3 in 1 move and move the two discs on peg 2 to peg 3 in 3 moves. Our total number of moves is then 3 + 1 + 3 = 7. For four discs we know we can move the top three discs to peg 2 in 7 moves. We then move the largest disc to peg 3 in 1 move. Finally we move the three discs on peg 2 to peg 3 in 7 moves giving a total of 7+1+7 = 15 moves. All is well at this point.

By the method above we 'know' that with five discs we would need 15+1+15 = 31 moves. Again it is what we need.

Now if we could move n discs in $2^n - 1$ moves, how many moves will it require to move $n + 1$ discs? By the method described above we could move the top n discs from peg 1 to peg 2 in $2^n - 1$ moves, the largest disc from peg 1 to peg 3 in 1 move, and then move the n discs on peg 2 to peg 3 in $2^n - 1$ moves. The total number of moves is then $2^n - 1 + 1 + 2^n - 1 = 2(2^n) - 1 = 2^{n+1} - 1$ moves.

Thus, if our 'rule' holds for n discs, it also holds for $n + 1$ discs.

But we know the rule holds for 1 disc and so it holds for 2 discs. We know it holds for 2 discs, so it holds for 3 discs. It holds for 3 discs, so it holds for 4 discs, etc.

We have thus established that the rule holds for as many discs as we care to use. We have gone further than simply inducing a rule, we have 'deduced' one that holds for ***all*** values of n.

This is the essence of a proof by ***mathematical induction***. Despite its name, the technique of mathematical induction is a deductive method, not an inductive one.

> *The Principle of Mathematical Induction*
>
> If a proposition, P(n), concerning integers n is true when $n = m$, and if the proposition P(k+1) is true whenever P(k) is true where k is an integer which is greater than or equal to m, then the proposition P(n) is true for all integers $n \geq m$.

Chapter 18

Example Prove that $1^2 + 2^2 + 3^2 + \cdots + n^2 = \frac{1}{6}n(n+1)(2n+1)$.

Step 1: Let P(n) be the proposition: $1^2 + 2^2 + 3^2 + \cdots + n^2 = \frac{1}{6}n(n+1)(2n+1)$.

Step 2: P(1) is true since $1^2 = \frac{1}{6}(1)(1+1)(2\times 1+1) = \frac{1}{6}(1)(2)(3) = 1$

Step 3: Assume that P(k) is true for some integer $k \geq 1$.
i.e. $1^2 + 2^2 + 3^2 + \cdots + k^2 = \frac{1}{6}k(k+1)(2k+1)$.

Step 4: Now, $1^2 + 2^2 + 3^2 + \cdots + k^2 + (k+1)^2$
$= \frac{1}{6}k(k+1)(2k+1) + (k+1)^2$ [by hypothesis, step 3]
$= \frac{1}{6}(k+1)\{k(2k+1) + 6(k+1)\}$
$= \frac{1}{6}(k+1)\{2k^2 + 7k + 6\}$
$= \frac{1}{6}(k+1)(k+2)(2k+3)$
$= \frac{1}{6}(k+1)([k+1]+1)(2[k+1]+1)$.

Thus P(k + 1) is true whenever P(k) is true or P(k) $\Rightarrow$ P(k + 1).

Step 5: Therefore by the principle of mathematical induction P(n) is true for all integers $n \geq 1$.

Example Prove that $3^{3n} + 1$ is divisible by 7 for all positive *odd* integers n.

Let P(n) be the proposition: $3^{3n} + 1$ is divisible by 7.
P(1) is true since $3^3 + 1 = 28$ is divisible by 7.
Assume P(k) is true for some integer $k \geq 1$.
i.e. $3^{3k} + 1$ is divisible by 7 or $3^{3k} + 1 = 7p$, $p \in \mathbb{Z}$.
Then $3^{3(k+2)} + 1 = 3^6 3^{3k} + 1$
$= 729(7p - 1) + 1$ [by hypothesis]
$= 729(7p) - 728$
$= 7(729p - 104)$
which is divisible by 7 since $(729p - 104) \in \mathbb{Z}$.

Thus P(k) $\Rightarrow$ P(k + 2) and so P(n) is true for n = 1, 3, 5, 7, ... i.e., for all positive *odd* integers n.

Example   Prove that $n! > 2^n$ for all integers $n \geq 4$.

Let P(n) be the proposition: $n! > 2^n$.
Then P(4) is true since $4! = 24$, $2^4 = 16$ and $24 > 16$.

Assume that P(k) is true for some integer $k \geq 4$.
i.e. $k! > 2^k$.
Then $(k+1)! = (k+1)k!$
$> (k+1)2^k$
$> 2 \times 2^k$ since $k + 1 > 2$ when $k \geq 4$
$= 2^{k+1}$.

Thus P(k) $\Rightarrow$ P($k+1$) and so P(n) is true for all integers $n \geq 4$.

Exercise 18.1

1. Use the principle of mathematical induction to prove that for all integers $n \geq 1$, $1 + 2 + 3 + \cdots + n = \frac{1}{2}n(n+1)$.

2. The sequence $\{a_n\}$ is defined recursively by $a_1 = 1$, $a_2 = 1$ and for all $n \geq 1$, $a_{n+2} = a_{n+1} + a_n$. Use the principle of mathematical induction to prove that
    (a) $a_1 + a_2 + a_3 + \cdots + a_n = a_{n+2} - 1$;   (b) $a_1 + a_3 + a_5 + \cdots + a_{2n-1} = a_{2n}$.

3. If $A = \begin{pmatrix} \cos \alpha & -\sin \alpha \\ \sin \alpha & \cos \alpha \end{pmatrix}$, prove by mathematical induction that for all integers $n \geq 1$, $A^n = \begin{pmatrix} \cos n\alpha & -\sin n\alpha \\ \sin n\alpha & \cos n\alpha \end{pmatrix}$.

4. Use the principle of mathematical induction to prove that each of the following statements is true for all integers $n \geq 1$:
    (a) $1 + 3 + 5 + \cdots + (2n-1) = n^2$;
    (b) $1 + 2 + 2^2 + 2^3 + \cdots + 2^{n-1} = 2^n - 1$;
    (c) $1 + 4 + 7 + \cdots + (3n-2) = \frac{1}{2}n(3n-1)$;
    (d) $1^2 + 3^2 + 5^2 + \cdots + (2n-1)^2 = \frac{1}{3}n(4n^2 - 1)$;
    (e) $\dfrac{1}{(1)(2)} + \dfrac{1}{(2)(3)} + \dfrac{1}{(3)(4)} + \cdots + \dfrac{1}{n(n+1)} = \dfrac{n}{n+1}$;

(f) $1+\frac{1}{2}+\frac{1}{4}+\cdots+\frac{1}{2^{n-1}}=2-\frac{1}{2^{n-1}}$;

(g) $a+(a+d)+(a+2d)+\cdots+(a+[n-1]d)=\frac{1}{2}n(2a+[n-1]d)$;

(h) $a+ar+ar^2+\cdots+ar^{n-1}=\frac{a(r^n-1)}{r-1}$, $r \neq 1$;

(i) $1^3+2^3+3^3+\cdots+n^3=\left[\frac{1}{2}n(n+1)\right]^2$;

(j) $\frac{1}{(1)(3)}+\frac{1}{(3)(5)}+\frac{1}{(5)(7)}+\cdots+\frac{1}{(2n-1)(2n+1)}=\frac{n}{2n+1}$;

(k) $\frac{1^2}{(1)(3)}+\frac{2^2}{(3)(5)}+\frac{3^2}{(5)(7)}+\cdots+\frac{n^2}{(2n-1)(2n+1)}=\frac{n(n+1)}{2(2n+1)}$;

(l) $(1^2)(2)+(2^2)(2^2)+(3^2)(2^3)+\cdots+(n^2)(2^n)=(n^2-2n+3)(2^{n+1})-6$;

(m) $6+24+60+\cdots+n(n+1)(n+2)=\frac{1}{4}n(n+1)(n+2)(n+3)$;

(n) $1^5+2^5+3^5+\cdots+n^5=\frac{1}{12}n^2(n+1)^2(2n^2+2n-1)$.

5. Use the principle of mathematical induction to prove that $\frac{d}{dx}(x^n)=nx^{n-1}$ for all positive integers n. [You may assume that $\frac{d}{dx}(x)=1$.]

6. Use the principle of mathematical induction to prove that
$\frac{d^n}{dx^n}(xe^x)=(x+n)e^x$ for all $n \in \mathbb{Z}^+$.

7. Use the principle of mathematical induction to prove that
$\frac{d^n}{dx^n}(\ln x)=(-1)^n\frac{(n-1)!}{x^n}$ for all $n \in \mathbb{Z}^+$.

8. Use the principle of mathematical induction to prove that each of the following propositions is true for all integers $n \geq 1$:
   (a) n^2+n is divisible by 2 ;
   (b) $n(n+1)(n+2)$ is divisible by 3 ;
   (c) 4^n-1 is divisible by 3 ;
   (d) $7^{2n}-1$ is divisible by 48 ;
   (e) $3^{3n}-1$ is divisible by 13 ;
   (f) $4^n+(-1)^{n-1}$ is divisible by 5 ;
   (g) $3^{3n}+2^{n+2}$ is divisible by 5 ;

- (h) $3^{4n+2} + 2^{6n+3}$ is divisible by 17 ;
- (i) $2^{4n+3} + 3^{3n+1}$ is divisible by 11.

9. (a) Simplify $a(a^n - b^n) + b^n(a - b)$.
   (b) Prove that if a and b are distinct integers, then $a^n - b^n$ is divisible by $a - b$ for all positive integers n.

10. (a) Simplify $a^2(a^n + b^n) - b^n(a^2 - b^2)$.
    (b) Prove that if a and b are distinct integers, then $a^n + b^n$ is divisible by $a + b$ for all *odd* positive integers n.

11. Use the principle of mathematical induction to prove that the following propositions are true for all positive integers n:
    (a) $2^n \geq n + 1$;
    (b) $3^n \geq 2n + 1$;
    (c) $2n^3 - 3n^2 + n + 1 > 0$;
    (d) $2n^3 - 4n^2 + 5n - 3 \geq 0$.

12. (a) Prove that $4x^2 \geq (x+1)^2$ for $x \geq 1$.
    (b) Use the principle of mathematical induction to prove that $4^n \geq 3n^2$ for all positive integers n.

13. (a) Find the positive values of x for which $2x^2 > (x+1)^2$.
    (b) Use the principle of mathematical induction to prove that $4^n > n^4$ for all positive integers $n \geq 5$.

14. Prove that $(1+a)^n \geq 1 + na$ given that $a > -1$, for all positive integers n. (Bernoulli's inequality).

15. Prove that for all integers $n \geq 2$, $\dfrac{1}{n+1} + \dfrac{1}{n+2} + \dfrac{1}{n+3} + \cdots + \dfrac{1}{2n} > \dfrac{13}{24}$.

16. Prove that $\dfrac{1}{1+x} + \dfrac{2}{1+x^2} + \dfrac{4}{1+x^4} + \cdots + \dfrac{2^n}{1+x^{(2^n)}} = \dfrac{1}{x-1} + \dfrac{2^{n+1}}{1-x^{(2^{n+1})}}$ for all $x \neq \pm 1$.

493

Chapter 18

17. The sequence $\{a_n\}$ is defined by $a_1 = 1, a_2 = 1$ and $a_{n+1} = a_n + a_{n-1}$ for all integers $n \geq 2$. Let Q be the matrix $\begin{pmatrix} 0 & 1 \\ 1 & 1 \end{pmatrix}$.

    Prove that $Q^n = \begin{pmatrix} a_{n-1} & a_n \\ a_n & a_{n+1} \end{pmatrix}$ for all integers $n \geq 2$.

18. (a) If a and b are positive numbers, prove that $\dfrac{a+b}{2} \geq \sqrt{ab}$.

    (b) If a_1 and a_2 are positive numbers, prove that $\dfrac{a_1}{a_2} + \dfrac{a_2}{a_1} \geq 2$.

    (c) If $a_1, a_2, a_3, \cdots, a_n$ are positive numbers, prove that
    $$(a_1 + a_2 + a_3 + \cdots + a_n)\left(\dfrac{1}{a_1} + \dfrac{1}{a_2} + \dfrac{1}{a_3} + \cdots + \dfrac{1}{a_n}\right) \geq n^2.$$

19. Prove that $\displaystyle\sum_{k=1}^{n} ka^k = \dfrac{a}{(1-a)^2}\{na^{n+1} - (n+1)a^n + 1\}$ for $a \neq 1$.

    What is the correct formula when $a = 1$?

20. Consider the sequence $\{a_n\}$ where $a_1 = 5$ and $a_{n+1} = 1 + 2a_n$ for $n \geq 1$. Use the principle of mathematical induction to prove that $a_n = 3(2^n) - 1$.

*21. Prove by mathematical induction that
$$n + 2(n-1) + 3(n-2) + \cdots + n = \tfrac{1}{6}n(n+1)(n+2).$$

*22. For any n distinct points on a circle, prove that the straight lines joining all pairs of points divide the region inside the circle into a maximum of $1 + \tfrac{1}{24}n(n-1)(n^2 - 5n + 18)$ regions.

*23. (a) If the proposition P(n) concerning positive integers n is such that P(n + 2) is true whenever both P(n + 1) and P(n) are true, what must also be done to prove that P(n) is true for all n?

    (b) Consider the sequence $\{a_n\}$ defined by $a_1 = 1, a_2 = 1$ and $a_{n+1} = a_n + a_{n-1}$ for all integers $n \geq 2$. Prove that $a_n = \dfrac{1}{\sqrt{5}}\{\alpha^n - \beta^n\}$ where α and β are the roots of the equation $x^2 - x - 1 = 0$, $\alpha > \beta$.

18.2 Making and Proving Conjectures

Example In the series
$$S_n = (1)(2)(3) + (2)(3)(4) + (3)(4)(5) + \cdots + n(n+1)(n+2)$$
the partial sums S_1, S_2, S_3 have the following forms:

$S_1 = (1)(2)(3) \qquad\qquad = \dfrac{(1)(2)(3)(4)}{4}$

$S_2 = (1)(2)(3) + (2)(3)(4) \qquad\qquad = \dfrac{(2)(3)(4)(5)}{4}$

$S_3 = (1)(2)(3) + (2)(3)(4) + (3)(4)(5) \qquad = \dfrac{(3)(4)(5)(6)}{4}.$

Guess a formula for S_n and prove your guess.

Guess: $S_n = \tfrac{1}{4}n(n+1)(n+2)(n+3)$.

Let P(n) be the proposition: $S_n = \tfrac{1}{4}n(n+1)(n+2)(n+3)$.

P(1) is true by conjecture.
Assume P(k) is true for some integer $k \geq 1$. That is
$(1)(2)(3) + (2)(3)(4) + \cdots + k(k+1)(k+2) = \tfrac{1}{4}k(k+1)(k+2)(k+3)$.
Then $(1)(2)(3) + (2)(3)(4) + \cdots + k(k+1)(k+2) + (k+1)(k+2)(k+3)$
$= \tfrac{1}{4}k(k+1)(k+2)(k+3) + (k+1)(k+2)(k+3)$
$= \tfrac{1}{4}(k+1)(k+2)(k+3)(k+4)$.
Therefore P(k) $\Rightarrow$ P(k + 1) and so P(n) is true for all integers $n \geq 1$.

Exercise 18.2

1. What is the greatest positive integer divisor of $3^{2n} - 1$ if n is a positive integer? Use the principle of mathematical induction to prove your answer.

2. Let $S_n = \dfrac{1}{2!} + \dfrac{2}{3!} + \dfrac{3}{4!} + \cdots + \dfrac{n}{(n+1)!}$.
   (a) Find S_1, S_2, S_3 and S_4.
   (b) Guess a formula for S_n.
   (c) Prove your guess is correct.

3. How many diagonals has a convex n-gon? Prove your conjecture.

Chapter 18

4. Consider the matrix $A = \begin{pmatrix} 0 & 1 \\ -1 & 2 \end{pmatrix}$.

   (a) Find the matrices A^2, A^3 and A^4.
   (b) Guess a formula for A^n where n is a positive integer.
   (c) Prove your conjecture using the principle of mathematical induction.

5. We define S_n for all positive integers $n \geq 1$ as the sum $S_n = \sum_{r=1}^{n} r(r!)$.

   (a) Find S_1, S_2, S_3, S_4 and S_5.
   (b) Guess a formula for S_n.
   (c) Prove your guess is correct for all integers $n \geq 1$.

6. If $(2 - \sqrt{3})^n = a_n - b_n\sqrt{3}$ for all positive integers n, where a_n and b_n are integers, show that $a_{n+1} = 2a_n + 3b_n$ and $b_{n+1} = a_n + 2b_n$.
   Calculate $a_n^2 - 3b_n^2$ for $n = 1, 2$ and 3.
   Hence guess a formula for $a_n^2 - 3b_n^2$ and prove your guess is true for all positive integers n.

7. The sum of n terms of the sequence $\left\{ \dfrac{1}{(1)(3)(5)}, \dfrac{1}{(3)(5)(7)}, \dfrac{1}{(5)(7)(9)}, \ldots \right\}$
   is denoted by S_n. Find the values of x_1, x_2, x_3, x_4 for which $S_n = \frac{1}{12} - \frac{1}{4}x_n$. Hence guess a formula for S_n and prove this formula is correct using the principle of mathematical induction.

Required Outcomes

After completing this chapter, a student should be able to:
- define the terms 'proposition', 'deduction', 'induction', 'hypothesis' and 'conjecture'.
- formulate a conjecture on the basis of the results of a few substitution instances.
- prove your conjecture using the principle of mathematical induction.

19 Polynomials

19.1 Addition, Multiplication and the Division Process

Definition A ***polynomial*** is a sum of a finite number of terms of the form kx^n where k is any number, and n is a non-negative integer.

Some examples of polynomials are: $x^2 + 2x - 4$, $3x - 1$, 5 ($= 5x^0$) and $\pi x^5 - 2\sqrt{3}x^3$.

Some examples of non-polynomials are: x^{-1}, $\sqrt{x}$, $\sin x$ and $\dfrac{3}{2x+1}$.

If we write $P(x) = a_n x^n + a_{n-1} x^{n-1} + \cdots + a_2 x^2 + a_1 x + a_0$, where $a_0, a_1, a_2, \cdots, a_n$ are all constants, then $P(x)$ is a polynomial.

If $a_n \neq 0$, n is called the ***degree*** of $P(x)$ ($n = \deg P$); each expression of the form $a_k x^k$ is called a ***term***; a_k is called the ***coefficient*** of the term in x^k; the coefficient a_n of the highest power of x is called the ***leading coefficient***; a_0 is called the ***constant term***.

Example $P(x) = 6x^2 - 2x^3 + 5$ is a polynomial of degree 3. The leading coefficient is –2; the coefficient of the term in x is 0; the constant term is 5.

The Zero Polynomial

If $P(x)$ is a polynomial for which $P(x) = 0$ *for all x*, then $P(x)$ is called the ***zero polynomial***.

The zero polynomial has *no degree*.

Equality

The polynomials $P(x) = a_n x^n + a_{n-1} x^{n-1} + \cdots + a_2 x^2 + a_1 x + a_0$, $a_n \neq 0$,

and $Q(x) = b_m x^m + b_{m-1} x^{m-1} + \cdots + b_2 x^2 + b_1 x + b_0$, $b_m \neq 0$,

are ***equal*** iff $m = n$ and $a_i = b_i$ for $i = 0, 1, 2, \ldots, n$.

Chapter 19

Thus two polynomials are equal iff they have the same degree and all corresponding coefficients are equal.

If two polynomials $P(x)$, $Q(x)$ are equal we write $P(x) \equiv Q(x)$. We use the symbol '$\equiv$' to distinguish equality of two polynomials from a simple equation connecting the polynomials. In the first case, $P(x)$ and $Q(x)$ take the same values for *all values of x*. In the second case, $P(x)$ and $Q(x)$ take the same values for a finite number of values of x.

The polynomials $P(x) = x^3$ and $Q(x) = x^3$ are equal polynomials, $P(x) \equiv Q(x)$, but the polynomials $P(x) = x^3$ and $Q(x) = x^2$ are not equal but $P(x) = Q(x)$ for $x = 0$ and $x = 1$ only.

Addition

Addition of polynomials simply involves "collecting like terms", i.e., terms of the same degree.

Example   Find the sum of the polynomials $P(x) = 3x^2 + x^3 + 5$ and $Q(x) = 3x - 3x^2 + 2$.

$$P(x) + Q(x) = x^3 + (3-3)x^2 + 3x + (5+2) = x^3 + 3x + 7.$$

Multiplication

Polynomials are multiplied in the usual way – each term of the first polynomial is multiplied by each term of the second polynomial and like terms are added.

Example   Find the product of the polynomials given in the previous example.

$$\begin{aligned} P(x)Q(x) &= (3x^2 + x^3 + 5)(3x - 3x^2 + 2) \\ &= 9x^3 - 9x^4 + 6x^2 + 3x^4 - 3x^5 + 2x^3 + 15x - 15x^2 + 10 \\ &= -3x^5 - 6x^4 + 11x^3 - 9x^2 + 15x + 10. \end{aligned}$$

Example   Find the values of the real numbers a and b if
$$a(x-2) + b(x-1) \equiv 5x - 8.$$

Method 1

Polynomials $a(x-2)+b(x-1) \equiv (a+b)x-(2a+b)$ and $5x-8$ are equal iff their respective coefficients are equal.
Thus, $a+b=5$,
and $2a+b=8$.
Solving these equations gives $a=3$, $b=2$.

Method 2

Since the polynomials $a(x-2)+b(x-1)$ and $5x-8$ are equal for all values of x, they are equal for any two particular values of x.
Putting $x=1$ in each gives: $-a=5-8$ or $a=3$.
Putting $x=2$ in each gives: $b=10-8$ or $b=2$.

Division

We can divide one polynomial by another in a fashion similar to our method for long division of decimal numbers.

Example Divide $2x^4 - 3x^2 + x + 2$ by $x^2 - 2x - 1$.

```
                           2x²  +   4x  +  7
         _____
x² - 2x - 1 | 2x⁴       -   3x²  +   x  +  2
              2x⁴  - 4x³ -  2x²
              _____
                     4x³ -   x²  +   x
                     4x³ -  8x²  -  4x
                     _____
                            7x²  +  5x  +  2
                            7x²  - 14x  -  7
                            _____
                                    19x  +  9
```

Thus $\dfrac{2x^4 - 3x^2 + x + 2}{x^2 - 2x - 1} \equiv 2x^2 + 4x + 7 + \dfrac{19x+9}{x^2 - 2x - 1}$

or $2x^4 - 3x^2 + x + 2 \equiv (2x^2 + 4x + 7)(x^2 - 2x - 1) + 19x + 9$.

Since all terms in the same power of x fall into the same column, we can present the work in the example above in the following way:

Chapter 19

```
                    2    4    7
   1  -2  -1 | 2  0  -3   1    2
               2  -4  -2
               ─────────
                  4  -1    1
                  4  -8   -4
                  ─────────
                     7    5    2
                     7  -14   -7
                     ─────────
                         19    9
```

This is known as the *method of detached coefficients*.

The Division Process

Consider the polynomials $P(x)$ and $D(x)$ with deg $D \le$ deg P. When $P(x)$ is divided by $D(x)$ there exists a *unique* quotient $Q(x)$ and a *unique* remainder $R(x)$ such that

$$P(x) \equiv Q(x)D(x) + R(x)$$

provided deg $R(x) <$ deg $D(x)$.

Example   Find the quotient and remainder when $P(x) = 3x^5 - 4x^3 - 4x^2 + 1$ is divided by $D(x) = x^3 - 2x^2 + 3$.

```
                             3    6    8
  1  -2  0  3 | 3   0  -4  -4   0    1
                3  -6   0   9
                ────────────
                    6  -4 -13    0
                    6 -12   0   18
                    ────────────
                        8 -13  -18    1
                        8 -16    0   24
                        ────────────
                            3  -18  -23
```

The quotient is $Q(x) = 3x^2 + 6x + 8$; the remainder is $R(x) = 3x^2 - 18x - 23$.

Exercise 19.1

1. If $P(x) = 3x^2 - 2x^3 + 1$ and $Q(x) = 3x^2 - 2x - 2$ find $P(x) + Q(x)$ and $P(x)Q(x)$.

2. Give an example of two polynomials $P(x)$ and $Q(x)$ such that deg $P = m$, deg $Q = n$ where $m \ge n$ and
   (a)   deg$(P + Q) = m$;   (b)   deg$(P + Q) < m$.

Polynomials

3. Polynomials $P(x)$ and $Q(x)$ have degrees m and n respectively. What can you say about the degree of
   (a) $P(x)Q(x)$; (b) $P(x) + Q(x)$?

4. Write down the degree, leading coefficient and constant term in each of the following polynomials:
   (a) $5x^2 - 3x^3 - 4$;
   (b) $6x - 2x^4$;
   (c) $6 - 3x - 5x^4$;
   (d) $(3x - 2x^3 - 5)(2x^2 + 4x - 3)$.

5. Multiply out each of the following:
   (a) $(2x^3 + 3x - 4)(3x + 5)$; (b) $(2x^3 - 3x^2 - 2x + 1)(4x^2 + 6x - 3)$;
   (c) $(2x^2 - 3x + 1)^2$; (d) $(x^2 - x - 2)^3$.

6. Find the coefficients of the terms in x^2 and x^3 in the expansion of $(x^3 - x^2 + 2x - 3)^2$.

7. (a) Find the coefficient of x in the expansion of
   $(1 + x)(1 + 2x)(1 + 3x)(1 + 4x)$.

   (b) Find the coefficient of x in the expansion of
   $(1 + x)(1 + 2x)(1 + 3x) \ldots (1 + nx)$.

8. In each of the following find the quotient and remainder when the first polynomial is divided by the second polynomial:
   (a) $2x - 5$, $x + 3$;
   (b) $2x^2 + 4x - 3$, $2x - 1$;
   (c) $2x^3 - 2x^2 + 2x - 1$, $x - 2$;
   (d) $x^4 - 3x^2 + 4x + 5$, $x^2 - 2x + 1$;
   (e) $3x^4 - 2x^2 - 3$, $x^3 + 2x$;
   (f) $3x^5 - 2x^3 + x^2$, $(x - 1)(x + 2)$;
   (g) $x^5 - 1$, $(x - 1)(x + 1)$;
   (h) $2x^4 - 3x^3 + 2$, $1 - 2x - x^2$.

9. Given that the following equalities are true for all values of x, find the values of the real numbers a and b in each case:
   (a) $2x^3 + 3x^2 - x - 1 = (x + 2)(2x^2 - x + a) + b$;
   (b) $x^4 - x^3 + 2x^2 + 3 = (x^2 - 2x - 5)(x^2 + x + 9) + ax + b$;
   (c) $3x^3 - 2x^2 + 1 = (x^2 + 2)(ax + b) + 5 - 6x$;
   (d) $2x^4 - 9x^2 + 12x + 5 = (2x^2 - 4x + 3)(x^2 + 2x - 2) + ax + b$.

10. Find the values of the real numbers a and b if, for all real x,
    (a) $7x + 1 = a(x + 1) + b(2x - 1)$;
    (b) $8x + 6 = a(2x + 1) + b(2x - 1)$;

501

(c) $a(2x-3)+b(3x+1)=11$;

(d) $x^2+x-6=a(x-1)(x-2)+b(x-2)(x-3)$.

11. Find, if possible, real numbers a and b for which the following pairs of polynomials are equal for all x:

(a) x^2+x+4, $a(x-1)(x+1)+b(x+2)(x-3)$;

(b) $x^2+14x-4$, $ax(x+2)+b(x-1)(x-3)$;

(c) $4x^2-6x+8$, $a(x-1)(x-2)+b(x+1)(x+2)$;

(d) x^2-3x-8, $a(x^2+1)+b(x+1)(x-2)$.

12. Find the values of the real numbers a, b and c such that the following pairs of polynomials are equal for all x:

(a) $10x-14$, $ax(x-1)+b(x-1)(x-2)+c(x-2)(x-3)$;

(b) $5-2x^2$, $a(x+1)(x+2)+b(x+2)(x+3)+c(x+3)(x+4)$;

(c) 16, $ax(x-3)+(bx+c)(x+1)$;

(d) $3x^2+3x+7$, $a(x-1)(x+1)+b(x+2)(x-2)+cx(x+3)$.

19.2 The Remainder and Factor Theorems

The Remainder Theorem

> When a *polynomial P(x)* is divided by $x-a$ until the remainder R is independent of x, then $R=P(a)$.

Proof Let $Q(x)$ be the quotient when $P(x)$ is divided by $x-a$. Then by the division process

$$P(x) \equiv (x-a)Q(x) + R$$

Putting $x=a$ gives $P(a) = (a-a)Q(a) + R$.
Thus $R = P(a)$.

The Factor Theorem

> The polynomial $P(x)$ has $x-a$ as a factor iff $P(a) = 0$.

Proof (1) If $P(a) = 0$ then when $P(x)$ is divided by $x-a$, the remainder is zero (remainder theorem).
Therefore $x-a$ is a factor of $P(x)$.

(2) If $x - a$ is a factor of $P(x)$ then there exists a polynomial $Q(x)$ such that $P(x) = (x - a)Q(x)$ for all x.
Putting $x = a$ gives $P(a) = 0$.

Corollary If $P(x)$ is a polynomial, $P(x) - P(a)$ is divisible by $x - a$.

Proof Let $Q(x)$ be the quotient when $P(x)$ is divided by $x - a$.
Then by the division process and the remainder theorem,
$P(x) = (x - a)Q(x) + P(a)$ for all x.
Thus $P(x) - P(a) = (x - a)Q(x)$ for all x.
Hence $P(x) - P(a)$ is divisible by $x - a$.

Note: The following statements concerning a polynomial $P(x)$ are equivalent:
(1) $P(a) = 0$;
(2) $x - a$ is a *factor* of $P(x)$;
(3) $x = a$ is a *root* of the equation $P(x) = 0$;
(4) a is a *zero* of $P(x)$.

Example Find the value of the real number a if $P(x) = x^3 + ax^2 + 3x - 1$ leaves a remainder of 1 when divided by $x - 2$.

When $P(x)$ is divided by $x - 2$ the remainder is $P(2)$.
Therefore $P(2) = 1$.
Thus $8 + 4a + 6 - 1 = 1$ giving $a = -3$.

Example When a polynomial $P(x)$ is divided by $(x - 1)(x - 2)$ the remainder is $2x + 1$. Find the remainders when $P(x)$ is divided separately by $x - 1$ and $x - 2$.

Let $Q(x)$ be the quotient when $P(x)$ is divided by $(x - 1)(x - 2)$.
Then $P(x) \equiv (x - 1)(x - 2)Q(x) + 2x + 1$.
Now $P(1) = (1 - 1)(1 - 2)Q(1) + 2 + 1 = 3$ and
$P(2) = (2 - 1)(2 - 2)Q(2) + 4 + 1 = 5$.
Therefore the required remainders are 3, 5 respectively.

Example When a polynomial $P(x)$ is divided by $x + 1$ the remainder is 4. When $P(x)$ is divided by $x - 2$ the remainder is 1. Find the remainder when $P(x)$ is divided by $(x + 1)(x - 2)$.

Let $Q(x)$ be the quotient and $ax + b$ the remainder when $P(x)$ is divided by $(x + 1)(x - 2)$.
Then $P(x) \equiv (x + 1)(x - 2)Q(x) + ax + b$.
But $P(-1) = 4$ and $P(2) = 1$ giving $-a + b = 4$ and $2a + b = 1$.
Solving these equations gives $a = -1$, $b = 3$.
Hence the required remainder is $3 - x$.

Chapter 19

Example Prove that $4^{3n}-1$ is divisible by 7 for all positive integers n.

Let $f(x) = x^n$ where n is a positive integer.
Then $f(x)$ is a polynomial.
Therefore $f(x) - f(a)$ is divisible by $x - a$ (corollary).
i.e. $f(64) - f(1)$ is divisible by $64 - 1 = 63$.
i.e. $64^n - 1$ is divisible by 63.
Thus $4^{3n} - 1$ is divisible by 63 and therefore by 7.

Example When a polynomial $P(x)$ is divided by $2x^2 - 3x + 1$ the quotient is $Q(x)$ and the remainder is $5x - 4$. When $Q(x)$ is divided by $x - 2$ the remainder is -2. Prove that $x - 2$ is a factor of $P(x)$.

By the division process $P(x) \equiv (2x^2 - 3x + 1)Q(x) + 5x - 4$ and $Q(2) = -2$.
Thus $P(2) = (8 - 6 + 1)Q(2) + 10 - 4 = 3 \times (-2) + 6 = 0$.
Therefore $x - 2$ is a factor of $P(x)$.

Exercise 19.2

1. Find the value of k if $f(x) = 2x^3 + kx^2 - 5x + 4$
   (a) leaves a remainder of 3 when divided by $x - 1$;
   (b) leaves a remainder of -2 when divided by $x + 2$;
   (c) leaves a remainder of 1 when divided by $2x + 1$;
   (d) leaves a remainder of 7 when divided by $2x - 3$.

2. Find the value of k in each of the following if $x - 2$ is a factor of the given polynomial:
   (a) $2x^3 - 3x^2 + kx + 6$;
   (b) $x^4 + 2x^3 + kx - 20$;
   (c) $kx^3 - 4x^2 - kx - 2$;
   (d) $3x^4 + kx^3 + kx^2 - 7x - 10$.

3. Find k in each of the following if $x - k$ is a factor of the given polynomial:
   (a) $x^3 - kx^2 + kx - 4$;
   (b) $x^3 + kx^2 + 4x - 6k$.

4. Show that the linear expression is a factor of the polynomial $f(x)$ in each of the following:
   (a) $x + 1$, $f(x) = 4x^3 - 5x^2 - 7x + 2$;
   (b) $2x - 1$, $f(x) = 2x^3 - 3x^2 + 7x - 3$;
   (c) $3x + 2$, $f(x) = 6x^3 + 7x^2 + 5x + 2$;
   (d) $2x + k$, $f(x) = 2x^3 - kx^2 + (4 - k^2)x + 2k$.

5. If the polynomial $f(x)$ leaves a remainder of $5x + 1$ when divided by $(x-2)(x+3)$, find the remainders when $f(x)$ is divided by
   (a) $x - 2$;   (b) $x + 3$.

6. The polynomial $f(x)$ leaves a remainder of $3x - 2$ when divided by $x^2 - x - 2$. Find the remainder when $f(x)$ is divided by $x + 1$.

7. The polynomial $f(x)$ leaves a remainder of -4 when divided by $x + 1$ and a remainder of 8 when divided by $x - 3$. Find the remainder when $f(x)$ is divided by $(x + 1)(x - 3)$.

8. When the polynomial $P(x)$ is divided by $2x^2 + x + 3$, $Q(x)$ is the quotient and $4x - 1$ is the remainder. If $Q(x)$ leaves a remainder of 1 when divided by $x + 2$, prove that $x + 2$ is a factor of $P(x)$.

9. Prove that for all positive integers n,
   (a) $4^n - 1$ is divisible by 3;   (b) $2^{3n} - 1$ is divisible by 7;
   (c) $5^{2n} - 1$ is divisible by 3 and 8;   (d) $3^{3n} - 1$ is divisible by 13.

10. Prove that for all odd positive integers n,
    (a) $5^n + 1$ is divisible by 6;   (b) $4^{2n} + 1$ is divisible by 17.

11. Prove that for all even positive integers n,
    (a) $7^n - 1$ is divisible by 8;   (b) $5^{2n} - 1$ is divisible by 13.

12. Find the value of a if the polynomial $f(x) = x^3 - ax^2 + x - 6$ is divisible by $x - a - 1$.

13. Show that $x + 1$ is a factor of the polynomial $P(x) = ax^3 + bx^2 + bx + a$ for all non-zero a and b, and find the ratio $a : b$ if $(x+1)^2$ is a factor of $P(x)$.

14. (a) If α is a zero of both polynomials $P(x)$ and $Q(x)$, prove that α is also a zero of the polynomial $P(x) - Q(x)$.
    (b) Find the real number m if the polynomials $mx^3 - 2x^2 - 12x + 8$ and $mx^3 + 14x^2 - 32$ have a common zero.

15. Find the values of a and b for which the zeros of the polynomial $x^2 - 2x + a$ are also two of the zeros of the polynomial $2x^3 + bx^2 + 6x - 2$.

505

Chapter 19

16. Find the values of a and b for which
$$9x^4 + 24x^3 + 7x^2 - 12x - 4 \equiv (3x^2 + ax)^2 - (bx+2)^2.$$
Hence solve the equation $9x^4 + 24x^3 + 7x^2 - 12x - 4 = 0$.

*17. When polynomial $f(x)$ is divided by $x - a$ the quotient is $q(x)$. Show that
  (a) when $q(x)$ is divided by $x - b$ where $b \neq a$, the remainder is
  $$\frac{f(b)-f(a)}{b-a};$$

  (b) when $f(x)$ is divided by $(x-a)(x-b)$ where $b \neq a$, the remainder is $(x-a)\left(\dfrac{f(b)-f(a)}{b-a}\right) + f(a)$;

  (c) when $f(x)$ is divided by $(x-a)^2$ the remainder is $(x-a)f'(a) + f(a)$.

*18. Prove that $(x-a)^2$ is a factor of the polynomial $f(x)$ iff $f(a)=0$ and $f'(a)=0$.

*19. Prove that real and unequal values of a and b may be found to satisfy the identity $x^3 - 3abx - (a^3+b^3) \equiv x^3 - 3px - 12$ provided $p^3 < 36$. By considering the turning points of the graph of $f(x) = x^3 - 3px - 12$ prove that the inequality $p^3 < 36$ is also the condition that the equation $x^3 - 3px - 12 = 0$ shall have only one real root and show that $x = a + b$ is the root. Reduce the expression $y^3 - 6y^2 + 3y - 2$ to the form $x^3 - 3px - 12$ by the substitution $y = x + k$ and find the real root of the equation $y^3 - 6y^2 + 3y - 2 = 0$.

19.3 Contracted (Synthetic) Division

The following is a method which may be used to determine the quotient when a polynomial is divided by a linear polynomial $x - k$.

Let $P(x) = a_n x^n + a_{n-1} x^{n-1} + \cdots + a_2 x^2 + a_1 x + a_0$ be any polynomial and let
$Q(x) = b_{n-1} x^{n-1} + b_{n-2} x^{n-2} + \cdots + b_2 x^2 + b_1 x + b_0$ be the quotient when $P(x)$ is divided by $x - k$.

Then $P(x) \equiv (x-k)Q(x) + P(k)$ (remainder theorem),
$$\equiv (x-k)(b_{n-1}x^{n-1} + b_{n-2}x^{n-2} + \cdots + b_2x^2 + b_1x + b_0) + P(k)$$
$$\equiv b_{n-1}x^n + (b_{n-2} - kb_{n-1})x^{n-1} + \cdots + (b_0 - kb_1)x + P(k) - kb_0.$$

Equating coefficients gives:
$$a_n = b_{n-1}$$
$$a_{n-1} = b_{n-2} - kb_{n-1}$$
$$a_{n-2} = b_{n-3} - kb_{n-2}$$
$$\vdots \quad \vdots \quad \vdots$$
$$a_2 = b_1 - kb_2$$
$$a_1 = b_0 - kb_1$$
$$a_0 = P(k) - kb_0.$$

Thus the coefficients of $Q(x)$ can be found as follows:
$$b_{n-1} = a_n$$
$$b_{n-2} = a_{n-1} + kb_{n-1}$$
$$b_{n-3} = a_{n-2} + kb_{n-2}$$
$$\vdots \quad \vdots \quad \vdots$$
$$b_1 = a_2 + kb_2$$
$$b_0 = a_1 + kb_1.$$

Finally, the remainder when $P(x)$ is divided by $x - k$, $P(k)$, can be found from
$$P(k) = a_0 + kb_0.$$

We set this work out in the following manner:

k	a_n	a_{n-1}	a_{n-2}	......	a_2	a_1	a_0
	0	kb_{n-1}	kb_{n-2}	......	kb_2	kb_1	kb_0
	b_{n-1}	b_{n-2}	b_{n-3}	......	b_1	b_0	$P(k)$

The procedure is to follow the arrows from left to right. Each element of the second row (except the first) is found by multiplying each element in the third row (in the column on the immediate left) by k. Each element of the third row is found by adding the corresponding elements of the first and second rows. The remainder is the last element in the third row.

Example Find the quotient and remainder when $2x^3 - 3x^2 + x + 2$ is divided by $x - 2$.

Chapter 19

$$\begin{array}{c|cccc} 2 & 2 & -3 & 1 & 2 \\ & 0 & 4 & 2 & 6 \\ \hline & 2 & 1 & 3 & | 8 \end{array}$$

Therefore the quotient is $2x^2 + x + 3$ and the remainder is 8.

Example Find the quotient and remainder when $6x^3 - x^2 - 4x + 4$ is divided by $2x + 1$.

$$\begin{array}{c|cccc} -\frac{1}{2} & 6 & -1 & -4 & 4 \\ & 0 & -3 & 2 & 1 \\ \hline & 6 & -4 & -2 & | 5 \end{array}$$

Here $6x^3 - x^2 - 4x + 4 \equiv (x + \frac{1}{2})(6x^2 - 4x - 2) + 5$
$\equiv (2x + 1)(3x^2 - 2x - 1) + 5$.

Therefore the quotient is $3x^2 - 2x - 1$ and the remainder is 5.

Note: The coefficients of the quotient are found by dividing the coefficients in the table by 2 (the coefficient of x in the divisor) and the remainder is found as in the first example.

Example Find the quotient and remainder when $x^4 + x^3 - 4x^2 - 14$ is divided by $(x - 2)(x + 3)$.

$$\begin{array}{c|ccccc} (x-2)\ 2 & 1 & 1 & -4 & 0 & -14 \\ & 0 & 2 & 6 & 4 & 8 \\ (x+3)\ -3 & 1 & 3 & 2 & 4 & |-6 \\ & 0 & -3 & 0 & -6 & \\ \hline & 1 & 0 & 2 & |-2 & \end{array}$$

Therefore $x^4 + x^3 - 4x^2 - 14 \equiv (x - 2)(x^3 + 3x^2 + 2x + 4) - 6$
$\equiv (x - 2)\{(x + 3)(x^2 + 2) - 2\} - 6$
$\equiv (x - 2)(x + 3)(x^2 + 2) - 2(x - 2) - 6$
$\equiv (x - 2)(x + 3)(x^2 + 2) - 2x - 2$.

Thus the quotient is $x^2 + 2$ and the remainder is $-2x - 2$.

Note: The quotient can be read directly from the last line in the table but the remainder is found by multiplying the "second remainder" by the first divisor and then adding the "first remainder". The remainder in the previous example is $-2(x-2) - 6 = -2x - 2$.

Also note that if the order of the divisors is changed, the quotient is obviously the same and the remainder is $-2(x+3) + 4 = -2x - 2$, again the same as before.

Example Express $x^3 - 4x^2 + 6x + 2$ as a polynomial in powers of $x - 1$.

Therefore
$$\begin{aligned} x^3 - 4x^2 + 6x + 2 &\equiv (x-1)(x^2 - 3x + 3) + 5 \\ &\equiv (x-1)\{(x-1)(x-2) + 1\} + 5 \\ &\equiv (x-1)^2(x-2) + (x-1) + 5 \\ &\equiv (x-1)^2\{(x-1) - 1\} + (x-1) + 5 \\ &\equiv (x-1)^3 - (x-1)^2 + (x-1) + 5. \end{aligned}$$

Note: The coefficients are read directly from the table beginning from the bottom left and moving up diagonally to the top right.

Exercise 19.3

1. Find the quotient and remainder when
   (a) $x^3 - 2x^2 + 3x + 3$ is divided by $x - 1$;
   (b) $2x^3 - 2x - 1$ is divided by $x - 2$;
   (c) $3x^3 + x^2 - 24x - 1$ is divided by $x + 3$;
   (d) $x^4 + x^3 + x^2 + 4x - 6$ is divided by $x + 2$.

2. Find the quotient and remainder when
   (a) $4x^3 + 4x^2 - x - 6$ is divided by $2x - 1$;
   (b) $3x^3 + x^2 - 8x + 9$ is divided by $3x - 2$;
   (c) $2x^4 + 7x^3 + 4x^2 - 11x - 2$ is divided by $2x + 3$;
   (d) $9x^4 - x^2 + 11x - 1$ is divided by $3x + 2$.

509

Chapter 19

3. Find the quotient and remainder when
   (a) $2x^3 - 5x^2 + x + 9$ is divided by $(x + 1)(x - 2)$;
   (b) $x^4 - 15x^2 + 8$ is divided by $(x + 2)(x + 3)$;
   (c) $x^5 - 5x^4 + 7x^3 - x^2 - 1$ is divided by $x(x - 3)$;
   (d) $2x^4 - 5x^3 + 2x^2 - 2x + 5$ is divided by $(x - 2)^2$.

4. Find the quotient and remainder when $x^4 - 5x^3 + 4x^2 + x + 2$ is divided by
   (a) $x^2 - x - 2$; (b) $x^2 + x + 1$.

5. Express
   (a) $2x^3 - x^2 + x + 1$ as a polynomial in powers of $x - 2$;
   (b) $3x^4 + x + 1$ as a polynomial in powers of $x + 1$;
   (c) $x^5 - x^3 - 1$ as a polynomial in powers of $x - 1$;
   (d) $2x^4 - 3x + 2$ as a polynomial in powers of $x + 3$.

6. If $(x - k)^2$ is a factor of $f(x) = x^3 + 3px + q$, show that $p = -k^2$ and find a similar expression for q in terms of k. Hence show that $4p^3 + q^2 = 0$ and find the other factor of $f(x)$.

7. Find the values of p if the polynomial $f(x) = 2x^3 - 5x^2 - 4x + p$ has a factor of the form $(x - k)^2$. For each of these values of p factorise $f(x)$ as a product of linear factors.

8. Find the largest positive integer value of n for which $(x - 2)^n$ is a factor of $x^5 - 6x^4 + 8x^3 + 16x^2 - 48x + 32$.

9. Show that the straight line $y = 2kx + 1$ meets the curve $y = x^3 + kx^2 - 6x + 5$ at $x = 2$ for all values of k, and find the value of k for which the line is a tangent to the curve.

10. Find the value of k for which $(x - k)^2$ is a factor of $x^5 + 40kx - 128$.

11. (a) Express $8x^3 + 4x^2 + 4x + 7$ in powers of $2x + 1$.
    (b) Find the quotient and remainder when $2x^5 - 3x^4 + x^3 + 2x^2 + 1$ is divided by $(x - 1)(x + 1)(x + 2)$.

19.4 Polynomial Equations with Integer Coefficients

Consider the polynomial $P(x) = a_n x^n + a_{n-1} x^{n-1} + \cdots + a_2 x^2 + a_1 x + a_0$ where the numbers $a_0, a_1, a_2, a_3, \cdots, a_n$ are all integers and $a_n \neq 0$.

(1) If $x = k$ is a root of the equation $P(x) = 0$ then $(x - k)$ is a factor of $P(x)$ and so k must be a factor of a_0.

(2) If $x = p/q$ (p and q are integers) is a rational root of the equation $P(x) = 0$ then $(qx - p)$ is a factor of $P(x)$ and so p must be a factor of a_0 and q must be a factor of a_n.

Example Find the real roots of the equation $x^3 + 2x^2 - 5x - 6 = 0$.

If $f(x) = x^3 + 2x^2 - 5x - 6$ then the possible (integral) roots of $f(x) = 0$ are $x = \pm 1, \pm 2, \pm 3, \pm 6$.
Now $f(-1) = -1 + 2 + 5 - 6 = 0$ and so $x + 1$ is a factor of $f(x)$.
Thus $f(x) \equiv (x+1)(x^2 + x - 6) \equiv (x+1)(x-2)(x+3)$.
Therefore the real roots of $f(x) = 0$ are $x = -1, 2, -3$.

Example Find the real roots of the equation $3x^3 + 5x^2 + 14x + 8 = 0$.

If $f(x) = 3x^3 + 5x^2 + 14x + 8$ then the possible (real) linear factors of $f(x)$ are $x \pm 1, x \pm 2, x \pm 4, x \pm 8, 3x \pm 1, 3x \pm 2, 3x \pm 4, 3x \pm 8$.
Now (perhaps after a long search) $f(-\frac{2}{3}) = -\frac{8}{9} + \frac{20}{9} - \frac{28}{3} + 8 = 0$ and $3x + 2$ is a factor of $f(x)$.
Thus $f(x) \equiv (3x+2)(x^2 + x + 4)$.
But the roots of $x^2 + x + 4 = 0$ are both non-real since $\Delta < 0$.
Therefore the only real root of $f(x) = 0$ is $x = -\frac{2}{3}$.

Exercise 19.4

1. Find all the real roots of the following equations:
   (a) $x^3 - 2x^2 - x + 2 = 0$;
   (b) $x^3 + 3x^2 - 4x - 12 = 0$;
   (c) $x^3 - 3x^2 + 4 = 0$;
   (d) $2x^3 + 9x^2 + 7x - 6 = 0$;
   (e) $2x^3 + x^2 - 7x - 6 = 0$;
   (f) $x^3 - 2x^2 - 2x - 3 = 0$.

Chapter 19

2. Find all the real zeros of the following polynomials:
   (a) $4x^3 + 12x^2 - x - 3$;
   (b) $6x^3 + 5x^2 - 8x - 3$;
   (c) $18x^3 + 15x^2 - x - 2$;
   (d) $2x^4 + 3x^3 + 12x^2 - 22x - 60$.

3. Solve the following equations:
   (a) $x^3 - 4x^2 + 3x + 2 = 0$;
   (b) $x^3 - 3x^2 - 3x + 1 = 0$;
   (c) $2x^3 + 3x^2 - 23x - 12 = 0$;
   (d) $2x^4 - x^3 - 11x^2 + 7x + 6 = 0$.

4. Find all the real roots of the following equations:
   (a) $x^2(x+1) = 5x + 2$;
   (b) $(x+2)(x^2+1) = 3x + 5$;
   (c) $4x^3 + 6 = 4x^2 + 11x$;
   (d) $4x^3 = -(x+1)$;
   (e) $x^2 - 4 = x^2(x+2)$;
   (f) $(x^2+2)^2 = 2x(x+1)^2 + 4x + 13$;
   (g) $x^4 = 6 - 5x$;
   (h) $4x^2 - 9 = (x^2 - 6)(2x+3)$;
   (i) $x^4 - 13x^2 + 36 = 0$;
   (j) $x^4 + 4x^3 + 6x^2 + 4x = 15$.

5. If the polynomial $f(x)$ is divided by $x^2 - x - 6$, the remainder is $7x - 8$.
   (a) Find the remainder when $f(x)$ is divided by $x + 2$.
   (b) If $f(x) = 2x^4 + ax^3 - 20x^2 - 27x + b$,
       (i) find the values of a and b and so check your answer to part (a) for this polynomial ;
       (ii) solve the equation $f(x) = 7x - 8$.

6. Let $f(x) = x^6 + 2x^5 - 4x^4 - 6x^3 + 4x^2 + 2x - 1$ and $g(x) = x^3 + 2x^2 - x - 2$.
   (a) Find the zeros of $g(x)$.
   (b) Show that $x^3 g(x - 1/x) \equiv f(x)$ and find the zeros of $f(x)$.

19.5 Relations between the Zeros and Coefficients of a Quadratic Polynomial (Optional)

Let α and β be the zeros of the quadratic polynomial $ax^2 + bx + c$ $(a \neq 0)$.
Then $ax^2 + bx + c \equiv a(x - \alpha)(x - \beta) \equiv a\left[x^2 - (\alpha + \beta)x + \alpha\beta\right]$.

Thus $\alpha + \beta = -\dfrac{b}{a}$ and $\alpha\beta = \dfrac{c}{a}$.

That is the **sum** of the zeros is $-\dfrac{b}{a}$ and the **product** of the zeros is $\dfrac{c}{a}$.

Polynomials

Conversely, a quadratic polynomial which has zeros whose sum is S and whose product is P is $x^2 - Sx + P$.

Example   Find the sum and product of the zeros of $2x^2 - 4x + 3$.

The sum of the zeros is $-\dfrac{b}{a} = -\dfrac{-4}{2} = 2$ and the product is $\dfrac{c}{a} = \dfrac{3}{2}$.

Example   Find a quadratic polynomial with integer coefficients which has as its zeros   (a) 3 and -2;   (b) $\tfrac{1}{2}$ and $\tfrac{1}{4}$;   (c) $3 \pm 2\sqrt{2}$.

(a)   The sum of the zeros is 1 and the product is -6.
A suitable quadratic is $x^2 - x - 6$.

Or a suitable quadratic is $(x-3)(x+2) \equiv x^2 - x - 6$.

(b)   The sum of the zeros is $\tfrac{3}{4}$ and the product is $\tfrac{1}{8}$.
A suitable quadratic is $8(x^2 - \tfrac{3}{4}x + \tfrac{1}{8}) \equiv 8x^2 - 6x + 1$.

Or a suitable quadratic is $(2x-1)(4x-1) \equiv 8x^2 - 6x + 1$.

(c)   The sum of the zeros is $(3 + 2\sqrt{2}) + (3 - 2\sqrt{2}) = 6$ and the product is $(3+2\sqrt{2})(3-2\sqrt{2}) = 3^2 - (2\sqrt{2})^2 = 9 - 8 = 1$.
A suitable quadratic is $x^2 - 6x + 1$.

Or a suitable quadratic is $\left(x - \left[3 + 2\sqrt{2}\right]\right)\left(x - \left[3 - 2\sqrt{2}\right]\right)$ which can be simplified to $x^2 - 6x + 1$ after a considerable amount of algebra.

Example   If α and β are the zeros of $2x^2 - 4x - 3$, find a quadratic which has zeros   (a) $\alpha - 2$ and $\beta - 2$;   (b) α^2 and β^2.

The sum of the zeros is 2 and the product of the zeros is $-\tfrac{3}{2}$.
Therefore $\alpha + \beta = 2$ and $\alpha\beta = -\tfrac{3}{2}$.

(a)   Let $S = (\alpha - 2) + (\beta - 2) = \alpha + \beta - 4 = -2$.
Let $P = (\alpha - 2)(\beta - 2) = \alpha\beta - 2(\alpha + \beta) + 4 = -\tfrac{3}{2} - 2(2) + 4 = -\tfrac{3}{2}$.
A suitable quadratic is $2(x^2 + 2x - \tfrac{3}{2}) \equiv 2x^2 + 4x - 3$.

513

Chapter 19

(b) Let $S = \alpha^2 + \beta^2 = (\alpha+\beta)^2 - 2\alpha\beta = 4+3 = 7$.

Let $P = \alpha^2\beta^2 = (\alpha\beta)^2 = \frac{9}{4}$.

A suitable quadratic is $4(x^2 - 7x + \frac{9}{4}) \equiv 4x^2 - 28x + 9$.

Example If α and β are zeros of $2x^2 - x - 5$, find the values of

(a) $\alpha^2 + \beta^2$; (b) $\dfrac{1}{\alpha} + \dfrac{1}{\beta}$; (c) $\alpha^3 + \beta^3$.

Here $\alpha + \beta = \frac{1}{2}$ and $\alpha\beta = -\frac{5}{2}$.

(a) $\alpha^2 + \beta^2 = (\alpha+\beta)^2 - 2\alpha\beta = \left(\frac{1}{2}\right)^2 - 2(-\frac{5}{2}) = \frac{1}{4} + 5 = \frac{21}{4}$

(b) $\dfrac{1}{\alpha} + \dfrac{1}{\beta} = \dfrac{\alpha+\beta}{\alpha\beta} = \dfrac{1/2}{-5/2} = -\dfrac{1}{5}$

(c) $\alpha^3 + \beta^3 = (\alpha+\beta)^3 - 3\alpha\beta(\alpha+\beta) = (\frac{1}{2})^3 - 3(-\frac{5}{2})(\frac{1}{2}) = \frac{1}{8} + \frac{15}{4} = \frac{31}{8}$

Exercise 19.5

1. Find a quadratic polynomial which has as its zeros
   (a) 4 and -1 ; (b) 3 and $-\frac{2}{3}$; (c) $\pm\sqrt{5}$;
   (d) $1 \pm \sqrt{3}$; (e) $-2 \pm \sqrt{5}$; (f) $\frac{1}{2}(-\sqrt{2} \pm 3\sqrt{2})$.

2. If α and β are the zeros of the polynomial $x^2 + 2x - 5$, find a polynomial which has as its zeros
   (a) $\alpha + 1$ and $\beta + 1$; (b) $2\alpha + 3$ and $2\beta + 3$;
   (c) $\dfrac{1}{\alpha}$ and $\dfrac{1}{\beta}$; (d) α^2 and β^2 ;
   (e) $\alpha + \dfrac{1}{\beta}$ and $\beta + \dfrac{1}{\alpha}$; (f) $\alpha + \dfrac{1}{\alpha}$ and $\beta + \dfrac{1}{\beta}$;
   (g) α^3 and β^3 ; (h) $\alpha^2 + \beta$ and $\beta^2 + \alpha$.

3. If α and β are the roots of the equation $2x^2 - x - 4 = 0$, find an equation whose roots are
   (a) $\alpha - 2$ and $\beta - 2$; (b) $3\alpha - 1$ and $3\beta - 1$;
   (c) $\dfrac{\alpha}{\beta}$ and $\dfrac{\beta}{\alpha}$; (d) $\alpha + \dfrac{1}{\beta}$ and $\beta + \dfrac{1}{\alpha}$;

(e) $\alpha - \dfrac{2}{\alpha}$ and $\beta - \dfrac{2}{\beta}$;   (f) $\alpha^2 + 1$ and $\beta^2 + 1$;

(g) α^3 and β^3 ;   (h) $\alpha^2 + \dfrac{1}{\alpha}$ and $\beta^2 + \dfrac{1}{\beta}$.

4. Find the values of p if the polynomial $3x^2 - px + 2$
   (a)   has -2 as one of its zeros ;
   (b)   has zeros whose sum is twice their product ;
   (c)   has zeros whose squares have a sum of $\tfrac{4}{9}$;
   (d)   has zeros whose difference is $\tfrac{1}{3}$.

5. Find the values of m if one of the zeros of $x^2 + m(x+1)$ is three times the other.

6. Find the values of c if the polynomial $2x^2 + c(x-2)$
   (a)   has 4 as one of its zeros ;
   (b)   has zeros whose sum is half their product ;
   (c)   has zeros whose squares have a sum of 5 ;
   (d)   has zeros whose cubes have a sum of -32.

7. Consider the polynomial $x^2 - p(2x-1) + 1$.
   (a)   If 3 is a zero, what is the other zero?
   (b)   If the zeros are equal, what are they?
   (c)   If the sum of the squares of the zeros is 10, what is the sum of their reciprocals?

8. Find the relationship between p and q such that the zeros of the polynomial $x^2 + px + q$ are in the ratio $3 : 1$?

9. Given that the equation $cx^2 - (1+c)x + 3c + 2 = 0$ has roots such that their sum is equal to twice their product, find c and the two roots.

10. The zeros α and β of the quadratic polynomial $x^2 - 3kx + k^2$ satisfy the condition $\alpha^2 + \beta^2 = \tfrac{7}{4}$. Find the possible values of k.

Chapter 19

Required Outcomes

After completing this chapter, a student should be able to:
- distinguish a polynomial from other functions.
- divide one polynomial by another using 'long' division or 'contracted' division.
- use the division process to express a polynomial in terms of the quotient, divisor and remainder.
- state and prove the remainder theorem.
- use the corollary to the remainder theorem in certain 'divisibility' problems.
- solve polynomial equations up to degree four using the remainder theorem.

20 Complex Numbers

20.1 Addition, Multiplication and Division

It has been known for centuries that the solution of the quadratic equation $ax^2 + bx + c = 0$ is $x = \dfrac{-b \pm \sqrt{b^2 - 4ac}}{2a}$ and that this solution is meaningless whenever $b^2 - 4ac < 0$ since the square root of a negative real number does not exist (in the domain of real numbers). While searching for a general solution of the cubic equation $ax^3 + bx^2 + cx + d = 0$ at the beginning of the sixteenth century, the Italian mathematician dal Ferro found that the solution of the simpler cubic equation $x^3 + Ax = B$ can be expressed in the form
$x = \left\{ \dfrac{B}{2} + \sqrt{\dfrac{B^2}{4} + \dfrac{A^3}{27}} \right\}^{1/3} + \left\{ \dfrac{B}{2} - \sqrt{\dfrac{B^2}{4} + \dfrac{A^3}{27}} \right\}^{1/3}$. However, when applying this formula it was sometimes found that real solutions existed even when $\dfrac{B^2}{4} + \dfrac{A^3}{27} < 0$. For example, the equation $x^3 - x = 0$ has the real solutions $x = 0$, ± 1. But for $A = -1$ and $B = 0$, $\dfrac{B^2}{4} + \dfrac{A^3}{27} = -\dfrac{1}{27}$ and dal Ferro's formula appeared to be unusable.

Mathematicians were then forced into the invention of an *imaginary* number $i = \sqrt{-1}$ and so began the development of the invaluable branch of mathematics known as *complex numbers*.

Definition A *complex number* is a number of the form $z = a + ib$ where a and b are real numbers and $i^2 = -1$.

a is called the **real part** of z and we write $a = \operatorname{Re} z$.
b is called the **imaginary part** of z and we write $b = \operatorname{Im} z$.

[*Note* that the imaginary part of z is **real**.]

If $b = 0$, then $z = a$ is **real**. Thus every real number is also a complex number.
If $a = 0$ and $b \neq 0$, then $z = ib$ is called **pure imaginary**.

Chapter 20

From the definition of i we find that $i^2 = -1$, $i^3 = -i$, $i^4 = 1$, $i^5 = i$, etc.
In fact, if n is an integer, $i^{4n} = 1$, $i^{4n+1} = i$, $i^{4n\pm 2} = -1$, $i^{4n-1} = -i$.

Equality

Two complex numbers $z_1 = a_1 + ib_1$ and $z_2 = a_2 + ib_2$ are *equal*, $z_1 = z_2$, iff $a_1 = a_2$ *and* $b_1 = b_2$.

That is two complex numbers are equal iff their real parts are equal and their imaginary parts are equal.

Example Find the real numbers a and b if $a + ib = 3 - 2i$.

Since the complex numbers $a + ib$ and $3 - 2i$ are equal, their real parts are equal, i.e. $a = 3$, and their imaginary parts are equal, i.e. $b = -2$.

Addition

If $z_1 = a_1 + ib_1$ and $z_2 = a_2 + ib_2$ are two complex numbers, then the sum of z_1 and z_2 is given by $z_1 + z_2 = (a_1 + a_2) + i(b_1 + b_2)$.

That is when two complex numbers are added, the sum is a complex number whose real part is the sum of the real parts of the given complex numbers and whose imaginary part is the sum of the imaginary parts of the given complex numbers.

Example Find the sum of the complex numbers $5 + 4i$, $-1 + 2i$ and 7.

Adding real parts gives $5 - 1 + 7 = 11$. Adding imaginary parts gives $4 + 2 + 0 = 6$. Therefore the required sum is $11 + 6i$.

Note: Addition is simply a matter of *collecting like terms*.

Multiplication

If $z_1 = a_1 + ib_1$ and $z_2 = a_2 + ib_2$ are two complex numbers, then the product of z_1 and z_2 is given by $z_1 z_2 = (a_1 a_2 - b_1 b_2) + i(a_1 b_2 + a_2 b_1)$.

This may appear to be a complicated operation, but in fact we carry out multiplication of complex numbers in exactly the same way we multiply two binomial expressions.

Thus $(a_1 + ib_1)(a_2 + ib_2) = a_1a_2 + ia_1b_2 + ia_2b_1 + i^2b_1b_2$
$= a_1a_2 + ia_1b_2 + ia_2b_1 - b_1b_2$ since $i^2 = -1$
$= (a_1a_2 - b_1b_2) + i(a_1b_2 + a_2b_1)$

which is our definition.

Example Evaluate the product $(3 - 4i)(5 + 2i)$.

Here $(3 - 4i)(5 + 2i) = 15 + 6i - 20i - 8i^2 = 23 - 14i$.

Example Write each of the following in the form $a + ib$ where a and b are real:

(a) $(4 - 3i)^2$;
(b) $(6 + i)(6 - i)$;
(c) $(3 + 2i)(5 - 4i) + (2 - 3i)^2 + 8$.

(a) $(4 - 3i)^2 = 16 - 24i + 9i^2 = 7 - 24i$
(b) $(6 + i)(6 - i) = 36 - i^2 = 37 + 0i$
(c) $(3 + 2i)(5 - 4i) + (2 - 3i)^2 + 8$
$= 15 - 12i + 10i - 8i^2 + 4 - 12i + 9i^2 + 8$
$= 26 - 14i$

Complex Conjugate The *complex conjugate* of the complex number $z = a + ib$ is denoted and defined by $z^* = a - ib$.

That is the complex conjugate of any complex number is found by changing the sign of the imaginary part.

Note: If a complex number and its conjugate are equal, $z = z^*$, then z is real.

Example Prove that the sum and product of any complex number and its conjugate are both real.

Let $z = a + ib$ where a and b are real numbers. Then $z^* = a - ib$.
Thus $z + z^* = (a + a) + i(b - b) = 2a$ which is real, and
$zz^* = (a + ib)(a - ib) = a^2 - (ib)^2 = a^2 + b^2$ which is real (and ≥ 0).

Summarising gives: $z + z^* = 2\,\text{Re}\,z$ and $zz^* = (\text{Re}\,z)^2 + (\text{Im}\,z)^2 \geq 0$.

Division of Complex Numbers

We make use of the fact that the product of two complex conjugates is real when we wish to express the quotient of two complex numbers in the form $a + ib$.

519

Chapter 20

Example Write $\dfrac{3+2i}{4-i}$ in the form $a+ib$ where a and b are real numbers.

$$\dfrac{3+2i}{4-i} = \dfrac{3+2i}{4-i} \times \dfrac{4+i}{4+i} = \dfrac{10+11i}{17} = \dfrac{10}{17} + \dfrac{11}{17}i$$

Example Find the real numbers x, y such that $(x+iy)(3-2i) = 6-17i$.

Method 1 $(x+iy)(3-2i) = (3x+2y) + i(-2x+3y) = 6-17i$ if and only if $3x + 2y = 6$ and $-2x + 3y = -17$.
Solving these equations gives $x = 4$, $y = -3$.

Method 2 $(x+iy)(3-2i) = 6-17i$

$$\Rightarrow x + iy = \dfrac{6-17i}{3-2i} = \dfrac{(6-17i)(3+2i)}{(3-2i)(3+2i)} = \dfrac{52-39i}{13} = 4 - 3i$$

Thus $x = 4$ and $y = -3$.

Example Find a quadratic whose zeros are $3 \pm 2i$.

The sum of the zeros is 6 and the product is 13.
Therefore a suitable quadratic is $z^2 - 6z + 13$.

Example Solve the equation $z^2 - 10z + 26 = 0$.

From the quadratic formula $z = \dfrac{10 \pm \sqrt{100-104}}{2} = \dfrac{10 \pm 2i}{2} = 5 \pm i$.

Example Find the real numbers x and y such that $(x+iy)^2 = 5 - 12i$.

$(x+iy)^2 = x^2 - y^2 + 2xyi = 5 - 12i$ iff $x^2 - y^2 = 5$ .... (i)
and $2xy = -12$ .... (ii).

From (ii) we find $y = -\dfrac{6}{x}$ .... (iii).

Substitute (iii) into (i): $x^2 - \dfrac{36}{x^2} = 5$

$$x^4 - 5x^2 - 36 = 0$$
$$(x^2 - 9)(x^2 + 4) = 0$$
$$x^2 = 9 \text{ or } x^2 = -4.$$

But x is real and so $x = \pm 3$, and from (iii) $y = \mp 2$.
Thus $x = 3$, $y = -2$ or $x = -3$, $y = 2$.

Note: The following problems are essentially the same:
(1) Find the values of the real numbers x, y if $(x+iy)^2 = 5-12i$.
(2) Solve the equation $z^2 = 5-12i$.
(3) Find the square roots of $5 - 12i$.

The answers are:
(1) $x = 3, y = -2$ or $x = -3, y = 2$
(2) $z = \pm(3 - 2i)$
(3) The square roots of $5 - 12i$ are $\pm(3 - 2i)$.

[*Note:* Quite obviously these answers can be obtained most easily by using a graphic display calculator. However students should be aware of the algebraic techniques required.]

Example Solve the equation $z^2 - iz = 4 + 2i$.

Rearranging the equation gives $z^2 - iz - (4+2i) = 0$.

Therefore $z = \dfrac{i \pm \sqrt{-1 - 4(-4-2i)}}{2} = \dfrac{i \pm \sqrt{15 + 8i}}{2}$.

Using the method outlined in the previous example, or using a graphic display calculator, we find that the square roots of $15 + 8i$ are $\pm(4+i)$.

Thus $z = \dfrac{i \pm (4+i)}{2} = \dfrac{4 + 2i}{2}, \dfrac{-4}{2} = 2 + i, -2$.

Example Express each of the following quadratics as a product of two linear factors:

(a) $z^2 + 25$; (b) $z^2 + 4z + 5$; (c) $z^2 + z + 1$.

(a) $z^2 + 25 = z^2 - (5i)^2 = (z-5i)(z+5i)$

(b) $z^2 + 4z + 5 = z^2 + 4z + 4 + 1 = (z+2)^2 - i^2 = (z+2-i)(z+2+i)$

Alternative Method:

The zeros of $z^2 + 4z + 5$ are $\dfrac{-4 \pm \sqrt{16-20}}{2} = -2 \pm i$.

Thus $z^2 + 4z + 5 = (z - [-2+i])(z - [-2-i]) = (z+2-i)(z+2+i)$.

(c) $z^2 + z + 1 = z^2 + z + \tfrac{1}{4} + \tfrac{3}{4} = (z + \tfrac{1}{2})^2 - (\tfrac{1}{2}i\sqrt{3})^2$

$= (z + \tfrac{1}{2} - \tfrac{1}{2}i\sqrt{3})(z + \tfrac{1}{2} + \tfrac{1}{2}i\sqrt{3})$

Chapter 20

Example Solve the equation $\dfrac{z+i}{z-i} = 2+3i$.

$\dfrac{z+i}{z-i} = 2+3i \Rightarrow z+i = (z-i)(2+3i) = (2+3i)z + 3 - 2i$

That is $(1+3i)z = -3+3i$ and so $z = \dfrac{-3+3i}{1+3i}$

$\qquad = \dfrac{(-3+3i)(1-3i)}{(1+3i)(1-3i)}$

$\qquad = \dfrac{6+12i}{10}$

$\qquad = \dfrac{3}{5} + \dfrac{6}{5}i.$

Example Solve: (a) $\quad z^4 + 13z^2 + 36 = 0$; (b) $\quad z^4 + 8z^2 + 36 = 0$.

(a) $\quad z^4 + 13z^2 + 36 = 0 \Rightarrow (z^2+4)(z^2+9) = 0 \Rightarrow z = \pm 2i, \pm 3i$

(b) $\quad z^4 + 8z^2 + 36 = 0 \Rightarrow z^4 + 12z^2 + 36 - 4z^2 = 0$

$\qquad \Rightarrow (z^2+6)^2 - (2z)^2 = 0$

$\qquad \Rightarrow (z^2 - 2z + 6)(z^2 + 2z + 6) = 0$

$\qquad \Rightarrow z = 1 \pm i\sqrt{5}, -1 \pm i\sqrt{5}.$

Exercise 20.1

1. Write each of the following in the form $a + ib$ where $a, b \in \mathbb{R}$:
    (a) $i^{12} + i^5$;
    (b) $2i^3(3i-1)$;
    (c) $(3+2i)(2-i)$;
    (d) $(5-i)(-2+3i)$;
    (e) $(4+3i)(4-3i)$;
    (f) $(7-i)(3-2i)$;
    (g) $(3-5i)^2$;
    (h) $(2-i)^3$.

2. Write each of the following in the form $a + ib$ where $a, b \in \mathbb{R}$:
    (a) $\dfrac{10}{3+i}$;
    (b) $\dfrac{26}{3-2i}$;
    (c) $\dfrac{1}{2+i}$;
    (d) $\dfrac{5-i}{1-i}$;
    (e) $\dfrac{7+i}{2+i}$;
    (f) $\dfrac{3+4i}{4-3i}$;
    (g) $\dfrac{3+i}{(2-i)^2}$;
    (h) $\dfrac{4i(1-2i)}{(1-i)^2}$;
    (i) $\left(\dfrac{2+i}{2-i}\right)^2$;

522

(j) $\dfrac{1-i}{3+i} + \dfrac{3+i}{1-i}$; (k) $\dfrac{2+i}{1+2i} - \dfrac{i}{3-i}$; (l) $\left(\dfrac{3-i}{2-i}\right)^2 - \dfrac{5i}{1+3i}$.

3. If $z = 5 - 2i$ and $w = 3 + i$, express each of the following in the form $a + ib$ where $a, b \in \mathbb{R}$:
   (a) z^2 ;
   (b) zw^* ;
   (c) z^*w ;
   (d) $\dfrac{z}{w}$;
   (e) $(z + w^*)^*$;
   (f) $\dfrac{10z^*}{w}$.

4. Find the real numbers x and y if:
   (a) $x + iy = \dfrac{17}{-4+i}$;
   (b) $x + iy = \dfrac{2+3i}{2-3i}$;
   (c) $(x + iy)(2 - 3i) = 21 + i$;
   (d) $x(2 - 3i) + y(4 + 5i) = 2 - 14i$;
   (e) $(x + iy)^2 = 3 + 4i$;
   (f) $(x + iy)^2 = -7 - 24i$;
   (g) $(x + iy)^2 = \dfrac{50}{4+3i}$;
   (h) $(x + iy)^2 = 4 + 3i$.

5. In each of the following find a quadratic which has zeros:
   (a) $\pm 3i$;
   (b) $\pm 6i$;
   (c) $1 \pm 2i$;
   (d) $-2 \pm 3i$;
   (e) $5 \pm 4i$;
   (f) $\dfrac{1}{2 \pm i}$.

6. Solve each of the following equations:
   (a) $z^2 + 2z + 5 = 0$;
   (b) $z^2 - 6z + 10 = 0$;
   (c) $z^2 - 12z + 37 = 0$;
   (d) $z^2 + z + 1 = 0$;
   (e) $z^2 + z\sqrt{3} + 1 = 0$;
   (f) $4z^2 - 4z + 5 = 0$;
   (g) $3z^2 - 18z + 28 = 0$;
   (h) $2z^2 + 4z + 3 = 0$.

7. Solve each of the following equations:
   (a) $\dfrac{z-i}{z+i} = 3$;
   (b) $\dfrac{z+2i}{z-2i} = 5$;
   (c) $\dfrac{2z+1}{2z-1} = 1+i$;
   (d) $\dfrac{z-2i}{z+2i} = 1-i$;
   (e) $\dfrac{2z+i}{2z-i} = 2+3i$;
   (f) $\dfrac{3z+1}{3z+i} = \dfrac{12-i}{10}$.

8. Express each of the following quadratics as a product of two linear factors:
   (a) $z^2 + 1$;
   (b) $z^2 + 10$;
   (c) $z^2 + 2z + 5$;
   (d) $z^2 - 6z + 13$;
   (e) $z^2 - z + 3$;
   (f) $z^2 - 3iz - 2$.

Chapter 20

9. Solve the following equations given that at least one root of each is pure imaginary:
    (a) $z^3 - 2z^2 + z - 2 = 0$;
    (b) $z^4 + 2z^3 + 9z^2 + 8z + 20 = 0$.

10. Find all complex numbers z for which
    (a) $z = (z*)^2$;
    (b) $z^2 = iz$;
    (c) $z = iz*$.

11. Prove that for all complex numbers z and w
    (a) $(z+w)* = z* + w*$;
    (b) $(zw)* = z*w*$.

12. Prove that if the sum and product of two complex numbers are both real, then either
    (a) the two complex numbers are both real ; or
    (b) the two complex numbers are conjugates.

20.2 Zeros of a Polynomial with Real Coefficients

Definition A polynomial whose coefficients are all real is called a *real polynomial*.

Theorem The *non-real* zeros of a *real* polynomial must occur in conjugate pairs.
[The proof of this theorem is not required at this stage.]

Example Find the values of the real numbers a and b if $2 + 3i$ is a zero of the polynomial $z^3 + az + b$.

Since the polynomial $z^3 + az + b$ is real, $2 \pm 3i$ are both zeros.
Now $2 \pm 3i$ are zeros of $z^2 - 4z + 13$ which is a factor of $z^3 + az + b$.
Hence $z^3 + az + b = (z^2 - 4z + 13)(z + 4) = z^3 - 3z + 52$.
Thus $a = -3$ and $b = 52$.

Example Find the values of the real numbers a and b if $1 - i$ is a zero of the polynomial $f(z) = z^4 - 10z^2 + az + b$ and find all the zeros of $f(z)$.

Since $f(z)$ is real $1 \pm i$ are both zeros of $f(z)$.
Now $1 \pm i$ are zeros of $z^2 - 2z + 2$ which is a factor of $f(z)$.
Let $f(z) = (z^2 - 2z + 2)(z^2 + pz + q)$, where $p, q \in \mathbb{R}$
$= z^4 + (p-2)z^3 + (-2p+q+2)z^2 + (2p-2q)z + 2q$.

Equating coefficients gives:
$$p - 2 = 0 \quad \ldots\ldots\ldots(i)$$
$$-2p + q + 2 = -10 \quad \ldots\ldots\ldots(ii)$$
$$2p - 2q = a \quad \ldots\ldots\ldots(iii)$$
$$2q = b \quad \ldots\ldots\ldots(iv).$$

From (i): $p = 2$.
From (ii): $q = -8$.
From (iii): $a = 20$.
From (iv): $b = -16$.

Thus $f(z) = (z^2 - 2z + 2)(z^2 + 2z - 8) = (z^2 - 2z + 2)(z + 4)(z - 2)$ and all the zeros are $1 \pm i$, -4 and 2.

[*Note:* This technique is known as *the method of undetermined coefficients.*]

Exercise 20.2

1. In each of the following, show that the given complex number is a zero of the given polynomial:

   (a) $2 + i$, $\quad f(z) = z^3 - 6z^2 + 13z - 10$;

   (b) $1 - 3i$, $\quad f(z) = z^3 + 6z + 20$;

   (c) $-1 + 2i$, $\quad f(z) = 2z^3 - z^2 - 25$;

   (d) $3 - i\sqrt{3}$, $\quad f(z) = 3z^3 - 17z^2 + 30z + 12$;

   (e) $2 + 3i$, $\quad f(z) = z^4 - z^2 + 44z + 26$;

   (f) $-\frac{1}{2}(1 + i\sqrt{3})$, $\quad f(z) = 2z^4 + z^3 - 2z - 1$.

2. Find all the zeros of each polynomial in Question 1.

3. Find the values of the real numbers a and b if the given complex number is a zero of the given polynomial:

   (a) $1 + i$, $\quad f(z) = z^3 + az + b$;

   (b) $3 - i$, $\quad f(z) = z^3 + az^2 + 16z + b$;

   (c) $-2 + i$, $\quad f(z) = z^3 + az^2 + bz - 15$;

   (d) $i - 4$, $\quad f(z) = 2z^3 + az^2 + 26z + b$;

   (e) $2 + i$, $\quad f(z) = z^4 + az^2 + bz + 15$;

   (f) $5 - 2i$, $\quad f(z) = 2z^4 - 22z^3 + az^2 - 68z + b$.

4. Solve the equation $f(z) = 0$ for each polynomial in Question 3.

Chapter 20

5. Find a polynomial with integer coefficients and least possible degree which has amongst its zeros
   (a) 2 and $1 + 2i$;
   (b) $-\frac{1}{2}$ and $2 - i$;
   (c) $1 + i$ and $3 + 2i$;
   (d) $2 - 3i$ and $4 + i$;
   (e) $\frac{1}{2}(1 - i\sqrt{3})$ and $\frac{1}{2}(1 + i\sqrt{2})$;
   (f) -1, $3 + i$ and $5 - 2i$.

6. In each of the following solve the equation $f(z) = 0$ given that $z = 2 - i$ is one solution:
   (a) $f(z) = 2z^3 - 5z^2 - 2z + 15$;
   (b) $f(z) = z^3 - 6z^2 + (12 + i)z - 3(3 + i)$;
   (c) $f(z) = z^3 - 2z^2 + (7 + 2i)z - 6(2 - i)$;
   (d)* $f(z) = z^3 - 2 + 11i$.

7. The real polynomial $p(z)$ has degree 3. Given that $p(1 + 2i) = 0$, $p(2) = 0$ and $p(0) = 20$, write $p(z)$ in the form $az^3 + bz^2 + cz + d$.

8. The real polynomial $p(z)$ has degree 4. Given that $p(3 - i) = 0$, $p(2i) = 0$ and $p(0) = 20$, write $p(z)$ in the form $az^4 + bz^3 + cz^2 + dz + e$.

9. Find a polynomial with least possible degree which has -2 and $1 + 3i$ as two of its zeros.

*10. Show that $z = i - 3$ is one root of the equation $z^3 = -18 + 26i$ and find the other two roots.

20.3 Geometrical Representation of a Complex Number – Modulus

Corresponding to each complex number $z = x + iy$ there is a point $P(x, y)$ in a rectangular coordinate plane. Thus $P(2, -3)$ corresponds to the complex number $2 - 3i$, $Q(0, 2)$ corresponds to the complex number $2i$ and $R(-3, 0)$ corresponds to the complex number -3.

Every point corresponding to a real number lies on the *x*-axis which is called the *real axis*; every point corresponding to a pure imaginary number lies on the *y*-axis which is called the *imaginary axis*.

Diagrams which display complex numbers in this way are called *Argand diagrams*.

Complex Numbers

Example Plot the points corresponding to each of the following complex numbers on an Argand diagram: $3 + 4i$, $-1 + 3i$, $-4 - 2i$, 3, $-5i$.

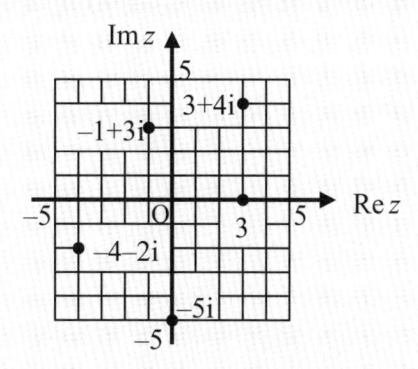

Addition of Complex Numbers in the Argand Plane

Let P and Q represent the complex numbers $z_1 = x_1 + iy_1$ and $z_2 = x_2 + iy_2$ respectively. Complete the parallelogram OPRQ where O is the origin. Then R represents the complex number $z_1 + z_2$.

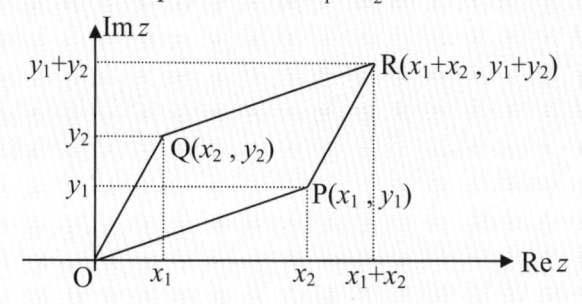

Example If P, Q represent the complex numbers z, w respectively, explain with the aid of a diagram how to find the points R, S, T and U which represent the complex numbers z^*, $z + w$, $-w$ and $z - w$ respectively.

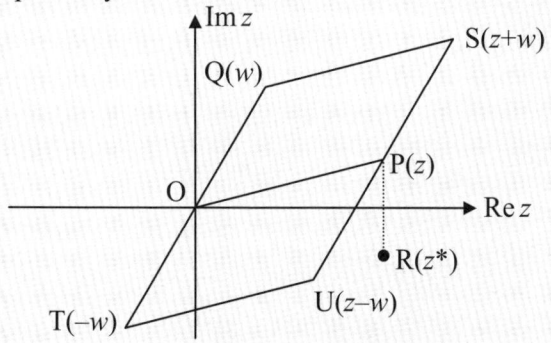

527

Chapter 20

If R is the image of P under a reflection in the x-axis, then R represents the complex number z*. Complete the parallelogram OPSQ. Then S represents the complex number z + w. If T is the image of Q under a rotation about O through an angle of 180°, then T represents the complex number –w. Complete the parallelogram OPUT. Then U represents the complex number z – w.

The Modulus of a Complex Number

Definition The *modulus* of the complex number $z = x + iy$ is denoted and defined by $|z| = \sqrt{x^2 + y^2}$.

Note: (1) $|z|$ is a real number.
    (2) $|z| \geq 0$ for all z with $|z| = 0$ iff $z = 0$.
    (3)   The definition is consistent with the definition of the modulus of a real number: if $y = 0$, $z = x$ is real and $|z| = |x| = \sqrt{x^2}$.

Geometrically $|z|$ measures the distance from the origin to the point P which represents z in the Argand plane.

Some Properties of $|z|$

(1) $|z|^2 = zz*$ (2) $|zw| = |z||w|$ (3) $\left|\dfrac{z}{w}\right| = \dfrac{|z|}{|w|}$

(4) $|z + w| \leq |z| + |w|$ (5) $|z - w| \geq |z| - |w|$

Proof   (1)   Let $z = x + iy$ then $z* = x - iy$.
Thus $zz* = x^2 + y^2 = |z|^2$.

(2) Let $w = u + iv$ and let z be as in part (1).
Then $zw = (xu - yv) + i(xv + yu)$.
Thus $|zw| = \sqrt{(xu - yv)^2 + (xv + yu)^2}$
$= \sqrt{x^2u^2 - 2xyuv + y^2v^2 + x^2v^2 + 2xyuv + y^2u^2}$
$= \sqrt{x^2(u^2 + v^2) + y^2(u^2 + v^2)}$
$= \sqrt{(x^2 + y^2)(u^2 + v^2)}$
$= \sqrt{x^2 + y^2}\sqrt{u^2 + v^2}$
$= |z||w|$.

(3) $\dfrac{1}{w} = \dfrac{w^*}{ww^*} = \dfrac{w^*}{|w|^2}$ {from (1)}

Thus $\left|\dfrac{1}{w}\right| = \dfrac{|w^*|}{|w|^2} = \dfrac{|w|}{|w|^2} = \dfrac{1}{|w|}$ and so $\left|\dfrac{z}{w}\right| = \left|z\left(\dfrac{1}{w}\right)\right| = |z|\left|\dfrac{1}{w}\right| = \dfrac{|z|}{|w|}$.

(4) $|z+w|^2 = (z+w)(z+w)^*$ {from (1)}
$= (z+w)(z^*+w^*)$
$= zz^* + ww^* + zw^* + z^*w$
$= |z|^2 + |w|^2 + 2|z||w| - \{2|z||w| - zw^* - z^*w\}$
$= (|z|+|w|)^2 - \{2|z||w| - [zw^* + (zw^*)^*]\}$
$= (|z|+|w|)^2 - \{2|z||w^*| - 2\operatorname{Re}(zw^*)\}$
$= (|z|+|w|)^2 - 2\{|zw^*| - \operatorname{Re}(zw^*)\}$.

Now $|\alpha| \geq \operatorname{Re}\alpha$ for all complex numbers α and so $\alpha - \operatorname{Re}\alpha \geq 0$. Therefore $|zw^*| - \operatorname{Re}(zw^*) \geq 0$. This gives $|z+w|^2 \leq (|z|+|w|)^2$, and since both $|z+w|$ and $|z|+|w|$ are positive, $|z+w| \leq |z|+|w|$.

(5) Put $z = z - w$ in (4).
This gives $|z-w+w| \leq |z-w| + |w|$ and then $|z-w| \geq |z| - |w|$ as required.

Example Find the modulus of each of the following complex numbers:
(a) $2+i$; (b) $-4+3i$; (c) $4i$;
(d) $\tfrac{1}{2}(1+i\sqrt{3})$; (e) -2 ; (f) $6(1+i)$;
(g) $(3-2i)^2$; (h) $\dfrac{5}{3-4i}$; (i) $\dfrac{2+i}{1-3i}$.

(a) $|2+i| = \sqrt{2^2+1^2} = \sqrt{5}$
(b) $|-4+3i| = \sqrt{(-4)^2+3^2} = 5$
(c) $|4i| = \sqrt{0^2+4^2} = 4$
(d) $\left|\tfrac{1}{2}(1-i\sqrt{3})\right| = \tfrac{1}{2}|1-i\sqrt{3}| = \tfrac{1}{2}\sqrt{1^2+(-\sqrt{3})^2} = 1$
(e) $|-2| = 2$

Chapter 20

(f) $|6(1+i)| = 6|1+i| = 6\sqrt{1^2+1^2} = 6\sqrt{2}$

(g) $|(3-2i)^2| = |3-2i|^2 = 3^2 + (-2)^2 = 13$

(h) $\left|\dfrac{5}{3-4i}\right| = \dfrac{|5|}{|3-4i|} = \dfrac{5}{\sqrt{3^2+(-4)^2}} = \dfrac{5}{5} = 1$

(i) $\left|\dfrac{2+i}{1-3i}\right| = \dfrac{|2+i|}{|1-3i|} = \dfrac{\sqrt{2^2+1^2}}{\sqrt{1^2+(-3)^2}} = \dfrac{\sqrt{5}}{\sqrt{10}} = \dfrac{1}{\sqrt{2}}$

Example If $\left|\dfrac{z+2}{z}\right| = 2$ and P represents z in the Argand plane, show that P lies on a circle and find the centre and radius of this circle.

Let $z = x + iy$ where $x, y \in \mathbb{R}$.

Then $\left|\dfrac{z+2}{z}\right| = 2$

$\Rightarrow$ $\dfrac{|z+2|}{|z|} = 2$

$\Rightarrow$ $|z+2| = 2|z|$

$\Rightarrow$ $|z+2|^2 = 4|z|^2$

$\Rightarrow$ $(x+2)^2 + y^2 = 4(x^2 + y^2)$

$\Rightarrow$ $x^2 + 4x + 4 + y^2 = 4x^2 + 4y^2$

$\Rightarrow$ $3x^2 + 3y^2 - 4x = 4$ which is the equation of a circle with centre at $\left(\tfrac{2}{3}, 0\right)$ and with radius of length $\sqrt{\tfrac{4}{3} + \left(\tfrac{2}{3}\right)^2 + 0^2} = \tfrac{4}{3}$.

Square Roots of a Complex Number using Modulus

Earlier in these notes we were able to find the square roots of any complex number, for example $-40 - 42i$, by setting $x + iy = \sqrt{-40 - 42i}$ for real x, y, squaring and equating both real and imaginary parts to provide two simultaneous equations: $x^2 - y^2 = -40$ and $2xy = -42$. These equations were solved via a quadratic equation in x^2 or y^2.

An alternative method equates real parts and the squares of the moduli and avoids solving a quadratic equation as follows:

Equating real parts gives $x^2 - y^2 = -40$ as before.

Equating the squares of the moduli gives $x^2 + y^2 = \sqrt{(-40)^2 + (-42)^2} = 58$.

Adding gives $2x^2 = 18$ so $x = \pm 3$. Subtracting gives $2y^2 = 98$ so $y = \pm 7$.
All we need to do now is to consider imaginary parts, $2xy = -42$ which gives $y = -7$ when $x = 3$ and $y = 7$ when $x = -3$.

Therefore the square roots of $-40 - 42i$ are $\pm(3 - 7i)$.

[Clearly, one can use a graphic display calculator to find the square roots without knowing any algebraic method.]

It is not immediately clear that every complex number has square roots. The method described earlier would fail if the quadratic in x^2 has two negative roots or had complex roots. In fact this cannot happen since finding the square roots of $a + ib$ leads to the equation $4x^4 - 4ax^2 - b^2$. Here $\Delta = 16a^2 + 16b^2 \geq 0$ for all a, b, and the product of the roots is $-\frac{1}{4}b^2 < 0$ so there are two real roots one of which is positive (and the other negative).

Exercise 20.3

1. Find the modulus of each of the following complex numbers:
   (a) $2 + 2i$;
   (b) $1 - 2i$;
   (c) $-3 + i\sqrt{3}$;
   (d) $-3i$;
   (e) 0;
   (f) 5;
   (g) $-5 - 12i$;
   (h) $\frac{1}{2}(\sqrt{3} + i)$;
   (i) $3\sqrt{3} - 9i$;
   (j) $\dfrac{1}{1+i}$;
   (k) $\dfrac{13}{3 - 2i}$;
   (l) $28 + 21i$;
   (m) $39 - 52i$;
   (n) $\dfrac{2 + 5i}{3 - 4i}$;
   (o) $(3 + 4i)(\sqrt{2} - i\sqrt{2})$.

2. If $z = 3 - i$ and $w = 1 + 2i$, calculate:
   (a) $|z|$;
   (b) $|w|$;
   (c) $|2z^* + i|$;
   (d) $|z^* + w^*|$;
   (e) $|w^2|$;
   (f) $|3z^3|$;
   (g) $|zw|$;
   (h) $\left|\dfrac{z}{w}\right|$;
   (i) $\left|\dfrac{z + w}{z^* - w^*}\right|$.

3. Find the modulus of the complex number $1 - i\tan\theta$ in simplest form.

4. Show that the relation $|z + w| < |z| + |w|$ is satisfied if $z = 3 + i$ and $w = 2 - i$.

Chapter 20

5. Give an example of two complex numbers z, w for which $|z+w|=|z|+|w|$. If P represents z and Q represents w in the Argand plane, what are the relative positions of P and Q when $|z+w|=|z|+|w|$?

6. If $\left|\dfrac{z-1}{z+1}\right|=1$, prove that the real part of z is zero.

7. If $\left|\dfrac{z+i}{z-i}\right|=1$, prove that z is real.

8. Given that $\left|\dfrac{2z+1}{z}\right|=1$, show that the point P which represents z in the Argand plane lies on a circle, and find the centre and radius of this circle.

9. If $|z|=1$, prove that

    (a) $z^* = \dfrac{1}{z}$;

    (b) $|z-w|=|z*w-1|$ for any complex number w.

10. Prove that $z + \dfrac{|z|^2}{z} = 2\operatorname{Re} z$.

11. Prove that $|z+w|^2 = |z|^2 + |w|^2 + 2\operatorname{Re}(zw^*)$ for all complex numbers z, w.

12. Prove that $|z+w|^2 + |z-w|^2 = 2|z|^2 + 2|w|^2$ for all complex numbers z, w.

13. Let P represent the complex number $z = \cos\theta + i\sin\theta$ for $0 < \theta < \tfrac{1}{2}\pi$. Show that $z^2 = \cos 2\theta + i\sin 2\theta$. Let Q represent z^2. Show that P and Q each lie on the unit circle. On a diagram, plot the relative positions of z, z^2, $-z^2$ and $1-z^2$. Hence show that $|1-z^2| = 2\sin\theta$.

14. Use the alternative method described in Section 12.3 to find the square roots of each of the following complex numbers:
    (a) i; (b) −i; (c) −5 + 12i; (d) 7 − 24i.

15. Solve the equation $(1+i)z^2 + (3-2i)z - (21-7i) = 0$.

532

20.4 Argument

Definition Let $z = x + iy$ be any complex number. If there is a *real number* θ such that $\cos\theta = \dfrac{x}{r}$ and $\sin\theta = \dfrac{y}{r}$ where $r = |z|$, then θ is called *an argument* of z. We write $\theta = \arg z$.

Geometrically, if P represents z, θ is the radian measure of the angle between the vector $\overrightarrow{OP}$ and the positive x-axis.

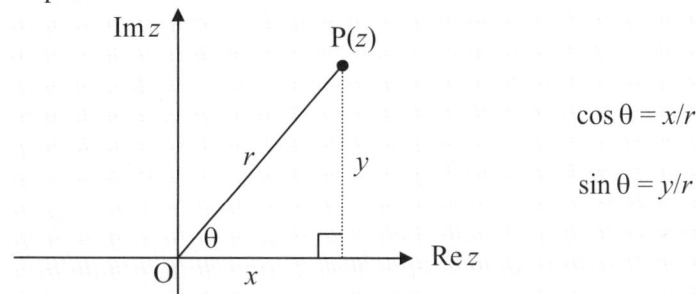

$\cos\theta = x/r$

$\sin\theta = y/r$

Since $\cos(\theta + 2\pi n) = \cos\theta$ and $\sin(\theta + 2\pi n) = \sin\theta$ for all integers n, each complex number has an infinite number of arguments any two of which differ by an integral multiple of 2π.

Principal Argument

Definition If θ is an argument of z and $-\pi < \theta \leq \pi$, then θ is called *the principal argument* of z. We denote the principal argument of z by Arg z.

When we refer to 'the argument of z' we shall always mean the principal argument.

Example Find the argument of each of the following complex numbers:
$1+i$, $2i$, -1, $-1+i\sqrt{3}$, $9\sqrt{3}-9i$, $-3-4i$.

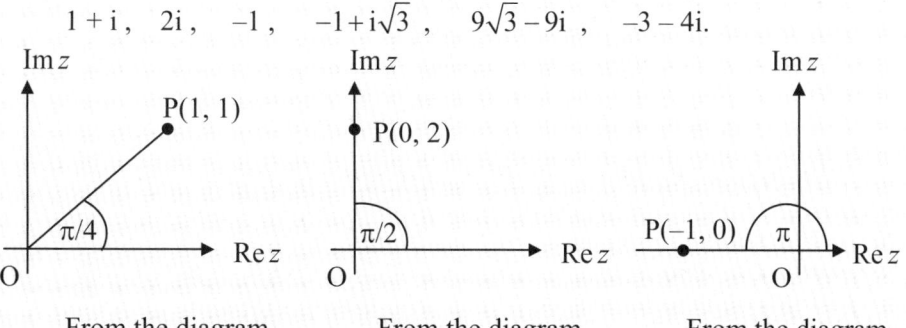

From the diagram, $\operatorname{Arg}(1+i) = \tfrac{1}{4}\pi$.

From the diagram, $\operatorname{Arg}(2i) = \tfrac{1}{2}\pi$.

From the diagram, $\operatorname{Arg}(-1) = \pi$.

533

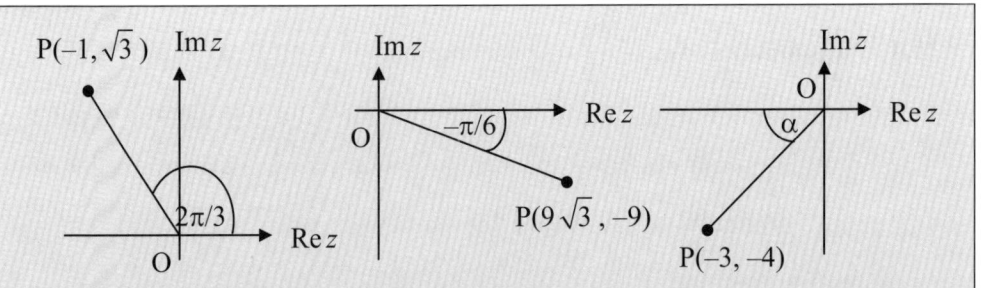

Chapter 20

P(−1, √3) Im z

2π/3

Re z

Im z

−π/6 → Re z

P(9√3, −9)

Im z

α → Re z

P(−3, −4)

From the diagram,
$\text{Arg}(-1 + i\sqrt{3}) = \frac{2}{3}\pi$.

From the diagram,
$\text{Arg}(9\sqrt{3} - 9i) = -\frac{1}{6}\pi$.

From the diagram,
$\text{Arg}(-3 - 4i)$
$= -\pi + \alpha$
$= -2.21$.

Exercise 20.4

1. Find the argument of the following complex numbers:
   (a) $2 + 2i$;
   (b) $1 + i\sqrt{3}$;
   (c) $3\sqrt{3} - 9i$;
   (d) $-i$;
   (e) 4;
   (f) $-\sqrt{3} + i$;
   (g) $6 - 2i\sqrt{3}$;
   (h) $-3 - 3i$;
   (i) $\frac{1}{4} + \frac{1}{4}i\sqrt{3}$.

2. Find the argument of each of the following complex numbers:
   (a) $4 + 3i$;
   (b) $-1 + 2i$;
   (c) $3 - 2i$;
   (d) $-4 - 5i$;
   (e) $3 + i$;
   (f) $1 - 4i$;
   (g) $-2 - 5i$;
   (h) $-3 + 7i$;
   (i) $-\sqrt{2} - 2i$.

3. Let z be the complex number $1 + 2i$ and let Arg $z = \theta$. If Arg$(z^2) = \alpha$, prove that $\alpha = 2\theta$.

4. Show that θ is an argument of the complex number $z = \cos\theta + i\sin\theta$.

5. Let $z = \cos\theta + i\sin\theta$ where $0 < \theta < \frac{1}{2}\pi$. Show that $z^2 = \cos 2\theta + i\sin 2\theta$ and that $|z| = |z^2| = 1$.

   Plot the points representing the complex numbers z, $-z^2$ and $1 - z^2$ on an Argand diagram and use the diagram to show that Arg$(1 - z^2) = \theta - \frac{1}{2}\pi$.

*6. If P represents the complex number z and P lies on the circle centre $(1, 0)$ with radius 1, show that $\arg(z - 1) = \arg(z^2)$. What is the complete set of points in the Argand plane for which $\arg(z - 1) = \arg(z^2)$?

20.5 The Polar Form of a Complex Number

If $z = x + iy$ is any complex number such that $|z| = r$ and $\arg z = \theta$, then $z = r(\cos\theta + i\sin\theta)$.

The form $z = r(\cos\theta + i\sin\theta)$ is called the *polar* or *modulus-argument* form of z while $z = x + iy$ is called the *Cartesian* form.

Note: cis θ ("sis theta") is a common abbreviation for $\underline{\cos}\theta + \underline{i}\underline{\sin}\theta$.

Example Express $z = 2\operatorname{cis}\frac{3}{4}\pi$ in Cartesian form.

$$z = 2\operatorname{cis}\tfrac{3}{4}\pi = 2\left(\cos\tfrac{3}{4}\pi + i\sin\tfrac{3}{4}\pi\right) = 2\left(-\tfrac{1}{2}\sqrt{2} + i\tfrac{1}{2}\sqrt{2}\right) = -\sqrt{2} + i\sqrt{2}.$$

Example Express $z = 2 - 2i\sqrt{3}$ in polar form.

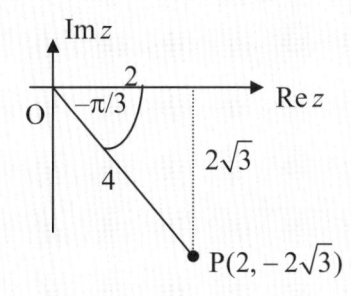

From the diagram, $|z| = 4$ and $\operatorname{Arg} z = -\tfrac{1}{3}\pi$.

Hence, $z = 4\left(\cos(-\tfrac{1}{3}\pi) + i\sin(-\tfrac{1}{3}\pi)\right)$.

Note: The following should be memorised:
$$\operatorname{cis} 0 = 1,\quad \operatorname{cis}(\tfrac{1}{2}\pi) = i,\quad \operatorname{cis}\pi = -1,\quad \operatorname{cis}(-\tfrac{1}{2}\pi) = -i.$$

Multiplication and Division of Complex Numbers in Polar Form

$z_1 = r_1 \operatorname{cis}\theta_1$ and $z_2 = r_2 \operatorname{cis}\theta_2 \Rightarrow z_1 z_2 = r_1 r_2 \operatorname{cis}(\theta_1 + \theta_2)$ and $\dfrac{z_1}{z_2} = \dfrac{r_1}{r_2}\operatorname{cis}(\theta_1 - \theta_2)$.

Proof
$$\begin{aligned}
z_1 z_2 &= r_1(\cos\theta_1 + i\sin\theta_1) \times r_2(\cos\theta_2 + i\sin\theta_2) \\
&= r_1 r_2 ([\cos\theta_1 \cos\theta_2 - \sin\theta_1 \sin\theta_2] + i[\sin\theta_1 \cos\theta_2 + \cos\theta_1 \sin\theta_2]) \\
&= r_1 r_2 (\cos[\theta_1 + \theta_2] + i\sin[\theta_1 + \theta_2]) \\
&= r_1 r_2 \operatorname{cis}(\theta_1 + \theta_2).
\end{aligned}$$

Chapter 20

$$\frac{z_1}{z_2} = \frac{r_1(\cos\theta_1 + i\sin\theta_1)}{r_2(\cos\theta_2 + i\sin\theta_2)}$$

$$= \frac{r_1}{r_2}\left(\frac{(\cos\theta_1 + i\sin\theta_1)(\cos\theta_2 - i\sin\theta_2)}{(\cos\theta_2 + i\sin\theta_2)(\cos\theta_2 - i\sin\theta_2)}\right)$$

$$= \frac{r_1}{r_2}\left(\frac{(\cos\theta_1\cos\theta_2 + \sin\theta_1\sin\theta_2) + i(\sin\theta_1\cos\theta_2 - \cos\theta_1\sin\theta_2)}{\cos^2\theta_2 + \sin^2\theta_2}\right)$$

$$= \frac{r_1}{r_2}(\cos[\theta_1 - \theta_2] + i\sin[\theta_1 - \theta_2])$$

$$= \frac{r_1}{r_2}\operatorname{cis}(\theta_1 - \theta_2).$$

Thus the modulus of a product of two complex numbers is equal to the product of the moduli of the numbers (as we have already seen), and an argument of the product is equal to the sum of the arguments of the numbers.

Also the modulus of a quotient of two complex numbers is equal to the quotient of the moduli of the numbers (as we have already seen), and an argument of the quotient is equal to the difference between the arguments of the numbers. [Argument of numerator minus argument of denominator.]

Example Multiply the complex numbers $z = 1 + i$ and $w = \sqrt{3} + i$ in both Cartesian and polar forms and hence find surd expressions for $\sin\frac{5}{12}\pi$ and $\cos\frac{5}{12}\pi$.

$z = 1 + i = \sqrt{2}\operatorname{cis}\frac{1}{4}\pi$ and $w = \sqrt{3} + i = 2\operatorname{cis}\frac{1}{6}\pi$

Therefore $zw = 2\sqrt{2}\operatorname{cis}(\frac{1}{4}\pi + \frac{1}{6}\pi) = 2\sqrt{2}\operatorname{cis}\frac{5}{12}\pi$ and

$zw = (1+i)(\sqrt{3}+i) = (\sqrt{3}-1) + i(\sqrt{3}+1)$.

Equating real and imaginary parts gives:
$2\sqrt{2}\cos\frac{5}{12}\pi = \sqrt{3} - 1$ and $2\sqrt{2}\sin\frac{5}{12}\pi = \sqrt{3} + 1$.

Thus $\sin\frac{5}{12}\pi = \frac{\sqrt{3}+1}{2\sqrt{2}} = \frac{1}{4}(\sqrt{6}+\sqrt{2})$ and $\cos\frac{5}{12}\pi = \frac{\sqrt{3}-1}{2\sqrt{2}} = \frac{1}{4}(\sqrt{6}-\sqrt{2})$.

Example Write $\dfrac{(-1+i)(\sqrt{3}-3i)}{-3+i\sqrt{3}}$ in polar form.

Complex Numbers

$$\frac{(-1+i)(\sqrt{3}-3i)}{-3+i\sqrt{3}} = \frac{(\sqrt{2}\operatorname{cis}\frac{3}{4}\pi)(2\sqrt{3}\operatorname{cis}-\frac{1}{3}\pi)}{2\sqrt{3}\operatorname{cis}\frac{5}{6}\pi}$$

$$= \sqrt{2}\operatorname{cis}\left(\tfrac{3}{4}\pi - \tfrac{1}{3}\pi - \tfrac{5}{6}\pi\right)$$

$$= \sqrt{2}\operatorname{cis}-\tfrac{5}{12}\pi.$$

Example If P represents the complex number z in the Argand plane, explain how to find the points representing the complex numbers iz and $-iz$.

If $z = r\operatorname{cis}\theta$ then $iz = (\operatorname{cis}\tfrac{1}{2}\pi)(r\operatorname{cis}\theta) = r\operatorname{cis}(\theta+\tfrac{1}{2}\pi)$.

Thus iz has a modulus equal to the modulus of z and an argument which is $\tfrac{1}{2}\pi$ more than the argument of z.

Thus the point representing iz is found by rotating P anticlockwise about the origin through an angle $\tfrac{1}{2}\pi$ or 90°.

Similarly $-iz = (\operatorname{cis}-\tfrac{1}{2}\pi)(r\operatorname{cis}\theta) = r\operatorname{cis}(\theta-\tfrac{1}{2}\pi)$.

Thus the point representing $-iz$ is found by rotating P clockwise about the origin through an angle $\tfrac{1}{2}\pi$ or 90°.

Exercise 20.5

1. Express each of the following complex numbers in Cartesian form:
   (a) $2\left(\cos\tfrac{1}{3}\pi + i\sin\tfrac{1}{3}\pi\right)$;
   (b) $\sqrt{2}\left(\cos-\tfrac{1}{4}\pi + i\sin-\tfrac{1}{4}\pi\right)$;
   (c) $3\left(\cos\tfrac{1}{2}\pi + i\sin\tfrac{1}{2}\pi\right)$;
   (d) $4(\cos\pi + i\sin\pi)$;
   (e) $\cos-\tfrac{5}{6}\pi + i\sin-\tfrac{5}{6}\pi$;
   (f) $2\left(\cos\tfrac{1}{6}\pi - i\sin\tfrac{1}{6}\pi\right)$;
   (g) $4\operatorname{cis}\tfrac{3}{4}\pi$;
   (h) $3\operatorname{cis}-\tfrac{2}{3}\pi$;
   (i) $\operatorname{cis}-\tfrac{1}{3}\pi$;
   (j) $6\operatorname{cis}\tfrac{1}{6}\pi$.

2. Express each of the following complex numbers in polar form:
   (a) 4;
   (b) $2i$;
   (c) -2;
   (d) $-5i$;
   (e) $\sqrt{3}+i$;
   (f) $2-2i$;
   (g) $-2+2i\sqrt{3}$;
   (h) $-4-4i$;
   (i) $-\sqrt{3}+3i$.

3. Express each of the following complex numbers in polar form:
   (a) $3+4i$;
   (b) $5+2i$;
   (c) $-2+i$;
   (d) $-4-3i$;
   (e) $2-3i$;
   (f) $-15-8i$.

Chapter 20

4. In each of the following express zw and $\dfrac{z}{w}$ in both Cartesian and polar forms:
   (a) $z = 1+i$, $w = -\sqrt{3}+i$;
   (b) $z = 2 - 2i\sqrt{3}$, $w = \text{cis}\frac{2}{3}\pi$;
   (c) $z = 6\,\text{cis}\frac{1}{6}\pi$, $w = 2\,\text{cis}-\frac{5}{6}\pi$;
   (d) $z = \cos\frac{1}{6}\pi - i\sin\frac{1}{6}\pi$, $w = -1+i\sqrt{3}$.

5. Multiply the complex numbers $z = \text{cis}\frac{1}{4}\pi$ and $w = \text{cis}\frac{1}{3}\pi$ in both polar and cartesian forms and hence obtain surd expressions for $\sin\frac{7}{12}\pi$ and $\cos\frac{7}{12}\pi$.

6. If $z = \text{cis}\,\theta$, prove that   (a) $z^* = \text{cis}(-\theta)$;   (b) $1/z = z^*$.

7. A square OABC in the complex plane is lettered anticlockwise with O the origin and A representing the complex number z. What complex numbers are represented by B and C?

8. The points P and Q represent the complex numbers z and w in the Argand plane and O is the origin. What can you say about the relative positions of P and Q if
   (a) $|z| = |w|$;
   (b) $z = iw$;
   (c) $\text{Arg}\,z = \text{Arg}\,w$;
   (d) $z + w = 0$;
   (e) $w = (1-i)z$;
   (f) $w = \dfrac{z}{1+i}$?

9. In the diagram, triangles OAP and OQR are similar. The point P represents z, the point Q represents w, and point A represents 1.

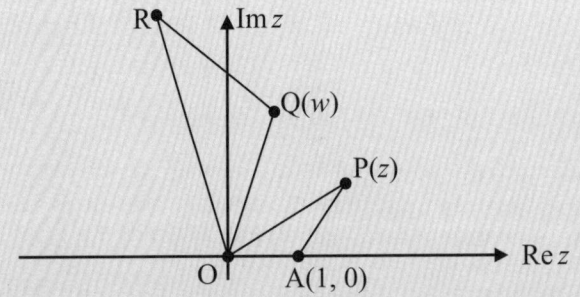

   Show that R represents zw.

*10. Points A, B, C and D represent the complex numbers z_1, z_2, z_3 and z_4 respectively. If ABCD is a square lettered anticlockwise, show that $z_3 = (1+i)z_2 - iz_1$, and find a similar expression for z_4 in terms of z_1 and z_2.

20.6 De Moivre's Theorem

Theorem For all integers n, $(\cos\theta + i\sin\theta)^n = \cos n\theta + i\sin n\theta$.

Proof Let P(n) be the proposition: $(\cos\theta + i\sin\theta)^n = \cos n\theta + i\sin n\theta$.

Then P(1) is true since $(\cos\theta + i\sin\theta)^1 = \cos\theta + i\sin\theta$.

Assume P(k) is true for some integer $k \geq 1$.

i.e. $(\cos\theta + i\sin\theta)^k = \cos k\theta + i\sin k\theta$.

Then $(\cos\theta + i\sin\theta)^{k+1}$

$= (\cos\theta + i\sin\theta)^k (\cos\theta + i\sin\theta)$
$= (\cos k\theta + i\sin k\theta)(\cos\theta + i\sin\theta)$
$= (\cos k\theta \cos\theta - \sin k\theta \sin\theta) + i(\sin k\theta \cos\theta + \cos k\theta \sin\theta)$
$= \cos(k\theta + \theta) + i\sin(k\theta + \theta)$
$= \cos(k+1)\theta + i\sin(k+1)\theta$.

Thus P(k) $\Rightarrow$ P(k + 1) and so P(n) is true for all positive integers n.

Let $m = -n$ where n is a positive integer. Then m is a negative integer.

Now $(\cos\theta + i\sin\theta)^m = (\cos\theta + i\sin\theta)^{-n}$

$= \dfrac{1}{(\cos\theta + i\sin\theta)^n}$

$= \dfrac{1}{\cos n\theta + i\sin n\theta}$ (from first part of proof)

$= \dfrac{1}{\cos n\theta + i\sin n\theta} \times \dfrac{\cos n\theta - i\sin n\theta}{\cos n\theta - i\sin n\theta}$

$= \dfrac{\cos n\theta - i\sin n\theta}{\cos^2 n\theta + \sin^2 n\theta}$

$= \cos n\theta - i\sin n\theta$
$= \cos(-n\theta) + i\sin(-n\theta)$
$= \cos m\theta + i\sin m\theta$.

Thus P(n) is true for all negative integers n.

Finally P(0) is true since $(\cos\theta + i\sin\theta)^0 = 1 = \cos 0 + i\sin 0$ and so P(n) is true for all integers n.

Example Express $\left(\frac{1}{2} + \frac{1}{2}i\sqrt{3}\right)^{10}$ in the form $a + ib$ where a and b are real.

Chapter 20

$$\left(\tfrac{1}{2}+\tfrac{1}{2}i\sqrt{3}\right)^{10}=\left(\cos\tfrac{1}{3}\pi+i\sin\tfrac{1}{3}\pi\right)^{10}=\cos\tfrac{10}{3}\pi+i\sin\tfrac{10}{3}\pi=-\tfrac{1}{2}-\tfrac{1}{2}i\sqrt{3}.$$

Example Expand $(\cos\theta+i\sin\theta)^3$ in two ways:
(1) using de Moivre's theorem ; (2) using the binomial theorem.
Hence prove: (a) $\cos 3\theta = 4\cos^3\theta - 3\cos\theta$;
(b) $\sin 3\theta = 3\sin\theta - 4\sin^3\theta$;
(c) $\tan 3\theta = \dfrac{3\tan\theta - \tan^3\theta}{1-3\tan^2\theta}$.

(1) $(\cos\theta+i\sin\theta)^3 = \cos 3\theta + i\sin 3\theta$.

(2) $(\cos\theta+i\sin\theta)^3$
$= \cos^3\theta + 3\cos^2\theta(i\sin\theta) + 3\cos\theta(i\sin\theta)^2 + (i\sin\theta)^3$
$= \cos^3\theta + 3i\cos^2\theta\sin\theta - 3\cos\theta\sin^2\theta - i\sin^3\theta$
$= (\cos^3\theta - 3\cos\theta\sin^2\theta) + i(3\cos^2\theta\sin\theta - \sin^3\theta)$.

(a) Equating real parts gives:
$\cos 3\theta = \cos^3\theta - 3\cos\theta\sin^2\theta$
$= \cos^3\theta - 3\cos\theta(1-\cos^2\theta)$
$= 4\cos^3\theta - 3\cos\theta$.

(b) Equating imaginary parts:
$\sin 3\theta = 3\cos^2\theta\sin\theta - \sin^3\theta$
$= 3(1-\sin^2\theta)\sin\theta - \sin^3\theta$
$= 3\sin\theta - 4\sin^3\theta$.

(c) From parts (a) and (b):
$\tan 3\theta = \dfrac{\sin 3\theta}{\cos 3\theta}$
$= \dfrac{3\sin\theta - 4\sin^3\theta}{4\cos^3\theta - 3\cos\theta}$
$= \dfrac{\dfrac{3\sin\theta}{\cos^3\theta} - \dfrac{4\sin^3\theta}{\cos^3\theta}}{\dfrac{4\cos^3\theta}{\cos^3\theta} - \dfrac{3\cos\theta}{\cos^3\theta}}$

Complex Numbers

$$= \frac{3\tan\theta \sec^2\theta - 4\tan^3\theta}{4 - 3\sec^2\theta}$$

$$= \frac{3\tan\theta(1 + \tan^2\theta) - 4\tan^3\theta}{4 - 3(1 + \tan^2\theta)}$$

$$= \frac{3\tan\theta - \tan^3\theta}{1 - 3\tan^2\theta}.$$

The General Solution of the Equation $z^n = \alpha$

Consider the equation $z^n = \alpha$ where α is a complex number and n is a positive integer. Let $\alpha = s\operatorname{cis}\phi$ and let $z = r\operatorname{cis}\theta$.

Then $r^n \operatorname{cis} n\theta = s\operatorname{cis}\phi$ and so $r^n \cos n\theta = s\cos\phi$ and $r^n \sin n\theta = s\sin\phi$.

Squaring and adding gives: $r^{2n}\left(\cos^2 n\theta + \sin^2 n\theta\right) = s^2\left(\cos^2\phi + \sin^2\phi\right)$ which simplifies to $r^{2n} = s^2$ or $r = s^{1/n}$ ($r > 0$).

Therefore $\cos n\theta = \cos\phi$ and $\sin n\theta = \sin\phi$

$\Rightarrow n\theta = \phi \ (+2\pi k, k \in \mathbb{Z})$

$\Rightarrow \theta = \dfrac{\phi}{n}\left(+\dfrac{2\pi k}{n}\right) = \dfrac{\phi}{n}, \dfrac{\phi}{n} \pm \dfrac{2\pi}{n}, \dfrac{\phi}{n} \pm \dfrac{4\pi}{n}, \ldots$.

Thus all n solutions of $z^n = \alpha$ have modulus $|\alpha|^{1/n}$ and successive arguments differ by $2\pi/n$. Therefore on an Argand diagram, the n solutions are equally spaced around the circle with centre at the origin and radius $|\alpha|^{1/n}$.

Example Solve the equation $z^4 = i$.

$z^4 = i = \operatorname{cis}\tfrac{1}{2}\pi$ and so

$z = \operatorname{cis}\left(\dfrac{\pi}{8} \pm \dfrac{2k\pi}{4}\right)(k \in \mathbb{Z}) = \operatorname{cis}\theta$ where $\theta = -\tfrac{7}{8}\pi, -\tfrac{3}{8}\pi, \tfrac{1}{8}\pi, \tfrac{5}{8}\pi$.

Example Find the five fifth roots of 32.

We need to solve the equation $z^5 = 32 = 32\operatorname{cis}0$.

Then $z = 32^{1/5}\operatorname{cis}\left(\dfrac{0}{5} + \dfrac{2k\pi}{5}\right)(k \in \mathbb{Z}) = 2\operatorname{cis}\theta$ where $\theta = 0, \pm\tfrac{2}{5}\pi, \pm\tfrac{4}{5}\pi$.

541

Chapter 20

Note: The non-real solutions in the previous example occur in conjugate pairs. We should expect this since the polynomial $z^5 - 32$ is real. But in the example before this, the non-real solutions do not occur in conjugate pairs since the polynomial $z^4 - i$ is not real.

Quadratic Polynomials with Conjugate Zeros

Given a pair of conjugate complex numbers, a real quadratic polynomial can always be found with these conjugates as its zeros.

Let $r \operatorname{cis} \theta$ and $r \operatorname{cis}(-\theta)$ be two complex conjugate zeros of a quadratic polynomial. The sum of these zeros is $2r\cos\theta$ and the product is r^2. Therefore a suitable quadratic is
$$z^2 - (2r\cos\theta)z + r^2.$$

Example Write down a real quadratic which has $2\operatorname{cis}\pm\tfrac{1}{3}\pi$ as its zeros.

Here $r = 2$ and $\theta = \tfrac{1}{3}\pi$ and so a suitable quadratic is
$$z^2 - (2r\cos\theta)z + r^2 = z^2 - (4\cos\tfrac{1}{3}\pi)z + 4 = z^2 - 2z + 4.$$

Example Solve the equation $z^6 + 1 = 0$ and by grouping the solutions into three pairs of conjugates, express $z^6 + 1$ as a product of 3 real quadratics.

$z^6 + 1 = 0$ and so $z^6 = -1 = \operatorname{cis}\pi$.

The solutions are $z = \operatorname{cis}\left(\tfrac{1}{6}\pi \pm \dfrac{2k\pi}{6}\right) (k \in \mathbb{Z}) = \operatorname{cis}\pm\tfrac{1}{6}\pi, \operatorname{cis}\pm\tfrac{1}{2}\pi, \operatorname{cis}\pm\tfrac{5}{6}\pi$.

Now $\operatorname{cis}\pm\tfrac{1}{6}\pi$ are the zeros of $z^2 - (2\cos\tfrac{1}{6}\pi)z + 1 = z^2 - z\sqrt{3} + 1$,

$\operatorname{cis}\pm\tfrac{1}{2}\pi$ are the zeros of $z^2 - (2\cos\tfrac{1}{2}\pi)z + 1 = z^2 + 1$,

$\operatorname{cis}\pm\tfrac{5}{6}\pi$ are the zeros of $z^2 - (2\cos\tfrac{5}{6}\pi)z + 1 = z^2 + z\sqrt{3} + 1$.

Therefore $z^6 + 1 = (z^2 - z\sqrt{3} + 1)(z^2 + 1)(z^2 + z\sqrt{3} + 1)$.

Example If $z = r\operatorname{cis}\theta$, show that (a) $z^n + z^{-n} = 2\cos n\theta$;
(b) $z^n - z^{-n} = 2i\sin n\theta$.

Hence prove that
$$\cos^5\theta = \tfrac{1}{16}\cos 5\theta + \tfrac{5}{16}\cos 3\theta + \tfrac{5}{8}\cos\theta, \text{ and}$$
$$\sin^5\theta = \tfrac{1}{16}\sin 5\theta - \tfrac{5}{16}\sin 3\theta + \tfrac{5}{8}\sin\theta.$$

(a) Since $z = r\operatorname{cis}\theta$, $z^n = \operatorname{cis} n\theta$ and $z^{-n} = \operatorname{cis}(-n\theta)$.

Thus $z^n + z^{-n} = \cos n\theta + i\sin n\theta + \cos n\theta - i\sin n\theta = 2\cos n\theta$.

(b) Also $z^n - z^{-n} = \cos n\theta + i\sin n\theta - \cos n\theta + i\sin n\theta = 2i\sin n\theta$.

Consider $\left(z + \dfrac{1}{z}\right)^5 = z^5 + 5z^4 \dfrac{1}{z} + 10z^3 \dfrac{1}{z^2} + 10z^2 \dfrac{1}{z^3} + 5z \dfrac{1}{z^4} + \dfrac{1}{z^5}$

$= \left(z^5 + \dfrac{1}{z^5}\right) + 5\left(z^3 + \dfrac{1}{z^3}\right) + 10\left(z + \dfrac{1}{z}\right).$

Using the result in part (a) gives:

$(2\cos\theta)^5 = 2\cos 5\theta + 5(2\cos 3\theta) + 10(2\cos\theta)$

$\Rightarrow \quad 32\cos^5\theta = 2\cos 5\theta + 10\cos 3\theta + 20\cos\theta$

$\Rightarrow \quad \cos^5\theta = \dfrac{1}{16}\cos 5\theta + \dfrac{5}{16}\cos 3\theta + \dfrac{5}{8}\cos\theta.$

Consider $\left(z - \dfrac{1}{z}\right)^5 = \left(z^5 - \dfrac{1}{z^5}\right) - 5\left(z^3 - \dfrac{1}{z^3}\right) + 10\left(z - \dfrac{1}{z}\right).$

Using the result of part (b) gives:

$(2i\sin\theta)^5 = 2i\sin 5\theta - 5(2i\sin 3\theta) + 10(2i\sin\theta)$

$\Rightarrow \quad 32i\sin^5\theta = 2i\sin 5\theta - 10i\sin 3\theta + 20i\sin\theta$

$\Rightarrow \quad \sin^5\theta = \dfrac{1}{16}\sin 5\theta - \dfrac{5}{16}\sin 3\theta + \dfrac{5}{8}\sin\theta.$

Exercise 20.6

1. Write each of the following in the form $a + ib$ where a and b are real:

   (a) $(\sqrt{3} + i)^6$; (b) $(1-i)^{10}$; (c) $\left(\dfrac{1 - i\sqrt{3}}{2}\right)^8$;

   (d) $\left(\dfrac{1+i}{\sqrt{2}}\right)^{12}$; (e) $(1 + i\sqrt{3})^9$; (f) $(2 + 2i)^5$.

2. Solve each of the following equations:

   (a) $z^3 = i$; (b) $z^4 = 1$; (c) $z^5 = 1$;
   (d) $z^5 = i$; (e) $z^4 = -4$; (f) $z^3 = 8i$;
   (g) $z^6 = i$; (h) $z^5 = -32$; (i) $z^4 = 8 + 8i\sqrt{3}$.

3. Find all the zeros of $z^{10} + 1$ which have negative imaginary parts.

Chapter 20

4. In each of the following find the zeros of the given polynomial and express this polynomial as a product of real quadratics:
   (a) $z^4 + 1$;
   (b) $z^8 + 1$;
   (c) $z^4 + 16$;
   (d) $z^6 + 64$;
   (e) $z^4 + 8$;
   (f) $z^{10} + 1$.

5. Find:
   (a) the square roots of $2i$;
   (b) the square roots of $3 + 4i$;
   (c) the 3 cube roots of -1 ;
   (d) the 4 fourth roots of i ;
   (e) the 5 fifth roots of -1 ;
   (f) the 6 sixth roots of $-i$.

6. Use de Moivre's theorem and the binomial theorem to express $\cos 4\theta$ in terms of $\cos\theta$.

7. Let $w = \cos\frac{2}{5}\pi + i\sin\frac{2}{5}\pi$.
   (a) Show that $1, w, w^2, w^3, w^4$ are the 5 zeros of $z^5 - 1$ where $z \in \mathbb{C}$.
   (b) By factorising $z^5 - 1$, or otherwise, prove that
   $$1 + w + w^2 + w^3 + w^4 = 0.$$
   (c) Find a real quadratic polynomial whose zeros are
   $$w + w^4 \text{ and } w^2 + w^3.$$
   (d) Hence show that $\cos\frac{2}{5}\pi = \frac{1}{4}(\sqrt{5} - 1)$.

8. Find the positive integers m for which $(\sqrt{3} + i)^m - (\sqrt{3} - i)^m = 0$.

9. If $z = 1, w_1, w_2, \cdots, w_{n-1}$ are the n roots of $z^n = 1$, find the equation which has $z = w_1, w_2, \cdots, w_{n-1}$ as its roots.

10. (a) Prove that $1 + \operatorname{cis}\theta + \operatorname{cis}2\theta + \operatorname{cis}3\theta + \cdots + \operatorname{cis}n\theta = \dfrac{1 - \operatorname{cis}(n+1)\theta}{1 - \operatorname{cis}\theta}$ for all positive integers n.

    (b) Evaluate $1 + \operatorname{cis}\frac{2}{7}\pi + \operatorname{cis}\frac{4}{7}\pi + \operatorname{cis}\frac{6}{7}\pi + \operatorname{cis}\frac{8}{7}\pi + \operatorname{cis}\frac{10}{7}\pi + \operatorname{cis}\frac{12}{7}\pi$.

11. (a) If $w = z + z^{-1}$ prove that
    (i) $z^2 + z^{-2} = w^2 - 2$;
    (ii) $z^4 + z^3 + z^2 + z + 1$
    $= z^2(w^2 + w + 1)$
    $= \left(z^2 + \tfrac{1}{2}z[1 + \sqrt{5}] + 1\right)\left(z^2 + \tfrac{1}{2}z[1 - \sqrt{5}] + 1\right)$.

544

(b) Show that the roots of $z^4 + z^3 + z^2 + z + 1 = 0$ are the four non-real roots of $z^5 = 1$.

(c) Deduce that $\cos 72° = \frac{1}{4}(\sqrt{5} - 1)$ and $\cos 36° = \frac{1}{4}(\sqrt{5} + 1)$.

12. (a) Find the solutions of the equation $\sin 5\theta = \sin 4\theta$ for $-\pi < \theta \leq \pi$.

(b) By considering $(\cos\theta + i\sin\theta)^5$, or otherwise, express $\dfrac{\sin 5\theta}{\sin \theta}$ in terms of $\cos\theta$ ($\sin\theta \neq 0$).

(c) Assuming, without proof, that $\dfrac{\sin 4\theta}{\sin \theta} = 8\cos^3\theta - 4\cos\theta$, show that when $\sin\theta \neq 0$,

$$\frac{\sin 5\theta - \sin 4\theta}{\sin \theta} = 16\cos^4\theta - 8\cos^3\theta - 12\cos^2\theta + 4\cos\theta + 1.$$

(d) By writing $x = 2\cos\theta$, show that the roots of the equation

$$x^4 - x^3 - 3x^2 + 2x + 1 = 0$$

are $1, 2\cos\frac{1}{9}\pi, 2\cos\frac{5}{9}\pi$ and $2\cos\frac{7}{9}\pi$.

*13. (a) Find the 5 fifth roots of 1.

(b) Show that if a non-zero complex number satisfies $\left|\dfrac{z+1}{z-1}\right| = 1$ then z is pure imaginary.

(c) If $\dfrac{z+1}{z-1} = \cos\theta + i\sin\theta$ where $\theta \neq 0$, show that $z = -i\cot\frac{1}{2}\theta$.

(d) Use the above results to solve the equation $\left(\dfrac{z+1}{z-1}\right)^5 = 1$.

(e) If z is a root of the equation in part (d), show that $z^4 + 2z^2 + 0.2 = 0$.

(f) Using the results in parts (d) and (e), write $z^4 + 2z^2 + 0.2$ as a product of two real quadratics.

(g) Hence show that $\tan^2\frac{1}{5}\pi \tan^2\frac{2}{5}\pi = 5$ and $\tan^2\frac{1}{5}\pi + \tan^2\frac{2}{5}\pi = 10$.

545

Chapter 20

Required Outcomes

After completing this chapter, a student should be able to:
- add, subtract, multiply and divide complex numbers in Cartesian form.
- find a real quadratic factor of a real polynomial given one non-real zero.
- convert a complex number to polar form from Cartesian form, and vice versa.
- multiply and divide complex numbers in polar form.
- describe the geometry of the complex plane corresponding to conjugate, addition and subtraction of complex numbers.
- state that a rotation of a complex number in the complex plane about the origin through 90° is equivalent to multiplication of that complex number by ±i.
- state and prove de Moivre's theorem for positive integers n.
- use de Moivre's theorem to solve equations of the form $z^n = \alpha$.

21 Probability Density Functions

21.1 Mean and Variance

If X is a continuous random variable, it is not a simple task to give meaning to the probability that X assumes a single value. If X can take any value between a and b and $P(X = c)$ is non-zero for all $c \in [a, b]$, then the sum of the infinite number of such probabilities would itself be infinite. Therefore we must assume that $P(X = c) = 0$. Clearly we need to define a more practical model.

Let X be a continuous random variable. Consider a function $f(x)$ which takes only non-negative values such that the area between the graph of $f(x)$ and the x-axis from $x = a$ to $x = b$ is equal to the probability that X takes a value which lies between a and b.

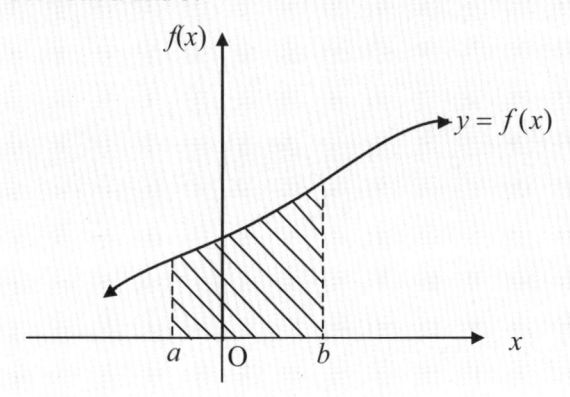

Then $f(x)$ is defined by

$$P(a \leq X \leq b) = \int_a^b f(x)\, dx.$$

Now as the sum of all possible probabilities must be 1, the total area under the curve must be 1. That is

$$\int_{-\infty}^{\infty} f(x)\, dx = 1.$$

For a discrete random variable X,

$$E(X) = \mu = \sum x f(x)$$
$$Var(X) = \sigma^2 = \sum (x-\mu)^2 f(x) = \sum x^2 f(x) - \mu^2.$$

547

Chapter 21

To obtain similar formulae for the mean and variance of a continuous random variable X, let us consider the probability that the value of X lies between x and $x + \delta x$. Provided that δx is small, this probability is given approximately by $f(x)\delta x$. Thus we let

$$E(X) = \mu = \lim_{\delta x \to 0} \sum x f(x) \, \delta x = \int_{-\infty}^{\infty} x f(x) \, dx, \text{ and}$$

$$\text{Var}(X) = \sigma^2 = \int_{-\infty}^{\infty} (x-\mu)^2 f(x) \, dx = \int_{-\infty}^{\infty} x^2 f(x) \, dx - \mu^2.$$

Example A continuous random variable X takes values between 0 and 1 and has a probability density function $f(x) = kx(1-x)$.

(a) Find the value of k.
(b) Calculate the mean and standard deviation of X.

(a) $\int_0^1 kx(1-x) \, dx = 1$

$\Rightarrow \quad k \int_0^1 (x - x^2) \, dx = 1$

$\Rightarrow \quad k \left[\tfrac{1}{2}x^2 - \tfrac{1}{3}x^3 \right]_0^1 = 1$

$\Rightarrow \quad k \left[\tfrac{1}{2} - \tfrac{1}{3} \right] = 1$

$\Rightarrow \quad k = 6.$

(b) $E(X) = \int_0^1 x f(x) \, dx$

$= 6 \int_0^1 x^2 (1-x) \, dx$

$= 6 \int_0^1 (x^2 - x^3) \, dx$

$= 6 \left[\tfrac{1}{3}x^3 - \tfrac{1}{4}x^4 \right]_0^1$

$= 6 \left[\tfrac{1}{3} - \tfrac{1}{4} \right]$

$= 0.5.$

Note: This should not be surprising since the probability density function is symmetrical about $x = \tfrac{1}{2}$.

$E(X^2) = \int_0^1 x^2 f(x) \, dx$

548

Probability Density Functions

$$= 6\int_0^1 (x^3 - x^4)\, dx$$

$$= 6\left[\tfrac{1}{4}x^4 - \tfrac{1}{5}x^5\right]_0^1$$

$$= 6\left[\tfrac{1}{4} - \tfrac{1}{5}\right]$$

$$= 0.3.$$

$\text{Var}(X) = E(X^2) - [E(X)]^2 = 0.3 - 0.25 = 0.05.$

$\text{SD}(X) = \sqrt{0.05} = 0.224.$

Uniform Distributions

One of the simplest types of continuous probability distribution is the rectangular or uniform distribution. A continuous random variable X is said to be uniformly distributed over the interval $[a, b]$ if its probability density function $f(x)$ has a constant non-zero value for $a \le x \le b$ and is zero elsewhere.

Example   Find the mean and variance of the continuous random variable X which is uniformly distributed over the interval $[0, 3]$.

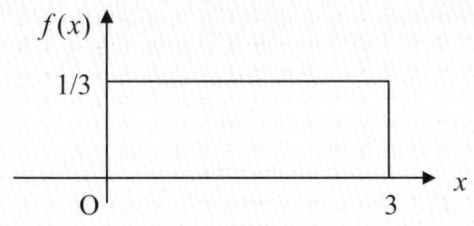

Since the total area under the graph of $f(x)$ must be 1,

$$f(x) = \begin{cases} \tfrac{1}{3} & \text{for } 0 \le x \le 3, \\ 0 & \text{elsewhere.} \end{cases}$$

Thus $E(X) = \int_0^3 x \times \tfrac{1}{3}\, dx = \left[\tfrac{1}{6}x^2\right]_0^3 = 1.5,$

and $\text{Var}(X) = \int_0^3 x^2 \times \tfrac{1}{3}\, dx - 1.5^2 = \left[\tfrac{1}{9}x^3\right]_0^3 - 1.5^2 = 3 - 2.25 = 0.75.$

Exercise 21.1

1. Find the mean and variance of the continuous random variable X which is uniformly distributed over the interval
   (a) $[0, 1]$; (b) $[2, 6]$;
   (c) $[0, b]$; (d) $[a, b]$.

549

Chapter 21

2. The radius of a circle is a random variable which is uniformly distributed between 10 cm and 15 cm.
   (a) Find the probability that the circumference lies between 50 cm and 60 cm.
   (b) Find the probability that the area of the circle exceeds 500 cm^2.

3. Triangle ABC is right-angled at B and AC = 10 cm. If BC = X cm and X is a random variable uniformly distributed between 6 cm and 8 cm, find the probability that the length of AB exceeds 7.5 cm.

4. A continuous random variable X takes values between 0 and 2 with a probability density function $f(x) = k(2x+1)$.
   Find: (a) the value of k; (b) the mean of X;
   (c) the standard deviation of X.

5. The continuous random variable X has a probability density function of the form
$$f(x) = \begin{cases} \frac{1}{4}x(3x-2) & 1 \le x \le 2 \\ 0 & \text{otherwise.} \end{cases}$$
   Find $P(|X - \mu| < \sigma)$.

6. A random variable X has a probability density function
$$f(x) = \begin{cases} mx^n & 0 \le x \le 1 \\ 0 & \text{otherwise.} \end{cases}$$
   If the mean of the distribution is $\frac{4}{5}$, find m and n.

7. A continuous random variable X has a probability density function defined by
$$f(x) = \begin{cases} \dfrac{k}{1+x} & 0 \le x \le 1 \\ 0 & \text{elsewhere.} \end{cases}$$

   (a) Find the value of k.

   (b) Find E(X) and Var(X).

21.2 Median and Mode

Mode

The mode or modal value of a continuous random variable X with probability density function $f(x)$ is the value of x for which $f(x)$ takes a maximum value. Thus the mode of a distribution is the x-coordinate of the maximum point on the graph of $y = f(x)$.

Example   The continuous random variable X has a probability density function given by $f(x) = \frac{3}{4}x^2(2-x)$, $0 \leq x \leq 2$. Find the mode of the distribution.

$f(x) = \frac{3}{4}x^2(2-x) = \frac{3}{4}(2x^2 - x^3)$
$f'(x) = \frac{3}{4}(4x - 3x^2) = \frac{3}{4}x(4-3x)$
$f'(x) = 0$ when $x = 0$, $x = \frac{4}{3}$ with the maximium turning point at $x = \frac{4}{3}$.
Therefore the mode is $\frac{4}{3}$.

Median

The median of a continuous random variable X with probability density function $f(x)$ is the value of m such that

$$\int_{-\infty}^{m} f(x)\,dx = \int_{m}^{\infty} f(x)\,dx = 0.5.$$

Thus the line $x = m$ divides the area under the graph of $f(x)$ into two equal areas.

Example   A continuous random variable X has a probability density function $f(x) = \frac{3}{26}(x+1)^2$, $0 \leq x \leq 2$. Find the median of the distribution.

Let the median be m.
Then $\int_{0}^{m} f(x)\,dx = \frac{1}{2}$

$\Rightarrow$ $\frac{3}{26}\int_{0}^{m}(x+1)^2\,dx = \frac{1}{2}$

$\Rightarrow$ $\left[\frac{1}{3}(x+1)^3\right]_{0}^{m} = \frac{13}{3}$

$\Rightarrow$ $(m+1)^3 - 1 = 13$

$\Rightarrow$ $m = \sqrt[3]{14} - 1 \approx 1.41$.

551

Chapter 21

Exercise 21.2

1. For each of the following probability density functions of the random variable X, find the mode.
   (a) $f(x) = \frac{3}{4}x(2-x)$, $0 \leq x \leq 2$;
   (b) $f(x) = \frac{15}{64}x^2(4-x^2)$, $0 \leq x \leq 2$.

2. A random variable X has a probability density function
$$f(x) = \begin{cases} kx(4-x)^2 & 0 \leq x \leq 4 \\ 0 & \text{otherwise.} \end{cases}$$
   Find the value of k and calculate the mean and mode of X.

3. A random variable X has a probability density function
$$f(x) = \begin{cases} \frac{3}{4}(x+1)^2 & -1 \leq x \leq 0 \\ \frac{3}{4} & 0 < x \leq 1. \end{cases}$$
   Calculate the median of X.

4. A random variable X has a probability density function $f(x) = kx^2$, $1 \leq x \leq 2$.
   Find (a) the value of the constant k ;
   (b) the mean, μ ;
   (c) the median, m ;
   (d) $P(1.2 < X < 1.5)$.

5. A continuous random variable X is distributed at random between the values 2 and 3 and has a probability density function $f(x) = 6/x^2$. Find the median value of X.

6. A continuous random variable X has a probability density function
$$f(x) = \begin{cases} ax^2 + bx & 0 \leq x \leq 2 \\ 0 & \text{elsewhere.} \end{cases}$$
   The mean of the distribution is $\frac{11}{8}$. Find:
   (a) the values of a and b ;
   (b) the variance of X ;
   (c) the median value of X.

7. (a) Use your GDC to evaluate $\int_0^\pi x\cos x \, dx$.

   (b) A continuous random variable X has a probability density function

   $$f(x) = \begin{cases} k(1+\cos x) & 0 \leq x \leq \pi \\ 0 & \text{otherwise.} \end{cases}$$

   Find the *exact* value of k and calculate the mean, μ, and median, m, of the distribution.

8. The probability density function of the random variable X is defined by
   $$f(x) = \frac{1}{1+x} \text{ for } 0 \leq x \leq k \text{ and } f(x) = 0 \text{ elsewhere.}$$

   Find: (a) the value of k;
   (b) the mean of X;
   (c) the standard deviation of X;
   (d) the median of X.

9. A continuous random variable X has probability density function

   $$f(x) = \begin{cases} kx(9-x^2) & \text{for } 0 \leq x \leq 3 \\ 0 & \text{elsewhere.} \end{cases}$$

   Find: (a) the value of k; (b) the mean;
   (c) the mode; (d) the median.

10. A random variable X has a probability density function $f(x) = ke^{-x}$, $2 \leq x \leq 3$.

    (a) Show that $k = \dfrac{e^3}{e-1}$.
    (b) Calculate the mean, μ.
    (c) Calculate the standard deviation, σ.
    (d) Calculate the median, m, of the distribution.
    (e) Evaluate $P(X > 2.5)$.
    (f) Evaluate $P(\mu - \sigma < X < \mu + \sigma)$.

Chapter 21

> **Required Outcomes**
>
> After completing this chapter, a student should be able to:
> - calculate the mean, median and mode of a continuous random variable whose probability density function is known.
> - calculate the variance and standard deviation of a continuous random variable.

22 Inverse Trigonometric Functions

22.1 The Inverse Sine, Inverse Cosine and Inverse Tangent Functions

The Inverse Sine Function

The function $f : x \mapsto \sin x$ with domain $\left[-\frac{1}{2}\pi, \frac{1}{2}\pi\right]$ and range $[-1, 1]$ is one-to-one and so its inverse function exists.
We write $f^{-1} : x \mapsto \arcsin x$ or $f^{-1} : x \mapsto \sin^{-1} x$.

Note: Do not confuse $\sin^{-1} x$ with $(\sin x)^{-1} = \csc x$; the inverse and reciprocal functions are clearly different.

The domain of $\arcsin x$ is $[-1, 1]$ and the range is $\left[-\frac{1}{2}\pi, \frac{1}{2}\pi\right]$.

Example $\arcsin \frac{1}{2} = \frac{1}{6}\pi$, $\arcsin(-1) = -\frac{1}{2}\pi$, but $\arcsin 2$ *does not exist* since 2 does not belong to the domain of $\arcsin x$.

The graph of $y = \arcsin x$ is the reflection of the graph of $y = \sin x$ in the line $y = x$.

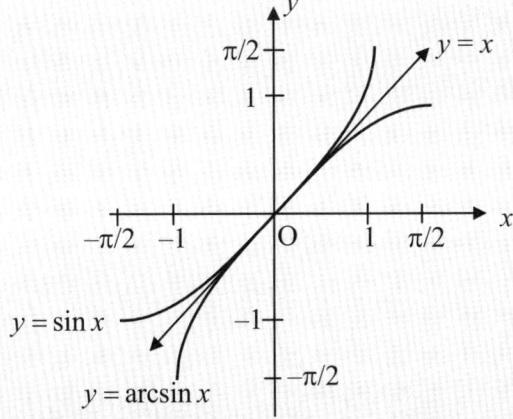

The Inverse Cosine Function

The function $f : x \mapsto \cos x$ with domain $[0, \pi]$ and range $[-1, 1]$ is one-to-one and so its inverse function exists.
We write $f^{-1} : x \mapsto \arccos x$ or $f^{-1} : x \mapsto \cos^{-1} x$.

Chapter 22

As with the inverse sine function, we must not confuse the inverse cosine function $\cos^{-1} x$ with the reciprocal cosine function $\sec x = (\cos x)^{-1}$.
The domain of $\arccos x$ is $[-1, 1]$ and the range is $[0, \pi]$.

Example $\arccos 1 = 0$, $\arccos\left(-\frac{1}{2}\sqrt{3}\right) = \frac{5}{6}\pi$, but $\arccos\sqrt{2}$ does not exist since $\sqrt{2}$ does not belong to the domain of $\arccos x$.

The graph of $y = \arccos x$ is the reflection of the graph of $y = \cos x$ in the line $y = x$.

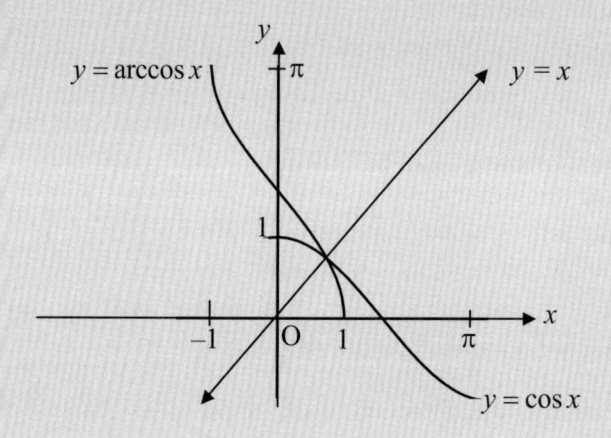

The Inverse Tangent Function

The function $f : x \mapsto \tan x$ with domain $\left]-\frac{1}{2}\pi, \frac{1}{2}\pi\right[$ and range $\mathbb{R}$ is one-to-one and so its inverse function exists.
We write $f^{-1} : x \mapsto \arctan x$ or $f^{-1} : x \mapsto \tan^{-1} x$.

The domain of $\arctan x$ is $\mathbb{R}$ and the range is $\left]-\frac{1}{2}\pi, \frac{1}{2}\pi\right[$.
Once again we must not confuse the inverse tangent function $\tan^{-1} x$ with the reciprocal tangent function $\cot x = (\tan x)^{-1}$.

Example $\arctan\sqrt{3} = \frac{1}{3}\pi$, $\arctan(-1) = -\frac{1}{4}\pi$, and indeed $\arctan x$ exists for *all* real values of x since the domain of $\arctan x$ is $\mathbb{R}$.

The graph of $y = \arctan x$ is the reflection of the graph of $y = \tan x$ in the line $y = x$.

Inverse Trigonometric Functions

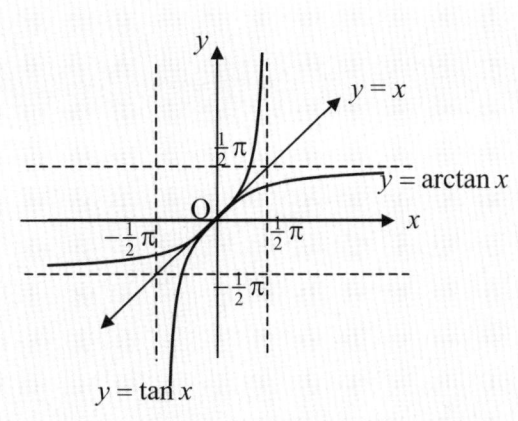

Example (a) For which numbers a is it true that $\tan(\arctan a) = a$?

(b) For which numbers b is it true that $\arctan(\tan b) = b$?

(a) $\tan(\arctan a) = a$ is true for all a since $\arctan a$ is defined to be the number whose tangent is a.

(b) $\arctan(\tan b) = b$ is only true for $-\tfrac{1}{2}\pi < b < \tfrac{1}{2}\pi$ since *all* values of $\arctan x$ lie in the interval $\left]-\tfrac{1}{2}\pi, \tfrac{1}{2}\pi\right[$.

Example Evaluate each of the following where possible:

(a) $\sin\left(\arcsin\tfrac{1}{2}\right)$;   (b) $\sin(\arccos 0.6)$;

(c) $\tan\left(\arctan\tfrac{1}{4}\pi\right)$;   (d) $\arctan\left(\tan\tfrac{1}{4}\pi\right)$;

(e) $\cos\left(\arctan\left(-\sqrt{3}\right)\right)$;   (f) $\arctan\left(\tan\tfrac{3}{4}\pi\right)$.

(a) $\sin\left(\arcsin\tfrac{1}{2}\right) = \sin\tfrac{1}{6}\pi = \tfrac{1}{2}$.

(b) $\sin(\arccos 0.6) = \sin\theta$ where $\cos\theta = 0.6$, $0 \le \theta \le \pi$.
$\sin\theta = 0.8$, $0 \le \theta \le \pi$, and so $\sin(\arccos 0.6) = 0.8$.

(c) $\tan\left(\arctan\tfrac{1}{4}\pi\right) = \tfrac{1}{4}\pi$ since $\tan(\arctan x) = x$ for all x.

(d) $\arctan\left(\tan\tfrac{1}{4}\pi\right) = \arctan 1 = \tfrac{1}{4}\pi$.

(e) $\cos\left(\arctan\left(-\sqrt{3}\right)\right) = \cos\left(-\tfrac{1}{3}\pi\right) = \tfrac{1}{2}$.

(f) $\arctan\left(\tan\tfrac{3}{4}\pi\right) = \arctan(-1) = -\tfrac{1}{4}\pi$.

Example Prove that $4\arctan\tfrac{1}{5} - \arctan\tfrac{1}{239} = \tfrac{1}{4}\pi$.

557

Chapter 22

Let $x = \arctan\frac{1}{5}$ then $\tan x = \frac{1}{5}$.

$\Rightarrow \quad \tan 2x = \dfrac{2\tan x}{1-\tan^2 x} = \dfrac{\frac{2}{5}}{1-\frac{1}{25}} = \dfrac{5}{12}$

$\Rightarrow \quad \tan 4x = \dfrac{2\tan 2x}{1-\tan^2 2x} = \dfrac{\frac{5}{6}}{1-\frac{25}{144}} = \dfrac{120}{119}$

Let $y = \arctan\frac{1}{239}$ then $\tan y = \frac{1}{239}$ and $4\arctan\frac{1}{5} - \arctan\frac{1}{239} = 4x - y$.

Now $\tan(4x - y) = \dfrac{\tan 4x - \tan y}{1 + \tan 4x \tan y}$

$= \dfrac{\frac{120}{119} - \frac{1}{239}}{1 + \left(\frac{120}{119}\right)\left(\frac{1}{239}\right)}$

$= \dfrac{(120)(239) - 119}{(119)(239) + 120}$

$= \dfrac{(119)(239) + 239 - 119}{(119)(239) + 120}$

$= 1.$

Therefore $4\arctan\frac{1}{5} - \arctan\frac{1}{239} = \frac{1}{4}\pi$.

Note: Both x and y, $x > y$, are positive and less than $\frac{1}{4}\pi$ and so $4x - y$ is positive and less than π. Thus even though $\tan\frac{5}{4}\pi = 1$, $4x - y \neq \frac{5}{4}\pi$.

Exercise 22.1

1. Evaluate where possible:
   (a) $\arcsin\frac{1}{2}$;
   (b) $\arccos 0$;
   (c) $\arctan(-1)$;
   (d) $\arcsin\left(-\frac{1}{2}\sqrt{3}\right)$;
   (e) $\arccos\left(-\frac{1}{2}\right)$;
   (f) $\arctan\left(-\sqrt{3}\right)$;
   (g) $\arccos\left(\cos\frac{1}{4}\pi\right)$;
   (h) $\arcsin\left(\sin\frac{3}{4}\pi\right)$;
   (i) $\arctan\left(\tan\frac{3}{4}\pi\right)$;
   (j) $\arcsin(\sin 0.2)$;
   (k) $\arctan(\tan(-\pi))$;
   (l) $\arccos(\cos(-\pi))$;
   (m) $\arcsin\left(\cos\frac{1}{4}\pi\right)$;
   (n) $\arccos\left(\sin\frac{1}{3}\pi\right)$;
   (o) $\arccos\left(\tan\frac{1}{3}\pi\right)$;
   (p) $\cos\left(\arcsin\frac{3}{5}\right)$;
   (q) $\sin\left(\arccos\left(-\frac{1}{2}\right)\right)$;
   (r) $\cos\left(\arcsin\left(-\frac{1}{2}\right)\right)$;
   (s) $\sin\left(2\arccos\frac{3}{5}\right)$;
   (t) $\cos\left(2\arctan\frac{3}{4}\right)$;
   (u) $\cos\left(2\arcsin\frac{1}{3}\right)$;
   (v) $\tan(2\arctan 2)$;
   (w) $\tan\left(2\arcsin\frac{5}{13}\right)$;
   (x) $\sin\left(2\arccos\frac{1}{5}\right)$.

Inverse Trigonometric Functions

2. (a) For which numbers a is it true that $\arcsin(\sin a) = a$?
   (b) For which numbers b is it true that $\sin(\arcsin b) = b$?

3. (a) For which numbers a is it true that $\arccos(\cos a) = a$?
   (b) For which numbers b is it true that $\cos(\arccos b) = b$?

4. Using the identity $\cos(\frac{1}{2}\pi - \theta) = \sin\theta$, show that $\arcsin x + \arccos x = \frac{1}{2}\pi$ for all $x \in [-1, 1]$.

5. Use $\tan 2\theta = \dfrac{2\tan\theta}{1-\tan^2\theta}$ to show that $2\arctan a = \arctan\left(\dfrac{2a}{1-a^2}\right)$ for suitable a. Specify the allowable values of a, and indicate the necessary modifications in the formula for other values of a.

6. Prove that
   (a) $\arctan\frac{1}{2} + \arctan\frac{1}{3} = \frac{1}{4}\pi$;
   (b) $2\arctan\frac{1}{3} + \arctan\frac{1}{7} = \frac{1}{4}\pi$.

22.2 The Derivatives of the Inverse Trigonometric Functions

Consider the function $y = \arcsin x$, $-1 < x < 1$, $-\frac{1}{2}\pi < y < \frac{1}{2}\pi$, then $\sin y = x$. Differentiating this with respect to x gives

$$\cos y \frac{dy}{dx} = 1$$

$$\Rightarrow \quad \frac{dy}{dx} = \frac{1}{\cos y}$$

$$= \frac{1}{\sqrt{1-\sin^2 y}} \quad \left(\text{since } \cos y > 0 \text{ for } -\tfrac{1}{2}\pi < y < \tfrac{1}{2}\pi\right)$$

$$= \frac{1}{\sqrt{1-x^2}} \quad (-1 < x < 1).$$

Therefore $\quad \boxed{\dfrac{d}{dx}(\arcsin x) = \dfrac{1}{\sqrt{1-x^2}} \quad (|x| < 1).}$

Consider the function $y = \arccos x$, $-1 < x < 1$, $0 < y < \pi$, then $\cos y = x$. Differentiating this with respect to x gives

559

Chapter 22

$$-\sin y \frac{dy}{dx} = 1$$

$$\Rightarrow \qquad \frac{dy}{dx} = \frac{-1}{\sin y}$$

$$= \frac{-1}{\sqrt{1-\cos^2 y}} \quad (\text{since } \sin y > 0 \text{ for } 0 < y < \pi)$$

$$= \frac{-1}{\sqrt{1-x^2}} \quad (-1 < x < 1).$$

Therefore $\quad \boxed{\dfrac{d}{dx}(\arccos x) = \dfrac{-1}{\sqrt{1-x^2}} \quad (|x|<1).}$

Consider the function $y = \arctan x$, $x \in \mathbb{R}$, $-\tfrac{1}{2}\pi < y < \tfrac{1}{2}\pi$, then $\tan y = x$.
Differentiating this with respect to x gives

$$\sec^2 y \frac{dy}{dx} = 1$$

$$\Rightarrow \qquad \frac{dy}{dx} = \frac{1}{\sec^2 y}$$

$$= \frac{1}{1+\tan^2 y}$$

$$= \frac{1}{1+x^2} \quad (x \in \mathbb{R}).$$

Therefore $\quad \boxed{\dfrac{d}{dx}(\arctan x) = \dfrac{1}{1+x^2} \quad (x \in \mathbb{R}).}$

Example Differentiate each of the following with respect to x:

(a) $\arctan(x^2+1)$; (b) $x \arcsin 2x$; (c) $\sin(\arccos 4x)$.

(a) $\dfrac{d}{dx}\left(\arctan(x^2+1)\right) = \dfrac{2x}{1+(x^2+1)^2}$

(b) $\dfrac{d}{dx}(x \arcsin 2x) = \arcsin 2x + x\left(\dfrac{2}{\sqrt{1-(2x)^2}}\right) = \arcsin 2x + \dfrac{2x}{\sqrt{1-4x^2}}$

(c) $\dfrac{d}{dx}(\sin(\arccos 4x)) = \cos(\arccos 4x) \dfrac{-4}{\sqrt{1-(4x)^2}} = \dfrac{-16x}{\sqrt{1-16x^2}}$

Inverse Trigonometric Functions

Note: Since $0 \leq \arccos 4x \leq \pi$, $\sin(\arccos 4x) \geq 0$ and

$\sin(\arccos 4x) = \sqrt{1-16x^2}$. Differentiating this expression gives the same result:

$$\frac{d}{dx}\sqrt{1-16x^2} = \frac{1}{2}(1-16x^2)^{-1/2}(-32x) = \frac{-16x}{\sqrt{1-16x^2}}.$$

Exercise 22.2

1. Differentiate each of the following with respect to x:
   (a) $\arcsin 3x$;
   (b) $\arccos(x+1)$;
   (c) $\arctan 2x$;
   (d) $\arcsin\sqrt{2x+1}$;
   (e) $\arccos(x^2-1)$;
   (f) $\arctan(x^2)$.

2. Differentiate each of the following with respect to x:
   (a) $\sin(\arctan x)$;
   (b) $\ln(\arcsin x)$;
   (c) $x \arctan(x^2)$.

3. For the function $f(x) = \arctan 2x$,
   (a) sketch the graph of $y = f(x)$;
   (b) evaluate $f(\tfrac{1}{2})$;
   (c) solve the equation $f(x) = 1$;
   (d) calculate the gradient of the curve at $x = 3$;
   (e) find the equation of the tangent to the curve at $x = -\tfrac{1}{2}$.

4. For the function $f(x) = \arcsin(\tfrac{1}{2}x)$,
   (a) sketch the graph of $y = f(x)$;
   (b) evaluate $f(-1)$;
   (c) solve the equation $f(x) = -\tfrac{1}{2}$;
   (d) calculate the gradient of the curve at $x = \sqrt{3}$;
   (e) find the equation of the tangent to the curve at the origin.

5. (a) Find the coordinates of the points on the curve $y = \arctan x$ where the tangent is parallel to $x = 2y$.
   (b) Find the coordinates of the points on the curve $y = \arcsin x$ where the tangent is parallel to $y = 2x$.

*6. (a) Given that $y = \dfrac{1}{x}\sqrt{1-x^2}$ $(-1 < x < 1, \ x \neq 0)$, prove that $\dfrac{dy}{dx}$ is always negative and sketch the graph of y.

561

Chapter 22

(b) Given that $z = \sqrt{1-x^2} \arcsin x$ $(-1 < x < 1)$, find $\dfrac{dz}{dx}$. Using your sketch in part (a), or otherwise, determine the number of turning points on the graph of z, and sketch this graph.

*7. Let $y = \arcsin\left(\sqrt{x}\right)$ where $0 \le x \le 1$.

(a) Show that $\dfrac{dy}{dx} \ge 1$ for $0 < x < 1$, and sketch the graph of y.

(b) By considering your sketch in part (a), show that
$$\int_0^1 \arcsin\sqrt{x}\,dx + \int_0^{\pi/2} \sin^2 y\,dy = \frac{1}{2}\pi.$$
Hence evaluate $\int_0^1 \arcsin\sqrt{x}\,dx$.

22.3 Integrals Involving Inverse Trigonometric Functions

From $\dfrac{d}{dx}(\arcsin x) = \dfrac{1}{\sqrt{1-x^2}}$, we have $\int \dfrac{1}{\sqrt{1-x^2}}\,dx = \arcsin x + c$ $(|x| < 1)$.

Also, from $\dfrac{d}{dx}(\arctan x) = \dfrac{1}{1+x^2}$, we have $\int \dfrac{1}{1+x^2}\,dx = \arctan x + c$ $(x \in \mathbb{R})$.

If $y = \arcsin\dfrac{x}{a}$, $a > 0$, then $\dfrac{dy}{dx} = \left(\dfrac{1}{a}\right)\dfrac{1}{\sqrt{1 - \left(\dfrac{x}{a}\right)^2}} = \dfrac{1}{\sqrt{a^2 - x^2}}$ $(|x| < a)$.

Therefore $\boxed{\int \dfrac{1}{\sqrt{a^2 - x^2}}\,dx = \arcsin\left(\dfrac{x}{a}\right) + c \quad (-a < x < a,\ a > 0).}$

If $y = \dfrac{1}{a}\arctan\left(\dfrac{x}{a}\right)$, $a > 0$, then $\dfrac{dy}{dx} = \left(\dfrac{1}{a^2}\right)\dfrac{1}{1 + \left(\dfrac{x}{a}\right)^2} = \dfrac{1}{a^2 + x^2}$ $(x \in \mathbb{R})$.

Therefore $\boxed{\int \dfrac{1}{a^2 + x^2}\,dx = \dfrac{1}{a}\arctan\left(\dfrac{x}{a}\right) + c \quad (x \in \mathbb{R},\ a > 0).}$

Inverse Trigonometric Functions

Example Integrate $\dfrac{1}{4+9x^2}$ with respect to x.

$$\int \dfrac{dx}{4+9x^2} = \dfrac{1}{9}\int \dfrac{dx}{\left(\tfrac{2}{3}\right)^2 + x^2} = \left(\dfrac{1}{9}\right)\dfrac{3}{2}\arctan\dfrac{3x}{2} + c = \dfrac{1}{6}\arctan\dfrac{3x}{2} + c$$

Example Evaluate $\displaystyle\int_{1}^{\sqrt{3}} \dfrac{dx}{\sqrt{4-x^2}}$.

$$\int_{1}^{\sqrt{3}} \dfrac{dx}{\sqrt{4-x^2}} = \left[\arcsin\dfrac{x}{2}\right]_{1}^{\sqrt{3}} = \arcsin\dfrac{\sqrt{3}}{2} - \arcsin\dfrac{1}{2} = \tfrac{1}{3}\pi - \tfrac{1}{6}\pi = \tfrac{1}{6}\pi.$$

Example Integrate $\dfrac{1}{\sqrt{7-6x-x^2}}$ with respect to x.

$$\int \dfrac{dx}{\sqrt{7-6x-x^2}} = \int \dfrac{dx}{\sqrt{16-(x+3)^2}} = \arcsin\left(\dfrac{x+3}{4}\right) + c \quad (-7 < x < 1)$$

Exercise 22.3

1. Integrate each of the following with respect to x:

   (a) $\dfrac{3}{\sqrt{9-x^2}}$; (b) $\dfrac{2}{\sqrt{16-x^2}}$; (c) $\dfrac{10}{\sqrt{25-x^2}}$;

   (d) $\dfrac{1}{\sqrt{1-4x^2}}$; (e) $\dfrac{6}{\sqrt{4-9x^2}}$; (f) $\dfrac{4}{\sqrt{2-3x^2}}$.

2. Integrate each of the following with respect to x:

   (a) $\dfrac{2}{4+x^2}$; (b) $\dfrac{6}{9+x^2}$; (c) $\dfrac{4}{16+x^2}$;

   (d) $\dfrac{3}{9+4x^2}$; (e) $\dfrac{1}{1+9x^2}$; (f) $\dfrac{2}{3+4x^2}$.

3. Find an antiderivative of each of the following:

   (a) $\dfrac{1}{\sqrt{4-(x+1)^2}}$; (b) $\dfrac{2}{\sqrt{1-(x-3)^2}}$; (c) $\dfrac{1}{\sqrt{3-2x-x^2}}$;

   (d) $\dfrac{1}{4+(x-3)^2}$; (e) $\dfrac{4}{10-2x+x^2}$; (f) $\dfrac{2}{4x^2+16x+25}$.

Chapter 22

4. Evaluate:
   (a) $\int_0^{3/2} \dfrac{dx}{\sqrt{9-x^2}}$;
   (b) $\int_{-5}^{-1} \dfrac{dx}{\sqrt{7-6x-x^2}}$;
   (c) $\int_0^2 \dfrac{2\,dx}{4+x^2}$;
   (d) $\int_{1/2}^{\sqrt{3}/2} \dfrac{2\,dx}{4x^2+1}$.

5. Find an antiderivative of each of the following:
   (a) $\dfrac{3}{3x+1}$;
   (b) $\dfrac{3}{3x^2+1}$;
   (c) $\dfrac{3x}{3x^2+1}$.

6. Find an antiderivative of each of the following:
   (a) $\dfrac{2}{1-4x}$;
   (b) $\dfrac{2}{\sqrt{1-4x^2}}$;
   (c) $\dfrac{2x}{\sqrt{1-4x^2}}$.

7. Find an antiderivative of each of the following:
   (a) $\dfrac{4x-1}{\sqrt{1-x^2}}$;
   (b) $\dfrac{3x+4}{x^2+4}$;
   (c) $\dfrac{2x-3}{4x^2+9}$;
   (d) $\dfrac{3x+2}{\sqrt{2-3x^2}}$;
   (e) $\left(4x - \dfrac{1}{\sqrt{x^2+1}}\right)^2$;
   (f) $\left(3 + \dfrac{2x}{\sqrt{x^2+4}}\right)^2$.

8. Evaluate:
   (a) $\int_0^2 \dfrac{x+1}{4+3x^2}\,dx$;
   (b) $\int_2^{2\sqrt{3}} \dfrac{(1+x)^2}{4+x^2}\,dx$.

Required Outcomes

After completing this chapter, a student should be able to:
- differentiate the inverse trigonometric functions arcsin x, arccos x and arctan x.
- sketch graphs of the inverse trigonometric functions.
- integrate functions of the forms
$$\dfrac{f'(x)}{\sqrt{a^2-[f(x)]^2}}, \quad -a < f(x) < a, \text{ and } \dfrac{f'(x)}{[f(x)]^2+a^2}.$$

23 Further Integration

23.1 Integration by Substitution

If $F(x)$ is an antiderivative of $f(x)$, then $F'(x) = f(x)$ and $F(x) = \int f(x)\,dx$.

If u is a function of x, then by the chain rule, $\dfrac{d}{dx}F(u) = F'(u)\dfrac{du}{dx} = f(u)\dfrac{du}{dx}$.

Hence $F(u)$ is an antiderivative of $f(u)\dfrac{du}{dx}$ (with respect to x).

Thus $F(u) = \int \left(f(u)\dfrac{du}{dx} \right) dx$.

But $F(u) = \int f(u)\,du$.

Therefore $\boxed{\int \left(f(u)\dfrac{du}{dx} \right) dx = \int f(u)\,du}$.

This is referred to as the *substitution formula*.

Example Integrate $2x\sqrt{x^2+1}$ with respect to x.

$$\begin{aligned}
\int 2x\sqrt{x^2+1}\,dx &= \int \sqrt{u}\,\dfrac{du}{dx}\,dx \quad (u = x^2+1) \\
&= \int u^{1/2}\,du \quad \text{(substitution formula)} \\
&= \dfrac{2}{3}u^{3/2} + c \\
&= \dfrac{2}{3}(x^2+1)^{3/2} + c.
\end{aligned}$$

The substitution formula can also be used to integrate certain functions where the type of substitution is not at first evident.

Example Find $\int \dfrac{x}{\sqrt{x-3}}\,dx \quad (x > 3)$.

565

Chapter 23

$$\int \frac{x}{\sqrt{x-3}}\, dx = \int \frac{u+3}{\sqrt{u}} \frac{dx}{du}\, du \quad (u = x-3,\ x = u+3 \text{ and } \frac{dx}{du} = 1)$$

$$= \int \left(u^{1/2} + 3u^{-1/2}\right) du$$

$$= \frac{2}{3} u^{3/2} + 6u^{1/2} + c$$

$$= \frac{2}{3}(x-3)^{3/2} + 6(x-3)^{1/2} + c.$$

Definite Integrals by Substitution

The rule is
$$\int_a^b f(x)\, dx = \int_{u(a)}^{u(b)} \left(f(u) \frac{du}{dx} \right) dx.$$

Mathematicians often use "*differentials*" when evaluating integrals by substitution. Thus if we use the substitution $u = x^2 + 1$ we write $du = 2x\, dx$ using the differentials "du" and "dx", instead of writing $\frac{du}{dx} = 2x$.

Example   Evaluate $\int_1^4 \frac{dx}{x + \sqrt{x}}$ using the substitution $u = \sqrt{x}$.

If $u = \sqrt{x}$, $x = u^2$ and $dx = 2u\, du$.
When $x = 1$, $u = 1$ and when $x = 4$, $u = 2$.

Therefore
$$\int_1^4 \frac{dx}{x + \sqrt{x}} = \int_1^2 \frac{2u\, du}{u^2 + u}$$

$$= \int_1^2 \frac{2}{u+1}\, du \quad (u \neq 0)$$

$$= 2\bigl[\ln|u+1|\bigr]_1^2$$

$$= 2(\ln 3 - \ln 2)$$

$$= \ln \tfrac{9}{4}.$$

Example   Calculate the area bounded by the curve $y = \dfrac{4}{1 + \sqrt{x}}$, the x-axis and the lines $x = 1$ and $x = 4$.

The required area is given by $A = \int_1^4 \dfrac{4}{1+\sqrt{x}} \, dx$.

Put $u = \sqrt{x}$ so that $x = u^2$ giving $dx = 2u \, du$.
When $x = 1, u = 1$ and when $x = 4, u = 2$.

Therefore $A = \int_1^2 \dfrac{4}{1+u}(2u) \, du$

$= \int_1^2 \left(8 - \dfrac{8}{u+1}\right) du$

$= \left[8u - 8\ln|u+1|\right]_1^2$

$= 16 - 8\ln 3 - 8 + 8\ln 2$

$= 8 - 8\ln 1.5 \; (\approx 4.76)$.

Exercise 23.1

1. Integrate each of the following functions using the suggested substitution:

   (a) $\dfrac{x}{\sqrt{1-x}}, \; u = 1-x$;
   (b) $(x-2)\sqrt{2x-1}, \; u = 2x-1$;

   (c) $\dfrac{2x+3}{(2x-1)^2}, \; u = 2x-1$;
   (d) $\sqrt{\dfrac{1-x}{x}}, \; x = \sin^2\theta$;

   (e) $\dfrac{1}{e^x + 4e^{-x}}, \; u = e^x$;
   (f) $\dfrac{1}{(x^2+4)^{3/2}}, \; x = 2\tan u$;

   (g) $\dfrac{e^x}{e^x + e^{-x}}, \; u = e^x$;
   (h) $\dfrac{\sec^2 x}{\sqrt{4 - \tan^2 x}}, \; u = \tan x$.

2. Evaluate each of the following integrals using the suggested substitution:

   (a) $\int_1^2 \dfrac{e^x}{e^x - 1} \, dx, \; u = e^x$;
   (b) $\int_{-2}^2 \sqrt{4 - x^2} \, dx, \; x = 2\sin\theta$;

   (c) $\int_4^5 \dfrac{x \, dx}{(x-3)^2}, \; u = x - 3$;
   (d) $\int_0^1 \dfrac{dx}{(x^2+1)^2}, \; x = \tan\theta$;

   (e) $\int_1^2 \dfrac{x+1}{(2x-1)^2} \, dx, \; u = 2x - 1$;
   (f) $\int_4^8 \dfrac{\sqrt{x-4}}{x} \, dx, \; u = \sqrt{x-4}$.

3. By means of the substitution $x = a\sec\theta$, find $\displaystyle\int_{a\sqrt{2}}^{2a} \dfrac{dx}{x^3 \sqrt{x^2 - a^2}}$.

Chapter 23

4. By means of the substitution $x = 2\cos^2\theta + 3\sin^2\theta$, evaluate
$$\int_{2\frac{1}{4}}^{2\frac{1}{2}} \frac{dx}{\sqrt{(3-x)(x-2)}}.$$

5. Evaluate $\displaystyle\int_1^4 \frac{dx}{\sqrt{x}\left(x+2\sqrt{x}+2\right)}$ by means of the substitution $x = u^2$.

23.2 Integration by Parts.

From the product rule we have $\dfrac{d}{dx}(u(x)v(x)) = u'(x)v(x) + u(x)v'(x)$.

By rewriting this in a different form we have
$$u(x)v'(x) = \frac{d}{dx}(u(x)v(x)) - u'(x)v(x).$$

Integrating each side with respect to x gives:

$$\boxed{\int u(x)v'(x)\,dx = u(x)v(x) - \int u'(x)v(x)\,dx.}$$ **Integration by Parts**

This formula expresses one integral in terms of another. The idea is to arrange the integrand in the form of a product $u(x)v'(x)$ such that the second integrand $u'(x)v(x)$ can be integrated by the standard procedures already outlined in this book.

Example   Integrate $x\sin x$ with respect to x.

We choose $u(x) = x$ and $v'(x) = \sin x$ giving $u'(x) = 1$ and $v(x) = -\cos x$.
The integrand $x\sin x$ is not one of the standard ones, but the second integrand is now $u'(x)v(x) = -\cos x$ which is a standard one and can easily be integrated.

$$\int x\sin x\,dx = x(-\cos x) - \int(-\cos x)\,dx$$
$$= -x\cos x + \sin x + c.$$

Note: If we choose $u(x) = \sin x$ and $v'(x) = x$, then the second integrand becomes $\tfrac{1}{2}x^2\cos x$ which is one step further removed from a standard form than the original integrand.

568

Further Integration

Example Find $\int \ln x \, dx$.

Choose $u(x) = \ln x$ and $v'(x) = 1$.

Then $u'(x) = \dfrac{1}{x}$ and $v(x) = x$.

Thus $\int \ln x \, dx = x \ln x - \int \dfrac{1}{x}(x) \, dx$
$= x \ln x - x + c$.

Example Find $\int e^{2x} \sin 3x \, dx$.

At first sight it does not appear that integration by parts could possibly simplify the integrand here since both derivatives and integrals of e^{2x} are multiples of e^{2x}, and both derivatives and integrals of $\sin 3x$ are multiples of either $\cos 3x$ or $\sin 3x$. This ensures that the second integrand is no closer to a standard form than the original.

It turns out, however, that if we integrate by parts *twice*, the original integral can be expressed in terms of itself and can then be determined without actually arriving at an integrand in standard form.

Let $I = \int e^{2x} \sin 3x \, dx$.

Choose $u(x) = e^{2x}$ and $v'(x) = \sin 3x$, so $u'(x) = 2e^{2x}$ and $v(x) = -\tfrac{1}{3}\cos 3x$.

Then $I = -\tfrac{1}{3}e^{2x} \cos 3x + \int \tfrac{2}{3} e^{2x} \cos 3x \, dx$.

Now we use integration by parts again, this time with the second integrand.
Choose $u(x) = e^{2x}$ and $v'(x) = \cos 3x$, so $u'(x) = 2e^{2x}$ and $v(x) = \tfrac{1}{3}\sin 3x$.

Now $I = -\tfrac{1}{3}e^{2x} \cos 3x + \tfrac{2}{3}\left\{\tfrac{1}{3}e^{2x} \sin 3x - \int \tfrac{2}{3} e^{2x} \sin 3x \, dx\right\}$

$= -\tfrac{1}{3}e^{2x} \cos 3x + \tfrac{2}{9}e^{2x} \sin 3x - \tfrac{4}{9}I + c$

$\tfrac{13}{9}I = -\tfrac{1}{3}e^{2x} \cos 3x + \tfrac{2}{9}e^{2x} \sin 3x + c$

$I = \tfrac{1}{13}e^{2x}\{2\sin 3x - 3\cos 3x\} + c_1$.

*[The following method is not required by this course but illustrates the use of complex numbers in the solution of certain types of *real* problems.

We also need Euler's formula for complex numbers namely $e^{i\theta} = \cos\theta + i\sin\theta$.

569

Firstly $e^{2x}(\cos 3x + i\sin 3x) = e^{2x}e^{3ix} = e^{(2+3i)x}$ and the imaginary part of the integral of $e^{(2+3i)x}$ is the imaginary part of the integral of $e^{2x}(\cos 3x + i\sin 3x)$.

Now $\int e^{(2+3i)x}\, dx = \dfrac{1}{2+3i} e^{(2+3i)x} + c$

$\qquad = \tfrac{1}{13}(2-3i)\{e^{2x}\cos 3x + ie^{2x}\sin 3x\} + c$.

Thus $\int e^{2x}\sin 3x\, dx = \operatorname{Im}\left(\tfrac{1}{13}(2-3i)\{e^{2x}\cos 3x + ie^{2x}\sin 3x\}\right) + c$

$\qquad = \tfrac{1}{13}e^{2x}\{2\sin 3x - 3\cos 3x\} + c$.

In addition, we may take the integrals of both real parts to obtain
$\int e^{2x}\cos 3x\, dx = \tfrac{1}{13}e^{2x}\{2\cos 3x + 3\sin 3x\} + c$ – 'two for one' in a shorter time!]

Exercise 23.2

1. Find an antiderivative of each of the following using integration by parts:
   (a) $x\cos 2x$; (b) xe^{3x} ; (c) $x\sin 4x$;
   (d) $x^2\ln x$; (e) $(2x+3)e^{2x}$; (f) xe^{-2x}.

2. Integrate each of the following using integration by parts:
   (a) $\ln(3x)$; (b) $\arcsin x$; (c) $\arctan x$.

3. Evaluate each of the following:
   (a) $\int_0^{\pi/2} x\sin 2x\, dx$; (b) $\int_0^{\pi/2} (x-1)\cos 3x\, dx$;
   (c) $\int_1^2 x\ln x\, dx$; (d) $\int_{\sqrt{3}}^2 \dfrac{\sqrt{4-x^2}}{x^2}\, dx$.

4. Integrate each of the following using integration by parts:
   (a) $x^2\sin 2x$; (b) $x^2 e^{-3x}$; (c) $x\ln(x^2+1)$;
   (d) $e^x\cos x$; (e) $e^{2x}\sin 2x$; (f) $e^{-3x}\cos 2x$.

5. The integral I_n is defined by $I_n = \int_0^{\frac{1}{2}\pi} \cos^n\theta\, d\theta$, $n \geq 2$. By writing $\cos^n\theta$ as $\cos^{n-1}\theta\cos\theta$ and using integration by parts, show that $I_n = \dfrac{n-1}{n} I_{n-2}$.

   Hence evaluate (a) $\int_0^{\frac{1}{2}\pi} \cos^8\theta\, d\theta$; (b) $\int_0^{\frac{1}{2}\pi} \cos^6\theta\sin^2\theta\, d\theta$.

6. (a) Derive the formula $\int x^n e^{ax}\, dx = \frac{1}{a} x^n e^{ax} - \frac{n}{a}\int x^{n-1} e^{ax}\, dx$, $n \geq 1$.

   (b) Find $\int x^4 e^{3x}\, dx$.

7. (a) Using integration by parts, show that

   (i) $\int e^{ax} \sin bx\, dx = -\frac{1}{b} e^{ax} \cos bx + \frac{a}{b} \int e^{ax} \cos bx\, dx$;

   (ii) $\int e^{ax} \cos bx\, dx = \frac{1}{b} e^{ax} \sin bx - \frac{a}{b} \int e^{ax} \sin bx\, dx$.

   (b) Using the results in part (a), show that
   $$\int e^{ax} \sin bx\, dx = \frac{e^{ax}}{a^2 + b^2}(a \sin bx - b \cos bx) + c$$
   and find a similar expression for $\int e^{ax} \cos bx\, dx$.

23.3 Partial Fractions

Definition A *rational function* $R(x)$ is one which can be expressed in the form
$$R(x) = \frac{P(x)}{Q(x)} \text{ where } P(x) \text{ and } Q(x) \text{ are polynomials.}$$

If, in addition, the degree of P is less than the degree of Q, then R is called *proper*. All other rational functions are said to be *improper*.

It is always possible to express the indefinite integral of a proper rational function in terms of elementary functions. Thus, if $R(x)$ is a proper rational function, $\int R(x)\, dx$ can be expressed in terms of other rational functions, logarithms of linear and/or quadratic polynomials, and functions of the form $\arctan(ax + b)$.

Example (a) $\int \frac{x^2}{(x^3+1)^2}\, dx = c - \frac{1}{3(x^3+1)}$

[a rational function]

(b) $\int \frac{dx}{x+3} = \ln|x+3| + c$

[a logarithm of a linear polynomial function]

Chapter 23

(c) $\int \dfrac{x}{x^2+1}\,dx = \dfrac{1}{2}\ln(x^2+1)+c$

[a logarithm of a quadratic polynomial function]

(d) $\int \dfrac{dx}{1+(x+2)^2} = \arctan(x+2)+c$

[a function of the form $\arctan(ax+b)$]

Expressing a Rational Function in Partial Fraction Format

A proper rational function is said to be resolved into its partial fractions if it is expressed as a sum of rational functions of the form

$$\dfrac{A}{(x-a)^n} \quad \text{and/or} \quad \dfrac{Ax+B}{(ax^2+bx+c)^n} \quad (n \in \mathbb{Z}^+),$$

where the quadratic ax^2+bx+c is positive definite (i.e., its discriminant is negative).

In our work we will be concerned only with the first form for $n = 1, 2$ and only with the second form for $n = 1$. Thus we will consider only proper rational functions which can be expressed as a sum of rational functions of the forms

$$\dfrac{A}{x-a},\ \dfrac{A}{(x-a)^2} \quad \text{and} \quad \dfrac{Ax+B}{ax^2+bx+c}.$$

If the rational function is improper, we must first divide out until the remainder has degree less than that of the denominator.

Example Integrate $\dfrac{x^3-2}{x^2+1}$ with respect to x.

By long division we find $\dfrac{x^3-2}{x^2+1} = x - \dfrac{x+2}{x^2+1}$.

Therefore $\int \dfrac{x^3-2}{x^2+1}\,dx = \int\left(x - \dfrac{x+2}{x^2+1}\right)dx$

$= \int\left(x - \dfrac{x}{x^2+1} - \dfrac{2}{x^2+1}\right)dx$

$= \tfrac{1}{2}x^2 - \tfrac{1}{2}\ln(x^2+1) - 2\arctan x + c.$

Further Integration

Only three situations may occur in this course:

1. If the denominator contains a linear factor $(x - a)$, it produces a term of the form $\dfrac{A}{x-a}$ in the partial fraction format.

2. If the denominator contains a repeated linear factor $(x-a)^2$, it produces a sum of terms of the form $\dfrac{A}{x-a} + \dfrac{B}{(x-a)^2}$ in the partial fraction format.

3. If the denominator contains a positive definite quadratic factor $ax^2 + bx + c$, it produces a term of the form $\dfrac{Ax+B}{ax^2+bx+c}$ in the partial fraction format.

Example Decompose $\dfrac{x-2}{x(x+2)}$ into its partial fractions.

The denominator contains two linear factors x and $x + 2$ and so the partial fraction format must contain the terms $\dfrac{A}{x}$ and $\dfrac{B}{x+2}$.

Let $\dfrac{x-2}{x(x+2)} = \dfrac{A}{x} + \dfrac{B}{x+2} = \dfrac{A(x+2) + Bx}{x(x+2)}$.

Thus $x - 2 = A(x + 2) + Bx$ for all x.
Put $x = 0$: $-2 = 2A \Rightarrow A = -1$.
Put $x = -2$: $-4 = -2B \Rightarrow B = 2$.

Therefore $\dfrac{x-2}{x(x+2)} = \dfrac{2}{x+2} - \dfrac{1}{x}$.

Example Decompose $\dfrac{x+1}{(x-1)^2}$ into its partial fractions.

Since the denominator contains a repeated linear factor $(x-1)^2$, the partial fraction format must contain the terms $\dfrac{A}{x-1}$ and $\dfrac{B}{(x-1)^2}$.

Let $\dfrac{x+1}{(x-1)^2} = \dfrac{A}{x-1} + \dfrac{B}{(x-1)^2} = \dfrac{A(x-1)+B}{(x-1)^2}$.

Chapter 23

Then $x + 1 = A(x - 1) + B$ for all x.
Clearly $A = 1$ and $B - A = 1$ giving $B = 2$.

Therefore $\dfrac{x+1}{(x-1)^2} = \dfrac{1}{x-1} + \dfrac{2}{(x-1)^2}$.

Example Decompose $\dfrac{1}{x(x^2+1)}$ into its partial fractions.

Since the denominator contains a linear factor x and a positive definite quadratic factor $x^2 + 1$, the partial fraction format must contain the terms $\dfrac{A}{x}$ and $\dfrac{Bx+C}{x^2+1}$.

Let $\dfrac{1}{x(x^2+1)} = \dfrac{A}{x} + \dfrac{Bx+C}{x^2+1} = \dfrac{A(x^2+1) + x(Bx+C)}{x(x^2+1)}$.

Then $1 = A(x^2+1) + x(Bx+C)$ for all x.
Put $x = 0$: $1 = A$.
Put $x = i$: $1 = i(Bi + C) = -B + Ci$ giving $B = -1$ and $C = 0$.

Therefore $\dfrac{1}{x(x^2+1)} = \dfrac{1}{x} - \dfrac{x}{x^2+1}$.

Note: The method used in the previous example sustituting $x = i$ to find **both** B and C at the same time can be quite useful, as here, but should not be overdone. However, this is simply another way in which complex numbers may be used to solve real problems with less work.

Exercise 23.3

1. Decompose each of the following into partial fractions:
   (a) $\dfrac{2}{(x-1)(x+1)}$;
   (b) $\dfrac{x}{(x+2)(x+1)}$;
   (c) $\dfrac{x-3}{x(2-x)}$;
   (d) $\dfrac{2x-3}{(x+1)(x-2)}$;
   (e) $\dfrac{7}{(2x-1)(x+3)}$;
   (f) $\dfrac{7}{(2x+3)(1-4x)}$.

2. Decompose each of the following into partial fractions:
   (a) $\dfrac{4}{x^2+2x}$;
   (b) $\dfrac{x+10}{x^2-4}$;
   (c) $\dfrac{4(x+4)}{x^2+6x+5}$;

(d) $\dfrac{5x-2}{x^2-x-2}$; (e) $\dfrac{6}{2x^2+x-1}$; (f) $\dfrac{1}{6x^2-5x+1}$.

3. Decompose each of the following into partial fractions:

(a) $\dfrac{x}{(x+1)^2}$; (b) $\dfrac{3x}{(x-2)^2}$; (c) $\dfrac{3x+2}{x(x+1)^2}$; (d) $\dfrac{8x^2-19x}{(x-2)^2(x+1)}$.

4. Decompose each of the following into partial fractions:

(a) $\dfrac{2-x}{x(x^2+1)}$; (b) $\dfrac{x-4}{(2x+1)(x^2+2)}$;

(c) $\dfrac{6x^2-x+5}{(x-2)(2x^2+1)}$; (d) $\dfrac{4x^2+13}{(2x+3)(2x^2-3x+2)}$.

5. Integrate each of the following functions with respect to x:

(a) $\dfrac{x-10}{x^2-4}$; (b) $\dfrac{2x}{(x-3)^2}$; (c) $\dfrac{3(x-6)}{x(x^2+9)}$;

(d) $\dfrac{2x^2-3x}{(x+2)(x^2+3)}$; (e) $\dfrac{1}{(x-3)(x-2)}$; (f) $\dfrac{5x}{3-14x+8x^2}$.

6. Evaluate each of the following:

(a) $\displaystyle\int_{-2}^{0} \dfrac{dx}{(3-x)(2-x)}$; (b) $\displaystyle\int_{-\frac{1}{2}}^{0} \dfrac{dx}{(4-2x)(1+x)}$;

(c) $\displaystyle\int_{0}^{2} \dfrac{x\,dx}{x^2+4x+3}$; (d) $\displaystyle\int_{1}^{2} \dfrac{dx}{(x^2+1)(x+1)}$.

7. Find the area bounded by the curve $y = \dfrac{3}{4-x^2}$ and the line $y = 1$.

8. Integrate each of the following with respect to x:

(a) $\dfrac{3x}{(x-2)^2}$; (b) $\dfrac{7x-2}{(2x-4)^2}$; (c) $\left(x-\dfrac{1}{x-2}\right)^2$.

9. Integrate each of the following functions with respect to x:

(a) $\dfrac{4x+7}{2x^2-3x-2}$; (b) $\dfrac{3x-2}{2x^2-11x+12}$;

(c) $\dfrac{4(x+1)}{4x^2-1}$; (d) $\dfrac{4+5x}{6+x-2x^2}$.

575

10. Evaluate each of the following:

   (a) $\int_0^{\frac{1}{2}} \dfrac{1+3x}{1-x^2}\,dx$;   (b) $\int_2^3 \dfrac{4x-1}{2x^2-x-1}\,dx$;

   (c) $\int_0^{\frac{1}{2}} \dfrac{12x-1}{2+x-6x^2}\,dx$;   (d) $\int_1^2 \dfrac{dx}{x^3+x}$.

11. Find the area enclosed by the curve $y = \dfrac{13x}{6-5x-6x^2}$, the x-axis and the lines $x = -1$ and $x = \tfrac{1}{2}$.

12. For the function $f(x) = \dfrac{x^3 - 28x - 32}{x^2 + 5x + 6}$, find

   (a) $f''(x)$;
   (b) $\int f(x)\,dx$.

23.4 Differential Equations

A differential equation is a relation which involves one or more derivatives of an unspecified function y of x. The relation may involve y itself, given functions of x, and constants.

The equations $\dfrac{d^2y}{dx^2} = \sin 2x$, $\dfrac{dy}{dx} = 3y$, and $(x^2+1)\dfrac{dy}{dx} = 3y$ are examples of differential equations which could be encountered in this course.

Differential Equations of the Form $\dfrac{dy}{dx} = f(x)$

The problem of finding a function y of x when we know its derivative $\dfrac{dy}{dx} = f(x)$ and its value y_0 at a particular point x_0 is called an *initial value problem*.

This problem is solved by first finding a general antiderivative of f, $y = F(x) + c$, and then using the *initial condition* that $y = y_0$ when $x = x_0$ to find the constant of integration c.

The general antiderivative $y = F(x) + c$ is called the **general solution** of the differential equation. A **particular solution** is found by using the initial condition to find the value of c.

Further Integration

Example The acceleration due to gravity near the earth's surface is 9.8 m s^{-2}. If an object is dropped from rest at a height of 1000 m, find the time taken for the object to reach the ground. (Ignore air resistance.)

$$a = 9.8 \implies \frac{dv}{dt} = 9.8$$

Integrating gives $v = 9.8t + c_1$ and the initial condition that the object is dropped *from rest* ($v = 0$ when $t = 0$) gives $c_1 = 0$.

Therefore $v = 9.8t \implies \frac{ds}{dt} = 9.8t$ where s is the displacement from the initial position.

Integrating once more gives $s = 4.9t^2 + c_2$ and the second initial condition ($s = 0$ when $t = 0$) gives $c_2 = 0$.

Therefore $s = 4.9t^2$.

The object reaches the ground when $s = 1000$ and then $t = \sqrt{\frac{1000}{4.9}} \approx 14.3$.

Therefore the required time is 14.3 s.

Since the techniques required to solve such differential equations have already been discussed in previous chapters, we will not be concerned with them here.

The remainder of this current chapter will be concerned with differential equations with separable variables.

Separable Differential Equations

Definition A (first order) *separable differential equation* is one which can be written in the form

$$g(y)\frac{dy}{dx} = f(x).$$

It is convenient to write this equation in the form

$$g(y)\, dy = f(x)\, dx$$

where the variables x and y are *separated* so that x appears only on the right and y appears only on the left. By integrating both sides of this relation we obtain

$$\int g(y)\, dy = \int f(x)\, dx + c.$$

577

Chapter 23

Example Find the general solution of the differential equation $y\dfrac{dy}{dx}+x=0$.

By separating variables we have $y\,dy = -x\,dx$.

By integrating both sides we find $\tfrac{1}{2}y^2 = -\tfrac{1}{2}x^2 + c_1$ or $x^2 + y^2 = c$.

Note: The solution represents a family of concentric circles.

A special type of separable differential equation has the form

$$\dfrac{dA}{dt} = kA \quad (k \text{ is a constant}),$$

that is, the time rate of change of the quantity A at any given time t is proportional to the amount present at that time.

To solve this we separate variables

$$\dfrac{dA}{A} = k\,dt$$

and then integrate

$$\ln A = kt + c.$$

From this equation we obtain $A = e^{(kt+c)} = e^c e^{kt}$ and with the initial condition that $A = A_0$ when $t = 0$ we find that $A_0 = e^c$ or $A = A_0 e^{kt}$.

This type of separable equation arises naturally in problems dealing with money invested at compound interest, the decay of radioactive materials and the growth of populations, to mention a few. In each of these situations we assume that the rate of change of the amount is proportional to the current amount, i.e., $\dfrac{dA}{dt} = kA$. When it is known that A decreases with time, it is convenient to choose the constant of proportionality as $-k$ where $k > 0$.

Note: In the case of the decay of radioactive substances, the term *half-life* is used to mean the time taken for half the original amount to disappear.

Example Find the half-life of a radioactive substance if 50 g decays to 49 g in 10 years.

Let $\dfrac{dA}{dt} = -kA \quad (k > 0)$.

Then $\displaystyle\int \dfrac{dA}{A} = -k \int dt$

$\Rightarrow \quad \ln A = -kt + c$

$\Rightarrow \quad A = A_0 e^{-kt}$.

Now $A_0 = 50$ and $A = 49$ when $t = 10$

$\Rightarrow \quad 49 = 50 e^{-10k}$

$\Rightarrow \quad e^{-10k} = \dfrac{49}{50} = 0.98$

$\Rightarrow \quad e^{-k} = (0.98)^{1/10}$. (The value of k itself is not always required.)

$\Rightarrow \quad A = 50(0.98)^{t/10}$.

For the half-life, $A = 25$ g

$\Rightarrow \quad 25 = 50(0.98)^{t/10}$

$\Rightarrow \quad \tfrac{1}{2} = (0.98)^{t/10}$

$\Rightarrow \quad \dfrac{t}{10} \log(0.98) = \log \tfrac{1}{2}$ (Any logarithm base may be used here.)

$\Rightarrow \quad t = \dfrac{10 \log \tfrac{1}{2}}{\log(0.98)} \approx 343$.

Therefore the half-life is 343 years.

There are other examples where the rate of change of some amount is proprtional to the difference between the amount present and a fixed amount. An example of this is the rate of cooling of the contents of a cup of coffee is proportional to the difference between the temperature of the coffee and the "ambient" (constant) temperature of the surroundings. This is known as *Newton's Law of Cooling*.

Example A cup of coffee at an initial temperature of 100°C is placed in a room whose constant temperature is 20°C. If the temperature of the coffee after 5 minutes is 84°C, find
(a) the temperature of the coffee after 10 minutes ;
(b) the time taken for the coffee to cool to a temperature of 21°C.

579

From Newton's law of cooling we know that if T °C is the temperature of the coffee after t minutes, then $\dfrac{dT}{dt} = -k(T-20)$ $k > 0$.

Separating variables gives $\dfrac{dT}{T-20} = -k\,dt$.

Integrating gives $\ln(T-20) = -kt + c$

$\Rightarrow \quad T - 20 = c_1 e^{-kt}$ (Here c_1 is *not* the initial temperature!)

$\Rightarrow \quad T = 20 + c_1 e^{-kt}$.

When $t = 0$, $T = 100$ and so $c_1 = 80$.

This gives $T = 20 + 80e^{-kt}$.

When $t = 5$, $T = 84$ and so $84 = 20 + 80e^{-5k}$ $\Rightarrow$ $e^{-k} = \left(\frac{64}{80}\right)^{1/5} = (0.8)^{1/5}$.

Thus $T = 20 + 80(0.8)^{t/5}$.

(a) When $t = 10$, $T = 20 + 80(0.8)^2 = 71.2$.

So after 10 minutes, the temperature of the coffee would be 71.2°C.

(b) When $T = 21$, $80(0.8)^{t/5} = 1$

$\Rightarrow \quad (0.8)^{t/5} = \tfrac{1}{80}$

$\Rightarrow \quad \dfrac{t}{5} \log 0.8 = -\log 80$

$\Rightarrow \quad t = \dfrac{-5\log 80}{\log 0.8} \approx 98.2$.

Therefore the time taken is 98.2 minutes.

Exercise 23.4

1. Find the general solution of each of the following differential equations:

  (a) $\dfrac{dy}{dx} = 2xy$;   (b) $\dfrac{dy}{dx} = 1 + y^2$;

  (c) $\dfrac{dy}{dx} = 2y$;   (d) $x\dfrac{dy}{dx} = 2y$;

  (e) $(2y+9)\dfrac{dy}{dx} = x(y+4)$;   (f) $(x^2+1)\dfrac{dy}{dx} + y^2 + 1 = 0$;

(g) $\dfrac{dy}{dx} = \dfrac{2\sqrt{y}}{x}$;

(h) $\dfrac{dy}{dx} = 2(y-10)$;

(i) $2\dfrac{dy}{dx} = y\tan x$;

(j) $y\dfrac{dy}{dx} = \sin^2 x$.

2.   Solve each of the following initial value problems:

(a) $y^2 \dfrac{dy}{dx} = x^2$, $y = 2$ when $x = 3$;

(b) $\dfrac{dy}{dx} = \dfrac{y}{x}$, $y = 6$ when $x = 2$;

(c) $xy\dfrac{dy}{dx} = 1 + y^2$, $y = 0$ when $x = 2$;

(d) $e^x y \dfrac{dy}{dx} = 1$, $y = 0$ when $x = 0$.

3. Find all the curves in the xy-plane such that
    (a) the tangent at each point (x, y) passes through the origin ;
    (b) the normal at each point (x, y) passes through the origin ;
    (c) the tangent at each point (x, y) intersects the x-axis at $(x - 1, 0)$;
    (d) the normal at each point (x, y) intersects the x-axis at $(x + 1, 0)$.

4. A particle is moving along the x-axis such that its velocity and acceleration at any time t s are given by $v(t)$ m s^{-1} and $a(t)$ m s^{-2}. If $a(t) = -0.2v(t)$ and the initial velocity is 10 m s^{-1}, find the velocity after 2 seconds and the distance travelled in that time.

5. A projectile is slowed in a resisting medium so that the time rate of change of its velocity is proportional to its instantaneous velocity. That is, $\dfrac{dv}{dt} = -kv$ where k is a positive constant. If the initial velocity of 100 m s^{-1} is reduced to 10 m s^{-1} in 2 seconds, find the value of k and the distance penetrated by the projectile in this time.

6. (a) If the number of radioactive nuclei, A, in a given sample decays at a rate which is proportional to the number of radioactive nuclei present, we know that $\dfrac{dA}{dt} = -kA$ where k is a positive constant. Show that the half-life of the sample is equal to $\dfrac{\ln 2}{k}$.

Chapter 23

(b) Radioactive carbon-14 with a half-life of 5700 years is used in the determination of the age of material which contains carbon-14 nuclei. Find the age of a sample if 80% of the original radioactive carbon-14 nuclei still remain.

7. A bowl of soup at an initial temperature of 90°C is brought into a room whose temperature is 25°C. Two minutes later the temperature of the soup is 80°C. How long will it take for the temperature of the soup to reach 70°C?

8. When \$$A$ is invested at r % per annum and interest is paid continously, then $\dfrac{dA}{dt} = \dfrac{rA}{100}$. Find r if \$1000 amounts to \$1648.72 in 10 years with interest paid continuously.

9. A thermometer which reads 16°C is brought into a room whose temperature is 25°C. One minute later the reading is 20°C. How long will it take for the reading to reach 24°C?

10. Calculate the annual interest rate if an amount of money doubles in value in 5 years when interest is paid continuously.

11. Radon-222 gas has a half-life of 3.85 days. How long will it take the radon to fall to 80% of its original value?

*12. A bowl of water at a temperature of 60°C was placed in a refrigerator. After 10 minutes the temperature of the water had dropped to 52°C and after a further 10 minutes, to 45°C. What was the temperature of the refrigerator?

23.5 Equations Reducible to Separable Form – Homogeneous Equations

Certain first-order differential equations are not separable but can be made separable by a change of variables.

A first order equation is called *homogeneous* if it can be written in the form $\dfrac{dy}{dx} = F\left(\dfrac{y}{x}\right)$ for some function F. Such an equation can be made separable by putting $\dfrac{y}{x} = v$ or $y = vx$ as follows:

582

If $y = vx$ then $\dfrac{dy}{dx} = x\dfrac{dv}{dx} + v$ and so $x\dfrac{dv}{dx} + v = F(v)$.

Now separating variables and integrating gives $\displaystyle\int \dfrac{dv}{F(v) - v} = \int \dfrac{dx}{x}$.

We can finally replace v by y/x to find the general solution of the original equation.

Example   Find the general solution of the following differential equation:
$$2xy\dfrac{dy}{dx} = y^2 - x^2.$$

Firstly we divide both sides by x^2 giving $2\left(\dfrac{y}{x}\right)\dfrac{dy}{dx} = \left(\dfrac{y}{x}\right)^2 - 1$ which confirms that the equation is homogeneous.

Putting $y = vx$ where $\dfrac{dy}{dx} = x\dfrac{dv}{dx} + v$ gives $2v\left(x\dfrac{dv}{dx} + v\right) = v^2 - 1$ which can be written in the form $\dfrac{2v\, dv}{1 + v^2} = -\dfrac{dx}{x}$.

Now integrating both sides gives $\ln(1 + v^2) = -\ln x + \ln k = \ln\left(\dfrac{k}{x}\right)$ where k is an arbitrary constant.

Thus $1 + v^2 = \dfrac{k}{x}$ and replacing v with y/x and multiplying both sides by x^2 gives $x^2 + y^2 = kx$ which is the required general solution.

Example   Solve the differential equation $xy\dfrac{dy}{dx} = 2y^2 + 4x^2$, $(x \geq \sqrt{2})$, given that $y = 4$ when $x = 2$.

Dividing by x^2 gives $\left(\dfrac{y}{x}\right)\dfrac{dy}{dx} = 2\left(\dfrac{y}{x}\right)^2 + 4$ and substituting $v = \dfrac{y}{x}$ gives the equation $v\left(x\dfrac{dv}{dx} + v\right) = 2v^2 + 4$ which can be written $xv\dfrac{dv}{dx} = v^2 + 4$.

Then $\displaystyle\int \dfrac{v}{v^2 + 4}\, dv = \int \dfrac{dx}{x}$ $\Rightarrow$ $\tfrac{1}{2}\ln(v^2 + 4) = \ln x + \ln k$ $(x > 0)$

$\Rightarrow$ $\sqrt{v^2 + 4} = kx$

583

Chapter 23

$$\Rightarrow \quad v^2 + 4 = k^2 x^2$$
$$\Rightarrow \quad y^2 + 4x^2 = k^2 x^4 = cx^4 \text{ where } c = k^2.$$

Thus $y^2 = x^2(cx^2 - 4)$ or $y = x\sqrt{cx^2 - 4}$.

But $y = 4$ when $x = 2$ and so $4 = 2\sqrt{4c - 4} \Rightarrow c = 2$.

Therefore the required solution is $y = x\sqrt{2x^2 - 4}$.

Sometimes we can reduce equations to separable form with other substitutions which may or may not be obvious. Substitutions which are not obvious will be given.

Example Use the substitution $v = y - x$ to find the general solution of the differential equation $\dfrac{dy}{dx} = \dfrac{y - x + 2}{y - x + 1}$.

$v = y - x$ and so $\dfrac{dv}{dx} = \dfrac{dy}{dx} - 1$ which gives $\dfrac{dv}{dx} = \dfrac{v + 2}{v + 1} - 1 = \dfrac{1}{v + 1}$.

Separating variables and integrating gives $\int (v+1)\,dv = \int dx$

$$\Rightarrow \quad \tfrac{1}{2}v^2 + v = x + c_1$$
$$\Rightarrow \quad v^2 + 2v = 2x + c$$
$$\Rightarrow \quad (y - x)^2 + 2(y - x) = 2x + c$$
$$\Rightarrow \quad x^2 - 2xy + y^2 - 4x + 2y = c.$$

Note: We can make y the subject of this equation as follows:
$(y - x)^2 + 2(y - x) = 2x + c$
$\Rightarrow \quad (y - x)^2 + 2(y - x) + 1 = 2x + c_1 \qquad (c_1 = c + 1)$
$\Rightarrow \quad (y - x + 1)^2 = 2x + c_1$
$\Rightarrow \quad y = x - 1 \pm \sqrt{2x + c_1}$ where the choice of sign can be determined if a value of y is known for a given value of x.

If $y = 5$ when $x = 5$, then $y = x - 1 + \sqrt{2x - 9}$ but if $y = 1$ when $x = 5$ then we find that $y = x - 1 - \sqrt{2x - 1}$.

Exercise 23.5

1. Find the general solution of each of the following differential equations:

   (a) $x\dfrac{dy}{dx} = x + y$;

   (b) $x\dfrac{dy}{dx} = 3x + 2y$;

   (c) $x\dfrac{dy}{dx} = x + 2y$;

   (d) $x\dfrac{dy}{dx} = 3y - x$;

   (e) $xy\dfrac{dy}{dx} = x^2 + y^2$;

   (f) $y\dfrac{dy}{dx} = x + y + \dfrac{y^2}{x}$;

   (g) $x\dfrac{dy}{dx} = y - xe^{-y/x}$;

   (h) $x\dfrac{dy}{dx} = y(1 + \ln y - \ln x)$;

   (i) $y\dfrac{dy}{dx} = 2x - y$;

   (j) $\dfrac{dy}{dx} = \dfrac{x + 2y}{3x + 2y}$.

2. Solve each of the following differential equations:

   (a) $xy\dfrac{dy}{dx} = x^2 + 2y^2$, if $y = 3$ when $x = -1$;

   (b) $(x^2 + 2y^2)\dfrac{dy}{dx} = xy$, if $y = 2$ when $x = 4$;

   (c) $x^2\dfrac{dy}{dx} = y^2 + 5xy + 4x^2$, if $y = -\dfrac{5}{2}$ when $x = 1$;

   (d) $\dfrac{dy}{dx} = \dfrac{y + x}{y - x}$, if $y = 3$ when $x = 1$.

*3. Use the given substitution to find the general solution of each of the following differential equations:

   (a) $\dfrac{dy}{dx} = (y - x)^2$, $(v = y - x)$;

   (b) $\dfrac{dy}{dx} = \tan(x + y) - 1$, $(v = x + y)$;

   (c) $x\dfrac{dy}{dx} = e^{-xy} - y$, $(v = xy)$;

   (d) $(x^2 - 2y)\dfrac{dy}{dx} = 2x(2x^2 + y)$, $\left(v = y/x^2\right)$.

585

Chapter 23

*4. Solve each of the following differential equations using the substitution given in parentheses:

(a) $\dfrac{dy}{dx} = \dfrac{y-x}{y-x-1}$, where $y = 2$ when $x = 2$, $(v = y - x)$;

(b) $\dfrac{dy}{dx} + y^2 x = 2xy$, where $y = 1$ when $x = 0$, $(v = 1/y)$;

(c) $\dfrac{dy}{dx} = \dfrac{1 - 2y - 4x}{1 + y + 2x}$, where $y = 2$ when $x = 1$, $(v = y + 2x)$;

(d) $\dfrac{dy}{dx} = (x + e^y - 1)e^{-y}$, where $y = 1$ when $x = 0$, $(v = x + e^y)$.

Required Outcomes

After completing this chapter, a student should be able to:
- integrate by substitution and by parts.
- express rational functions in terms of their partial fractions.
- solve first-order separable differential equations.
- solve problems involving growth and decay.
- solve homogeneous linear differential equations using the substitution $y = vx$.
- solve non-homogeneous linear differential equations which can be converted to separable form by a given substitution.

24 Matrices and Transformations in the Plane

24.1 Linear Transformations in the Plane

Definition A *linear transformation in the plane* is a transformation which maps each point P(x, y) in the plane to a unique point P'(x', y') also in the plane such that
$$x' = ax + by$$
$$y' = cx + dy$$
where a, b, c and d are constants.

The original point P is called the *object* point and the resultant point P' is called the *image* point. We say that P *maps to* P' under the transformation.

If we denote a given transformation by *T*, we write
$$T(x, y) = (x', y') \text{ or } T : (x, y) \rightarrow (x', y').$$

Example *T* is a transformation in the plane such that $T(x, y) = (2x+3y, 3x-y)$.
(a) Find the image of the point P(3, –2) under *T*. i.e. Find *T*(P).
(b) Find the coordinates of the point which maps to (4, –5) under transformation *T*.

(a) The equations of *T* are: $x' = 2x + 3y$ and $y' = 3x - y$.
Therefore $T(P) = (2 \times 3 + 3 \times (-2), 3 \times 3 - (-2)) = (0, 11)$.

(b) Let P(x, y) map to (4, –5).
Then $2x + 3y = 4$ and $3x - y = -5$.
Solving these equations simultaneously gives $x = -1$ and $y = 2$.
Therefore the point (–1, 2) maps to (4, –5) under *T*.

The equations of the linear transformation *T* in the plane
$$x' = ax + by$$
$$y' = cx + dy$$

can be written in matrix form: $\begin{pmatrix} x' \\ y' \end{pmatrix} = \begin{pmatrix} a & b \\ c & d \end{pmatrix} \begin{pmatrix} x \\ y \end{pmatrix}$.

The 2 × 2 matrix $T = \begin{pmatrix} a & b \\ c & d \end{pmatrix}$ is called *the matrix of the transformation*.

Chapter 24

Example   Find the matrix M of the linear transformation in the plane whose equations are $x' = 3x - y$ and $y' = 2y$.

The equations are $x' = 3x + (-1)y$
and $ y' = 0x + 2y.$

Therefore $M = \begin{pmatrix} 3 & -1 \\ 0 & 2 \end{pmatrix}.$

Example   Find the matrix M of the linear transformation in the plane which maps $A(2, 0)$ to $A'(2, 4)$ and $B(-1, 2)$ to $B'(7, -4)$.

Let the equations of the transformation be $\quad x' = ax + by$
$and\quad y' = cx + dy.$
Then $2a + 0b = 2$ and $2c + 0d = 4$ since $(2, 0)$ maps to $(2, 4)$.
Also $-a + 2b = 7$ and $-c + 2d = -4$ since $(-1, 2)$ maps to $(7, -4)$.
Solving these pairs of simultaneous equations gives $a = 1$, $b = 4$, $c = 2$, $d = -1$ and $\begin{pmatrix} 1 & 4 \\ 2 & -1 \end{pmatrix}$ is the required matrix.

Alternatively:

Since $(2, 0)$ maps to $(2, 4)$ and $(-1, 2)$ maps to $(7, -4)$, then $M\begin{pmatrix} 2 \\ 0 \end{pmatrix} = \begin{pmatrix} 2 \\ 4 \end{pmatrix}$
and $M\begin{pmatrix} -1 \\ 2 \end{pmatrix} = \begin{pmatrix} 7 \\ -4 \end{pmatrix}$ or $M\begin{pmatrix} 2 & -1 \\ 0 & 2 \end{pmatrix} = \begin{pmatrix} 2 & 7 \\ 4 & -4 \end{pmatrix}.$

Solving the equation gives $M = \frac{1}{4}\begin{pmatrix} 2 & 7 \\ 4 & -4 \end{pmatrix}\begin{pmatrix} 2 & 1 \\ 0 & 2 \end{pmatrix} = \begin{pmatrix} 1 & 4 \\ 2 & -1 \end{pmatrix}.$

Example   Find the image of the line $y = 2x + 1$ under a transformation in the plane with matrix $\begin{pmatrix} 1 & 2 \\ 2 & 3 \end{pmatrix}.$

Under a linear transformation the image of a straight line will always be a straight line. Now $A(1, 3)$ and $B(2,5)$ lie on $y = 2x + 1$. The images of A and B under this transformation are $A'(7, 11)$ and $B'(12, 19)$. The equation of $(A'B')$ is $8x - 5y = 1$ which the equation of the image of $y = 2x + 1$.

Alternatively: The point $P(a, 2a + 1)$ lies on $y = 2x + 1$ for all a.
$\begin{pmatrix} 1 & 2 \\ 2 & 3 \end{pmatrix}\begin{pmatrix} a \\ 2a+1 \end{pmatrix} = \begin{pmatrix} 5a+2 \\ 8a+3 \end{pmatrix}$ so $P' = (5a + 2, 8a + 3).$

Matrices and Transformations of the Plane

If $x' = 5a + 2$ and $y' = 8a + 3$, then $8x' - 5y' = 1$ (by elimination of a) and so the image of $y = 2x + 1$ is $8x - 5y = 1$.

Since $\begin{pmatrix} a & b \\ c & d \end{pmatrix}\begin{pmatrix} 1 & 0 \\ 0 & 1 \end{pmatrix} = \begin{pmatrix} a & b \\ c & d \end{pmatrix}$, then $(1, 0)$ maps to (a, c) and $(0, 1)$ maps to (b, d) under the transformation whose matrix is $\begin{pmatrix} a & b \\ c & d \end{pmatrix}$. Thus whenever the images of $(1, 0)$ and $(0, 1)$ are known, the matrix of the transformation can be written down at once, the first column containing the coordinates of the image of $(1, 0)$ and the second column containing the coordinates of the image of $(0, 1)$.

Example Find the matrix of the transformation in the plane which maps $(1, 0)$ to $(3, -2)$ and $(0, 1)$ to $(-4, 1)$.

The matrix is $\begin{pmatrix} 3 & -4 \\ -2 & 1 \end{pmatrix}$.

Note: The convention used throughout this book will be that the same symbol will represent both a transformation and its matrix.

Theorem When an object (a closed geometrical figure) in a plane maps to its image under a linear transformation in the plane T, the ratio $\dfrac{\text{area of image}}{\text{area of object}}$ is always equal to $|\det T|$.

Proof Consider the unit square OABC in the diagram below under the transform-ation T whose matrix is $T = \begin{pmatrix} a & b \\ c & d \end{pmatrix}$.

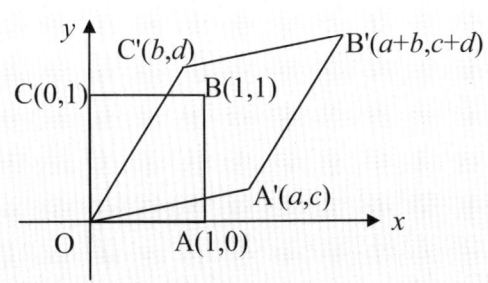

The coordinates of the vertices of the image are found:

$\begin{pmatrix} a & b \\ c & d \end{pmatrix}\begin{pmatrix} 1 & 1 & 0 \\ 0 & 1 & 1 \end{pmatrix} = \begin{pmatrix} a & a+b & b \\ c & c+d & d \end{pmatrix}$ {(0, 0) always maps to itself}.

589

Notice that the square becomes a parallelogram under a linear transformation. This is a particular case of the general rule that the image of a parallelogram is another parallelogram since parallel lines remain parallel under a linear transformation. The proof of this is left as an exercise for the reader.

$$\begin{aligned}
\text{Now the area of OA'B'C'} &= |\overrightarrow{OA'} \times \overrightarrow{OC'}| \\
&= \left\| \begin{matrix} i & j & k \\ a & c & 0 \\ b & d & 0 \end{matrix} \right\| \\
&= |(ad - bc)k| \\
&= |ad - bc| \\
&= |\det T|.
\end{aligned}$$

Thus $\dfrac{\text{area of image}}{\text{area of object}} = |\det T|$.

When a geometrical figure is reflected in a straight line, we say that the image has a 'sense' opposite to that of the object. In the following diagram, triangle A'B'C' is the image of the triangle ABC under a reflection in line ℓ.

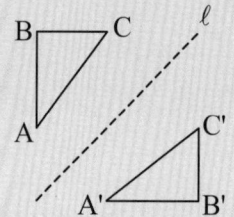

Triangle ABC has a 'clockwise sense' in that moving from A to B and then to C is a clockwise motion. Triangle A'B'C', on the other hand, has an 'anticlockwise sense'.

Theorem If det $T > 0$, then 'sense' is conserved under the linear transformation T.

Proof Consider the unit square OABC and its image as in the diagram on the previous page. Let θ be the angle of the anticlockwise rotation about the origin which maps $\overrightarrow{OA'}$ to $\overrightarrow{OC'}$. 'Sense' is clearly conserved if $0 < \theta < \pi$.
That is if $\sin \theta > 0$.
Now $\overrightarrow{OA'} \times \overrightarrow{OC'} = (ad - bc)k = (\det T)k$, as shown previously.
This vector points in the direction of the positive z-axis (up out of the page) if $ad - bc > 0$. As $\overrightarrow{OA'} \times \overrightarrow{OC'} = (OA)(OC)\sin \theta u$ where u is a unit vector whose direction is that of k if $\sin \theta > 0$, then 'sense' is conserved if det $T > 0$.

Matrices and Transformations of the Plane

Neither a rotation nor a reflection changes the area of any closed geometrical figure. A reflection, but not a rotation, changes the 'sense'. Therefore the determinant of the matrix of a rotation is +1 and that of a reflection is −1.

Exercise 24.1

1. Find the matrix of each of the following linear transformations in the plane:
    (a) a reflection in the x-axis ;
    (b) a reflection in the y-axis ;
    (c) an anticlockwise rotation about O through 90° ;
    (d) a rotation about O through 180° ;
    (e) a clockwise rotation about O through 90° ;
    (f) a reflection in the line $y = x$;
    (g) a reflection in the line $y = -x$;
    (h) a rotation about O through 360°.

2. Find the matrix of the linear transformation in the plane which maps
    (a) (1, 0) to (3, 2) and (0, 1) to (5, −1) ;
    (b) (1, 0) to (−1, 2) and (2, 1) to (−1, 3) ;
    (c) (4, 1) to (13, −5) and (−1, 2) to (−1, 8) ;
    (d) (1, 1) to (5, 0) and (−2, 1) to (−1, 3) ;
    (e) (−2, −1) to (5, −4) and (3, 2) to (−8, 5) ;
    (f) i to $3j$ and j to $-2i$;
    (g) $j - i$ to $2i + 3j$ and $2j$ to $3i - 2j$.

3. Consider the linear transformation in the plane T where
    $$T : (x, y) \rightarrow (2x - y, 3x + 2y).$$

    (a) Write down the matrix of the transformation.
    (b) Find the image of (3, −1) under T.
    (c) Find $T(-1, -2)$.
    (d) Find the image of the line $x + 2y = 6$ under T.
    (e) Plot the unit square and its image under T on the same set of axes. Show that the area of this image is $|\det T|$.

4. Find the image of the following under the linear transformation T where
    $T = \begin{pmatrix} 4 & -2 \\ 3 & -1 \end{pmatrix}$.

    (a) the x-axis ;   (b) the line $y = 3x$;
    (c) the line $y = 2x - 3$;   (d) the line $y = x$;
    (e) the line $2y = 3x$;   (f) the unit square.

591

Chapter 24

24.2 Rotations, Reflections, Stretches, Dilations and Shears

A Rotation about the Origin

In the diagram at the top of the next page, the coordinates of A' are $(\cos\theta, \sin\theta)$ from the definition of $\sin\theta$ and $\cos\theta$.

Similarly the coordinates of B' are
$$\left(\cos\left[\tfrac{1}{2}\pi+\theta\right],\ \sin\left[\tfrac{1}{2}\pi+\theta\right]\right) = (-\sin\theta,\ \cos\theta).$$

Thus under an anticlockwise rotation about the origin through angle θ, $(1, 0)$ maps to $(\cos\theta, \sin\theta)$ and $(0, 1)$ maps to $(-\sin\theta, \cos\theta)$.

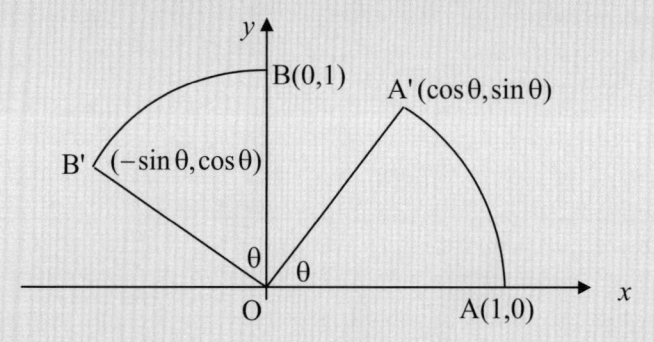

Therefore the matrix of an anticlockwise rotation about the origin through angle θ is $\begin{pmatrix} \cos\theta & -\sin\theta \\ \sin\theta & \cos\theta \end{pmatrix}$.

The determinant of this matrix is $\cos^2\theta + \sin^2\theta = 1$ so areas and sense are conserved under rotations as we already know.

Example Find the matrix of an anticlockwise rotation about the origin through 120°.

The required matrix is $\begin{pmatrix} \cos 120° & -\sin 120° \\ \sin 120° & \cos 120° \end{pmatrix} = \dfrac{1}{2}\begin{pmatrix} -1 & -\sqrt{3} \\ \sqrt{3} & -1 \end{pmatrix}$.

Example Describe the transformation whose matrix is $\begin{pmatrix} 0.6 & -0.8 \\ 0.8 & 0.6 \end{pmatrix}$.

This matrix is $\begin{pmatrix} \cos\theta & -\sin\theta \\ \sin\theta & \cos\theta \end{pmatrix}$ where $\cos\theta = 0.6$ and $\sin\theta = 0.8$.

Therefore the required transformation is an anticlockwise rotation about the origin through an angle $\arctan\frac{4}{3}$ ($\approx 53.1°$).

Example Find the image of
(a) the point (5, 12);
(b) the line $x + 2y = 5$, under an anticlockwise rotation about the origin through an angle of $\arctan\frac{5}{12}$.

(a) If $\theta = \arctan\frac{5}{12}$, $\cos\theta = \frac{12}{13}$, $\sin\theta = \frac{5}{13}$ and the matrix is $\frac{1}{13}\begin{pmatrix} 12 & -5 \\ 5 & 12 \end{pmatrix}$.

Now $\frac{1}{13}\begin{pmatrix} 12 & -5 \\ 5 & 12 \end{pmatrix}\begin{pmatrix} 5 \\ 12 \end{pmatrix} = \begin{pmatrix} 0 \\ 13 \end{pmatrix}$ and so the image of (5, 12) is (0, 13).

(b) Also $\frac{1}{13}\begin{pmatrix} 12 & -5 \\ 5 & 12 \end{pmatrix}\begin{pmatrix} 5-2a \\ a \end{pmatrix} = \frac{1}{13}\begin{pmatrix} 60-29a \\ 25+2a \end{pmatrix}$ and so the image of $P(5-2a, a)$ is $P'\left(\frac{1}{13}[60-29a], \frac{1}{13}[25+2a]\right)$ which lies on the line $2x + 29y = 65$. This is the required image.

Example T is an anticlockwise rotation in the plane about the origin which maps (3, 4) to (5, 0). Find the angle of the rotation.

Let the equations of the rotation be
$x' = cx - sy$
$y' = sx + cy$.
Then $5 = 3c - 4s$ and $0 = 3s + 4c$ which gives $c = 0.6$ and $s = -0.8$.
Therefore the required angle is $360° - \arctan\frac{4}{3} \approx 306.9°$.

Alternatively:

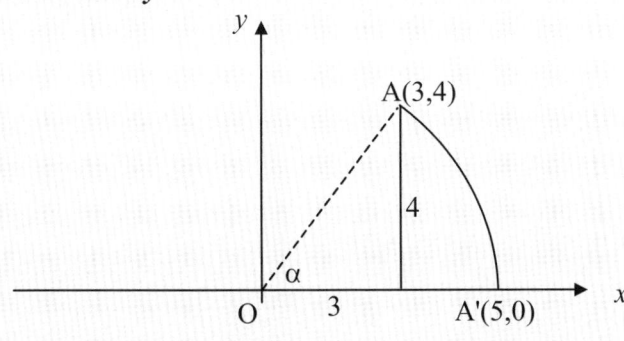

From the diagram, $\tan\alpha = \frac{4}{3}$ and so $\alpha = 53.1°$.
Therefore the required angle = $360° - 53.1° = 306.9°$.

593

Chapter 24

A Reflection in the Line $y = mx$ where $m = \tan\theta$

Under a reflection in the line $y = mx$ where $m = \tan\theta$, the point $A(1, 0)$ is mapped to A'. Since $OA' = OA$ and angle $AOA' = 2\theta$, then $A' = (\cos 2\theta, \sin 2\theta)$.

The point $B(0, 1)$ is mapped to B'. The angle $AOB' = 2\theta - \frac{1}{2}\pi$ and so $B' = \left(\cos[2\theta - \frac{1}{2}\pi], \sin[2\theta - \frac{1}{2}\pi]\right) = (\sin 2\theta, -\cos 2\theta)$.

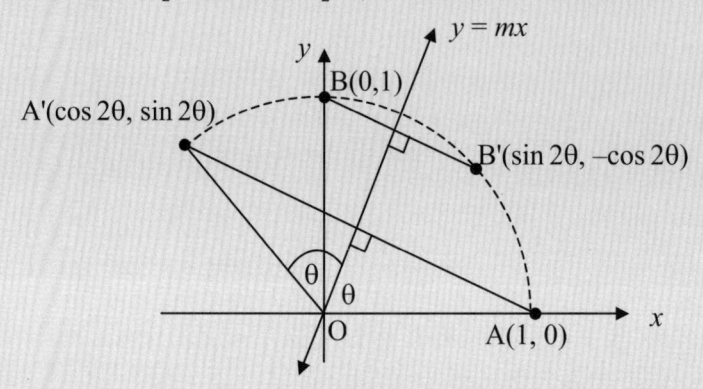

Therefore the matrix of a reflection in $y = mx = x\tan\theta$ is $\begin{pmatrix} \cos 2\theta & \sin 2\theta \\ \sin 2\theta & -\cos 2\theta \end{pmatrix}$.

The determinant of this matrix is $-\cos^2 2\theta - \sin^2 2\theta = -1$ so area is conserved but sense is not under a reflection as we already know. [The object has been 'flipped over'.]

Using the identities $\cos 2\theta = \dfrac{1 - \tan^2\theta}{1 + \tan^2\theta}$ and $\sin 2\theta = \dfrac{2\tan\theta}{1 + \tan^2\theta}$, the matrix of a reflection in $y = mx = x\tan\theta$ can be written in the form $\begin{pmatrix} \dfrac{1 - m^2}{1 + m^2} & \dfrac{2m}{1 + m^2} \\ \dfrac{2m}{1 + m^2} & -\dfrac{1 - m^2}{1 + m^2} \end{pmatrix}$.

Example Find the matrix of a reflection in the line $y = x\sqrt{3}$.

Here $\tan\theta = \sqrt{3}$, $\theta = 60°$, $\cos 2\theta = -\frac{1}{2}$ and $\sin 2\theta = \frac{1}{2}\sqrt{3}$.

Thus the required matrix is $\dfrac{1}{2}\begin{pmatrix} -1 & \sqrt{3} \\ \sqrt{3} & 1 \end{pmatrix}$.

Alternatively:

For a reflection in $y = x\sqrt{3}$, $m = \sqrt{3}$, $\dfrac{1-m^2}{1+m^2} = -\dfrac{1}{2}$, $\dfrac{2m}{1+m^2} = \dfrac{\sqrt{3}}{2}$.

It follows that the required matrix is $\dfrac{1}{2}\begin{pmatrix} -1 & \sqrt{3} \\ \sqrt{3} & 1 \end{pmatrix}$.

Example Find the matrix of a reflection in the line $y = 2x$.

In this case, the second method described in the previous example is easier to use than the first since the exact value of θ is not known. But $\dfrac{1-m^2}{1+m^2} = -\dfrac{3}{5}$ and $\dfrac{2m}{1+m^2} = \dfrac{4}{5}$ for $m = 2$ giving the matrix $\begin{pmatrix} -0.6 & 0.8 \\ 0.8 & 0.6 \end{pmatrix}$.

Example Describe the transformation whose matrix is $\begin{pmatrix} 0.8 & 0.6 \\ 0.6 & -0.8 \end{pmatrix}$.

The matrix is of the form $\begin{pmatrix} \cos 2\theta & \sin 2\theta \\ \sin 2\theta & -\cos 2\theta \end{pmatrix}$ with $\cos 2\theta = \dfrac{4}{5}$ and $\sin 2\theta = \dfrac{3}{5}$. This is clearly the matrix of a reflection.

To find the axis of the reflection, we find the set of ***invariant points*** (i.e. points which are not moved by the reflection).

Definition A point is ***invariant*** if it is its own image, i.e., $P(x, y) = P'(x', y')$.

Returning to the example:
The equations for this transformation are
$$x' = 0.8x + 0.6y$$
$$y' = 0.6x - 0.8y.$$

For invariant points, $x = 0.8x + 0.6y$, which gives $y = \tfrac{1}{3}x$.

Therefore the axis of the reflection is $y = \tfrac{1}{3}x$. [***Check:*** $y = 0.6x - 0.8y$ which simplifies to $y = \tfrac{1}{3}x$.]

Chapter 24

A Stretch Parallel to a Coordinate Axis

(1) Parallel to the *x*-axis

The equations of this transformation are
$$x' = kx$$
and $\quad y' = y.$

The matrix is $\begin{pmatrix} k & 0 \\ 0 & 1 \end{pmatrix}.$

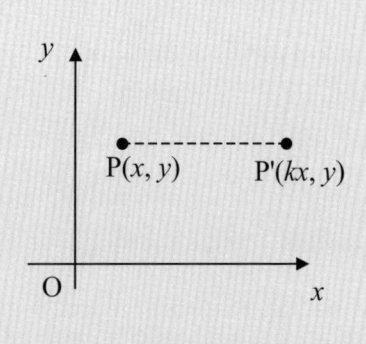

(2) Parallel to the *y*-axis

The equations of this transformation are
$$x' = x$$
and $\quad y' = ky.$

The matrix is $\begin{pmatrix} 1 & 0 \\ 0 & k \end{pmatrix}.$

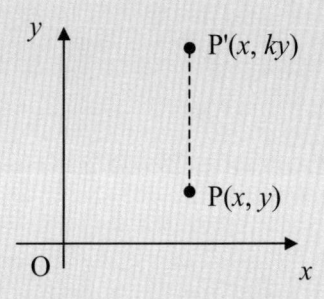

(3) A combination of stretches parallel to the the coordinate axes
The equations of this transformation are
$$x' = kx$$
$$y' = my.$$

The matrix is $\begin{pmatrix} k & 0 \\ 0 & m \end{pmatrix}.$

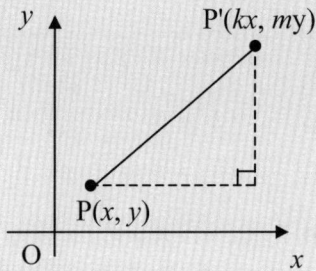

A Dilation (Enlargement), Centre O, Scale Factor *k*

From the diagram, $x' = kx$ and $y' = ky.$

The matrix is therefore $M = \begin{pmatrix} k & 0 \\ 0 & k \end{pmatrix}.$

Note that $\det M = k^2$ and so $\dfrac{\text{area of image}}{\text{are of object}} = k^2.$

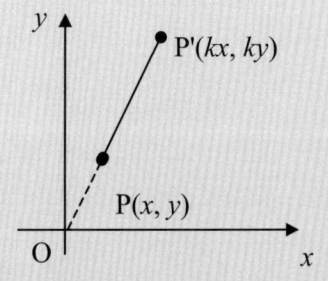

This transformation is clearly equivalent to a combination of stretches parallel to the coordinate axes with a common scale factor.

Matrices and Transformations of the Plane

A Shear with Shear Factor k

(1) Parallel to the x-axis

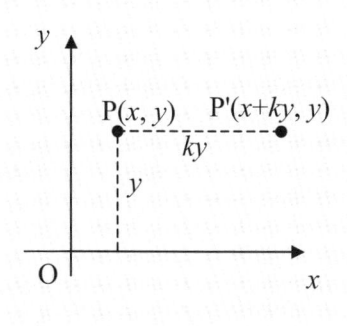

A shear parallel to the x-axis is a transformation which translates each point of the plane parallel to the x-axis through a distance which is proportional to the object's distance from the x-axis. The constant of proportionality k is called the **shear factor**. Under such a transformation, all points on the x-axis (the shear line) are invariant and points above the x-axis are moved to the right for positive k and to the left for negative k. Points below the x-axis are moved in the opposite directions.

The equations of the transformation are $x' = x + ky$
and $y' = y$.

Therefore the matrix of the shear in the x-axis is $M = \begin{pmatrix} 1 & k \\ 0 & 1 \end{pmatrix}$.

Note that $\det M = 1$ and so area is conserved under a shear.

(2) Parallel to the y-axis

In a similar way, it can be shown that the equations of a shear in the y-axis are

$x' = x$
and $y' = kx + y$.

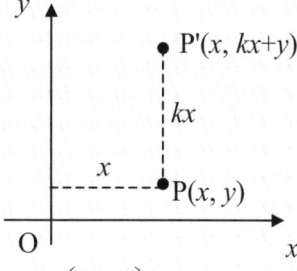

Therefore, the matrix of a shear in the y-axis is $M = \begin{pmatrix} 1 & 0 \\ k & 1 \end{pmatrix}$.

Here $\det M = 1$ (area is conserved) and points to the right of the y-axis are moved up if k is positive and down if k is negative, with points to the left of the y-axis moving in the opposite directions. Points on the y-axis (shear line) are invariant.

Chapter 24

Combining Transformations

If S and T are two transformations, then ST is that transformation which is equivalent to first performing T and then performing S on the result of T.

Example Describe the transformation in the plane which is equivalent to an anticlockwise rotation about O through 60° followed by a reflection in the line $y = x\sqrt{3}$.

The matrix of the rotation is $\begin{pmatrix} \cos 60° & -\sin 60° \\ \sin 60° & \cos 60° \end{pmatrix} = \frac{1}{2}\begin{pmatrix} 1 & -\sqrt{3} \\ \sqrt{3} & 1 \end{pmatrix}$, and

the matrix of the reflection is $\begin{pmatrix} \cos 120° & \sin 120° \\ \sin 120° & -\cos 120° \end{pmatrix} = \frac{1}{2}\begin{pmatrix} -1 & \sqrt{3} \\ \sqrt{3} & 1 \end{pmatrix}$.

The correct matrix product is $\frac{1}{4}\begin{pmatrix} -1 & \sqrt{3} \\ \sqrt{3} & 1 \end{pmatrix}\begin{pmatrix} 1 & -\sqrt{3} \\ \sqrt{3} & 1 \end{pmatrix} = \frac{1}{2}\begin{pmatrix} 1 & \sqrt{3} \\ \sqrt{3} & -1 \end{pmatrix}$

which is the matrix of a reflection in $x = \frac{1}{2}x + \frac{1}{2}\sqrt{3}y$ (using $x' = x$ to find the set of invariant points). This gives $y = \frac{1}{3}\sqrt{3}x$, the required transformation being a reflection in this line.

Example Find the matrix A^6 when $A = \frac{1}{2}\begin{pmatrix} \sqrt{3} & -1 \\ 1 & \sqrt{3} \end{pmatrix}$.

A is the matrix of an anticlockwise rotation about O through 30°.
Therefore A^6 is the matrix of an anticlockwise rotation about O through $6 \times 30° = 180°$.

The required matrix is $\begin{pmatrix} \cos 180° & -\sin 180° \\ \sin 180° & \cos 180° \end{pmatrix} = \begin{pmatrix} -1 & 0 \\ 0 & -1 \end{pmatrix}$.

Exercise 24.2

1. Find the matrix of each of the following linear transformations in the plane:
   (a) an anticlockwise rotation about O through 150°;
   (b) a reflection in the line $y = \frac{1}{2}x$;
   (c) an enlargement, centre O, scale factor 3;
   (d) a shear in the x-axis with shear factor 2;
   (e) an anticlockwise rotation about O through $\arctan \frac{1}{2}$;
   (f) a reflection in the line $2x + 3y = 0$;
   (g) a clockwise rotation about O through 120°;
   (h) a shear in the y-axis with shear factor -2.

Matrices and Transformations of the Plane

2. The triangle ABC has vertices A(−1, 1), B(1, 2) and C(−2, 3).
   (a) Find the coordinates of A', B' and C ', the images of A, B and C respectively under the linear transformation which maps (1, 0) to (2, 3) and (0, 1) to (−1, 1).

   (b) Find the coordinates of A", B" and C", the images of A', B' and C' respectively under the linear transformation $x" = x' − y'$
   $y" = x' + 2y'$.

   (c) Write down the matrix of the linear transformation which maps A, B and C to A", B" and C" respectively.

   (d) If the order of the transformations is reversed, what would be the final images of A, B and C? Write down the matrix of this transformation.

   (e) Are transformations commutative under the operation 'follows'?

3. If $A = \begin{pmatrix} \cos\theta & -\sin\theta \\ \sin\theta & \cos\theta \end{pmatrix}$ find, without actually multiplying matrices, A^2, A^3, A^n and A^{-1}.

4. Find the smallest positive integer value of n for which $A^n = I$ if
$A = \frac{1}{2}\begin{pmatrix} 1 & -\sqrt{3} \\ \sqrt{3} & 1 \end{pmatrix}$.

5. Describe the transformations in the plane whose matrices are:

   (a) $\begin{pmatrix} 1 & 0 \\ 0 & -1 \end{pmatrix}$;
   (b) $\begin{pmatrix} 0 & 1 \\ -1 & 0 \end{pmatrix}$;
   (c) $\frac{1}{5}\begin{pmatrix} -3 & -4 \\ 4 & -3 \end{pmatrix}$;

   (d) $\frac{1}{5}\begin{pmatrix} 3 & -4 \\ -4 & -3 \end{pmatrix}$;
   (e) $\begin{pmatrix} \frac{1}{2} & 0 \\ 0 & \frac{1}{2} \end{pmatrix}$;
   (f) $\begin{pmatrix} 1 & 4 \\ 0 & 1 \end{pmatrix}$;

   (g) $\begin{pmatrix} 2 & 0 \\ 0 & 3 \end{pmatrix}$;
   (h) $\frac{1}{13}\begin{pmatrix} 5 & 12 \\ 12 & -5 \end{pmatrix}$;
   (i) $\frac{1}{17}\begin{pmatrix} 15 & -8 \\ 8 & 15 \end{pmatrix}$;

   (j) $\begin{pmatrix} 1 & 0 \\ -3 & 1 \end{pmatrix}$;
   (k) $\begin{pmatrix} -2 & 0 \\ 0 & 1 \end{pmatrix}$;
   (l) $\begin{pmatrix} 1 & -1 \\ 1 & 1 \end{pmatrix}$.

6. Prove that an anticlockwise rotation about O through an angle θ followed by an anticlockwise rotation in the plane about O through angle α is equivalent to a single anticlockwise rotation about O through angle $\theta + \alpha$.

599

Chapter 24

7. Show that a reflection in the line $y = x\tan\theta$ followed by a reflection in the line $y = x\tan\alpha$ is equivalent to an anticlockwise rotation about O through angle $2(\alpha - \theta)$.

8. Describe the linear transformation in the plane which maps (12, 5) to (0, –13) and (1, –5) to (5, 1).

9. Find the matrix of the anticlockwise rotation of the plane about O which maps the point (3, 4) to the point (–5, 0). What is the angle of this rotation?

10. The following are transformation matrices. By inspection, write down their inverses giving reasons for your answers.

   (a) $\begin{pmatrix} 1 & 3 \\ 0 & 1 \end{pmatrix}$; (b) $\begin{pmatrix} 2 & 0 \\ 0 & 2 \end{pmatrix}$; (c) $\begin{pmatrix} 1 & 0 \\ 4 & 1 \end{pmatrix}$;

   (d) $\dfrac{1}{2}\begin{pmatrix} 1 & -\sqrt{3} \\ \sqrt{3} & 1 \end{pmatrix}$; (e) $\dfrac{1}{5}\begin{pmatrix} 3 & 4 \\ 4 & -3 \end{pmatrix}$; (f) $\dfrac{1}{\sqrt{2}}\begin{pmatrix} -1 & -1 \\ 1 & -1 \end{pmatrix}$.

11. Write down the matrix of the single transformation for each of the following:
   (a) a reflection in the x-axis followed by a reflection in the line $y = x\sqrt{3}$;
   (b) a reflection in the line $y = x$ followed by a clockwise rotation about O through 90° ;
   (c) a reflection in $y = 2x$ followed by a reflection in $y = -x$;
   (d) an anticlockwise rotation about O through 60° followed by a reflection in the line $x + y\sqrt{3} = 0$;
   (e) an anticlockwise rotation about O through 45° followed by an enlargement centre O scale factor 2 ;
   (f) an anticlockwise rotation about O through 90° followed by a shear in the y-axis with shear factor 2.

12. Describe the transformation $T : (x, y) \to (3x - y, x + 3y)$.
   [*Hint:* Consider the unit square.]

13. Find the image of
   (a) the point (3, 1) ;
   (b) the line $x + y = 4$,
   under a reflection in the line $y = 2x$.

600

14. By calculating the products of the matrices involved, express each of the following composite transformations more simply:
   (a) a reflection in the y-axis followed by an anticlockwise rotation about O through 240°;
   (b) an anticlockwise rotation about O through 120° followed by a reflection in the y-axis;
   (c) a reflection in the line $x = y\sqrt{3}$ followed by an anticlockwise rotation about O through 240°;
   (d) a reflection in the y-axis followed by a reflection in $x + y\sqrt{3} = 0$;
   (e) a reflection $y = x\sqrt{3}$ followed by a reflection in $y = -x\sqrt{3}$;
   (f) a reflection in $y = 2x$ followed by a reflection in $2y + x = 0$;
   (g) a clockwise rotation about O through an angle $\arctan\frac{1}{3}$ followed by a reflection in the line $y = x$.

15. Find the equations of the linear transformation $T : (x, y) \to (x', y')$ defined by
   (a) a reflection in $y = 3x$ followed by an anticlockwise rotation about O through 90°;
   (b) an anticlockwise rotation about O through 90° followed by a reflection in $y = 3x$.

   Under each of the above transformations, find the images of
   (i) the line $2x + 3y = 6$;
   (ii) the circle with equation $x^2 + y^2 - 2x - 4y = 0$.

16. The triangle ABC with vertices A(–2, 3), B(1, 1) and C(2,4) is transformed by a shear S with matrix $\begin{pmatrix} 1 & 2 \\ 0 & 1 \end{pmatrix}$, followed by an anticlockwise rotation R about O with matrix $\begin{pmatrix} 0 & -1 \\ 1 & 0 \end{pmatrix}$. Sketch the final image and write down the matrix of the combined transformations. Are the transformations commutative under 'follows'? There is a shear S_1 which can be applied to the final image, followed by a rotation R^{-1} to map the final image to the original triangle. Identify S_1 and find its matrix. Verify that $R^{-1}S_1 = S^{-1}R^{-1}$.

601

Chapter 24

17. Under the transformation which maps i to $3i$ and j to $3j$, find the image of
    (a) the line $y = 2x - 3$;
    (b) the circle $x^2 + y^2 - 2x + 2y = 5$;
    (c) the parabola $y = x^2 - 2x + 1$.
    State the nature of the transformation.

18. (a) If $A = \begin{pmatrix} \cos\theta & -\sin\theta \\ \sin\theta & \cos\theta \end{pmatrix}$ where $\theta = \dfrac{\pi}{4}$, compute the matrix $A^{-1}BA$ where B is the matrix of the transformation defined by $x' = 3x$, $y' = y$.
    (b) Describe geometrically the transformation whose matrix is B.

19. Show that the matrix $\begin{pmatrix} a & -b \\ b & a \end{pmatrix}$, where a and b are both positive, represents an enlargement centre O followed by a suitable anticlockwise rotation about O. Find the scale factor of the enlargement and the angle of the rotation.

24.3 The Image of the Curve $y = f(x)$

The equation of the image of the curve $y = f(x)$ under

(a) a translation with vector $\begin{pmatrix} a \\ b \end{pmatrix}$ is $y = f(x-a) + b$;

(b) a reflection in the line $x = a$ is $y = f(2a - x)$;

(c) a reflection in the line $y = b$ is $y = 2b - f(x)$;

(d) a rotation of 180° about the point (a, b) is $y = 2b - f(2a - x)$.

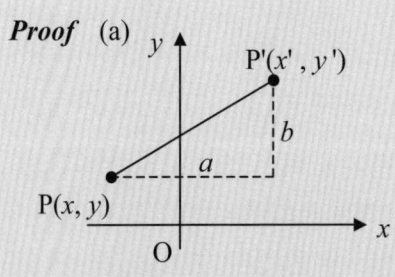

Proof (a) If P'(x', y') is the image of P(x, y) under a translation with vector $\begin{pmatrix} a \\ b \end{pmatrix}$, then

$x' = x + a$ and $y' = y + b$.

Rewriting gives $x = x' - a$ and $y = y' - b$.

Therefore the curve $y = f(x)$ becomes $y' - b = f(x' - a)$ which is the condition that P'(x', y') lies on the curve $y - b = f(x - a)$.

Thus the image of $y = f(x)$ is $y = f(x - a) + b$.

(b) If P'(x', y') is the image of P(x, y) under a reflection in the line $x = a$, then A(a, y) is the midpoint of [PP'].

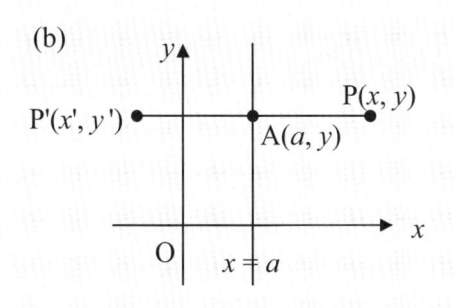

Thus $\dfrac{x+x'}{2} = a$ and $y = y'$.

Rewriting gives $x = 2a - x'$ and $y = y'$.

Therefore the curve $y = f(x)$ becomes $y' = f(2a - x')$ which is the condition that P'(x', y') lies on the curve $y = f(2a - x)$.

Thus the image of $y = f(x)$ is $y = f(2a - x)$.

(c) If P'(x', y') is the image of P(x, y) under a reflection in the line $y = b$, then B(x, b) is the midpoint of [PP'].

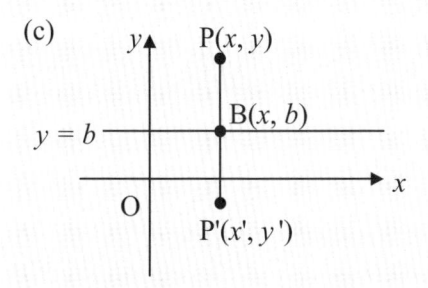

Thus $x = x'$ and $\dfrac{y + y'}{2} = b$.

Rewriting gives $x = x'$ and $y = 2b - y'$.

Thus the image of $y = f(x)$ becomes $2b - y' = f(x')$ which is the condition that P'(x', y') lies on the curve $y = 2b - f(x)$.

Thus the image of $y = f(x)$ is $y = 2b - f(x)$.

(d) If P'(x', y') is the image of P(x, y) under a rotation of 180° about the point X(a, b), then X is the midpoint of $[PP']$.

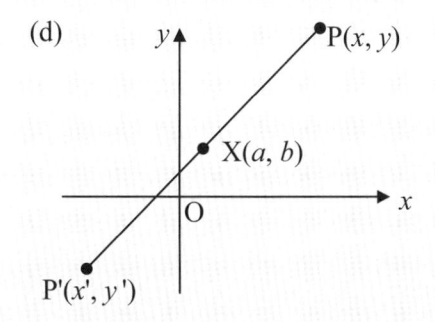

Thus $\dfrac{x+x'}{2} = a$ and $\dfrac{y+y'}{2} = b$.

Rewriting gives
$x = 2a - x'$ and $y = 2b - y'$.

Thus the image of $y = f(x)$ becomes $2b - y' = f(2a - x')$ which is the condition that P'(x', y') lies on the curve $y = 2b - f(2a - x)$.

Thus the image of $y = f(x)$ is $y = 2b - f(2a - x)$.

Chapter 24

Example The curve $y = x^3$ is translated with vector $\begin{pmatrix} 3 \\ -2 \end{pmatrix}$ and then the image is reflected in the line $y = -2$. Find the equation of the final image.

Under the translation the image is $y + 2 = (x - 3)^3$.
The equations of the reflection are $x' = x$, $y' = -4 - y$ or $x = x'$, $y = -4 - y'$.
Therefore the final image is $-4 - y + 2 = (x - 3)^3$ or $y = (3 - x)^3 - 2$.

Exercise 24.3

1. Under what translation will the curve $y = (x + 2)^2 - 1$ become $y = x^2$?

2. Give the vector of the translation which maps $y = x^2$ onto $y = x^2 + 2x - 3$?

3. Find the equation of the image of the curve $y = x^2 - 4x - 2$ under a translation with vector $\begin{pmatrix} -2 \\ 6 \end{pmatrix}$.

4. The curve $y = 2x^3 - 6x$ is translated with vector $\begin{pmatrix} 1 \\ -4 \end{pmatrix}$ and then reflected in the y-axis. Find the equation of the final image.

5. Find the equation of the final image of the curve $y = -x^2 + 2x + 3$ under a reflection in $y = 2$ followed by a translation with vector $\begin{pmatrix} -1 \\ 0 \end{pmatrix}$.

6. The curve $y = 2x^3 - 6x - 6$ is translated with vector $\begin{pmatrix} 1 \\ 2 \end{pmatrix}$. Find the equation of the image and without using a GDC, sketch the curve of the image and then the original curve on the same set of axes.

7. Find the equation of the image of the curve $y = x^3 + 3x + 1$ under a rotation of 180° about the origin. Name another transformation which is equivalent to this one.

8. Find the equation of the image of the curve $y = 3x - x^2 - x^3$ under a reflection in the line $x = 1$.

9. Are the transformations
   (a)   a reflection in the line $x = a$, and
   (b)   a rotation of $180°$ about the point (b, c)
   commutative under 'follows'?

10. Show that under a rotation of $180°$ about the point $(1, 3)$, the curve $y = 5x - 3x^2 + x^3$ is its own image.

11. Rotate the curve $y = 2x^3 + 12x^2 + 15x$ through $180°$ about its point of inflexion. Give the equation of the final image.

12. Rotate the curve $y = x^3 - 6x^2 + 12x - 5$ through $180°$ about the point $(2, 3)$ and then translate the image with vector $\begin{pmatrix} -2 \\ -3 \end{pmatrix}$. Find the equation of the final image.

13.

$y = g(x)$ $\qquad$ $f(x) = (x+1)(x-3)^2 + 2$

$(-1, 2)$

   (a)   Describe the transformations required to map $f(x)$ to $g(x)$, given that the point $(-1, 2)$ maps to the origin.

   (b)   Write $g(x)$ in the form $ax^3 + bx^2 + cx + d$.

Chapter 24

Required Outcomes

After completing this chapter, a student should be able to:
- write down the matrix corresponding to a linear transformation whose equations are given.
- find the images of points, lines and curves under a given transformation.
- find the matrix of a rotation about the origin, a reflection in any line through the origin, stretches parallel to the coordinate axes, dilations and shears.
- describe a transformation whose matrix is known.
- find points which are invariant under a given transformation.
- combine transformations by multiplying their matrices.

PART 3
HIGHER LEVEL OPTIONS

25 Sets, Relations and Groups

25.1 Sets

The concept of a set is basic to all of mathematics and mathematical applications. The first section of this chapter deals with the language of sets and the binary operations defined on sets.

The concepts and notation of sets, elements of sets, subsets, the universal set and the null set have all been defined in Chapter 13 covering the core material and need not be repeated here. The set operations, however, are so critical to this work that they are reproduced below.

Union The *union* of two sets, A and B, is denoted by $A \cup B$ and consists of all the elements which are members of either A or B or both A and B. i.e. $A \cup B = \{x \mid x \in A \text{ or } x \in B\}$.

Intersection The *intersection* of sets A and B is denoted by $A \cap B$ and consists of all those elements which belong to both A and B. i.e. $A \cap B = \{x \mid x \in A \text{ and } x \in B\}$.

Difference The *difference* of A and B is denoted by $A - B$ and consists of those elements which belong to A but not to B, i.e. belong to A only.
i.e. $A - B = \{x \mid x \in A, x \notin B\}$.

Complement The complement of A is denoted by A' and consists of all those elements in the universe which do not belong to A.
i.e. $A' = \{x \mid x \in U, x \notin A\}$.

Note that $A - B = A \cap B'$.

Definition Two sets, A and B, are said to be *disjoint* if they do not have any common elements. Thus A and B are disjoint if $A \cap B = \emptyset$.

Venn diagrams are used to illustrate the relationships between sets. We use a rectangle to represent the universal set and often circles within this rectangle to represent other sets.

Chapter 25

$A \cup B$ is shaded

$A \cap B$ is shaded

$A - B$ is shaded

A' is shaded

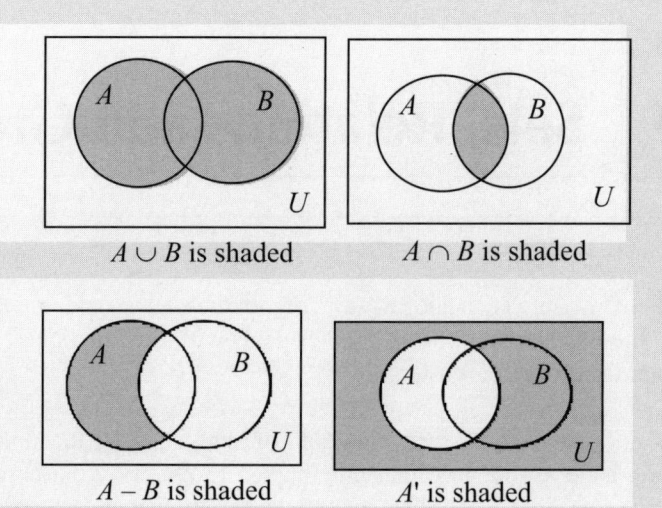

Laws of the Algebra of Sets

Associative Laws $(A \cup B) \cup C = A \cup (B \cup C)$
$(A \cap B) \cap C = A \cap (B \cap C)$

Commutative Laws $A \cup B = B \cup A$
$A \cap B = B \cap A$

Distributive Laws $A \cup (B \cap C) = (A \cup B) \cap (A \cup C)$
$A \cap (B \cup C) = (A \cap B) \cup (A \cap C)$

De Morgan's Laws $(A \cup B)' = A' \cap B'$
$(A \cap B)' = A' \cup B'$

These laws can be verified using Venn diagrams, but such verifications are not considered to be rigorous proofs.

Example Illustrate the first of de Morgan's laws using a Venn diagram.

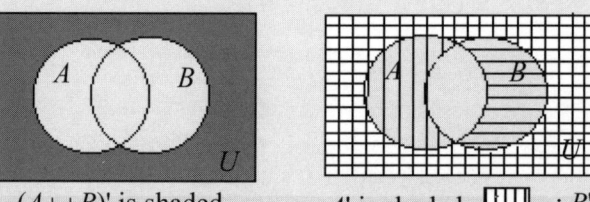

$(A \cup B)'$ is shaded

A' is shaded |||| ; B' is shaded ≡

$A' \cap B'$ is shaded ▦

From the above diagrams, it can be seen that $(A \cup B)' = A' \cap B'$.

Sets, Relations and Groups

A rigorous proof of the previous law could be something like:

Let $a \in (A \cup B)'$.
Then $a \notin A \cup B$ and so $a \notin A$ and $a \notin B$.
Therefore $a \in A'$ and $a \in B'$
$\Rightarrow a \in A' \cap B'$.
Hence $(A \cup B)' \subseteq A' \cap B'$.

Let $a \in A' \cap B'$.
Then $a \in A'$ and $a \in B'$.
Therefore $a \notin A$ and $a \notin B$
$\Rightarrow a \notin A \cup B$ and so $a \in (A \cup B)'$.
Hence $A' \cap B' \subseteq (A \cup B)'$.

Therefore $(A \cup B)' = A' \cap B'$ which is the required result.

The Order of a Set

The *order* of a finite set A is denoted by $|A|$ and is equal to the number of elements in it.

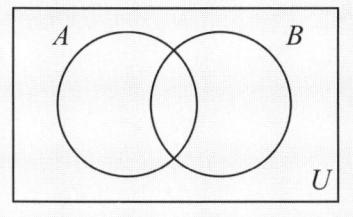

We can see from the diagram on the left that the sum $|A| + |B|$ includes the number of elements in $A \cap B$ twice. Thus $|A \cup B| = |A| + |B| - |A \cap B|$.

Note that union and intersection may be interchanged in this equation. Thus we also have $|A \cap B| = |A| + |B| - |A \cup B|$.

Example In a class of 25 students, 19 study geography and 14 study history and all students study at least one of these subjects. How many students study both geography and history?

Let G and H represent the set of all students who study geography and history respectively. Then $|G| = 19$, $|H| = 14$, $|G \cup H| = 25$.

Then $|G \cap H| = |G| + |H| - |G \cup H|$
$= 19 + 14 - 25$
$= 8$.

Thus 8 students study both geography and history.

Example In a survey of 100 households, 59 read newspaper X and 71 read newspaper Y. What can be said about the number of households who read both papers?

Let X and Y represent the sets of households reading those newspapers.
Then $|X \cap Y| = |X| + |Y| - |X \cup Y|$
$= 59 + 71 - |X \cup Y|$
$= 130 - |X \cup Y|$.

611

Chapter 25

But $|X \cup Y|$ cannot exceed 100, i.e., $|X \cup Y| \leq 100$, so $|X \cap Y| \geq 30$.

Also, the number of elements in $X \cap Y$ cannot exceed the number of elements in X (or the number of elements in Y). Thus $|X \cap Y| \leq 59$.

Thus at least 30, but no more than 59, households read both newspaper X and newspaper Y.

Partitions of a Set

A *partition* of a set X divides X into non-overlapping subsets.
A collection S of non-empty subsets of X is said to be a *partition* of X if every element of X belongs to exactly one member of S.

Example The set $S = \{ \{1, 3, 5\}, \{2, 4\}, \{6\} \}$ is a partition of the set $X = \{1, 2, 3, 4, 5, 6\}$ since each element of X belongs to exactly one set in S.

Exercise 25.1

1. What relation must hold between sets A and B if
   (a) $A \cap B = A$;   (b) $A \cup B = A$;
   (c) $(A \cap B)' = B'$ (d) $(A \cup B)' = B'$?

2. List the subsets of each of the following sets:
   (a) $\{1\}$;   (b) $\{a, b\}$;
   (c) $\{a, b, c\}$;   (d) $\{2, 3, 4, 5\}$.

3. (a) Using the results of Question 2, conjecture the total number of subsets of a set X with n elements.

   *(b) Prove your conjecture using mathematical induction, or otherwise.

4. For the sets given in Question 2, list all the partitions of each.

5. (a) Use Venn diagrams to verify the associative law for union:
   $$(A \cup B) \cup C = A \cup (B \cup C).$$

   (b) Give a rigorous proof of this law.

6. (a) Use Venn diagrams to verify the associative law for intersection:
   $$(A \cap B) \cap C = A \cap (B \cap C).$$

   (b) Give a rigorous proof of this law.

Sets, Relations and Groups

7. (a) Use Venn diagrams to verify the distributive laws:
   (i) $A \cap (B \cup C) = (A \cap B) \cup (A \cap C)$;
   (ii) $A \cup (B \cap C) = (A \cup B) \cap (A \cup C)$.

   (b) Give rigorous proofs of these laws.

8. (a) Use Venn diagrams to verify the second of de Morgan's laws:
   $(A \cap B)' = A' \cup B'$.

   *(b) Give a rigorous proof of this law.

9. (a) Use Venn diagrams to verify that
   $(A - B) \cup (B - A) = (A \cup B) - (A \cap B)$.

   *(b) Use de Morgan's laws to prove the result in part (a).

25.2 Cartesian Products and Binary Relations

The concept of an ordered pair (x, y) where x is a member of one set, A, and y is a member of another set, B, is very common in mathematics. For the coordinates (x, y) of a point in the plane, $x \in \mathbb{R}$ and $y \in \mathbb{R}$. For the ordered pairs (x, y) for which $y = \sin x$, x could be any member of the set of all angles, and y any real number between -1 and 1 inclusive. The rational number p/q could be represented by the ordered pair (p, q) where p is any integer and q is any non-zero integer.

Definition The **Cartesian product** of two sets, A and B, is the set of all ordered pairs (x, y) such that $x \in A$ and $y \in B$. We denote the Cartesian product of A and B by $A \times B$.

Thus $A \times B = \{ (x, y) \mid x \in A, y \in B \}$.

Because the Cartesian plane consists of the set of points (x, y) where $x \in \mathbb{R}$ and $y \in \mathbb{R}$, we may denote it by $\mathbb{R} \times \mathbb{R}$, or $\mathbb{R}^2$.

Example Let $A = \{1, 2, 4\}$ and $B = \{2, 3\}$. List the members of the Cartesian product $A \times B$.

$A \times B = \{ (1, 2), (1, 3), (2, 2), (2, 3), (4, 2), (4, 3) \}$.

Example If $X = \{1, 2, 3\}$ and $Y = \{a, b\}$, then
$X \times Y = \{(1, a), (1, b), (2, a), (2, b), (3, a), (3, b)\}$
$Y \times X = \{(a, 1), (a, 2), (a, 3), (b, 1), (b, 2), (b, 3)\}$

Chapter 25

$$X \times X = \{(1, 1), (1, 2), (1, 3), (2, 1), (2, 2), (2, 3), (3, 1), (3, 2), (3, 3)\}$$
$$Y \times Y = \{(a, a), (a, b), (b, a), (b, b)\}$$

This example shows that, in general, $X \times Y \neq Y \times X$.

But note that $|X \times Y| = |X| \times |Y|$.

Binary Relations

A **binary relation** R from a set X to a set Y is a subset of the Cartesian product $X \times Y$. If $(x, y) \in R$, we write xRy and say that **x is related to y**. When $X = Y$, we call R a binary relation **on X**.

The set $\{x \in X \mid (x, y) \in R \text{ for some } y \in Y\}$ is called the **domain** of R.

The set $\{y \in Y \mid (x, y) \in R \text{ for some } x \in X\}$ is called the **range** of R.

Example   Let $X = \{2, 3, 4, 5\}$ and $Y = \{5, 6, 7, 8, 9, 10\}$. Define a relation R from X to Y by $(x, y) \in R$ if x divides y. Find the domain and range of R.

$R = \{(2, 6), (2, 8), (2, 10), (3, 6), (3, 9), (4, 8), (5, 5), (5, 10)\}$

The domain of $R = \{2, 3, 4, 5\}$ and the range of $R = \{5, 6, 8, 9, 10\}$.

Example   Let R be a relation on a set $X = \{1, 2, 3, 4\}$ defined by $(x, y) \in R$ provided $x < y$ and $x, y \in X$. Find the domain and range of R.

$R = \{(1, 2), (1, 3), (1, 4), (2, 3), (2, 4), (3, 4)\}$

The domain of $R = \{1, 2, 3\}$ and the range of $R = \{2, 3, 4\}$.

Definition   A relation R on a set X is said to be **reflexive** if $(x, x) \in R$ for all $x \in X$.

Example   Show that the relation R defined on set $X = \{1, 2, 3, 4\}$ by $(x, y) \in R$ provided $x \leq y$ for all $x, y \in X$ is reflexive.

Since $x \leq x$, $(x, x) \in R$ for all $x \in X$ and so R is reflexive on X.

Example   Is the relation R defined on $X = \{a, b, c, d\}$ given by
$$R = \{(a, a), (b, b), (b, c), (c, c), (c, d)\}$$
reflexive?

Since $d \in X$ but $(d, d) \notin R$, then R is not reflexive on X.

Sets, Relations and Groups

Definition A relation R on a set X is said to be ***symmetric*** if whenever $(x, y) \in R$, then $(y, x) \in R$ for all $x, y \in X$.

Example The relation $R = \{(a, a), (a, b), (b, a), (b, c), (c, b), (d, d)\}$ defined on $X = \{a, b, c, d\}$ is symmetric since whenever $(x, y) \in R$, $(y, x) \in R$.

Example The relation R defined on $X = \{1, 2, 3\}$ by $(x, y) \in R$ provided $x \geq y$ whenever $x, y \in X$, is not symmetric since $(2, 1) \in R$ but $(1, 2) \notin R$.

Definition A relation R defined on a set X is said to be ***transitive*** if whenever both (x, y) and $(y, z) \in R$, then $(x, z) \in R$ for all $x, y, z \in X$.

Example Show that the relation R defined on $X = \{1, 2, 3\}$ by $(x, y) \in R$ provided $x \leq y$ for all $x, y \in X$, is transitive.

To verify that R is transitive on X, we must list all pairs of the form (x, y) and $(y, z) \in R$ and then show that in every case, $(x, z) \in R$.

Pairs of the Form $(x, y), (y, z)$	(x, z)
(1, 1), (1, 1)	(1, 1)
(1, 1), (1, 2)	(1, 2)
(1, 1), (1, 3)	(1, 3)
(1, 2), (2, 2)	(1, 2)
(1, 2), (2, 3)	(1, 3)
(1, 3), (3, 3)	(1, 3)
(2, 2), (2, 2)	(2, 2)
(2, 2), (2, 3)	(2, 3)
(2, 3), (3, 3)	(2, 3)
(3, 3), (3, 3)	(3, 3)

Since each element in the right-hand column is a member of R, then R is transitive.

Note: If $x = y$, then (x, y) and (y, z) belong to R implies that (x, x) and (x, z) belong to R and so (x, z) automatically belongs to R.

If $y = z$, then (x, y) and (y, z) belong to R implies that (x, y) and (y, y) belong to R and so $(x, z) = (x, y)$ automatically belongs to R.

Thus we do not need to verify the transitive property for pairs of elements (x, y) and (y, z) of R whenever $x = y$ or $y = z$.

Thus in the previous example we can eliminate the cases $x = y$ and $y = z$ leaving only the pair $(1, 2), (2, 3) \in R$ and $(1, 3) \in R$ to establish that R is transitive on X.

Chapter 25

Exercise 25.2

1. Consider the relation R defined on $X = \{1, 2, 3, 4, 5, 6\}$ by $(x, y) \in R$ if 3 divides $x - y$.
   (a) List the elements of R.
   (b) Find the domain and range of R.
   (c) Is R reflexive, symmetric and/or transitive?

2. Consider the relation R defined on $X = \{1, 2, 3, 4, 5, 6\}$ by $(x, y) \in R$ if $y - x = 1$.
   (a) List the elements of R.
   (b) Find the domain and range of R.
   (c) Is R reflexive, symmetric and/or transitive?

3. Consider the relation R defined on $X = \{1, 2, 3, 4, 5, 6\}$ by $(x, y) \in R$ if $x + y \leq 6$.
   (a) List the elements of R.
   (b) Find the domain and range of R.
   (c) Is R reflexive, symmetric and/or transitive?

4. Consider the relation R defined on $\mathbb{R}^+$ by $(x, y) \in R$ if $x = y$. Determine whether R is reflexive, symmetric and/or transitive.

5. Consider the relation R defined on $\mathbb{R}^+$ by $(x, y) \in R$ if $x > y$. Determine whether R is reflexive, symmetric and/or transitive.

6. Consider the relation R defined on $\mathbb{R}^+$ by $(x, y) \in R$ if $y = x^2$. Determine whether R is reflexive, symmetric and/or transitive.

7. Consider the relation R defined on $\mathbb{R}^+$ by $(x, y) \in R$ if $x \geq y$. Determine whether R is reflexive, symmetric and/or transitive.

8. Consider the relation R defined on $\mathbb{Z}^+$ by $(x, y) \in R$ if the greatest common divisor of x and y is 1. Determine whether R is reflexive, symmetric and/or transitive.

9. Consider the relation R defined on $\mathbb{Z}^+$ by $(x, y) \in R$ if 2 divides $y - x$. Determine whether R is reflexive, symmetric and/or transitive.

10. Find a relation R defined on $X = \{1, 2, 3, 4, 5\}$ which is
    (a) reflexive but neither symmetric nor transitive ;
    (b) neither reflexive nor symmetric but transitive.
    (c) reflexive and symmetric but not transitive.

25.3 Equivalence Relations

Theorem Let S be a partition of set X. Define xRy to mean that for some set $Y \in S$, both x and y belong to Y. Then R is reflexive, symmetric and transitive.

Proof Let $x \in X$. Since S is a partition of X, then x belongs to some member Y of S and so xRx and R is reflexive.

Suppose that xRy. Then both x and y belong to some set $Y \in S$. Since both x and y belong to Y, yRx and so R is symmetric.

Suppose xRy and yRz. Then both x and y belong to some set $Y \in S$ and both y and z belong to some set $Z \in S$. But y belongs to just one set in S and so $Y = Z$ and so both x and z belong to Y giving xRz which proves that R is transitive.

Example Consider the partition $S = \{ \{1, 2\}, \{3, 4\}, \{5\} \}$ of the set $X = \{1, 2, 3, 4, 5\}$. Let xRy mean that x and y belong to the same set in S. Then $R = \{(1, 1), (1, 2), (2, 1), (2, 2), (3, 3), (3, 4), (4, 3), (4, 4), (5, 5)\}$.

$(1, 1), (2, 2), (3, 3), (4, 4), (5, 5)$ all belong to R and so R is reflexive.

$(1, 2), (2, 1), (3, 4), (4, 3)$ are the only elements of R with differing components and so R is symmetric.

Since no pairs of elements (x, y) and (y, z) where $x \neq y$ and $y \neq z$ exist in R, we do not need to check for transitivity.

Note: For set S and relation R from the theorem above, if $Y \in S$, then members of Y can be considered to be equivalent as far as the relation R is concerned. Thus relations which are reflexive, symmetric and transitive are called ***equivalence relations***.

Definition A relation that is reflexive, symmetric and transitive on a set X is called ***an equivalence relation on X***.

Example Let $X = \{1, 2, 3, 4, 5, 6\}$. Define xRy to mean that $x - y$ is even. Show that R is an equivalence relation on X.

Firstly, $x - x = 0$ which is even and so $(x, x) \in R$ for all $x \in X$.
Therefore R is reflexive.

Chapter 25

Secondly, if $x - y$ is even, so is $y - x$ and so R is symmetric.

Thirdly, if $x - y$ and $y - z$ are both even, then their sum is even. Thus $x - z$ is even and R is transitive

Therefore R is an equivalence relation on X.

Example Let $X = \{1, 2, 3, 4, 5, 6\}$. Define R to mean that $(x, y) \in R$ provided $x \leq y$ whenever $x, y \in X$. Explain why R is not an equivalence relation on X.

If xRy then $x \leq y$ and so $y \geq x$. Thus y is not related to x ($y \leq x$ is not true). Therefore R is not symmetric and R is not an equivalence relation on X.

[*Note:* $x \leq x$ so R is reflexive; if $x \leq y$ and $y \leq z$ then $x \leq z$ and so R is transitive.]

Equivalence Classes

Given an equivalence relation on a set X, we can partition X by grouping related members of X together. Elements related to one another may be thought of as equivalent.

Theorem Let R be an equivalence relation on a set X. For each $a \in X$ let

$$[a] = \{ x \in X \mid xRa \}.$$

Then $S = \{ [a] \mid a \in X \}$ is a partition of X.

Note: The sets $[a]$ defined by this theorem are called ***equivalence classes of X given by the relation R***.

Example Consider the equivalence relation $R = \{(1, 1), (1, 2), (2, 1), (2, 2), (3, 3), (3, 4), (4, 3), (4, 4), (5, 5)\}$ on the set $X = \{1, 2, 3, 4, 5\}$. Find all the equivalence classes.

The equivalence class $[1]$ contains all the elements $x \in X$ for which $(x, 1) \in R$. Thus $[1] = \{1, 2\}$. Continuing in this way we obtain the following equivalence classes: $[1] = [2] = \{1, 2\}$, $[3] = [4] = \{3, 4\}$ and $[5] = \{5\}$.

Example Let R be a relation defined on $\mathbb{Z}^+$ by xRy if $x - y$ is divisible by 3. Show that R is an equivalence relation on $\mathbb{Z}^+$ and find the equivalence classes.

Sets, Relations and Groups

$x - x = 0$ which is divisible by 3 and so R is reflexive.
If $x - y$ is divisible by 3 then $y - x$ is divisible by 3 and so R is symmetric.
If $x - y$ and $y - z$ are both divisible by 3 then their sum, $x - z$, is divisible by 3 and so R is transitive.

Therefore R is an equivalence relation on $\mathbb{Z}^+$.

The equivalence classes are: $[1] = \{1, 4, 7, 10, \ldots \}$, $[2] = \{2, 5, 8, 11, \ldots \}$ and $[3] = \{3, 6, 9, 12, \ldots \}$.

{These equivalence classes contain the positive integers which leave remainders of 1, 2, 0 respectively when divided by 3.}

Exercise 25.3

1. Consider the set $X = \{1, 2, 3, 4, 5, 6\}$. For each of the following relations, determine which are equivalence relations on X. If the relation is an equivalence relation, list the equivalence classes:
   (a) $R = \{(1, 1), (1, 3), (1, 5), (3, 1), (3, 3), (3, 5), (5, 1), (5, 3), (5, 5),$
       $(2, 2), (2, 4), (4, 2), (4, 4), (6, 6) \}$;
   (b) $R = \{(1, 1), (2, 2), (3, 3), (4, 4), (5, 5), (6, 6) \}$;
   (c) $R = \{(x, y) \mid x - y \text{ is divisible by 3} \}$;
   (d) $R = \{(x, y) \mid x + y \text{ is divisible by 4} \}$.

2. Define a relation R on $\mathbb{Z}^+ \times \mathbb{Z}^+$ by $(a, b)R(c, d)$ if $a + d = b + c$. Prove that R is an equivalence relation on $\mathbb{Z}^+ \times \mathbb{Z}^+$.

3. Define a relation R on $\mathbb{Z}^+ \times \mathbb{Z}^+$ by $(a, b)R(c, d)$ if $ad = bc$. Prove that R is an equivalence relation on $\mathbb{Z}^+ \times \mathbb{Z}^+$.

4. Define a relation R on $\mathbb{Z} \times \mathbb{Z}^+$ by $(a, b)R(c, d)$ if and only if $ad = bc$.
   (a) Prove that R is an equivalence relation on $\mathbb{Z} \times \mathbb{Z}^+$.
   (b) Show how R partitions $\mathbb{Z} \times \mathbb{Z}^+$ and describe the equivalence classes.

5. Define the relation R on the set $\mathbb{Z}$ by xRy if and only if $xy \geq 0$. Determine whether or not R is an equivalence relation on $\mathbb{Z}$.

6. Determine which of the following relations is an equivalence relation. Describe the partition arising from each equivalence relation.
   (a) xRy on $\mathbb{R}$ if $x \geq y$;
   (b) xRy on $\mathbb{R}$ if $|x - y| \leq 2$;
   (c) xRy on $\mathbb{Z}^+$ if x and y have the same number of decimal digits.

619

25.4 Matrices of Relations

A matrix is a very convenient way to represent a relation from set X to set Y (if X and Y contain just a few members each). We label the rows of the matrix with the elements of X and the columns with elements of Y. Then the ijth entry in the matrix is a '1' if iRj and a zero otherwise.

Example Write down the matrix of the relation R from $\{a, b, c\}$ to $\{d, e, f, g\}$ given that $R = \{(a, e), (a, f), (a, g), (b, f), (b, g), (c, f), (c, g)\}$.

The matrix is:

$$\begin{array}{c} \\ a \\ b \\ c \end{array} \begin{array}{cccc} d & e & f & g \\ \left(\begin{array}{cccc} 0 & 1 & 1 & 1 \\ 0 & 0 & 1 & 1 \\ 0 & 0 & 1 & 1 \end{array} \right. & & & \left. \begin{array}{c} \\ \\ \\ \end{array} \right) \end{array}.$$

Example Find the matrix of the relation from $\{2, 3, 4\}$ to $\{5, 6, 7, 8\}$ defined by xRy provided x is a factor of y.

2 is a factor of 6 and 8; 3 is a factor of 6; 4 is a factor of 8.

Therefore the matrix is

$$\begin{array}{c} \\ 2 \\ 3 \\ 4 \end{array} \begin{array}{cccc} 5 & 6 & 7 & 8 \\ \left(\begin{array}{cccc} 0 & 1 & 0 & 1 \\ 0 & 1 & 0 & 0 \\ 0 & 0 & 0 & 1 \end{array} \right. & & & \left. \begin{array}{c} \\ \\ \\ \end{array} \right) \end{array}.$$

Example Find the matrix of the relation R defined on $\{1, 2, 3, 4\}$ by xRy if $x - y$ is even.

The matrix is

$$\begin{array}{c} \\ 1 \\ 2 \\ 3 \\ 4 \end{array} \begin{array}{cccc} 1 & 2 & 3 & 4 \\ \left(\begin{array}{cccc} 1 & 0 & 1 & 0 \\ 0 & 1 & 0 & 1 \\ 1 & 0 & 1 & 0 \\ 0 & 1 & 0 & 1 \end{array} \right. & & & \left. \begin{array}{c} \\ \\ \\ \\ \end{array} \right) \end{array}.$$

Note: The matrix of a relation R on a set X is always square.

Sets, Relations and Groups

The matrix A of a relation R on a set X can be used to determine whether or not R is an equivalence relation on X.

1. Since R is reflexive if xRx for all $x \in X$, then the elements on the leading diagonal of A must all be '1s' if R is reflexive on X.
2. Since R is symmetric if yRx whenever xRy, then $a_{ij} = a_{ji}$. That is, the matrix must be symmetrical about the leading diagonal.
3. The condition on the matrix A under which R is transitive on X is not obvious. It can be shown, however, that if the matrix A^2 has zeros in exactly the same positions as does the matrix A, then R is transitive on X.

Example Let $X = \{a, b, c, d, e, f\}$ and let R be a relation on X defined by the matrix

$$A = \begin{pmatrix} 1 & 0 & 1 & 0 & 1 & 0 \\ 0 & 1 & 0 & 1 & 0 & 0 \\ 1 & 0 & 1 & 0 & 1 & 0 \\ 0 & 1 & 0 & 1 & 0 & 0 \\ 1 & 0 & 1 & 0 & 1 & 0 \\ 0 & 0 & 0 & 0 & 0 & 1 \end{pmatrix}.$$

(a) Prove that R is an equivalence relation on X.
(b) Give the partition of X corresponding to R.

(a) Since A has only '1s' on the leading diagonal, R is reflexive.
Since A is symmetric about the leading diagonal, R is symmetric.
Since the matrix

$$A^2 = \begin{pmatrix} 3 & 0 & 3 & 0 & 3 & 0 \\ 0 & 2 & 0 & 2 & 0 & 0 \\ 3 & 0 & 3 & 0 & 3 & 0 \\ 0 & 2 & 0 & 2 & 0 & 0 \\ 3 & 0 & 3 & 0 & 3 & 0 \\ 0 & 0 & 0 & 0 & 0 & 1 \end{pmatrix}$$

has zeros in the same positions as the zeros in A, R is transitive. Therefore, R is an equivalence relation on X.

(b) The partition of X corresponding to R can be read from the matrix A^2. This partition is $S = \{\{a, c, e\}, \{b, d\}, \{f\}\}$.

Chapter 25

Example Let R be a relation defined on $X = \{1, 2, 3, 4\}$ by the following

matrix $A = \begin{pmatrix} 1 & 1 & 0 & 0 \\ 1 & 1 & 1 & 0 \\ 0 & 1 & 1 & 0 \\ 0 & 0 & 0 & 1 \end{pmatrix}$. Is R is an equivalence relation on X?

A has all '1s' on its leading diagonal and is symmetrical about that diagonal. Thus R is both reflexive and symmetric on X. However

$$A^2 = \begin{pmatrix} 2 & 2 & 1 & 0 \\ 2 & 3 & 2 & 0 \\ 1 & 2 & 2 & 0 \\ 0 & 0 & 0 & 1 \end{pmatrix}$$

which has non-zero entries in the first row-third column and the third row-first column where A has zero entries. Thus R is not transitive on X and therefore not an equivalence relation on X. [$(1, 2), (2, 3) \in R$, but $(1, 3) \notin R$]

Exercise 25.4

1. In each of the following, find the matrix of the relation R from X to Y:
   (a) $R = \{(1, a), (1, b), (2, b), (3, a), (3, c), (4, b), (4, c)\}$, $X = \{1, 2, 3, 4\}$, $Y = \{a, b, c\}$;
   (b) $R = \{(1, 1), (1, 2), (2, 2), (3, 1), (4, 1), (4, 2)\}$, $X = \{1, 2, 3, 4, 5\}$, $Y = \{1, 2\}$;
   (c) $R = \{(a, 2), (a, 4), (b, 1), (b, 3), (c, 1), (c, 2), (c, 3)\}$, $X = \{a, b, c\}$, $Y = \{1, 2, 3, 4\}$;
   (d) $R = \{(p, a), (p, b), (q, a), (q, c)\}$, $X = \{p, q\}$, $Y = \{a, b, c\}$.

2. In each of the following, find the matrix of the relation R on X:
   (a) $R = \{(1, 1), (1, 2), (1, 3), (2, 1), (2, 2), (3, 2)\}$, $X = \{1, 2, 3\}$;
   (b) $R = \{(x, y) \mid x + y = 4\}$, $X = \{1, 2, 3, 4\}$;
   (c) $R = \{(x, y) \mid x \geq y\}$, $X = \{1, 2, 3, 4, 5\}$;
   (d) $R = \{(x, y) \mid x - y = 1\}$, $X = \{1, 2, 3, 4, 5, 6\}$.

3. A relation R on $X = \{a, b, c, d\}$ has matrix

$$\begin{array}{c} \\ a \\ b \\ c \\ d \end{array} \begin{pmatrix} a & b & c & d \\ 1 & 0 & 1 & 0 \\ 0 & 0 & 0 & 0 \\ 1 & 0 & 1 & 0 \\ 0 & 0 & 0 & 1 \end{pmatrix}.$$

Determine whether R is reflexive, symmetric and/or transitive.

Sets, Relations and Groups

4. Explain why the relation R on $X = \{1, 2, 3, 4, 5, 6\}$ with matrix

$$\begin{array}{c} \\ 1 \\ 2 \\ 3 \\ 4 \\ 5 \\ 6 \end{array} \begin{pmatrix} 1 & 2 & 3 & 4 & 5 & 6 \\ 1 & 1 & 0 & 0 & 0 & 1 \\ 1 & 1 & 0 & 0 & 0 & 1 \\ 0 & 0 & 1 & 1 & 0 & 0 \\ 0 & 0 & 1 & 1 & 0 & 0 \\ 0 & 0 & 0 & 0 & 1 & 0 \\ 1 & 1 & 0 & 0 & 0 & 1 \end{pmatrix}$$

is an equivalence relation and give a partition of X corresponding to R.

5. Determine whether or not the relation R on $X = \{a, b, c, d, e\}$ given by the matrix

$$A = \begin{pmatrix} 1 & 1 & 0 & 0 & 0 \\ 1 & 1 & 1 & 0 & 0 \\ 0 & 1 & 1 & 0 & 0 \\ 0 & 0 & 0 & 1 & 0 \\ 0 & 0 & 0 & 0 & 1 \end{pmatrix}$$

is an equivalence relation.

25.5 Functions – Injections, Surjections and Bijections

Definition A ***function*** is a relation in which no two different ordered pairs have the same first member.

More formally, a ***function***, f with domain A and codomain B is a subset of the Cartesian product $A \times B$ such that
(a) for every $x \in A$, there exists at least one $y \in B$ such that $(x, y) \in f$;
(b) if $(x, y_1) \in f$ and $(x, y_2) \in f$, then $y_1 = y_2$.

Example Show that the relation $f = \{(x, 3x - 2) \mid x \in \mathbb{R}\} \subset \mathbb{R}^2$ is a function.

Consider any $x \in \mathbb{R}$. There exists (exactly) one value of $(3x - 2) \in \mathbb{R}$, and if $(x, y_1) \in f$ and $(x, y_2) \in f$, then $y_1 = 3x - 2$ and $y_2 = 3x - 2$ giving $y_1 = y_2$. Therefore f is a function.

Note: For subsets of the Cartesian plane, no vertical line (parallel to the y-axis) may cross the graph of a function at more than one point.

Chapter 25

Definition Two functions $f_1 : A_1 \to B_1$ and $f_2 : A_2 \to B_2$ are said to be equal, $f_1 = f_2$, if $A_1 = A_2$, $B_1 = B_2$, and if $x \in A_1$, $f_1(x) = f_2(x)$.

Thus equal functions have the same domain and codomain, and have equal values for each element of the domain.

Injections

Definition A function f is said to be ***injective*** (or ***one-to-one***) if $f(x_1) = f(x_2)$ implies that $x_1 = x_2$. (We also say that f is an ***injection***.)

The graph of an injection cannot be cut more than once by a horizontal line (parallel to the x-axis).

Example The function $f : x \mapsto x^2$ with domain $\mathbb{R}$ and codomain $\mathbb{R}$, is not an injection since $f(-2) = f(2)$ but obviously $-2 \neq 2$. However, the function $f : x \mapsto x^2$ with domain $[0, \infty[$ and codomain $\mathbb{R}$ is an injection since if $f(x) = f(y)$, $x = y$.

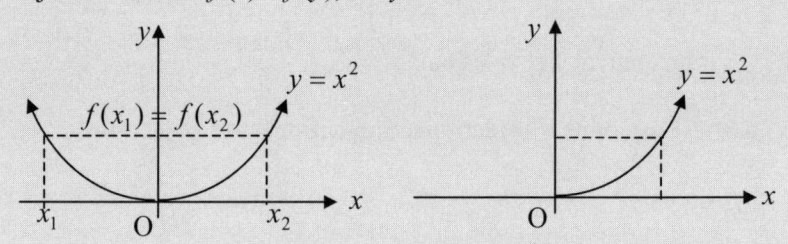

Surjections

Definition A function $f : A \to B$ is said to be ***surjective*** (or ***onto***) if any element $b \in B$ is the image of some element $a \in A$. (We also say that f is a ***surjection***.)

For a general function $f : A \to B$ we say that f maps A *into* B, but for a surjection we say that f maps A *onto* B.

Note: Given that $f : A \to B$ is an injection, then if the equation $f(x) = k$, $k \in B$, has a solution, $x \in A$, then this solution is unique, but it need not have a solution at all. However, if f is a surjection, then the equation $f(x) = k$ has a solution, $x \in A$, for all $k \in B$, but this solution need not be unique.

Sets, Relations and Groups

Theorem The function f with domain A and codomain B is a surjection *if and only if* (written *iff*) the range is equal to the codomain. [The range of a function should already be familiar to the reader.]

Proof
(i) If the range, R, is equal to the codomain, B, then for any $b \in B$, $b \in R$. Thus there exists an element $a \in A$ for which $f(a) = b$. Hence f is a surjection.

(ii) If f is a surjection, then for any element $b \in B$ (the codomain), there exists at least one element $a \in A$ for which $f(a) = b$. Hence $b \in R$, the range of f, and so $R = B$.

This proves the theorem.

Example Prove that the function $f: \mathbb{R} \to \mathbb{R}$, defined by $f(x) = 3x - 2$, is both an injection and a surjection.

Let b be any real number, then the equation $f(x) = b$ has a unique solution $x = \frac{1}{3}(b + 2)$ which is real.
Therefore f is both an injection and a surjection.

Example Prove that $f: \mathbb{R} \to \mathbb{R}$, defined by $f(x) = x^2$ is neither an injection nor a surjection.

(i) Consider the element $4 \in \mathbb{R}$, the codomain. Then the equation $f(x) = 4$ has two solutions $x = -2, 2 \in \mathbb{R}$, the domain.
Therefore f is not an injection.

(ii) Consider an element $-4 \in \mathbb{R}$, the codomain. Then the equation $f(x) = -4$ has no solution $x \in \mathbb{R}$, the domain.
Therefore f is not a surjection.

Example Show that the function $f: \mathbb{R} \to \mathbb{R}$, defined by $f(x) = x^3 + 3x^2$, is a surjection but not an injection.

(i) Consider the element $0 \in \mathbb{R}$, the codomain. Then, the equation $f(x) = 0$ has two solutions $x = 0, -3 \in \mathbb{R}$, the domain.
Therefore f is not an injection.

(ii) The function f is a surjection since for any $b \in \mathbb{R}$ (the codomain), the equation $x^3 + 3x^2 = b$ has at least one real solution. A graph of the function can be used to confirm this.

Chapter 25

Example Show that the function $f: \mathbb{R} \to \mathbb{R}$, defined by $f(x) = 2^x$, is an injection, but not a surjection.

Let b be any real number. Then the equation $2^x = b$ has a unique solution $x = \log_2 b$ if b is positive, and no solution at all if b is not positive. Therefore f is an injection but not a surjection.

Note: The function $f: \mathbb{R} \to \mathbb{R}^+$, defined by $f(x) = 2^x$ is both an injection and a surjection. Once again, a graph of the function will confirm this.

Bijections

Definition A function $f: A \to B$ is said to be **bijective** if and only if it is both injective and surjective. (We also say that f is a **bijection**.)

Example Show that the function $f: x \mapsto x^3$ is a bijection from $\mathbb{R}$ onto $\mathbb{R}$.

(i)   If b is any real number, then the equation $x^3 = b$ has at most one real solution $x = \sqrt[3]{b}$, and so f is an injection.

(ii) If b is any real number, then the equation $x^3 = b$ has exactly one real solution $x = \sqrt[3]{b}$, and so f is a surjection.

Therefore f is a bijection from $\mathbb{R}$ onto $\mathbb{R}$.

Definition A set A is said to be **countable** if there exists a bijection $f: \mathbb{N} \to A$, where $\mathbb{N}$ is the set of natural numbers $\{0, 1, 2, 3, \ldots\}$.

Example Show that the set of even numbers is countable.

Let A be the set of even numbers $\{\ldots, -4, -2, 0, 2, 4, \ldots\}$.
Consider a function $f: \mathbb{N} \to A$, defined by $f(x) = \begin{cases} x, & \text{if } x \text{ is even,} \\ -(x+1), & \text{if } x \text{ is odd.} \end{cases}$

The function f is a bijection since the equation $f(x) = b$, where b is any even number, has exactly one solution $x = b$ if b is non-negative or $x = -(b+1)$ if b is negative.

Therefore the set of even numbers is countable.

Sets, Relations and Groups

Exercise 25.5

1. Decide whether the following functions are injections and/or surjections:
   (a) $f: \mathbb{R} \to \mathbb{R}$, defined by $f(x) = 2 - x$;
   (b) $f: \mathbb{R} \to \mathbb{R}$, defined by $f(x) = \sin x$;
   (c) $f: \mathbb{R} \to \mathbb{R}$, defined by $f(x) = x^3$;
   (d) $f: \mathbb{R}^+ \to \mathbb{R}$, defined by $f(x) = \log x$;
   (e) f with domain $[0, 2\pi]$, and codomain $\mathbb{R}$, defined by $f(x) = \cos x$;
   (f) f with domain $\mathbb{R}$, codomain $]0, 2]$ and defined by $f(x) = 2/(1+x^2)$;
   (g) f with domain $[0, \infty[$, codomain $\mathbb{R}$ and defined by $f(x) = \sqrt{x}$;
   (h) $f: \mathbb{R} \to \mathbb{R}$, defined by $f(x) = x(x-2)$;
   (i) f with domain $[-2, 2]$, codomain $[0, 2]$, defined by $f(x) = \sqrt{4-x^2}$;
   (j) f with domain $[0, \pi]$, codomain $[0, 1]$ and defined by $f(x) = \sin x$;
   (k) $f: \mathbb{R} \to \mathbb{R}^+$, defined by $f(x) = 1 + 2^x$;
   (l) $f: \mathbb{R} \to \left]-\tfrac{1}{2}\pi, \tfrac{1}{2}\pi\right[$, defined by $f(x) = \arctan x$.

2. Find the largest domain for which the function defined by $f(x) = 1 - x^2$ is an injection. For the domain chosen, give the codomain required to ensure that f is also a surjection.

3. Define a function f, which is both an injection and a surjection, for which
   (a) the domain is $\mathbb{R}^+$ and the codomain is $\mathbb{R}^+$;
   (b) the domain is $[0, \infty[$ and the codomain is $[0, \infty[$;
   (c) the domain is $]-1, 1[$ and the codomain is $\mathbb{R}$.

4. Which of the following functions are bijections? If a function is not a bijection, explain why this is so.
   (a) $f: \mathbb{R} \to \mathbb{R}$, defined by $f(x) = 5x + 4$;
   (b) $f: \mathbb{R} \to \mathbb{R}$, defined by $f(x) = x^2 + 1$;
   (c) $f: \mathbb{R} \to \mathbb{R}$, defined by $f(x) = x^3 + 1$;
   (d) $f: \mathbb{R} \to \mathbb{R}$, defined by $f(x) = x^3 - x$;
   (e) f with domain and codomain $[\tfrac{1}{2}, \infty[$, defined by $f(x) = 2x^2 - 2x + 1$;
   (f) f with domain $[0, 2\pi]$, codomain $[-1, 1]$, defined by $f(x) = \cos \tfrac{1}{2} x$;
   (g) $f: \mathbb{R} \to \mathbb{R}$, defined by $f(x) = 2^{-x}$;
   (h) $f: \mathbb{R}^+ \to \mathbb{R}$, defined by $f(x) = \log_2 x$;

627

Chapter 25

(i) $f: \mathbb{R}^+ \to \mathbb{R}^+$, defined by $f(x) = 4/x$;

(j) f with domain and codomain $]1, \infty[$, defined by $f(x) = x/(x-1)$.

5. Prove that the following functions are bijections:
   (a) $f: \mathbb{R} \to \mathbb{R}$, defined by $f(x) = x^3$;
   (b) $f: \mathbb{R} \to \mathbb{R}^+$, defined by $f(x) = 10^x$;
   (c) f with domain $[-\frac{1}{2}\pi, \frac{1}{2}\pi]$, codomain $[-1, 1]$, defined by $f(x) = \sin x$;
   (d) f with domain $[0, \infty[$, codomain $]0, 1]$, defined by $f(x) = 1/(1+x^2)$.

6. Show that each of the following sets is countable:
   (a) $\{1, 2, 3, 4, 5, \ldots\}$;
   (b) $\{1, 3, 5, 7, 9, \ldots\}$;
   (c) $\{1, 2, 4, 8, 16, 32, \ldots\}$;
   (d) $\mathbb{Z}$.

25.6 Binary Operators and Closure

Definition A ***binary operator*** combines two 'elements' together to give a unique third 'element'.

Examples of binary operators on real numbers include addition and multiplication. In the case of transformations of the plane, the normal binary operator is 'follows' in the sense that one transformation follows another to give a single third transformation. For example, if *A* is a reflection in the *x*-axis, ***B*** is a reflection in the *y*-axis and ***C*** is a rotation of 180° about the origin, then *A* 'follows' ***B*** is equal to ***C*** in the sense that a reflection in the *x*-axis following a reflection in the *y*-axis is equivalent to a single rotation of 180° about the origin..

Closure of a Set with Respect to an Operation

Definition A set *S* is ***closed*** under a binary operator ∗ if, whenever *a* and *b* are in *S*, *a* ∗ *b* is also in *S*.

The set of non-zero real numbers is closed under multiplication but not under addition since the product of any two non-zero real numbers is another (unique) non-zero real number, but the sum of any two non-zero real numbers is not necessarily another non-zero real number. For example, the sum of the two non-zero real numbers 6 and −6 is 0 which is not a non-zero real number.

The set of numbers of the form *a.bc*, where *a* is a natural number, . is a decimal point and *b* and *c* are natural numbers between 0 and 9 inclusive, is closed under addition but not under multiplication.

Sets, Relations and Groups

Exercise 25.6

1. Which of the following sets are closed under addition? The set of all
   - (a) even integers ;
   - (b) odd integers ;
   - (c) positive integers ;
   - (d) negative integers ;
   - (e) non-zero integers ;
   - (f) prime numbers ;
   - (g) rational numbers ;
   - (h) positive rational numbers ;
   - (i) negative rational numbers ;
   - (j) non-zero rational numbers ;
   - (k) irrational numbers ;
   - (l) real numbers ;
   - (m) positive real numbers ;
   - (n) negative real numbers ;
   - (o) multiples of 3 ;
   - (p) numbers of the form $a + b\sqrt{2}$, $a, b \in \mathbb{Q}$.

2. Which of the sets in Question 1 are closed under multiplication?

3. Which of the following subsets of $\mathbb{N}$ are closed under addition?
   - (a) $\mathbb{Z}^+$;
   - (b) $\{3, 4, 5, 6, \ldots\}$;
   - (c) $\{1, 3, 4, 5, 6, \ldots\}$;
   - (d) $\{4, 7, 8, 9, 10, \ldots\}$;
   - (e) $\{0\}$;
   - (f) $\{1\}$;
   - (g) $\{0, 2, 4, 6, 8, \ldots\}$;
   - (h) $\{1, 3, 5, 7, 9, \ldots\}$;
   - (i) $\{2^m \mid m \in \mathbb{N}\}$;
   - (j) $\{6m \mid m \in \mathbb{N}\}$.

4. Which of the sets in Question 3 are closed under multiplication?

5. Which of the following subsets of $\mathbb{R}$ are closed under multiplication?
   - (a) $\{1, 2\}$;
   - (b) $\{0, 1, 4, 9, 16, \ldots\}$;
   - (c) $\{6m \pm 1 \mid m \in \mathbb{Z}^+\}$;
   - (d) $\{4, 6, 8, 9, 10, 12, \ldots\}$ – the set of all composite numbers ;
   - (e) $\{n \mid n \in \mathbb{N}, 101 \text{ is not a divisor of } n\}$;
   - (f) $\{n \mid n \in \mathbb{N}, 91 \text{ is not a divisor of } n\}$.

6. Show that there is only one subset of $\mathbb{N}$ which is closed under subtraction.

7. Show that there is only one subset of $\mathbb{N}$ which is closed under division.

8. Find three finite subsets of $\mathbb{N}$ which are closed under multiplication.

9. Find the smallest subset of $\mathbb{Z}$ which is closed under subtraction and contains the number 2.

10. Find the smallest subset of $\mathbb{Z}$ which is closed under multiplication and contains the number -1.

Chapter 25

11. Find two subsets of $\mathbb{Z}$ which are closed under division.

12. Which of the following subsets of $\mathbb{Q}$ are closed under addition?
    (a) $\{q \mid q \in \mathbb{Q}, q < -2\}$;
    (b) $\{q \mid q \in \mathbb{Q}, q < 0\}$;
    (c) $\{q \mid q \in \mathbb{Q}, q < 2\}$;
    (d) $\{q \mid q \in \mathbb{Q}, q \neq 0\}$;
    (e) $\{q \mid q \in \mathbb{Q}, q^2 > 1\}$;
    (f) $\{q \mid q \in \mathbb{Q}, q^2 < 1\}$;
    (g) $\{q \mid q \in \mathbb{Q}, 0 < q < q^2\}$;
    (h) $\{n(2^{-k}) \mid n, k \in \mathbb{Z}^+\}$.

13. Let $Q(\sqrt{2}) = \{a + b\sqrt{2} \mid a, b \in \mathbb{Q}\}$. Show that $Q(\sqrt{2})$ is closed under three of the four operations $+, -, \times, \div$. Discuss the closure of the set of all non-zero members of $Q(\sqrt{2})$ under these four operations.

14. Prove that if A is a subset of $\mathbb{R}$ which is closed under addition, then $B = \{-a \mid a \in A\}$ is also closed under addition. Is B closed under subtraction?

25.7 The Associative Law

Definition The set S is *associative* under the binary operation $*$ if, for all a, b and c in S, $a*(b*c) = (a*b)*c$.

Example The set $\mathbb{R}$ is associative under both addition and multiplication, and so is any subset of $\mathbb{R}$, but $\mathbb{R}$ is not associative under either subtraction or division since
$a + (b + c) = (a + b) + c$ for all real a, b and c;
$a \times (b \times c) = (a \times b) \times c$ for all real a, b and c;
$a - (b - c) \neq (a - b) - c$, in general, and
$a \div (b \div c) \neq (a \div b) \div c$, in general.

Example Show that $\mathbb{R}$ is associative under the binary operator $*$ where $a*b = 2ab + a + b$ for all real a, b.

If a, b and c are real,
$$\begin{aligned} a*(b*c) &= a*(2bc + b + c) \\ &= 2a(2bc + b + c) + a + (2bc + b + c) \\ &= 4abc + 2ab + 2ac + 2bc + a + b + c. \end{aligned}$$
Also,
$$\begin{aligned} (a*b)*c &= (2ab + a + b)*c \\ &= 2(2ab + a + b)c + (2ab + a + b) + c \\ &= 4abc + 2ab + 2ac + 2bc + a + b + c. \end{aligned}$$
Clearly $a*(b*c) = (a*b)*c$ and so $\mathbb{R}$ is associative under $*$.

630

Sets, Relations and Groups

Exercise 25.7

1. Under which of the following operations is $\mathbb{R}$ associative?
   (a) $a*b = a+b-4$;
   (b) $a*b = a+b-4ab$;
   (c) $a*b = 2a+b$;
   (d) $a*b = ab+a+b$;
   (e) $a*b = \sqrt{a^2+b^2}$;
   (f) $a*b = |a+b|$.

2. Consider the set $S = \{(x, y) \mid x, y \in \mathbb{R}\}$ under the operation $*$ where
   $(x_1, y_1) * (x_2, y_2) = (x_1 x_2 - y_1 y_2, x_1 y_2 + x_2 y_1)$.
   Is S associative under $*$?

3. Let $A = \{a, b, c\}$ and let $*$ be defined on A by means of the following 'multiplication table'.

*	a	b	c
a	a	b	c
b	b	c	a
c	c	a	b

   [The table is constructed so that the value of $a*b$ is shown in the same row as a and the same column as b.]

   (a) Verify that $a*(b*c) = (a*b)*c$.
   (b) How many separate checks are needed to test the associative law for $*$ on A?
   (c) Can you see any property of the operation that will enable you to reduce the labour required to check the associative law? How many checks are now required?

25.8 The Identity Element

Definition A set A under a binary operation $*$ is said to contain **an identity element** e if, for all a in A, $a*e = e*a = a$.

Example $\mathbb{R}, \mathbb{Q}, \mathbb{N}$ and $\mathbb{Z}$ all have zero as the identity element with respect to addition and 1 as the identity element with respect to multiplication.

Example The set of non-negative even numbers $\{0, 2, 4, 6, \ldots\}$ has 0 as the identity under addition but no identity under multiplication. The set of positive odd numbers $\{1, 3, 5, 7, \ldots\}$ has 1 as the identity under multiplication but no identity under addition.

631

Chapter 25

Example   Find the identity element of $\mathbb{R}$ with respect to the operation $*$ defined on $\mathbb{R}$ by $a * b = a + b - 2ab$ for all a, b in $\mathbb{R}$.

Let e be the identity, if it exists.
Then
$$a * e = a$$
$$a + e - 2ae = a$$
$$e(1 - 2a) = 0$$
$\Rightarrow \quad e = 0 \text{ or } a = \tfrac{1}{2}$.

Hence, if $a * e = a$ for all a in $\mathbb{R}$, it follows that $e = 0$.

We must now check that $e = 0$ is in fact the identity element.
$$a * 0 = a + 0 - 2a(0) = a \quad \text{and} \quad 0 * a = 0 + a - 2(0)a = a.$$
Therefore 0 is the identity element.

Theorem   The identity element is unique.

Proof   Let e and e' both be identity elements for the operation $*$ defined on a set A. Then $\quad e * e' = e'$ since e is an identity element, and
$$e * e' = e \quad \text{since } e' \text{ is an identity element.}$$
Hence $e' = e$ and so the identity element is unique.

Exercise 25.8

1. Determine the identity element (if it exists) for the operation $*$ defined on $\mathbb{R}$, if for a, b in $\mathbb{R}$
   (a) $a * b = a + b + 3$;   (b) $a * b = 3ab$;
   (c) $a * b = ab - a - b$;   (d) $a * b = a + b - ab$.

2. Find the identity element, if it exists, for the following structures:
   (a)   Real numbers under $*$ where $a * b = 2a + ab + 2b$.
   (b)   Set $S = \{ (x, y) \mid x, y \in \mathbb{R} \}$ under $*$ where
   $$(x_1, y_1) * (x_2, y_2) = (x_1 x_2 - y_1 y_2, x_1 y_2 + x_2 y_1)$$
   for all $(x_1, y_1), (x_2, y_2)$ in S.
   (c)   The set of all subsets of a given universal set U under union.
   (d)   The set of all subsets of a given universal set U under intersection.
   (e)   Set $S = \{2, 4, 6, 8\}$ under $*$ where $a * b$ is the least positive remainder when ab is divided by 10.

3. Consider the set $S = \{2, 4, 6, 8, 10, 12\}$ under $*$ where $a * b$ is the least positive remainder when ab is divided by 14. Draw up a multiplication table for S under $*$ and hence determine the identity element.

Sets, Relations and Groups

4. Consider the set $S = \left\{ \begin{pmatrix} a & b \\ c & d \end{pmatrix} \middle| \ a, b, c, d \in \mathbb{R} \right\}$ under the operator $*$ where

$$\begin{pmatrix} a & b \\ c & d \end{pmatrix} * \begin{pmatrix} e & f \\ g & h \end{pmatrix} = \begin{pmatrix} ae + bg & af + bh \\ ce + dg & cf + dh \end{pmatrix} \text{ for all } \begin{pmatrix} a & b \\ c & d \end{pmatrix}, \begin{pmatrix} e & f \\ g & h \end{pmatrix} \text{ in } S.$$

Find the identity element for $*$ on S.

5. In each of the following, a set S and an operation $*$ on S are defined by a 'multiplication' table. Find the identity element in each, if it exists.

(a)

*	a	b	c
a	b	c	a
b	c	a	b
c	a	b	c

(b)

*	a	b	c	d
a	b	d	a	c
b	c	a	d	b
c	a	b	c	d
d	d	c	b	a

What property of a multiplication table determines the existence of an identity element?

25.9 Inverse Elements

Definition Let A be a set with binary operation $*$ and with an identity element e. The element a' in A is said to be an inverse of the element a in A if

$$a * a' = a' * a = e.$$

Clearly, if a' is an inverse of a then a is an inverse of a'.

Example The inverse of a real number a under addition is its negative $-a$ since $a + (-a) = (-a) + a = 0$ where 0 is the identity element under addition.

The inverse of the non-zero number a under multiplication is its reciprocal a^{-1} since $a \times a^{-1} = a^{-1} \times a = 1$ where 1 is the identity element under multiplication. [0 has no inverse under multiplication.]

Notation: In all that follows, we will generally use the symbol a^{-1} to represent the inverse of the element a under **all** operations defined on a set S. In most cases this must not be confused with the reciprocal of the real number a.

Example Find the inverse of the real number a with respect to the operation $*$ defined on $\mathbb{R}$ by $a * b = ab + a + b$ for all a, b in $\mathbb{R}$.

633

Chapter 25

First we must find the identity element.
Let e be the identity element, if it exists.
Then $a*e = a$ and so $ae + a + e = a$.
Thus $e(a+1) = 0$ which gives $e = 0$ or $a = -1$.
But $a*e = a$ for all a in $\mathbb{R}$ and so $e = 0$.
We must now check that 0 is the identity: $a*0 = a(0) + a + 0 = a$
$$0*a = (0)a + 0 + a = a.$$
Therefore 0 is indeed the identity.

Now let a^{-1} be the inverse of a if it exists.
Then $\quad a*a^{-1} = e$
$\Rightarrow \quad aa^{-1} + a + a^{-1} = 0$
$\Rightarrow \quad a^{-1}(a+1) = -a$
$\Rightarrow \quad a^{-1} = \dfrac{-a}{a+1}$ provided $a \neq -1$.

Finally we must check that this is indeed the inverse of a ($\neq -1$).
$$a*\left(\dfrac{-a}{a+1}\right) = a\left(\dfrac{-a}{a+1}\right) + a + \dfrac{-a}{a+1} = \dfrac{-a^2 + a^2 + a - a}{a+1} = 0.$$

Therefore the inverse of a is $\dfrac{-a}{a+1}$ provided $a \neq -1$.
If $a = -1$, there is no inverse.

Exercise 25.9

1. Find the inverse of the real number c with respect to the operation $*$ defined on $\mathbb{R}$ by each of the following:
   (a) $a*b = a + b + 4$;
   (b) $a*b = 2ab$;
   (c) $a*b = 2ab + a + b$;
   (d) $a*b = ab + 2a + 2b$;
   (e) $a*b = a + b - 3ab$;
   (f) $a*b = 2ab + 10 - 4a - 4b$.

2. For the set $S = \{(x, y) \mid x, y \in \mathbb{R}\}$ under the operation $*$ defined by
   $$(x_1, y_1) * (x_2, y_2) = (x_1 x_2 - y_1 y_2, x_1 y_2 + x_2 y_1)$$
   for all $(x_1, y_1), (x_2, y_2)$ in S, we have shown that $(1, 0)$ is the identity element. What is the inverse of the element (p, q)? Which elements in S, if any, do not have inverses?

3. Prove that for every set with a binary operator defined on it, the identity element (if it exists) has a unique inverse.

Sets, Relations and Groups

4. Consider the set $S = \{1, 2, 3, 4\}$ under the binary operation $*$ where $a * b$ is the smallest positive remainder when ab is divided by 5. Find the inverse of each member of S with respect to $*$.

5. Each of the following tables defines an operation $*$ on a set. Find the identity element and the inverse of each element.

(a)

*	a	b	c	d
a	b	a	d	c
b	a	b	c	d
c	d	c	a	b
d	c	d	b	a

(b)

*	a	b	c	d
a	b	c	a	c
b	c	d	b	c
c	a	b	c	d
d	d	c	d	a

6. (a) Find the inverse of each member of $\{1, 2, 3, 4, 5, 6, 7, 8, 9, 10\}$ if $a * b$ means the least positive remainder when ab is divided by 11.
   (b) As for part (a) but using $\{2, 4, 6, 8\}$ and division by 10.
   (c) As for part (a) but using $\{3, 6, 9, 12, 15, 18\}$ and division by 21.
   (d) Which members of the set of all natural numbers less than or equal to 14 have inverses under the operation $*$ defined on the set by taking $a * b$ to be the least non-negative remainder when the product is divided by 15? What are their inverses?

*7. (a) Simplify (i) $(x+d)(x^2 - dx + d^2)$;
   (ii) $(\sqrt[3]{2}+1)(\sqrt[3]{4} - \sqrt[3]{2} + 1)$.
   (b) Prove that the set $\{a\sqrt[3]{4} + b\sqrt[3]{2} + c \mid a, b, c \in \mathbb{Q}\}$ is closed under multiplication. Find the inverse of $\sqrt[3]{2} + 1$ and also that of $\sqrt[3]{2} - 1$.

25.10 Residue Classes

The set of integers $\{ \ldots, -4, -1, 2, 5, 8, 11, \ldots \}$ each member of which has the same remainder (2) when divided by 3, is called a ***residue class***, modulo 3. There are three different residue classes, modulo 3, corresponding to the three possible remainders 0, 1, 2 on division by 3. These classes are:

$\{ \ldots, -6, -3, 0, 3, 6, 9, \ldots \}$;
$\{ \ldots, -5, -2, 1, 4, 7, 10, \ldots \}$;
$\{ \ldots, -4, -1, 2, 5, 8, 11, \ldots \}$.

Similarly there are two residue classes, modulo 2. They are the set of all odd numbers and the set of all even numbers. It is clear that for each positive integer m, there are m different residue classes, modulo m, and every integer is a member of one and only one of these classes.

Chapter 25

Definition   Two integers, a and b, which are in the same residue class, modulo m, are said to be **congruent, modulo m**. This relation is written
$$a \equiv b \pmod{m},$$
and means that
$\qquad\qquad a - b$ is a multiple of m ;
or $\qquad a - b = mq$ where q is an integer ;
or $\qquad a = b + mq$.

Theorem   Suppose $a \equiv b \pmod{m}$ and $c \equiv d \pmod{m}$.
Then   (i) $a + c \equiv b + d \pmod{m}$;
(ii) $ac \equiv bd \pmod{m}$.

Proof   (i)   For some integers q_1 and q_2 we have
$a = b + mq_1$ and $c = d + mq_2$.
Therefore $a + c = b + mq_1 + d + mq_2$
$= b + d + m(q_1 + q_2)$
$= b + d + mq_3$ (q_3 is an integer)
$\equiv b + d \pmod{m}$.

(ii)   Also $ac = (b + mq_1)(d + mq_2)$
$= bd + m(q_1 d + q_2 b + mq_1 q_2)$
$= bd + mq_4$ (q_4 is an integer)
$\equiv bd \pmod{m}$.

This theorem means that, as far as the congruence relation is concerned, one member of a residue class is as good as another. Thus we speak of

(a) the residue class 4 (mod 7), meaning the solution set of the congruence $x \equiv 4 \pmod{7}$;

(b) the sum of the residue classes 4 and 2 (mod 7), meaning the solution set of the congruence $x \equiv 4 + 2 \pmod{7}$ [i.e., { ... , –1, 6, 13, 20, ... }] ;

(c) the product of the residue classes 4 and 2 (mod 7), meaning the solution set of the congruence $x \equiv 4 \times 2 \pmod{7}$ [i.e., { ... , –6, 1, 8, 15, ... }].

In naming a residue class, we can use any member of it, but generally we choose to use its smallest non-negative member.

Example   Find the remainder when $54 \times 69 \times 137$ is divided by 7.

Sets, Relations and Groups

$54 \equiv 5$, $69 \equiv 6$ and $137 \equiv 4 \pmod{7}$
Therefore $54 \times 69 \times 137 \equiv 5 \times 6 \times 4 \equiv 2 \times 4 \equiv 1 \pmod{7}$.
Thus the required remainder is 1.

Example $\{t_n\}$ is a sequence of integers such that $t_1 = 6$ and for every $n \geq 1$, $t_{n+1} = t_n^2 + 4$. Find the least n for which t_n is a multiple of 13.

n	t_n
1	6
2	$6^2 + 4 \equiv 1 \pmod{13}$
3	$1^2 + 4 \equiv 5 \pmod{13}$
4	$5^2 + 4 \equiv 3 \pmod{13}$
5	$3^2 + 4 \equiv 0 \pmod{13}$

Therefore the least value of n is 5.

[Check? $t_1 = 6$, $t_2 = 40$, $t_3 = 1604$, $t_4 = 2\,572\,820$,
$t_5 = 6\,619\,402\,752\,404 = 13 \times 509\,184\,827\,108$]

Exercise 25.10

1. Prove that for all integers a, b and c,
   (a) $100a + 10b + c \equiv a + b + c \pmod{9}$;
   (b) $100a + 10b + c \equiv a - b + c \pmod{11}$.
   State and prove similar results for $a_0 + 10a_1 + 10^2 a_2 + 10^3 a_3 + \cdots + 10^n a_n$.

2. Find the digit x if $5\,38x\,239$ is divisible by (a) 9; (b) 11.

3. Let $t_1 = 4$ and $t_{n+1} = t_n^2 + t_n + 2$ for all $n \geq 1$. Show that for every n, either $t_n \equiv 1 \pmod{7}$ or $t_1 \equiv 4 \pmod{7}$.

4. Let $t_1 = 3$ and $t_{n+1} = t_n^3 + 2$ for all $n \geq 1$. Show that for every n, the smallest positive remainder when t_n is divided by 9 is either 1, 2, or 3.

5. Let $t_1 = 2$, $t_2 = 3$ and $t_{n+2} = t_{n+1} + t_n$ for all $n \geq 1$. Find the smallest value of n for which t_n is a multiple of 19.

6. Find the smallest positive integer x that is a solution of the congruence $ax \equiv 1 \pmod{11}$ for each integer value of a from 1 to 10 inclusive.

Chapter 25

7. Repeat Question 6 using the congruence $ax \equiv 1 \pmod{10}$ and all the numbers a such that $1 \leq a \leq 9$ for which a solution x exists.

*8. Let p be a prime number and q a positive integer, not a multiple of p. Show that no two of the numbers $q, 2q, 3q, \ldots, (p-1)q$ are congruent, modulo p. Deduce that the congruence $ax \equiv 1 \pmod{p}$ must have a solution x.

25.11 Permutations

Definition Let $A = \{1, 2, 3, \ldots, n\}$. A ***permutation*** p of A is a function whose domain and range are both A, i.e., the set of values of p is the whole of A.

Like any numerical function with finite domain, p may be defined by means of a table of values:

k	$p(k)$
1	$p(1)$
2	$p(2)$
3	$p(3)$
•	•
•	•
n	$p(n)$

It is customary to write this table horizontally rather than vertically for obvious reasons:

$$p = \begin{pmatrix} 1 & 2 & 3 & \cdots & k & \cdots & n \\ p(1) & p(2) & p(3) & \cdots & p(k) & \cdots & p(n) \end{pmatrix},$$

where the numbers on the second row are precisely those of the first, possibly in a different order.

Example If $A = \{1, 2, 3, 4\}$, some permutations of set A might be

$$a = \begin{pmatrix} 1 & 2 & 3 & 4 \\ 2 & 3 & 4 & 1 \end{pmatrix}, \quad b = \begin{pmatrix} 1 & 2 & 3 & 4 \\ 2 & 3 & 1 & 4 \end{pmatrix}, \quad c = \begin{pmatrix} 1 & 2 & 3 & 4 \\ 2 & 1 & 3 & 4 \end{pmatrix},$$

$$d = \begin{pmatrix} 1 & 2 & 3 & 4 \\ 2 & 1 & 4 & 3 \end{pmatrix}, \quad e = \begin{pmatrix} 1 & 2 & 3 & 4 \\ 1 & 2 & 3 & 4 \end{pmatrix}.$$

Sets, Relations and Groups

Note: The symbol $\begin{pmatrix} 1 & 2 & 3 & 4 \\ 1 & 4 & 1 & 3 \end{pmatrix}$ does ***not*** represent a permutation as 2 is not in the range of values given by this table. Also, $a = \begin{pmatrix} 2 & 3 & 4 & 1 \\ 3 & 4 & 1 & 2 \end{pmatrix} = \begin{pmatrix} 4 & 2 & 3 & 1 \\ 1 & 3 & 4 & 2 \end{pmatrix}$, etc. Thus the information is read vertically, not horizontally.

The symbol a here denotes a function for which the usual notation can be used, e.g., $a(2) = 3$, $a(4) = 1$, just as we write $\sin(\pi/6)=0.5$.

Definition We denote by S_n the set of all permutations of $\{1, 2, 3, \ldots, n\}$.

Example The set of all permutations of $\{1, 2, 3\}$ is
$$\begin{pmatrix} 1 & 2 & 3 \\ 1 & 2 & 3 \end{pmatrix}, \begin{pmatrix} 1 & 2 & 3 \\ 1 & 3 & 2 \end{pmatrix}, \begin{pmatrix} 1 & 2 & 3 \\ 3 & 2 & 1 \end{pmatrix}, \begin{pmatrix} 1 & 2 & 3 \\ 2 & 1 & 3 \end{pmatrix}, \begin{pmatrix} 1 & 2 & 3 \\ 2 & 3 & 1 \end{pmatrix}, \begin{pmatrix} 1 & 2 & 3 \\ 3 & 1 & 2 \end{pmatrix}.$$

Theorem S_n has $n!$ members.

Proof Let p be one of the permutations. Then $p(1)$ may have any one of n possible values. To each such value, there are $n - 1$ possible values for $p(2)$ since $p(1)$ may not be used twice. Thus there are $n(n - 1)$ possible pairs $(p(1), p(2))$. To each of these there are $n - 2$ choices for $p(3)$. Having assigned values to $p(1), p(2), p(3)$, there remains $n - 3$ choices for $p(4)$, and so on. Thus the total number of possible ways of defining p is the product $n(n - 1)(n - 2) \ldots (3)(2)(1) = n!$, as required.

Composition of Permutations

The composition of permutations is their composition as functions, but for notational convenience we write pq instead of $p * q$ or $p \circ q$. To compute pq, where p and q are permutations on the set $\{1, 2, 3, 4\}$ given by

$$p = \begin{pmatrix} 1 & 2 & 3 & 4 \\ 2 & 3 & 4 & 1 \end{pmatrix} \quad \text{and} \quad q = \begin{pmatrix} 1 & 2 & 3 & 4 \\ 2 & 1 & 3 & 4 \end{pmatrix}$$

we argue as follows: $pq(1) = p(q(1)) = p(2) = 3$; $\quad pq(2) = p(q(2)) = p(1) = 2$;
$\quad\quad\quad\quad\quad\quad\quad\quad\quad pq(3) = p(q(3)) = p(3) = 4$; $\quad pq(4) = p(q(4)) = p(4) = 1$.

Thus $pq = \begin{pmatrix} 1 & 2 & 3 & 4 \\ 3 & 2 & 4 & 1 \end{pmatrix}$.

Chapter 25

Definition The permutation $e = \begin{pmatrix} 1 & 2 & \ldots & k & \ldots & n \\ 1 & 2 & \ldots & k & \ldots & n \end{pmatrix}$ of $A = \{1, 2, \ldots, n\}$ is called the *identity permutation* of A.

Thus $pe = ep = p$ for all permutations p of A.

Exercise 25.11

1. Label the six permutations of the set $\{1, 2, 3\}$ as $p_1 = \begin{pmatrix} 1 & 2 & 3 \\ 1 & 2 & 3 \end{pmatrix}$, $p_2 = \begin{pmatrix} 1 & 2 & 3 \\ 1 & 3 & 2 \end{pmatrix}$, $p_3 = \begin{pmatrix} 1 & 2 & 3 \\ 3 & 2 & 1 \end{pmatrix}$, $p_4 = \begin{pmatrix} 1 & 2 & 3 \\ 2 & 1 & 3 \end{pmatrix}$, $p_5 = \begin{pmatrix} 1 & 2 & 3 \\ 2 & 3 & 1 \end{pmatrix}$, $p_6 = \begin{pmatrix} 1 & 2 & 3 \\ 3 & 1 & 2 \end{pmatrix}$ and construct a 'multiplication' table under the composition of permutations. [The entry in the ith row and jth column will be the composition $p_i p_j$ or p_i 'follows' p_j.]

   (a) State the identity and give the inverse of each permutation.
   (b) Simplify each of the following:

   (i) p_5^3;
   (ii) $p_2 p_3 p_4$;
   (iii) $p_4 p_3 p_2$;
   (iv) $p_3 p_5^{-1} p_4$;
   (v) $(p_6 p_5 p_4)^{-1}$;
   (vi) $p_4^2 p_5^3$.

2. With the six permutations given in Question 1, find the permutations x which satisfy each of the following equations:

   (a) $p_2 x = p_1$;
   (b) $x p_5 = p_6$;
   (c) $p_2 x p_3 = p_4$;
   (d) $p_5 x p_5^{-1} = p_6$;
   (e) $x^2 = p_1$;
   (f) $x^3 = p_1$.

3. In S_5 let $e = \begin{pmatrix} 1 & 2 & 3 & 4 & 5 \\ 1 & 2 & 3 & 4 & 5 \end{pmatrix}$ and find

   (a) a permutation x such that x, x^2, x^3, x^4 are all distinct and $x^5 = e$;
   (b) a permutation y such that $y \neq e$ but $y^2 = e$.

4. A shuffle of 6 cards is described by the permutation $\begin{pmatrix} 1 & 2 & 3 & 4 & 5 & 6 \\ 3 & 6 & 5 & 1 & 2 & 4 \end{pmatrix}$.

   Find the number of times this shuffle has to be carried out consecutively in order to replace the cards in their original order.

25.12 Cyclic Notation (Optional)

The tabular format for writing down permutations is quite 'clumsy'. A much more economical notation is the 'cyclic notation'.

As an example consider the permutation $p = \begin{pmatrix} 1 & 2 & 3 & 4 \\ 4 & 2 & 1 & 3 \end{pmatrix}$ in S_4. Since p maps 1 to 4, 4 to 3, 3 to 1 and 2 to itself, p may be represented cyclically as follows:

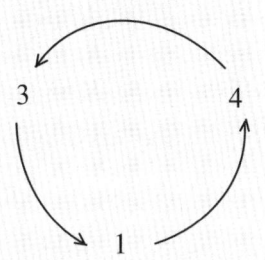

We denote this by (143), i.e., $1 \rightarrow 4 \rightarrow 3 \rightarrow 1$, and since 2 is not mentioned, $2 \rightarrow 2$. This is the most economical way of writing p that is possible since each element is written once and the fixed element 2 is not written down at all. We have to write 3 numbers instead of 8.

Definition Let $k_1, k_2, \cdots, k_r$ be distinct members of the set $A = \{1, 2, \ldots, n\}$ in the given order. The cycle $c = (k_1 k_2 \cdots k_r)$ is the permutation of A such that
(a) $c(k_i) = k_{i+1}$ for $i = 1, 2, \ldots, r-1$ and $c(k_r) = k_1$; and
(b) $c(j) = j$ for any $j \in A$ distinct from $k_1, k_2, \cdots, k_r$.

Note: The number r is known as the **length** of the cycle.

Each member of the cycle is taken to the next on the right and the last is taken to the first thereby 'closing' the cycle.

The cycle (143) may equally well be written (431) or (314), but (413) represents a different permutation. A peculiarity of this notation is that the symbol (143) could represent permutations of S_4, S_5, S_6, etc. Thus to avoid ambiguity, the set A should be specified. A cycle of length 1, written (1) in general, represents the identity permutation of A.

Example For the set $A = \{1, 2, 3, 4\}$, let $p = (132)$ and $q = (234)$. Find the permutation pq.

Chapter 25

q maps 1 to 1 and p maps 1 to 3 ;
q maps 2 to 3 and p maps 3 to 2 ;
q maps 3 to 4 and p maps 4 to 4 ;
q maps 4 to 2 and p maps 2 to 1.

Therefore the required permutation is (134) or $\begin{pmatrix} 1 & 2 & 3 & 4 \\ 3 & 2 & 4 & 1 \end{pmatrix}$.

Exercise 25.12

1. The following cycles are permutations of the set {1, 2, 3, 4, 5}. Write these cycles in tabular form.
   (a) (23) ;   (b) (123) ;   (c) (1524) ;
   (d) (54321) ;   (e) (1) ;   (f) (45) ;
   (g) (124) ;   (h) (1523) ;   (i) (15423).

2. Express the following permutations of {1, 2, 3, 4, 5} in tabular form:
   (a) (12)(45) ;   (b) (23)(23) ;   (c) (12)(13) ;
   (d) (12)(13)(14) ;   (e) (123)(453) ;   (f) (123)(45).

3. For each of the following, find the least positive power to which it must be raised in order to obtain the identity permutation:
   (a) (1) ;   (b) (24) ;   (c) (153) ;
   (d) (1432) ;   (e) (12345) ;   (f) (12)(34) ;
   (g) (123)(234) ;   (h) (23)(145) ;   (i) (1234)(345).

*4. (a) Let $p = (123 \ldots m)$ and $q = (m+1\ m+2 \ldots n)$ be two cycles with common domain A. If $a \in A$ show that $pq(a) = qp(a)$.

   (b) Two cycles $(abc \ldots i)$ and $(jkl \ldots r)$ are said to be **disjoint** if no number appears in both cycles. Show that if p and q represent disjoint cycles, then they commute, i.e., $pq = qp$. By considering the example $p = (1234)$ and $q = (1432)$, show that the converse is not true, i.e., it is possible for two cycles to have numbers in common and yet still commute.

*5. Show that if 1, a, b, c are 4 different numbers, then $(1a)(1b)(1c) = (1cba)$. Extend this to the product $(12)(13)(14) \ldots (1n)$.

*6. A cycle of length 2 is called a **transposition**. Prove that any cycle of length r can be expressed as a product of $r - 1$ transpositions.

Sets, Relations and Groups

25.13 Groups

Definition A ***group*** is a set of elements G together with a binary operator $*$ which satisfies the following axioms:
  (a) G is ***closed*** under $*$.
      Whenever a, b are in G, $a * b$ is in G.
  (b) G is ***associative*** under $*$.
      For all a, b, c in G, $a * (b * c) = (a * b) * c$.
  (c) There exists an ***identity*** element e in G such that for all a in G, $a * e = e * a = a$.
  (d) Every element a in G has a corresponding ***inverse*** element a^{-1} in G such that $a * a^{-1} = a^{-1} * a = e$.

Note: In general, the most difficult axiom to prove is the associative axiom. To simplify the procedure, you may assume that the following sets are associative under the given operation.

- any subset of the real numbers under addition and multiplication;
- any subset of the complex numbers under addition and multiplication of complex numbers;
- any set of square matrices under matrix addition and multiplication;
- any set of residue classes (mod m) under addition or multiplication (mod m);
- any set of permutations under composition of permutations;
- any set of transformations of the plane under 'follows';
- any set of symmetries (of the square etc.) under 'follows';
- any set of functions under composition of functions.

Example Show that the set $\mathbb{Z}$ of integers is a group under addition.

  (a) $\mathbb{Z}$ is closed under addition since the sum of any two integers is also an integer.
  (b) $\mathbb{Z}$ is a subset of $\mathbb{R}$ and so must be associative under addition.
  (c) 0 is the identity.
  (d) If a is an integer, then so is $(-a)$ and $a + (-a) = (-a) + a = 0$. Therefore $(-a)$ is the inverse of a for all integers a.

Hence the set $\mathbb{Z}$ is a group under addition.

Example Show that the set $S = \{2, 4, 6, 8\}$ of residue classes, modulo 10, is a group under multiplication.

643

Chapter 25

The 'multiplication table' is as follows:

*	2	4	6	8
2	4	8	2	6
4	8	6	4	2
6	2	4	6	8
8	6	2	8	4

(a) Each entry in the table is a member of S. Therefore S is closed under multiplication.
(b) The associative rule can be assumed for residue classes under multiplication.
(c) The third row is the same as the row at the top and the third column is the same as the column on the left. Therefore 6 is the identity element.
(d) Since the identity (6) appears once in every row and every column, every element has an inverse. The inverses of 2, 4, 6, 8 are respectively 8, 4, 6, 2.

Thus S is a group under multiplication of residue classes (mod 10).

Exercise 25.13

1. Decide which of the following are groups:
   (a) $\mathbb{N}$ under addition ;
   (b) $\mathbb{N}$ under multiplication ;
   (c) $\mathbb{Z}$ under addition ;
   (d) $\mathbb{Z}$ under multiplication ;
   (e) $\mathbb{Q}$ under addition ;
   (f) $\mathbb{Q}$ under multiplication ;
   (g) $\mathbb{R}$ under addition ;
   (h) $\mathbb{R}$ under multiplication.

2. Decide which of the following are groups:
   (a) $\{1, -1\}$ under multiplication ;
   (b) $\{a + b\sqrt{2} \mid a, b \in \mathbb{Q}\}$ under addition ;
   (c) $\{a + b\sqrt{2} \mid a, b \in \mathbb{Q}; a, b \text{ not both zero}\}$ under multiplication ;
   (d) $\{2^m \mid m \in \mathbb{Z}\}$ under multiplication ;
   (e) $\{5m + 1 \mid m \in \mathbb{Z}\}$ under multiplication ;
   (f) $\{f_1(x), f_2(x), f_3(x), f_4(x)\}$ under composition of functions where $f_1(x) = x$, $f_2(x) = -x$, $f_3(x) = 1/x$, $f_4(x) = -1/x$;
   (g) $\mathbb{Z}$ under * where $a * b = a + b + 4$ for all a, b in $\mathbb{Z}$;
   (h) the set G of all real numbers except 1 under * where for all a, b in G, $a * b = 2(a-1)(b-1) + 1$
   (i) the set of all complex numbers with modulus 1 under multiplication.

Sets, Relations and Groups

3. Which of the following are groups under addition? For each reject, state which of the group axioms do not hold.
   (a) The set of all even integers.
   (b) The set of all odd integers.
   (c) The set of all multiples of three.
   (d) The set of all rational numbers.

4. Find the smallest multiplicative group of complex numbers which contains
   (a) $\frac{1}{2}(1+i\sqrt{3})$;
   (b) $\frac{1}{2}\sqrt{2}(1-i)$.

5. Find the identity element e and the inverse a^{-1} of a for an operation $*$ defined on $\mathbb{R}$ by $a * b = a + b + 2$. Is $\mathbb{R}$ a group under this operation?

6. Which of the group axioms are satisfied by the operation $*$ defined on $\mathbb{R}$ by $a * b = a(b+2)$?

7. Show that the following sets of residue classes, modulo m, are groups under multiplication:
   (a) $\{1, 2, 3, 4\}$, $m = 5$;
   (b) $\{1, 5, 7, 11\}$, $m = 12$;
   (c) $\{1, 3, 7, 9\}$, $m = 10$;
   (d) $\{1, 7, 9, 15\}$, $m = 16$.

8. Let $e = (1)$, $a = (12)(34)$, $b = (13)(24)$ and $c = (14)(23)$ be four permutations with domain $\{1, 2, 3, 4\}$. Show that they form a group under composition.

9. Under what condition does the set of all non-zero residue classes, modulo m, form a group under multiplication, modulo m?

10. Consider the set G of non-zero residue classes, modulo 11. Which of the following subsets of G forms a group under multiplication?
    (a) $\{1, 3, 4, 5, 9\}$;
    (b) $\{1, 3, 5, 7, 8\}$.

11. Let $A = \{(x, y) \mid x, y \in \mathbb{R},\ x, y \text{ not both zero}\}$, and let
    $$(x_1, y_1) * (x_2, y_2) = (x_1 x_2 - y_1 y_2,\ x_1 y_2 + x_2 y_1)$$
    for all (x_1, y_1), (x_2, y_2) in A. Prove that $(A, *)$ is a group.

12. Let G be the set of all real numbers excluding -1 under $*$ where
    $$a * b = ab + a + b$$
    for all a, b in G. Prove that $(G, *)$ is a group.

25.14 Properties of a Group

Theorem (Cancellation) Let G be a group under $*$. Let a, b, x be in G. If $a * x = b * x$ then $a = b$.

Proof
$$a * x = b * x$$
$$\Rightarrow (a * x) * x^{-1} = (b * x) * x^{-1} \quad \text{[closure and inverse axioms]}$$
$$\Rightarrow a * (x * x^{-1}) = b * (x * x^{-1}) \quad \text{[associative axiom]}$$
$$\Rightarrow a * e = b * e \quad \text{[inverse axiom]}$$
$$\Rightarrow a = b \quad \text{[identity axiom]}.$$

Theorem The inverse of an element of a group is unique.

Proof Suppose a' and a'' are two elements of the group, both inverses of the element a.
Then $a' * a = e$ [inverse axiom]
and $a'' * a = e$ [inverse axiom]
$\Rightarrow a' = a''$ [cancellation].
Therefore the inverse of a is unique.

Theorem Let a, b, x be elements of a group G under $*$. If $a * x = b$, then $x = a^{-1} * b$.

Proof
$$a * x = b$$
$$a^{-1} * (a * x) = a^{-1} * b \quad \text{[closure axiom]}$$
$$(a^{-1} * a) * x = a^{-1} * b \quad \text{[associative axiom]}$$
$$e * x = a^{-1} * b \quad \text{[inverse axiom]}$$
$$x = a^{-1} * b \quad \text{[identity axiom]}.$$

Theorem If $(G, *)$ is a group, then for any x, y in G, $(x * y)^{-1} = y^{-1} * x^{-1}$.

Proof
$$(x * y) * (y^{-1} * x^{-1})$$
$$= \{x * (y * y^{-1})\} * x^{-1} \quad \text{[associative axiom]}$$
$$= \{x * e\} * x^{-1} \quad \text{[inverse axiom]}$$
$$= x * x^{-1} \quad \text{[identity axiom]}$$
$$= e \quad \text{[inverse axiom]}.$$

Similarly, $(y^{-1} * x^{-1}) * (x * y) = e$.
Therefore $(x * y)^{-1} = y^{-1} * x^{-1}$ [inverse axiom].

Sets, Relations and Groups

Definition The *order* of an element x of a group $(G, *)$ is the smallest positive integer n such that $x^n = e$ (the identity).

Note: We write $x * x * x * \ldots * x$ (to n factors) as x^n.

Example Find the order of the element 3 in the group of residue classes $\{0, 1, 2, 3, 4\}$ modulo 5 under addition of residue classes modulo 5.

0 is the identity and $3^2 = 1$ $(3 + 3 = 1, \bmod 5)$, $3^3 = 4$, $3^4 = 2$, $3^5 = 0$. Therefore 3 has order 5.

Exercise 25.14

1. Let e be the identity element of a group $(G, *)$.
   Prove that   (a) $e^{-1} = e$;   (b) $(x^{-1})^{-1} = x$ for any x in G.

2. If x, y, z are elements of a group $(G, *)$, write $(x * y * z)^{-1}$ without brackets.

3. In conventional algebra, if $x^2 = x$ then $x = 0$ or $x = 1$. How many solutions has the equation $x^2 = x$ in the algebra of groups? Find a group which has more than one element x such that $x^3 = x$.

4. Find the smallest group of transformations of the plane under 'follows' which contains
   (a) $T_1 : (x, y) \mapsto (-x, y)$ and $T_2 : (x, y) \mapsto (x, -y)$;
   (b) $R : (x, y) \mapsto (-y, x)$.

5. Let $(G, *)$ be a group with three members e, a, b. Which member of G is $a * b$? Show that $a * b = b * a$.

6. Let $(G, *)$ be a group with four members e, a, b, c. Which members of G could be $a * b$? (Give examples.) Show that $a * b = b * a$.

7. Use the properties of a group to simplify each of the following:
   (a) $(xyx^{-1})^2$;   (b) $(xyx^{-1})^3$;   (c) $(xyx^{-1})^4$;
   (d) $(xyx^{-1})^{-1}$;   (e) $(xyx^{-1})^{-2}$;   (f) $(xyx^{-1})^{-3}$.

8. Given that a, b, c are elements of a group, simplify each of the following:
   (a) $(a^{-1})^{-1}$;   (b) $(ab)^{-1} a^2$;   (c) $(a^{-1}b^{-1})^{-1}$;
   (d) $a(ba)^{-2}(ab)^{-1}$;   (e) $(abc)^2 (bc)^{-1}$;   (f) $c(abc)^{-1} ab$;
   (g) $bc(abd)^{-1} a$;   (h) $(ab)^{-1}(abc)^2 (bc)^{-1}$.

647

Chapter 25

9. Find the orders of
    (a) the elements of the group G of non-zero residue classes modulo 7 under multiplication;
    (b) the function $f(x) = 1/(1-x)$ as an element of a group under function composition;
    (c) the transformation in the plane $T: (x, y) \mapsto (x-y, x)$ under 'follows'.

10. (a) Consider the group of residue classes mod 5 under addition. Show that every element of the group except for 0 has the same order.
    (b) Consider the group of residue classes modulo n under addition. For what n is it true that all the elements of the group except for 0 have the same order?

11. Let x be an element of order 3 in a group. Express each of the following in the form x^n where n is the smallest possible positive integer:
    (a) x^7;   (b) x^{11};   (c) x^{-1};   (d) x^{-2}.

12. Let x with order 2 and y with order 3 be two members of a group.
    (a) Find the order of xy if $xy = yx$.   (b) Find the order of xy if $xy = yx^2$.
    (c) Simplify $(xy)^{-2}$ if $xy = yx$.   (d) Simplify $(xy)^{-2}$ if $xy = yx^2$.
    (e) If $xy = yx^2$, prove that $yx = x^2 y$.

13. Let a be an element of order 3 and b be any other element of the same group. What is the order of bab^{-1}?

14. Suppose a and b are two elements of a group with orders 4 and 2 respectively.
    (a) If the order of ab is 2, prove that
        (i) $ab = ba^3$;
        (ii) $ba = a^3 b$.
    (b) If $ab = ba$, prove that the order of ab is 4.

15. If a, b are elements of a group such that ab has order 2, show that ba has order 2.

16. If a, b and ab, all with order 2, are elements of a group, show that $ab = ba$.

25.15 Subgroups, Cyclic Groups and Isomorphism

Definition   The *order* of a group is the number of elements in it. If the group has an infinite number of elements its order is said to be infinite. We denote the order of a group G by $|G|$.

Sets, Relations and Groups

Definition A group $(G, *)$ is said to be ***Abelian (commutative)*** if $x * y = y * x$ for all x, y in G.

Example A group G is such that $x^2 = e$ for all elements x in G where e is the identity element in G. Show that every element of G is its own inverse. Hence, by considering the inverse of xy, where x and y are in G, show that G is Abelian.

Since $x^2 = xx = e$, then $x^{-1} = x$.
As $(xy)^{-1} = y^{-1}x^{-1} = yx$ and $(xy)^{-1} = xy$, then $xy = yx$ and G is Abelian.

Definition Let $(G, *)$ be a group. If H is a subset of G and $(H, *)$ is a group, then $(H, *)$ is said to be a ***subgroup*** of $(G, *)$.

Note: Every subgroup of G must contain the identity element of G. Thus any group G contains at least 2 subgroups, viz., $\{e\}$ and G. Since these subgroups are obvious, we distinguish all other subgroups as ***proper***.

Theorem A subset H of a group $(G, *)$ is a subgroup of G iff
(a) H is closed under $*$;
(b) the identity e of G is in H ;
(c) for all $a \in H$, $a^{-1} \in H$.
[There is no need to check the associative axiom since any subset of $(G, *)$ ***must*** be associative under the operation $*$.]

Cyclic Groups

Definition A group G is said to be ***cyclic*** if there exists an element $g \in G$ such that the sequence $e, g, g^2, g^3, \cdots$ contains all the members of G. The element g is called a ***generator*** of the group.

Example Show that the group of residue classes $\{1, 2, 3, 4, 5, 6\}$ under multiplication modulo 7 is cyclic and find all its generators.

$3^2 = 2, 3^3 = 6, 3^4 = 4, 3^5 = 5, 3^6 = 1$ and so powers of 3 produce all the elements of the group. Therefore the group is cyclic.

$2^2 = 4, 2^3 = 1, 2^4 = 2, \cdots$ and so powers of 2 produce the elements 1, 2 and 4 only. Therefore 2 is not a generator.

Chapter 25

Similarly $4^2 = 2, 4^3 = 1, 4^4 = 4, \cdots$ so 4 is not a generator but $5^2 = 4, 5^3 = 6, 5^4 = 2, 5^5 = 3, 5^6 = 1$ and so 5 is a second generator. [This should not be surprising since 5 is the inverse of 3 under the operation.]

Theorem If G is any finite group, not necessarily cyclic, and $g \in G$, then the first member of the sequence $e, g, g^2, g^3, \cdots$ to be repeated will be e.

Proof Assume that the first power of g to be repeated is g^m and that this is a repeat of an earlier member $g^n \neq e$.

Then $g^m = g^n$ where $m > n$.

Now $\left(g^{-1}\right)^n g^m = \left(g^{-1}\right)^n g^n \Rightarrow g^{m-n} = e$ and so e has already appeared in the list which contradicts our assumption.

Therefore the first member of the sequence $e, g, g^2, g^3, \cdots$ to be repeated will be e.

Note: The order of e is always 1, and a group G is cyclic if and only if it contains an element whose order is $|G|$.

Definition If a is an element of a group G, then the set of all elements of G of the form a^k where $k \in \mathbb{Z}$, is a subgroup of G.

This subgroup is called the **cyclic subgroup of G generated by a**.

The order of the element a is the same as the order of the cyclic subgroup of G which it generates.

We denote the cyclic subgroup of G generated by a by $\langle a \rangle$.

Example Consider the group of non-zero residue classes (mod 7) under multiplication. The group $\langle 2 \rangle = \langle 4 \rangle = \{1, 2, 4\} = \{e (= g^3), g, g^2\}$ for $g = 2$ (or 4) is a cyclic subgroup of the original group generated by 2 (or 4).

Theorem Every subgroup of a cyclic group of order n is also cyclic; and furthermore, its order is a divisor of n.

Example Make a list of all the proper cyclic subgroups of a cyclic group of order 8 generated by g.

Sets, Relations and Groups

Since the order of a subgroup of any cyclic group divides the order of the group, the order of any subgroup of the cyclic group of order 8 is 1, 2, 4 or 8.

Therefore the order of the generator of any proper cyclic subgroup is 2 or 4. Now g^4 has order 2 and both g^2 and g^6 have order 4.

The proper subgroups are $\{g^4, g^8(=e)\}$ and $\{g^2, g^4, g^6, g^8(=e)\}$.

The result of the previous theorem is simply an example of a more general result given by a theorem due to Lagrange.

Theorem (Lagrange)
    If H is a subgroup of a finite group G, then $|H|$ divides $|G|$.

The proof of this theorem is beyond the scope of this book.

Example   Prove that if a is a member of a group G, then the order of a divides the order of G.

The order of a is equal to the order of the cyclic subgroup of G generated by a. By the theorem of Lagrange, the order of a divides the order of G.

Example   If the order of the group G is n, then each element a of G satisfies the relation $a^n = e$.

If the order of a is m, then by the result of the previous example, m divides n, i.e., $n = km$ for some $k \in \mathbb{Z}^+$.

Therefore $a^m = e \Rightarrow (a^m)^k = e$
                          $\Rightarrow a^{mk} = e$
                          $\Rightarrow a^n = e$.

Example   Prove that if the order of a group is prime, then the group is cyclic.

Consider a group G whose order is p, a prime number.
Let g be any member of G, $g \neq e$ (e is the identity).
Now since the order of any element of a group must divide the order of the group (Lagrange), g must be a divisor of p (other than 1). That is, $g = p$ and so g generates the whole group. Therefore G is cyclic.

Example   If e (the identity), a and b are 3 members of a non-cyclic group of order 4, find the fourth member of the group given that this group is known to exist.

Chapter 25

Since the group is non-cyclic, there is no element of order 4.
But since the order of each element must divide the order of the group, a and b each have order 2. Thus $a^2 = b^2 = e$.
Now ab must belong to the group since the group is closed, but ab is not e, a or b since

$$ab = e \implies a = b^{-1} = b \quad \text{(contradiction since } b = b^{-1}\text{)};$$
$$ab = a \implies b = e \quad \text{(contradiction)};$$
$$ab = b \implies a = e \quad \text{(contradiction)}.$$

Therefore the fourth member of the group is ab.

From the result of the previous example, it is clear that there can be only 2 different groups of order 4:

1. The cyclic group generated by $g - \{e, g, g^2, g^3\}$.
2. The non-cyclic group, called a Klein 4-group $- \{e, a, b, ab\}$.

The group tables have the following forms:

Cyclic group of order 4

*	e	a	b	c
e	e	a	b	c
a	a	b	c	e
b	b	c	e	a
c	c	e	a	b

Klein 4-group

*	e	a	b	c
e	e	a	b	c
a	a	e	c	b
b	b	c	e	a
c	c	b	a	e

$[b = a^2, c = a^3, e = a^4]$

Isomorphism

Definition

Let $(G, *)$ and $(H, \circ)$ be groups. If there exists a bijection $\phi : G \to H$ such that $\phi(g_1) = h_1$, $\phi(g_2) = h_2$ and $\phi(g_1 * g_2) = h_1 \circ h_2$, then the groups $(G, *)$ and $(H, \circ)$ are said to be *isomorphic*.

Note: Two finite groups are clearly isomorphic if their 'group tables' have exactly the same structure.

Example Show that the group $\{1, 5, 7, 11\}$ of residue classes, modulo 12, under $*$ (multiplication (mod 12)) is isomorphic to the group of four congruence transformations (or symmetries) of a rhombus under $\circ$, (composition of transformations).

Sets, Relations and Groups

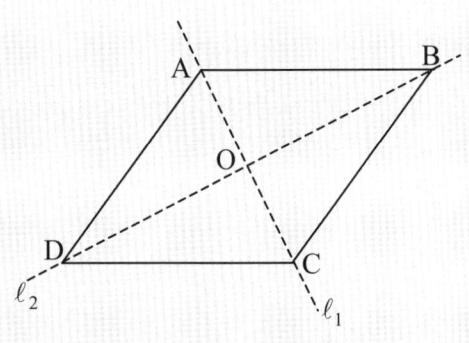

Let the symmetries of ABCD be:
I = Rotation about O through 0°
R = Rotation about O through 180°
M_1 = Reflection in line ℓ_1
M_2 = Reflection in line ℓ_2
[∘ is equivalent to 'follows']

For example, $M_1 \circ R$ (ABCD) = M_1 (CDAB) = CBAD = M_2 (ABCD), and so $M_1 \circ R = M_2$. [Here the vertices are labelled clockwise from the top left.]

The group table is:

∘	I	R	M_1	M_2
I	I	R	M_1	M_2
R	R	I	M_2	M_1
M_1	M_1	M_2	I	R
M_2	M_2	M_1	R	I

The group table for the residue classes is:

*	1	5	7	11
1	1	5	7	11
5	5	1	11	7
7	7	11	1	5
11	11	7	5	1

These groups are clearly isomorphic since the tables are identical in structure under the mapping I → 1, R → 5, M_1 → 7 and M_2 → 11.

Example Let $\mathbb{C}^*$ be the set of non-zero complex numbers under multiplication of complex numbers and G the set of non-zero matrices of real numbers of the form $\begin{pmatrix} a & b \\ -b & a \end{pmatrix}$ under multiplication of matrices. Show that $\mathbb{C}^*$ and G form groups under the given operations and prove that these groups are isomorphic.

1.1 Since the product of 2 non-zero complex numbers is another non-zero complex number, $\mathbb{C}^*$ is closed under multiplication.
1.2 Any subset of the set of complex numbers is associative under multiplication and so $\mathbb{C}^*$ is associative under multiplication.
1.3 The non-zero complex number $1 + 0i$ is the identity element.

653

Chapter 25

1.4 The inverse of the non-zero complex number $a + bi$ is
$$\frac{a}{a^2+b^2} + \frac{-b}{a^2+b^2}i$$
which is a non-zero complex number.
Hence $\mathbb{C}^*$ is a group under multiplication of complex numbers.

2.1 Since the matrix $\begin{pmatrix} a & b \\ -b & a \end{pmatrix}$ cannot be zero, $\left| \begin{pmatrix} a & b \\ -b & a \end{pmatrix} \right| = a^2 + b^2 \neq 0$.

Let $a_1, b_1, a_2, b_2 \in \mathbb{R}$ such that $\begin{pmatrix} a_1 & b_1 \\ -b_1 & a_1 \end{pmatrix}$ and $\begin{pmatrix} a_2 & b_2 \\ -b_2 & a_2 \end{pmatrix}$ are non-zero matrices. Then their determinants are non-zero and so their product is a matrix with non-zero determinant.

Also $\begin{pmatrix} a_1 & b_1 \\ -b_1 & a_1 \end{pmatrix}\begin{pmatrix} a_2 & b_2 \\ -b_2 & a_2 \end{pmatrix} = \begin{pmatrix} a_1a_2 - b_1b_2 & a_1b_2 + b_1a_2 \\ -[a_1b_2 + a_2b_1] & a_1a_2 - b_1b_2 \end{pmatrix}$ which is

of the form $\begin{pmatrix} a_3 & b_3 \\ -b_3 & a_3 \end{pmatrix}$ where $a_3 = a_1a_2 - b_1b_2$ and $b_3 = a_1b_2 + b_1a_2$.

Hence G is closed under matrix multiplication.

2.2 Any subset of the set of 2×2 matrices is associative under matrix multiplication and so G is associative under matrix multiplication.

2.3 $\begin{pmatrix} 1 & 0 \\ 0 & 1 \end{pmatrix} \in G$ is the identity.

2.4 Each element of G has an inverse (since its determinant cannot be zero) $= \begin{pmatrix} a/(a^2+b^2) & -b/(a^2+b^2) \\ b/(a^2+b^2) & a/(a^2+b^2) \end{pmatrix}$ which is of the form $\begin{pmatrix} c & d \\ -d & c \end{pmatrix}$

where $a/(a^2+b^2) = c$ and $-b/(a^2+b^2) = d$ and so every element of G has an inverse in G.

Therefore G is a group under matrix multiplication.

Consider the mapping $\phi : \mathbb{C}^* \to G$ defined by $\phi(a+bi) = \begin{pmatrix} a & b \\ -b & a \end{pmatrix}$.

ϕ is clearly a bijection and
$$\begin{aligned}
\phi((a_1+b_1 i)(a_2+b_2 i)) &= \phi((a_1 a_2 - b_1 b_2) + i(a_1 b_2 + a_2 b_1)) \\
&= \begin{pmatrix} a_1 a_2 - b_1 b_2 & a_1 b_2 + a_2 b_1 \\ -[a_1 b_2 + a_2 b_1] & a_1 a_2 - b_1 b_2 \end{pmatrix} \\
&= \begin{pmatrix} a_1 & b_1 \\ -b_1 & a_1 \end{pmatrix}\begin{pmatrix} a_2 & b_2 \\ -b_2 & a_2 \end{pmatrix} \\
&= \phi(a_1+b_1 i)\phi(a_2+b_2 i).
\end{aligned}$$
Therefore the groups are isomorphic.

Sets, Relations and Groups

Exercise 25.15

1. Make a list of all the proper subgroups of a cyclic group of order 12 and generator g.

2. If g is a generator of a cyclic group of order n, state which powers of g are also generators of G when
   (a) $n = 4$;
   (b) $n = 5$;
   (c) $n = 6$;
   (d) $n = 12$.

3. Consider the group $M_p = \{1, 2, 3, \ldots, p-1\}$ under the operation of multiplication, modulo p, where p is a prime number. Taking in turns the values $p = 3, 5, 7, 11, 13, 17$, show that each group is cyclic and verify (for these cases) that either 2 or 3 is a generator.

4. Find all the proper subgroups of the group with the following multiplication table:

*	a	b	c	d	e	f
a	b	e	d	f	a	c
b	e	a	f	c	b	d
c	f	d	e	b	c	a
d	c	f	a	e	d	b
e	a	b	c	d	e	f
f	d	c	b	a	f	e

5. Show that the group C_n of rotations of the plane about a point C through angles which are integer multiples of $2\pi/n$, under 'follows', is isomorphic to the group $\mathbb{Z}_n$ of integers under addition modulo n.

6. Let G_1 be the group of symmetries of an equilateral triangle under 'follows';
   G_2 be the group S_3 of permutations under composition;
   G_3 be the group of rotations of a regular hexagon under 'follows'.
   State with reasons which two of the above groups are isomorphic to each other, and which group is isomorphic to neither of the others.

7. Let S be the set of all matrices of real numbers of the form $\begin{pmatrix} a & -b \\ b & a \end{pmatrix}$ with not both a and b equal to zero.
   (a) Show that S is closed under matrix multiplication.
   (b) Assuming that S is associative under matrix multiplication, show that S forms an Abelian group under matrix multiplication.

655

Chapter 25

(c) Show that $T = \left\{ \begin{pmatrix} a & 0 \\ 0 & a \end{pmatrix} \mid a \neq 0 \right\}$ is a subgroup of S.

(d) Show that T is isomorphic to the group of non-zero real numbers under multiplication.

8. Consider the three points $(0, 2)$, $(\sqrt{3}, 1)$ and $(\sqrt{3}, -1)$ in the plane. Let S be the set of rotations of the plane which send any one of these points to any other (including the possibility of sending a point to itself).

(a) Show that S contains exactly 5 elements and that S does **not** form a group under the operation of composition of rotations.
(b) Find a rotation such that, if it is added to S, the resultant set does form a group under composition of rotations. Show that this resultant group is cyclic and find a generator for it.

9. Write out the 'multiplication' table for the set of all non-zero residue classes (mod 7) under multiplication (mod 7) and hence show that this set and operator form a group. Consider the set of rotations by angles of $n\pi/3$ ($n \in \mathbb{Z}$) of a regular hexagon about its centre. Show that this set forms a group under composition of rotations which is isomorphic to the multiplicative group of non-zero residue classes modulo 7.

10. State the axioms for a group G with binary operator $*$. Prove from the axioms:

(a) every element a of G has a unique inverse a^{-1};
(b) if a, b are any two elements of G, then $(a * b)^{-1} = b^{-1} * a^{-1}$.

Prove that if $G = \{a, b, c, d\}$, $a * c = d$ and $d * c = a$, then b is the identity of G. Hence construct the two multiplication tables for G.

11. What is meant by a "cyclic group"? Give an example of an infinite cyclic group. Consider each of the following statements. If it is true, prove it; if it is false, give a counter example:
(a) Every group of order three is cyclic.
(b) A cyclic group has no proper subgroup.

12. (a) Let $\mathbb{Z}_n$ be the set of integers modulo n under addition (mod n). Show that $\mathbb{Z}_n$ is cyclic.
(b) How many distinct elements are generators of $\mathbb{Z}_n$ when
(i) $n = 3$; (ii) $n = 4$; (iii) $n = 5$;
(iv) $n = 6$; (v) $n = 7$; (vi) $n = 20$?

Sets, Relations and Groups

13. Let $(G, *)$ and $(H, \circ)$ be two groups. Consider the Cartesian product $G \times H$ under the binary operation Δ defined by
$$(g_1, h_1)\Delta(g_2, h_2) = (g_1 * g_2, h_1 \circ h_2).$$
   (a) Show that $(G \times H, \Delta)$ is a group.
   (b) Taking the group $\mathbb{Z}_n$ defined in Question 8, state the order of $\mathbb{Z}_3 \times \mathbb{Z}_4$ and evaluate $(2, 3) \Delta (1, 3)$.
   (c) Show that $\mathbb{Z}_2 \times \mathbb{Z}_3$ is cyclic and list all possible generators.
   (d) Determine whether the following are cyclic:
        (i) $\mathbb{Z}_2 \times \mathbb{Z}_2$; (ii) $\mathbb{Z}_2 \times \mathbb{Z}_4$; (iii) $\mathbb{Z}_3 \times \mathbb{Z}_4$.
   (e) How many elements of $\mathbb{Z}_2 \times \mathbb{Z}_4$ have order (i) 2; (ii) 4?

14. Establish an explicit isomorphism between the six roots of $z^6 = 1$ under multiplication and the set $\{1, 2, 4, 5, 7, 8\}$ under multiplication (mod 9).

15. Let $A = \{z \mid z \in \mathbb{C} \text{ and } z^3 = 1\}$ and define $f(z) = z^2$ for all $z \in A$.
   (a) Prove that A forms a group under multiplication of complex numbers.
   (b) Prove that f is an isomorphism from $(A, \times)$ onto $(A, \times)$.

16. Let $S = \left\{ x \mid x = a + b\sqrt{2}\,; a, b \in \mathbb{Q}, a^2 - 2b^2 \neq 0 \right\}$.
   (a) Prove that $(S, \times)$ is a group where $\times$ is real number multiplication.
   (b) For $x = a + b\sqrt{2}$, define $f(x) = a - b\sqrt{2}$. Prove that f is an isomorphism from $(S, \times)$ onto $(S, \times)$.

17. Prove that the groups $(\mathbb{R}, +)$ and $(\mathbb{R}^+, \times)$ are isomorphic.

18. Let $\phi : G \to H$ be an isomorphism from $(G, *)$ to $(H, \circ)$. Let e be the identity element of G and e' that of H. Let $\phi(a) = a'$.
   Prove:
   (a) $\phi(e) = e'$;
   (b) $\phi(a^{-1}) = (a')^{-1}$.

657

Chapter 25

> **Required Outcomes**
>
> After completing this chapter, a student should be able to:
> - prove various set relations using de Morgan's laws or justify them using a Venn diagram.
> - determine whether or not a given binary relation is an equivalence relation.
> - determine the equivalence classes of a given equivalence relation.
> - use the matrix of a relation to determine whether it is an equivalence relation.
> - determine whether or not a function is an injection, surjection or bijection.
> - determine whether a given set together with a binary operator forms a group.
> - determine whether or not a given group is a subgroup of another.
> - decide whether or not a given group is cyclic.
> - use Lagrange's theorem in a variety of situations.
> - determine whether or not two given groups are isomorphic.

26 Analysis and Approximation

26.1 Infinite Series – Tests for Convergence

The sum of terms of a sequence is an ***infinite series***. That is, if $\{u_n\}_1^\infty$ is a sequence, the expression $\sum_{n=1}^\infty u_n = u_1 + u_2 + u_3 + \cdots$ is an infinite series. How can we assign meaning to this definition? We certainly cannot add infinitely many numbers.

Consider the series $\sum_{k=1}^\infty \frac{1}{2^{k-1}} = 1 + \frac{1}{2} + \frac{1}{4} + \frac{1}{8} + \cdots$. Let S_n be the sum of the first n terms. Then from our work in geometric series we have $S_n = \frac{1-\left(\frac{1}{2}\right)^n}{1-\frac{1}{2}} = 2 - \frac{1}{2^{n-1}}$. The sequence $\{S_n\} = \{1, 1\frac{1}{2}, 1\frac{3}{4}, \cdots, 2 - \frac{1}{2^{n-1}}, \cdots\}$, called the sequence of ***partial sums***, clearly converges to 2. Thus $\lim_{n\to\infty} S_n = 2$ and we say that the sum of the infinite series $\sum_{k=1}^\infty \frac{1}{2^{k-1}} = 1 + \frac{1}{2} + \frac{1}{4} + \frac{1}{8} + \cdots$ is 2.

Definition Let $\sum_{k=1}^\infty u_k$ be a given series and let $S_n = \sum_{k=1}^n u_k$ be the ***n-th partial sum***. If $S = \lim_{n\to\infty} S_n$ exists, then we say that S is the ***sum*** of the infinite series. Thus $S = \lim_{n\to\infty} \sum_{k=1}^n u_k = \sum_{k=1}^\infty u_k$.

If the sum S exists, the series is called ***convergent***. If the sum S does not exist, the series is called ***divergent***.

Formulae like the one for the sum of a convergent geometric series are not common. Indeed, in the following work we will mostly be concerned with the convergence or divergence of a series rather than with the limit itself (if it exists) which can be very difficult to find.

Chapter 26

Another series for which the sum can be found exactly is the following "telescoping series".

Example Find the sum of the series $\sum_{n=1}^{\infty} \dfrac{1}{n(n+1)}$.

$$\sum_{k=1}^{\infty} \dfrac{1}{k(k+1)}$$
$$= \sum_{k=1}^{\infty}\left(\dfrac{1}{k}-\dfrac{1}{k+1}\right)$$
$$= \lim_{n\to\infty}\left(\sum_{k=1}^{n}\left(\dfrac{1}{k}-\dfrac{1}{k+1}\right)\right)$$
$$= \lim_{n\to\infty}\left(\left(1-\dfrac{1}{2}\right)+\left(\dfrac{1}{2}-\dfrac{1}{3}\right)+\left(\dfrac{1}{3}-\dfrac{1}{4}\right)+\cdots+\left(\dfrac{1}{n-1}-\dfrac{1}{n}\right)+\left(\dfrac{1}{n}-\dfrac{1}{n+1}\right)\right)$$
$$= \lim_{n\to\infty}\left(1-\dfrac{1}{n+1}\right) \text{ since the sum of all the interior pairs is zero,}$$
$$= 1.$$

Therefore $\sum_{n=1}^{\infty} \dfrac{1}{n(n+1)} = 1$.

Theorem For any convergent series $\sum_{n=1}^{\infty} u_n$, $\lim_{n\to\infty} u_n$ must be zero.

Proof Let $S_n = \sum_{k=1}^{n} u_k$ and since the series is convergent there exists a number S such that $\lim_{n\to\infty} S_n = S$.

But $u_n = S_n - S_{n-1}$ and so $\lim_{n\to\infty} u_n = \lim_{n\to\infty} S_n - \lim_{n\to\infty} S_{n-1} = S - S = 0$.

Note however that it is not true that if $\lim_{n\to\infty} u_n = 0$, then $\sum_{n=1}^{\infty} u_n$ is convergent.

The condition for convergence of the infinite series $\sum_{n=1}^{\infty} u_n$, namely $\lim_{n\to\infty} u_n = 0$, is *necessary* but *not sufficient*.

To illustrate this we consider the following theorem.

Analysis and Approximation

Theorem The *harmonic series* $\sum_{n=1}^{\infty} \frac{1}{n}$ is divergent.

Proof Let $S_n = \sum_{k=1}^{n} \frac{1}{k}$.

Then
$$S_{2n} - S_n = \left(1 + \frac{1}{2} + \frac{1}{3} + \cdots + \frac{1}{n} + \frac{1}{n+1} + \cdots + \frac{1}{2n}\right) - \left(1 + \frac{1}{2} + \frac{1}{3} + \cdots + \frac{1}{n}\right)$$

$$= \frac{1}{n+1} + \frac{1}{n+2} + \frac{1}{n+3} + \cdots + \frac{1}{2n}$$

$$\geq \frac{1}{2n} + \frac{1}{2n} + \frac{1}{2n} + \cdots + \frac{1}{2n}$$

$$= n\left(\frac{1}{2n}\right)$$

$$= \frac{1}{2}.$$

Now if $\lim_{n\to\infty} S_n = S$, then $\lim_{n\to\infty} S_{2n} = S$ and so $\lim_{n\to\infty} (S_{2n} - S_n) = 0$ which is impossible since $S_{2n} - S_n \geq \frac{1}{2}$ for all n.

Therefore the harmonic series is divergent.

Condition for Convergence of a Series of Non-negative Terms

Definition A sequence $\{S_n\}$ such that $S_n \leq S_{n+1}$ for all n is said to be *non-decreasing*.

Definition A sequence $\{S_n\}$ is *bounded above* if there exists a number N such that $S_n \leq N$ for all n. (In this case, N is called an "*upper bound*" of the sequence.)

If $u_n \geq 0$ for all n, the series $\sum_{n=1}^{\infty} u_n$ converges if and only if the sequence of its partial sums is bounded above.

Chapter 26

Example Prove that the series $\sum_{n=1}^{\infty} \frac{1}{n^2}$ is convergent.

Let $S_n = 1 + \frac{1}{2^2} + \frac{1}{3^2} + \cdots + \frac{1}{n^2}$.

Then for all n, $S_n < 1 + \frac{1}{(1)(2)} + \frac{1}{(2)(3)} + \cdots + \frac{1}{(n-1)n}$ and so $S_n < 2 - \frac{1}{n}$ from the example on page 658.

Therefore $\{S_n\}$ is bounded above by 2 and so the series is convergent.

(In fact, $\sum_{n=1}^{\infty} \frac{1}{n^2} = \frac{\pi^2}{6}$ although the proof is beyond the scope of this book.)

Example Prove that the series $\sum_{n=0}^{\infty} \frac{1}{n!}$ is convergent.

Let $S_n = 1 + \frac{1}{1!} + \frac{1}{2!} + \frac{1}{3!} + \cdots + \frac{1}{n!}$.

Then $S_n \leq 1 + 1 + \frac{1}{2} + \frac{1}{2^2} + \cdots + \frac{1}{2^n} < 1 + \sum_{n=0}^{\infty} \frac{1}{2^n} = 1 + \frac{1}{1-\frac{1}{2}} = 3$.

Therefore the sequence of partial sums is bounded above by 3 and so the series converges.

Note: The series does not converge to 3 even though the sequence of partial sums is bounded above by 3. We will find in a later section that the series converges to e.

Comparison Test for Series of Non-negative Terms

Let $\sum a_n$ be an infinite series of non-negative terms.

(1) $\sum a_n$ converges if there exists a convergent series $\sum b_n$ such that $a_n \leq b_n$ for all $n > N$, where N is some positive integer.

(2) $\sum a_n$ diverges if there exists a divergent series $\sum c_n$ such that $a_n \geq c_n$ for all $n > N$, where N is some positive integer.

Analysis and Approximation

Example Use the comparison test to show that the series $\sum_{n=1}^{\infty} \frac{2n+1}{n}\left(\frac{1}{2}\right)^n$ is convergent.

$\frac{2n+1}{n}\left(\frac{1}{2}\right)^n \leq 3\left(\frac{1}{2}\right)^n$ and $\sum_{n=1}^{\infty} 3\left(\frac{1}{2}\right)^n$ is a convergent geometric series converging to 6.

Therefore the series $\sum_{n=1}^{\infty} \frac{2n+1}{n}\left(\frac{1}{2}\right)^n$ converges by comparison.

Example Use the comparison test to show that the series $\sum_{n=1}^{\infty} \frac{1}{2n-1}$ is divergent.

$\frac{1}{2n-1} > \frac{1}{2n} = \frac{1}{2}\left(\frac{1}{n}\right)$ and $\sum_{n=1}^{\infty} \frac{1}{n}$ is divergent (the harmonic series).

Therefore the series $\sum_{n=1}^{\infty} \frac{1}{2n-1}$ is divergent by comparison.

The Limit Comparison Test

If the series $\sum a_n$ and $\sum b_n$ are two series of non-negative terms, and $\lim_{n \to \infty} \frac{a_n}{b_n}$ is finite, then both series converge or both diverge.

Example Decide whether the series $\sum_{n=1}^{\infty} \frac{n+1}{n^3+1}$ converges or diverges.

Let $a_n = \frac{n+1}{n^3+1}$ and $b_n = \frac{1}{n^2}$, then $\lim_{n \to \infty} \frac{a_n}{b_n} = \lim_{n \to \infty} \frac{n^3+n^2}{n^3+1} = \lim_{n \to \infty} \frac{1+\frac{1}{n}}{1+\frac{1}{n^3}} = 1$.

Since $\sum_{n=1}^{\infty} \frac{1}{n^2}$ converges, $\sum_{n=1}^{\infty} \frac{n+1}{n^3+1}$ converges by the limit comparison test.

Example Decide whether the series $\sum_{n=1}^{\infty} \frac{n+1}{2n^2+1}$ converges or diverges.

663

Chapter 26

Let $a_n = \dfrac{n+1}{2n^2+1}$ and $b_n = \dfrac{1}{n}$, then $\lim\limits_{n\to\infty} \dfrac{a_n}{b_n} = \lim\limits_{n\to\infty} \dfrac{n^2+n}{2n^2+1} = \lim\limits_{n\to\infty} \dfrac{1+\frac{1}{n}}{2+\frac{1}{n^2}} = \dfrac{1}{2}$.

Since $\sum\limits_{n=1}^{\infty} \dfrac{1}{n}$ diverges, then so does $\sum\limits_{n=1}^{\infty} \dfrac{n+1}{2n^2+1}$ by the limit comparison test.

For comparison, we know the following:

(1) $\sum \dfrac{1}{n}$ diverges.

(2) $\sum ar^n$ converges for $|r|<1$ and diverges for $|r|\geq 1$.

(3) $\sum \dfrac{1}{n^2}$ converges.

(4) $\sum \dfrac{1}{n!}$ converges.

Exercise 26.1

1. (a) Find the numbers A, B and C for which $\dfrac{2}{n(n^2-1)} = \dfrac{A}{n-1} + \dfrac{B}{n} + \dfrac{C}{n+1}$.

    (b) Find $\sum\limits_{n=2}^{\infty} \dfrac{2}{n(n^2-1)}$.

2. Consider the series $\sum\limits_{n=1}^{\infty} nr^n$. Show that if S_n is the sum of the first n terms, then $S_n - rS_n = \left(\sum\limits_{i=1}^{n} r^i\right) - nr^{n+1}$ and hence prove that the series converges to $\dfrac{r}{(1-r)^2}$ provided $|r|<1$.

3. Decide which of the following series converge and which diverge:

    (a) $\sum\limits_{n=1}^{\infty} \dfrac{1}{3^n}$; (b) $\sum\limits_{n=1}^{\infty} \dfrac{2n}{n+2}$; (c) $\sum\limits_{n=1}^{\infty} \dfrac{1}{\sqrt{n}}$;

    (d) $\sum\limits_{n=1}^{\infty} \dfrac{3n}{n^2+1}$; (e) $\sum\limits_{n=1}^{\infty} \dfrac{4}{n^2+1}$; (f) $\sum\limits_{n=1}^{\infty} \dfrac{2}{n+\sqrt{n}}$;

    (g) $\sum\limits_{n=1}^{\infty} \dfrac{2n}{n^2-n+1}$; (h) $\sum\limits_{n=1}^{\infty} \dfrac{\cos^2 n}{3^n}$; (i) $\sum\limits_{n=1}^{\infty} \dfrac{(\ln n)^2}{n^3}$;

Analysis and Approximation

(j) $\displaystyle\sum_{n=1}^{\infty}\frac{5}{n+4}$; (k) $\displaystyle\sum_{n=1}^{\infty}\frac{2^n}{n}$; (l) $\displaystyle\sum_{n=1}^{\infty}\frac{3}{5^n}$;

(m) $\displaystyle\sum_{n=1}^{\infty}\frac{n+2}{2n^2+n}$; (n) $\displaystyle\sum_{n=1}^{\infty}\frac{1}{e^{2n}+1}$; (o) $\displaystyle\sum_{n=2}^{\infty}\frac{\sqrt{n}}{\ln n}$;

(p) $\displaystyle\sum_{n=1}^{\infty}\frac{1}{n^2(2^n)}$; (q) $\displaystyle\sum_{n=1}^{\infty}\frac{\sqrt{n}}{n^2+1}$; (r) $\displaystyle\sum_{n=1}^{\infty}n\left(\frac{1}{2}\right)^n$.

26.2 The Ratio and nth–Root Tests

Let $\sum u_n$ be a series with positive terms and let $\displaystyle\lim_{n\to\infty}\frac{u_{n+1}}{u_n}=L$, then

(1) the series converges if $L<1$;
(2) the series diverges if $L>1$;
(3) the series may or may not converge if $L=1$ (i.e., the test is inconclusive).

To illustrate the fact that the ratio test is inconclusive when $L=1$, consider the series $\displaystyle\sum_{n=1}^{\infty}\frac{1}{n}$ and $\displaystyle\sum_{n=1}^{\infty}\frac{1}{n^2}$. The first (harmonic) series is known to diverge and the second is known to converge.

For the first series $\displaystyle\lim_{n\to\infty}\frac{u_{n+1}}{u_n}=\lim_{n\to\infty}\frac{n}{n+1}=1$,

and for the second series $\displaystyle\lim_{n\to\infty}\frac{u_{n+1}}{u_n}=\lim_{n\to\infty}\left(\frac{n}{n+1}\right)^2=1$.

Example Decide, if possible, whether or not the series $\displaystyle\sum_{n=1}^{\infty}\frac{3^n+1}{5^n}$ converges or diverges by using the ratio test.

Let $u_n=\dfrac{3^n+1}{5^n}$ then $\dfrac{u_{n+1}}{u_n}=\left(\dfrac{3^{n+1}+1}{5^{n+1}}\right)\left(\dfrac{5^n}{3^n+1}\right)=\dfrac{1}{5}\left(\dfrac{3^{n+1}+1}{3^n+1}\right)=\dfrac{1}{5}\left(\dfrac{3+3^{-n}}{1+3^{-n}}\right)$.

Therefore $\displaystyle\lim_{n\to\infty}\frac{u_{n+1}}{u_n}=\frac{3}{5}<1$ and so the series is convergent.

Example Decide whether or not the series $\displaystyle\sum_{n=1}^{\infty}\frac{(n!)^2}{(2n)!}$ converges or diverges by using the ratio test.

665

Let $u_n = \dfrac{(n!)^2}{(2n)!}$.

Then $\dfrac{u_{n+1}}{u_n} = \dfrac{((n+1)!)^2}{(2(n+1))!} \times \dfrac{(2n)!}{(n!)^2} = \dfrac{(n+1)(n+1)}{(2n+2)(2n+1)} = \dfrac{n+1}{2(2n+1)} = \dfrac{1+\frac{1}{n}}{2(2+\frac{1}{n})}$.

Thus $\lim\limits_{n\to\infty} \dfrac{u_{n+1}}{u_n} = \dfrac{1}{4} < 1$ and so the series converges.

Example   Show that the ratio test cannot be used to determine whether or not the series $\sum\limits_{n=1}^{\infty} \dfrac{1}{\sqrt{n}}$ is convergent.

Let $u_n = \dfrac{1}{\sqrt{n}}$.

Then $\lim\limits_{n\to\infty} \dfrac{u_{n+1}}{u_n} = \lim\limits_{n\to\infty} \left(\dfrac{n}{n+1}\right)^{1/2} = 1$.

Therefore the ratio test is inconclusive.

The *n*th–Root Test

Let $\sum\limits_{n=1}^{\infty} u_n$ be a series of positive terms and let $\lim\limits_{n\to\infty} \sqrt[n]{u_n} = L$. Then

(1) the series converges if $L < 1$;
(2) the series diverges if $L > 1$;
(3) the test is inconclusive if $L = 1$.

A result which is quite useful when applying the *n*th-root test to many series is

$$\lim_{n\to\infty} \sqrt[n]{n} = 1.$$

As before, we can illustrate the fact that the test is inconclusive when $L = 1$ by considering the series $\sum\limits_{n=1}^{\infty} \dfrac{1}{n}$ and $\sum\limits_{n=1}^{\infty} \dfrac{1}{n^2}$.

In the first case $\lim\limits_{n\to\infty} \sqrt[n]{u_n} = \lim\limits_{n\to\infty} \dfrac{1}{\sqrt[n]{n}} = 1$,

and in the second case $\lim\limits_{n\to\infty} \sqrt[n]{u_n} = \lim\limits_{n\to\infty} \dfrac{1}{\left(\sqrt[n]{n}\right)^2} = 1$.

Analysis and Approximation

Example Determine whether or not the series $\sum_{n=1}^{\infty} \dfrac{n}{3^n}$ converges.

$$\lim_{n\to\infty} \sqrt[n]{\dfrac{n}{3^n}} = \lim_{n\to\infty} \dfrac{\sqrt[n]{n}}{3} = \dfrac{1}{3} < 1 \text{ and so the given series is convergent.}$$

Example Determine whether or not the series $\sum_{n=1}^{\infty} \dfrac{3^n}{n^2 2^n}$ is convergent.

$$\lim_{n\to\infty} \sqrt[n]{\dfrac{3^n}{n^2 2^n}} = \lim_{n\to\infty} \dfrac{3}{2\left(\sqrt[n]{n}\right)^2} = \dfrac{3}{2} > 1 \text{ and so the series does not converge.}$$

Exercise 26.2

1. Use either the ratio test or the nth-root test to determine which of the following series converge and which diverge:

    (a) $\sum_{n=1}^{\infty} \dfrac{n^4}{3^n}$;
    (b) $\sum_{n=1}^{\infty} \dfrac{n^5}{5^n}$;
    (c) $\sum_{n=1}^{\infty} \dfrac{2^{3n}}{3^{2n}}$;

    (d) $\sum_{n=1}^{\infty} \dfrac{n!}{2^n}$;
    (e) $\sum_{n=1}^{\infty} \dfrac{n!}{(2n+1)!}$;
    (f) $\sum_{n=1}^{\infty} \dfrac{(n+1)(n+2)}{n!}$.

2. What can be deduced from the ratio test about the following series?

    (a) $\sum_{n=1}^{\infty} \dfrac{n^2}{n^3+1}$;
    (b) $\sum_{n=1}^{\infty} \dfrac{n^3}{2^n}$;
    (c) $\sum_{n=1}^{\infty} \dfrac{\sqrt{n}}{2n+1}$.

3. Use either the ratio test or the nth-root test to determine which of the following series converge and which diverge:

    (a) $\sum_{n=1}^{\infty} \left(\dfrac{n+1}{2n}\right)^n$;
    (b) $\sum_{n=1}^{\infty} \dfrac{10^n}{n!}$;
    (c) $\sum_{n=1}^{\infty} \dfrac{(n!)^n}{n^{2n}}$;

    (d) $\sum_{n=1}^{\infty} n! e^{-n}$;
    (e) $\sum_{n=1}^{\infty} \dfrac{(2n)!}{6^n (n!)^2}$;
    (f) $\sum_{n=1}^{\infty} \dfrac{n^3 2^n}{3^n}$;

    (g) $\sum_{n=1}^{\infty} \dfrac{2^n}{n^3+1}$;
    (h) $\sum_{n=1}^{\infty} \dfrac{(n+1)(n+2)}{3^n}$; (i) $\sum_{n=1}^{\infty} \dfrac{n}{(\ln n)^n}$.

Chapter 26

4. Show that the ratio test cannot be used to determine whether or not the series $\sum_{n=1}^{\infty} \frac{4^n (n!)^2}{(2n)!}$ converges or diverges and use an alternative method to show that it in fact diverges.

5. Show that neither the ratio test nor the nth-root test can be used to determine whether or not the series $\sum_{n=1}^{\infty} \frac{1}{n^p}$ converges or diverges.

26.3 Improper Integrals and the Integral Test

Consider the function $f(x)$ which is continuous for all $x \geq 0$. Then the integral $\int_0^{\infty} f(x)\, dx$ is called an *improper* integral.

Now consider $F(a) = \int_0^a f(x)\, dx$.

If $\lim_{a \to \infty} F(a)$ exists and is equal to L, then the integral $\int_0^{\infty} f(x)\, dx$ is said to *converge to L*.

If $\lim_{a \to \infty} F(a)$ does not exist, then the integral $\int_0^{\infty} f(x)\, dx$ is said to be *divergent*.

Note: The lower limit of the integral need not be zero. If it is k, then $f(x)$ must be continuous for all $x \geq k$.

Example Show that the integral $\int_1^{\infty} \frac{1}{x^2}\, dx$ converges and that the integral $\int_1^{\infty} \frac{1}{x}\, dx$ diverges.

$\lim_{a \to \infty} \int_1^a \frac{1}{x^2}\, dx = \lim_{a \to \infty} \left[-\frac{1}{x} \right]_1^a = \lim_{a \to \infty} \left(1 - \frac{1}{a} \right) = 1$ and so $\int_1^{\infty} \frac{1}{x^2}\, dx$ converges (to 1).

$\lim_{a \to \infty} \int_1^a \frac{1}{x}\, dx = \lim_{a \to \infty} \left[\ln x \right]_1^a = \lim_{a \to \infty} (\ln a)$ which does not exist and so $\int_1^{\infty} \frac{1}{x}\, dx$ diverges.

Analysis and Approximation

The Integral Test

If $f(x)$ is a continuous, positive, decreasing function of x for all $x \geq 1$, and $u_n = f(n)$ for all positive integers n, then the series $\sum_{n=1}^{\infty} u_n$ and the integral $\int_0^{\infty} f(x)\, dx$ both converge or both diverge.

Example Use the integral test to show that the harmonic series is divergent and the series $\sum_{n=1}^{\infty} \dfrac{1}{n^2}$ is convergent.

Let $u_n = f(n) = \dfrac{1}{n}$. Then $f(x) = \dfrac{1}{x}$ is clearly continuous, positive and decreasing for all $x \geq 1$.

Now from the previous example, $\int_1^{\infty} \dfrac{1}{x}\, dx$ diverges and so the harmonic series, $\sum_{n=1}^{\infty} \dfrac{1}{n}$, also diverges.

Let $u_n = f(n) = \dfrac{1}{n^2}$. Then $f(x) = \dfrac{1}{x^2}$ is clearly continuous, positive and decreasing for all $x \geq 1$.

Now from the previous example, $\int_1^{\infty} \dfrac{1}{x^2}\, dx$ converges and so the series $\sum_{n=1}^{\infty} \dfrac{1}{n^2}$ also converges.

The p–Series

If p is a real constant, then the series $\sum_{n=1}^{\infty} \dfrac{1}{n^p}$ is called *the p-series*.

Theorem The p-series converges if $p > 1$ and diverges if $p \leq 1$.

Proof Let $f(x) = \dfrac{1}{x^p}$.

669

Chapter 26

(1) $p > 1$

Clearly $f(x)$ is continuous, positive and decreasing for all $x \geq 1$.

Now $\int_1^\infty \dfrac{1}{x^p}\,dx = \lim_{a \to \infty} \int_1^a x^{-p}\,dx$

$\qquad = \lim_{a \to \infty} \left[\dfrac{1}{1-p} x^{1-p} \right]_1^a$

$\qquad = \dfrac{1}{1-p} \lim_{a \to \infty} \left(a^{1-p} - 1 \right)$

$\qquad = \dfrac{1}{1-p} \lim_{a \to \infty} \left(\dfrac{1}{a^{p-1}} - 1 \right)$

$\qquad = \dfrac{1}{1-p}(0 - 1)$

$\qquad = \dfrac{1}{p-1}.$

Since $\int_1^\infty \dfrac{1}{x^p}\,dx$ converges then so does the series $\sum_{n=1}^\infty \dfrac{1}{n^p}$.

(2) $p = 1$

The series $\sum_{n=1}^\infty \dfrac{1}{n}$ (the harmonic series) is known to diverge and so does the p-series.

(3) $p < 1$

$\sum_{n=1}^\infty \dfrac{1}{n^p} = \dfrac{1}{1^p} + \dfrac{1}{2^p} + \dfrac{1}{3^p} + \cdots \cdots > \dfrac{1}{1} + \dfrac{1}{2} + \dfrac{1}{3} + \cdots \cdots$.

Thus $\sum_{n=1}^\infty \dfrac{1}{n^p}$ diverges by comparison with the harmonic series.

Therefore the p-series converges for $p > 1$ and diverges for $p \leq 1$.

Example Determine whether or not the series $\sum_{n=1}^\infty \dfrac{2}{\sqrt{n}}$ converges.

The p-series with $p = \tfrac{1}{2}$, $\sum_{n=1}^\infty \dfrac{1}{n^{1/2}}$, diverges and so $\sum_{n=1}^\infty \dfrac{2}{\sqrt{n}}$ diverges.

Analysis and Approximation

Exercise 26.3

1. Use the integral test to determine which of the following series converge and which diverge:

   (a) $\sum_{n=1}^{\infty} \dfrac{1}{2n-1}$; (b) $\sum_{n=2}^{\infty} \dfrac{1}{n \ln n}$; (c) $\sum_{n=1}^{\infty} \dfrac{n}{\sqrt{n^2+25}}$;

   (d) $\sum_{n=1}^{\infty} \dfrac{1}{1+n^2}$; (e) $\sum_{n=1}^{\infty} \dfrac{1}{n(n+1)}$; (f) $\sum_{n=1}^{\infty} \dfrac{\arctan n}{1+n^2}$;

   (g) $\sum_{n=1}^{\infty} n e^{-n^2}$; (h) $\sum_{n=2}^{\infty} \dfrac{1}{n(\ln n)^3}$; (i) $\sum_{n=2}^{\infty} \dfrac{1}{n(\ln n)^p}$, $p > 1$.

2. (a) If $u_n = \dfrac{2n}{n^2+2}$, show that $u_{n+1} < u_n$ for all positive integers n.

   (b) Show that the conditions needed to apply the integral test to establish whether or not the series $\sum_{n=1}^{\infty} \dfrac{2n}{n^2+1}$ is convergent, are satisfied.

   (c) Determine whether or not the series in part (b) is convergent.

3. Use the p-series test to decide which of the following series are convergent and which are divergent:

   (a) $\sum_{n=1}^{\infty} \dfrac{\sqrt{n}}{n^2+1}$; (b) $\sum_{n=1}^{\infty} \dfrac{1}{n\sqrt{n}}$; (c) $\sum_{n=2}^{\infty} \dfrac{n^{3/2}}{n-1}$;

   (d) $\sum_{n=1}^{\infty} \dfrac{n^{2/3}}{n^2+1}$; (e) $\sum_{n=1}^{\infty} \dfrac{n}{n^4+1}$; (f) $\sum_{n=1}^{\infty} \dfrac{1}{n\sqrt[n]{n}}$.

26.4 Alternating Series and Absolute Convergence

Definition A series in which the terms are alternately positive and negative is called *an alternating series*.

Leibniz's Theorem

The alternating series $\sum_{n=1}^{\infty} (-1)^{n+1} u_n = u_1 - u_2 + u_3 - u_4 + \cdots$ converges if all three of the following conditions hold:

(1) u_i is positive for all $i = 1, 2, 3, \ldots$;

(2) $u_{n+1} \leq u_n$ for all n ;

(3) $u_n \to 0$ as $n \to \infty$.

Chapter 26

Example Show that the alternating series $\sum_{n=1}^{\infty}\frac{(-1)^{n+1}}{n} = \frac{1}{1} - \frac{1}{2} + \frac{1}{3} - \frac{1}{4} + \cdots$
(the alternating harmonic series), converges.

Let $u_n = \frac{1}{n}$.

Then (1) u_i is positive for all $i = 1, 2, 3, 4, \ldots$.

(2) $\frac{1}{n+1} \leq \frac{1}{n}$ for all n and so $u_{n+1} \leq u_n$ for all n.

(3) $u_n \to 0$ as $n \to \infty$.

Therefore by Leibniz's theorem, the alternating harmonic series converges.

The Alternating Series Estimation Theorem

If the alternating series $\sum_{n=1}^{\infty}(-1)^{n+1}u_n$ converges to L and

$$S_n = u_1 - u_2 + u_3 - \cdots + (-1)^{n+1}u_n,$$

then S_n approximates the sum L with an error $\varepsilon_n = L - S_n$ such that $|\varepsilon_n| < u_{n+1}$, the numerical value of the first unused term. Also the difference $L - S_n$ has the same sign as the first unused term.

Example Estimate the magnitude of the error involved in using the sum of the first seven terms to approximate the sum L of the geometric series $\sum_{n=1}^{\infty}(-1)^{n+1}\frac{1}{2^n}$, and confirm that $L - S_7$ has the same sign as the eighth term of the series.

Firstly we have $\sum_{n=1}^{\infty}(-1)^{n+1}\frac{1}{2^n} = \frac{\frac{1}{2}}{1-(-\frac{1}{2})} = \frac{1}{3} = L$.

The eighth term of the series is $-\frac{1}{256}$ so $|\varepsilon_7| < \frac{1}{256}$.

Also $L - S_7 = \frac{1}{3} - \frac{u_1(1-(-\frac{1}{2})^7)}{1-(-\frac{1}{2})} = \frac{1}{3} - \frac{1}{3}\left(1 + \frac{1}{2^7}\right) = -\frac{1}{3 \times 2^7} < 0$ and so $L - S_7$ has the same sign as the eighth term.

Note: $|\varepsilon_7| = \frac{1}{3 \times 2^7} = \frac{1}{384} < \frac{1}{256}$, the numerical value of the eighth term.

Analysis and Approximation

Absolute Convergence

A series $\sum_{n=1}^{\infty} u_n$ is ***absolutely convergent*** if the corresponding series of absolute values, $\sum_{n=1}^{\infty} |u_n|$, converges.

Example Show that the geometric series $\sum_{n=0}^{\infty} \left(-\frac{2}{3}\right)^n$ is absolutely convergent.

If $u_n = \left(-\frac{2}{3}\right)^n$ then $|u_n| = \left(\frac{2}{3}\right)^n$ and the series $\sum_{n=0}^{\infty} |u_n|$ is geometric with $r = \frac{2}{3} < 1$ and so converges.

Therefore $\sum_{n=0}^{\infty} \left(-\frac{2}{3}\right)^n$ is absolutely convergent.

Definition A series which converges but does not converge absolutely is said to ***converge conditionally*** or is said to be ***conditionally convergent***.

Example Show that the alternating harmonic series, $\sum_{n=1}^{\infty} \frac{(-1)^{n+1}}{n}$, is conditionally convergent.

We have already seen in Section 26.4 that the alternating harmonic series is convergent and in Section 26.1 that the harmonic series is divergent.

Therefore the alternating harmonic series is conditionally convergent.

The Absolute Convergence Theorem

If $\sum_{n=1}^{\infty} |u_n|$ converges, then $\sum_{n=1}^{\infty} u_n$ also converges.

That is every absolutely convergent series converges.

Example Show that $\sum_{n=1}^{\infty} \frac{(-1)^{n+1}}{n^2}$ converges.

Chapter 26

The given series is absolutely convergent since $\sum_{n=1}^{\infty} \frac{1}{n^2}$ converges.

Therefore the series $\sum_{n=1}^{\infty} \frac{(-1)^{n+1}}{n^2}$ converges.

Example Determine whether or not the series $\sum_{n=1}^{\infty} (-1)^{n+1} \frac{\cos n}{n^2}$ converges.

$$\sum_{n=1}^{\infty} \frac{|\cos n|}{n^2} = \frac{|\cos 1|}{1^2} + \frac{|\cos 2|}{2^2} + \frac{|\cos 3|}{3^2} + \cdots \leq 1 + \frac{1}{2^2} + \frac{1}{3^2} + \cdots = \sum_{n=1}^{\infty} \frac{1}{n^2}$$

since $|\cos n| \leq 1$ for all n.

Since $\sum_{n=1}^{\infty} \frac{1}{n^2}$ converges then $\sum_{n=1}^{\infty} (-1)^{n+1} \frac{\cos n}{n^2}$ converges absolutely and is therefore convergent.

Exercise 26.4

1. Determine which of the following alternating series converge:

   (a) $\sum_{n=1}^{\infty} \frac{(-1)^{n+1}}{2^n}$;

   (b) $\sum_{n=1}^{\infty} (-1)^{n+1} n^{-3/2}$;

   (c) $\sum_{n=1}^{\infty} (-1)^n \frac{3}{n^2 + 1}$;

   (d) $\sum_{n=1}^{\infty} (-1)^{n+1} \frac{n+2}{n+1}$;

   (e) $\sum_{n=2}^{\infty} \frac{(-1)^{n+1}}{\ln n}$;

   (f) $\sum_{n=1}^{\infty} (-1)^{n+1} \frac{2n+1}{n^2}$;

   (g) $\sum_{n=1}^{\infty} (-1)^{n+1} \frac{n^2}{n^3 + 1}$;

   (h) $\sum_{n=1}^{\infty} (-1)^{n+1} n e^{-n}$.

2. Show that the series $\sum_{n=1}^{\infty} \frac{(-1)^{n+1}}{n^p}$ converges for $p > 0$, converges absolutely for $p > 1$ and converges conditionally for $0 < p \leq 1$.

3. For each of the following series, find an upper bound for the error if the sum of the first 5 terms is used as an approximation to the sum of the series:

   (a) $\sum_{n=1}^{\infty} (-1)^{n+1} \frac{2}{n}$;

   (b) $\sum_{n=1}^{\infty} (-1)^{n+1} \frac{1}{(n+1)(n+2)}$;

(c) $\sum_{n=1}^{\infty}(-1)^n 3n^{-2}$; (d) $\sum_{n=1}^{\infty}(-1)^n \dfrac{6}{n^3}$;

(e) $\sum_{n=1}^{\infty}(-1)^{n+1} \dfrac{3}{2^n}$; (f) $\sum_{n=1}^{\infty}(-1)^{n+1} \dfrac{1}{(n+2)^3}$.

4. Find the sum of each of the following series correct to three decimal places:

(a) $\sum_{n=1}^{\infty}(-1)^{n+1}\left(\dfrac{2}{3}\right)^n$; (b) $\sum_{n=1}^{\infty}(-1)^{n+1}\dfrac{3}{2^n}$;

(c) $\sum_{n=1}^{\infty}(-1)^{n+1}\dfrac{1}{(3n+2)^3}$; (d) $\sum_{n=1}^{\infty}(-1)^{n+1}\dfrac{1}{n!}$;

(e) $\sum_{n=1}^{\infty}(-1)^{n+1}\dfrac{1}{n 2^n}$; (f) $\sum_{n=1}^{\infty}(-1)^{n+1}\dfrac{1}{3n^4}$.

5. Which of the following series converge absolutely, which converge conditionally and which diverge?

(a) $\sum_{n=1}^{\infty}(-1)^{n+1}\dfrac{1}{n\sqrt{n}}$; (b) $\sum_{n=1}^{\infty}(-1)^{n+1}(0.99)^n$;

(c) $\sum_{n=1}^{\infty}(-1)^n \dfrac{2n}{n^3+2}$; (d) $\sum_{n=1}^{\infty}\dfrac{(-1)^n}{\sqrt{n}}$;

(e) $\sum_{n=1}^{\infty}\dfrac{(-1)^n}{3n+2}$; (f) $\sum_{n=1}^{\infty}(-1)^{n+1}\dfrac{2n+1}{2n+3}$;

(g) $\sum_{n=1}^{\infty}(-1)^n n^2 \left(\dfrac{2}{3}\right)^n$; (h) $\sum_{n=1}^{\infty}(-1)^n \dfrac{\arctan n}{1+n^2}$;

(i) $\sum_{n=1}^{\infty}(-1)^{n+1}\left(\dfrac{n+2}{2n}\right)^n$; (j) $\sum_{n=1}^{\infty}\dfrac{(-1)^{n+1}}{n^2+1}$;

(k) $\sum_{n=1}^{\infty}\dfrac{\sin\frac{1}{2}(2n-1)\pi}{n^{1.2}}$; (l) $\sum_{n=1}^{\infty}(-1)^n \left(\sqrt{n+2}-\sqrt{n}\right)$.

26.5 Power Series

Definition A series of the form

$$\sum_{n=0}^{\infty} c_n(x-a)^n = c_0 + c_1(x-a) + c_2(x-a)^2 + \cdots\cdots$$

where a and the coefficients $c_0, c_1, c_2, \cdots\cdots$ are constants, is called a *power series* in $(x-a)$ or a power series *centred on a*.

675

Chapter 26

With a power series, one of three possibilities must occur.

(1) The power series may converge for all values of x.
(2) The power series may diverge for all x except $x = a$.
(3) There exists a number $r > 0$ such that the series converges absolutely for $|x-a| < r$ and diverges if $|x-a| > r$. In this case there are several possibilities when $x = a - r$ or $x = a + r$.

In the third case, the set of possible values of x for which the series converges is called **the interval of convergence**. This interval may be open, $]a-r, a+r[$, half-open, $]a-r, a+r]$ or $[a-r, a+r[$, or closed, $[a-r, a+r]$. In all of these cases, the number r is called **the radius of convergence**.

In the first case, the radius of convergence is $r = \infty$; in the second case, the radius of convergence is $r = 0$.

Example Consider the power series $\sum_{n=0}^{\infty} x^n = 1 + x + x^2 + x^3 + \cdots$.

Find the values of x for which this series converges.

The series is geometric series with $r = x$ and so is convergent for $|r| < 1$ or $-1 < x < 1$, and divergent for $x > 1$ and for $x < -1$.

When $x = 1$, the series is $1 + 1 + 1 + 1 + \ldots$ which is clearly divergent.

When $x = -1$, the series is $1 - 1 + 1 - 1 + \ldots$ which is also divergent.

Note: When the series converges, it converges to $\dfrac{1}{1-x}$.

That is $\dfrac{1}{1-x} = 1 + x + x^2 + x^3 + \cdots$ for $-1 < x < 1$.

Example For which values of x does the power series $\sum_{n=1}^{\infty} (-1)^{n+1} \dfrac{x^n}{n}$ converge?

$u_n = (-1)^{n+1} \dfrac{x^n}{n}$ and $\lim_{n \to \infty} \left| \dfrac{u_{n+1}}{u_n} \right| = \lim_{n \to \infty} \left| \dfrac{x^{n+1}}{n+1} \times \dfrac{n}{x^n} \right| = \lim_{n \to \infty} \dfrac{n}{n+1} |x| = |x|$ (ratio test), or $\lim_{n \to \infty} \sqrt[n]{\left| \dfrac{x^n}{n} \right|} = \lim_{n \to \infty} \dfrac{|x|}{\sqrt[n]{n}} = |x|$ (nth-root test).

Therefore the series is absolutely convergent for $|x| < 1$ (and divergent for $|x| > 1$).

676

Analysis and Approximation

When $x = 1$ the series is $1 - \frac{1}{2} + \frac{1}{3} - \frac{1}{4} + \cdots$ which is the alternating harmonic series which converges.

When $x = -1$ the series is $-1 - \frac{1}{2} - \frac{1}{3} - \frac{1}{4} - \cdots$ which is the negative of the harmonic series which diverges.

Therefore the given power series converges for $-1 < x \leq 1$.

Example Find the interval of convergence and the radius of convergence of the power series $\sum_{n=1}^{\infty} (-1)^{n+1} \left(\frac{2x+3}{5} \right)^n$.

The series is geometric with common ratio $r = -\left(\frac{2x+3}{5} \right)$.

Therefore the series is convergent for $\left| \frac{2x+3}{5} \right| < 1$ or $\left| x + \frac{3}{2} \right| < \frac{5}{2}$, i.e., for

$-5 < 2x + 3 < 5 \implies -4 < x < 1$.

Therefore the interval of convergence is $]-4, 1[$ and the radius of convergence is $\frac{5}{2}$.

Example Find the radius of convergence of the power series $\sum_{n=0}^{\infty} \frac{x^n}{n!}$.

$u_n = \frac{x^n}{n!}$ and $\lim_{n \to \infty} \left| \frac{u_{n+1}}{u_n} \right| = \lim_{n \to \infty} \left| \frac{x^{n+1}}{(n+1)!} \times \frac{n!}{x^n} \right| = \lim_{n \to \infty} \frac{|x|}{n+1} = 0$ for all x.

Therefore the radius of convergence is $r = \infty$.

Example Find the radius of convergence of the power series $\sum_{n=0}^{\infty} n!(x-1)^n$.

$u_n = n!(x-1)^n$ and $\lim_{n \to \infty} \left| \frac{u_{n+1}}{u_n} \right| = \lim_{n \to \infty} \left| \frac{(n+1)!(x-1)^{n+1}}{n!(x-1)^n} \right| = \lim_{n \to \infty} (n+1)|x-1| = 0$

for $x = 1$ only.

Therefore the radius of convergence is $r = 0$.

677

Chapter 26

Exercise 26.5

1. For each of the following power series, determine the values of x for which the series converges absolutely and the radius of convergence:

   (a) $\sum_{n=0}^{\infty}(x+2)^n$;
   (b) $\sum_{n=0}^{\infty} n(3x+1)^n$;

   (c) $\sum_{n=0}^{\infty}(-1)^n(2x+3)^n$;
   (d) $\sum_{n=0}^{\infty}\dfrac{(x-1)^n}{2^n}$;

   (e) $\sum_{n=0}^{\infty}(-1)^n\dfrac{x^n}{n!}$;
   (f) $\sum_{n=0}^{\infty}\dfrac{x^n}{\sqrt{n^2+1}}$;

   (g) $\sum_{n=0}^{\infty}(-1)^n\dfrac{x^{2n+1}}{(2n+1)!}$;
   (h) $\sum_{n=0}^{\infty}\dfrac{(n+1)(x+5)^n}{3^n}$;

   (i) $\sum_{n=0}^{\infty} n!(x-3)^n$;
   (j) $\sum_{n=1}^{\infty}(-1)^{n+1}\dfrac{(n+1)x^n}{n!}$.

2. If a and b are positive integers, find the radius of convergence of the power series $\sum_{n=1}^{\infty}\dfrac{(n+a)!}{n!(n+b)!}x^n$.

3. Determine both the radius of convergence and the interval of convergence for each of the following power series:

   (a) $\sum_{n=0}^{\infty}\dfrac{(2x+3)^n}{4^n}$;
   (b) $\sum_{n=1}^{\infty} n^3 x^n$;

   (c) $\sum_{n=0}^{\infty}\dfrac{x^n}{n^2+2}$;
   (d) $\sum_{n=0}^{\infty}\dfrac{n^2 x^n}{2^n}$;

   (e) $\sum_{n=1}^{\infty}\dfrac{x^n}{2^n\sqrt{n}}$;
   (f) $\sum_{n=1}^{\infty}(-1)^{n+1}\dfrac{(2x)^n}{n3^n}$.

4. If $\sum_{n=1}^{\infty} u_n$ is an absolutely convergent series, prove that the power series $\sum_{n=1}^{\infty} u_n x^n$ is absolutely convergent for $|x|\le 1$.

5. Prove that if the radius of convergence of the power series $\sum_{n=1}^{\infty} u_n x^n$ is r, then the radius of convergence of $\sum_{n=1}^{\infty} u_n x^{2n}$ is $\sqrt{r}$.

26.6 The Mean Value Theorem

Rolle's Theorem

Consider the function $y = f(x)$ for which $f(a) = f(b) = 0$. If $f(x)$ is continuous for $a \leq x \leq b$ and differentiable for $a < x < b$, then there exists at least one value of c between a and b for which $f'(c) = 0$.

Proof

The maximum or minimum value of a continuous function on a given interval $[a, b]$ is either $f(a)$, $f(b)$ or $f(c)$ where c is between a and b and $f'(c) = 0$ or $f'(c)$ does not exist.

Since $y = f(x)$ is differentiable for all x between a and b, $f'(c)$ must exist and so $f'(c) = 0$.

If either the maximum or minimum occurs at $x = c$, $a < c < b$, $f'(c) = 0$.

If both maximum and minimum occur at $x = a$ and $x = b$, then $f(x) = 0$ for all $a \leq x \leq b$ and $f'(c) = 0$ for all c between a and b.

This proves the theorem.

Rolle's theorem makes considerable sense geometrically. Consider the following graphical examples:

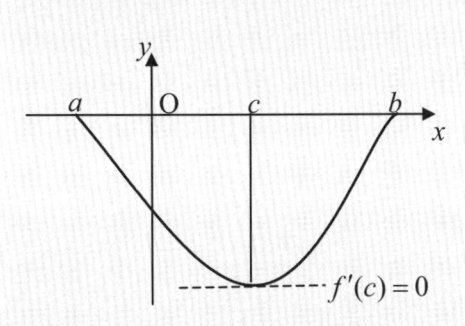

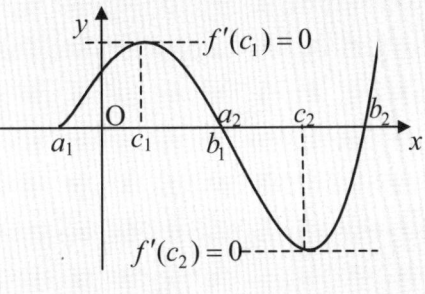

If a smooth curve crosses the x-axis in two places, $x = a$ and $x = b$, there is at least one point $x = c$ between $x = a$ and $x = b$ where the tangent to the curve is horizontal.

The need for the condition in Rolle's theorem concerning the differentiability of $f(x)$ in the interval $a < x < b$ is illustrated by the the graph of $f(x) = |x| - 1$.

Chapter 26

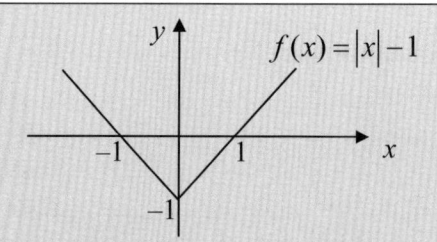

[Graph showing $f(x) = |x| - 1$ with V-shape, vertices at $(-1, 0)$, $(0, -1)$, $(1, 0)$.]

Here, $f(-1) = f(1) = 0$ and $f(x)$ is continuous for $-1 \le x \le 1$, but the graph does not have a horizontal tangent anywhere since $f(x)$ is not differentiable at $x = 0$.

The Mean Value Theorem

If $y = f(x)$ is continuous for all $x \in [a, b]$ and differentiable for all $x \in \,]a, b[$, then there exists at least one number $c \in \,]a, b[$ for which

$$\frac{f(b) - f(a)}{b - a} = f'(c).$$

Proof

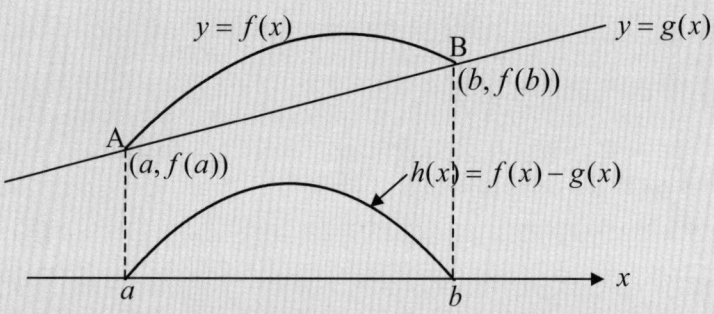

[Diagram showing curve $y = f(x)$ with points $A(a, f(a))$ and $B(b, f(b))$, line $y = g(x)$ through A and B, and function $h(x) = f(x) - g(x)$ below.]

Let $A(a, f(a))$ and $B(b, f(b))$ be any two points on a smooth curve $y = f(x)$. The equation of the line (AB) is $y = g(x)$ where

$$g(x) - f(a) = \frac{f(b) - f(a)}{b - a}(x - a)$$

since the gradient is $\dfrac{f(b) - f(a)}{b - a}$ and the point $A(a, f(a))$ lies on it.

Consider the function $h(x) = f(x) - g(x)$ which gives the distance between the curve $y = f(x)$ and the straight line (AB). Then

$$h(x) = f(x) - f(a) - \frac{f(b) - f(a)}{b - a}(x - a).$$

680

Analysis and Approximation

Now $h(x)$ satisfies the conditions needed for Rolle's theorem to hold. That is $h(a) = h(b) = 0$ and $h(x)$ is continuous for all $x \in [a, b]$ and differentiable for all $x \in]a, b[$.

Thus there exists a number $c \in]a, b[$ for which $h'(c) = 0$.

Now $h'(x) = f'(x) - \dfrac{f(b) - f(a)}{b - a}$ and $h'(c) = f'(c) - \dfrac{f(b) - f(a)}{b - a} = 0$.

Therefore $f'(c) = \dfrac{f(b) - f(a)}{b - a}$ as required.

Note: The condition of the mean value theorem that does not require differentia-bility at the end-points is illustrated by the function $f(x) = \sqrt{1 - x^2}$ defined on $[-1, 1]$.

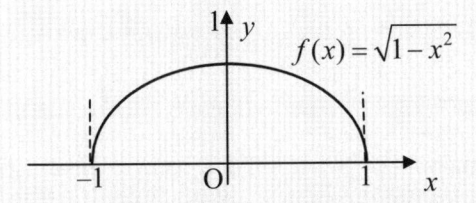

Here $f'(-1)$ and $f'(1)$ do not exist (the tangents at the end-points are vertical), but $f'(0) = \dfrac{f(1) - f(-1)}{1 - (-1)} = 0$ and $0 \in]-1, 1[$.

Example Apply the mean value theorem to the function $f(x) = x^3$, $0 \le x \le 2$, to find a value of c, $c \in]0, 2[$, for which $f'(c)$ is equal to the gradient of the straight line joining $(0, 0)$ to $(2, 8)$.

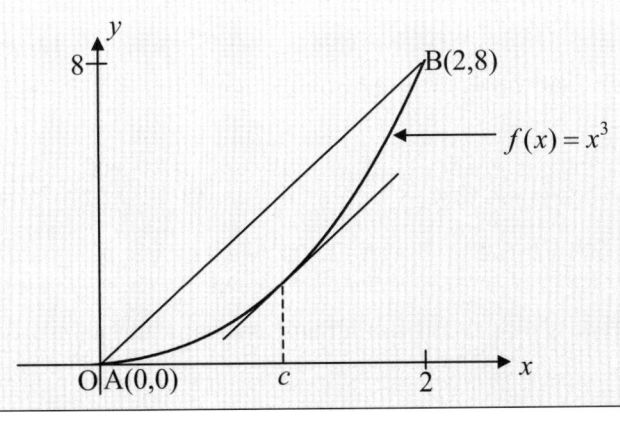

681

Chapter 26

$f(x)$ is continuous on $[0, 2]$ and differentiable on $]0, 2[$. Then there exists a number $c \in]0, 2[$ such that $f'(c) = \dfrac{f(2) - f(0)}{2 - 0}$ and so $3c^2 = \dfrac{8 - 0}{2} = 4$.

Therefore $c^2 = \frac{4}{3}$ and so $c = \frac{2}{\sqrt{3}}, c \in]0, 2[$.

Example A car starting from rest takes 12 seconds to travel 288 metres. Show that at some time the speed of the car was exactly 24 ms^{-1}.

Let $s = f(t)$ be the distance travelled by the car in the first t seconds. Then $f(t)$ is continuous on $[0, 12]$, differentiable on $]0, 12[$, and $f(0) = 0$ and $f(12) = 288$.

By the mean value theorem there exists a number $c \in]0, 12[$ such that $f'(c) = \dfrac{f(12) - f(0)}{12 - 0} = \dfrac{288}{12} = 24$. Thus at some time $t = c$ where c is between 0 and 12, the speed, $f'(c)$, of the car was exactly 24 ms^{-1}.

The Mean Value Theorem for Definite Integrals

If the function f is continuous on $[a, b]$, then at some point $c \in [a, b]$,

$$f(c) = \dfrac{1}{b-a} \int_a^b f(x)\, dx.$$

The proof is beyond the scope of this book.

Example Consider the function $f(x) = \ln x$ for $x \in [1, 1 + \frac{1}{n}]$ where n is a positive integer. Use the mean value theorem, or otherwise, to show that $n \ln\left(1 + \dfrac{1}{n}\right) \leq 1$.

Firstly, $f(x)$ is continuous on $[1, 1+\frac{1}{n}]$ and differentiable on $]1, 1+\frac{1}{n}[$. Therefore there exists a number $c \in]1, 1+\frac{1}{n}[$ for which

$$f'(c) = \dfrac{f\left(1+\frac{1}{n}\right) - f(1)}{\left(1+\frac{1}{n}\right) - 1} = n \ln\left(1 + \dfrac{1}{n}\right).$$

Now $f'(x) = \dfrac{1}{x}$ and $\dfrac{1}{x} \leq 1$ for $x \in [1, 1+\frac{1}{n}]$.

Thus $f'(c) \leq 1$ for $c \in]1, 1+\frac{1}{n}[$ and so $n \ln\left(1 + \dfrac{1}{n}\right) \leq 1$ as required.

Analysis and Approximation

Using the integral form of the mean value theorem with $f(x) = \dfrac{1}{x}$ defined on $[1, 1+\tfrac{1}{n}]$, there exists a value of $c \in [1, 1+\tfrac{1}{n}]$ such that

$$f(c) = \dfrac{1}{\left(1+\tfrac{1}{n}\right)-1} \int_1^{1+\tfrac{1}{n}} \dfrac{1}{x}\,dx = n[\ln x]_1^{1+\tfrac{1}{n}} = n\ln\left(1+\dfrac{1}{n}\right).$$

Now $\dfrac{1}{c} \leq 1$ for $c \in [1, 1+\tfrac{1}{n}]$ and so $n\ln\left(1+\dfrac{1}{n}\right) \leq 1$ as required.

Consider the following geometrical proof.

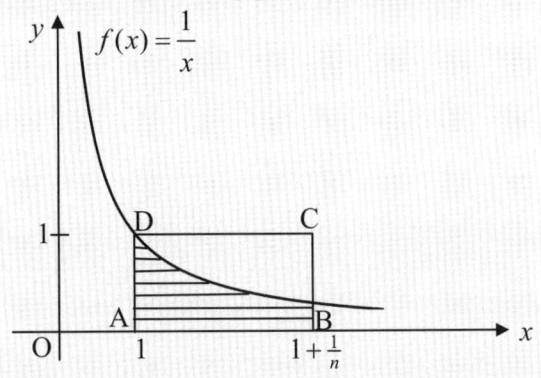

The shaded area = $\displaystyle\int_1^{1+\tfrac{1}{n}} \dfrac{1}{x}\,dx <$ area of rectangle ABCD $= \dfrac{1}{n}$.

Therefore $\ln\left(1+\dfrac{1}{n}\right) - \ln 1 < \dfrac{1}{n} \leq \dfrac{1}{n}$ and so $n\ln\left(1+\dfrac{1}{n}\right) \leq 1$ as required.

Exercise 26.6

1. In each of the following, find a value of c in the interior of the given interval $[a, b]$ such that $f'(c) = \dfrac{f(b)-f(a)}{b-a}$:

   (a) $f(x) = x^2$, $[1, 3]$;
   (b) $f(x) = x^2 + 2x + 1$, $[-1, 1]$;
   (c) $f(x) = \sqrt{x}$, $[1, 9]$;
   (d) $f(x) = x^{2/3}$, $[0, 1]$;
   (e) $f(x) = \dfrac{1}{x}$, $[1, 2]$;
   (f) $f(x) = \dfrac{x+2}{x}$, $[1, 2]$.

2. The displacement from the origin (in metres) of a particle moving along the x-axis at time t seconds is given by $x = f(t)$. If $f(t)$ satisfies the conditions of the mean value theorem for $t \in [a, b]$, $0 \leq a \leq b$, prove that at

683

Chapter 26

some time between $t = a$ and $t = b$, the instantaneous velocity of the particle is exactly equal to its average velocity over the whole interval.

3. Consider the function $f(x) = x^4 - x^3 + 2x^2 - 2x$. Use Rolle's theorem to prove that the equation $4x^3 - 3x^2 + 4x - 2 = 0$ has at least one real root between 0 and 1.

4. Use Rolle's theorem to prove that the equation $2x^3 - 7x + 3 = 0$ has at least one real root between 1 and 2.

5. Use Rolle's theorem to prove that if c is any constant, the equation $2x^3 + 5x + c = 0$ cannot have more than one real root.

6. Suppose that $f(x)$ is continuous on the interval $[0, 6]$ and differentiable on $]0, 6[$. If $f(0) = -8$ and $|f'(x)| \leq 10$ for $0 < x < 6$, show that $-68 \leq f(x) \leq 52$ for $0 \leq x \leq 6$.

7. Suppose that $f(x)$ is continuous on the interval $[2, 5]$ and differentiable on $]2, 5[$. If $f(2) = 3$ and $|f'(x)| \leq 4$ for $2 < x < 5$, find the bounds for $f(x)$ on the interval $[2, 5]$.

8. Use Rolle's theorem to prove that between any two zeros of a polynomial $a_n x^n + a_{n-1} x^{n-1} + \cdots + a_2 x^2 + a_1 x + a_0$ there is a zero of the polynomial $n a_n x^{n-1} + (n-1) a_{n-1} x^{n-2} + \cdots + 2 a_2 x + a_1$.

9. Use the mean value theorem to prove that $\frac{1}{9} < \sqrt{66} - 8 < \frac{1}{8}$. {*Hint:* Consider the function $f(x) = \sqrt{x}$ in the interval $[64, 66]$.}

10. Apply the mean value theorem to the function $f(x) = \ln x$ on the interval $[e, 2.8]$ to show, without the use of a calculator, that $1 < \ln 2.8 < \frac{2.8}{e}$.

11. Use the mean value theorem to show, without the use of a calculator, that $\frac{\pi}{4} + \frac{5}{61} < \arctan 1.2 < \frac{\pi}{4} + \frac{1}{10}$.

12. Use the mean value theorem to prove that for any numbers a and b
    (a) $|\sin b - \sin a| \leq |b - a|$;
    (b) $|\sin a \cos a - \sin b \cos b| \leq |a - b|$.

26.7 Taylor and Maclaurin Series

A Taylor series is a special case of a power series.

Definition The power series

$$\sum_{k=0}^{\infty} \frac{f^{(k)}(a)}{k!}(x-a)^k$$

$$= f(a) + f'(a)(x-a) + \frac{f''(a)}{2!}(x-a)^2 + \cdots + \frac{f^{(n)}(a)}{n!}(x-a)^n + \cdots$$

where $f(x)$ has derivatives of all orders on some open interval containing a, is called the *Taylor series of f at a*.

The special case when $a = 0$:

$$\sum_{k=0}^{\infty} \frac{f^{(k)}(0)}{k!}x^k = f(0) + xf'(0) + x^2\frac{f''(0)}{2!} + \cdots + x^n\frac{f^{(n)}(0)}{n!} + \cdots$$

is called the *Maclaurin series of f*.

Example Find the Maclaurin series for $f(x) = \ln(1+x)$, and find the radius and interval of convergence of this series.

$$f(x) = \ln(1+x), \quad f'(x) = \frac{1}{1+x}, \quad f''(x) = \frac{-1}{(1+x)^2}, \quad f'''(x) = \frac{2!}{(1+x)^3}, \quad \ldots,$$

$$f^{(n)}(x) = \frac{(-1)^{n+1}(n-1)!}{(1+x)^n}.$$

Therefore $f(0) = 0$, $f'(0) = 1$, $f''(0) = -1$, $f'''(0) = 2!$, $f^{(4)}(0) = -3!$, $\ldots$, $f^{(n)}(0) = (-1)^{n+1}(n-1)!$.

The required series is $\displaystyle\sum_{n=0}^{\infty} (-1)^{n+1} x^n \frac{(n-1)!}{n!}$

$$= x - \frac{x^2}{2} + \frac{x^3}{3} - \frac{x^4}{4} + \cdots + (-1)^{n+1}\frac{x^n}{n} + \cdots.$$

Let $u_n = (-1)^{n+1}\dfrac{x^n}{n}$ which gives $\displaystyle\lim_{n\to\infty}\left|\frac{u_{n+1}}{u_n}\right| = |x|\lim_{n\to\infty}\left(\frac{n}{n+1}\right) = |x|$, so the radius of convergence is 1.

685

Chapter 26

When $x = 1$, $u_n = \dfrac{(-1)^n}{n}$ and the series converges.

When $x = -1$, $u_n = -\dfrac{1}{n}$ and the series diverges.

Therefore the interval of convergence is $]-1, 1]$.

Example Find the Taylor series generated by $f(x) = \dfrac{1}{x}$ at $x = 1$ and find the values of x for which this series converges to $\dfrac{1}{x}$.

$$f'(x) = -\frac{1}{x^2}, \; f''(x) = \frac{2!}{x^3}, \; f'''(x) = -\frac{3!}{x^4}, \; \cdots, \; f^{(n)}(x) = (-1)^n \frac{n!}{x^{n+1}}.$$

Thus $f(1) = 1$, $f'(1) = -1$, $f''(1) = 2$, $f'''(1) = -3!$, $\cdots$, $f^{(n)}(1) = (-1)^n n!$.

Therefore the required Taylor series is

$$\sum_{n=0}^{\infty} (-1)^n (x-1)^n = 1 - (x-1) + (x-1)^2 - (x-1)^3 + \cdots + (-1)^n (x-1)^n + \cdots.$$

This series is geometric with $r = 1 - x$ and it converges for $-1 < 1 - x < 1$ or $0 < x < 2$. The sum of the series is $\dfrac{u_1}{1-r} = \dfrac{1}{1-(1-x)} = \dfrac{1}{x}$.

Thus the Taylor series generated by $f(x) = \dfrac{1}{x}$ at $x = 1$ converges to $\dfrac{1}{x}$ whenever $0 < x < 2$.

Taylor Polynomials

Any power series representation of a function is unique. Therefore any power series representation of a function $f(x)$ at $x = a$ must be the function's Taylor series at $x = a$.

Let $f(x) = c_0 + c_1(x-a) + c_2(x-a)^2 + c_3(x-a)^3 + \cdots$ be a power series representation of function $f(x)$.

Then
$$f'(x) = c_1 + 2c_2(x-a) + 3c_3(x-a)^2 + 4c_4(x-a)^3 + \cdots$$
$$f''(x) = 2!c_2 + (3)(2)c_3(x-a) + (4)(3)(x-a)^2 + \cdots$$
$$f'''(x) = 3!c_3 + (4)(3)(2)c_4(x-a) + (5)(4)(3)c_5(x-a)^2 + \cdots$$
$$f^{(4)}(x) = 4!c_4 + (5)(4)(3)(2)c_5(x-a) + \cdots \quad \text{etc,.}$$

Analysis and Approximation

Hence $f(a) = c_0$, $f'(a) = c_1$, $f''(a) = 2!c_2$, $f'''(a) = 3!c_3$, $\cdots$, $f^{(n)}(a) = n!c_n$

and so $f(x) = \sum_{n=0}^{\infty} \dfrac{f^{(n)}(a)}{n!}(x-a)^n$ which is the Taylor series for the function $f(x)$ at $x = a$.

Definition If $f(x)$ has derivatives of all orders on an open interval $]a-r, a+r[$, then for all $n \geq 0$, the Taylor polynomial of order n generated by f at $x = a$ is the polynomial

$$P_n(x) = f(x) + f'(a)(x-a) + \dfrac{f''(a)}{2!}(x-a)^2 + \cdots + \dfrac{f^{(n)}(a)}{n!}(x-a)^n.$$

Example Find the Taylor polynomials of orders $2n+1$ and $2n+2$ generated by the function $f(x) = \sin x$ at $x = 0$.

$f'(x) = \cos x$, $f''(x) = -\sin x$, $f'''(x) = -\cos x$, $f^{(4)}(x) = \sin x$, etc,.

Thus $f(0) = 0$, $f'(0) = 1$, $f''(0) = 0$, $f'''(0) = -1$, $f^{(4)}(0) = 0$, $f^{(5)}(0) = 1$, etc,.

Therefore the Taylor polynomial of order $2n+1$ generated by $f(x) = \sin x$ at $x = 0$ is $P_{2n+1}(x) = 0 + x + 0x^2 - \dfrac{x^3}{3!} + 0x^4 + \dfrac{x^5}{5!} + \cdots + (-1)^{n+1} \dfrac{x^{2n+1}}{(2n+1)!}$, and

the Taylor polynomial of order $2n+2$ generated by $f(x) = \sin x$ at $x = 0$

is $P_{2n+2}(x) = 0 + x + 0x^2 - \dfrac{x^3}{3!} + 0x^4 + \dfrac{x^5}{5!} + \cdots + (-1)^{n+1} \dfrac{x^{2n+1}}{(2n+1)!} + 0x^{2n+2}$.

[*Note:* These polynomials are identical.]

Example Find the Taylor polynomials of orders 1, 2, 3 generated by $f(x) = e^x$ at $x = 0$. Sketch graphs of $f(x)$ and the three Taylor polynomials on the same set of coordinate axes for $-1.2 \leq x \leq 1.2$.

$f^{(n)}(x) = e^x$ and $f^{(n)}(0) = 1$ for $n = 0, 1, 2, 3$ ($f^{(0)}(x)$ denotes the function).

Therefore the Taylor polynomials of orders 1, 2, 3 are:

$$P_1(x) = 1 + x, \quad P_2(x) = 1 + x + \tfrac{1}{2}x^2 \text{ and } P_3(x) = 1 + x + \tfrac{1}{2}x^2 + \tfrac{1}{6}x^3.$$

The graphs of these functions follow at the top of the next page.

Chapter 26

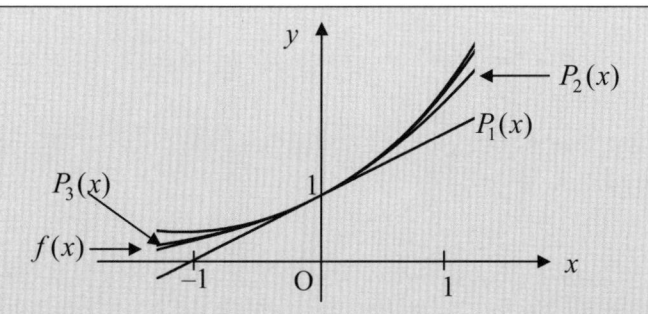

Taylor's Formula

If we replace x with $x + a$ in the Taylor series $f(x) = \sum_{k=0}^{\infty} \frac{f^{(k)}(a)}{k!}(x-a)^k$, we

obtain $f(x+a) = \sum_{k=0}^{\infty} \frac{f^{(k)}(a)}{k!} x^k = f(a) + xf'(a) + \frac{x^2}{2!}f''(a) + \frac{x^3}{3!}f'''(a) + \cdots$.

If $f(x)$ has derivatives of all orders defined on some open interval containing a, then for each positive integer n and for each x in the interval,

$$f(x+a) = f(a) + xf'(a) + \frac{x^2}{2!}f''(a) + \cdots + \frac{x^n}{n!}f^{(n)}(a) + R_n(x)$$

where $R_n(x) = \frac{x^{n+1}}{(n+1)!} f^{(n+1)}(c)$ for some c between a and $a + x$ (excluding the end-points).

The function $R_n(x)$ is called the *remainder of order n* or simply the *error term*.

$R_n(x)$ gives the error involved when we use the Taylor polynomial of order n as an approximation for the function $f(x+a)$ in a given open interval. If $R_n(x) \to 0$ as $n \to \infty$ for all x in the given interval, we say that the Taylor series *converges* to $f(x+a)$ on the interval.

Example Show that the Maclaurin series for $f(x) = e^x$ represents the function for all values of x.

From the previous example we find that the required Maclaurin series is

$$f(x) = e^x = \sum_{n=0}^{\infty} \frac{x^n}{n!}.$$

688

Analysis and Approximation

If we truncate this series after the term in x^n the error term is

$$R_n(x) = \frac{f^{(n+1)}(c)}{(n+1)!} x^{n+1} = \frac{e^c}{(n+1)!} x^{n+1}$$

where c is between 0 and x.

If $x > 0$, $0 < c < x$ and so $e^c < e^x$.

Therefore multiplying both sides by the positive number $\dfrac{x^{n+1}}{(n+1)!}$ gives

$$0 < \frac{e^c}{(n+1)!} x^{n+1} < \frac{e^x}{(n+1)!} x^{n+1} \text{ or } 0 < R_n(x) < \frac{e^x}{(n+1)!} x^{n+1} \quad \ldots\ldots\ldots (*).$$

From the second example on page 675 we found that the series $\displaystyle\sum_{n=0}^{\infty} \frac{x^n}{n!}$ is convergent for all values of x and so $\displaystyle\lim_{n \to \infty} \frac{x^{n+1}}{(n+1)!} = 0$ and

$\displaystyle\lim_{n \to \infty} e^x \frac{x^{n+1}}{(n+1)!} = 0$. Therefore $\displaystyle\lim_{n \to \infty} R_n(x) = 0$ from (*).

If $x < 0$, $x < c < 0$ and so $0 < e^x < e^c < e^0$ or $0 < e^c < 1$ .... (**)

(i) If $x^{n+1} > 0$ and we multiply (**) throughout by $\dfrac{x^{n+1}}{(n+1)!}$ (>0) we

obtain $0 < e^c \dfrac{x^{n+1}}{(n+1)!} < \dfrac{x^{n+1}}{(n+1)!}$ or $0 < R_n(x) < \dfrac{x^{n+1}}{(n+1)!}$ and we find $\displaystyle\lim_{n \to \infty} R_n(x) = 0$.

(ii) If $x^{n+1} < 0$ and we multiply (**) throughout by $\dfrac{x^{n+1}}{(n+1)!}$ (<0), we

obtain $\dfrac{x^{n+1}}{(n+1)!} < e^c \dfrac{x^{n+1}}{(n+1)!} < 0$ and once again we find $\displaystyle\lim_{n \to \infty} R_n(x) = 0$.

If $x = 0$, $\displaystyle\sum_{n=0}^{\infty} \frac{x^n}{n!} = 1 + x + \frac{x^2}{2!} + \frac{x^3}{3!} + \cdots = 1 = e^0$.

Therefore $\displaystyle\sum_{n=0}^{\infty} \frac{x^n}{n!} = e^x$ for all values of x.

689

Chapter 26

Example Show that the Taylor series for $f(x) = \cos x$ at $x = a$ represents the function for all values of x.

The required Taylor series is

$$\cos(x+a) = \cos a - (\sin a)x - (\cos a)\frac{x^2}{2!} + (\sin a)\frac{x^3}{3!} + \cdots .$$

We must show that $\lim\limits_{n \to \infty} R_n(x) = \lim\limits_{n \to \infty} \dfrac{f^{(n+1)}(c)}{(n+1)!} x^{n+1} = 0$.

Now $\left|f^{(n+1)}(c)\right| \leq 1$ since $f^{(n+1)}(c)$ is one of $\sin c$, $-\sin c$, $\cos c$ or $-\cos c$.

Thus $0 < \left|R_n(x)\right| \leq \dfrac{|x|^{n+1}}{(n+1)!}$.

Since $\lim\limits_{n \to \infty} \dfrac{x^{n+1}}{(n+1)!} = 0$ for all x, then $\lim\limits_{n \to \infty} R_n(x) = 0$.

Therefore the Taylor series represents the function for all values of x.

Example Find a Maclaurin series for e^{2x}.

The Maclaurin series for e^x is $\sum\limits_{n=0}^{\infty} \dfrac{x^n}{n!}$ and so the Maclaurin series for e^{2x}

is $\sum\limits_{n=0}^{\infty} \dfrac{(2x)^n}{n!}$.

Example Find a Maclaurin series for $x \cos x$.

The Maclaurin series for $\cos x$ is $\sum\limits_{n=0}^{\infty} (-1)^n \dfrac{x^{2n}}{(2n)!}$.

Thus the Maclaurin series for $x \cos x$ is $x \sum\limits_{n=0}^{\infty} (-1)^n \dfrac{x^{2n}}{(2n)!} = \sum\limits_{n=0}^{\infty} (-1)^n \dfrac{x^{2n+1}}{(2n)!}$.

The Remainder Estimation Theorem

If we can find positive numbers M and r such that $\left|f^{(n+1)}(c)\right| \leq Mr^{n+1} \dfrac{|x|^{n+1}}{(n+1)!}$ for all c between a and $a + x$, then the remainder term $R_n(x)$ in Taylor's theorem satisfies

Analysis and Approximation

$$|R_n(x)| \leq Mr^{n+1}\frac{|x|^{n+1}}{(n+1)!}$$

and so the series converges to $f(a+x)$.

Example Calculate cos 1 with an error less than 10^{-6}.

The Maclaurin series for $\cos x$ is $\sum_{n=0}^{\infty}(-1)^n \frac{x^{2n}}{(2n)!}$.

Therefore $\cos 1 = 1 - \frac{1}{2!} + \frac{1}{4!} - \frac{1}{6!} + \cdots + \frac{(-1)^n}{(2n)!} + R_n(1)$

and $|R_n(1)| = \left| f^{(n+1)}(c)\frac{1}{(2n)!} \right| \leq \frac{1}{(2n)!}$ for some c between 0 and 1.

By trial-and-error we find that $\frac{1}{10!} < 10^{-6} < \frac{1}{8!}$.

Therefore the value of cos 1 with an error less than 10^{-6} is

$$1 - \frac{1}{2!} + \frac{1}{4!} - \frac{1}{6!} + \frac{1}{8!} - \frac{1}{10!} \approx 0.540\,302.$$

[A calculator gives $\cos 1 = 0.540\,302\,3...$ which confirms our answer.]

Example Calculate the value of $\sqrt{e}$ correct to 4 decimal places using the Maclaurin series for $f(x) = e^x$.

The Maclaurin series for $f(x) = e^x$ is $\sum_{n=0}^{\infty} \frac{x^n}{n!}$.

We require the error term to have a numerical value which is less than 5×10^{-5} to ensure a value correct to 4 decimal places.

Now $R_n(x) = \frac{e^c}{(n+1)!}x^{n+1}$ where c is between 0 and x.

$$\left|R_n(\tfrac{1}{2})\right| < \frac{e^{1/2}}{2^{n+1}(n+1)!} < \frac{2}{2^{n+1}(n+1)!} = \frac{1}{2^n(n+1)!} \text{ since } e^{1/2} < 2.$$

Here $\frac{1}{2^4 5!} \approx 5.2 \times 10^{-4}$ and $\frac{1}{2^5 6!} \approx 2.2 \times 10^{-5}$ and so $P_5(\tfrac{1}{2})$ will give us a value for $\sqrt{e}$ which is correct to four decimal places.

Thus $\sqrt{e} \approx 1 + \tfrac{1}{2} + \tfrac{1}{8} + \tfrac{1}{48} + \tfrac{1}{384} + \tfrac{1}{3840} \approx 1.648\,7$.

691

Chapter 26

Example Find the values of x for which the polynomial $P_2(x) = 1 - \frac{1}{2}x^2$ gives a value for $\cos x$ with an error which is less than 10^{-3}.

The Maclaurin series for $\cos x$ is $\sum_{n=0}^{\infty} (-1)^n \frac{x^{2n}}{(2n)!}$ which is an alternating series for all non-zero x. Therefore we can apply the alternating series estimation theorem to show that the error is no greater than the numerical value of the third term, that is $|\varepsilon_3| < \frac{x^4}{4!} = \frac{1}{24} x^4$.

Thus $x^4 < 24 \times 10^{-3} = 0.024$ and so $|x| < 0.393$.

[**Note:** Since the third term of the series is $\frac{1}{24} x^4$ which is positive ($x \neq 0$), the alternating series estimation theorem provides us with information not obtainable from the remainder estimation theorem, i.e. the polynomial $P_2(x)$ gives an **underestimate** of the value of $\cos x$.]

Also, if we apply the remainder estimation theorem to the above problem we have $\cos x = 1 - \frac{1}{2}x^2 + R_2(x)$ where $R_2(x) \leq \frac{1}{6}|x|^3$ which is nowhere near as good as the error found using the alternating series estimation theorem. However, the Taylor polynomial of order 3 is the same as that of order 2 since

$$\cos x = 1 - \frac{1}{2}x^2 + 0x^3 + R_3(x)$$

and the remainder estimation theorem then gives $|R_3(x)| \leq \frac{1}{24} x^4$ which is the same as that found using the alternating series estimation theorem.

The Maclaurin series $\ln(1+x) = \sum_{n=0}^{\infty} (-1)^{n+1} \frac{x^n}{n}$ which converges for $-1 < x \leq 1$, should be memorised.

The following Maclaurin series which represent the given functions for all values of x, should also be memorised:

$$e^x = \sum_{n=0}^{\infty} \frac{x^n}{n!}$$

$$\sin x = \sum_{n=0}^{\infty} (-1)^n \frac{x^{2n+1}}{(2n+1)!}$$

$$\cos x = \sum_{n=0}^{\infty} (-1)^n \frac{x^{2n}}{(2n)!}.$$

Exercise 26.7

1. Find the Maclaurin series up to the term in x^3 for each of the following:
   (a) $(1+x)^{1/2}$;   (b) $(1-x)^{-2}$;   (c) $(1+x)^{3/2}$.

2. Find the Taylor series for
   (a) $2x^3 + x^2 - 2x + 1$ in powers of $x - 1$;
   (b) x^4 in powers of $x + 2$.
   Check your results algebraically.

3. Calculate the value of e^{-1} correct to 4 decimal places using the Maclaurin series for the function $f(x) = e^x$.

4. Calculate the value of $\ln 1.2$ correct to 6 decimal places using the Maclaurin series for the function $f(x) = \ln(1+x)$.

5. Find the Taylor series for each function at $x = a$.
   (a) $f(x) = e^x,\ a = 1$;
   (b) $f(x) = \cos x,\ a = \frac{1}{2}\pi$;
   (c) $f(x) = \ln(1+x),\ a = 2$;
   (d) $f(x) = \frac{1}{x},\ a = 1$.

6. Using the Maclaurin series for $\sin x$, $\cos x$, e^x and $\ln(1+x)$, find the Maclaurin series for:
   (a) $x^2 \sin x$;   (b) $\cos 2x$;   (c) $\cos^2 x$;
   (d) $\sin^2 x$;   (e) e^{-x} ;   (f) $\ln(1+x^2)$.

7. (a) Find the Taylor polynomial of order 3 at $x = 9$ for the function $f(x) = \sqrt{x}$.
   (b) Estimate the size of $|R_3(x)|$ for $9 < x \leq 10$.
   (c) Calculate the value of $\sqrt{10}$ correct to as many decimal places as are justified by using the polynomial found in part (a).

8. (a) Find the Taylor series for $\sin x$ at $x = \frac{1}{4}\pi$.
   (b) Show that the series found in part (a) represents the function $\sin x$ for all values of x.
   (c) Calculate the value of $\sin 46°$ with an error less than 10^{-6}.

Chapter 26

26.8 Differentiation and Integration of Power Series

Theorem Let $\sum_{n=0}^{\infty} c_n x^n$ be a power series whose radius of convergence is $r > 0$. If $f(x) = \sum_{n=0}^{\infty} c_n x^n$, then $f'(x)$ exists for all x in the open interval $]-r, r[$ and $f'(x) = \sum_{n=1}^{\infty} n c_n x^{n-1}$.

[The proof is beyond the scope of this book.]

Example Find the radius and interval of convergence of each of the power series $\sum_{n=0}^{\infty} \frac{x^{n+1}}{(n+1)^2}$ and $\sum_{n=0}^{\infty} \frac{x^n}{n+1}$.

Let $f(x) = \sum_{n=0}^{\infty} u_n = \sum_{n=0}^{\infty} \frac{x^{n+1}}{(n+1)^2}$.

Then $\lim_{n \to \infty} \left| \frac{u_{n+1}}{u_n} \right| = \lim_{n \to \infty} \left| \frac{(n+1)^2 x^{n+2}}{(n+2)^2 x^{n+1}} \right| = \lim_{n \to \infty} |x| \left(\frac{1 + \frac{1}{n}}{1 + \frac{2}{n}} \right)^2 = |x|$ and so the radius of convergence is $r_1 = 1$.

When $x = 1$, the series is $\sum_{n=0}^{\infty} \frac{1}{(n+1)^2}$ which is convergent, and when $x = -1$, the series is $\sum_{n=0}^{\infty} \frac{(-1)^{n+1}}{(n+1)^2}$ which is absolutely convergent.

Therefore the interval of convergence is $[-1, 1]$.

Let $g(x) = \sum_{n=0}^{\infty} v_n = \sum_{n=0}^{\infty} \frac{x^n}{n+1}$.

Then $\lim_{n \to \infty} \left| \frac{v_{n+1}}{v_n} \right| = \lim_{n \to \infty} \left| \frac{(n+1) x^{n+1}}{(n+2) x^n} \right| = \lim_{n \to \infty} |x| \left(\frac{1 + \frac{1}{n}}{1 + \frac{2}{n}} \right) = |x|$ and so the radius of convergence is $r_2 = 1$.

Analysis and Approximation

When $x = 1$, the series is $\sum_{n=0}^{\infty} \frac{1}{n+1}$ which is divergent, and when $x = -1$, the series is $\sum_{n=0}^{\infty} \frac{(-1)^n}{n+1}$ which is convergent.

Therefore the interval of convergence is $[-1, 1[$.

Since $g(x)$ is obtained by differentiating $f(x)$ term-by-term, this example illustrates the fact that even though f and f' have the same radius of convergence, they do not necessarily have the same interval of convergence.

Example Find a power series representation of the function $f(x) = \frac{1}{(1-x)^2}$.

The series $\sum_{n=0}^{\infty} x^n$ is geometric with $u_1 = 1$ and $r = x$. It therefore converges to $\frac{1}{1-x}$ for $|x| < 1$.

Thus $\frac{1}{1-x} = 1 + x + x^2 + x^3 + \cdots + x^n + \cdots$ for $|x| < 1$.

Differentiating term-by-term we get

$$\frac{1}{(1-x)^2} = 1 + 2x + 3x^2 + 4x^3 + \cdots + nx^{n-1} + \cdots \quad \text{for } |x| < 1$$

which is a power series representation of the required function.

Theorem Let $\sum_{n=0}^{\infty} c_n x^n$ be a power series whose radius of convergence is $r > 0$.

If $f(x) = \sum_{n=0}^{\infty} c_n x^n$, then f is integrable on every closed subinterval of $]-r, r[$, and the integral of f is evaluated by integrating the given power series term-by-term. Hence if $x \in]-r, r[$,

$$\int_0^x f(t)\, dt = \sum_{n=0}^{\infty} \frac{c_n}{n+1} x^{n+1}$$

and the radius of convergence of the resulting series is r.

[The proof is beyond the scope of this book.]

Chapter 26

Example Find a power series in x for $\int_0^x \frac{\sin t}{t}\,dt$.

We know that

$$\sin x = \sum_{n=0}^{\infty}(-1)^n \frac{x^{2n+1}}{(2n+1)!} = x - \frac{x^3}{3!} + \frac{x^5}{5!} - \cdots + (-1)^n \frac{x^{2n+1}}{(2n+1)!} + \cdots \text{ for all } x.$$

Therefore $\dfrac{\sin x}{x} = \sum_{n=0}^{\infty}(-1)^n \dfrac{x^{2n}}{(2n+1)!} = 1 - \dfrac{x^2}{3!} + \dfrac{x^4}{5!} - \cdots + (-1)^n \dfrac{x^{2n}}{(2n+1)!} + \cdots$

for all $x \neq 0$.

Thus $\int_0^x \dfrac{\sin t}{t}\,dt = x - \dfrac{x^3}{3(3!)} + \dfrac{x^5}{5(5!)} - \cdots + (-1)^n \dfrac{x^{2n+1}}{(2n+1)(2n+1)!} + \cdots$.

Example Find a power series representation for $\arctan x$.

From a result found previously, that is

$$1 + x + x^2 + x^3 + \cdots + x^n + \cdots = \frac{1}{1-x} \text{ for } |x| < 1,$$

we obtain

$$\frac{1}{1+x^2} = 1 + (-x^2) + (-x^2)^2 + \cdots + (-x^2)^n + \cdots$$

$$= 1 - x^2 + x^4 - x^6 + \cdots + (-1)^n x^{2n} + \cdots, \text{ for } |x| < 1.$$

Now $\int_0^x \dfrac{dt}{1+t^2} = \arctan x = x - \dfrac{1}{3}x^3 + \dfrac{1}{5}x^5 - \cdots + \dfrac{(-1)^n}{2n+1}x^{2n+1} + \cdots, \ |x| < 1.$

Example Evaluate $\int_0^{0.5} e^{-t^2}\,dt$ correct to five decimal places.

$$e^x = \sum_{n=0}^{\infty} \frac{x^n}{n!} \text{ for all } x \text{ and so } e^{-x^2} = \sum_{n=0}^{\infty} \frac{(-x^2)^n}{n!} = \sum_{n=0}^{\infty}(-1)^n \frac{x^{2n}}{n!} \text{ for all } x.$$

Now $\int_0^x e^{-t^2}\,dt = \sum_{n=0}^{\infty}(-1)^n \dfrac{x^{2n+1}}{n!(2n+1)}$ and $\int_0^{0.5} e^{-t^2}\,dt = \sum_{k=0}^{\infty}(-1)^k \dfrac{0.5^{2k+1}}{k!(2k+1)}$.

This is a convergent alternating series and so the error, ε_n, created by truncating the series after the nth term is such that

$$|\varepsilon_n| < \left|\frac{0.5^{2n+3}}{(n+1)!(2n+3)}\right| = \frac{1}{2^{2n+3}(n+1)!(2n+3)}.$$

Analysis and Approximation

Now $|\varepsilon_3| \approx 9.04 \times 10^{-6}$ and $|\varepsilon_4| \approx 3.7 \times 10^{-7}$ and so to obtain a value of the integral correct to five decimal places we need to add the first 5 terms of the series.

Therefore $\int_0^{0.5} e^{-t^2} dt \approx 0.5 - \dfrac{0.5^3}{3} + \dfrac{0.5^5}{2!(5)} - \dfrac{0.5^7}{3!(7)} + \dfrac{0.5^9}{4!(9)} = 0.46128$.

Exercise 26.8

1. Using term-by-term differentiation of the power series for $\sin x$, show that we obtain the power series for $\cos x$.

2. Using term-by-term differentiation of the power series for $\cos x$, show that we obtain the power series for $-\sin x$.

3. Using term-by-term differentiation of the power series for $f(x) = e^x$, show that $f'(x) = e^x$.

4. Use the result $\dfrac{1}{(1-x)^2} = \sum\limits_{n=1}^{\infty} nx^{n-1}$ for $|x| < 1$ to find a power series representation of $\dfrac{1}{(1-x)^3}$.

5. Use the result $\dfrac{1}{1-x} = \sum\limits_{n=0}^{\infty} x^n$ for $|x| < 1$ to find a power series representation of $\dfrac{1}{(1+x)^2}$.

6. (a) Find a power series representation of e^{x^2}.
   (b) Find a power series representation of xe^{x^2}.

7. Find a power series representation of $e^x \sin x$ up to the term in x^3.

8. By multiplication, obtain the first few terms of the Maclaurin series for $\sin x \cos x$. Show how the whole series may be obtained from the Maclaurin series for $\sin 2x$.

9. Obtain the Maclaurin polynomial of degree 4 for $\cos^2 x$ by
   (a) multiplication;
   (b) using the identity $\cos^2 x = \tfrac{1}{2}(1 + \cos 2x)$.

Chapter 26

10. If $f(x) = \sum_{n=0}^{\infty} (-1)^n \dfrac{x^n}{2^{2n}}$, find $f'(0.5)$ correct to three decimal places.

11. Find a power series in x for

 (a) $\displaystyle\int_0^x \sin t^2 \, dt$; (b) $\displaystyle\int_0^x \dfrac{1}{2-t} \, dt$, $0 < x < 2$.

12. In each of the following, use a power series to estimate the value of the integral correct to four decimal places:

 (a) $\displaystyle\int_0^{0.2} \arctan x \, dx$; (b) $\displaystyle\int_0^{0.2} x e^{-x^2} \, dx$;

 (c) $\displaystyle\int_0^{0.2} \ln(1+x) \, dx$; (d) $\displaystyle\int_2^3 \dfrac{dx}{4-x}$.

13. (a) Find a power series representation of $\dfrac{1}{1-x^2}$.

 (b) Find a power series representation of $\ln\left(\dfrac{1+x}{1-x}\right)$ using term-by-term integration of the series in part (a).

14. (a) Find a power series of xe^x.

 (b) Show that $\displaystyle\int_0^1 xe^x \, dx = 1$.

 (c) By integrating term-by-term the series found in part (a), show that
 $$\sum_{n=0}^{\infty} \dfrac{1}{n!(n+2)} = \dfrac{1}{2}.$$

15. Let $f'(x) = \sum_{n=0}^{\infty} (-1)^n \dfrac{(x-1)^n}{n!}$.

 (a) Find a power series representation of $\displaystyle\int_1^x f'(t) \, dt$.

 (b) Evaluate $f(\tfrac{5}{4})$ to an accuracy of four decimal places.

26.9 The Definite Integral as a Limit of a Sum

The following results will be useful in the following section:

(1) $\displaystyle\sum_{i=1}^{n} i = \dfrac{n}{2}(n+1)$

698

(2) $\sum_{i=1}^{n} i^2 = \frac{n}{6}(n+1)(2n+1)$

(3) $\sum_{i=1}^{n} i^3 = \left[\frac{n}{2}(n+1)\right]^2$

Each can be proved by mathematical induction, but this is left as an exercise for the reader.

Consider the function $y = f(x)$ which is continuous and non-negative on the interval $[a, b]$. Let this interval be divided into n sub-intervals of equal length h. Thus $h = \frac{b-a}{n}$. Denote the end-points of these intervals by

$x_0 = a, \ x_1 = a + h, \ \cdots, \ x_i = a + ih, \ \cdots, \ x_{n-1} = a + (n-1)h, \ x_n = a + nh$.

The ith sub-interval is denoted by $[x_{i-1}, x_i]$. Since $f(x)$ is continuous on $[a, b]$ it is continuous on each closed sub-interval.

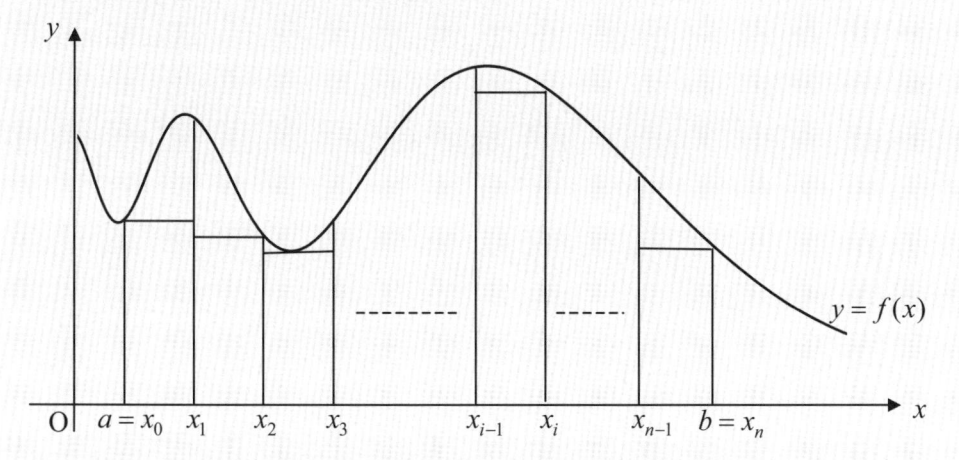

Let $c_i \in [x_{i-1}, x_i]$ be such that $f(c_i)$ is the minimum value of f on $[x_{i-1}, x_i]$. Consider n rectangles each of width h and altitude $f(c_i)$ (see diagram above).

Let the sum of the areas of these rectangles be L_n, then $L_n = \sum_{i=1}^{n} hf(c_i)$.

Let $d_i \in [x_{i-1}, x_i]$ be such that $f(d_i)$ is the maximum value of f on $[x_{i-1}, x_i]$.

Consider n rectangles each of width h and altitude $f(d_i)$ (see diagram at the top of the next page).

Chapter 26

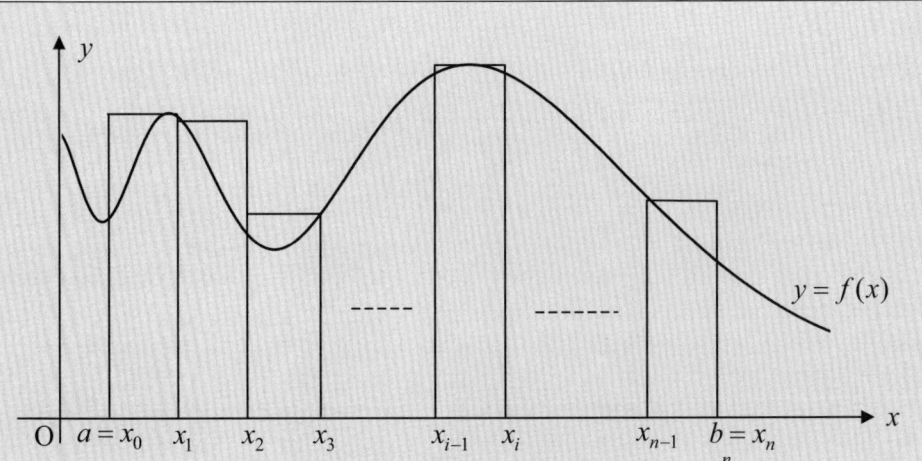

Let the sum of the areas of these rectangles be U_n, then $U_n = \sum_{i=1}^{n} hf(d_i)$.

Clearly the area A between the curve $y = f(x)$ and the x-axis from $x = a$ to $x = b$ is such that $L_n \leq A \leq U_n$.

As the number of sub-intervals increases, we assert (without proof) that both L_n and U_n approach A.

Thus $\quad A = \lim_{n \to \infty} \sum_{i=1}^{n} hf(c_i) = \lim_{n \to \infty} \sum_{i=1}^{n} hf(d_i) \quad$ or

$$\int_a^b f(x)\, dx = \sum_{i=1}^{\infty} hf(c_i) = \sum_{i=1}^{\infty} hf(d_i).$$

Example   Evaluate $\int_1^2 x^2\, dx$ by calculating the value of $\lim_{n \to \infty} U_n$.

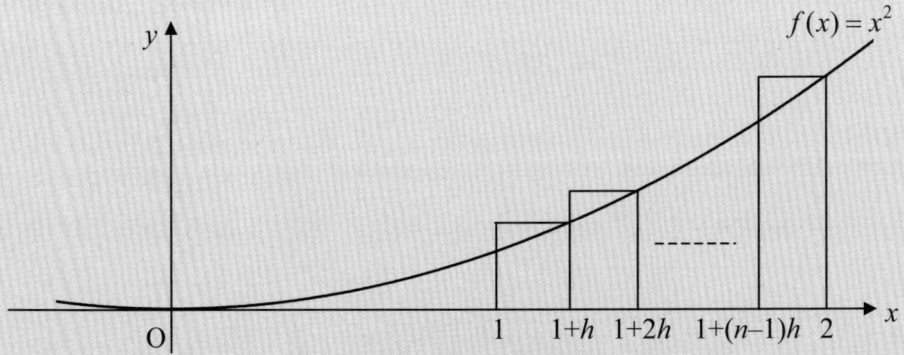

Here $f(x) = x^2$, $a = x_0 = 1$, $b = x_n = 2$, $d_i = x_i = 1 + ih = 1 + \dfrac{i}{n}$ and so

700

$$U_n = \sum_{i=1}^{n} \left(\frac{1}{n}\right)\left(1+\frac{i}{n}\right)^2$$

$$= \frac{1}{n}\left(1+\frac{1}{n}\right)^2 + \frac{1}{n}\left(1+\frac{2}{n}\right)^2 + \frac{1}{n}\left(1+\frac{3}{n}\right)^2 + \cdots + \frac{1}{n}\left(1+\frac{n-1}{n}\right)^2 + \frac{1}{n}(2)^2$$

$$= \frac{1}{n^3}\left((n+1)^2 + (n+2)^2 + (n+3)^2 + \cdots + (n+[n-1])^2 + (n+n)^2\right)$$

$$= \frac{1}{n^3}\left(\frac{2n}{6}(2n+1)(4n+1) - \frac{n}{6}(n+1)(2n+1)\right) \quad \text{[result (2) on page 699]}$$

$$= \frac{2n+1}{6n^2}(8n+2-n-1)$$

$$= \frac{(2n+1)(7n+1)}{6n^2}$$

$$= \frac{7}{3} + \frac{3}{2n} + \frac{1}{6n^2}.$$

Now $\lim_{n \to \infty} U_n = \frac{7}{3}$ and so $\int_1^2 x^2 \, dx = \frac{7}{3}$.

Exercise 26.9

1. Find the area of the region enclosed by the curve $f(x) = 4 - x$ and the coordinate axes by finding $\lim_{n \to \infty} L_n$.

2. Find the area enclosed by the curve $f(x) = 4 - x^2$ and the x-axis by finding $\lim_{n \to \infty} U_n$.

3. Find the area enclosed by the curve $f(x) = x^3$ and the x-axis from $x = -1$ to $x = 2$ by using $\lim_{n \to \infty} L_n$ twice, once between $x = -1$ and $x = 0$ and then between $x = 0$ and $x = 2$.

4. Find the area enclosed by the curve $f(x) = mx$ $(m > 0)$, the x-axis and the lines $x = a$, $x = b$ $(0 < a < b)$ using $\lim_{n \to \infty} U_n$.

26.10 Numerical Integration

In attempting to evaluate $\int_a^b f(x)\,dx$, we may find an anti-derivative of $f(x)$ very hard, or even impossible, to find. A numerical approximation would be quite useful in such situations. Since we can interpret the integral as an area, geometric techniques can be considered.

The Trapezium Rule

Consider the function $y = f(x)$ which is positive and continuous for all $x \in [a,b]$. We divide the interval $[a, b]$ into n strips all of width $h = \dfrac{b-a}{n}$. The integral $\int_a^b f(x)\,dx$ gives the area between the curve and the x-axis from $x = a$ to $x = b$. We estimate the area of each strip by joining the points on the curve at the ends of the strip with a straight line. Thus the area of each strip can be approximated by finding the area of a trapezium.

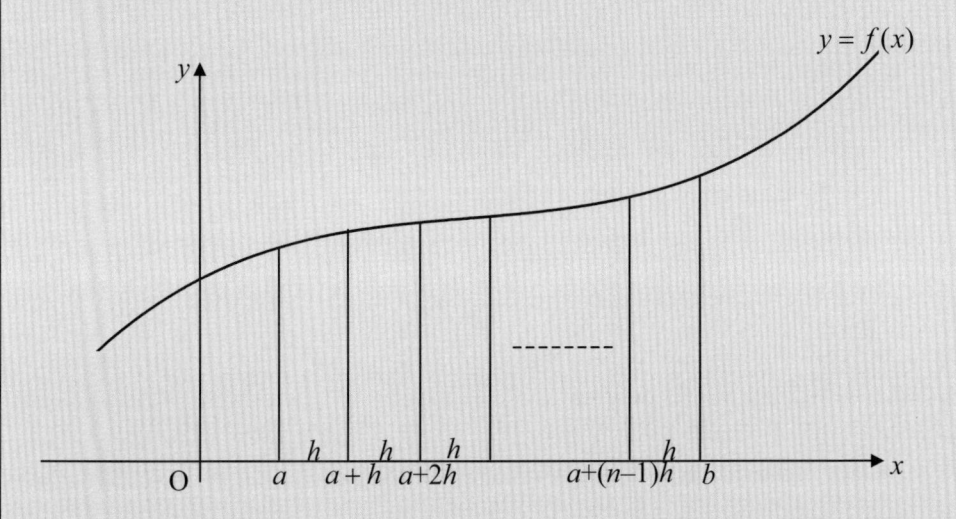

Let $y_r = f(a + rh)$ for $r = 1, 2, 3, \ldots, n-1$, and $y_0 = a$, $y_n = b$. Then the area of the rth strip is approximately
$$A_r = \tfrac{1}{2}(y_{r-1} + y_r)h.$$

Analysis and Approximation

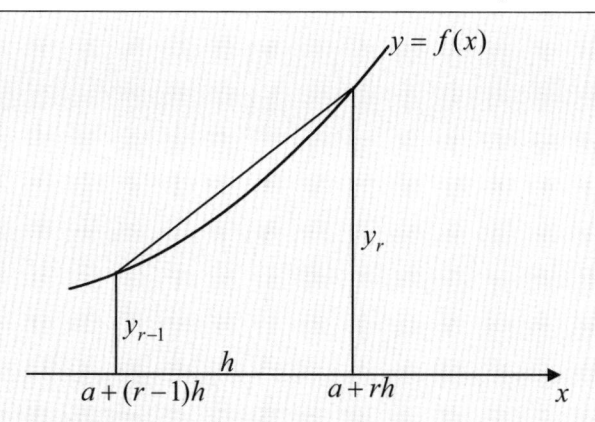

The total area of all the strips, A_T, is then given by

$$A_T = \tfrac{1}{2}h(y_0 + y_1) + \tfrac{1}{2}h(y_1 + y_2) + \tfrac{1}{2}h(y_2 + y_3) + \cdots + \tfrac{1}{2}h(y_{n-1} + y_n)$$
$$= \tfrac{1}{2}h(y_0 + 2y_1 + 2y_2 + \cdots + 2y_{n-1} + y_n).$$

Example Estimate the value of $\int_1^2 x^2 \, dx$ using the trapezium rule with four equal sub-intervals.

Here, $f(x) = x^2$, $a = 1$, $b = 2$ and $h = \tfrac{1}{4}$.

$$\int_1^2 x^2 \, dx \approx \tfrac{1}{2}h\left(f(1) + 2f(\tfrac{5}{4}) + 2f(\tfrac{3}{2}) + 2f(\tfrac{7}{4}) + f(2)\right)$$
$$= \tfrac{1}{8}\left(1 + \tfrac{25}{8} + \tfrac{9}{2} + \tfrac{49}{8} + 4\right)$$
$$= \tfrac{75}{32} \quad (\approx 2.34).$$

[The exact answer is $\tfrac{7}{3} \approx 2.33$.]

Theorem Let the function f be continuous on the closed interval $[a, b]$, and let f' and f'' both exist on $[a, b]$. If $\varepsilon_T = \int_a^b f(x) \, dx - T$ where T is the approximate value of $\int_a^b f(x) \, dx$ found using the trapezium rule, then there exists some number $c \in]a, b[$ such that

$$\varepsilon_T = -\frac{(b-a)h^2}{12} f''(c).$$

[The proof is beyond the scope of this book.]

Chapter 26

In the preceding example, $f(x) = x^2$ and so $f''(x) = 2$ for all x in $]1, 2[$. Thus the error term is $\varepsilon_T = -\dfrac{(2-1)(\frac{1}{4})^2}{12}(2) = -\dfrac{1}{96}$ and $\int_1^2 x^2 \, dx = \dfrac{75}{32} - \dfrac{1}{96} = \dfrac{7}{3}$, the exact answer.

The exact error can not always be calculated, but in most cases, a close approximation may be found.

Example Find an approximate value of $\int_1^3 \dfrac{1}{x} \, dx$ using the trapezium rule with five equal subintervals. Find an approximate value for the error and use a graph to explain why this error is negative, i.e., the estimated value of the integral is greater than the exact value of $\ln 3 \approx 1.10$.

Here $f(x) = \dfrac{1}{x}$, $a = 1$, $b = 3$ and $h = 0.4$.

$\int_1^3 \dfrac{1}{x} \, dx \approx \dfrac{1}{2}(0.4)\left(f(1) + 2[f(1.4) + f(1.8) + f(2.2) + f(2.6)] + f(3)\right) \approx 1.11$.

$f(x) = \dfrac{1}{x}$, $f'(x) = -\dfrac{1}{x^2}$, $f''(x) = \dfrac{2}{x^3}$ and $\varepsilon_T = -\dfrac{2(0.4)^2}{12}f''(c) = -\dfrac{0.16}{12}\left(\dfrac{2}{c^3}\right)$.

An approximate value of the error is $-\dfrac{2 \times 0.4^2}{12}\left(\dfrac{2}{1^3}\right) = -0.053$.

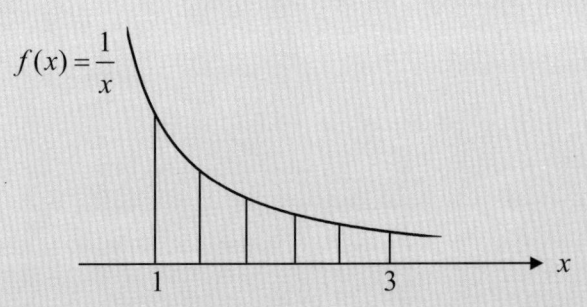

$f(x) = \dfrac{1}{x}$

Each straight line joining the points on the graph at the beginning and end of each interval lies ***above*** the graph and the sum of the areas of the 5 trapeziums is therefore greater than the area between the x-axis and the curve.

Analysis and Approximation

Simpson's Rule

In this method two strips are taken at a time. Therefore the interval $[a, b]$ must always be divided into and even number of sub-intervals. When applying Simpson's rule, instead of joining pairs of points on the curve with straight lines, we select three points, find a parabola which passes through them and evaluate the area under the parabola.

Consider the positive function $y = f(x)$ from $x = -h$ to $x = h$ with two strips of width h one from $x = -h$ to $x = 0$ and the other from $x = 0$ to $x = h$.
Let $y_0 = f(-h)$, $y_1 = f(0)$ and $y_2 = f(h)$.

Let the parabola passing through $(-h, y_0)$, $(0, y_1)$ and (h, y_2) have the equation $y = ax^2 + bx + c$.

The area under this parabola is

$$\int_{-h}^{h} (ax^2 + bx + c)\, dx = \left[\frac{ax^3}{3} + \frac{bx^2}{2} + cx \right]_{-h}^{h} = \frac{2ah^3}{3} + 2ch.$$

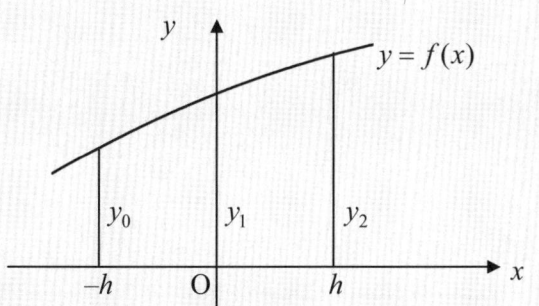

Point $(-h, y_0)$ lies on the parabola so $y_0 = ah^2 - bh + c$ ....(1).
Point $(0, y_1)$ lies on the parabola so $y_1 = c$ .... (2).
Point (h, y_1) lies on the parabola so $y_2 = ah^2 + bh + c$ .... (3).
Equation (1) – 2 × Equation (2) + Equation (3) gives: $y_0 - 2y_1 + y_2 = 2ah^2$.
Substituting this into the formula for the area given above gives:

$$\text{area under the parabola} = \frac{h}{3}(y_0 - 2y_1 + y_2) + 2hy_1 = \frac{h}{3}(y_0 + 4y_1 + y_2).$$

Adding the areas of all pairs of strips gives a total area A_S where

$$A_S = \tfrac{1}{3}h(y_0 + 4y_1 + y_2) + \tfrac{1}{3}h(y_2 + 4y_3 + y_4) + \cdots + \tfrac{1}{3}h(y_{n-2} + 4y_{n-1} + y_n)$$
$$= \tfrac{1}{3}h(y_0 + 4y_1 + 2y_2 + 4y_3 + 2y_4 + \cdots + 2y_{n-2} + 4y_{n-1} + y_n).$$

705

Chapter 26

Example Using Simpson's rule with 4 sub-intervals, estimate the value of the definite integral $\int_0^1 \frac{4\,dx}{1+x^2}$ correct to 3 decimal places.

Here $f(x) = \frac{4}{1+x^2}$, $a = 0$, $b = 1$ and $h = 0.25$.

Therefore $\int_0^1 \frac{4\,dx}{1+x^2}$ = $\frac{1}{12}(f(0) + 4f(0.25) + 2f(0.5) + 4f(0.75) + f(1))$

$= \frac{1}{12}(4 + 4(3.7647) + 2(3.2) + 4(2.56) + 2)$

$= 3.142$.

[The exact value is $\pi = 3.14159265\ldots$ and the estimated value is $3.1415686\ldots$ and so our estimate is actually correct to 4 decimal places.]

Theorem Let the function f be continuous on the closed interval $[a, b]$, and let f', f'', f''' and $f^{(4)}$ all exist on $[a, b]$. If $\varepsilon_S = \int_a^b f(x)\,dx - S$ where S is the approximate value of $\int_a^b f(x)\,dx$ found using Simpson's rule, then there exists some number $c \in]a, b[$ such that

$$\varepsilon_S = -\frac{(b-a)h^4}{180} f^{(4)}(c).$$

[The proof is beyond the scope of this book.]

Example Use Simpson's rule with six sub-intervals to find an approximate value for $\int_1^4 \frac{dx}{x}$. Give the answer correct to three decimal places.

Find also an upper bound for the absolute value of the error in the estimate. How many sub-intervals are needed to ensure that the estimate is accurate to five decimal places?

Here $a = 1$, $b = 4$, $f(x) = \frac{1}{x}$ and $h = 0.5$.

$\int_1^4 \frac{dx}{x} \approx \frac{1}{6}(f(1) + 4f(1.5) + 2f(2) + 4f(2.5) + 2f(3) + 4f(3.5) + f(4))$

$= \frac{1}{6}\left(1 + \frac{8}{3} + 1 + \frac{8}{5} + \frac{2}{3} + \frac{8}{7} + \frac{1}{4}\right)$

$= 1.386$.

706

Analysis and Approximation

$$f(x) = \frac{1}{x}, \quad f'(x) = -\frac{1}{x^2}, \quad f''(x) = \frac{2}{x^3}, \quad f'''(x) = -\frac{6}{x^4}, \quad f^{(4)}(x) = \frac{24}{x^5}$$

For $1 \le c \le 4$, $\dfrac{3}{128} \le f^{(4)}(c) \le 24$ and so $|\varepsilon_S| \le \dfrac{3(\frac{1}{2})^4}{180}(24) = 0.025$.

Thus an upper bound for the absolute value of the error is 0.025.

We require the smallest *even* value of n for which $|\varepsilon_S| < 5 \times 10^{-6}$.

Then $\left| \dfrac{3(\frac{3}{n})^4}{180}(24) \right| < 5 \times 10^{-6}$

$\Rightarrow \quad n^4 > \dfrac{24 \times 3^5}{5 \times 180} \times 10^6$

$\Rightarrow \quad n > 50.4$.

The least n is therefore $n = 52$.

Note: Although we have found the least number of sub-intervals needed according to the error estimation theorem, this may not necessarily be the actual number required since the optimum value of $f^{(4)}(c)$ may indeed be less than 24. However, $n = 52$ will **guarantee** that the estimate is accurate to five decimal places.

Exercise 26.10

1. (i)   In each of the following, use the trapezium rule to estimate the given integral using the number of sub-intervals indicated. Give each answer correct to three decimal places.

   (a) $\int_0^1 x^2 \, dx$, $n = 4$;   (b) $\int_0^\pi \sin x \, dx$, $n = 6$;

   (c) $\int_0^1 \dfrac{dx}{\sqrt{1+3x^2}}$, $n = 5$;   (d) $\int_0^1 e^{-x^2} \, dx$, $n = 4$.

   (ii) In each of the following, use Simpson's rule to estimate the given integral using the number of sub-intervals indicated. Give each answer correct to three decimal places.

   (a) $\int_0^\pi x \sin x \, dx$, $n = 4$;   (b) $\int_0^1 \dfrac{dx}{x^2 + x + 2}$, $n = 6$;

   (c) $\int_0^1 \sqrt{1-x^2} \, dx$, $n = 6$;   (d) $\int_0^{\pi/2} \sqrt{\cos x} \, dx$, $n = 8$.

Chapter 26

2. (i) Use the trapezium rule to estimate each integral using the data provided.

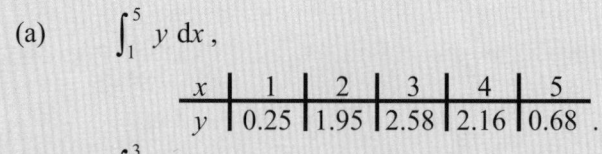

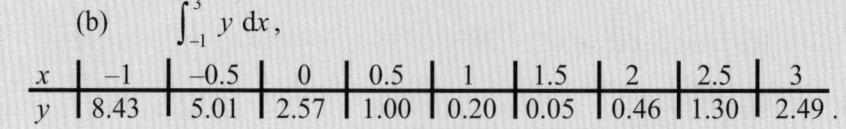

(a) $\int_1^5 y\, dx$,

x	1	2	3	4	5
y	0.25	1.95	2.58	2.16	0.68

(b) $\int_{-1}^3 y\, dx$,

x	−1	−0.5	0	0.5	1	1.5	2	2.5	3
y	8.43	5.01	2.57	1.00	0.20	0.05	0.46	1.30	2.49

(ii) (a) Using the same data as in part (i)(a), use Simpson's rule to estimate the given integral.

(b) Using the same data as in part (i)(b), use Simpson's rule to estimate the given integral.

3. (i) In each of the following, use the trapezium rule to estimate the given integral using the number of sub-intervals indicated. Calculate the maximum error in your calculation.

(a) $\int_0^1 x^3\, dx$, $n=4$; (b) $\int_1^3 \dfrac{dx}{x+1}$, $n=5$;

(c) $\int_0^{\pi/2} \cos x\, dx$, $n=6$; (d) $\int_1^2 e^x\, dx$, $n=8$.

(ii) In each of the following, use Simpson's rule to estimate the given integral using the number of sub-intervals indicated. Calculate the maximum error in your calculation.

(a) $\int_0^1 x^3\, dx$, $n=4$; (b) $\int_1^2 \dfrac{dx}{x+2}$, $n=6$;

(c) $\int_0^{\pi} \sin x\, dx$, $n=6$; (d) $\int_0^1 e^{-x}\, dx$, $n=8$.

4. Using the error estimate for the trapezium rule, determine the number of sub-intervals needed to ensure that the calculated value of each integral is correct to five decimal places.

(a) $\int_0^1 x^4\, dx$; (b) $\int_0^{\pi/2} \sin x\, dx$;

(c) $\int_1^2 \dfrac{dx}{x+3}$; (d) $\int_{-1}^1 e^x\, dx$.

Analysis and Approximation

5. Using the error estimate for Simpson's rule, determine the number of sub-intervals needed to ensure that the calculated value of each integral from Question 4 is correct to five decimal places.

6. Use the error estimate formula to explain why Simpson's rule gives the *exact* value for the integral $\int_a^b f(x)\,dx$ if $f(x)$ is a polynomial of degree less than four, even if only two sub-intervals are used.

7. The velocity of a car travelling in a straight line is measured at one-second intervals. Estimate the total distance travelled in the first eight seconds.

t (s)	0	1	2	3	4	5	6	7	8
v (m s^{-1})	5	6.4	8.2	10.6	13.6	17.5	22.4	28.8	36.9

8. The standardised normal probability density function is defined by $\Phi(z) = \dfrac{1}{\sqrt{2\pi}} e^{-z^2/2}$. Use Simpson's rule with four sub-intervals to estimate the probability that a normal random variable X with mean 5 and standard deviation 0.5 takes a value between 4 and 5.5.

9. A small river is 12 m wide at a given point. The depths at intervals of 2 m are measured and the results are as follows:

Distance from one Bank (m)	0	2	4	6	8	10	12
Depth (m)	0	1.8	4.3	4.5	2.8	1.4	0

If the river is flowing at the given point at 1 km hr^{-1}, estimate the volume of water passing this point in one minute.

26.11 Numerical Solutions of Equations

The aim of this section is to provide methods for the solution of equations of the form $f(x) = 0$. In all that follows, a sketch graph of $y = f(x)$ should always be produced. The general aim is to find values of x, say $x = x_1$ and $x = x_2$, such that the root we seek is between x_1 and x_2 and $f(x_1)$ and $f(x_2)$ are of opposite sign. Difficulties may arise if $f(x)$ is not continuous or if there are two solutions very close together. A sketch of the graph will always help to minimise such difficulties.

Chapter 26

Linear Interpolation

If we know that there is one and only one root $x = k$ of the equation $f(x) = 0$ between $x = x_1$ and $x = x_2$, we can obtain an approximate value for k by joining the points $P(x_1, y_1)$ and $Q(x_2, y_2)$ with a straight line. If this line meets the x-axis at $x = x_3$, then x_3 is the approximation to k we seek.

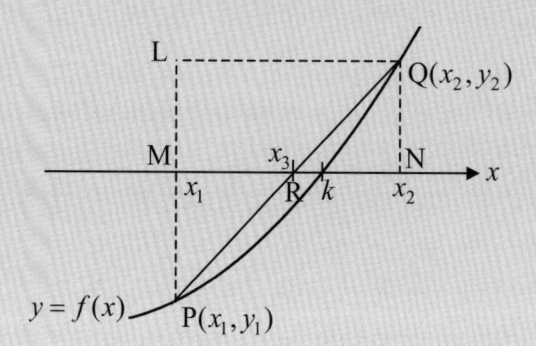

Triangles PMR, PLQ are similar and so $\dfrac{MR}{LQ} = \dfrac{PM}{PL} \Rightarrow \dfrac{x_3 - x_1}{x_2 - x_1} = \dfrac{-y_1}{y_2 - y_1}$

$$\Rightarrow x_3 - x_1 = \dfrac{(x_1 - x_2)y_1}{y_2 - y_1}$$

$$\Rightarrow x_3 = \dfrac{x_1 y_2 - x_2 y_1}{y_2 - y_1}.$$

Just how close x_3 is to k depends on how close x_1 and x_2 are together and how close the curve $y = f(x)$ between $x = x_1$ and $x = x_2$ is to a straight line.

We can continue this process, obtaining successive values $x_3, x_4, x_5, \cdots$ which converge to k.

If $f(x_3)$ has the same sign as $f(x_1)$, then $x_3 < k < x_2$; if $f(x_3)$ has the same sign as $f(x_2)$, then $x_1 < k < x_3$. In the first case we apply linear interpolation using the points (x_3, y_3) and (x_2, y_2); in the second case we apply linear interpolation using the points (x_1, y_1) and (x_3, y_3).

Example Show that there is a root of the equation $3x^2 - 6x + 1$ which lies between $x = 1$ and $x = 2$. Use linear interpolation to find this root correct to two decimal places.

Let $y = 3x^2 - 6x + 1$.

Here $x_1 = 1, x_2 = 2, y_1 = -2, y_2 = 1$.

Since y_1, y_2 have opposite signs, there is a root of the equation between 1 and 2.

x_1	x_2	x_3	$f(x_3)$
1	2	1.667	−0.667
1.667	2	1.800	−0.08
1.800	2	1.815	−0.008
1.815	2	1.816	−0.0008
1.816	2	1.816	

Therefore the required root is 1.82 (2 decimal places).

The Newton-Raphson Method

The Newton-Raphson method for estimating a root of the equation $f(x) = 0$ is an iterative process. That is, given an initial approximation, x_0, to a root k, we obtain a sequence of approximations $x_1, x_2, x_3, \cdots$ converging to k, stopping only when the present approximation is sufficiently accurate for our purposes.

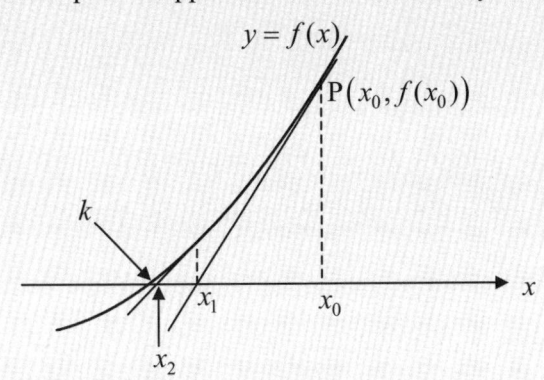

Let x_0 be an approximation to the root $x = k$ of the equation $f(x) = 0$. The equation of the tangent to the curve $y = f(x)$ at $P(x_0, f(x_0))$ is

$$xf'(x_0) - y = x_0 f'(x_0) - f(x_0).$$

This tangent meets the x-axis at $(x_1, 0)$ where $x_1 f'(x_0) - 0 = x_0 f'(x_0) - f(x_0)$ which gives

$$x_1 = x_0 - \frac{f(x_0)}{f'(x_0)}.$$

Chapter 26

From the graph we see that x_1 is clearly closer to k than x_0 is.

We can continue in this way to find a sequence of approximations $x_1, x_2, x_3, \ldots$ which converges to k and where

$$x_{n+1} = x_n - \frac{f(x_n)}{f'(x_n)}, \quad n = 0, 1, 2, 3, \ldots .$$

The choice of the first approximation is critical. In an attempt to use the Newton-Raphson formula to find the root $x = k_1$ of the equation $f(x) = 0$ in the following diagram, the choice of x_0 as the first approximation leads to a sequence which converges to k_2 instead of k_1.

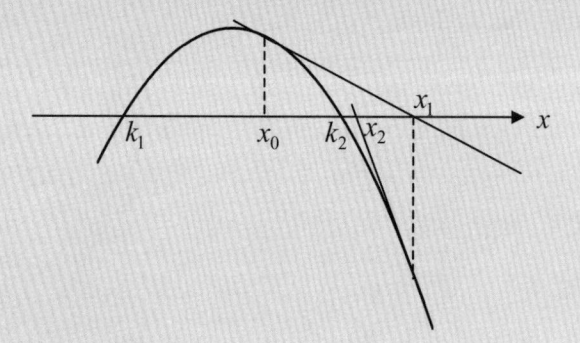

In the following diagram, $f'(x_0) = 0$ and so the method fails since the tangent at $P(x_0, f(x_0))$ does not meet the x-axis at all.

In any case, x_1 can not be calculated using the Newton-Raphson formula since $\dfrac{f(x_0)}{f'(x_0)}$ does not exist when $f'(x_0) = 0$.

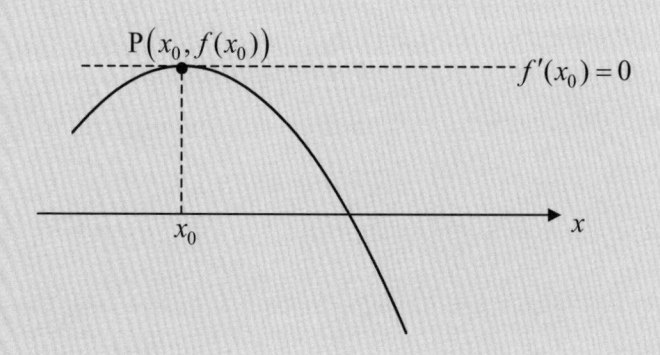

712

Analysis and Approximation

How can we make sure that our choice of the first approximation is going to provide convergence to the required root? As a general 'rule', if $|x_0 - k| < 0.1$ and there is no turning point on the graph of $y = f(x)$ between $x = x_0$ and $x = k$, the vast majority of equations will lend themselves to solution by means of the Newton-Raphson method.

Example   Use the Newton-Raphson method to find the root of the equation $e^{-x} = x - 1$ correct to five decimal places.

From the following sketch, we select $x_0 = 1.3$ as our first approximation.

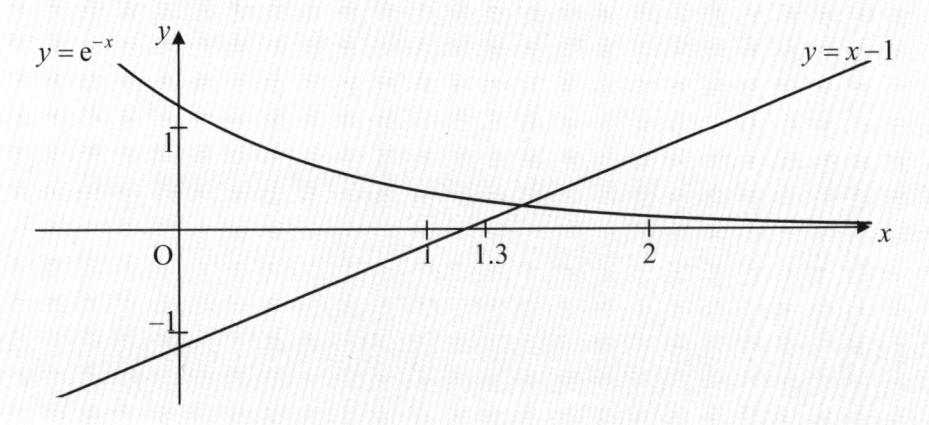

The required root is the solution of the equation $f(x) = 0$ where the function $f(x) = e^{-x} - x + 1$.

Now $x_{n+1} = x_n - \dfrac{f(x_n)}{f'(x_n)} = x_n - \dfrac{e^{-x_n} - x_n + 1}{-e^{-x_n} - 1} = x_n + 1 - \dfrac{x_n}{e^{-x_n} + 1}$.

Repeated applications give the following table of approximations:

n	x_n
1	1.3
2	1.2784145
3	1.2784645
4	1.2784645

Therefore the required root is $x = 1.27846$ (5 decimal places).

Chapter 26

Fixed-Point Iteration

In applying this method we re-write the equation $f(x)=0$ in the form $x = F(x)$ and then consider the iterative formula

$$x_{n+1} = F(x_n).$$

After choosing x_0, the sequence $x_1, x_2, x_3, \cdots$ will converge to a root of $x = F(x)$ *if it converges at all*.

Consider the following graphs of $y = F(x)$ and $y = x$. The required solution to the equation $x = F(x)$ is $x = k$.

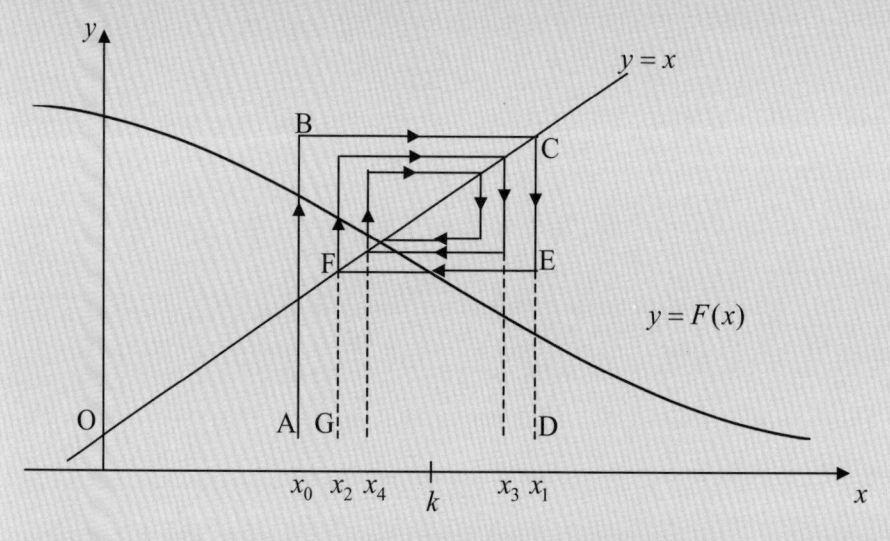

Now, since $x_1 = F(x_0)$ and $F(x_0) = AB = CD = OD$, D has an x-coordinate x_1. Continuing we have $x_2 = F(x_1)$ which is the 'height' of the curve $y = F(x)$ at $x = x_1$ and $DE = GF = OG$. Therefore G has an x-coordinate x_2. The sequence of approximations $x_1, x_2, x_3, \cdots$ then converges to k.

The diagram above and those on the next page suggest the following:

(1) there is two-sided convergence if $-1 < F'(x) < 0$ near the root;
(2) there is one-sided convergence if $0 < F'(x) < 1$ near the root;
(3) there is two-sided divergence if $F'(x) < -1$ near the root;
(4) there is one-sided divergence if $F'(x) > 1$ near the root.

These results will be proved in the next section of the text.

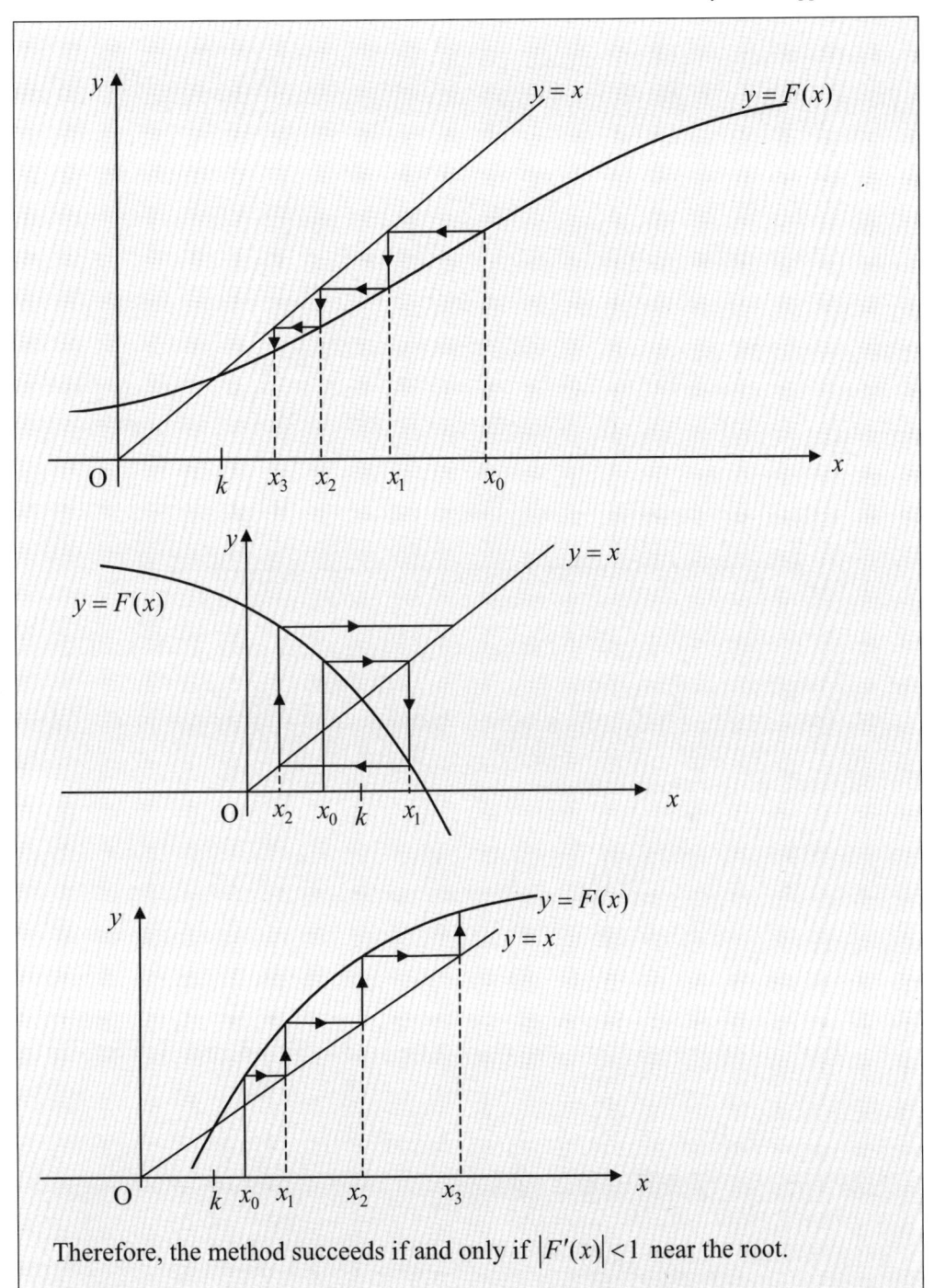

Therefore, the method succeeds if and only if $|F'(x)| < 1$ near the root.

Example Use a fixed-point iteration to find the real root of the equation $x^3 + 2x - 2 = 0$ correct to three decimal places.

Chapter 26

The equation $x^3 + 2x - 2 = 0$ can be written as $x(x^2 + 2) = 2$ or $x = \dfrac{2}{x^2 + 2}$.

Let $F(x) = \dfrac{2}{x^2 + 2}$ then $F'(x) = \dfrac{-4x}{(x^2 + 2)^2}$.

The graphs of $y = F(x)$ and $y = x$ are as follows:

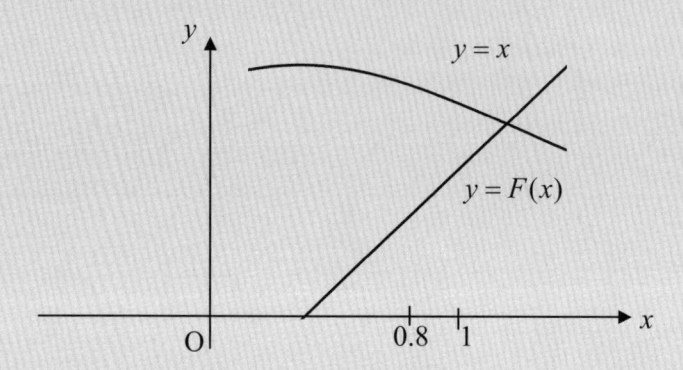

$|F'(0.8)| = 0.46 < 1$ and so the method converges and since $F'(0.8) < 0$ the convergence is two-sided.

Let $x_0 = 0.8$. The following table gives the successive approximations.

n	x_n
1	0.7576
2	0.7770
3	0.7681
4	0.7722
5	0.7703
6	0.7712
7	0.7708

Since there is two-sided convergence, 0.7712 is larger than the required root and 0.7708 is smaller.

Therefore the required root is 0.771 (3 decimal places).

Note how slowly this method converges in comparison with the Newton-Raphson method:

$f(x) = x^3 + 2x - 2$ and so $f'(x) = 3x^2 + 2$

$$x_{n+1} = x_n - \dfrac{x_n^3 + 2x_n - 2}{3x_n^2 + 2}, \quad x_0 = 0.8$$

n	x_n
1	0.7714
2	0.7709
3	0.7709

Thus after only 3 iterations (instead of 7), the root = 0.771 (3 decimal places).

Convergence of Iterative Solutions to $f(x)=0$

Consider the fixed-point method in which $x_{n+1} = F(x_n)$ and $x = k$ is the required root (i.e., $k = F(k)$ or $f(k) = 0$).

Let ε_n be the error in x_n so that $x_n = k + \varepsilon_n$.
Then $\qquad x_{n+1} = F(x_n)$

$$\Rightarrow \quad k + \varepsilon_{n+1} = F(k+\varepsilon_n) = F(k) + \varepsilon_n F'(k) + \frac{\varepsilon_n^2}{2!} F''(k) + \cdots \quad \text{(Taylor's theorem)}.$$

But $k = F(k)$ and so $\varepsilon_{n+1} = \varepsilon_n F'(k) + \dfrac{\varepsilon_n^2}{2!} F''(k) + \cdots$.

If we neglect terms in ε_n^2 and higher powers of ε_n, we have $\varepsilon_{n+1} \approx \varepsilon_n F'(k)$.

Now, if $|F'(k)| \geq 1$, then the absolute values of the errors remain the same or they increase, and the method does not converge. Also if $F'(k)$ is negative, the errors alternate in sign and we get two-sided convergence ($-1 < F'(k) < 0$) or two-sided divergence ($F'(k) \leq -1$). Finally, if $F'(k)$ is positive, the errors have the same sign and we get one-sided convergence ($0 < F'(k) < 1$) or one-sided divergence ($F'(k) > 1$).

Convergence of the form $\varepsilon_{n+1} \approx K\varepsilon_n$ ($K \neq 0$) is called *linear convergence*, and the errors approach zero in an approximate geometric progression.

If $F'(k) = 0$, we obtain from Taylor's theorem, $\varepsilon_{n+1} \approx \dfrac{\varepsilon_n^2}{2!} F''(k)$.

Convergence of the form $\varepsilon_{n+1} \approx K\varepsilon_n^2$ ($K \neq 0$) is called *second-order convergence*.

If $|K|$ is small, each iterate using second-order convergence should be approximately accurate to twice as many decimal places as its predecessor. Thus second-order convergence is very fast.

Chapter 26

The Newton-Raphson method always provides second-order convergence.

$$x_{n+1} = x_n - \frac{f(x_n)}{f'(x_n)}$$

$$\Rightarrow k + \varepsilon_{n+1} = k + \varepsilon_n - \frac{f(x_n)}{f'(x_n)}$$

$$\Rightarrow \varepsilon_{n+1} = \varepsilon_n - \frac{f(x_n)}{f'(x_n)}$$

$$\Rightarrow \varepsilon_{n+1} = \frac{1}{f'(x_n)}\{\varepsilon_n f'(x_n) - f(x_n)\}$$

$$= \frac{1}{f'(x_n)}\{\varepsilon_n f'(k+\varepsilon_n) - f(k+\varepsilon_n)\}$$

$$= \frac{1}{f'(x_n)}\left\{\varepsilon_n\left[f'(k) + \varepsilon_n f''(k) + \frac{\varepsilon_n^2}{2!}f'''(k) + \cdots\right]\right.$$

$$\left. -\left[f(k) + \varepsilon_n f'(k) + \frac{\varepsilon_n^2}{2!}f''(k) + \cdots\right]\right\}.$$

Neglecting terms in ε_n^3 and higher powers of ε_n and remembering that $f(k) = 0$, we have $\varepsilon_{n+1} \approx \frac{1}{f'(x_n)}\left\{\frac{\varepsilon_n^2}{2}f''(k)\right\} = \left\{\frac{f''(k)}{2f'(x_n)}\right\}\varepsilon_n^2 = K\varepsilon_n^2$ and so convergence is at least second-order.

Exercise 26.11

- 1. Show that the equation $x^3 + x^2 - 9x + 23 = 0$ has a root between -5 and -4. Use a single linear interpolation to find an approximation for this root.

2. (a) Prove that the equation $x^3 + x - 1 = 0$ has exactly one real root, $x = k$.
   (b) Find two integers a, b ($a < b$) between which k lies.
   (c) Use linear interpolation to show that $\frac{7}{11} < k < 1$.
   (d) Use the Newton-Raphson method to find k to 5 decimal places.

3. Show that the equation $24x^3 - 18x^2 + 1 = 0$ has three real roots all between -1 and 1. Use the Newton-Raphson method to find all three roots to 4 decimal places.

4. Show that the equation $x^4 - 5x + 2 = 0$ has a root, k, between 1 and 2. Show also that each of the following equations corresponds to the equation $x^4 - 5x + 2 = 0$. Of these equations, determine which will converge to k given a suitable starting value x_0, and use these to find k to 3 decimal places.

(a) $x = \dfrac{x^4 + 2}{5}$; (b) $x = \sqrt[4]{5x - 2}$; (c) $x = x^4 - 4x + 2$;

(d) $x = \dfrac{5x - 2}{x^3}$; (e) $x = \dfrac{2}{5 - x^3}$; (f) $x = \dfrac{25x - x^4 - 2}{20}$.

5. (a) Show that the equation $x^3 + x^2 + 5 = 0$ has a root k which lies between -3 and -2.

(b) Use part (a), or otherwise, to show that the equation $5x^3 + x + 1 = 0$ has a root between $-\frac{1}{2}$ and $-\frac{1}{3}$.

(c) Use the Newton-Raphson method with $x_0 = -\frac{1}{2}$ to find the root of $5x^3 + x + 1 = 0$ correct to 6 decimal places.

(d) Use the result of part (c) to find the root of $x^3 + x^2 + 5 = 0$ correct to 5 decimal places.

6. Show that if x_n is an approximation for $\sqrt{N}$, then a better approximation x_{n+1} found by using the Newton-Raphson method is $x_{n+1} = \dfrac{1}{2}\left(x_n + \dfrac{N}{x_n}\right)$.

Use this method to find $\sqrt{3}$ correct to 5 decimal places.

7. Use the Newton-Raphson method to find $\sqrt[3]{10}$ and $\sqrt[4]{5}$ correct to 6 decimal places.

8. Use the Newton-Raphson method to solve the following equations correct to 6 decimal places:

(a) $e^{-x} = \ln x$; (b) $\dfrac{1}{1 + x^2} = x$; (c) $\sin x = 3x - 1$;

(d) $e^x = 2x^2$, $2 < x < 3$.

9. (a) Sketch the graphs of $y = x$ and $y = \tan x$ for $0 \le x < \frac{3}{2}\pi$.

(b) Find an estimate for the positive root of the equation $\tan x = x$, $0 \le x < \frac{3}{2}\pi$.

(c) Using this estimate as the initial approximation, find the smallest positive root of the equation $\tan x = x$ correct to six decimal places.

10. Show that there is a root k of the equation $6x^3 - 65x^2 + 51x - 7 = 0$ between 0 and $\frac{1}{2}$. Use $x_0 = \frac{1}{2}$ as the initial approximation to this root and apply the Newton-Raphson method to find the next four iterates x_1, x_2, x_3, x_4. Explain why these iterates do *not* approach k.

Required Outcomes

After completing this chapter, a student should be able to:
- determine whether a given series converges using various tests which include comparison, limit comparison, the ratio test, the nth–root test, the integral test and the p–series test.
- determine whether an alternating series converges using Leibniz's theorem.
- use the alternating series estimation theorem to determine the accuracy of a given partial sum.
- determine whether a given series is absolutely convergent or conditionally convergent.
- determine the radius of convergence and the interval of convergence of a given power series.
- use Rolle's theorem and the mean value theorem in a variety of contexts.
- find the Taylor and Maclaurin series and polynomials for a variety of functions.
- use the remainder estimation theorem to determine the accuracy of a Taylor polynomial.
- find Taylor and Maclaurin series for compound functions by differentiating or integrating other Taylor and Maclaurin series.
- use the trapezium rule and Simpson's rule to estimate areas under curves.
- calculate the number of sub-intervals needed for a given accuracy when the trapezium rule or Simpson's rule is used.
- solve equations using linear interpolation, the Newton-Raphson method or fixed-point iteration.

PART 4
ANSWERS

Answers

Exercise 1.1

1. (i) $-2, \sqrt{441}, 1.\bar{9}$ (ii) $-2, \frac{3}{7}, \sqrt{441}, \sqrt{6\frac{1}{4}}, 0.8, 0.\bar{8}, 1.\bar{9}, \dfrac{4-\sqrt{32}}{\sqrt{18}-3}$

   (iii) $\sqrt{5}, \dfrac{1}{\pi}, \dfrac{1}{\sqrt{3}-1}, \pi^2$

3. $\dfrac{1}{25}, \dfrac{11}{9}, \dfrac{7}{2}, \dfrac{23}{99}, \dfrac{617}{5000}, \dfrac{1234}{9999}, \dfrac{2}{7}, \dfrac{1}{1}, \dfrac{7342}{5555}$

Exercise 1.2

1. (a) $x<4$ (b) $x\leq -6$ (c) $x>\frac{1}{6}$ (d) $x\geq -5$ (e) $x>-1.4$ (f) $x>-1$
   (g) $x>2$ (h) $x>\frac{3}{2}$ (i) $x<2$ (j) $x<1$ (k) $x>-\frac{1}{2}$ (l) $x>\frac{4}{3}$
3. (a) $x>0$ (b) $x<-13$ (c) No x (d) $-3<x<-1$

Exercise 1.3

1. (a) $\sqrt{15}$ (b) $\sqrt{xy^2}$ (c) $\sqrt{90}$ (d) $\sqrt[3]{20}$
3. (a) $3\sqrt{2}$ (b) $10\sqrt{2}$ (c) $5\sqrt{3}$ (d) $6\sqrt{2}$ (e) $15\sqrt{2}$ (f) $0.7\sqrt{2}$
5. (a) 30 (b) 63 (c) 726 (d) $6-\sqrt{15}$ (e) $10(\sqrt{2}-\sqrt{5})$ (f) $6(2-\sqrt{2})$
7. (a) $\dfrac{\sqrt{3}}{3}$ (b) $\dfrac{2\sqrt{6}}{3}$ (c) $\dfrac{\sqrt{3}}{6}$ (d) $\sqrt{2}+1$ (e) $2(\sqrt{5}-\sqrt{2})$ (f) $3\sqrt{2}-2\sqrt{3}$
   (g) $5+2\sqrt{6}$ (h) $\frac{1}{8}(5+3\sqrt{5})$ (i) $\frac{1}{5}(7-3\sqrt{6})$
9. (a) $a=\frac{5}{13}, b=\frac{2}{13}$ (b) $a=4, b=2$ or $a=-3, b=-\frac{8}{3}$

Exercise 1.4

1. (a) 12 (b) 17 (c) No x (d) $-\frac{7}{2}$ (e) $\frac{13}{3}$ (f) $\frac{2}{3}$
3. (a) 7 (b) $\frac{3}{2}$ (c) 3, 18 (d) $-\frac{2}{9}$ (e) $-\frac{23}{24}$ (f) 1

Exercise 1.5

1. (a) a^6 (b) b^9 (c) 1 (d) a^2 (e) $\dfrac{1}{a}$ (f) $\dfrac{x}{y}$
3. (a) 8 (b) 8 (c) $\frac{1}{3}$ (d) 0.008 (e) 1000 (f) $|x|$
5. (a) ± 2 (b) 3 (c) No x (d) 9 (e) ± 8 (f) $\pm\frac{1}{8}$ (g) 1024 (h) 0.2 (i) $\pm\frac{1}{27}$
6. (a) 1, 16 (b) 0, 4 (c) 0, 2 (d) $-2, \frac{1}{2}$ (e) $\pm 1, \pm 16\sqrt{2}$ (f) 4 (g) 9
   (h) $\pm\frac{27}{8}, \pm\frac{8}{27}$

723

Answers

Exercise 1.6
1. (a) $\log_3 9 = 2$ (b) $\log_2 \frac{1}{4} = -2$ (c) $\log_q p = 3$ (d) $\log_{\frac{1}{2}} y = x$ (e) $\log_k \ell = 3$
   (f) $\log_5 q = -p$
3. (a) 5 (b) 1.5 (c) 1.09 (d) 1.5 (e) –3 (f) 1.05
5. (a) $y = 2x^3$ (b) $y = 5/\sqrt{x}$ (c) $y = 100x^3$ (d) $y \approx 5x^{1/4}$
7. (a) 10 years (b) 6.02 years (c) 10.8 years (d) 26.0 years
9. 12.9°C
11. (a) 20.8 years (b) 138 years
13. (a) 3.33 seconds (b) 4.72 seconds (c) 13.3 seconds
15. $t = 0.2\sqrt{\ell}$
17. $k = 8, n = 1.2$

Exercise 2.1
1. (a) $-\frac{5}{4}$ (b) –3 (c) $\frac{4}{3}$ (d) $\frac{3}{2}$ (e) 2 (f) $-\frac{11}{5}$ (g) $\frac{5}{7}$ (h) $-\frac{5}{9}$ (i) –3 (j) —
   (k) 0 (l) $\frac{3}{7}$
3. (a) $2x - 3y = 1$ (b) $x - y = 3$ (c) $3x + 4y = 12$ (d) $x = -2$ (e) $y = 1$
   (f) $x + y = a + b$
5. (a) $3x - 2y = 5$ (b) $x + 3y + 10 = 0$ (c) $4x + 5y + 15 = 0$ (d) $5x + 3y = 41$
7. $\left(\frac{21}{16}, \frac{39}{8}\right)$
9. $\left(-1, -\frac{4}{3}\right)$

Exercise 2.2
1. (a) 3 (b) 2 (c) 2 (d) $\sqrt{5}$ (e) $2\sqrt{5}$ (f) $\frac{1}{2}\sqrt{10}$
3. (a) $\frac{10}{\sqrt{34}}$ (b) $\frac{7}{\sqrt{58}}$ (c) 3 (d) $\frac{27}{2\sqrt{2}}$ (e) $\frac{5}{\sqrt{2}}$ (f) $\frac{21}{2\sqrt{5}}$
5. (a) $21\frac{1}{2}$ (b) 9 (c) 26 (d) $\frac{125}{8}$ (e) $22\frac{1}{2}$
7. $9 \pm 5\sqrt{13}$
9. $2, 14\frac{1}{2}$
11. (a) $x + y > 8$, $x - y > -4$, $7x + y < 44$ (b) $y = 6$, $2x + y = 14$ (c) (4, 6)

Exercise 2.3
1. (a) (–1, –4), $x = -1$ (b) (2, –1), $x = 2$ (c) (–1, 3), $x = -1$ (d) (2, 9), $x = 2$
   (e) $\left(-\frac{1}{2}, -4\frac{1}{4}\right)$, $x = -\frac{1}{2}$ (f) $\left(1\frac{1}{2}, -\frac{1}{4}\right)$, $x = 1\frac{1}{2}$ (g) $\left(2\frac{1}{2}, 6\frac{1}{4}\right)$, $x = 2\frac{1}{2}$
   (h) $\left(-1\frac{1}{2}, 5\frac{1}{4}\right)$, $x = -1\frac{1}{2}$ (i) $\left(\frac{1}{2}, -4\frac{3}{4}\right)$, $x = \frac{1}{2}$ (j) $\left(-\frac{1}{3}, 6\frac{1}{3}\right)$, $x = -\frac{1}{3}$
3. 81 cm^2
5. 1250 m^2

Answers

Exercise 2.4
1. (a) −2, −3 (b) −3, −4 (c) −5, −4 (d) 2, 5 (e) 1, 5 (f) −2, 1 (g) −7, 2 (h) −7, 3 (i) 5, −2 (j) 2, −1
3. (a) 0, 9 (b) $\pm\frac{5}{2}$ (c) −1, −2 (d) −1, −3 (e) 1, $\frac{1}{2}$ (f) $5\pm\sqrt{5}$
   (g) $\frac{1}{10}\left(1\pm\sqrt{21}\right)$ (h) $\frac{1}{4}\left(3\pm\sqrt{17}\right)$ (i) $-3\pm\sqrt{11}$ (j) $-6, \frac{3}{2}$
4. (a) $\pm 1, \pm\frac{3}{2}$ (b) $\pm 2, \pm\sqrt{\frac{2}{3}}$ (c) $-1, \frac{1}{2}$ (d) 16 (e) 8, −27 (f) 1, 2, $-\frac{2}{3}, -\frac{5}{3}$

Exercise 2.5
1. (a) $\Delta = 29$; $\frac{1}{2}\left(-3\pm\sqrt{29}\right)$ (b) $\Delta = 33$; $\frac{1}{4}\left(1\pm\sqrt{33}\right)$ (c) $\Delta = -20$; no solution
   (d) $\Delta = 28$; $\frac{1}{3}\left(-1\pm\sqrt{7}\right)$ (e) $\Delta = 0$; $\frac{3}{2}$ (f) $\Delta = 196$; $-\frac{1}{5}, -3$
   (g) $\Delta = 12$; $\frac{1}{3}\left(-3\pm\sqrt{3}\right)$ (h) $\Delta = 1$; $-1, -\frac{2}{3}$ (i) $\Delta = 0$; $\frac{5}{2}$
   (j) $\Delta = -23$; no solution
5. P, Q, R are midpoints of AB, AC, BC; maximum area = 7 cm^2
7. 18 ha
9. 0, 16
11. 12 cm^2

Exercise 2.6
3. (a) $f(x) = 3 - (x-2)^2$ (b) $f(x) = 2(x+3)^2$ (c) A right shift of 1, a stretch parallel to the y-axis with stretch constant 2 and an upward shift of 3.
   (d) A left shift of 2, a stretch parallel to the x-axis with stretch constant 2, a reflection in the x-axis and an upward shift of 1.

Exercise 2.7
1. (a) P = (2, 6), Q = (10, 2) (b) P = $\left(\frac{8}{5}, \frac{31}{5}\right)$, Q = (16, −1)
   (c) P = $\left(\frac{2}{3}, \frac{20}{3}\right)$, Q = (−26, 20) (d) P = $\left(-\frac{1}{2}, \frac{29}{4}\right)$, Q = $\left(-5, \frac{19}{2}\right)$
3. (a) (−5, −8) (b) (4, 0) (c) $\left(\frac{23}{5}, -\frac{4}{5}\right)$
5. P = (−2, 1), Q = (6, 5), R = (0, −4)
7. 1 : 3 externally

Exercise 2.8
1. (a) $(x-3)^2 + (y-1)^2 = 16$ (b) $(x+1)^2 + (y+2)^2 = 25$
   (c) $(x+3)^2 + (y+2)^2 = 2\frac{1}{4}$ (d) $(x+3)^2 + y^2 = 8$
   (e) $(x-2)^2 + (y-2)^2 = 25$ (f) $(x-5)^2 + (y+4)^2 = 2$
   (g) $(x+3)^2 + (y-2)^2 = 4$ (h) $(x+5)^2 + (y+4)^2 = 25$

Answers

(i) $(x-2)^2 + (y+\frac{3}{2})^2 = 45$ (j) $x^2 + (y+4)^2 = 6\frac{1}{4}$
(k) $(x-3)^2 + (y-1)^2 = 20$ (l) $(x+1)^2 + y^2 = 25$
(m) $(x-4)^2 + (y+1)^2 = 4$ (n) $(x+5)^2 + (y+3)^2 = 20$

3. (a) external, (3, 3), $x + 2y = 9$ (b) internal, (8, 4), $x = 8$
(c) internal, (10, –1), $3x - y = 31$ (d) external, $(\frac{5}{4}, \frac{5}{4})$, $4x + 8y = 15$

5. (a) (2, 1) (b) (7, –2) (c) (–1, 7) (d) (4, 2)

7. (a) ± 13; $(\pm 2, \pm 3)$ (b) 14, –6; (6, –4), (2, 4) (c) –2.4; $(\frac{14}{13}, \frac{8}{13})$

Exercise 3.1

1. (a) $a = 34.6$, $c = 20$, $C = 30°$ (b) $A = 25°$, $b = 38.8$, $c = 23.7$
(c) $A = 88.7°$, $B = 31.3°$, $a = 23.1$ (d) $B = 130.5°$, $C = 19.5°$, $b = 22.8$
(e) $B = 101.4°$, $C = 48.6°$, $b = 19.6$ or $B = 18.6°$, $C = 131.4°$, $b = 6.38$

3. 1.52 km
5. 343 m
7. 44.0 m
9. 9.78 cm
11. 48.6°
    $A = 49.9°$, $C = 95.1°$, $c = 52.1$ cm or $A = 130.1°$, $C = 14.9°$, $c = 13.4$ cm
13. $a = b \sin A$ or $a = b \tan A$
15. 1190 m

Exercise 3.2

1. (a) $b = 13.7$ cm, $A = 49.1°$, $C = 70.9°$
   (b) $b = 5.21$ m, $A = 35.5°$, $C = 27.5°$
   (c) $A = 23.7°$, $B = 29.7°$, $C = 126.6°$
   (d) $A = 38.1°$, $B = 81.1°$, $C = 60.8°$

3. (a) 86.4° (b) 95.5°
5. 2.87 km
7. (a) 21.8°, 38.2°, 120° (b) 41.4°, 55.8°, 82.8°
9. (b) 28.5°, 65.6°, 85.9° (c) 7 (d) $\dfrac{x^2 - 6x - 7}{2(x+1)(2x+1)}$, 2 (e) 6

11. 237 m

Exercise 4.1

1.

	θ	$\sin \theta$	$\cos \theta$	$\tan \theta$
(a)	0°	0	1	0
(b)	90°	1	0	—
(c)	45°	$\frac{1}{2}\sqrt{2}$	$\frac{1}{2}\sqrt{2}$	1

Answers

| (d) | 120° | $\frac{1}{2}\sqrt{3}$ | $-\frac{1}{2}$ | $-\sqrt{3}$ |
| (e) | −210° | $\frac{1}{2}$ | $-\frac{1}{2}\sqrt{3}$ | $-\frac{1}{3}\sqrt{3}$ |

3. (a) $\cos\theta = \pm 0.6$, $\tan\theta = \pm 4/3$ (b) $\sin\theta = \pm\frac{1}{2}\sqrt{3}$, $\tan\theta = \pm\sqrt{3}$
   (c) $\sin\theta = \pm 2/\sqrt{5}$, $\cos\theta = \mp 1/\sqrt{5}$
5. (a) $\sin\theta$ (b) $\cos\theta$ (c) $2\cos\theta$ (d) $2\sin\theta$ (e) $2\cos\theta$
7. (a) Any 4 multiples of 180°. (b) Any 4 multiples of 90°.
   (c) Any 4 **odd** multiples of 90°.

Exercise 4.2
1. 23.6°, 156.4°
3. −116.6°, 63.4°, 243.4°

Exercise 4.3
1. (a) 180° (b) 90° (c) 120° (d) 108° (e) 220° (f) 75° (g) 45.8° (h) 70.5°
3.

$\theta°$	0°	30°	45°	60°	90°	120°	135°	150°	180°
x^c	0^c	$\frac{1}{6}\pi^c$	$\frac{1}{4}\pi^c$	$\frac{1}{3}\pi^c$	$\frac{1}{2}\pi^c$	$\frac{2}{3}\pi^c$	$\frac{3}{4}\pi^c$	$\frac{5}{6}\pi^c$	π^c
$\sin x$	0	$\frac{1}{2}$	$\frac{1}{2}\sqrt{2}$	$\frac{1}{2}\sqrt{3}$	1	$\frac{1}{2}\sqrt{3}$	$\frac{1}{2}\sqrt{2}$	$\frac{1}{2}$	0
$\cos x$	1	$\frac{1}{2}\sqrt{3}$	$\frac{1}{2}\sqrt{2}$	$\frac{1}{2}$	0	$-\frac{1}{2}$	$-\frac{1}{2}\sqrt{2}$	$-\frac{1}{2}\sqrt{3}$	−1
$\tan x$	0	$\frac{1}{3}\sqrt{3}$	1	$\sqrt{3}$	—	$-\sqrt{3}$	−1	$-\frac{1}{3}\sqrt{3}$	0

7. (a) $\sin(2x+40°)$ (b) $-\cos 4x$ (c) 1 (d) $\cos(2x+\frac{1}{2}\pi)$ (e) $\cos^2\frac{1}{2}x$
   (f) $1-\sin 2x$
9. $-1, \frac{7}{9}$
11. (a) $2\cos x$ (b) $-\tan x$ (c) $\cos 2x$

Exercise 4.5
7. (a) π (b) π (c) π (d) 2π (e) π (f) 2π

Exercise 4.8
1. (a) $\frac{5}{3}\pi$ cm (b) $\frac{10}{3}\pi$ cm (c) 2.58 m (d) 41.4 cm
3. (a) 8.21 cm (b) $\frac{8}{3}\pi$ cm
5. (a) 120 cm (b) $\frac{100}{9}$ cm, $\frac{25}{9}\sqrt{65}$ cm
7. 10.8 m²
9. (a) 12.9 cm (b) 14.0 cm (c) 20.6 cm²

Answers

13. 11.4 cm^2

Exercise 4.9
1. (a) $\frac{2}{3}\pi, \frac{4}{3}\pi$ (b) $\frac{1}{3}\pi, \frac{2}{3}\pi$ (c) 2.68, 5.82 (d) 2.21, 5.36 (e) no x
 (f) 1.82, 4.46 (g) 0.841, 2.30, 3.98, 5.44 (h) $\frac{1}{4}\pi, \frac{3}{4}\pi, \frac{5}{4}\pi, \frac{7}{4}\pi$
 (i) 0, π, $\frac{7}{6}\pi$, $\frac{11}{6}\pi$ (j) 0.464, 1.11, 3.61, 4.25 (k) 1.23, π, 5.05
 (l) $\frac{1}{4}\pi$, 2.90, $\frac{5}{4}\pi$, 6.04
3. (a) 10°, 50°, 130°, 170°, 250°, 290° (b) 90°, 270°
 (c) 54.2°, 144.2°, 234.2°, 324.2° (d) 21.1°, 81.1°, 141.1°, 201.1°, 261.1°, 321.1°
 (e) 47.5°, 107.5°, 227.5°, 287.5° (f) 43.4°, 91.6°, 223.4°, 271.6°
 (g) 21.1°, 81.1°, 141.1°, 201.1°, 261.1°, 321.1°
 (h) 26.2°, 93.8°, 146.2°, 213.8°, 266.2°, 333.8° (i) 45°, 135°, 225°, 315°
 (j) 90°, 270°

Exercise 4.10
1. (a) 1 (b) $\sec^2 \frac{1}{4} A$ (c) 1 (d) $\csc^2 \theta$ (e) 1 (f) 1 (g) 1 (h) 2
 (i) $\cos^2 A$ (j) $\sin^2 2B$ (k) $\tan^2 \theta$ (l) $-\cot^2 A$

Exercise 4.11
3. (a) $\sin 2x \cos y + \cos 2x \sin y$ (b) $\sin 3B \cos 40° + \cos 3B \sin 40°$
 (c) $\sin 2A \cos 2B + \cos 2A \sin 2B$ (d) $\sin x \cos 2y - \cos x \sin 2y$
 (e) $\frac{1}{2}\sqrt{3} \sin \theta - \frac{1}{2} \cos \theta$ (f) $\cos x \cos \frac{1}{2} y - \sin x \sin \frac{1}{2} y$
 (g) $\cos 3A \cos 3B - \sin 3A \sin 3B$ (h) $\frac{1}{2}\sqrt{3} \cos A + \frac{1}{2} \sin A$
 (i) $\frac{1}{2} \cos A - \frac{1}{2}\sqrt{3} \sin A$ (j) $\cos B \cos C + \sin B \sin C$
 (k) $\sin B$ (l) $\frac{1}{2} \cos 2A - \frac{1}{2}\sqrt{3} \sin 2A$ (m) $\frac{1}{2}\sqrt{3} \cos x + \frac{1}{2}\sin x$
 (n) $2 \sin x \cos x$ (o) $\cos^2 x - \sin^2 x$
5. (a) $-\frac{24}{25}$ (b) -1 (c) 0
7. (a) $\frac{1}{4}(\sqrt{2} - \sqrt{6})$ (b) $\frac{1}{4}(\sqrt{6} - \sqrt{2})$ (c) $\frac{1}{4}(\sqrt{6} - \sqrt{2})$ (d) $\frac{1}{4}(\sqrt{6} + \sqrt{2})$

Exercise 4.12
3. (a) 1 (b) $\cot(A - B)$ (c) $\frac{1}{3}\sqrt{3}$ (d) $\tan(\frac{1}{4}\pi + A)$
 (e) $\tan(\frac{1}{4}\pi + A)$ (f) $\tan 2A$

Exercise 4.13
3. (a) $\frac{1}{2}$ (b) $\frac{1}{2}\sqrt{2}$ (c) $\frac{1}{2}\sqrt{2}$ (d) $\frac{1}{4}$ (e) $\frac{1}{3}\sqrt{3}$ (f) 2
 (g) 1 (h) $\frac{1}{4}$ (i) $\frac{1}{8}$ (j) $\frac{1}{2}\sqrt{3}$

Answers

5. (a) $\pm\frac{1}{3}\sqrt{6}$, $\pm\frac{1}{3}\sqrt{3}$ (b) $\pm\frac{1}{4}\sqrt{7}$, $\pm\frac{3}{4}$, $\pm\frac{1}{3}\sqrt{7}$ (e) -1, 0.98

Exercise 4.14
1. (a) $\sqrt{13}\sin(x+0.983)$, $\pm\sqrt{13}$ (b) $\sqrt{10}\sin(x+0.322)$, $\pm\sqrt{10}$
   (c) $\sqrt{29}\sin(x+1.19)$, $\pm\sqrt{29}$
3. (a) 0, 4.07 (b) 1.79, 3.42 (c) 0.432, 2.32 (d) 1.08
5. (a) 0, 0.464, π, 3.61 (b) 1.25, 3.00, 4.39, 6.14
   (c) 0.519, 2.03, 3.66, 5.18 (d) 0.561, 1.99, 3.70, 5.13

Exercise 4.15
5. (a) $\sqrt{3}$ (b) $\sqrt{3}$
7. (a) $\frac{1}{4}\sqrt{6}$ (b) 0 (c) 0 (d) 0

Exercise 4.16
1. (a) $\sin(A+B)+\sin(A-B)$ (b) $\sin(2x+y)-\sin(2x-y)$
   (c) $\cos 96° + \cos 8°$ (d) $\cos 15° - \cos 35°$
   (e) $\frac{1}{2}\cos 4x + \frac{1}{2}\cos 2x$ (f) $\frac{1}{2}\cos 2x - \frac{1}{2}\cos 10x$
   (g) $\frac{1}{2}\sin 5x - \frac{1}{2}\sin x$ (h) $\frac{1}{2}\sin 7x - \frac{1}{2}\sin 3x$
5. (a) $\frac{27}{25}$ (b) $\frac{1}{10}(\sqrt{5}-4)$
7. (a) $\frac{1}{6}\pi$, $\frac{1}{2}\pi$, $\frac{7}{6}\pi$, $\frac{3}{2}\pi$ (b) $\frac{1}{3}\pi$, $\frac{2}{3}\pi$, $\frac{4}{3}\pi$, $\frac{5}{3}\pi$

Exercise 5.1
1. (a) function (b) function (c) function (d) not a function
3. (a) $\mathbb{R}$ (b) $[0,\infty[$ (c) $]1,\infty[$ (d) $\mathbb{R}$ (e) $[-1,5]$ (f) $[0,1]$ (g) $[0,3]$
   (h) $\mathbb{R}^+$ (i) $]-\infty,0]$ (j) $]0,1]$ (k) $]0,1]$ (l) $[0,2[$
5. (a) $\{x|x\in\mathbb{R}, x\geq 1 \text{ or } x\leq -1\}$, $\{y|y\in\mathbb{R}, y\geq 0\}$
   (b) $\{x|x\in\mathbb{R}, x\neq 0\}$, $\{y|y\in\mathbb{R}, y>0\}$
   (c) $\{x|x\in\mathbb{R}, x\neq 2\}$, $\{y|y\in\mathbb{R}, y\neq 0\}$
   (d) $\{x|x\in\mathbb{R}, x\neq 1\}$, $\{y|y\in\mathbb{R}, y\neq 1\}$
   (e) $\{x|x\in\mathbb{R}, x\neq 1, x\neq 2\}$, $\{y|y\in\mathbb{R}, y\leq -4, y>0\}$
   (f) $\mathbb{R}$, $\{y|y\in\mathbb{R}, 0<y\leq 4\}$

Exercise 5.2
1. (a) 2 (b) 2 (c) 3 (d) 1 (e) 2 (f) 3
3. $f\circ g: x \mapsto \sqrt{2x-1}$; Domain = $[\frac{1}{2},\infty[$; Range = $[0,\infty[$
   $g\circ f: x \mapsto 2\sqrt{x}-1$; Domain = $[0,\infty[$; Range = $[-1,\infty[$

729

Answers

5. (a) $h: x \mapsto \frac{1}{2}(x+1)$ (b) $h: x \mapsto 5-2x$ (c) $h: x \mapsto 3x+4$
   (d) $h: x \mapsto \frac{2}{3}(1-6x)$
7. $f^2: x \mapsto x$
9. $\{x \mid x \in \mathbb{R}, x \neq -d/c\}$

Exercise 5.3
1. (a) 1–1 (b) Not 1–1 (c) 1–1 (d) 1–1 (e) Not 1–1 (f) Not 1–1
   (g) 1–1 (h) Not 1–1 (i) Not 1–1 (j) 1–1 (k) 1–1 (l) 1–1
3. (a) $f^{-1}(x) = \frac{1}{3}x$ (b) $f^{-1}(x) = 4-x$ (c) $f^{-1}(x) = \frac{1}{3}(x-2)$
   (d) $f^{-1}(x) = \sqrt[3]{x}$ (e) $f^{-1}(x) = 4/x$ (f) $f^{-1}(x) = \sqrt[3]{x-1}$
   (g) $f^{-1}(x) = \frac{1}{2-x}$ (h) $f^{-1}(x) = 1 - \frac{3}{x}$
5. (a) $f^{-1}(x) = \frac{1-x}{x}$ (b) $f^{-1}(x) = \frac{x+2}{x-1}$ (c) $f^{-1}(x) = \frac{1-x}{2(x+1)}$
   (d) $f^{-1}(x) = \frac{2x-3}{3x-2}$
7. $f^{-1}(x) = \sqrt{\frac{x-1}{2}+1}$

Exercise 6.1
1. (a) $\{3, 6, 9, \ldots, 30, \ldots\}$ (b) $\{-3, -1, 1, \ldots, 15, \ldots\}$
   (c) $\{1, 2, 4, \ldots, 512, \ldots\}$ (d) $\{\frac{3}{2}, \frac{5}{3}, \frac{7}{4}, \ldots, \frac{21}{11}, \ldots\}$
   (e) $\{0.1, 0.02, 0.003, \ldots, 10^{-9}, \ldots\}$ (f) $\{3, 1, 3, \ldots, 1, \ldots\}$
3. (a) $\{1, 1, 2, 3, 5, 8, 13, 21, 34, 55, \ldots\}$ (b) $\{1, 1, 2, 3, 5, \ldots\}$

Exercise 6.2
1. (a) arithmetic, $d = 4$ (b) arithmetic, $d = 3a$ (c) not arithmetic
   (d) arithmetic, $d = -6$ (e) not arithmetic (f) arithmetic, $d = b - a$
3. (a) $u_n = 4n + 1, u_{10} = 41$ (b) $u_n = 99 - 12n, u_{10} = -21$
   (c) $u_n = 0.6n + 1.7, u_{10} = 7.7$ (d) $u_n = 2 + 5x - 2xn, u_{10} = 2 - 15x$
5. (a) $u_1 = 5, d = 7$ (b) $u_1 = 17, d = -6$ (c) $u_1 = 24, d = 4$ (d) $u_1 = \frac{29}{2}, d = -\frac{1}{2}$
7. (a) 11, 17, 23 (b) 5, 8, 11, 14 (c) $-4, -1, 2, 5, 8$

Exercise 6.3
1. (a) 3775 (b) 1800 (c) 420 (d) 243 (e) $4a^3$ (f) $\frac{3}{2}a^2(13-9a)$
3. (a) $\{-8, -5, -2, \ldots, 25, \ldots\}$ (b) $\{27, 22, 17, \ldots, -28, \ldots\}$
   (c) $\{3\frac{1}{2}, 3, 2\frac{1}{2}, \ldots, -2, \ldots\}$ (d) $\{3, 9, 15, \ldots, 69, \ldots\}$
   (e) $\{11, 5, -1, \ldots, -55, \ldots\}$ (f) $\{-38, -35, -32, \ldots, -5, \ldots\}$

Answers

7. 765
9. 36 hours, 16.65 kilolitres
11. 37 years
13. {11, 15, 19, 23, ... }
15. (a) $\left[\tfrac{1}{2}n(n+1)\right]^2$ (b) $n^2(1+n^2)$

Exercise 6.4

1. (a) geometric, $r = 4$ (b) geometric, $r = -\tfrac{1}{2}$ (c) not geometric
   (d) geometric, $r = -1/a$ (e) geometric, $r = 8/9$ (f) geometric, $r = \sqrt{2}/2$
3. (a) $u_5 = 5(3^4)$, $u_{12} = 5(3^{11})$, $u_n = 5(3^{n-1})$
   (b) $u_5 = 2(-3)^4$, $u_{12} = 2(-3)^{11}$, $u_n = 2(-3)^{n-1}$
   (c) $u_5 = 27\left(\tfrac{2}{3}\right)^4$, $u_{12} = 27\left(\tfrac{2}{3}\right)^{11}$, $u_n = 27\left(\tfrac{2}{3}\right)^{n-1}$
   (d) $u_5 = x(-xy)^4 = x^5 y^4$, $u_{12} = x(-xy)^{11} = -x^{12}y^{11}$,
   $u_n = x(-xy)^{n-1} = (-y)^{n-1} x^n$
5. (a) $u_{28} \approx 1190$ (b) $u_{12} \approx 1380$

Exercise 6.5

1. (a) $4^{10} - 1$ (b) $10(1.2^{12} - 1)$ (c) $0.1(1 - 2^{20})$ (d) $-3(1 + 0.4^{15})$
3. $S_n = 6(1.5^n - 1)$, $n \geq 18$
5. 113
7. $75 800; $49 600
9. (b) $r = -1.5$

Exercise 6.6

1. (a) $8954.24   (b) $964.65
3. (a) $1279.32   (b) 8.23 years
5. (a) (i) $6513.53   (ii) $6888.06   (b) $2582.48
7. $502.02
9. (a) $263.80   (b) $4096.64

Exercise 6.7

1. (a) 2   (b) 40.5 (c) 51.2 (d) $8\tfrac{1}{3}$ (e) 526 (f) 2.89
3. (a) $\tfrac{5}{11}$ (b) $\tfrac{2081}{990}$ (c) $\tfrac{1577}{6660}$
5. 7.5 ; 10
7. 693
9. (a) 0.998   (b) 1900
11. $k > \tfrac{1}{2}$
15. (a) $x < 0$ (b) 149, 67

731

Answers

Exercise 7.1
1. 115

Exercise 7.2
1. (a) 8, 9, 6   (b) 194, 194, 195   (c) 0.725, 0.73, 0.73
3. (a) 3.925, 4   (b) 19.8, 18.4   (c) 31.0, 30.0

Exercise 7.3
1. (a) 53   (b) 21   (c) 58   (d) 78
3. (a) 46   (b) 28   (c) 135

Exercise 7.4
1. 24, 1.19
3. (a) 775, 120 250   (b) 157 cm, 5.89 cm
5. $2 - 3\bar{x}$, $9s^2$
7. 22.9°, 3.27°
9. 11, 14

Exercise 7.5
1. 27.1, 10.9   (b) 54.4, 18.0
3. 170, 10.0

Exercise 8.1
1. (a) 144   (b) 72
3. (a) 216   (b) 72   (c) 108
5. (a) 300   (b) 156   (c) 108
7. (a) 9000   (b) 3168
9. (a) 24   (b) 8   (c) 0
11. (a) 81   (b) 54

Exercise 8.2
1. (a) 210   (b) $\dfrac{1}{110}$ (c) 600   (d) 81
3. (a) $n-1$ (b) $(n+2)(n+1)$ (c) $n(n^2 + 3n + 1)$

Exercise 8.3
1. 120
3. 5040
5. 168
7. (a) 240   (b) 600
9. (a) 1000   (b) 720   (c) 990
11. (a) 103 680   (b) 34 560

Exercise 8.4
1. (a) 6 (b) 84 (c) 1820
3. 1140
5. 525
7. (a) 462 (b) 56 (c) 20
9. 1260
11. (a) 924 (b) 34 650
15. (a) 729 (b) 28
17. 10
19. 61 (including the given straight line)
21. 56
23. (a) 286 (b) 84

Exercise 8.5
1. (a) $x^4 + 4x^3y + 6x^2y^2 + 4xy^3 + y^4$
    (b) $a^7 - 7a^6b + 21a^5b^2 - 35a^4b^3 + 35a^3b^4 - 21a^2b^5 + 7ab^6 - b^7$
    (c) $64 + 192p^2 + 240p^4 + 160p^6 + 60p^8 + 12p^{10} + p^{12}$
    (d) $32h^5 - 80h^4k + 80h^3k^2 - 40h^2k^3 + 10hk^4 - k^5$
    (e) $x^3 + 3x + 3x^{-1} + x^{-3}$
    (f) $z^8 - 4z^6 + 7z^4 - 7z^2 + \frac{35}{8} - \frac{7}{4}z^{-2} + \frac{7}{16}z^{-4} - \frac{1}{16}z^{-6} + \frac{1}{256}z^{-8}$
3. $64x^5 + 160x^{-1} + 20x^{-7}$
5. 0, 1 (trivial) and 6
7. 2
9. 30.43168
11. (a) 560 (b) –590 625 (c) –720 (d) –448 (e) 1 966 080 (f) $-\frac{7}{144}$
13. $\pm\frac{2}{3}$
15. (a) 112 (b) 196 (c) –2214 (d) 140
19. (a) –14 (b) 26 (c) 352 (d) 10

Exercise 9.1
1. (a) $\begin{pmatrix} 5 & 0 \\ 5 & -5 \end{pmatrix}$ (b) $\begin{pmatrix} 1 & 3 \\ 4 & -1 \end{pmatrix}$ (c) $\begin{pmatrix} 5 & 5 \\ 0 & -5 \end{pmatrix}$ (d) $\begin{pmatrix} 5 & 5 \\ 0 & -5 \end{pmatrix}$
5. (a) $X = -\frac{2}{3}B + \frac{7}{3}C$ (b) $X = -\frac{6}{7}A + \frac{2}{7}B - \frac{3}{7}C$

Exercise 9.2
1. AB is 4×4; BA is 5×5

Answers

3. (a) $\begin{pmatrix} -6 & 6 \\ 5 & 9 \end{pmatrix}$ (b) $\begin{pmatrix} 1 & 0 \\ 0 & 1 \end{pmatrix}$ (c) $\begin{pmatrix} -8 & -11 & -14 \\ -8 & -13 & -18 \\ 1 & 2 & 3 \end{pmatrix}$ (d) $\begin{pmatrix} -4 & 5 \\ 16 & -12 \end{pmatrix}$

(e) (0) (f) $\begin{pmatrix} 6 & -2 & -4 \\ 9 & -3 & -6 \\ 3 & -1 & -2 \end{pmatrix}$ (g) $\begin{pmatrix} 2 & 4 & 16 & 14 \\ -5 & -2 & 0 & 5 \end{pmatrix}$ (h) $(-13 \quad 4)$

(i) $\begin{pmatrix} 2 & 16 \\ 3 & 12 \\ -12 & -16 \end{pmatrix}$ (j) $\begin{pmatrix} -22 & 32 & 57 \\ 16 & 25 & 24 \\ -34 & 23 & 54 \end{pmatrix}$ (k) $\begin{pmatrix} 1 & 0 & 0 \\ 0 & 1 & 0 \\ 0 & 0 & 1 \end{pmatrix}$ (l) $\begin{pmatrix} 30 \\ 10 \end{pmatrix}$

5. $\begin{pmatrix} 10 & -20 \\ 5 & -10 \end{pmatrix}$

7. $a = 2, b = -2$

9. (a) $a = 5, b = 1$ (b) $a = -1, b = 5$ (c) $a = -1, b = 4$ (d) $a = -2, b = 0$

11. $\begin{pmatrix} 0.3 \\ 2.6 \\ 1.2 \end{pmatrix}$ – percentages of silver, lead and zinc in the combined samples

15. (a) $\begin{pmatrix} 1 & 0 \\ 0 & 1 \end{pmatrix}$ (b) $\begin{pmatrix} 1 & 0 \\ 0 & 1 \end{pmatrix}$

17. $\begin{pmatrix} -3 & 4 \\ 4 & -5 \end{pmatrix}$

19. (a) $a = \pm 1, b = \mp 1$ (b) $a = \pm\sqrt{2}, b = \mp\sqrt{2}$ (c) $a = \dfrac{1 \pm \sqrt{5}}{2}, b = \dfrac{1 \mp \sqrt{5}}{2}$

Exercise 9.3
1. (a) 9 (b) 1 (c) 1 (d) –1 (e) –2 (f) –2 (g) 6 (h) 0
3. $-2, -1, 3$
5. $-\lambda^3 + \lambda^2 + \lambda - 1$
7. 1
9. (a) x^2 (b) x^m

Exercise 9.4
1. (a) $\begin{pmatrix} 3 & -2 \\ -7 & 5 \end{pmatrix}$ (b) $\begin{pmatrix} -8 & 3 \\ 11 & -4 \end{pmatrix}$ (c) $\begin{pmatrix} -2 & 3 \\ -3 & 5 \end{pmatrix}$ (d) $\begin{pmatrix} -2 & 2\frac{1}{2} \\ 1 & -1 \end{pmatrix}$

(e) $\begin{pmatrix} \frac{1}{3} & 0 \\ 2 & -1 \end{pmatrix}$ (f) $\begin{pmatrix} \cos\alpha & \sin\alpha \\ -\sin\alpha & \cos\alpha \end{pmatrix}$ (g) $\begin{pmatrix} \cos\alpha & \sin\alpha \\ \sin\alpha & -\cos\alpha \end{pmatrix}$

Answers

(h) $\begin{pmatrix} \frac{1-m^2}{1+m^2} & \frac{2m}{1+m^2} \\ \frac{2m}{1+m^2} & -\frac{1-m^2}{1+m^2} \end{pmatrix}$

3. $\dfrac{1}{128}$

5. $\begin{pmatrix} 1 & 0 \\ 0 & 1 \end{pmatrix}$, $\dfrac{1}{2}\begin{pmatrix} -1 & \sqrt{3} \\ -\sqrt{3} & -1 \end{pmatrix}$

7. (a) $\begin{pmatrix} -7 & -11 \\ 3 & 4 \end{pmatrix}$ (b) $\begin{pmatrix} 1 \\ 2 \end{pmatrix}$ (c) $\begin{pmatrix} 2 & 4 \\ -1 & -3 \end{pmatrix}$ (d) $(3 \ 1)$ (e) $\begin{pmatrix} 25 & 34 \\ -94 & -128 \end{pmatrix}$

9. (a) $A^{-1} = \tfrac{1}{7}(4I - A^2)$ (b) $\begin{pmatrix} 5 \\ -1 \\ 1 \end{pmatrix}$

11. $A^2 = I$, $A^{-1} = A$
13. $AB = I$, $A^{-1} = B$
15. $A^2 = I$, $A^{-1} = A$

17. $x = \begin{pmatrix} 3 \\ -2 \\ 4 \end{pmatrix}$

Exercise 9.5

1. (a) $x = 3$, $y = 5$ (b) $x = 2$, $y = 1$ (c) $x = 3$, $y = -2$ (d) $x = 3$, $y = 1$

3. $k \neq 4$; $x = \dfrac{7}{3}$, $y = -\dfrac{2}{3}$, $z = 0$

   If $k = 4$, the solution is $x = 3 - 2t$, $y = -2t$, $z = 3t - 1$ for all real t.

5. (a) $x = 3$, $y = 0$, $z = 1$ (b) $x = 13$, $y = -12$, $z = 16$
7. $x = 1 - 2k$, $y = 4k - 8$, $z = -k - 2$
9. (a) $x = -a - b + c$; $y = 10a - b - 2c$; $z = -6a + b + c$

   (b) $\begin{pmatrix} -1 & -1 & 1 \\ 10 & -1 & -2 \\ -6 & 1 & 1 \end{pmatrix}$

Exercise 9.6

1. (a) $x = 3$, $y = 1$: The lines meet at $(3, 1)$.
   (b) $x = 1\tfrac{1}{2}$, $y = \tfrac{1}{2}$: The lines meet at $(1\tfrac{1}{2}, \tfrac{1}{2})$.
   (c) No solution exists : The lines are parallel and distinct.

Answers

    (d) $x = \frac{1}{2}(1 + 3t)$, $y = t$ for all real t : The lines are coincident.
    (e) $x = 4 - 5t$, $y = t - 1$, $z = t$ for all real t : The planes meet in a line.
    (f) No solution exists : The planes are parallel and distinct.
    (g) $x = 3$, $y = 1$, $z = 1$: The planes meet at (3, 1, 1).
    (h) No solution exists : Two planes are parallel and distinct.
    (i) $x = t + 6$, $y = 1 - t$, $z = t$ for all real t : The planes meet in a line – two of the planes coincide.
    (j) No solution exists : The line of intersection of any two of the planes is parallel to and distinct from the third plane.
    (k) $x = 4 - 3t$, $y = 1 - 3t$, $z = t$ for all real t : The planes meet in a line.
    (l) No solution exists : The line of intersection of any two of the planes is parallel to and distinct from the third plane.
    (m) $x = 1$, $y = \frac{1}{3}$, $z = -\frac{4}{3}$: The planes meet at $(1, \frac{1}{3}, -\frac{4}{3})$.
    (n) No solution exists : The line of intersection of any two of the planes is parallel to and distinct from the third plane.
3. The line of intersection of any two planes is parallel to and distinct from the third plane.
5. $k = 3$ or $k = -1$: $x = \dfrac{2}{1+k}$, $y = \dfrac{2}{1+k}$, $k \neq 3$, $k \neq -1$
    If $k = 3$, an infinite number of solutions exist. The lines coincide.
    If $k = -1$, no solution exists. The lines are parallel and distinct.

Exercise 10.1
1. (a) $\to 0$ (b) $\to 0$ (c) $\to \frac{1}{2}$ (d) no limit   (e) $\to -2$ (f) $\to 0$
   (g) no limit   (h) $\to 1$ (i) no limit   (j) $\to 0$ (k) $\to -6$ (l) $\to 1$
3. (a) 8   (b) $-\frac{35}{4}$ (c) $\frac{3}{4}$ (d) 4   (e) $\frac{1}{8}\sqrt{2}$ (f) $\frac{1}{6}$
5. (a) $\frac{11}{5}$ (b) $\frac{2}{3}$ (c) $\frac{2}{11}$ (d) $\frac{1}{2}$

Exercise 10.2
1. $-4 - 2h$; -4 ; the gradient of the curve at P is -4
3. (a) 2   (b) -3 (c) 12   (d) -1 (e) -5 (f) 3
4. (a) $6x - y = 3$ (b) $4x - y = -5$ (c) $2x - y = -2$ (d) $y = 0$ (e) $x + y = 3$
   (f) $x + 2y = 5$ (g) $3x - 4y = 1$ (h) $x - 4y = -4$

Exercise 10.3
1. -6
3. (a) $4x + 3$ (b) $6x^2$ (c) $\dfrac{1}{(x+1)^2}$ (d) $-\dfrac{4}{x^3}$ (e) $\dfrac{2x^2 - 2x - 2}{(2x-1)^2}$ (f) $\dfrac{1}{2\sqrt{x}}$
5. (a) $3x - y = 1$ (b) $6x + y = 4$ (c) $4x + y = 8$ (d) $2x - 9y = -1$
   (e) $x + 4y = 11$ (f) $7x - 16y = 3$

Answers

Exercise 10.4
1. (a) $4x^3$ (b) $12x^2$ (c) $16x$ (d) $2x-4$ (e) $-3-6x^2$ (f) $12x^3-12x^2+5$
   (g) $6x-6x^2$ (h) $10x^4-9x^2$ (i) $-4-4x-15x^2$
3. (a) $2x+y=-6$ (b) $y=-\frac{4}{3}$
5. (a) $-\frac{3}{t^2}+\frac{8}{t^3}-\frac{15}{t^4}$ (b) $-4+\frac{18}{5t^4}$ (c) $-\frac{3}{2t^3}+\frac{4}{9t^4}$ (d) $11-20t$
   (e) $6\pi t^2-4\pi t+10\pi$ (f) $-\frac{4}{\pi t^3}+\frac{25}{t^2}$
7. (a) ± 8 (b) $-\frac{16}{3}$

Exercise 10.5
1. (a) $5, 0$ (b) $6x-6, 6$ (c) $6x^2-10x+4$, $12x-10$ (d) $3x^2+\frac{2}{x^2}$, $6x-\frac{4}{x^3}$
   (e) $\frac{3}{2x^{1/2}}-\frac{1}{x^{3/2}}$, $-\frac{3}{4x^{3/2}}+\frac{3}{2x^{5/2}}$ (f) $-\frac{6}{x^2}+\frac{6}{x^3}-\frac{12}{x^4}$, $\frac{12}{x^3}-\frac{18}{x^4}+\frac{48}{x^5}$
3. (a) Body changes direction at $t=2$.
   (b) Body does not change direction in the first 3 seconds.
5. 28 m
7. (a) $(19.6-9.8t)\,\text{ms}^{-1}$, $-9.8\,\text{ms}^{-2}$ (b) 2 s (c) 19.6 m (d) 0.586 s, 3.41 s

Exercise 10.6
1. (a) $(0, 3)$ – local maximum (b) $(-3, -27)$ – local minimum
       $(2, -1)$ – local minimum $(0, 0)$ – inflexion
   (c) $(2, 0)$ – inflexion   (d) $(-1, 1)$ – local maximum
                                          $(0, 0)$ – local minimum
                                          $(1, 1)$ – local maximum
   (e) $(0, 0)$ – inflexion   (f) $(0, 0)$ – inflexion
       $\left(\frac{9}{4}, -\frac{2187}{256}\right)$ – local minimum $(3, 108)$ – local maximum
                                          $(5, 0)$ – local minimum
3. $a=-3, b=-12$
5. (a) $a=-6, b=9$; $(3, 0)$ (b) $a=-3, b=0, c=5$; $y=9x+10$
   (c) $a=2, b=-3, c=-12, d=-20$; $\left(-\frac{1}{2}, 5\right)$

Exercise 10.7
1. (a) $3, -9$ (b) $\frac{32}{27}, 0$ (c) $\frac{32}{27}, -\frac{49}{27}$
3. (a) (i) $0, -3$ (ii) $(0, 0), (-2, 4)$ (iii) $(-1, 2)$
   (b) (i) $2, -\frac{1}{2}(5\pm\sqrt{21})$ (ii) $(-3, 25), (1, -7)$ (iii) $(-1, 9)$
   (c) (i) $0, -\frac{8}{3}$ (ii) $(-2, -16)$ (iii) $(0, 0), \left(-\frac{4}{3}, -\frac{256}{27}\right)$

737

Answers

(d) (i) 0, 2 (ii) (0, 0), (1, 1), (2, 0) (iii) $\left(1 \pm \tfrac{1}{3}\sqrt{3}, \tfrac{4}{9}\right)$

(e) (i) 0, ±4 (ii) $(\pm 2, -36\sqrt[3]{4})$, (0, 0) (iii) There are no points of inflexion.

(f) (i) 0, –3 (ii) $(-\tfrac{3}{4}, -2.04)$ (iii) (0, 0), $(\tfrac{3}{2}, 5.15)$

Exercise 10.8
1. 2 : 1
3. $\tfrac{8}{27}$ m^3
5. $\dfrac{20}{\sqrt{3\pi}} \approx 6.51$ cm
9. 320 m
11. (a) $\dfrac{10\pi}{4+\pi}$ (b) 10

Exercise 11.1
1. $(a + b) + c = \overrightarrow{AC} + \overrightarrow{CD} = \overrightarrow{AD}$
$a + (b + c) = \overrightarrow{AB} + \overrightarrow{BD} = \overrightarrow{AD}$

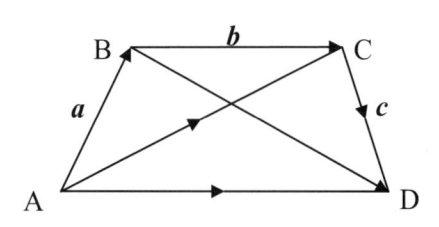

$\overrightarrow{AC} = \overrightarrow{AB} + \overrightarrow{BC} = \overrightarrow{AD} + \overrightarrow{DC}$
$a + b = b + a$

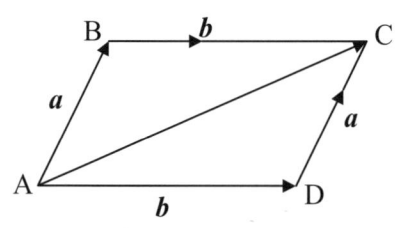

3. (a) $a + b$ (b) $a + b + c$ (c) $a + d$ (d) $d - b - c$
9. (a) $10\sqrt{5}$, 206.6° (b) $10\sqrt{5}$, 333.4° (c) 10, 330° (d) 29.1, 189.9°
11. a and b have the same direction

Exercise 11.2
4. $b - a$, $b - 2a$, $2(b - a)$, $2b - 3a$, $b - 2a$
5. (a) $\tfrac{1}{3}\overrightarrow{A} + \tfrac{2}{3}\overrightarrow{B}$ (b) $\tfrac{3}{7}\overrightarrow{A} + \tfrac{4}{7}\overrightarrow{B}$ (c) $\tfrac{3}{5}\overrightarrow{A} + \tfrac{2}{5}\overrightarrow{B}$ (d) $\tfrac{3}{2}\overrightarrow{A} - \tfrac{1}{2}\overrightarrow{B}$
 (e) $-\tfrac{2}{3}\overrightarrow{A} + \tfrac{5}{3}\overrightarrow{B}$ (f) $4\overrightarrow{A} - 3\overrightarrow{B}$
9. $\overrightarrow{OM} = \tfrac{1}{2}(a + b)$ $\overrightarrow{ON} = \tfrac{1}{3}(a + b + c)$

Exercise 11.3

1. (a) $\begin{pmatrix} 2 \\ -1 \end{pmatrix}, \sqrt{5}, \begin{pmatrix} 2/\sqrt{5} \\ -1/\sqrt{5} \end{pmatrix}$ (b) $\begin{pmatrix} 5/2 \\ 6 \end{pmatrix}, 6.5, \begin{pmatrix} 5/13 \\ 12/13 \end{pmatrix}$

   (c) $\begin{pmatrix} -5 \\ -2 \end{pmatrix}, \sqrt{29}, \begin{pmatrix} -5/\sqrt{29} \\ -2/\sqrt{29} \end{pmatrix}$

3. (a) $\begin{pmatrix} 0.6 \\ -0.8 \end{pmatrix}$ (b) $\begin{pmatrix} -5/13 \\ 12/13 \end{pmatrix}$ (c) $\begin{pmatrix} -1/\sqrt{2} \\ -1/\sqrt{2} \end{pmatrix}$ (d) $\begin{pmatrix} 3/\sqrt{13} \\ -2/\sqrt{13} \end{pmatrix}$

7. (a) right-angled isosceles (b) isosceles (c) equilateral
9. (a) (2, 3) (b) (1, 0) (c) (3, 7) (d) $(5, \frac{1}{2})$
13. $\frac{2}{3}$

Exercise 11.4

1. (a) 1 (b) 2 (c) 31 (d) 10 (e) −16 (f) 0 (g) 5 (h) 24 (i) 47
5. (a) $\frac{8}{5}$ (b) $\frac{9}{13}$ (c) $\frac{1}{2}\sqrt{2}$ (d) $-\frac{6}{25}$ (e) $\frac{6}{\sqrt{82}}$ (f) $-\frac{17}{\sqrt{10}}$
7. (a) 60 (b) 50 (c) 1 (d) 42
9. (−1, −6)
11. (a) 90°, 45°, 45° (b) 63.4°, 90°, 26.6° (c) 45°, 108.4°, 26.6°
    (d) 87.7°, 37.9°, 54.5°
19. $t = -7$ or $t = \frac{101}{7}$

Exercise 11.5

1. (a) $r = 3i + 2j + t(3i + j)$ (b) $r = 2i - j + t(i + j)$ (c) $r = (t - 4)i$
   (d) $r = 2i - 5j + t(-7i + 3j)$
3. (a) $x - 3y = -3$ (b) $x - y = 3$ (c) $y = 0$ (d) $3x + 7y = -29$
5. (a) $i + 2j$ (b) $3i - 2j$ (c) $2i - j$ (d) $4i + 3j$ (e) j (f) i
7. (a) $3x - 2y = -7$ (b) $4x + 5y = -23$ (c) $2x + y = 3$ (d) $3x - y = 12$
   (e) $y = -4$ (f) $x = 3$
9. (a) $3i - j$ (b) 13 m s^{-1} (c) 10.5 s

Exercise 11.6

1. (a) $2i + 7j - 3k$ (b) $2i - j - 9k$ (c) $4i - 6j - 21k$ (d) 7 (e) 5 (f) $\sqrt{62}$
3. (a) $(5, -2, 5)$ (b) $(\frac{15}{4}, \frac{1}{2}, \frac{5}{4})$ (c) $(\frac{13}{3}, -\frac{2}{3}, 3)$ (d) $(\frac{36}{7}, -\frac{16}{7}, \frac{38}{7})$
5. (a) parallelogram (b) rhombus (c) rectangle (d) square (e) rectangle
9. (1, 0, 0)
13. 2

Exercise 11.7

1. 0, k, $-j$, i, 0, 0
5. (a) 7 (b) 30 (c) 15 (d) $7\sqrt{3}$

Answers

9. (a) 0 (b) $-abc$

Exercise 11.8
1. (a) 17 (b) 14 (c) 7 (d) $3\frac{1}{2}$
3. (b) (i) coplanar (ii) not coplanar (iii) coplanar

Exercise 11.9
1. (a) $r = 2i - 3j + k + \lambda(i + 2j - 3k)$ (b) $r = i - j + \lambda(i + j + k)$
3. (a) $r = i + k + \lambda(i - j + 2k)$ (b) $r = 7j - 7k + \mu(i + 2j - k)$ (c) $-2i + 3j - 5k$
5. (AM): $r = a + \lambda(2a - b - c)$; (BD): $r = b + \mu(-a + 2b - c)$

Exercise 11.10
1. (a) $r = 2i + 3j + 4k + \lambda(2i - 3j + 2k) \mu(j + 2k)$
   (b) $r = -2k + \lambda(3i + 3j - k) + \mu(i - j + k)$
   (c) $r = -2i - j - 3k + \lambda(i + k) + \mu(2i + j + k)$
   (d) $r = 5i + j - 4k + \lambda(i - j + k) + \mu(3i - j - k)$
3. (a) $r = (1 - \lambda - \mu)(i + j) + \lambda(j + 4k) + \mu(i + 5k)$ (b) $r \cdot (4i + 5j + k) = 9$
   (c) $4x + 5y + z = 9$
5. (a) $r \cdot (3i - j - 2k) = 7$ (b) $r \cdot (2i + 3j - 4k) = 9$
7. $r \cdot (3i + 4j - k) = 7$
9. $i - j + k$, $5i + k$, $x - 4y - 5z + 27 = 0$
11. (a) $2x + 3y - 4z = 29$
    (b) $r = (1 - \lambda - \mu)(6i) - 3\lambda k + 3\mu(i + 2j)$ or $2x + y - 4z = 12$
    (c) $r = \lambda(5i + 2j - 7k) + \mu(-2i + 4j - 2k)$ or $x + y + z = 0$
    (d) $r = i + j - k + \lambda(2i + j + 2k) + \mu(5j + 4k)$ or $3x + 4y - 5z = 12$
    (e) $2x - y = 4$ (f) $5y + 2z = -11$ (g) $x = 3$
13. $x - 2y - z = 0$

Exercise 11.11
1. (a) 80.4° (b) 72.5° (c) 38.9°
3. (a) 26.4° (b) 58.9° (c) 11.1° (d) 40.2°

Exercise 11.12
1. (a) skew (b) intersect at $(5, 3, -1)$ (c) coincident (d) intersect at $(-5, 6, 4)$
   (e) parallel and distinct
3. $x - 5y + 3z = 0$
5. $4i - j - 2k$
7. (a) $3j$ (b) $-i + 2j + 3k$ (c) line is parallel to the plane (d) $5i + 4j - k$
   (e) line is in the plane (f) $\frac{5}{2}i + \frac{1}{2}j + k$
9. (a) $x = t + 1$, $y = -t$, $z = t$ (b) $x = 5 - 9t$, $y = 1 - 4t$, $z = t$
   (c) $x = 3 - 7t$, $y = 3 - t$, $z = 3t$ (d) $x = 2 - t$, $y = t - 1$, $z = t$
11. $x = 5t + 8$, $y = -7t - 7$, $z = t$, $(18, -21, 2)$

Answers

Exercise 11.13
1. (a) $\sqrt{2}$ (b) 3 (c) $\sqrt{10}$ (d) $\frac{1}{2}\sqrt{138}$
3. (a) 7, (1, 2, 3) (b) $2\sqrt{6}$, (2, –1, –1) (c) 6, (4, 0, 0) (d) $\frac{1}{3}\sqrt{42}$, $(\frac{5}{3}, \frac{4}{3}, \frac{11}{3})$
5. (a) $\sqrt{14}$ (b) 4 (c) $\sqrt{2}$ (d) $\frac{10}{\sqrt{6}}$ (e) $\frac{18}{\sqrt{35}}$ (f) $\frac{15}{\sqrt{17}}$
7. (a) $\frac{3}{2}\sqrt{11}$ (b) 4
9. (a) 48 (b) 11
11. (a) skew, 3 (b) intersect at (2, 1, –7) (c) skew, $3\sqrt{3}$ (d) skew, $\dfrac{7}{2\sqrt{59}}$
13. $2x + y - z = 10$, $\sqrt{6}$

Exercise 12.1
1. (a) $9(3x+1)^2$ (b) $20(5x-2)^3$ (c) $-15(1-3x)^4$ (d) $6x(x^2+2)^2$
   (e) $5(1-2x)(1+x-x^2)^4$ (f) $\dfrac{-2}{(2x+5)^2}$ (g) $\dfrac{8x}{(3-x^2)^2}$ (h) $\dfrac{-12x^2}{(x^3+1)^3}$
   (i) $\dfrac{90x}{(1-3x^2)^4}$ (j) $\dfrac{5}{2\sqrt{5x-4}}$ (k) $\dfrac{4x}{3(2x^2+5)^{2/3}}$ (l) $\dfrac{15x}{(12-5x^2)^{3/2}}$
3. (a) $20(\sqrt{2}+1)$ cm (b) 5
5. (a) (–1, 2) – min ; (–3, –2) – max   (b) (1, 2) – min
   (c) (0, –2) – max ; $(\frac{4}{3}, 6)$ – min   (d) (–1, 128) – max ; (1, 32) – min
7. (a) $6ax + (a-1)^3 y = 9a - 3$, $a = \frac{1}{3}$ (b) $a = 8$, $b = -6$
   (c) $8(2a+1)^3 x - y = (2a+1)^3(6a-1)$; $y = 0$ and $y = \frac{512}{27}x$

Exercise 12.2
1. (a) $3(2x+3) + 2(3x+2)$ (b) $2x(2x-1) + 2(x^2+1)$
   (c) $4x(1-3x) - 3(2x^2-1)$ (d) $6x(2x^3+5) + 6x^2(3x^2-2)$
   (e) $2\sqrt{2x^2+3} + \dfrac{4x^2}{\sqrt{2x^2+3}}$ (f) $6x\sqrt{3x^2+1} + \dfrac{3x(3x^2-1)}{\sqrt{3x^2+1}}$
   (g) $3(2x-1)^3 + 6(2x-1)^2(3x+1)$
   (h) $-4x(1-x^2)(1+x^2)^3 + 6x(1-x^2)^2(1+x^2)^2$
   (i) $3(2x+1)^{3/2} + 9x(2x+1)^{1/2}$ (j) $6(3x-2)^2(2x-3)^2 + 6(2x-3)^3(3x-2)$
   (k) $\dfrac{(3x+1)^{3/2}}{(2x-1)^{1/2}} + \frac{9}{2}(2x-1)^{1/2}(3x+1)^{1/2}$ (l) $-\sqrt{2-3x^2} - \dfrac{3x(1-x)}{\sqrt{2-3x^2}}$
3. (a) (–1, –1) – min   (b) (–3, –3) – min   (c) (6, 96) – max ; (14, 0) – min
5. (a) $t = 2$, $x = 4$; $t = 4$, $x = 0$ (b) $2 < t < 4$, 3 m s^{-1}

741

Answers

(c)

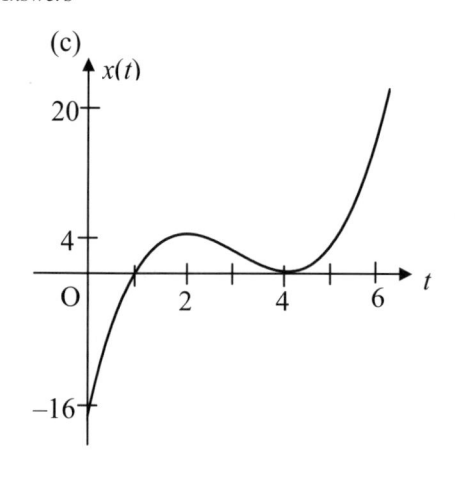

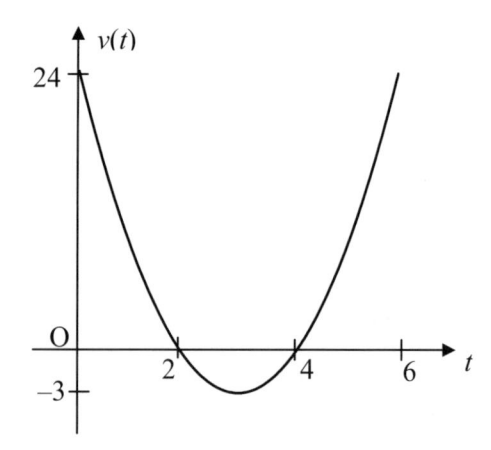

(d) (i) 44 m (ii) $\frac{22}{3}$ m s^{-1} (iii) 6 m s^{-1}

7. 12 cm × 12$\sqrt{3}$ cm

Exercise 12.3

1. (a) $\dfrac{2}{(x+2)^2}$ (b) $\dfrac{9}{(x+5)^2}$ (c) $\dfrac{6x(x+5)}{(2x+5)^2}$ (d) $\dfrac{-6x-7}{(2x-1)^3}$

   (e) $\dfrac{3-8x-3x^2}{(x^2+1)^2}$ (f) $\dfrac{2}{(x^2+2)^{3/2}}$ (g) $\dfrac{16x}{(x^2+4)^2}$ (h) $\dfrac{6-8x}{(2x+3)^4}$

   (i) $\dfrac{2x^2-8x-7}{(x-2)^2}$ (j) $\dfrac{x+1}{(2x+1)^{3/2}}$ (k) $\dfrac{-2}{\sqrt{2x+1}(2x-1)^{3/2}}$ (l) $\dfrac{21-\frac{3}{2}x}{(2x^2-x+7)^{3/2}}$

3. (a) (−1, −1) − min (b) (−2, −9) − max ; (2, $-\frac{1}{9}$) − min
   (c) (3, 2) − max ; (5, 6) − min
5. First month ; 7.5%
7. (a) $a = 9, b = -8$ (b) $a = 2, b = -5$ (c) $a = 3, b = 4, y = 3x + 28$
9. $(-1, \frac{1}{4})$ − max ; $(5, \frac{25}{16})$ − min
11. (c) $x = 15.75$ (d) 27.3 cm

Exercise 12.4

1. (a) $5\dfrac{dy}{dx}$; $6y\dfrac{dy}{dx}$; $\dfrac{1}{2\sqrt{y}}\dfrac{dy}{dx}$; $\dfrac{-1}{y^2}\dfrac{dy}{dx}$; $\dfrac{-6}{y^4}\dfrac{dy}{dx}$

   (b) $y+x\dfrac{dy}{dx}$; $2xy+x^2\dfrac{dy}{dx}$; $3y^2+6xy\dfrac{dy}{dx}$; $\dfrac{2xy-x^2\dfrac{dy}{dx}}{y^2}$; $12xy^3+18x^2y^2\dfrac{dy}{dx}$

(c) $2(5x-2y)\left(5-2\dfrac{dy}{dx}\right)$; $3(2x+3y)^2\left(2+3\dfrac{dy}{dx}\right)$; $\dfrac{2(x+y)-2x\left(1+\dfrac{dy}{dx}\right)}{(x+y)^2}$;

$\dfrac{1}{\sqrt{x^2+y^2}}\left(x+y\dfrac{dy}{dx}\right)$; $\dfrac{(x-2y)\left(1+2\dfrac{dy}{dx}\right)-(x+2y)\left(1-2\dfrac{dy}{dx}\right)}{(x-2y)^2}$

3. (a) $7x-y=13$, $x+7y=9$ (b) $9x-8y=6$, $8x+9y+43=0$
   (c) $9x+10y+8=0$, $10x-9y+29=0$ (d) $x+3y=3$, $3x-y+1=0$
5. (a) $(0, \pm 2)$, $(\pm 3, 0)$ (b) $(\tfrac{1}{2}, \tfrac{3}{4})$, $(-\tfrac{3}{2}, -\tfrac{1}{4})$
9. $-\tfrac{3}{16}$

Exercise 12.5

1. 60 cm^2 s^{-1}
3. 32π cm^2 s^{-1}
5. (a) $\tfrac{1}{4}$ cm s^{-1} (b) $\tfrac{1}{9}$ cm s^{-1}
7. $\tfrac{1}{2}$ m; $\tfrac{4}{15}$ m min^{-1}
9. $\tfrac{20}{73}\sqrt{73}$ km h^{-1}

Exercise 12.6

1. (a) $x=1, y=2$ (b) $x=0, y=0$ (c) $x=-\tfrac{5}{2}, y=-\tfrac{3}{2}$ (d) $x=1, x=2, y=0$
   (e) $x=1, x=2, y=0$ (f) $x=3, x=-1, y=1$ (g) $x=0, y=x+5$
   (h) $x=1, y=-\tfrac{1}{2}x-\tfrac{3}{2}$

Exercise 12.7

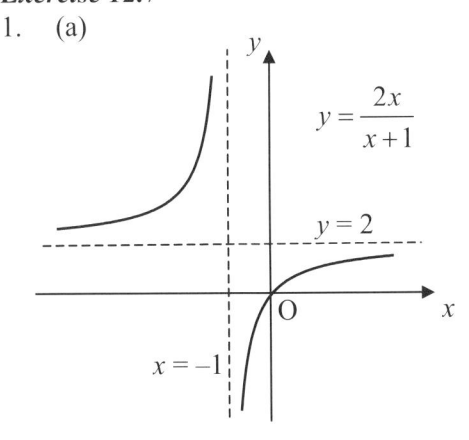

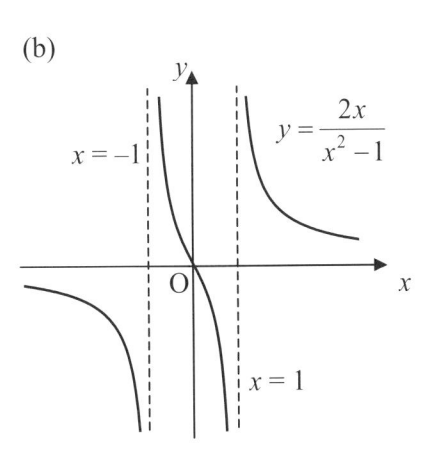

1. (a) [graph of $y=\dfrac{2x}{x+1}$ with asymptotes $x=-1$ and $y=2$]
   (b) [graph of $y=\dfrac{2x}{x^2-1}$ with asymptotes $x=-1$ and $x=1$]

743

Answers

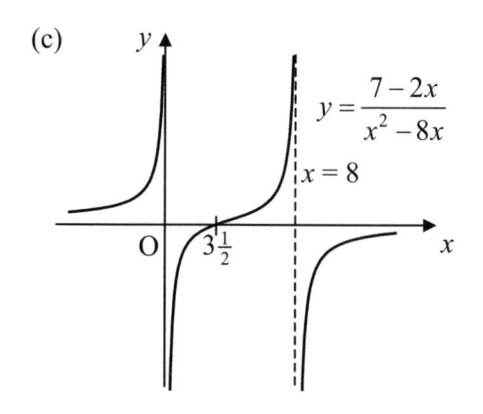

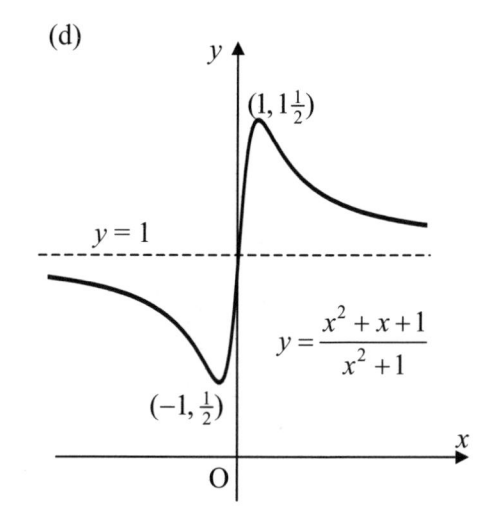

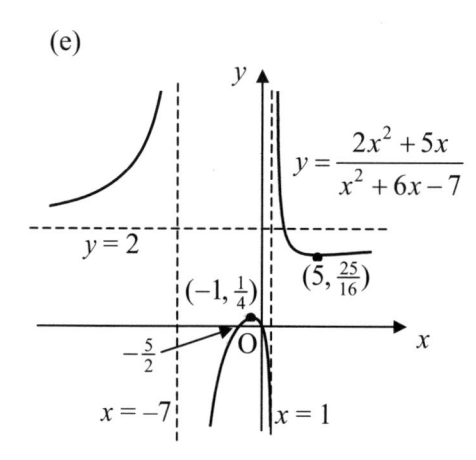

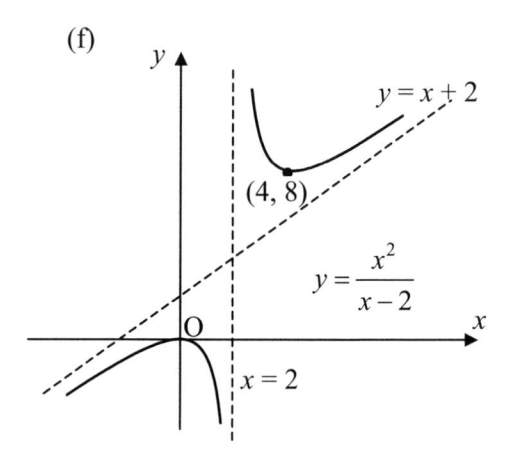

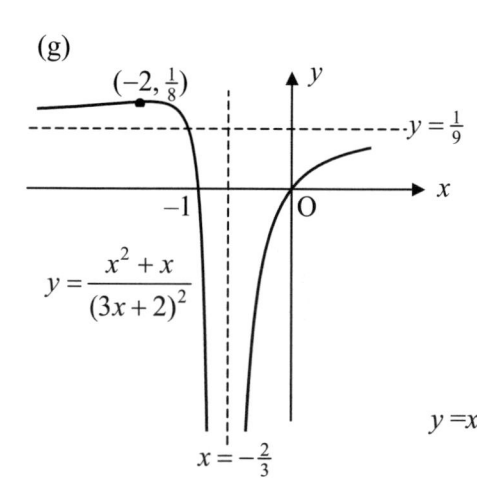

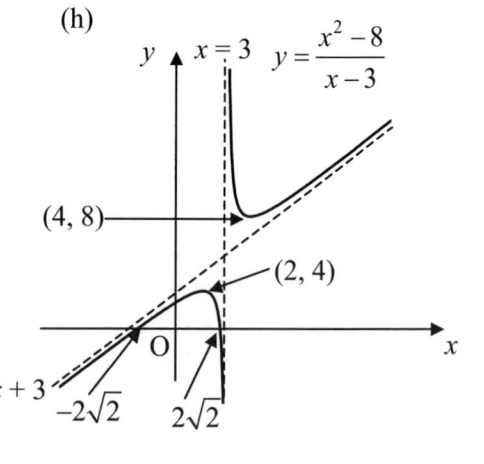

(c) $y = \dfrac{7-2x}{x^2-8x}$; asymptotes $x = 8$; intercept $3\tfrac{1}{2}$

(d) $y = \dfrac{x^2+x+1}{x^2+1}$; asymptote $y = 1$; turning points $(1, 1\tfrac{1}{2})$ and $(-1, \tfrac{1}{2})$

(e) $y = \dfrac{2x^2+5x}{x^2+6x-7}$; asymptotes $y = 2$, $x = -7$, $x = 1$; turning points $(-1, \tfrac{1}{4})$ and $(5, \tfrac{25}{16})$; intercept $-\tfrac{5}{2}$

(f) $y = \dfrac{x^2}{x-2}$; asymptotes $y = x+2$, $x = 2$; turning point $(4, 8)$

(g) $y = \dfrac{x^2+x}{(3x+2)^2}$; asymptotes $y = \tfrac{1}{9}$, $x = -\tfrac{2}{3}$; turning point $(-2, \tfrac{1}{8})$; intercept -1

(h) $y = \dfrac{x^2-8}{x-3}$; asymptotes $y = x+3$, $x = 3$; turning points $(4, 8)$ and $(2, 4)$; intercepts $-2\sqrt{2}$, $2\sqrt{2}$

744

(i)

$$y = \frac{1-x}{(x+1)^2}$$

points: $(3, -\frac{1}{8})$; asymptote $x = -1$

(j)

$$y = \frac{2}{(x+1)^2}$$

asymptote $x = -1$

(k)

$$y = \frac{2x+3}{x^2 + 2x + 3}$$

points: $(0, 1)$, $(-3, -\frac{1}{2})$, $-\frac{3}{2}$

(l)

$$y = \frac{4x^2 + 6x + 15}{x^2 + 4x + 3}$$

point $(-2, 1)$; asymptotes $y = 4$, $x = -3$, $x = -1$; point 5

3.

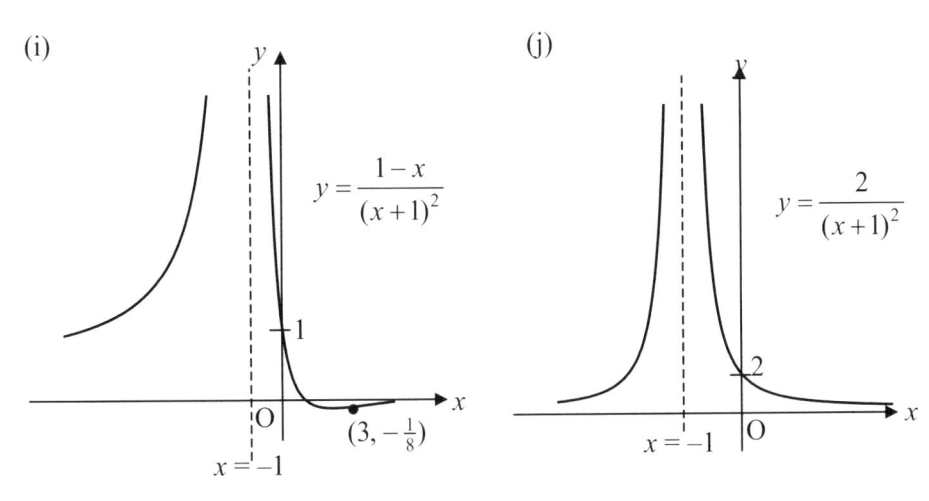

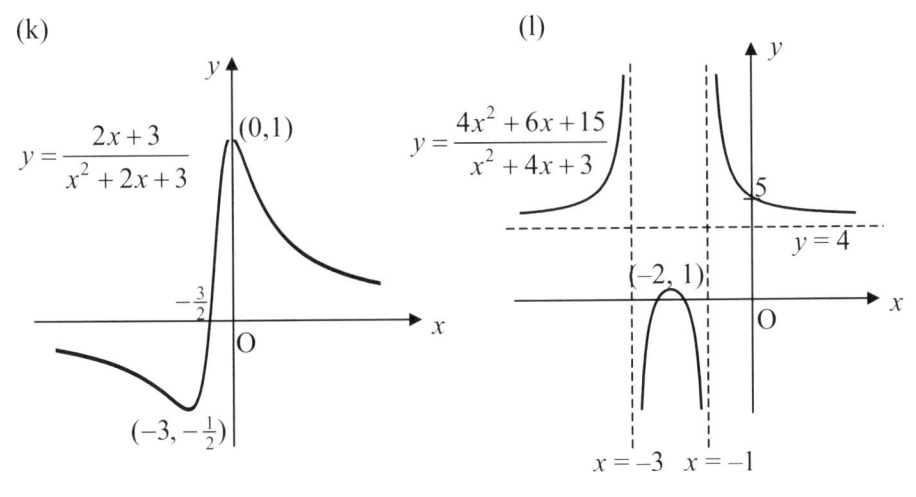

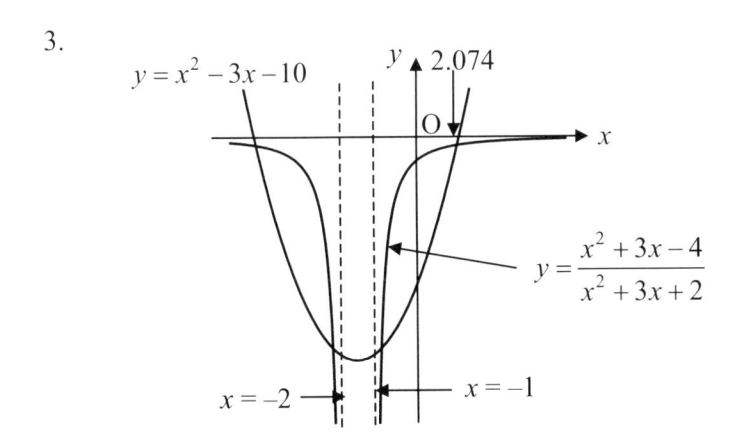

$y = x^2 - 3x - 10$

$$y = \frac{x^2 + 3x - 4}{x^2 + 3x + 2}$$

y-value 2.074; $x = -2$, $x = -1$

$x = 2.074$

745

Answers

5.

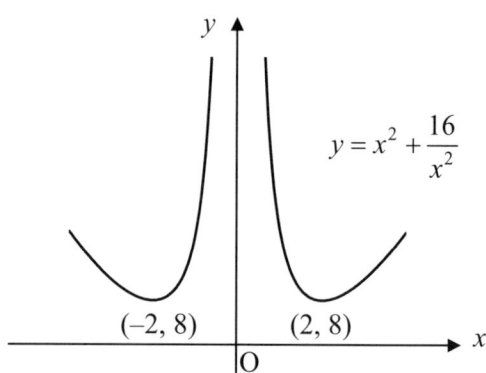

Minimum $y = 8$

Graph of $y = x^2 + \dfrac{16}{x^2}$ with minima at $(-2, 8)$ and $(2, 8)$.

Exercise 13.1
1. (a) $\{1, 3, 4, 5, 6, 7\}$ (b) $\{3, 5, 7\}$ (c) $\{4, 6\}$ (d) $\{1, 2, 3, 5, 7\}$ (e) $\{1\}$
   (f) $\{2\}$
3. (a) $\{3, 4, 5, 6\}$ (b) $\{1, 4, 9, 16, 25\}$ (c) $\{2, 3, 4, 5, 6\}$
   (d) $\{1, 11, 13, 143\}$ (e) $\{1\}$ (f) $\{1\}$
5.

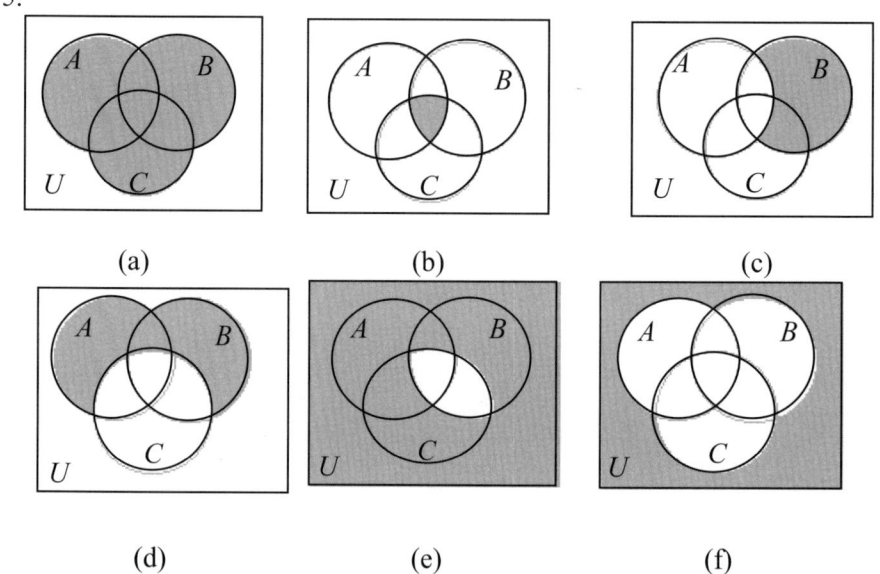

(a) (b) (c)

(d) (e) (f)

9. (a) 250 (b) 166 (c) 83 (d) 667

Exercise 13.2
1. (a) $\frac{1}{2}$ (b) $\frac{1}{2}$ (c) $\frac{1}{2}$
3. (a) $\frac{1}{4}$ (b) $\frac{3}{4}$ (c) $\frac{7}{24}$ (d) $\frac{11}{12}$

5.

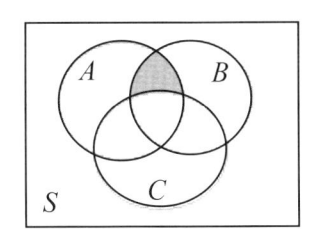

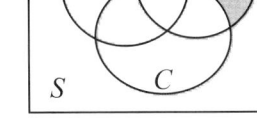

(a) $A \cap B \cap C'$ (b) $A' \cap B \cap C'$

7. (a) 0.5 (b) 0.14 (c) 0.07 (d) 0.43
9. (a) $\frac{1}{17}$ (b) $\frac{15}{34}$
11. (a) $\frac{1}{6}$ (b) $\frac{5}{126}$

Exercise 13.3
1. $\frac{13}{24}$
3. (a) 0.2 (b) 0.5 (c) 0.6
5. (a) $\frac{5}{8}$ (b) $\frac{23}{24}$ (c) $\frac{1}{8}$ (d) $\frac{3}{8}$
9. (a) $\frac{1}{9}$ (b) $\frac{5}{9}$
11. (a) $\frac{1}{216}$ (b) $\frac{5}{72}$ (c) $\frac{91}{216}$
13. $\frac{3}{4}$

Exercise 13.4
1. $\frac{2}{5}$
3. (a) $\frac{375}{4096}$ (b) $\frac{3}{28}$
5. (a) $\frac{67}{120}$ (b) $\frac{24}{53}$
7. (a) $\frac{1}{10}$ (b) $\frac{3}{10}$
9. $\frac{1}{48}$
11. $\frac{3}{13}$
13. 9

Exercise 13.5
1. (a) $\frac{3}{8}$ (b) $\frac{1}{4}$ (c) $\frac{1}{16}$ (d) $\frac{11}{16}$
3. $\frac{2048}{6561} \approx 0.312$
5. (a) 375 (b) 62.5 (c) 625
7. 0.00376
9. $\frac{896}{3^8} \approx 0.137$

Answers

11. (a) 0.2 (b) 0.2
13. (c) (i) 5 (ii) 7 or 8

Exercise 14.1

1. (a), (c), (f) represent discrete variables
3.

(a)
x	P(X = x)
0	1/27
1	2/9
2	4/9
3	8/27

(b)
x	P(X = x)
1	1/5
2	3/5
3	1/5

5. (a) 0.1 (b) 0.8 (c) 0.375
7.

x	P(X = x)	x	P(X = x)
0	0.6588	3	0.0017
1	0.2995	4	0.0000
2	0.0399		

9. $P(X = x) = \left(\frac{5}{6}\right)^{x-1}\left(\frac{1}{6}\right), x = 1, 2, 3, \ldots$

11.

x	P(X = x)
0	1/2
1	3/10
2	3/20
3	1/20

13.

x	P(X = x)
0	0.2373
1	0.3955
2	0.2637
3	0.0879
4	0.0146
5	0.0010

Exercise 14.2

1. (a) 1.9 (b) 2.4 (c) 0.4 (d) 9.23
3. 1
5. 0
7. 1

Exercise 14.3
1. (a) 2, 1   (b) 1, 0.756
3. (a) 3.2, 0.8   (b) 2.4, 0.8
5. 1.875, 0.709
7. 0.75, 0.887
9. (a) 20/3   (b) 60   (c) 40/3   (d) 200/3

Exercise 14.4
2. (a) 5, 2.5   (b) 20, 16   (c) 12, 9   (d) 90, 36
3. (a) 0.117   (b) 0.974
4. 0.763

Exercise 14.5
1. (a) 0.964   (b) 0.285
2. (a) 0.994   (b) 0.159   (c) 0.302
3. (a) 0.900   (b) –0.400   (c) 0.230   (d) –1.65
5. (a) 0.841   (b) 0.691   (c) 0.547   (d) 0.0455
7. (a) 0.106   (b) 0.0304
9. 0.122
11. (a) 2.90   (b) 0.365   (c) 0.484
13. (a) 0.444   (b) 11.5   (c) 0.00155

Exercise 14.6
1. (a) 20, 3.43   (b) 54, 15.2   (c) 103.6, 29.2
3. (a) 32.2, 13.9   (b) 54.0, 18.5
5. Y_1 and Y_2 are unbiased; Y_1 is the more efficient

Exercise 14.7
1. (27.9, 32.1), (27.6, 32.4), (27.1, 32.9)
3. (a) (24.5, 25.9)   (b) (3.37, 3.47)   (c) (142, 168)
5. (15.4, 16.6)
7. (a) 1080   (b) 1540   (c) 2160

Exercise 14.8
1. No evidence to support "$\mu = 25$".
3. Students have not underperformed.
5. (a) (19.03, 21.97)   (b) (18.57, 22.43)   (c) (18.03, 22.97)

Exercise 14.9
1. (a) independent   (b) not independent   (c) not independent
   (d) independent
3. Job satisfaction is independent of job.
5. There is evidence of an association between political association and age at the 5% level of significance.

749

Answers

Exercise 15.1

1. (a) $2x^2 - 5x$ (b) $2x^3 + 2x^2 + 3x$ (c) $\frac{1}{4}x^4 + \frac{1}{3}x^3 + \frac{1}{2}x^2$
   (d) $3x^3 - 12x^2 + 16x$ (e) $5x$ (f) $-\frac{8}{3}x^3 + 9x^2 - 9x$ (g) $\frac{2}{5}x^5 - x^3 - 5x$
   (h) $\frac{4}{5}x^5 - \frac{4}{3}x^3 + x$ (i) $\frac{1}{4}x^4 - 2x^3 + 6x^2 - 8x$

3. $y = 2x^3 - 3x^2 - 2x + 3$; $2x + y = 2$

5. (a) $y = 2x^2 - x + 3$ (b) $y = 3x + \dfrac{2}{x} + 8$ (c) $y = \frac{9}{2}x^2 - 8x^{3/2} + 4x - 20$
   (d) $y = \frac{1}{2}x^2 - \frac{9}{5}x^{5/3} + \frac{9}{4}x^{4/3} - x + \frac{8}{5}$

7. (a) $y = 2x^3 - x^2 + x - 4$ (b) $y = x^3 + x^2 - 2x$, $(0, 0)$, $(-2, 0)$

9. $y = x^3 - x^2 - x + 2$; local maximum at $\left(-\frac{1}{3}, \frac{59}{27}\right)$

11. $t = 0, x = 0$ and $t = 0.6, x = -0.744$

Exercise 15.2

1. (a) $\frac{1}{6}(2x+1)^3 + c$ (b) $-\frac{1}{20}(3-5x)^4 + c$ (c) $\frac{1}{20}(5x+2)^4 - 2x + c$
   (d) $3x^2 + \frac{1}{15}(1-3x)^5 + c$ (e) $x^3 + \frac{1}{4}(5-2x)^4 + c$ (f) $\frac{4}{5}x^5 - 4x^3 + 9x + c$
   (g) $2x^2 + x + (2-3x)^4 + c$ (h) $\frac{1}{6}(2x+3)^3 + \frac{9}{5}x^5 + 4x^3 + 4x + c$

3. (a) $\frac{1}{9}(6x+3)^{3/2}$ (b) $\frac{2}{25}(5x-2)^{5/2}$ (c) $\frac{3}{16}(8x+1)^{2/3}$ (d) $\frac{2}{3}(2-3x)^{-1/2}$
   (e) $\frac{1}{3}(2x+5)^{3/2}$ (f) $\frac{1}{4}(3x-8)^{4/3}$ (g) $4(2x+3)^{1/2}$ (h) $\frac{4}{3}(1-3x)^{-1/2}$

5. (a) $a = 6, b = 3$, $y = 6\sqrt{2x+1} - 3x + 1$ (b) $3x - y + 7 = 0$

Exercise 15.3

1. (a) $\frac{1}{4}(3x^2 - 1)^4 + c$ (b) $-\frac{1}{5}(1-x^2)^5 + c$ (c) $\frac{1}{9}(x^3 + 2)^3 + c$
   (d) $\frac{1}{16}(x^4 - 3)^4 + c$ (e) $\frac{3}{40}(2x^5 + 3)^4 + c$ (f) $-\frac{2}{9}(1-3x^2)^6 + c$
   (g) $-\frac{1}{21}(2-3x^7)^3 + c$ (h) $\frac{3}{32}(4x^4 - 2)^6 + c$ (i) $-\frac{5}{12}(3-x^3)^4 + c$
   (j) $-\frac{1}{6}(2x^{-1} - 3)^3 + c$ (k) $-\frac{1}{12}(3x^4 + 1)^{-1} + c$ (l) $\frac{1}{2}(x^{-2} + 1)^{-1} + c$
   (m) $\frac{1}{8}x^8 + \frac{4}{5}x^5 + 2x^2 + c$ (n) $\frac{1}{9}(x^3 + 2)^3 + c$ (o) $\frac{1}{10}x^{10} + \frac{4}{7}x^7 + x^4 + c$

3. $a = 1, b = -2, c = 3$; $\frac{1}{2}x^2 - 2x - \dfrac{3}{2(x^2+1)} + c$

Exercise 15.4

1. (a) 6 (b) $-\frac{32}{3}$ (c) $\frac{31}{3}$ (d) $-\frac{1}{8}$ (e) $\frac{4}{3}$ (f) $\frac{19}{3}$ (g) $\frac{56}{15}$ (h) $\frac{3}{4}$

3. (a) $\frac{15}{4}$ (b) 2 (c) 9 (d) $\frac{8}{3}$

5. (a) $\frac{9}{2}$ (b) $\frac{37}{12}$ (c) $\frac{32}{3}$ (d) $\frac{148}{3}$

Answers

7. (a) 3.14 (b) 1.10 (c) 10.5 (d) 2.28
9. $\frac{16}{15}a^5$
11. (a) $\frac{125}{6}a^3$ (b) $\frac{9}{8}a^3$ (c) $\frac{1}{2}a^4$
15. $\frac{2}{3}a^{3/2} - a + \frac{1}{3}$

Exercise 16.1

1. (a) $\to 3$ (b) no limit (c) $\to 3$ (d) $\to \frac{1}{3}$ (e) no limit (f) $\to 9$
3. (a) 2 (b) $\frac{1}{5}\sqrt{2}$ (c) $-\frac{1}{2}$ (d) $\frac{1}{32}$ (e) $\frac{2}{5}$ (f) $\frac{1}{20}$

Exercise 16.2

1. (a) $2\cos 2x - 2\sin 2x$ (b) $-2\sin x + 6\sin 2x$ (c) $1/\cos^2 x$
   (d) $\cos 3x - 3x\sin 3x$ (e) $-4\sin 2x - 4\sin^2 2x + 2\cos^2 2x$
   (f) $2/\cos^2 x + 6\sin 2x$ (g) $2x\cos 3x - 3x^2 \sin 3x$ (h) $\dfrac{1}{1+\cos x}$
   (i) $\dfrac{2}{1-\sin 2x}$ (j) $\dfrac{2x(3-\tan 2x + x/\cos^2 2x)}{(3-\tan 2x)^2}$

3. (a) $6x - y + 4 = 0$ (b) $x + y = \pi$ (c) $a = \frac{2}{3}$, $b = -2$

5. (a) (0.395, 13) – max ; (3.54, –13) – min
   (b) $\left(\frac{1}{6}\pi, \frac{3}{2}\sqrt{3}\right)$ – max ; $\left(\frac{5}{6}\pi, -\frac{3}{2}\sqrt{3}\right)$ – min ; $\left(\frac{3}{2}\pi, 0\right)$ – inflexion
   (c) (–1.89, 7) – max ; (1.25, 7) – max ; (–0.322, –3) – min ; (2.82, –3) – min
   (d) $\left(\frac{1}{6}\pi, \frac{1}{6}\pi - \frac{1}{2}\sqrt{3}\right)$ – min ; $\left(\frac{5}{6}\pi, \frac{5}{6}\pi + \frac{1}{2}\sqrt{3}\right)$ – max
   (e) (0, 0) – inflexion ; (π, –π) – min
   (f) $\left(\frac{1}{6}\pi, \sqrt{3}\right)$ – min ; $\left(\frac{5}{6}\pi, -\sqrt{3}\right)$ – max
   (g) $\left(-\frac{1}{3}\pi, -3\sqrt{3}\right)$ – min ; $\left(\frac{1}{3}\pi, 3\sqrt{3}\right)$ – max

7. (a) $\frac{1}{2}\cos\frac{1}{2}a$ (b) $-2\sin 2a$ (c) $3/\cos^2 3x$ (d) $1/(3\cos^2 \frac{1}{3}x)$
   (e) $2\cos 2x$ (f) $-\pi\sin \pi x$

9. (a) $3.44°\ \text{s}^{-1}$ (b) $-\frac{255}{8}\ \text{cm}^2\ \text{s}^{-1}$
11. $-6.41°\ \text{s}^{-1}$; $0.447\ \text{m s}^{-1}$

Exercise 16.3

1. (a) $-\frac{1}{3}\cos 3x + c$ (b) $4\cos\left(\frac{1}{2}\pi - \frac{1}{4}x\right) + c$ (c) $2\sin\frac{1}{2}x + c$
   (d) $\frac{2}{5}\sin 5x + c$ (e) $\frac{3}{2}\tan\frac{2}{3}x + c$ (f) $2\tan\frac{1}{2}(x-2) + c$
   (g) $\frac{1}{2}x^2 + \frac{2}{3}\cos 3x + c$ (h) $\frac{3}{2}\sin 2(x-1) - x^3 + c$ (i) $\frac{3}{2}\tan(2x - 3) + c$

751

Answers

3. (c) (i) $\frac{1}{2}x+\frac{1}{12}\sin 6x+c$ (ii) $\frac{1}{2}x-\frac{1}{8}\sin 4x+c$ (iii) $\frac{1}{3}x^3-\frac{1}{2}x-\frac{1}{2}\sin x+c$
   (iv) $\frac{7}{2}x-\frac{1}{20}\sin 10x-x^2+c$

5. (a) $-6\cos 3x \sin 3x$ (b) $-\frac{1}{6}\cos^2 3x+c$

7. (a) $y=-\cos 2x-\sin x+2$ (b) $y=-\frac{1}{8}\sin 4x+\frac{5}{2}x+4$
   (c) $y=-\frac{1}{2}\cos 2x-\sin x+4x+3$

Exercise 16.4

1. (a) $-2\csc 2x \cot 2x$ (b) $\frac{1}{2}\sec\frac{1}{2}x \tan\frac{1}{2}x$ (c) $-4\csc^2 4x$

3. (a) $3\sec^2 2x+2\sec 2x \tan 2x$ (b) $2\tan x \sec^2 x-4\sec^2 2x \tan 2x$
   (c) $-2\csc^2 2x \cot 2x+6\csc^2 3x \cot 3x$ (d) $-2\csc^2 x+4\csc^2 2x \cot 2x$
   (e) $\tan 2x \sec x \tan x+2\sec x \sec^2 2x$
   (f) $\frac{1}{2}\sec^2\frac{1}{4}(x-\pi)\tan\frac{1}{4}(x-\pi)+\frac{3}{2}\tan^2\frac{1}{2}(x-\pi)\sec^2\frac{1}{2}(x-\pi)$

Exercise 16.5

1. (a) $\frac{3}{2}\tan\frac{2}{3}x+c$ (b) $-2\cot\frac{1}{2}(x-2)+c$ (c) $\frac{1}{2}-\frac{1}{2}\sin x+c$
   (d) $\frac{1}{2}x+\frac{1}{12}\sin 2(1-3x)+c$ (e) $\frac{1}{2}x+\frac{1}{3}\sin\frac{3}{2}x+c$ (f) $\frac{1}{2}x+\frac{1}{2\pi}\sin \pi x+c$
   (g) $\frac{1}{3}\tan 3x-x+c$ (h) $\frac{3}{2}\tan\frac{2}{3}x-x+c$ (i) $-\frac{1}{5}\cot 5x-x+c$
   (j) $\frac{1}{4}\cot(\frac{2}{3}\pi-4x)-x+c$ (k) $-\frac{1}{3}\cos^3 x+c$ (l) $-\frac{1}{9}\cos^3 3x+c$
   (m) $2\sqrt{\sin x}+c$ (n) $-2\csc 2x+c$

3. (a) $\sin x+\frac{1}{2}\cos 2x+c$ (b) $-\frac{1}{8}\cos 4x-\frac{1}{2}\cos 2x+c$
   (c) $\frac{9}{2}x+\frac{4}{3}\cos 3x-\frac{1}{12}\sin 6x+c$ (d) $x+\frac{1}{4}\cos 4x+c$ (e) $-2\csc x+c$
   (f) $-\frac{1}{3}\sin 3x-\frac{1}{3}\tan 3x+c$ (g) $c-\sqrt{3-2\sin x}$ (h) $\dfrac{1}{12(1+2\cos 3x)^2}+c$
   (i) $\tan x-x+c$ (j) $\tan 2x+c$ (k) $c-\frac{1}{2}\cos(x^2)$ (l) $\frac{1}{3}(\tan(x^3)-x^3)+c$

5. (a) $\frac{3}{8}+\frac{1}{4}\sin 2x+\frac{1}{32}\sin 4x+c$
   (b) $\dfrac{d}{dx}(x\sin x)=\sin x+x\cos x$ $\int x\cos x\, dx=x\sin x-\cos x+c$
   (c) $\sin x-x\cos x+c$

7. $y=\sin x(1-2\cos x)$

   Area $=-\displaystyle\int_0^{\pi/3} y\, dx+\int_{\pi/3}^{\pi} y\, dx=2.5$

Answers

9. 15.3
11. $\frac{4}{3}$

Exercise 17.1
1. e = 2.718282 (12 terms)
2. (a) $3e^{3x+1} - 2e^{-2x}$ (b) $2e^{2-x}$ (c) $e^x + e^{-x}$ (d) $2e^{2x} - 2x - 2e^{-2x}$
   (e) $2e^{2x}(\sin 2x + \cos 2x)$ (f) $(2\cos 2x)e^{\sin 2x}$ (g) $xe^{3x}(2-3x)$
   (h) $\dfrac{12}{(e^{3x}+e^{-3x})^2}$ (i) $2x\exp(\cos 2x)[1 - x\sin 2x]$
3. (a) $\left(-1, -\dfrac{1}{e}\right)$ – min (b) $\left(-2, \dfrac{5}{e^2}\right)$ – max ; $(1, -e)$ – min (c) $(0, 2)$ – min
   (d) $(0, 6)$ – min (e) $(0, 0)$ – inflexion ; $\left(3, \dfrac{27}{e^3}\right)$ – max
   (f) $\left(\tfrac{3}{4}\pi, \tfrac{1}{2}\sqrt{2}e^{3\pi/4}\right)$ – max ; $\left(\tfrac{7}{4}\pi, -\tfrac{1}{2}\sqrt{2}e^{7\pi/4}\right)$ – min
5. (c) $m = 3, 4$
7. (a) 18.4° $(\arctan \tfrac{1}{3})$
9. (a) $(0, 1)$ – max ; $\left(\pm \tfrac{1}{2}\sqrt{2}, e^{-1/2}\right)$ – inflexions

Exercise 17.2
1. (a) $\dfrac{1}{x}$ (b) $\dfrac{1}{x+2}$ (c) $\dfrac{3}{3x+5}$ (d) $\dfrac{6x^2}{2x^3+3}$ (e) $3\cot 3x$ (f) $\dfrac{1}{\sin x \cos x}$
   (g) $3x^2 \ln 2x + x^2$ (h) $\dfrac{1}{x\ln x}$ (i) $\dfrac{2}{4x+3}$ (j) $\dfrac{1-2\ln x}{x^3}$ (k) $-\ln x - 1$
   (l) $-2\tan x$
3. 0.327
5. (a) $\ln 2, \ln 3$ (b) $0, \ln 6$ (c) $-\ln 2$ (d) ± 0.658
7. (a) $\dfrac{1}{x\ln 6}$ (b) $\dfrac{x}{\ln 10}(2\ln x + 1)$ (c) $-\dfrac{1}{x\ln 2}$
9. 6.31

Exercise 17.3
1. (a) $2e^x - 3e^{-x} + c$ (b) $2e^{2x} + 12e^{-x/2} + c$ (c) $20e^{0.2x} - xe^{0.2} + c$
   (d) $-e^{1-2x} + e^{1-3x} + c$ (e) $2x + \tfrac{1}{2}e^{-2x} + c$ (f) $-\tfrac{1}{2}e^{-2x} - 2e^{-x} + x + c$
3. $f(x) = e^{2x} - e^x + \tfrac{5}{2}$
5. 2.56
7. (a) $2k$ m s^{-1} (b) 1.39 s

Answers

Exercise 17.4
1. (a) $\frac{1}{4}(e^x+2)^4+c$ (b) $\exp(\sin x)+c$ (c) $\frac{3}{2}\exp(x^2+1)+c$
   (d) $\exp(\tan x)+c$ (e) $\dfrac{5^{2x}}{\ln 25}+\dfrac{3^x}{\ln 3}+c$ (f) $\dfrac{10^{x^2}}{\ln 100}+c$
   (g) $2\sqrt{e^x+1}+c$ (h) $\frac{1}{3}(e^x+e^{-x})+c$
3. 3.17

Exercise 17.5
1. (a) $\ln(x+3)+c$ (b) $\frac{1}{3}\ln(3x-2)+c$ (c) $-2\ln(x-1)+c$
   (d) $\frac{2}{3}\ln(2-3x)+c$
3. (a) $\ln 4$ (b) $\frac{1}{2}\ln 3$ (c) $\frac{1}{3}\ln 4$ (d) $\ln \frac{1}{2}$ (e) $2-\ln 1.5$ (f) $6-8\ln 3$
5. $\dfrac{5(1-x)}{(2x-1)(3-x)}$; $\frac{1}{10}\ln\frac{3}{16}$

Exercise 17.6
1. (a) $\frac{1}{4}\ln\left|2x^2+4x+1\right|+c$ (b) $\ln(2+\sin x)+c$ (c) $\frac{1}{3}\ln\left|\sin 3x\right|+c$
   (d) $\ln(1+e^x)+c$ (e) $\ln\left|\ln x\right|+c$ (f) $2\ln\left(\sqrt{x}+1\right)+c$
3. $\dfrac{d}{dx}\ln(\sec x+\tan x)=\sec x$; $\displaystyle\int \sec x\,dx=\ln\left|\sec x+\tan x\right|+c$
5. 1.60
7. $y=\frac{1}{2}\ln\frac{1}{2}-\frac{1}{2}\ln\left|\cos 2x\right|+c$

Exercise 18.2
1. 8
3. $\frac{1}{2}n(n-3)$
5. (a) 1, 5, 23, 119, 719 (b) $S_n=(n+1)!-1$
7. $\frac{1}{15},\frac{1}{35},\frac{1}{63},\frac{1}{99}$; $S_n=\dfrac{1}{12}-\dfrac{1}{4(2n+1)(2n+3)}$

Exercise 19.1
1. $6x^2-2x^3-2x-1$; $-6x^5+13x^4-2x^3-3x^2-2x-2$
3. (a) $\deg(P(x)Q(x))=m+n$ (b) $\deg(P(x)+Q(x))\le \max(m,n)$
5. (a) $6x^4+10x^3+9x^2+3x-20$ (b) $8x^5-32x^3+x^2+12x-3$
   (c) $4x^4-12x^3+13x^2-6x+1$ (d) $x^6-3x^5-3x^4+11x^3+6x^2-12x-8$
7. (a) $1+2+3+4=10$ (b) $1+2+3+\ldots+n=\frac{1}{2}n(n+1)$
9. (a) $a=1, b=-3$ (b) $a=23, b=48$ (c) $a=3, b=-2$ (d) $a=-2, b=11$

11. (a) $a = 2, b = -1$ (b) No a, b exist. (c) $a = 3, b = 1$ (d) $a = -2, b = 3$

Exercise 19.2
1. (a) 2 (b) 0 (c) –21 (d) $\frac{5}{3}$
3. (a) ±2 (b) $0, \pm 1$
5. (a) 11 (b) –14
7. $3x - 1$
13. $1 : 3$
15. $2, -5 \,;\; 1, -6$
19. $k = 2, p = 3, y = \sqrt[3]{3} + \sqrt[3]{9} + 2$

Exercise 19.3
1. (a) $x^2 - x + 2, 5$ (b) $2x^2 + 4x + 6, 11$ (c) $3x^2 - 8x, -1$
   (d) $x^3 - x^2 + 3x - 2, -2$
3. (a) $2x - 3, 2x + 3$ (b) $x^2 - 5x + 4, 10x - 16$ (c) $x^3 - 2x^2 + x + 2, 6x - 1$
   (d) $2x^2 + 3x + 6, 10x - 19$
5. (a) $2(x-2)^3 + 11(x-2)^2 + 21(x-2) + 15$
   (b) $3(x+1)^4 - 12(x+1)^3 + 18(x+1)^2 - 11(x+1) + 3$
   (c) $(x-1)^5 + 5(x-1)^4 + 9(x-1)^3 + 7(x-1)^2 + 2(x-1) - 1$
   (d) $2(x+3)^4 - 24(x+3)^3 + 108(x+3)^2 - 219(x+3) + 173$
7. $p = -\frac{19}{27}, f(x) = (x+\frac{1}{3})^2(2x-\frac{19}{3}); \; p = 12, f(x) = (x-2)^2(2x+3)$
9. –3
11. (a) $(2x+1)^3 - 2(2x+1)^2 + 3(2x+1) + 5$ (b) $2x^2 - 7x + 17 \,;\, 35 + 3x - 35x^2$

Exercise 19.4
1. (a) ±1, 2 (b) ±2, –3 (c) –1, 2 (d) $-2, -3, \frac{1}{2}$ (e) $-1, 2, -\frac{3}{2}$ (f) 3
3. (a) $2, 1 \pm \sqrt{2}$ (b) $-1, 2 \pm \sqrt{3}$ (c) $-\frac{1}{2}, -4, 3$ (d) $2, -\frac{1}{2}, \frac{1}{2}\left(-1 \pm \sqrt{13}\right)$
5. (a) –22 (b) (i) $a = 4, b = 4$ (ii) $-2, 3, \frac{1}{2}\left(-3 \pm \sqrt{13}\right)$

Exercise 19.5
1. (a) $x^2 - 3x - 4$ (b) $3x^2 - 7x - 6$ (c) $x^2 - 5$ (d) $x^2 - 2x - 2$
   (e) $x^2 + 4x - 1$ (f) $x^2 + x\sqrt{2} - 4$
3. (a) $2x^2 + 7x + 2$ (b) $2x^2 + x - 37$ (c) $8x^2 + 17x + 8$ (d) $4x^2 - x - 2$
   (e) $4x^2 - 4x + 1$ (f) $4x^2 - 25x + 37$ (g) $8x^2 - 25x - 64$ (h) $16x^2 - 64x + 31$
5. $0, \frac{16}{3}$

Answers

7. (a) 1 (b) $\frac{1}{2}(1+\sqrt{5})$ or $\frac{1}{2}(1-\sqrt{5})$ (c) 6 or $\frac{4}{3}$
9. $c = -\frac{3}{5}$, $x = \frac{1}{3}, -1$

Exercise 20.1
1. (a) $1+i$ (b) $6+2i$ (c) $8+i$ (d) $-7+17i$ (e) $25+0i$ (f) $19-17i$
   (g) $-16-30i$ (h) $2-11i$
3. (a) $21-20i$ (b) $13-11i$ (c) $13+11i$ (d) $\frac{13}{10}-\frac{11}{10}i$ (e) $8+3i$
   (f) $17+i$
5. (a) z^2+9 (b) z^2+36 (c) z^2-2z+5 (d) $z^2+4z+13$
   (e) $z^2-10z+41$ (f) $5z^2-4z+1$
7. (a) $-2i$ (b) $3i$ (c) $\frac{1}{2}-i$ (d) $4-2i$ (e) $\frac{3}{10}+\frac{3}{5}i$ (f) $2-i$
9. (a) $\pm i, 2$ (b) $\pm 2i, -1 \pm 2i$

Exercise 20.2
3. (a) $-2, 4$ (b) $-7, -10$ (c) $1, -7$ (d) $15, -17$ (e) $-8, 8$ (f) $79, 29$
5. (a) $z^3 - 4z^2 + 9z - 10$ (b) $2z^3 - 7z^2 + 6z + 5$ (c) $z^4 - 8z^3 + 27z^2 - 38z + 26$
   (d) $z^4 - 12z^3 + 62z^2 - 172z + 221$ (e) $4z^4 - 8z^3 + 11z^2 - 7z + 3$
   (f) $z^5 - 15z^4 + 83z^3 - 175z^2 + 16z + 290$
7. $p(z) = -2z^3 + 8z^2 - 18z + 20$
9. $z^2 + (1-3i)z - 2(1+3i)$

Exercise 20.3
1. (a) $2\sqrt{2}$ (b) $\sqrt{5}$ (c) $2\sqrt{3}$ (d) 3 (e) 0 (f) 5 (g) 13 (h) 1
   (i) $6\sqrt{3}$ (j) $\frac{1}{2}\sqrt{2}$ (k) $\sqrt{13}$ (l) 35 (m) 65 (n) $\frac{1}{5}\sqrt{29}$ (o) 10
3. $|\sec\theta|$
5. $z = 1+i$, $w = 2+2i$; O, P, Q are collinear with P and Q on the same side of O.
15. $3-i$, $\frac{7}{2}(-1+i)$

Exercise 20.4
1. (a) $\frac{1}{4}\pi$ (b) $\frac{1}{3}\pi$ (c) $-\frac{1}{3}\pi$ (d) $-\frac{1}{2}\pi$ (e) 0 (f) $\frac{5}{6}\pi$ (g) $-\frac{1}{6}\pi$
   (h) $-\frac{3}{4}\pi$ (i) $\frac{1}{3}\pi$

Exercise 20.5
1. (a) $1+i\sqrt{3}$ (b) $1-i$ (c) $3i$ (d) -4 (e) $-\frac{1}{2}\sqrt{3}-\frac{1}{2}i$ (f) $\sqrt{3}-i$
   (g) $-2\sqrt{2}+2i\sqrt{2}$ (h) $-\frac{3}{2}-\frac{3}{2}i\sqrt{3}$ (i) $\frac{1}{2}-\frac{1}{2}i\sqrt{3}$ (j) $3\sqrt{3}+3i$
3. (a) $5\operatorname{cis} 0.927$ (b) $\sqrt{29}\operatorname{cis} 0.381$ (c) $\sqrt{5}\operatorname{cis} 2.68$ (d) $5\operatorname{cis} -2.50$
   (e) $\sqrt{13}\operatorname{cis} -0.983$ (f) $17\operatorname{cis} -2.65$

Answers

5. $\sin\frac{7}{12}\pi = \frac{\sqrt{6}+\sqrt{2}}{4}$ $\cos\frac{7}{12}\pi = \frac{\sqrt{2}-\sqrt{6}}{4}$

7. $(1+i)z$, iz

Exercise 20.6

1. (a) $-64 + 0i$ (b) $0 - 32i$ (c) $-\frac{1}{2} - \frac{1}{2}i\sqrt{3}$ (d) $-1 + 0i$ (e) $-512 + 0i$
   (f) $-128 - 128i$

3. $\operatorname{cis} -\frac{1}{10}(2k+1)\pi$, $(k = 0, 1, 2, 3, 4)$

5. (a) $\pm(1+i)$ (b) $\pm(2+i)$ (c) $-1, \frac{1}{2}\pm\frac{1}{2}i\sqrt{3}$
   (d) $\operatorname{cis}\frac{1}{8}(4k-3)\pi$, $(k = 0, \pm 1, 2)$ (e) $-1, \operatorname{cis}\pm\frac{3}{5}\pi, \operatorname{cis}\pm\frac{1}{5}\pi$
   (f) $\operatorname{cis}\frac{1}{12}(4k-1)\pi$, $(k = 0, \pm 1, \pm 2, 3)$

7. (c) $z^2 + z - 1$

9. $z^{n-1} + z^{n-2} + z^{n-3} + \cdots + z^2 + z + 1 = 0$

13. (a) $1, \operatorname{cis}\pm\frac{2}{5}\pi, \operatorname{cis}\pm\frac{4}{5}\pi$ (d) $z = -i\cot\pm\frac{1}{5}\pi, -i\cot\pm\frac{2}{5}\pi$
    (f) $z^4 + 2z^2 + 0.2 = \left(z^2 + \cot^2\frac{1}{5}\pi\right)\left(z^2 + \cot^2\frac{2}{5}\pi\right)$

Exercise 21.1

1. (a) $\frac{1}{2}, \frac{1}{12}$ (b) $4, \frac{4}{3}$ (c) $\frac{1}{2}b, \frac{1}{12}b^2$ (d) $\frac{1}{2}(a+b), \frac{1}{12}(b-a)^2$
3. 0.307
5. 0.631
7. (a) 1.44 (b) $E(X) = 0.443$, $\operatorname{Var}(X) = 0.0827$

Exercise 21.2

1. (a) 1 (b) $\sqrt{2}$
3. $\frac{1}{3}$
5. 2.4
7. (a) -2 (b) $k = \frac{1}{\pi}$, $\mu = 0.934$, $m = 0.832$
9. (a) $\frac{4}{81}$ (b) $\frac{8}{5}$ (c) 1.73 (d) 1.62

Exercise 22.1

1. (a) $\frac{1}{6}\pi$ (b) $\frac{1}{2}\pi$ (c) $-\frac{1}{4}\pi$ (d) $-\frac{1}{3}\pi$ (e) $\frac{2}{3}\pi$ (f) $-\frac{1}{3}\pi$ (g) $\frac{1}{4}\pi$
   (h) $\frac{1}{4}\pi$ (i) $-\frac{1}{4}\pi$ (j) 0.2 (k) 0 (l) π (m) $\frac{1}{4}\pi$ (n) $\frac{1}{6}\pi$ (o) —
   (p) 0.8 (q) $\frac{1}{2}\sqrt{3}$ (r) $\frac{1}{2}\sqrt{3}$ (s) 0.96 (t) 0.28 (u) $\frac{7}{9}$ (v) $-\frac{4}{3}$
   (w) $\frac{120}{119}$ (x) $\frac{4}{25}\sqrt{6}$

3. (a) $0 \le a \le \pi$ (b) $-1 \le b \le 1$

Answers

5. $-1<a<1$ Modifications: (i) $a<-1$; $2\arctan a = \arctan\dfrac{2a}{1-a^2} - \pi$

(ii) $a=-1$; $2\arctan a = -\tfrac{1}{2}\pi$

(iii) $a=1$; $2\arctan a = \tfrac{1}{2}\pi$

(iv) $a>1$; $2\arctan a = \arctan\dfrac{2a}{1-a^2} + \pi$

Exercise 22.2

1. (a) $\dfrac{3}{\sqrt{1-9x^2}}$ $(-\tfrac{1}{3}<x<\tfrac{1}{3})$ (b) $\dfrac{-1}{\sqrt{1-(x+1)^2}}$ $(-2<x<0)$

(c) $\dfrac{2}{1+4x^2}$ (all x) (d) $\dfrac{1}{\sqrt{-2x(2x+1)}}$ $(-\tfrac{1}{2}<x<0)$

(e) $\dfrac{-2x}{\sqrt{1-(x^2-1)^2}}$ $(0<x<\sqrt{2})$ (f) $\dfrac{2x}{1+x^4}$ (all x)

3. (b) $\tfrac{1}{4}\pi$ (c) $\tfrac{1}{2}\tan 1$ (d) $\tfrac{2}{37}$ (e) $4x - 4y = \pi - 2$

5. (a) $(1, \tfrac{1}{4}\pi)$, $(-1, -\tfrac{1}{4}\pi)$ (b) $(\tfrac{1}{2}\sqrt{3}, \tfrac{1}{3}\pi)$, $(-\tfrac{1}{2}\sqrt{3}, -\tfrac{1}{3}\pi)$

7. (b) $\tfrac{1}{4}\pi$

Exercise 22.3

1. (a) $3\arcsin\tfrac{1}{3}x + c$ (b) $2\arcsin\tfrac{1}{4}x + c$ (c) $10\arcsin\tfrac{1}{5}x + c$

(d) $\tfrac{1}{2}\arcsin 2x + c$ (e) $2\arcsin\tfrac{3}{2}x + c$ (f) $\tfrac{4}{3}\sqrt{3}\arcsin\sqrt{\tfrac{3}{2}}x + c$

2. (a) $\arctan\tfrac{1}{2}x + c$ (b) $2\arctan\tfrac{1}{3}x + c$ (c) $\arctan\tfrac{1}{4}x + c$

(d) $\tfrac{1}{2}\arctan\tfrac{2}{3}x + c$ (e) $\tfrac{1}{3}\arctan 3x + c$ (f) $\tfrac{1}{3}\sqrt{3}\arctan\tfrac{2}{3}\sqrt{3}x + c$

3. (a) $\arcsin\tfrac{1}{2}(x+1) + c$ (b) $2\arcsin(x-3) + c$ (c) $\arcsin\tfrac{1}{2}(x+1) + c$

(d) $\tfrac{1}{2}\arctan\tfrac{1}{2}(x-3) + c$ (e) $\tfrac{4}{3}\arctan\tfrac{1}{3}(x-1) + c$ (f) $\tfrac{1}{3}\arctan\tfrac{2}{3}(x+2) + c$

5. (a) $\ln|3x+1|$ (b) $\sqrt{3}\arctan\sqrt{3}x$ (c) $\tfrac{1}{2}\ln(3x^2+1)$

7. (a) $-4\sqrt{1-x^2} - \arcsin x$ (b) $\tfrac{3}{2}\ln(x^2+4) + 2\arctan\tfrac{1}{2}x$

(c) $\tfrac{1}{4}\ln(4x^2+9) - \tfrac{1}{2}\arctan\tfrac{2}{3}x$ (d) $-\sqrt{2-3x^2} + \tfrac{2}{3}\sqrt{3}\arcsin\tfrac{1}{2}\sqrt{6}x$

(e) $\tfrac{16}{3}x^3 - 8\sqrt{x^2+1} + \arctan x$ (f) $13x + 12\sqrt{x^2+4} - 8\arctan\tfrac{1}{2}x$

Exercise 23.1

1. (a) $\tfrac{2}{3}(1-x)^{2/3} - 2(1-x)^{1/2} + c$ (b) $\tfrac{1}{10}(2x-1)^{5/2} - \tfrac{1}{2}(2x-1)^{3/2} + c$

(c) $\frac{1}{2}\ln|2x-1| - \frac{2}{2x-1} + c$ (d) $\arcsin\sqrt{x} + \sqrt{x-x^2} + c$

(e) $\frac{1}{2}\arctan\frac{1}{2}e^x + c$ (f) $\frac{x}{4\sqrt{4+x^2}} + c$ (g) $\frac{1}{2}\ln(e^{2x}+1) + c$

(h) $\arcsin\left(\frac{1}{2}\tan x\right) + c$

3. $\frac{1}{24a^3}\left(\pi + 3\sqrt{3} - 6\right)$

5. $2(\arctan 3 - \arctan 2)$

Exercise 23.2

1. (a) $\frac{1}{2}x\sin 2x + \frac{1}{4}\cos 2x$ (b) $\frac{1}{9}(3x-1)e^{3x}$ (c) $\frac{1}{16}\sin 4x - \frac{1}{4}x\cos 4x$

 (d) $\frac{1}{9}x^3(3\ln x - 1)$ (e) $(x+1)e^{2x}$ (f) $-\frac{1}{4}(2x+1)e^{-2x}$

3. (a) $\frac{1}{4}\pi$ (b) $\frac{1}{18}(4-3\pi)$ (c) $\ln 4 - \frac{3}{4}$ (d) $\frac{1}{3}\left(\pi + \sqrt{3}\right) - \arcsin 1$

5. (a) $\frac{35}{256}\pi$ (b) $\frac{5}{256}\pi$

7. (b) $\int e^{ax}\cos bx \, dx = \frac{e^{ax}}{a^2+b^2}(b\sin bx + a\cos bx) + c$

Exercise 23.3

1. (a) $\frac{1}{x-1} - \frac{1}{x+1}$ (b) $\frac{2}{x+2} - \frac{1}{x+1}$ (c) $\frac{1}{2(x-2)} - \frac{3}{2x}$

 (d) $\frac{5}{3(x+1)} + \frac{1}{3(x-2)}$ (e) $\frac{2}{2x-1} - \frac{1}{x+3}$ (f) $\frac{1}{2x+3} + \frac{2}{1-4x}$

3. (a) $\frac{1}{x+1} - \frac{1}{(x+1)^2}$ (b) $\frac{3}{x-2} + \frac{6}{(x-2)^2}$ (c) $\frac{2}{x} - \frac{2}{x+1} + \frac{1}{(x+1)^2}$

 (d) $\frac{5}{x-2} - \frac{2}{(x-2)^2} + \frac{3}{x+1}$

5. (a) $3\ln|x+2| - 2\ln|x-2| + c$ (b) $2\ln|x-3| - \frac{6}{x-3} + c$

 (c) $\ln(x^2+9) + \arctan\frac{1}{3}x - 2\ln|x| + c$ (d) $2\ln|x+2| - \sqrt{3}\arctan\frac{1}{3}\sqrt{3}x + c$

 (e) $\ln\left|\frac{x-3}{x-2}\right| + c$ (f) $\frac{3}{4}\ln|2x-3| - \frac{1}{8}\ln|4x-1| + c$

7. $2 - \frac{3}{2}\ln 3$

9. (a) $3\ln|x-2| - \ln|2x+1| + c$ (b) $2\ln|x-4| - \frac{1}{2}\ln|2x-3| + c$

 (c) $\frac{3}{2}\ln|2x-1| - \frac{1}{2}\ln|2x+1| + c$ (d) $-2\ln|x-2| - \frac{1}{2}\ln|2x+3| + c$

Answers

11. $\ln 13.5 - \frac{2}{3}\ln 5$

Exercise 23.4
1. (a) $y = ce^{x^2}$ (b) $y = \tan(x+c)$ (c) $y = y_0 e^{2x}$ (d) $y = cx^2$
   (e) $x^2 - 4y - \ln(y+4)^2 = c$ (f) $y = \dfrac{c-x}{1+cx}$ (g) $y = (\ln x + c)^2$
   (h) $y = 10 + ce^{2x}$ (i) $y^2 = c\sec x$ (j) $y^2 = x - \frac{1}{2}\sin 2x + c$
3. (a) $y = cx$ (b) $x^2 + y^2 = c$ (c) $y = ce^x$ (d) $y^2 = 2x + c$
5. $k = \frac{1}{2}\ln 10$; 78.2 m
7. A further 2.40 minutes ($\approx$ 2 minutes 24 seconds)
9. A further 2.74 minutes ($\approx$ 2 minutes 44 seconds)
11. 1.24 days

Exercise 23.5
1. (a) $y = x\ln(cx)$ (b) $y = x(cx - 3)$ (c) $y = x(cx - 1)$
   (d) $y = \frac{1}{2}x(cx^2 + 1)$ (e) $y^2 = 2x^2 \ln cx$ (f) $y = x\ln(c[x+y])$
   (g) $y = x\ln(c - \ln x)$ (h) $y = xe^{cx}$ (i) $(2x+y)^2(x-y) = c$
   (j) $(2y-x)^4 = c(x+y)$

2. (a) $x^2 + y^2 = 10x^4$ (b) $x^2 = 2y^2 \ln\left(\dfrac{e^2 y^2}{4}\right)$ (c) $y = -x\left(2 + \dfrac{1}{\ln x + 2}\right)$
   (d) $y^2 - 2xy - x^2 = 2$

3. (a) $y = x - 1 + \dfrac{2}{1 - ce^{2x}}$ (b) $y = \arcsin(ce^x) - x$ (c) $y = \dfrac{1}{x}\ln(x+c)$
   (d) $\arctan\left(\dfrac{y}{x^2}\right) = \ln c(x^4 + y^2)$

4. (a) $y = x + 1 - \sqrt{2x - 3}$ (b) $y = \dfrac{2}{1 + \exp(-x^2)}$ (c) $y = \sqrt{6x + 19} - 2x - 1$
   (d) $y = \ln(e^{x+1} - x)$

Exercise 24.1
1. (a) $\begin{pmatrix} 1 & 0 \\ 0 & -1 \end{pmatrix}$ (b) $\begin{pmatrix} -1 & 0 \\ 0 & 1 \end{pmatrix}$ (c) $\begin{pmatrix} 0 & -1 \\ 1 & 0 \end{pmatrix}$ (d) $\begin{pmatrix} -1 & 0 \\ 0 & -1 \end{pmatrix}$
   (e) $\begin{pmatrix} 0 & 1 \\ -1 & 0 \end{pmatrix}$ (f) $\begin{pmatrix} 0 & 1 \\ 1 & 0 \end{pmatrix}$ (g) $\begin{pmatrix} 0 & -1 \\ -1 & 0 \end{pmatrix}$ (h) $\begin{pmatrix} 1 & 0 \\ 0 & 1 \end{pmatrix}$

Answers

3. (a) $T = \begin{pmatrix} 2 & -1 \\ 3 & 2 \end{pmatrix}$ (b) (7, 7) (c) (0, –7) (d) $4x - 5y = -42$

Exercise 24.2

1. (a) $\frac{1}{2}\begin{pmatrix} -\sqrt{3} & -1 \\ 1 & -\sqrt{3} \end{pmatrix}$ (b) $\frac{1}{5}\begin{pmatrix} 3 & 4 \\ 4 & -3 \end{pmatrix}$ (c) $\begin{pmatrix} 3 & 0 \\ 0 & 3 \end{pmatrix}$ (d) $\begin{pmatrix} 1 & 2 \\ 0 & 1 \end{pmatrix}$

   (e) $\frac{1}{\sqrt{5}}\begin{pmatrix} 2 & -1 \\ 1 & 2 \end{pmatrix}$ (f) $\frac{1}{13}\begin{pmatrix} 5 & -12 \\ -12 & -5 \end{pmatrix}$ (g) $\frac{1}{2}\begin{pmatrix} -1 & \sqrt{3} \\ -\sqrt{3} & -1 \end{pmatrix}$ (h) $\begin{pmatrix} 1 & 0 \\ -2 & 1 \end{pmatrix}$

3. $A^2 = \begin{pmatrix} \cos 2\theta & -\sin 2\theta \\ \sin 2\theta & \cos 2\theta \end{pmatrix}$, $A^3 = \begin{pmatrix} \cos 3\theta & -\sin 3\theta \\ \sin 3\theta & \cos 3\theta \end{pmatrix}$, $A^n = \begin{pmatrix} \cos n\theta & -\sin n\theta \\ \sin n\theta & \cos n\theta \end{pmatrix}$

   $A^{-1} = \begin{pmatrix} \cos \theta & \sin \theta \\ -\sin \theta & \cos \theta \end{pmatrix}$

5. (a) Reflection in the *x*-axis.
   (b) Clockwise rotation about O through 90°.
   (c) Anticlockwise rotation about O through 126.9° (180° – arctan $\frac{4}{3}$).
   (d) Reflection in $x + 2y = 0$.
   (e) Enlargement, centre O, scale factor $\frac{1}{2}$.
   (f) Shear in the *x*-axis with shear constant 4.
   (g) Stretches parallel to the axes with scale factors 2 and 3.
   (h) Reflection in $2x - 3y = 0$.
   (i) Anticlockwise rotation about O through 28.1° (arctan $\frac{8}{15}$).
   (j) Shear in the *y*-axis with shear constant –3.
   (k) Stretch parallel to the *x*-axis with scale factor –2.
   (l) Anticlockwise rotation about O through 45° followed by an enlargement, centre O, scale factor $\sqrt{2}$ (or vice versa).

9. $\frac{1}{5}\begin{pmatrix} -3 & -4 \\ 4 & -3 \end{pmatrix}$, 126.9° (180° – arctan $\frac{4}{3}$).

11. (a) $\frac{1}{2}\begin{pmatrix} -1 & -\sqrt{3} \\ \sqrt{3} & -1 \end{pmatrix}$ (b) $\begin{pmatrix} 1 & 0 \\ 0 & -1 \end{pmatrix}$ (c) $\frac{1}{5}\begin{pmatrix} -4 & -3 \\ 3 & -4 \end{pmatrix}$

    (d) $\frac{1}{2}\begin{pmatrix} -1 & -\sqrt{3} \\ -\sqrt{3} & 1 \end{pmatrix}$ (e) $\sqrt{2}\begin{pmatrix} 1 & -1 \\ 1 & 1 \end{pmatrix}$ (f) $\begin{pmatrix} 0 & -1 \\ 1 & -2 \end{pmatrix}$

13. (a) (–1, 3) (b) $x + 7y = 20$

15. (a) $x' = \frac{1}{5}(-3x - 4y)$, $y' = \frac{1}{5}(-4x + 3y)$
    (i) $y = 18x + 30$ (ii) $(5x + 11)^2 + (5y - 2)^2 = 125$ or $5x^2 + 5y^2 + 22x - 4y = 0$

761

Answers

   (b) $x' = \frac{1}{5}(3x + 4y)$, $y' = \frac{1}{5}(4x - 3y)$
   (i) $y = 18x - 30$ (ii) $(5x - 11)^2 + (5y + 2)^2 = 125$ or $5x^2 + 5y^2 - 22x + 4y = 0$

17. (a) $y = 2x - 9$ (b) $x^2 + y^2 - 6x + 6y = 45$ (c) $y = \frac{1}{3}(x - 3)^2 = \frac{1}{3}x^2 - 2x + 3$
    Enlargement, centre O, scale factor 3.

19. $\sqrt{a^2 + b^2}$, $\arctan(b/a)$

Exercise 24.3

1. $\begin{pmatrix} 2 \\ 1 \end{pmatrix}$

3. $y = x^2$

5. $y = x^2$

7. $y = x^3 + 3x - 1$; translation with vector $\begin{pmatrix} 0 \\ -2 \end{pmatrix}$

9. No

11. $y = 2x^3 + 12x^2 + 15x$

13. (a) Reflection in y-axis followed by a translation with vector $\begin{pmatrix} -1 \\ -2 \end{pmatrix}$.

    (b) $g(x) = -x^3 - 8x^2 - 16x$

Exercise 25.1

1. (a) $A \subseteq B$ (b) $B \subseteq A$ (c) $B \subseteq A$ (d) $A \subseteq B$

3. (a) 2^n

Exercise 25.2

1. (a) $\{(1,1), (1,4), (2,2), (2,5), (3,3), (3,6), (4,1), (4,4), (5,2), (5,5), (6,3), (6,6)\}$
   (b) Domain = Range = $\{1, 2, 3, 4, 5, 6\}$ (c) R is reflexive, symmetric and transitive

3. (a) $\{(1,1), (1,2), (1,3), (1,4), (1,5), (2,1), (2,2), (2,3), (2,4), (3,1), (3,2), (3,3), (4,1), (4,2), (5,1)\}$
   (b) Domain = Range = $\{1, 2, 3, 4, 5\}$ (c) R is symmetric

5. R is transitive

7. R is reflexive and transitive

9. R is reflexive, symmetric and transitive

Exercise 25.3

1. (a) Equivalence relation: $[1] = \{1, 3, 5\}$, $[2] = \{2, 4\}$, $[6] = \{6\}$
   (b) Equivalence relation: $[a] = \{a\}$, $a = 1, 2, 3, 4, 5, 6$
   (c) Equivalence relation: $[1] = \{1, 4\}$, $[2] = \{2, 5\}$, $[3] = \{3, 6\}$

(d) Not an equivalence relation. e.g., $(1, 1) \notin R$
5. Not an equivalence relation. e.g., $(1, 0) \in R$, $(0, -1) \in R$ but $(1, -1) \notin R$

Exercise 25.4
1. (a)
$$\begin{array}{c|ccc} & a & b & c \\ \hline 1 & 1 & 1 & 0 \\ 2 & 0 & 1 & 0 \\ 3 & 1 & 0 & 1 \\ 4 & 0 & 1 & 1 \end{array}$$

(b)
$$\begin{array}{c|cc} & 1 & 2 \\ \hline 1 & 1 & 1 \\ 2 & 0 & 1 \\ 3 & 1 & 0 \\ 4 & 1 & 1 \\ 5 & 0 & 0 \end{array}$$

(c)
$$\begin{array}{c|cccc} & 1 & 2 & 3 & 4 \\ \hline a & 0 & 1 & 0 & 1 \\ b & 1 & 0 & 1 & 0 \\ c & 1 & 1 & 1 & 0 \end{array}$$

(d)
$$\begin{array}{c|ccc} & a & b & c \\ \hline p & 1 & 1 & 0 \\ q & 1 & 0 & 1 \end{array}$$

3. R is symmetric and transitive
5. R is not transitive on X

Exercise 25.5
1. (a) both (b) neither (c) both (d) both (e) neither (f) surjection
   (g) injection (h) neither (i) surjection (j) surjection (k) injection
   (l) both
3. (a) $f : x \mapsto \dfrac{1}{x}$ (b) $f : x \mapsto \sqrt{x}$ (c) $f : x \mapsto \tan\dfrac{\pi x}{2}$

Exercise 25.6
1. (a), (c), (d), (g), (h), (i), (l), (m), (n), (o), (p)
3. (a), (b), (d), (e), (g), (j)
5. (b), (c), (d), (e)
9. {even integers}
11. $\{1\}, \{1, -1\}$

Exercise 25.7
1. (a), (b), (d), (e)
3. (b) 27 (c) If any of d, e or f is a, then $(d * e) * f = d * (e * f)$. Only 8 checks.

Exercise 25.8
1. (a) -3 (b) $\tfrac{1}{3}$ (c) — (d) 0
3. $e = 8$
5. (a) c (b) —

Exercise 25.9
1. (a) $-8 - c$ (b) $\dfrac{1}{4c}$ $(c \neq 0)$ (c) $\dfrac{-c}{2c+1}$ $(c \neq -\tfrac{1}{2})$ (d) —
   (e) $\dfrac{c}{3c-1}$ $(c \neq \tfrac{1}{3})$ (f) $\dfrac{8c-15}{4(c-2)}$ $(c \neq 2)$

763

Answers

5. (a) $e = b$, $a^{-1} = a$, $b^{-1} = b$, $c^{-1} = d$, $d^{-1} = c$
   (b) $e = c$, $a^{-1} = b$, $b^{-1} = a$ or d, $c^{-1} = c$, $d^{-1} = b$

7. (a) (i) $x^3 + d^3$ (ii) 3 (b) $\left(\sqrt[3]{2}+1\right)^{-1} = \frac{1}{3}\sqrt[3]{4} - \frac{1}{3}\sqrt[3]{2} + \frac{1}{3}$,
   $\left(\sqrt[3]{2}-1\right)^{-1} = \sqrt[3]{4} + \sqrt[3]{2} + 1$

Exercise 25.10
5. 16
7. $a = 1, x = 1$; $a = 3, x = 7$; $a = 7, x = 3$; $a = 9, x = 9$

Exercise 25.11
1. (a) $e = p_1$, $p_1^{-1} = p_1$, $p_2^{-1} = p_2$, $p_3^{-1} = p_3$, $p_4^{-1} = p_4$, $p_5^{-1} = p_6$, $p_6^{-1} = p_5$
   (b) (i) p_1 (ii) p_3 (iii) p_3 (iv) p_6 (v) p_4 (vi) p_1

3. (a) $\begin{pmatrix} 1 & 2 & 3 & 4 & 5 \\ 2 & 3 & 4 & 5 & 1 \end{pmatrix}$ (b) $\begin{pmatrix} 1 & 2 & 3 & 4 & 5 \\ 2 & 1 & 3 & 4 & 5 \end{pmatrix}$

Exercise 25.12
1. (a) $\begin{pmatrix} 1 & 2 & 3 & 4 & 5 \\ 1 & 3 & 2 & 4 & 5 \end{pmatrix}$ (b) $\begin{pmatrix} 1 & 2 & 3 & 4 & 5 \\ 2 & 3 & 1 & 4 & 5 \end{pmatrix}$ (c) $\begin{pmatrix} 1 & 2 & 3 & 4 & 5 \\ 5 & 4 & 3 & 1 & 2 \end{pmatrix}$

   (d) $\begin{pmatrix} 1 & 2 & 3 & 4 & 5 \\ 5 & 1 & 2 & 3 & 4 \end{pmatrix}$ (e) $\begin{pmatrix} 1 & 2 & 3 & 4 & 5 \\ 1 & 2 & 3 & 4 & 5 \end{pmatrix}$ (f) $\begin{pmatrix} 1 & 2 & 3 & 4 & 5 \\ 1 & 2 & 3 & 5 & 4 \end{pmatrix}$

   (g) $\begin{pmatrix} 1 & 2 & 3 & 4 & 5 \\ 2 & 4 & 3 & 1 & 5 \end{pmatrix}$ (h) $\begin{pmatrix} 1 & 2 & 3 & 4 & 5 \\ 5 & 3 & 1 & 4 & 2 \end{pmatrix}$ (i) $\begin{pmatrix} 1 & 2 & 3 & 4 & 5 \\ 5 & 3 & 1 & 2 & 4 \end{pmatrix}$

3. (a) 1 (b) 2 (c) 3 (d) 4 (e) 5 (f) 2 (g) 3 (h) 6 (i) 12

Exercise 25.13
1. (c), (e) and (g) are groups
3. (a) group (b) not closed, no identity and no inverses (c) group
   (d) group
5. $e = -2$, $a^{-1} = -4 - a$; group
9. m must be prime

Exercise 25.14
3. $x^2 = x$ has only one solution: $x = e$
   The set of residue classes $\{2, 4, 6, 8\}$ under multiplication, modulo 10 has 2 elements x satisfying $x^3 = x$. They are $x = 6$ (the identity) and $x = 4$.
5. $a * b = e$
7. (a) xy^2x^{-1} (b) xy^3x^{-1} (c) xy^4x^{-1} (d) $xy^{-1}x^{-1}$ (e) $xy^{-2}x^{-1}$

764

(f) $xy^{-3}x^{-1}$

9. (a)

Element	1	2	3	4	5	6
Order	1	3	6	3	6	2

(b) 3   (c) 6

11. (a) x (b) x^2 (c) x^2 (d) x
13. 3

Exercise 25.15

1. $\{e, g^6\}$, $\{e, g^4, g^8\}$, $\{e, g^3, g^6, g^9\}$, $\{e, g^2, g^4, g^6, g^8, g^{10}\}$, $(e = g^{12})$

11. The even integers under addition – 2 is a generator.
    (a) True
    (b) False – The set of 4 rotations about the origin through 0°, 90°, 180° and 270° under 'follows' is cyclic and the rotation of 90° generates the group. The rotations through 0° and 180° under 'follows' is a proper subgroup.

13. (b) 12, (0, 2)   (c) (1, 1), (1, 2)   (d) (iii) is cyclic
    (e) (i) (0, 2), (1, 0), (1, 2) have order 2   (ii) (0, 1), (0, 3), (1, 1), (1, 3) have order 4

Exercise 26.1

1. (a) $A = 1, B = -2, C = 1$ (b) $\frac{1}{2}$
3. (a) converges   (b) diverges   (c) diverges   (d) diverges   (e) converges
   (f) diverges   (g) diverges   (h) converges   (i) converges   (j) diverges
   (k) diverges   (l) converges   (m) diverges   (n) converges   (o) diverges
   (p) converges   (q) converges   (r) converges (by result of Q2)

Exercise 26.2

1. (a) converges   (b) converges   (c) converges   (d) diverges   (e) converges
   (f) converges
3. (a) converges   (b) converges   (c) diverges   (d) diverges   (e) converges
   (f) converges   (g) diverges   (h) converges   (i) converges

Exercise 26.3

1. (a) diverges   (b) diverges   (c) diverges   (d) converges   (e) converges
   (f) converges   (g) converges   (h) converges   (i) converges
3. (a) convergent   (b) convergent   (c) divergent   (d) convergent
   (e) convergent   (f) convergent

Exercise 26.4

1. (a) converges   (b) converges   (c) converges   (d) diverges
   (e) converges   (f) converges   (g) converges   (h) converges

Answers

3. (a) $\frac{1}{3}$ (b) $\frac{1}{56}$ (c) $\frac{1}{12}$ (d) $\frac{1}{36}$ (e) $\frac{3}{64}$ (f) $\frac{1}{512}$

5. (a) converges absolutely (b) converges absolutely (c) converges absolutely
   (d) converges conditionally (e) converges conditionally (f) diverges
   (g) converges absolutely (h) converges absolutely (i) converges absolutely
   (j) converges absolutely (k) converges absolutely (l) converges absolutely

Exercise 26.5

1. (a) $-3 < x < -1$, $r = 1$ (b) $-\frac{2}{3} < x < 0$, $r = \frac{1}{3}$ (c) $-2 < x < -1$, $r = \frac{1}{2}$
   (d) $-1 < x < 3$, $r = 2$ (e) all x, $r = \infty$ (f) $-1 < x < 1$, $r = 1$ (g) all x, $r = \infty$
   (h) $-8 < x < -2$, $r = 3$ (i) $x = 3$, $r = 0$ (j) all x, $r = \infty$

3. (a) $-\frac{7}{2} < x < \frac{1}{2}$, $r = 2$ (b) $-1 < x < 1$, $r = 1$ (c) $-1 \leq x \leq 1$, $r = 1$
   (d) $-2 < x < 2$, $r = 2$ (e) $-2 \leq x < 2$, $r = 2$ (f) $-\frac{3}{2} \leq x \leq \frac{3}{2}$, $r = \frac{3}{2}$

Exercise 26.6

1. (a) 2 (b) 0 (c) 4 (d) $\frac{8}{27}$ (e) $\sqrt{2}$ (f) $\sqrt{2}$

7. $-9 \leq f(x) \leq 15$

Exercise 26.7

1. (a) $(1+x)^{1/2} \approx 1 + \frac{1}{2}x - \frac{1}{8}x^2 + \frac{1}{24}x^3$, $|x| < 1$
   (b) $(1-x)^{-2} \approx 1 + 2x + 3x^2 + 4x^3$, $|x| < 1$
   (c) $(1+x)^{3/2} \approx 1 + \frac{3}{2}x + \frac{3}{8}x^2 - \frac{1}{16}x^3$, $|x| < 1$

3. 0.3679

5. (a) $e^x = e \sum_{n=0}^{\infty} \frac{(x-1)^n}{n!}$ (b) $\cos x = \sum_{n=1}^{\infty} (-1)^n \frac{(x-\frac{1}{2}\pi)^{2n-1}}{(2n-1)!}$
   (c) $\ln(1+x) = \ln 3 + \sum_{n=1}^{\infty} \frac{(-1)^{n+1}}{n 3^n}(x-2)^n$ (d) $\frac{1}{x} = \sum_{n=0}^{\infty} (-1)^n (x-1)^n$

7. (a) $P_3(x) = 3 + \frac{1}{6}x - \frac{1}{216}x^2 + \frac{1}{3888}x^3$ (b) 0.00002   (c) $\sqrt{10} \approx 3.1623$

Exercise 26.8

5. $\frac{1}{(1+x)^2} = \sum_{n=0}^{\infty} (-1)^n (n+1)x^n$

7. $e^x \sin x = x + x^2 + \frac{1}{3}x^3 + \cdots$

9. (a) $\cos^2 x = \left(1 - \frac{1}{2}x^2 + \frac{1}{24}x^3 + \cdots\right)\left(1 - \frac{1}{2}x^2 + \frac{1}{24}x^3 + \cdots\right) = 1 - x^2 + \frac{1}{3}x^4 + \cdots$

   (b) $\frac{1}{2}(1 + \cos 2x) = \frac{1}{2} + \frac{1}{2}\left(1 - \frac{(2x)^2}{2} + \frac{(2x)^4}{24} + \cdots\right) = 1 - x^2 + \frac{1}{3}x^4 + \cdots$

11. (a) $\int_0^x \sin t^2 \, dt = \sum_{n=0}^{\infty} (-1)^n \frac{x^{4n+3}}{(4n+3)(2n+1)!}$

(b) $\int_0^x \frac{1}{2-t} \, dt = \sum_{n=1}^{\infty} \frac{x^n}{n2^n}$ $(0 < x < 2)$

13. (a) $\frac{1}{1-x^2} = \sum_{n=0}^{\infty} x^{2n}$, $|x|<1$ (b) $\ln\left(\frac{1+x}{1-x}\right) = 2\sum_{n=0}^{\infty} \frac{x^{2n+1}}{2n+1}$, $|x|<1$

15. (a) $\sum_{n=0}^{\infty} \frac{(-1)^n}{(n+1)!}(x-1)^{n+1}$ (b) 0.2212

Exercise 26.9
1. 8
3. 3.75

Exercise 26.10
1. (i) (a) 0.344 (b) 1.954 (c) 0.759 (d) 0.743
   (ii)(a) 3.149 (b) 0.368 (c) 0.778 (d) 1.191
3. (i) (a) 0.266, 0.032 (b) 0.696, 0.007 (c) 0.994, 0.009 (d) 4.677, 0.010
   (ii)(a) 0.25, 0 (b) 0.28768229, 4.2×10^{-7} (c) 2.0009, 0.0013
   (d) 0.63212, 1.4×10^{-6}
5. (a) 14 (b) 12 (c) 4 (d) 18
7. 128 m
9. 500 m^3

Exercise 26.11
1. −4.3
3. −0.2085, 0.3065, 0.6520
5. (c) −0.472 513 (d) −2.116 34
7. $\sqrt[3]{10} \approx 2.154\,435$ $\sqrt[4]{5} \approx 1.495\,349$
9. (a) [graph showing $y = x$ and $y = \tan x$ with intersection near 4.5, between $\pi/2$ and $3\pi/2$] (b) 4.5 (c) 4.493 409

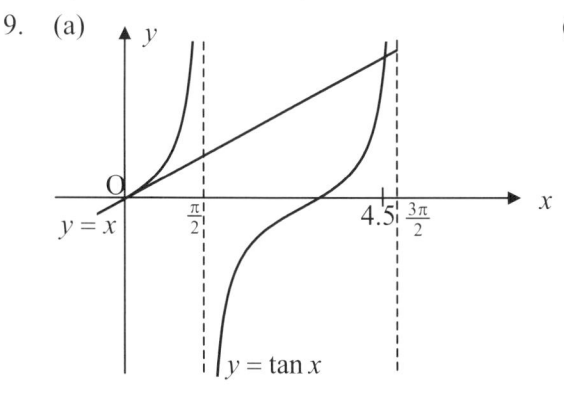